Engineering Principles of Combat Modeling and Distributed Simulation

Engineering Principles of Combat Modeling and Distributed Simulation

Edited by

Andreas Tolk

A John Wiley & Sons, Inc., Publication

Published by John Wiley & Sons, Inc., Hoboken, New Jersey.
Published simultaneously in Canada.

For general information on our other products and services or for technical support, please contact our Customer Care Department within the United States at (800) 762-2974, outside the United States at (317) 572-3993 or fax (317) 572-4002.

Wiley also publishes its books in a variety of electronic formats. Some content that appears in print may not be available in electronic formats. For more information about Wiley products, visit our web site at www.wiley.com.

Library of Congress Cataloging-in-Publication Data:

Tolk, Andreas.
　Engineering principles of combat modeling and distributed simulation / Andreas Tolk.
　　p. cm.
　Includes bibliographical references and index.
　Chapters 1-15 written by Andreas Tolk; chapters 16-32 written by various authors.
　ISBN 978-0-470-87429-5 (cloth)
　1. War games–Data processing. 2. Military art and science–Computer simulation. 3. Combat–Mathematical models. 4. Combat–Simulation methods. I. Title.
　U310.T63 2012
　355.4′80285–dc23
2011031418

Printed in the United States of America

10 9 8 7 6 5 4 3 2 1

Contents

Preface

Looking through the present volume, I am struck by how long such a book has been needed and how the fact of the book coming together is a tangible indication of how much the field of modeling and simulation has matured in the last 20 years. Computer models of combat have been a major element in force planning, training, and system development for roughly 40 years—give or take a decade and depending on the criteria used. For most of that period, however, the computer models were built individually by talented individuals and teams who plunged ahead with courage, innovation, and hard work, but without the benefit of a discipline to guide them or the technology to make modeling and simulation systematic and adaptive. None of them had been educated to do what they did.

Many of the early combat models were pure attrition models shaped by the role of massive firepower in World War II, mathematical methods familiar to that era's defense scientists, and what computers of the era could and could not do. Other models were much more "micro" in nature, representing the low level physics of combat. Over time, researchers moved toward more systemic treatments that included such critical functions as mobilizing reserves, transporting them to where war was to be fought, and fighting the mostly separate air, land, and sea battles. Some of the detailed models of the era became extraordinarily accurate in their representation of, for example, missile trajectories, accuracy, and effectiveness. By the late 1980s, integrated models were emerging that could deal with multi-theater conflict, better represent the air–land battle, and (in at least some work) represent adaptive alternative military strategies.

Despite such progress, and many substantial accomplishments, those of us helping the Department of Defense to review the state of combat modeling in the early 1990s were very troubled by the haphazard and often mysterious relationships among models, issues of validity (especially where models could not realistically be tested against hard data), the continued failure to incorporate the "soft factors" known to be crucial in warfare (e.g. the related effects of leadership, troop quality, morale, training, the second-class treatment of command and control), and the failure to deal well with uncertainty (which has been and remains a profound problem to this day). There were concerns as well about the severe limitations of the era's modeling and simulation methods, which were lagging what was possible technologically as computer science was yielding technologies with major implications for software engineering in the large (e.g. composable rather than monolithic systems), in specific methods (e.g. object-oriented programming and agent-based modeling) and for such then-still-visionary concepts as distributed interactive simulation and exceedingly accurate entity level simulations of battlefield operations such as were demonstrated by DARPA's path-breaking SIMNET program.

In imagining what might be possible in the way ahead, a recurring observation was that the field of combat modeling needed to be more professionalized: those building the increasingly complex and important simulations should have shared foundational knowledge of subject matter, the art and science of actually building simulations, and the technology that was allowing far more advanced modeling and simulation. At the time, there was no academic infrastructure for collecting or conveying such knowledge systematically. Not surprisingly, there were few dedicated professional journals and textbooks, and no real "community" of what the current book properly refers to as simulationists.

And now we come to the present. Many problems and challenges remain and I remain one of the more impatient of critics. However, what has been accomplished is stunning. Distributed war games and simulations are taken for granted as part of the way we do business in training and exercising. In actual war, command and control includes near-continuous interaction of component commanders in geographically dispersed locations who share incredible amounts of information and have a remarkable degree of shared awareness. Mission rehearsal can sometimes use the same simulations as for training and even weapon system analysis. The more privileged workers routinely bring to bear powerful, reasonably standardized software engineering methods for model composition and federation. To be sure, the best versions of all this are very different from what is often the norm, but a craft and profession has been emerging. The current book reflects this evolution. In one volume it combines a wealth of information relating to everything from foundational knowledge on representing the key elements of military operations; to technology for building powerful, adaptive, and interoperable simulations; to assuring that the simulation products relate well to the mindsets, needs, and language of military users. Remarkably, and as another indication of how far we have come, the book even contains material on professional ethics and good practices. Much of the book reflects Professor Tolk's experience building a coherent academic curriculum and teaching many of its related courses. Other chapters bring in the insights and experience of diverse experts from a number of organizations. The book is a milestone accomplishment for which the editor and authors should be congratulated.

PAUL K. DAVIS
The RAND Corporation
July, 2011

Contributors

Gnana K. Bharathy, PhD

Systems Engineering/ACASA, University of Pennsylvania, Philadelphia, Pennsylvania, USA

Jose (Joe) L. Bricio

Combat Direction Systems Activity Dam Neck, US Navy, Virginia Beach, Virginia, USA

Erdal Cayirci, PhD

Electrical Engineering and Computer Science, University of Stavanger, Stavanger, Norway

Salim Chemlal

Electrical and Computer Engineering, Old Dominion University, Norfolk, Virginia, USA

Andrew Collins, PhD

Virginia Modeling Analysis and Simulation Center, Old Dominion University, Norfolk, Virginia, USA

Paulo C. G. Costa, PhD

C4I Center and SEOR, George Mason University, Fairfax, Virginia, USA

Saikou Y. Diallo, PhD

Virginia Modeling Analysis and Simulation Center, Old Dominion University, Suffolk, Virginia, USA

Robert W. Franceschini, PhD

Science Applications International Corporation, Orlando, Florida, USA

Paul Gustavson, CTO

SimVentions Inc., Fredericksburg, Virginia, USA

Holly A. H. Handley, PhD, PE
Engineering Management and Systems Engineering, Old Dominion University, Norfolk, Virginia, USA

Heber Herencia-Zapana, PhD
National Institute of Aerospace, Hampton, Virginia, USA

Patrick T. Hester, PhD
Engineering Management and Systems Engineering, Old Dominion University, Norfolk, Virginia, USA

Robert C. Holcomb Jr
VT-MAK Technologies, Suffolk, Virginia, USA

Robert H. Kewley, PhD
Systems Engineering, United States Military Academy, West Point, New York, USA

Tag Gon Kim, PhD
Electrical Engineering, KAIST, Daejeon, Republic of Korea

J. David Lashlee, PhD, CMSP
Modeling and Simulation Coordination Office, OSD AT&L/ASD(R&E), Alexandria, Virginia, USA

Kathryn Blackmond Laskey, PhD
Systems Engineering and Operations Research, George Mason University, Fairfax, Virginia, USA

Margaret L. Loper, PhD
Information & Communications Laboratory, Georgia Tech Research Institute, Atlanta, Georgia, USA

Il-Chul Moon, PhD
Industrial and Systems Engineering, KAIST, Daejeon, Republic of Korea

J. Russell Noseworthy, PhD
TENA SDA, Science Application International Corporation Alexandria, Virginia, USA

S. K. Numrich, PhD, CMSP
Institute for Defense Analyses, Alexandria, Virginia, USA

Tuncer Ören, PhD

Electrical Engineering and Computer Science, University of Ottawa, Ottawa, Ontario, Canada

José J. Padilla, PhD

Virginia Modeling Analysis and Simulation Center, Old Dominion University, Suffolk, Virginia, USA

James Panagos

Gnosys Systems Incorporated, Providence, Rhode Island, USA

Mikel D. Petty, PhD

University of Alabama in Huntsville, Huntsville, Alabama, USA

P. M. Picucci, PhD

Institute for Defense Analyses, Alexandria, Virginia, USA

Edward T. Powell, PhD

Science Applications International Corp., Fairfax, Virginia, USA

William T. Richards

Virginia Modeling Analysis and Simulation Center, Old Dominion University, Suffolk, Virginia, USA

Roger D. Smith, PhD

Nicholson Center for Surgical Advancement, Florida Hospital, Orlando, Florida, USA

Joe Sorroche

Flight and Embedded Software/Simulation, Sandia National Laboratories, Albuquerque, New Mexico, USA

Andreas Tolk, PhD

Engineering Management and Systems Engineering, Old Dominion University, Norfolk, Virginia, USA

Charles D. Turnitsa

Virginia Modeling Analysis and Simulation, Center, Old Dominion University, Suffolk, Virginia, USA

Robert L. Wittman Jr, PhD

MITRE Corporation, Orlando, Florida, USA

Marc Wood, ME

Systems Engineering, United States Military, Academy, West Point, New York, USA

Levent Yilmaz, PhD

Computer Science and Software Engineering, Auburn University, Auburn, Alabama, USA

Biographies

Andreas **Tolk** is Professor for Engineering Management and Systems Engineering at the Old Dominion University in Norfolk, Virginia, USA. He received his PhD in Computer Science (1995) and MS in Computer Science (1988) from the University of the Federal Armed Forces, Germany.

Andreas Tolk joined the faculty of Old Dominion University in 2006, where he is instrumental in faculty involvement in the National Centers for Systems of Systems Engineering (NCSOSE) as well as in the Virginia Modeling Analysis and Simulation Center (VMASC). He was a Senior Research Scientist at VMASC from 2002 to 2006. He was Vice President for Land Weapon Systems with the German company IABG from 1998 to 2002. He was Project Manager for decision support systems and integration of M&S into command and control systems from 1995 to 1998 with the German company ESG. From 1983 to 1995, Andreas Tolk served as an Officer in the air defense branch of the German Army, followed by army reserve assignments from 1995 to 2002. He left the army as a Major of the Reserve.

Andreas Tolk's research focuses on model-based systems engineering, which includes research on modeling and simulation interoperability challenges in particular in the context of complex systems and system of systems. His research on simulation interoperability documented in more than 200 publications is internationally recognized by over 30 outstanding paper awards. In 2008, he received the Award for Excellence in Research by Old Dominion University's Frank Batten College for Engineering and Technology. In 2010, he received the first Technical Merit Award from the Simulation Interoperability Standards Organization (SISO). He contributed to textbooks on Agent-Directed Simulation and Systems Engineering, Conceptual Modeling for Discrete-Event Simulation, and Modeling and Simulation Fundamentals. He has edited books on Complex Systems in Knowledge-Based Environments and Intelligence-Based Systems Engineering. He is also on the editorial board of several journals.

Andreas Tolk is senior member of the Institute of Electrical and Electronics Engineers (IEEE) and the Society for Modeling and Simulation International (SCS). He is a member of the American Society for Engineering Management (ASEM), the Association for Computing Machinery (ACM) Special Interest Group Simulation (SIGSIM), the Military Operational Research Society (MORS), the National Defense Industrial Association (NDIA), and the Simulation Interoperability Standards Organization (SISO).

BIOGRAPHICAL SKETCHES OF CHAPTER AND ANNEX AUTHORS

Gnana K. Bharathy is a Researcher, Project Manager and Consultant at the University of Pennsylvania. His areas of research broadly include risk management, analytics, and modeling and simulation, particularly of social systems. He was educated at the University of Pennsylvania, the University of Canterbury, New Zealand, and the National Institute of Technology (formerly Regional Engineering College), Trichy, India. He also holds Project Management Professional certification and is a full member of the Institution of Engineers Australia (MIEAust).

Joe L Bricio is an Engineer in the US Navy, Combat Direction Systems Activity Dam Neck, Virginia Beach. He holds a Bachelor's Degree in Mechanical Engineering and a Master of Science in Modeling and Simulation, both from Old Dominion University. As a Navy Engineer, he is the Co-Chair of the Maritime Theater Missile Defense Forum Modeling and Simulation Working Group and NAVSEA Modeling and Simulation Forum Chair. Prior to the Navy, he worked both in large and small businesses, Project Scientist at Virginia Modeling Analysis and Simulation Center (VMASC), and the Center for Advanced Engineering Environments at NASA Langley as a Modeling and Simulation Engineer. He is currently a PhD student in Engineering Management and Systems Engineering at Old Dominion University.

Erdal Cayirci is currently Chief, Computer Assisted Exercise Support Branch in the NATO Joint Warfare Center in Stavanger, Norway, and also a faculty with the Electrical and Computer Engineering Department of University of Stavanger. He graduated from the Turkish Army Academy in 1986 and from the Royal Military Academy Sandhurst in 1989. He received his MS degree from Middle East Technical University and the PhD degree from Bogazici University in Computer Engineering in 1995 and 2000, respectively. He retired from the Turkish Army as a colonel in 2005. He served on the editorial board of *IEEE Transactions on Mobile Computing, Ad Hoc Networks* (Elsevier Science) and *ACM Kluwer Wireless Networks*. He received the 2002 IEEE Communications Society Best Tutorial Paper Award and the Excellence Award in ITEC 2006. He co-authored three books published by Wiley and Sons. His research interests include sensor networks, mobile communications, tactical communications, and military constructive simulation.

Salim Chemlal is a PhD candidate in the Electrical and Computer Engineering Department at Old Dominion University. He received Dual Bachelor Degrees in Electrical and Computer Engineering and an MS in Computer Engineering from Old Dominion University. His areas of interest include medical modeling and simulation, mobile applications, data mining, augmented reality and scientific visualization, and combat modeling and simulation. To date, his projects in these areas have led to several publications and outstanding awards.

Andy Collins is a Research Assistant Professor at the Virginia Modeling Analysis and Simulation Center (VMASC) where he applies his expertise in game theory and agent-based modeling and simulation to a variety of projects including foreclosure and entrepreneur modeling. He earned his PhD and MSc in Operational Research from the University of Southampton in the UK. He has spent the last 10 years, while conducting his PhD and as an Analyst for the UK's Ministry of Defence, applying game theory to a variety of practical problems.

Paulo C. G. Costa is a Research Associate Professor at the George Mason University (GMU) C4I Center and an Affiliate Professor of the Systems Engineering and Operations Research Department at GMU. He received a PhD and MSc from GMU and a BSc from the Brazilian Air Force Academy. He is a retired Air Force Officer from the Brazilian Air Force with expertise on integrating semantic technology and uncertainty management, and a pioneer in the field of probabilistic ontologies. He developed PR-OWL, a probabilistic ontology language, and is a key contributor to UnBBayes-MEBN, a probabilistic reasoning framework that implements PR-OWL. His research path also includes work as a W3C invited expert in the area of uncertainty reasoning, as a leading organizer of workshops and conferences on semantic technologies (e.g. URSW, STIDS), and as program committee member in diverse academic fields. Paulo Costa teaches courses on decision theory, decision support systems design, and models for probabilistic reasoning. He holds courtesy affiliations from the University of Brasilia and the Instituto Tecnológico da Aeronáutica (Brazil).

Saikou Y. Diallo is a Research Assistant Professor at the Virginia Modeling Analysis and Simulation Center (VMASC) of the Old Dominion University (ODU) in Norfolk, VA. He received his MS and PhD in Modeling and Simulation from ODU and currently leads the Interoperability Lab at VMASC. His research focus is on command and control to simulation interoperability, formal theories of M&S, web services and model-based data engineering. He participates in a number of M&S related organizations and conferences and is currently the Co-Chair of the Coalition Battle Management Language drafting group in the Simulation Interoperability Standards Organization.

Robert W. Franceschini is a Vice President, Chief Engineer, and Technical Fellow at SAIC, where he is a technology leader for modeling, simulation, and training. He earned a PhD in Computer Science from the University of Central Florida. His modeling and simulation research work, documented in over 50 research publications, includes multi-resolution simulation, course of action analysis, terrain reasoning, modes of human interaction with simulation, and simulation interoperability. He has proposed and directed over $19 million in modeling and simulation research and development projects.

Paul Gustavson is a Chief Technology Officer and Co-Founder of SimVentions Inc. He has over 22 years of experience including the design, development and integration of US DoD systems, simulations, standards, and software applications, and has authored numerous technical publications and tutorials on modeling and simulation, and software development. He supports the US M&S Coordination Office in identifying key metadata for M&S assets. He is an active leader within the Simulation Interoperability Standards Organization (SISO) involved in multiple standards efforts including the Base Object Model (BOM), Distributed

Simulation Engineering and Execution Process (DSEEP), and HLA Evolved. He is a co-author of several books, including "*C++Builder 6 Developer's Guide.*" He holds a Bachelor of Science degree in Computer Engineering from Old Dominion University and is a certified John Maxwell Team coach and speaker.

Holly Ann Heine Handley is an Assistant Professor in the Engineering Management and System Engineering Department at Old Dominion University in Norfolk, VA. Dr Handley applies systems engineering principles and experience in computational modeling to conduct research and perform analysis on challenging problems of complex organizational systems. Her education includes a BS in Electrical Engineering from Clarkson College (1984), an MS in Electrical Engineering from the University of California at Berkeley (1987) and an MBA from the University of Hawaii (1995). She received her PhD from George Mason University in 1999. Before joining ODU, Dr Handley worked as a Design Engineer for Raytheon Company (1984–1993) and as a Senior Engineer for the Pacific Science and Engineering Group (2002–2010), as well as in various academic research positions. Dr Handley is a Licensed Professional Electrical Engineer and a member of the Institute of Electrical and Electronic Engineers (IEEE) Senior Grade, the International Council on System Engineers (INCOSE) and Sigma Xi, the Scientific Research Society.

Heber Herencia-Zapana is a Research Scientist at the National Institute of Aerospace in Hampton, Virginia. He holds a PhD in Electrical and Computer Science Engineering from Old Dominion University. His topics of interest are formal methods research on aviation systems at NASA Langley Research Center, designing formal specifications of conflict detection and resolution of aircraft transportation system, proving theorems of formal specification of the detection and resolution of aircraft using the Prototype Verification System (PVS), and mathematical foundations of modeling and simulation.

Patrick T. Hester is an Assistant Professor of Engineering Management and Systems Engineering at Old Dominion University and a Principal Researcher at the National Centers for System of Systems Engineering. He received a PhD in Risk and Reliability Engineering at Vanderbilt University. His research interests include multi-attribute decision making under uncertainty, complex system governance, and decision making using modeling and simulation. He is a member of the Society for Modeling and Simulation International, Society for Judgment and Decision Making, and the International Society on Multiple Criteria Decision Making.

Robert C. Holcomb Jr is a Solution Architect for VT-MAK Technologies. He holds a MS in Modeling and Simulation from Old Dominion University and a BS in Software Engineering from the US Military Academy at West Point. He has significant practical experience supporting his company's worldwide distributed customers with modeling and simulation architectures, large scale software development, teaching development, and application of standards, including the integration of geospatial information of various formats into simulation systems.

Robert H. Kewley is a Professor and Department Head in the US Military Academy (USMA) Department of Systems Engineering. He is a Colonel in the US Army. His research interests include systems engineering and distributed simulation for integration of military systems. He has also done research in command and control systems. He has analytic experience as the Director of West Point's Operations Research Center and as an Analyst in the Center for Army Analysis. He has an MS in Industrial Engineering and a PhD in Decision Science and Engineering Systems, both from Rensselaer Polytechnic Institute.

Tag Gon Kim received his PhD in Computer Engineering with a specialization in Systems Modeling and Simulation from University of Arizona, Tucson, AZ, in 1988. He was an Assistant Professor at the Electrical and Computer Engineering, University of Kansas, Lawrence, KS, from 1989 to 1991. He joined the Electrical Engineering Department at the KAIST, Tajeon, Korea, in Fall 1991 and has been a full Professor in the Electrical Engineering and Computer Science Department since Fall 1998. He was the President of the Korea Society for Simulation (KSS) and the Editor-in-Chief for *Simulation: Transactions for Society for Computer Modeling and Simulation International* (SCS). He is a co-author of the textbook *Theory of Modelling and Simulation*, Academic Press, 2000. He has published about 200 papers in M&S theory and practice in international journals and conference proceedings. He is very active in research and education in defense modeling and simulation in Korea. He was/is a Technical Advisor for defense M&S in various Korean government organizations, including the Ministry of Defence, the Defence Agency for Technology and Quality, the Korea Institute for Defence Analysis, and the Agency for Defence Development. He developed a tools set, call DEVSimHLA, for HLA-compliant wargame models development, which has been used for the development of three military wargame models for the Navy, Air Force and Marine force in Korea. He is a Fellow of the SCS, a Senior Member of the Institute of Electrical and Electronics Engineers (IEEE) and Eta Kappa Nu.

J. David Lashlee is an Associate Director for Data at the US Department of Defense Modeling and Simulation Coordination Office in Alexandria, Virginia. He has 25 years of experience developing digital terrain databases for combat modeling and simulation applications. He has Graduate Degrees in Geographic Techniques, Physical Geography, and Earth Information Science and Technology. His dissertation research involved hyperspectral modeling of arid and tropical environments using laboratory, field, and imaging spectrometry data. His Postdoctorate study examined the appropriate use of digital terrain data for developmental and operational testing of Army battle command systems. In 2009, he served as Assistant Operations Officer at the Korea Battle Simulation Center during Exercise Key Resolve, a developmental assignment sponsored by the US Army Simulation Proponent Office. He is a Certified Mapping Scientist in both geographic information systems and remote sensing, a Level III Certified Member of the Acquisition Workforce, and a Certified Modeling and Simulation Professional (CSMP).

Kathryn Blackmond Laskey is an Associate Professor of Systems Engineering and Operations Research at George Mason University and Associate Director of the Center of Excellence in Command, Control, Communications, Computing and Intelligence (C4I Center). She received her PhD in Statistics and Public Affairs from Carnegie Mellon University in 1985, her MS degree in Mathematics from the University of Michigan in 1978, and her BS degree in Mathematics from the University of Pittsburgh in 1976. Laskey teaches and performs research on computational decision theory and evidential reasoning. Her research involves methods for representing knowledge in forms that can be processed by computers, extending traditional knowledge representation methods to represent uncertainty, eliciting knowledge from human experts, applying probability theory to draw conclusions from evidence arising from multiple sources.

Margaret L. Loper is the Chief Scientist for the Information and Communications Laboratory at the Georgia Tech Research Institute. She holds a PhD in Computer Science from the Georgia Institute of Technology, an MS in Computer Engineering from the University of Central Florida, and a BS in Electrical Engineering from Clemson University. Margaret's technical focus is parallel and distributed simulation, and she has published more than 50 papers as book chapters, journal contributions, or in conference proceedings. She is a senior member of the IEEE and ACM, and member of the Society for Modeling and Simulation. She is a founding member of the Simulation Interoperability Standards Organization (SISO) and received service awards for her work with the Distributed Interactive Simulation (DIS) and High Level Architecture (HLA) standards and the DIS/SISO transition. Her research contributions are in the areas of temporal synchronization, simulation testing, and simulation communication protocols.

Il-Chul Moon received his PhD in Computation, Organization, and Society from Carnegie Mellon University in 2008. He is an Assistant Professor at the Department of Industrial and Systems Engineering, KAIST, from 2011. His theoretic research interests include social network analysis, multi-agent modeling and simulation, game theory, machine learning, and artificial intelligence. His practical research interests include military C2 structure, counter-terrorism, and disaster modeling and simulation.

J. Russell Noseworthy received his PhD in Computer Systems Engineering from Rensselaer Polytechnic Institute in 1996. At Rensselaer, he researched distributed and real-time computing systems at the NASA Center for Intelligent Robotic Systems for Space Exploration. After working a year at Lockheed Martin's Distributed Processing Lab, he became employee number three of Object Sciences Corporation (OSC) in 1997. OSC, which he helped to make the second fastest growing technology company in Northern Virginia in 2004, was acquired by Science Application International Corporation in 2005. Since the year 2000, Dr Noseworthy has served as the Chief Software Engineer and the Development Team Leader for the TENA software development activity. For the three years prior to that, he performed the same role on the team that created the HLA

RTI-NG, a distributed simulation infrastructure. His experience helped to shape the evolution of work done with the HLA RTI-NG into TENA.

S. K. Numrich holds an AB, MA and PhD in Physics. Her experience includes underwater sound in the Arctic, fluid–structure interactions, parallel processing, modeling and simulation and virtual reality. Recent studies focused on irregular warfare, the impact of cultural awareness on military operations, and culturally "aware" modeling and simulation tools. She is a Research Staff Member at the Institute for Defense Analyses.

Tuncer Ören is a Professor Emeritus at the University of Ottawa in Canada. He has been involved in simulation since the early 1960s. He authored/co-authored over 450 publications and has been active in about 380 conferences and seminars held in 30 countries. In the 2000s alone, he has been a keynote, plenary, or invited speaker or was involved at honorary positions in over 50 conferences/seminars. In 2011, he was inducted to the Society for Modeling and Simulation International Hall of Fame, a Lifetime Achievement Award, with the following citation: "For his outstanding contributions to the field, particularly new M&S methodologies and synergies that increase the effectiveness and broaden the application of simulation, and a Code of Ethics that have significantly impacted the use of simulation throughout the world."

José J. Padilla is a Research Scientist with the Virginia Modeling, Analysis and Simulation Center (VMASC) at Old Dominion University, Suffolk, VA. He received his PhD in Engineering Management from Old Dominion University. He also holds a BSc in Industrial Engineering from la Universidad Nacional de Colombia, Medellín, Colombia, and a Master of Business Administration from Lynn University, Boca Raton, Florida. His research interest is on the nature of the processes of understanding and interoperability and their implications in human social culture behavior (HSCB) modeling.

James Panagos is the Founder and President of Gnosys Systems Incorporated, a small company applying artificial intelligence and other advanced software techniques for challenging command and control computer simulation problems. He is one of the pioneers in the early work on computer generated forces through the SIMNET project and was the lead in the first successfully fielded commercial computer generated forces system a few years later—the Leopard II tactical tank trainer. Subsequently, he has worked on many large Department of Defense initiatives such as Joint Precision Strike Demonstration, Synthetic Theater of War, Warfare Simulation, and the What-If Simulation System for Advanced Research and Development. He presently researches and develops techniques for virtual simulations in the areas of combat behaviors and multi-resolution modeling. He is also working on first-principles simulations of electron beams and microwave device design tools. He received his SM Degree from MIT in 1985.

Mikel D. Petty is the Director of the University of Alabama in Huntsville's Center for Modeling, Simulation, and Analysis and a Research Professor in both the Computer Science and the Industrial and Systems Engineering and Engineering Management Departments. Prior to joining UAH, he was Chief Scientist at

Old Dominion University's Virginia Modeling, Analysis, and Simulation Center and Assistant Director at the University of Central Florida's Institute for Simulation and Training. He received a PhD in Computer Science from the University of Central Florida in 1997. Dr Petty has worked in modeling and simulation research and development since 1990 in areas that include simulation interoperability and composability, human behavior modeling, multi-resolution simulation, and applications of theory to simulation. He has published over 160 research papers and has been awarded over $14 million in research funding. He served on a National Research Council committee on modeling and simulation, is a Certified Modeling and Simulation Professional, and is an Editor of the journals *SIMULATION* and *Journal of Defense Modeling and Simulation*. While at Old Dominion University he was the Dissertation Advisor to the First and Third students in the world to receive PhDs in Modeling and Simulation and is currently Coordinator of the M&S degree program at UA Huntsville.

P. M. Picucci is a Research Staff Member at the Institute for Defense Analyses. He holds an MA in National Security Studies from California State University, San Bernardino, and a PhD in Political Science from the University of Kansas. His primary research efforts have centered on non-traditional conflict (irregular warfare and terrorism) and the use of computerized content analysis for the study of Islamic terrorism.

Edward T. Powell is a Senior Scientist and Program Manager for SAIC. Currently, he is the Lead Architect for the Test and Training Enabling Architecture. After receiving his PhD in Astrophysics from Princeton University, he worked for the Lawrence Livermore National Laboratory performing simulation-based analysis. He moved to SAIC in 1994, and participated as Lead Architect in some of the most complex distributed simulation programs in the DoD, including the Joint Precision Strike Demonstration (JPSD), the Synthetic Theater of War (STOW), and the Joint Simulation System (JSIMS). He then worked in the intelligence community for two years on architectures for integrating large scale diverse ISR systems. He is working on expanding the applicability of TENA, and integrating multiple interoperability architecture approaches using ontology-based systems.

William T. Richards is a Project Scientist for the Virginia Modeling Analysis and Simulation Center in Suffolk, Virginia. He holds an MS and a BS from Christopher Newport University in Newport News, Virginia. He has worked in the field of modeling and simulation since 1995 concentrating in automated geospatial terrain database generation and software development. He has supported numerous military simulation exercises such as Cobra Gold, Unified Endeavor and Tandem Thrust. As a result he is familiar with several different simulation systems that require and use geospatial terrain data. He has also written a number of papers regarding the use of terrain and environmental data in simulations.

Roger D. Smith is the Chief Technology Officer for the Nicholson Center for Surgical Advancement at Florida Hospital. He previously served as the Chief Technology Officer for US Army Simulation, Training and Instrumentation (PEO

STRI) and Research Scientist for Texas A&M University. He is applying simulation and related technologies to surgical education and military training. He has published three books on simulation, created multiple commercial simulation courses, published over 150 papers and presentations on technical and management topics, and has served on the faculties of four universities. Dr Smith holds a BS in Applied Mathematics, MS in Statistics, Master's and Doctorate in Business Administration, and PhD in Computer Science.

Joe Sorroche is a Principal Member of the Technical Staff for Sandia National Laboratories. He currently works in the Flight and Embedded Software/Simulation Group supporting satellite systems design, modeling, and simulation. Previously, Mr Sorroche worked as a Senior Systems Engineer with Arctic Slope Regional Corporation Communications (ASRCC) for 15 years at the USAF Distributed Missions Operations Center (DMOC) for the System Architecture Group. He was the DMOC Engineering lead for JEFX, Blue Flag, and Virtual Flag exercises. He chaired the SISO TADIL TALES Product Support Group, the Link 11/11B Product Development Group, and was the SISO Liaison for the NATO Tactical Data Link Interoperability Testing Syndicate. Mr Sorroche has Bachelors and Masters of Science Degrees in Electrical Engineering from New Mexico State University. He is a member of the IEEE, and Tau Beta Pi and Eta Kappa Nu Honor Societies.

Charles D. Turnitsa is a Research Scientist with the Virginia Modeling, Analysis and Simulation Center (VMASC) at Old Dominion University, Suffolk, VA. He received his BS in Computer Science (1991) from Christopher Newport University (Newport News, Virginia), and his MS in Modeling and Simulation (2006) from ODU. He is a PhD candidate in the Modeling and Simulation program. He authored or co-authored multiple conference papers and book chapters on combat modeling and distributed simulation. He also supported teaching of Principles of Military Modeling and Simulation with focus on the history. Prior to his work at VMASC, he spent a decade doing research for the US Army and NASA in the areas of data interoperability, knowledge representation and modeling. Most of his life he has enjoyed the hobby of military wargaming, including activities such as publishing, organizing national events, and game design and development.

Robert L. Wittman Jr is a Principal in Modeling and Simulation for the MITRE Corporation in Orlando, Florida, and is the Chief Architect for the Army's OneSAF Simulation. He holds a PhD in Industrial Engineering/Interactive Simulation from the University of Central Florida, an MS in Software Engineering from the University of West Florida, and a BS in Computer Science from Washington State University. He has supported US DoD simulation development across the training, experimentation, analysis, and testing domains over the past 20 years. He continues to lead the evolution of the Military Scenario Definition Language (MSDL) international standard and has played critical roles in its initial development as Vice-Chair and Co-Editor on the Simulation Interoperability Standards Organization (SISO) MSDL Product Development Group.

Marc D. Wood served as an Aviation Officer in the US Army for 11 years in a variety of assignments, where he experienced the value of combat simulation as a training tool. After earning a Master of Engineering Degree in Industrial and Systems Engineering from Texas A&M University, he joined West Point's Department of Systems Engineering where he now teaches combat modeling and serves as an Analyst in the Operations Research Center.

Levent Yilmaz is an Associate Professor of Computer Science and Software Engineering and holds a joint appointment with the Industrial and Systems Engineering at Auburn University. He received his MS and PhD degrees from Virginia Tech. His research interests are in modeling and computer simulation, agent-directed simulation, and complex adaptive systems. He serves as the Editor-in-Chief of *Simulation: Transactions of the Society for Modeling and Simulation International* and is the founding organizer and General Chair of the annual Agent-Directed Simulation conference series.

Acknowledgments

$\mathbf{M}$y first thank you note must go to all chapter authors who contributed, in particular, to Part IV of this book. Most of them served in several roles, as they were not only chapter authors, but also peer reviewers and editorial supporters. The mutual support of all authors in improving the chapters of this book helped to make it become a compendium of knowledge capturing the state of the art in combat modeling and distributed simulation as simulation engineers should know it.

This book could not have been written without the help of many additional peer reviewers and expert discussions conducted in support of writing the various chapters. I therefore would like to thank all colleagues and friends who helped in contributing recommendations for improvement to at least one of the chapters. Your constructive criticism and innovative ideas are hopefully shining through. Particular thanks go to:

Robert Aaron, US Army Test and Evaluation Command

Khaldoon Al-Harthi, Carleton University

Lisa J. Bair, Weisel Science and Technology Corporation

Osman Balci, Virginia Tech

Catherine M. Banks, Old Dominion University

Jerry M. Couretas, Lockheed Martin Corporation

Paul K. Davis, RAND Corporation

Paul Fishwick, University of Florida

Johnny J. Garcia, SimIS Inc.

Randall B. Garrett, Northrop Grumman Technical Services

Dean S. Hartley III, Hartley Consulting

Jean-Louis Igarza, Antycip Simulation Inc.

Max Karlström, BAE Systems Sweden

Charles B. Keating, Old Dominion University

Thomas Kreitmair, NATO Consultation, Command and Control Agency (NC3A)

Staffan Löf, Pitch

Thomas W. Lucas, Naval Postgraduate School

Robert R. Lutz, Johns Hopkins University

Katherine L. Morse, Johns Hopkins University

Eckehard Neugebauer, Industrieanlagen Betriebsgesellschaft m.b.H. (IABG)

Karl-Heinz Neumann, Industrieanlagen Betriebsgesellschaft m.b.H. (IABG)

Q.A.H. (Mimi) Nguyen, NATO Consultation, Command and Control Agency (NC3A)

Michael Proctor, Institute for Simulation Technology

Paul F. Reynolds Jr, University of Virginia

Dave Taylor, Medium Extended Air Defense System (MEADS) International, Inc.

James A. Wall, Texas A&M University

My personal thanks go to **Kim B. Sibson** who read in minuscule detail through all my chapters making legions of corrections to ensure that my German roots and interesting interpretations of the English language will not get into the way of understanding the book. She also ensured that the writing style remained consistent over the long time of compiling and writing these chapters.

My personal thanks go furthermore to **Colonel Roy Van McCarty**. His military expertise and feedback on how to describe conceptual and technical challenges of combat modeling and distributed simulation better for master level students and practitioners in the field helped to improve the chapters.

Last but not least, I thank Haseen Khan and her team at Laserwords for the minuscule detail of the final editing process. Their contributions to creating the index, aligning the reference styles, and turning this book from a good idea into a professional product is very much appreciated.

Abbreviations

AAR	After Action Review
ABCS	Army Battle Command Systems
ABL	Airborne Laser
AC	Aircraft
ACASA	Ackoff Collaboratory for Advancement of Systems Approach
ACE	Adaptive Communication Environment
ACM	Association for Computing Machinery
ACSIS	Army C4ISR and Simulation Initialization System
ACTF-MRM	Army Constructive Training Federation Multi-Resolution Modeling
ADS	Agent Directed Simulation
AFDRG	Anglo French Defence Research Group
AFQT	Armed Forces Qualifying Test
AFV	Armored Fighting Vehicale
AI	Artificial Intelligence
AIDZ	Air Defense Identification Zone
AIEE	American Institute of Electrical Engineers
ALSP	Aggregate Level Simulation Protocol
AMC	Army Material Command
AMM	Army Mobility Model
AMSAA	Army Material Systems Analysis Activity
AMSO	Army M&S Office
AMT	Architecture Management Team
ANSI	American National Standards Institute
API	Application Programming Interface
APWG	ATCCIS Permanent Working Group
ARE	Acronym Rich Environment
ARPANET	Advanced Research Project Agency Network
ASCII	American Standard Code for Information Interchange
ASD	Assistant Secretary of Defense

ASOC	Aerospace Operation Center
ASVAB	Armed Services Vocational Aptitude Battery
AT&L	Acquisition, Technology, and Logistics
ATCCIS	Army Tactical Command and Control Information System
ATCCS	Army Tactical Command & Control Systems
ATDL	Army Tactical Data Link
AV	All View or All Viewpoint
AVT	Applied Vehicle Technology
AWACS	Airborne Warning and Control System
AWSIM	Air Warfare Simulation
BCMS	Battle Command Management Services
BDZ	Base Defense Zone
BFT	Blue Force Tracker
BIP	Battlefield Interoperability Protocol
BLOS	Beyond Line-of-Sight
BML	Battle Management Language
BN	Bayesian Network
BOM	Base Object Model
BPS	Bits per Second
BRIMS	Behavior Representation in Modeling and Simulation
C2	Command and Control
C2IS	Command and Control Information System
C3	Command, Control, and Communication (US)
C3	Command, Control, and Consultation (NATO)
C4	Command, Control, Communication, and Computers
C4I	Command, Control, Communication, Computers, and Intelligence
C4ISR	Command, Control, Communication, Computers, Intelligence, Surveillance, and Reconnaissance
CADM	Core Architecture Data Model
CAESAR	Computer Aided Evaluation of System Architectures
CAOC	Combined Aerospace Operations Center
CATT	Combined Arms Tactical Trainer
CAX	Computer Assisted Exercise
C-BML	Coalition-Battle Management Language
CBR	Chemical, Biological and Radiological
CBS	Corps Battle Simulation
CBT	Computer-based Training
CC	Component Commander
CCRP	Command and Control Research Program
CCRTS	Command and Control Research and Technology Symposium
CCTT	Close Combat Tactical Training
CEIT	Commanders Exercise Initialization Toolkit
CEP	Circular Error Probable
CEPA	Common European Priority Area

CES	Core Enterprise Services
CGF	Computer Generated Forces
CIS	Communication and Information System
CIWS	Closed-In Weapon System
CJTF	Combined Joint Task Force
CLCS	Company Level Constructive Simulation
CMNTS	Comments
CMO	Civil Military Operations
CMSD	Core Manufacturing Simulation Data
CNI	Communications, Navigation and Identification
CNN	Cable News Network
COA	Course of Action
COBP	Code of Best Practice
COE	Common Operating Environment
COI	Community of Interest
COIN	Counter Insurgency
COL	Colonel
COMPOEX	Conflict Modeling, Planning and Outcome Experimentation
COMSEC	Communications Security
CONCEN	Control Center
COPB	Code of Best Practice
CORBA	Common Object Request Broker Architecture
COTS	Commercial-Off-The-Shelf
CPOF	Command Post of the Future
CPT	Conditional Probability Table
CPU	Central Processing Unit
CRC	Control and Reporting Centre
CRM	Common Reference Model
CSI	Common System Interface
CSPI	COTS Simulation Package Interoperability
CTAPS	Contingency Theater Automated Planning System
CTPS	Combat Trauma Patient Simulation
CV	Capability Viewpoint
DAG	Directed Acyclic Graph
DARPA	Defense Advanced Research Projects Agency
DATTRF	Data Transfer
DCR	DOTMLPF Change Request
DDCA	Distributed Debrief Control Architecture
DDM	Data Distribution Management
DDMS	DoD Discovery Metadata Specification
DDS	Data Distribution Service
DEC	Digital Equipment Corporation
DEM	Data Exchange Mechanism, also Digital Elevation Model
DEVIL	Demonstrator for the Connection of Virtual and Live Simulation

DEVS	Discrete Event Systems Specification
DEVSML	DEVS Markup Language
DEVSSOA	DEVS Service Oriented Architecture
DGA	Direction Générale de l'Armement
DI	Data Initiative
DIAS	Dynamic Information Architecture System
DII	Defense Information Infrastructure
DIME	Diplomatic, Information, Military, and Economic
DIS	Distributed Interactive Simulation
DISA	Defense Information Systems Agency
DIV	Data and Information Viewpoint
DLC	Dynamic Link Compatibility
DM	Data Management
DMAO	DSEEP Multi-Architecture Overlay
DMIF	Dynamic Multi-User Information Fusion
DMSO	Defense Modeling and Simulation Office
DMT	Distributed Mission Training
DNDAF	Department of National Defence and the Canadian Forces Architecture Framework
DoD	Department of Defense
DoDAF	Department of Defense Architecture Framework
DOT&E	Director, Operational Test & Evaluation
DOTMLPF	Doctrine, Organization, Training, Material, Leadership, Personnel, and Facilities
DP	Dimensional Parameter
DR	Disaster Relief
DRM	Data Replication Mechanism
DS	Direct Support
DSB	Defense Science Board
DSCA	Defense Support to Civilian Authorities
DSEEP	Distributed Simulation Engineering and Execution Process
DTDMA	Distributed Time Division Multiple Access
DTED	Digital Terrain Elevation Data
DTRA	Defense Threat Reduction Agency
DUSA-OR	Deputy Under Secretary of the Army—Operations Research
EDAC	Error Detection and Correction
EDCS	Environmental Data Coding Standard
EDLC	Evolved DLC
EFFBD	Enhanced Functional Flow Block Diagram
EMBR	Enterprise Metacard Builder Resource
EMF	Exercise Management & Feedback
ENIAC	Electronic Numerical Integrator and Computer
ENM	EPLRS Network Manager
EOB	Electronic Order of Battle

EOI	Event of Interest
EP	Exercise Process
EPLRS	Enhanced Position Location Radio Set
ERC	Environmental Runtime Component
ERM	Entity Relationship Model
ESRI	Environmental Systems Research Institute
ETO	Exercise/Training Objectives
EU	EPLRS Unit
EXCEN	Exercise Center
EXCON	Exercise Control
EXDIR	Exercise Director
EXPLAN	Exercise Plan
FACC	Feature and Attribute Coding Catalogue
FACET	Framework for Addressing Cooperative Extended Transactions
FAO	Foreign Area Officer
FAX	Facsimile
FEAT	Federation Engineering Agreements Template
FEBA	Forward Edge of the Battle Area
FEDEP	Federation Development and Execution Process
FIT	Führung und Informations-Technologie
FMCW	Frequency Modulated Continuous-Wave
FOM	Federation Object Model
FPU	Forwarding Participating Unit
FRAGO	Fragmentary Order
FRU	Forwarding Reporting Unit
GCC	Geocentric Coordinate system
GCCS	Global Command and Control System
GDC	Geodetic Coordinate system
GES	GIG Enterprise Services
GFM	Global Force Management
GFMIEDM	Global Force Management Data Information Exchange Model
GIG	Global Information Grid
GIS	Geographic Information System
GLOBE	Global Leadership and Organizational Behavior Effectiveness
GM VV&A	General Model for VV&A
GOIS	Geo-referenced Environment, Object and Infrastructure Service
GOTS	Government-Off-the-Shelf
GPS	Global Positioning System
GPU	Graphics Processing Unit
GRASS	Geographic Resources Analysis Support System
GRIM	Guidance, Rationale, and Interoperability Modalities
GS	General Support
GSP	Goals, Standards, and Preferences
GSR	General Support Reinforcing

HA	Humanitarian Assistance
HBR	Human Behavior Representation
HF	High Frequency
HFM	Human Factors & Medicine
HICON	Higher Control Cell
HIDACS	High-Density Airspace Control Zone
HLA	High Level Architecture
HQ	Headquarter
HSCB	Human, Social, Cultural, and Behavioral
HTTP	Hypertext Transfer Protocol
IACM	Information Age Combat Model
IBS	Integrated Broadcast Service
IDA	Institute for Defense Analysis
IEC	International Electrotechnical Commission
IED	Improvised Explosion Device
IEEE	Institute of Electrical and Electronics Engineers
IEJU	Initial Entry JTIDS Unit
IER	Information Exchange Requirement
IFF	Identification Friend and Foe
IIDBT	Integrated Interactive Data Briefing Tool
IIE	Institute of Industrial Engineers
IJMS	International Journal of Modelling and Simulation
IMC	Information Management Committee
INCOSE	International Council on Systems Engineering
INFORMS	Institute for Operations Research and the Management Sciences
IO	International Organizations
IP	Internet Protocol
ISAAC	Irreducible Semi-Autonomous Adaptive Combat
ISAF	International Security Assistance Force
ISO	International Organization for Standardization
ISR	Intelligence, Surveillance and Reconnaissance
IST	Information Systems Technology
IT	Information Technology
ITEC	International Training and Education Conference
IVT	Interface Verification Tool
IW	Information Warfare
IWARS	Infantry Warrior Simulation
JC3IEDM	Joint Consultation, Command and Control Information Exchange Data Model
JCA	Joint Capability Area
JCAS	Joint Close Air Support
JCATS	Joint Conflict and Tactical Simulation
JDL	Joint Directors of Laboratories
JDMS	Journal of Defense Modeling and Simulation

JEF	Joint Experimental Federation
JFCOM	Joint Forces Command
JIPOE	Joint Intelligence Preparation of the Operational Environment
JLVC	Joint Live Virtual Constructive
JMETC	Joint Mission Environment Test Capability
JMRF	Joint Multi-resolution Federation
JMRM	Joint Multi-resolution Model
JNTC	Joint National Training Capability
JOC	Joint Operating Concept
JOPP	Joint Operation Planning Process
JP	Joint Publication
JPEG	Joint Photographic Experts Group
JS	Joint Staff
JSAF	Joint Semi-Automated Forces
JTC	Joint Technical Committee
JTDS	Joint Training Data System
JTIDS	Joint Tactical Information Distribution System
JTLS	Joint Theatre Level Simulation
JTRS	Joint Tactical Radio System
JTT	Joint Tactical Terminal
JU	JTIDS unit
JVMF	Joint Variable Message Format
JWID	Joint Warrior Interoperability Demonstrator
KISS	"Keep it short and simple," also "Keep it simple and stupid"
KSA	Knowledge, Skills, and Abilities
LAN	Local Area Network
LCIM	Levels of Conceptual Interoperability Model
LEP	Linear Error Proble
LIDAR	Light Detection And Ranging
LLC	Link Level COMSEC
LLTR	Low-Level Transit Route
LOCON	Lower Control Cell
LOS	Line-of-Sight
LPD	Local Probability Distribution
LPI	Low Probability Interference
LRC	Local RTI Component
LROM	Logical Range Object Model
LTDP	Long-Term Defense Plan
LVC	Live Virtual Constructive
M&S	Modeling and Simulation
MALO	Mission, Area, Level, and Operator
MAMID	Methodology for Analysis and Modeling of Individual Differences
MANA	Map Aware Non-uniform Automata
MASON	Multi-Agent Simulator Of Neighborhoods

MATREX	Modeling Architecture for Technology, Research, and Experimentation
MBDE	Model-based Data Engineering
MCE	Modular Control Equipment
MCI	MIP Common Interface
MDA	Model Driven Architecture
MEBN	Multi-entity Bayesian Network
MEL	Main Event List
MEM	Message Exchange Mechanism
METOC	Meteorology and Oceanography
MGRS	Military Grid Reference System
MHS	Message Handling Service
MIDS	Multi-Function Information Distribution System
MIL	Master Incident List
MIL-STD	Military Standard
MIP	Multilateral Interoperability Program
MISS	McLeod Institute of Simulation Sciences
MIT	Massachusetts Institute of Technology
MLRS	Multiple Launch Rocket System
MMOG	Massively Multiplayer Online Games
MOC	Maritime Operations Center
MoCE	Measures of C2 Effectiveness
MoD	Ministry of Defence
MoDAF	Ministry of Defence Architecture Framework
MOE	Measure of Effectiveness
MoFE	Measures of Force Effectiveness
MOM	Management Object Model
MOOTW	Military Operations other than War
MOP	Measure of Performance
MoPE	Measures of Policy Effectiveness
MOPP	Mission Oriented Protective Posture
MORS	Military Operations Research Society
MOS	Military Occupational Specialty
MOU	Memorandum of Understanding
MRCI	Modular Reconfigurable C4I Interface
MRM	Multi-Resolution Modeling
MRR	Minimum-Risk Route
MRT	Mission Rehearsal Training
MSC-DMS	M&S COI Discovery Metadata Specification
MSCO	Modeling and Simulation Coordination Office
MSDL	Military Scenario Definition Language
MSEC	Message Security Encryption Code
MSG	Modeling and Simulation Group
MSGID	Message Indentifyer

MSIAC	Modeling and Simulation Information Analysis Center
MSIS	Modeling and Simulation Information System
MSRR	Modeling and Simulation Resource Repository
MTF	Message Text Format
MTI	Moving Target Indicator
MTS	Message Transceiver Service
MTWS	Marine Tactical Warfare Simulation
MUAV	Multi-UAV
NAF	NATO Architecture Framework
NASA	National Aerospace Agency
NATO	North Atlantic Treaty Organization
NBC	Nuclear, Biological, and Chemical
NCES	Net-Centric Enterprise Services
NCO	Net-centric Operations
NCS	Net Control Station
NDA	NHQC3S Data Administration
NDRM	NATO Data Replication Mechanism
NEC	Network Enabled Capability
NECC	Net Enabled Command Capability
NGO	Non-Governmental Organization
NHQC3S	NATO Headquarters Consultation, Command and Control Systems
NILE	NATO Improved Link Eleven
NIMS	National Incident Management System
NIST	National Institute of Standards and Technology
NLP	Natural Language Processing
NMSG	NATO Modeling and Simulation Group
NOEM	National Operational Environment Model
NOS	Not Otherwise Specified
NPG	Network Participation Group
NRC	National Research Council
NRL	Naval Research Laboratory
NRT	Near-Real-Time
NTR	Network Time Reference
NWARS	National Wargaming System
NWDC	Navy Warfare Development Command
OA	Operational Analysis
OASES	Ocean, Atmosphere, and Space Environmental Services
OBS	Order of Battle Service
OCC	Ortony, Clore, and Collins
OCE	Officer Conducting the Exercise
OCS	Organic Communication Service
ODE	Officer Directing the Exercise
OIPT	Overarching Integrated Product Team
OMC	Object Model Compiler

OMG	Object Management Group
OMT	Object Model Template
OneSAF	One Semi-Automated Forces
ONR	Office of Naval Research
OOB	Order of Battle
OODA	Observe, Orient, Decide, and Act
OOS	OneSAF Operational Systems
OOTW	Operations other than War
OP	Operation
OPFOR	Opposing Force
OPNET	Operations Network
OPORD	Operational Order
OR	Operations Research
ORB	Object Request Broker
ORBAT	Order of Battle
ORCEN	OR Center
ORM	Other Reference Model
ORMT	Object Reference Model Template
OSE	Officer Specifying the Exercise
OSI	Open Systems Interconnection
OTH	Over the Horizon
OV	Operational View or Operational Viewpoint
OWL	Web Ontology Language
OWL-DL	OWL Description Logic
P&R	Personnel and Readiness
PA&E	Program, Analysis, and Evaluation
PACOM	Pacific Command
PADS	Principles of Advanced Distributed Simulation
PASS	Publish and Subscribe Service
PDG	Product Development Group
PDP	Programmed Data Processor
PDU	Protocol Data Unit
PEO	Program Executive Office
PEO-STRI	PEO for Simulation, Training, and Instrumentation
PfP	Partnership for Peace
PFS	Priority First Search
PIC	Public Information Center
PM	Project Manager
PMESII	Political, Military, Economic, Social, Information, and Infrastructure
PMF	Performance Moderator Function
PO	Probabilistic Ontology
POMC	Probabilistic Ontology Modeling Cycle
PPF	1. Platform Proto-Federation
PPI	Planned Position Indicator

PPLI	Precise Position Location Indicators
PSG	Product Support Group
PSI	Political Science-Identity
PSP	Playstation Portable
PU	Participating Unit
PV	Product Viewpoint
PVO	Private Volunteer Organizations
QIP	Quatroliteral Interoperability Protocol
RAM	Rolling Airframe Missile
RC	Response Cell
RDECOM	Research, Development and Engineering Command
RDF	Resource Description Framework
RDFS	RDF-Schema
RESA	Research, Evaluation, and System Analysis
REVVA	Referential for Verification, Validation, and Accreditation
RHI	Range Height Indicator
RIF	Rule Interchange Format
RLS	Real-Life Support
RMI	Remote Method Invocation
ROA	Restricted Operations Area
RPC	Remote Procedure Call
RPDM	Recognition Prime Decision Module
RPG	Recommended Practice Guide
RPR-FOM	Realtime Platform Reference FOM
RTA	Research and Technology Agency
RTB	Research and Technology Board
RTI	Runtime Infrastructure
RTO	Research and Technology Organization
RTP	Research Technology Program
RTT	Round Trip Timing
RU	Reporting Unit
RV	Random Variable
SAC	Standards Activity Committee
SADL	Situational Awareness Data Link
SAF	Semi-Automated Forces
SAGE	Semi Automatic Ground Environment
SAIC	Science Applications International Corporation
SAM	Surface-to-Air Missile
SANDS	Situation Awareness and Display
SAS	System Analysis & Studies
SBIR	Small Business Innovation Research
SCI	Systems Concepts & Integration
SCM	Simulation Conceptual Modeling
SCS	Society for Modeling and Simulation

SDO	Stateful Distributed Object
SDU	Secure Data Unit
SE	Synthetic Environment
SEAD	Suppression of Enemy Air Defense
SEAS	System Effectiveness Analysis Simulation
SEDEP	Synthetic Environment Development and Exploitation Process
SEDRIS	Synthetic Environment Data Representation and Interchange Specification
SET	Sensors & Electronics Technology
SIMCI	Simulation to C4I Interoperability
SIMNET	Simulator Networking
SISO	Simulation Interoperability Standards Organization
SITFOR	Situation Forces
SIW	Simulation Interoperability Workshop
SMART	Simulation and Modeling for Acquisition, Requirements, and Training
SME	Subject Matter Expert
SNC	System Network Controller
SOA	Service Oriented Architecture
SOAP	Simple Object Access Protocol
SOAR	State, Operator, and Results
SOM	Simulation Object Model
SOPES	Shared Operational Picture Exchange Services
SORASCS	Service Oriented Architecture for Socio-Cultural Systems
SPARQL	Simple Protocol and RDF Query Language
SPI	Simulation Publications Incorporated
SRF	Spatial Reference Frame
SRM	Spatial Reference Model
SRML	Scenario Reference Markup Language
SRTM	Shuttle Radar Topography Mission
SSBN	Situation-specific Bayesian Network
SSG	Standing Support Group
SSSB	Ship Shore Ship Buffer
SSTR	Secure, Stabilize, Transition, and Reconstruction
STANAG	Standardization Agreement
StdV	Standards Viewpoint
STF	SEDRIS Transmittal Format
STMS	Soldier Tactical Mission System
STTR	SBIR and Technology Transfer Research
SV	Systems View or Systems Viewpoint
SvcV	Services Viewpoint
SWRL	Semantic Web Rule Language
SysML	System Modeling Language
TA	Training Audience
TACC	Tactical Air Control Center

TACS	Theater Air Control System
TADIL	Tactical Digital Information Link
TADIXS	Tactical Data Information Exchange Subsystem
TADL	Tactical Data Link
TALES	Technical Advice and Lexicon for Enabling Simulation
TAO	Tactical Air Operations
TAOC	Tactical Air Operations Center
TBM	Theatre Ballistic Missile
TCP	Transmission Control Protocol
TDDS	TRAP Data Dissemination System
TDL	TENA Definition Language
TDMA	Time Division Multiple Access
TEK	Traffic Encryption Key
TENA	Test and Training Enabling Architecture
TIBS	Tactical Information Broadcast Service
TIDE	TENA Integrated Development Environment
TIN	Triangular Information Network
TLCS	Theater Level Constructive Simulation
TMI	Timing Master Initiator
ToA	Terms of Acceptance
TOE	Table of Equipment
ToVV	Terms of Verification and Validation
TPED	Task/Process/Exploit/Disseminate
TPPU	Task/Post/Process/Use
TRAC	TRADOC Analysis Center
TRADOC	Training and Doctrine Command
TRANSEC	Transport Security
TRAP	Tactical Related Applications
TRCE	TENA in a Resource Constrained Environment
TRIXS	Tactical Reconnaissance Intelligence Exchange System
TRMC	Test Resource Management Center
TSA	Transportation Security Administration
TSEC	Transmission Security Encryption Code
TT	Training Team
TTP	Tactics, Techniques, and Procedure
TV	Technical View
UAV	Unmanned Air Vehicles
UDDI	Universal Description, Discovery and Integration
UDP	User Datagram Protocol
UHF	Ultra High Frequency
UML	Unified Modeling Language
UMP-ST	Uncertainty Modeling Process for Semantic Technologies
UNITDES	Unit Description
UNITLOC	Unti Location

UPDM	Unified Profile for DoDAF/MODAF
URI	Uniform Resource Identifier
USAF	United States Air Force
USD A&T	Under Secretary of Defense for Acquisition and Technology
USGS	United States Geological Survey
USMC	United States Marine Corps
USMTF	United States Message Text Format
UTM	Universal Transverse Mercator
V&V	Verification and Validation
VHF	Very High Frequency
VIDSVC	Video Service
VMASC	Virginia Modeling Analysis and Simulation Center
VMF	Variable Message Format
VOB	Visitor Office Bureau
VOCSVC	Voice Service
VR	Virtual Reality
VRS	Vortex Ring State
VV&A	Validation, Verification, and Accreditation
WAN	Wide Area Network
WARNO	Warning Order
WEAG	Western European Armaments Group
WEAO	Western European Armaments Organisation
WEU	Western European Union
WEZ	Weapon Engagement Zone
WFZ	Weapon Free Zone
WRC	WEAG Research Cell
WSC	Winter Simulation Conference
WSDL	Web Service Defintion Language
WSMR	White Sands Missile Range
XML	Extensible Markup Language
XMSF	Extensible Modeling and Simulation Framework
XSD	XML Schema Definition

Chapter 1

Challenges of Combat Modeling and Distributed Simulation

Andreas Tolk

SIMULATION ENGINEERS AND THEIR MULTIPLE-VIEW CHALLENGES

There are many good books written on the topics of combat modeling and distributed simulation and most of them are used as references in this book, so why write another book on this topic? The reason is simple: while all other books in this domain successfully highlight special topics in detail none of them compiles the basic knowledge of all contributing fields that a simulation engineer needs to be aware of, in particular when he or she starts a job. To my knowledge, none of the existing books give the required holistic overview of combat modeling and distributed simulation needed to get a primary understanding of the challenges.

An editorial remark: in this book I will address the simulation engineer in the male form using he or his as pronouns—without implying that female engineers are less common or qualified; I just prefer to address the engineer simply as he instead of complex he/she combinations or the impersonal it.

There are good books that focus on various topics of combat modeling, but they do not look at what modeling decisions mean for the ensuing simulation task(s). Other books introduce the engineer to the specifics, such as specific Institute of Electrical and Electronics Engineers (IEEE) standards for simulation interoperability, but they assume that no modeling needs to be done by the engineer doing the integration, although it will be shown in this book that understanding the underlying models is pivotal to making sure that the resulting simulation federations are valid. Other books on validation and verification do not address the combat modeling tasks or the distributed simulation task in detail, but all three topics have to go hand in hand. Finally, the operational analysis community developed many valuable insights that can be reused by a simulation engineer but the applicability of their results to support combat modeling and distributed simulation is not dealt with explicitly, as simulation is perceived as just one powerful tool within a set of alternative tools. For the simulation engineer, this knowledge will be educational when confronted with such broader views that need to be supported by work.

In addition to all these technical aspects, the simulation engineer has to understand the customer: the soldiers and decision makers in the military domain. The various terms and their semantics of language need to be understood. An understanding of how soldiers fight and conduct their operations is required in order to understand how to support their training and their real-world operations. Tactical, operational, and strategic principles must also be understood in order to model entities and their behavior and relations in the situated battlefield. These operations grow in complexity and are conducted on a global scale. The same is true for training and support of operations, and the simulation engineer is the person who needs to ensure that the right systems are interconnected providing the appropriate capabilities in an orchestrated fashion. An understanding of the problem is needed along with the conceptualizations of supporting models, the implementation details of the resulting simulation systems, and the applicable standards supporting the task of integrating the systems to support the soldiers. Applicable solutions need to be identified as well as selecting the best set for the task, composing them into a coherent solution, and orchestrating their execution. All aspects will be addressed in the upcoming chapters and the state of the art and current research will be presented in the invited chapters written by internationally recognized experts of their special domains. In other words: the simulation engineer has to understand the *operational foundations* of the supported domain, the *conceptual foundations* required for the modeling, and the *technical foundations* of implementing and composing simulation systems.

The chapters of this book are oriented at education needs identified while teaching various courses on 'Engineering Principles of Combat Modeling and Distributed Simulation' and related topics in graduate programs and professional tutorials. They address the needs of graduate and postgraduate students in the engineering and computer science fields as well as scholars and practitioners in the field. To address the variety of challenges and recommended solutions, the book is divided in four parts: Foundations addresses the operational aspects, Combat Modeling addresses the conceptual aspects, Distributed Simulation addresses the technical aspects, and Advanced Topics.

- The *Foundations* section provides a consistent and comprehensive overview of relevant topics and recommended practices for graduate students and practitioners. It provides an initial understanding necessary to cope with challenges in the domain of combat modeling and distributed simulation.

- The *Combat Modeling* section focuses on the challenges of modeling the core processes of *move, shoot, look*, and *communicate* in the synthetic environment. The simulation engineer will learn the basics about modeling a synthetic battle sphere that is the conceptual basis for simulations. Modeling is the process of abstracting, theorizing, and capturing the resulting concepts and relations in a conceptual model on the abstraction level. This section educates the simulation engineer and provides the ability for understanding these concepts and their limitations.

- The *Distributed Simulation* section introduces the main challenges of advanced distributed simulation. Simulation is the process of specifying, implementing, and executing the models. Simulation resides on the implementation level. In particular when simulation systems are developed independently from each other, the simulation engineer has to know which of these systems can be composed and what standards are applicable. The basics of validation and verification will also be explained as well as how such systems can support the operational environment of the warfighter.

- The *Advanced Topics* section highlights new and current developments and special topic areas. Recognized experts in their domains contributed these chapters. These topics address the needs of advanced students and scholars. They can be used in advanced teaching and in-depth study of special topics, or as a source for scholarly work. I invited recognized experts of these various domains to provide their insight in these chapters.

   It should be pointed out that the chapters in Part IV of this book are written by invited *Subject Matter Experts* of the advanced topics. The views expressed in these chapters reflect the views of the authors alone, and do not necessarily reflect the views of any of their organizations. In particular they do not reflect the views of the US Government or its organizations. The views are based on education and experience of the individual experts. Neither the selection of advanced topics nor the selection within the chapters is meant to be complete or exclusive, but will give examples that can be extended.

When reading the invited expert chapters the reader will notice that there is some overlap between the chapters within Part IV as well as with the first three parts of the book. For examples, Chapters 18 and 25 look into attrition modeling that is introduced in Chapter 9, and Chapters 16 and 19 look at simulation interoperability standards that are introduced in Chapter 12. This redundancy results from the idea that the chapters in Part IV of the book were written to be assigned as independent introductions into the advanced topics and as such can stand as publications by themselves. Furthermore, the viewpoint of all contributions is slightly different, which will contribute to the diversity of the presentations and therefore to the diversity of the education of the simulation engineer. It not only reflects on the multitude of domains in this body of knowledge—the comprehensive and concise representation of concepts, terms, and activities needed that make up the professional combat modeling and distributed simulation domain representing the common understanding of relevant professionals and professional associations—it also reflects the diversity of opinions within the core domain and the contribution domains.

No other compendium addresses this broad variety of topics which are all important to a simulation engineer who has to support combat modeling and distributed simulation. It is hoped that this book will replace the collection of copies of selected book chapters and proceedings papers that I had to use in the last decade to teach. However, I am sure that I will never be able to get rid of additional material for the lectures, as the research on these related topics is rapidly evolving. Nonetheless, the core knowledge captured in this book should remain stable for some time to come.

In my lectures I observed that students sometimes gets lost after the first couple of sessions as so many different aspects are important that on first look do not seem to be related. What do methods of semantic web technology have to do with validation and verification? Why is the resolution of the terrain model important to kill probabilities of military systems? For new students in the domain, even the vocabulary used becomes a challenge, as many military terms are used that need to be understood.

The following sections will provide a sort of overview of the book: where to find information, why chapters are written, what is the common thread through the books, etc.

SELECTED CHALLENGES THE SIMULATION ENGINEER WILL ENCOUNTER—WHERE TO FIND WHAT IN THIS BOOK

What should the reader expect to find in the following chapters? The first three parts of the book are structured following the challenges a simulation engineer will face in the process of conducting work. They address the foundations (the basics necessary to conduct the job correctly), combat modeling (the processes of abstracting and theorizing), and distributed simulation (focusing on building federations and supporting the soldier and decision maker). The fourth part addresses topics that give a historic perspective of where we are and where we came from as a community, in-depth presentations of theory, methods, and solutions, and more.

Finally, two annexes provide starting points for simulation engineers that are looking for more information. The first annex enumerates professional organizations and societies—structured using the categories of government, industry, and academia. For each entry, a short description is given and a website address is provided where more information and contact addresses can be obtained. The second annex gives some examples of currently used simulation systems. Both lists are neither complete nor exclusive but are meant as starting points for more research.

As the author and editor of this book I obviously hope that all chapters will be helpful to the readers. For scholars, educators, and students, the following overview can serve as an introduction to the topics. For readers who use this book more as a handbook or compendium, the overview will help them find the appropriate chapters to read in detail.

Foundations

What contributes to the foundations of combat modeling and distributed simulation for simulation engineers? What needs to be addressed? Should the focus be on how to conceptualize and theorize and build a good combat model? Or should the emphasis be on software engineering and distributed systems that are necessary and essential to support the kind of worldwide simulation federations that contribute to the success stories of military simulation applications?

The approach chosen in this book is different. It focuses, in the foundations part, on methods and solutions that allow a better understanding of the challenges that the military user faces, as the problem(s) to be solved is pivotal. Without understanding the customer, neither models nor simulation can provide insight or solutions. Furthermore, the need to understand the general limits of the approach in comparison with alternatives is needed: applying combat modeling and distributed simulation for the sake of applying it because it is technically possible cannot be a driving factor to recommend or to choose a solution. Generally, recommending one particular solution although other good alternatives are available is unethical. These ideas established the foundations for this book: ethics, best practices for operational assessment, and the problem domain of military users. Figure 1.1 shows the foundations and the four contributing domains.

The Foundation part starts with Chapter 2, dealing with applicable codes of ethics developed and adopted by professional organizations. These codes provide the central guidelines for professional conduct as proposed by the IEEE, the Military Operations Research Society (MORS), and the Society for Modeling and Simulation (SCS). All applicable codes of ethics focus on the necessity to be honest about the research and its limitations. However, ethical conflicts can arise in myriad situations, in particular when engineering goals and business interests are in conflict. Furthermore, own research interests can lead to breaching professional ethics. Simulation engineers in the domain of combat modeling and distributed simulation must clearly understand that the research and solutions are applied to train soldiers and support decisions in the defense domain and therefore directly contribute to what can be a matter of life and death, not limited to the soldiers. If, for example, a flawed design occurs in the national missile defense system, hundreds or thousands of civilians can suffer the consequences of a missile attack. Bad training leads to bad decisions which lead to less protection of those that are dependent on soldiers. Combat modeling and distributed simulation is a serious business and requires highest ethical principles to be applied.

The simulation engineer must make every attempt to ensure that the best combat models and distributed simulation solutions are applied. In order to do

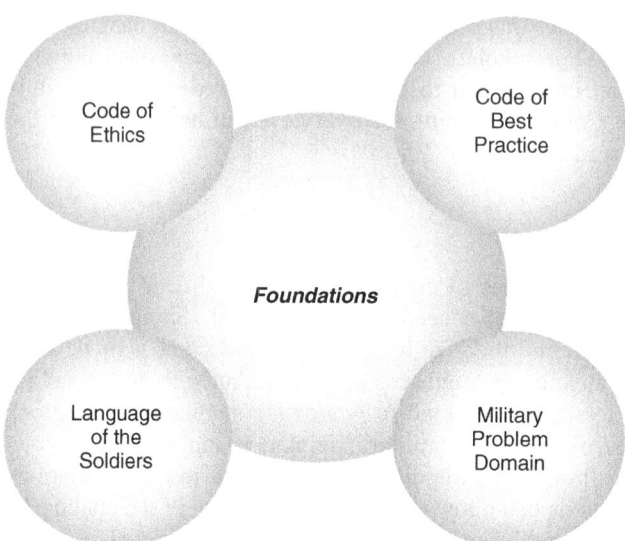

Figure 1.1   Foundations.

this, the simulation engineer needs to understand the big picture of the customer; best practices are needed to guide choices. The *NATO Code of Best Practice* (COBP) was written as a guideline for professional operational analysts on what to consider when setting up and conducting studies in general. It addresses simulation, but shows the application within an orchestrated set of tools of which simulation is just one of many. As such, the COBP addresses two important aspects. First, it shows the applicability constraints and limitations of simulations in comparison with other alternatives. Second, the best practices collected by a group of international experts in the domain of operational analysis are generally applicable, and as such applicable to studies supported by combat modeling and distributed simulation as well. How does one capture the sponsor's problem and come up with a solution strategy? How does one set up scenarios to capture applicable measures of merit of relevant human and organizational issues? How does one select the best methods and tools and obtain the necessary data? How does one reduce risks? Good practices and guidelines are provided and should be known and applied by all simulation engineers.

Chapter 3 introduces the simulation engineer to the "languages," or jargon, required to communicate in this field: those of the simulationist, the military simulationist, and the military customer. In *Terms and Application Domains*, the simulation engineer is introduced to the principles of modeling and simulation first. There are different modeling paradigms that can be used to model entities, behaviors, and relations in the battlefield. There are different domains that can be supported: the training of individual soldiers, groups of soldiers, or command posts; the evaluation of alternative courses of action in real operations; detection capability gaps in doctrine for possible future operations; procurement of new military equipment; and more. Simulationists have already developed special terms to address these concepts, but the military simulation community developed additional or alternative terms to address their areas of concern. On top of these, military language is full of terms with very special meaning the simulation engineer must know to efficiently communicate with subject matter experts. Among these are military hierarchy, the basics of weapon systems, and more in order to model required entities correctly. After studying the terms and concepts introduced in Chapter 4, the simulation engineer will be able to study papers on special topics without too many difficulties regarding the language used.

The last chapter of the first section, Chapter 5, builds a bridge between the foundational understanding and how to model the concepts. The chapter *Scenario Elements* gives an overview of the military entities, their behavior, and their relations within the situated environment. It puts the terms and concepts introduced in Chapter 4 into the context of combat modeling. The *principle of alignment and harmonization*—that is applied subsequently in the following chapters—is introduced and motivated. This principle requires the alignment and harmonization of what we model (represented concepts), the internal rules driving the behavior (decision logic), and the applied measure of merits defining the success (evaluation logic). It can be understood as follows: (1) if something is modeled, it should be used in the model; (2) if something is used, it needs to be measured and evaluated regarding how successfully the use contributed to the overall success; and (3) the overall success must be guided by the common purpose behind the simulation. The common purpose is solving the sponsor's problem by addressing the research question. If any of these aspects of this principle is violated, strange behavior and unintuitive results will result.

Equipped with the theory, the methods, and tools, the simulation engineer can now address the challenges of modeling entities to answer the needs of the sponsor.

Combat Modeling

Modeling is the purposeful abstraction and simplification of the perception of a real or imagined system with the intention to solve a sponsor's problem or to answer a research question. Combat modeling therefore purposefully abstracts and simplifies combat entities, their behaviors, activities, and interrelations to answer defense-related research questions. There cannot be a general model that answers all questions, and even if such a model could be constructed it would become more complex than the real thing, as it not only includes real systems but also imagined ones. How does one address such a challenge in a book on combat modeling and distributed simulation?

The way chosen in this book is captured in Figure 1.2. In five chapters, Chapter 6 to Chapter 10, the simulation engineer will learn how to model the core activities that can be found on every battlefield: *shooting*, *moving*, *looking*, and *communicating*. This traditional combat view is generalized into modeling intended effects of actions, movements on the battlefield including transportation

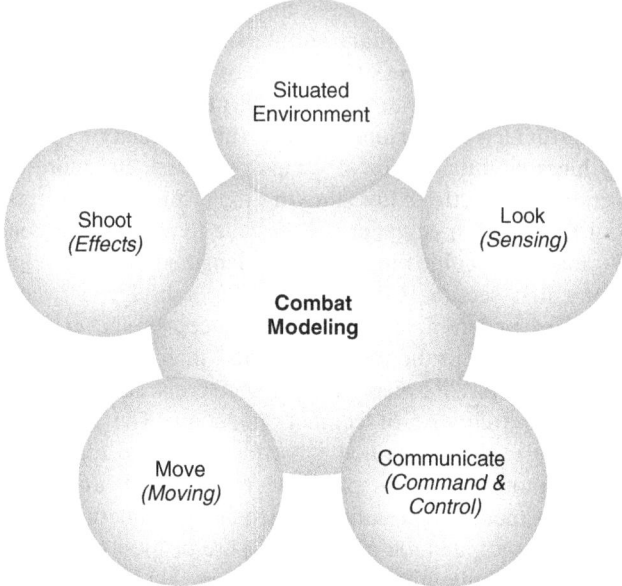

Figure 1.2   Combat modeling.

and mounted operations, modeling sensors that help the soldier to understand what is going on in proximity on the battlefield, and modeling the various aspects of command and control: generating orders and observing their execution by subordinated entities.

The part on *Combat Modeling* starts with analyzing the challenges to model the *Situated Environment*. All modeled combat entities are situated in their environment, the virtual battlefield, or the battle sphere. They perceive their environment including other entities, and map their perception to an internal representation based on its knowledge and goals. They communicate and act with other combat entities. The environment contains all objects, passive ones, like obstacles, as well as active ones. As such, modeling the environment is as important as modeling the entities and their core processes. Even more, the modeling choices for core processes are constrained by the model chosen for the environment. For example, if the environment models vegetation appropriately the influence of seasons on visibility can be evaluated: if the vegetation is leafy it may block visual contacts while other entities can be easily spotted through the branches of trees in the winter. Again, the principle of alignment of data and harmonization of processes comes into play: if something is important for the military operation, then it should be included, used, and evaluated. In particular when, in the third part of the book, several alternative models have to be combined the need for the simulation engineer to understand all these aspects and how they can be composed becomes obvious.

The first core process evaluated in more detail is *Modeling Movement.* The chapter introduces entity level models that deal with the movement of individual weapon systems as well as aggregate models that are used to model the movements of groups of systems. A tight connection with the modeling of the environment is not only obvious for land based systems, as clouds and thunderstorms can influence the flight paths of aircraft, and currents can constrain movement on and under water. The models use patches and grids; they use physical models for weapon systems and reference schemas for unit movement. The focus in this chapter lies on models used to move on land, but the ideas can be generalized for air and sea based movement as well.

The next of the core processes described is *modeling sensing*, answering the question: how does one look on the battlefield? Addressing this process increases in the light of new technological advances as well as new types of military operations that even include terms like information warfare (the fight for and with superior information), or even more important in this context, information superiority (knowing more than your adversaries). To gain and ensure situational awareness—knowing what is going on and how things unfold on the battlefield—becomes a main capability in modern warfare. Various types of sensors, such as acoustic, chemical, electromagnetic, thermal, and optical sensors, can contribute to perceiving the environment and the other entities as close to reality as possible. In real operations, military decision makers will be confronted with wrong, contradictory, and imprecise information; training situations need to prepare them realistically for this kind of challenge. Intelligence, surveillance, and reconnaissance operations contribute to similar requirements. And as before, the alignment and harmonization principle becomes pivotal: in order to sense special properties of an entity, each of these special properties needs to be modeled explicitly. If it is modeled explicitly, it needs to make a difference in the reconnaissance process. Examples are extra antennas on command and control vehicles that otherwise look like other combat vehicles. Similarly, weather influences are important as they may affect the sensors' need to be modeled in the environment. Furthermore, if a detail is important for the military decision process, it needs to be part of the perception, and hence needs to be observed by sensors, which requires that the respective things are modeled as properties of the entities.

Chapter 9 deals with the topics most traditional combat modeling books focus on, *Modeling Effects.* Effects on the battlefield are no longer limited to attrition, but modeling the outcomes of duels between weapon systems and battles between units is still a topic of major interest. After all, fighting against opposing forces is the main thing for which troops are trained and weapons are designed. On the weapon system level, direct and indirect fire are analyzed. Direct fire means that a more or less straight line between the shooter and the target describes how the bullet flies. Artillery systems and other ballistic weapons do not need to see the target and shoot at it straight. Their weapons follow a ballistic curve being described by the term indirect fire. Many models have been developed to keep up with the score, from game based point systems that count how often and where a target is hit to hit and kill probabilities that use real-world data derived from battles or from live fire ranges and exercises. Damage classes are used to find out what effect a hit has: a system can be totally destroyed or merely show any effect at all, depending on who the shooter is and who the target is. This chapter also deals with the famous Lanchester models of warfare that describe attrition of aggregated units in the form of differential equations. Other effects are only mentioned in this chapter. The reader is referred to other in-depth publications to deal with them in more detail. The chapter on modeling effects closes with an overview of Epstein's model of warfare and looks at some new approaches that take recent developments in agent directed simulation into account.

Finally, the second part of this book closes with a chapter on *Modeling of Communications, Command and Control*. It ties all the earlier chapters together as command and control is situated in the environment and commands the entities to shoot, move, observe, and communicate. Several models of command and control in military headquarters are discussed, as more and more simulation models have to come up with decisions based on available information where until recently human decision makers had to be involved. The better command and control is modeled the less military experts are needed to provide a realistic training environment. Furthermore, only when partners and opponents behave realistically can the resulting simulation systems be used in real operations to support the evaluation of alternative options. The chapter closes with some detailed observations on how messages are used in the real domain to communicate and how this can be used to model communication.

In summary, the second part on combat modeling will give the simulation engineer the necessary basic understanding of how military operations comprising moving forces, observing the battlefield, engaging the opponent, and communicating to optimize the engagement can be modeled. The simulation engineer will appreciate the principle of aligning the data and harmonizing the processes of all these models to ensure consistency and avoid wrong results and strange behavior, an exercise the simulation engineer will be able to use often when dealing with the topic of the following part of the book: composing a simulation federation out of several independently developed simulation systems.

Distributed Simulation

The third part of the book addresses the challenges of and existing methods, tools, solutions, and guidelines for composing several in general independently developed simulation systems into a federation. Providing homogeneous support of required capabilities by composing heterogeneous solutions that expose implemented functionality is conceptually demanding as well as technically sophisticated. Many of the conceptual tasks of aligning the data and harmonizing the processes can be supported by the engineering principles introduced in Part II. The focus of this part is the technical integration based on available methods and guidelines that are provided in particular for defense modeling. Many of these methods can be applied in slightly modified form for domains other than

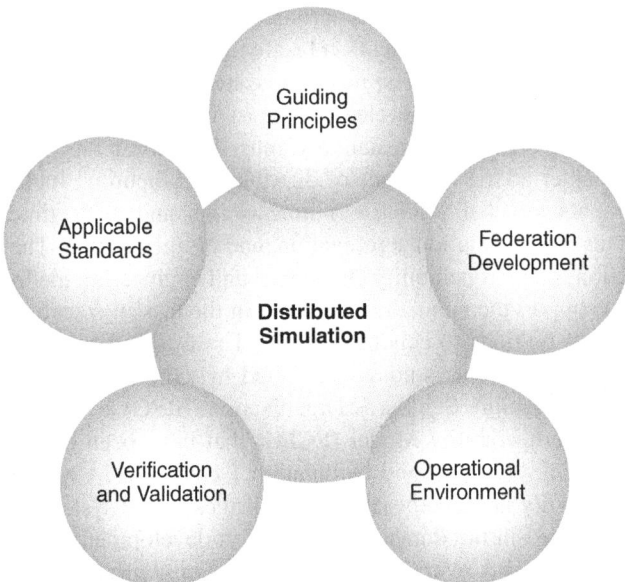

Figure 1.3 Distributed simulation.

the defense domain as well—and they will have to be applied when federations comprise tools from supporting domains, such as social science models in support of human, social, cultural, and behavioral modeling.

Part III deals with five core topics to focus on the technical foundations of combat modeling and distributed simulation, which are shown in Figure 1.3. First are the general principles of advanced distributed simulation that need to be assured in every simulation federation, no matter which simulation interoperability standard is applied or which architecture viewpoint is supported. Next are applicable standards that can support the federation development process, followed by several guidelines and best practices for the process of building a federation itself. The last two topics deal with verification and validation challenges and how to integrate simulation systems and simulation federations into the operational environment of the military user.

In Part II of this book, the conceptual foundations are laid to understand the modeling challenges of combat modeling and distributed simulation. Before diving into the technical aspects of distributed simulation, many of those aspects addressed and supported by software engineering principles, Chapter 11 on *Challenges of Distributed Simulation* puts the conceptual foundations into the context of identifying applicable simulation systems that can contribute to address the sponsor's need and evaluate the modeling question, selecting the best systems under the given constraints, composing the selected solution to provide the required capabilities as a coherent federation, and orchestrating their execution. The application of engineering principles helps avoid inconsistencies, anomalies, and unfair fight situations. Furthermore, the infrastructure that needs to be provided has to support the interoperability protocol and the information exchange model must fulfill three requirements: (1) all information exchange elements must be delivered to the correct simulation systems (effectiveness); (2) only the required information exchange elements must be delivered to the simulation systems (efficiency); and (3) the delivery must happen at the right time (correctness). The chapter addresses issues of distributed computing, looks into consistency requirements and challenges for entities, events, and states, time management, and addresses multiresolution, aggregation, and disaggregation. The chapter closes with a section on interoperability challenges and engineering methods that can be applied to support the simulation engineer in the task of selecting the best methods, tools, and solutions to conduct the job. These include general aspects on interoperability infrastructure studies conducted by experts under the lead of the Simulation Interoperability Standards Organization (SISO) as well as the Levels of Conceptual Interoperability Model (LCIM) that have been developed at the Virginia Modeling Analysis and Simulation Center (VMASC) to generally address interoperability and composability challenges in the system of systems environment as exposed in the domain of federation development.

Following this introduction and overview, Chapter 12, *Standards for Distributed Simulation*, focuses on the technical standards. Applicable guidelines, that also have been standardized, are dealt with in a later chapter in more detail. The view of this chapter is purposefully broad and not limited to the simulation interoperability standards predominantly applied in the defense domain. The vision of the SISO is to serve the global community of modeling and simulation (M&S) professionals, providing an open forum for the collegial exchange of ideas, the examination and advancement of M&S related technologies and practices, and the development of standards and other products that enable greater M&S capability, interoperability, credibility, reuse, and cost-effectiveness. Therefore, the chapter starts with an overview of past and current standardization activities. Some of these are dealt with in more detail in advanced topic chapters in Part IV of the book, such as base object models described in Chapter 19, the tactical data link simulation standards in Chapter 23, or the military scenario definition language in Chapter 24. Not all successfully applied engineering principles and resulting methods, tools, and solutions have been standardized. The Test and Training Enabling Architecture (TENA) described in Chapter 20 is an example. After this overview on SISO related activities, the chapter gives an introduction to other activities of the community—borrowing from the systems engineering community—and describes the basics and applicability in the context of combat modeling and distributed simulation of the Discrete Events Simulation Formalism (DEVS), the Unified Modeling Language (UML), the System Modeling Language (SysML), and the US Department of Defense Architecture Framework (DoDAF). DEVS and DoDAF are also topics of in-depth chapters in Part IV. The semantic web stack of supporting standards for web based distribution and alignment of interpretation of data closes the general overview. The chapter closes with describing the two IEEE Modeling and Simulation Interoperability Standards: the

IEEE 1278 Distributed Interactive Simulation (DIS), and the IEEE 1516 High Level Architecture (HLA).

Chapter 13 covers several of the methods and guidelines for *Modeling and Simulation Development and Preparation Processes.* Technical standards and solutions can address most challenges that arise from different infrastructures, networks, or communication protocols, but if the principle of aligning data and harmonizing processes is violated, they cannot help. Technical standards cannot be used to solve conceptual problems—conceptual solutions are needed instead. Supporting methods and guidelines have been developed in several professional organizations. One of the most general views is captured in a guidance document developed by the US Modeling and Simulation Coordination Office (MSCO) using modeling and simulation as a means within the problem solving process. This guideline distinguishes three use cases for M&S: (1) developing new M&S solutions; (2) adapting existing M&S solutions for new problems; and (3) constructing an M&S federation. In particular the third use case is supported in detail for the high level architecture standards by the standardized guideline for the Federation Development and Execution Process (FEDEP) and the Synthetic Environment Development and Exploitation Process (SEDEP). The generalization of these processes resulted in the most recent standards, the Distributed Simulation Engineering and Execution Process (DSEEP). All three processes—FEDEP, SEDEP, and DSEEP—are introduced and compared, as they have slightly different viewpoints on challenges to address and how to address them. The simulation engineer should know all three of them, as even if only one of them is used explicitly in support of the project, the alternative views can still be used to guide solutions based on a broader understanding of the problems. The chapter closes with a new project management method that in particular is useful in service-oriented distributed environments: the SCRUM methodology. SCRUM allows users to explicitly take the agility of the problem domain—the rapid and continuous change of operational constraints and resulting requirements—into account. In particular when new generations of distributed simulations are based on standards developed for the semantic web, SCRUM can be used to guide federation development processes in support of lifecycle use cases of federations instead of limited point solutions.

The next chapter, Chapter 14, addresses *Verification and Validation.* Ultimately, the application domains of combat modeling and distributed simulation are connected with the well-being of humans. Wrong decisions based on the application of wrong solutions or the rejection of correct advice as well as insufficient training of soldiers using insufficient training simulation systems will likely result in harm and even the loss of lives. The processes of verification and validation (V&V) should not be confused with quality control means of software engineering that address the implementation challenges. V&V mainly addresses the conceptual challenges. The chapter starts with an overview of the underlying academic groundwork showing that the objective of correct and suitable solutions may be out of reach, but that heuristics can be applied to create solutions that are at least credible. Credibility is based on the whole process integrating the various aspects. While validation ensures the accuracy of the process leading to a suitable model for the intended purpose, verification ensures the accuracy of the software engineering process and the information preserving transformation processes ensuring the use of the right scenario and the right data to complete these processes. As such, the NATO Code of Best Practice principles are utilized and the principle of aligned data and harmonized processes is supported as well, closing the circle. The following principles of V&V ensure good results: (1) V&V must be an integrated activity with model selection, development, and integration activities; (2) the intended purpose needs to be specified precisely; (3) sufficient knowledge about the system to be simulated is necessary; and (4) sufficient V&V tools and V&V experts are available. The Verification, Validation, and Accreditation (VV&A) Recommended Practice Guide (RPG) provides guidelines to support the simulation engineers to meet such requirements by defining roles, tasks, and products. Outside of the United States, the European effort Referential for Verification, Validation, and Accreditation (REVVA) is important to know.

The final chapter of Part III, Chapter 15, deals with challenges the simulation engineer has to solve when tasked with the *Integration of M&S Solutions into the Operational Environment*. The operational environment is the environment of the sponsor, which normally is the military decision maker or the soldier. The infrastructure used by soldiers to support their command and control of operations is often significantly different from the simulation infrastructure with which the simulation engineer is familiar. The chapter starts with several use cases that require the coupling of operational and simulation infrastructures, focusing on training, testing, and decision support. It describes the standards used in the operational environment, such as tactical messages and their format, the Common Operating Environment (COE), the Multilateral Interoperability Program (MIP), and the Global Information Grid (GIG). The chapter closes with current research work on frameworks that can facilitate the work of the simulation engineer. It also introduces the underlying ideas of current standardization efforts for a common language that helps to bridge the gap between different operational views within a coalition as well as between different implementation decisions in supported information infrastructures: the Coalition Battle Management Language (C-BML) that is intended to become the common unambiguous language used to command and control forces and equipment conducting military operations and to provide for situational awareness and a shared, common operational picture.

In summary, Part III of the book provides the simulation engineer with the technical foundations to construct federations and to identify the best practices and guidelines that help to ensure conceptual consistency. It helps with validation and verification methods and shows the simulation engineer how to integrate M&S and operation solutions to the benefit of the warfighter.

Advanced Topics

While the first three parts of this book are rooted in the material used to teach the engineering principles of combat modeling, Part IV on advanced topics comprises a number of invited chapters on special topics. These chapters either can be read in parallel to the first three parts of the book to provide alternative views and opinions, or they can be read as an independent part that summarizes the state of the art in the form of a compendium or handbook. For scholars and students, these chapters can become an introduction of a problem sub-domain that still exposes several gaps that are required to be closed. For faculty and teachers, the fourth part of the book can be used either to prepare a course for advanced students or in support of additional research assignments in parallel to the coursework captured in the first three parts.

Chapter 16 introduces the simulation engineer to the *History of Combat Modeling and Distributed Simulation.* In order to better understand where we are it is necessary to understand where we are coming from as a community. This chapter gives an overview of the history of combat models and simulation from board games and maneuvers to modern day worldwide distributed international training federations. It starts with early board games used to teach tactical understanding and ends with recent efforts to integrate live systems with simulators and simulation systems to provide global exercise capability. Only those who know their history are not doomed to repeat the same mistakes again, so there are many valuable lessons learned in this chapter.

Chapter 17 deals with *Serious Games, Virtual Worlds, and Interactive Digital Worlds.* The growing community of game developers does not always focus on engineering principles of combat modeling and simulation. However, modern simulation systems not only have their roots in games, today's simulation systems and computer games share technical, social, and economic domains. Common visualization tools and solutions decrease development costs and may even increase the credibility of solutions. This chapter gives an overview of commonalities, differences, and trends between combat modeling and distributed simulation and serious games and shows where mutual support of both communities is possible and desirable.

Chapter 18 provides a deeper introduction to two topics of special interest to combat modeling, namely game theoretic applications, in particular in the context of operational analysis, and Lanchestrian models of warfare. *Mathematical Applications for Combat Modeling* introduces more background and theory for Lanchester's models of warfare that are still dominating aggregated attrition models. In order to better support rational decision making, the application of game theoretic ideas can be applied. In particular when several individually developed and principally independent systems are federated into a system of systems, the optimum for the federated systems only sometimes results from individual optimal decisions within all contributing individual systems. As such, this chapter is a great enhancement to the general principles discussed in Chapter 9.

Chapter 19 addresses two standardization efforts of general interest and complements the description given in Chapter 12. This chapter on *High Level Architecture and Base Object Models* documents two success stories of the SISO. In the focus of this chapter are new standardization efforts and developments. The High Level Architecture (HLA) continues to be improved to meet the needs for distributed simulation better. Modular federation object models and better support of net-centric environments are two aspects that are currently discussed and recently integrated into the standard. In addition, new standard developments, like Base Object Models (BOM), support HLA as well as alternatives. This chapter gives a broad overview of how the standards on HLA and BOM were developed and where they are heading. It can serve as an independent introduction to the topic.

Chapter 20 deals with a very important topic that did not get enough attention in the first three parts of the book: the *Test and Training Enabling Architecture* (TENA). Developed under a joint interoperability initiative within the US Department of Defense, TENA has been designed with the goal of enabling interoperability among ranges, facilities, and simulations in a quick and cost-efficient manner, and fostering reuse of range resources and range system developments. This section will give a better overview of the TENA components and their application(s). As the application of TENA is not limited to the United States—many partner nations are using TENA and are contributing actively to its continuous improvements—and also not limited to testing—several training events were successfully supported by TENA—the simulation engineer should know philosophy, theory, methods, and solutions as discussed in this chapter.

Chapter 21 introduces the simulation engineer to *Combat Modeling using the DEVS Formalism.* Within the United States, the application of the Discrete Event Simulation (DEVS) formalism, as introduced in Chapter 12 of Part III, for combat modeling and distributed simulation was until recently nearly completely limited to academic efforts. In other nations, in particular in the Pacific domain, the use of DEVS to develop combat simulations for the armed forces is a well known procedure. Because DEVS defines system structures as well as behavior and uses atomic components to build up composites that finally represent the modeled systems, its application allows the application to combat models, as hierarchical structures and necessary capabilities for operations are well captured. The chapter describes a real-world application within the Korean armed forces.

Chapter 22 addresses a topic of growing interest, using *Geospatial Data for Combat Modeling.* Geospatial information is as important for military operations as it is for operational data. This chapter examines how to model and integrate geographic information system (GIS) data for combat modeling and prepare the distribution of data in distributed simulation environments. The operational community with command and control systems is facing similar challenges, so that common standards may be possible. The chapter introduces important standards and procedures as currently used within the GIS community and gives examples on how these approaches can be used to enrich M&S solutions.

The *Tactical Data Links* (TADL) for military command and command are standardized across NATO partners and other allies to support in particular air operations, but they are used in all services. They were addressed in Chapter 10, as they are important for modeling command and control, and in Chapter 15, as they can be used for integration purposes of M&S into the operational environment as well. This real-life command and control standard was also standardized in M&S standards to allow for their evaluation as well as their support for more realistic training. This chapter introduces the military view on TADL as well as the resulting standards and their application, as the simulation engineer must know them. Every military application that uses TADL, from air defense to missile operations, should evaluate the applicability of the ideas as captured in Chapter 23.

Chapter 24 introduces another successful standardization effort of interest to all simulation engineers, the *Military Scenario Definition Language* (MSDL). MSDL is a standard used to describe the initialization data for combat simulation systems. It allows several simulation systems to be initialized with the same consistent set of scenario data. It also provides a common language for heterogeneous authoritative data sources to provide their data to potential users. MSDL was introduced in Chapter 12 and mentioned also in Chapter 15. This invited chapter, however, gives a much more detailed technical overview of the XML schema and its use in support of scenario distribution and systems' initialization. MSDL is internationally used and supported.

Chapter 25 returns to the conceptual challenges and provides valuable information regarding various *Multi-Resolution Modeling Challenges.* Whenever models of different resolutions have to be federated, multi-resolution problems have to be solved. Several heuristics have been developed in the community to cope with the challenges. This field of research can look back to many successful and documented solutions that the simulation engineer should be aware of. Among other means, the chapter provides a tabular overview of important research results. The chapter is a very valuable in-depth add-on to the general principles introduced to the simulation engineer as an overview in Chapter 11. The examples can neither be complete nor exclusive, but they definitely are a great start for every simulation engineer to learn more.

Another problem addressed in principle several times in the first parts of the book is the fact of new operational scenarios. Today's military focus has moved away from the force-on-force battlefield of the past century and into the domain of irregular warfare and its companion, security, stability, transition and reconstruction missions. Knowing how to model moving, sensing, communicating, and shooting is still needed, but not sufficient. Chapter 26 addresses *New Challenges: Human, Social, Cultural, and Behavioral Modeling.* The success of this new type of operation depends on understanding social dynamics, tribal politics, social networks, religious influences and cultural mores. This chapter introduces the simulation engineer to this new world that is characterized by the absence of standards and established models and requires new approaches that the community so far has not yet agreed upon, and may conceptually never be able to do, as psycho-socio models do not follow a commonly accepted world view, such as physics-based attrition and movement models do.

Chapter 27 describes an approach that may help with the application of human, social, cultural, and behavioral models. It introduces *Agent-directed Simulation for Combat Modeling and Distributed Simulation.* The agent metaphor changed the way simulations are written and used in many domains, and combat modeling and distributed simulation is no exception. This chapter introduces the foundational ideas of agent-directed simulation and gives selected examples of agent applications. An agent can be understood as a software object with an attitude: while normal objects expose their methods to the user and execute the methods when invoked, an agent decides if and how the simulation engineer reacts to invocations. This makes the agent metaphor a powerful approach to represent human beings within simulations as well as to represent independent and individual systems with system of systems environment.

Another aspect that is introduced, in principle, in several places but in particular in Chapter 8 and in modeling sensing is the aspect of representing uncertain, incomplete, vague, and contradictory information. Chapter 28 introduces methods for *Uncertainty Representation and Reasoning for Combat Simulation Models*. This chapter is a little bit more challenging from the underlying mathematics but is suitable for graduate students of engineering disciplines or of computer science. It introduces Bayesian theory with Bayesian networks as well as multi-entity Bayesian networks and applies these ideas to build a vehicle identification Bayesian network. The theory is then generalized into the probabilistic argument and applicable related theories. The simulation engineer will find several useful methods and tools to better cope with uncertainty and vagueness in combat modeling and distributed simulation challenges.

Chapter 29 describes an engineering method developed at the Virginia Modeling Analysis and Simulation Center in more detail: *Model-Based Data Engineering* (MBDE). Models are individual abstractions of reality and will necessarily result in different scope, different structure, and different resolution. MBDE was developed to support developers of federations with a heuristic to identify shareable data and common concepts. In particular for scholars of data modeling theory, this chapter will be a valuable tool to identify shareable information between simulation solutions—or systems of the operational environment. If applicable data units have to be identified or a common information exchange model needs to be created from the beginning, MBDE heuristics help to avoid many common mistakes.

The multifaceted and multidisciplinary field of systems engineering and the emerging new discipline of system of systems engineering address several challenges identified for combat modeling and distributed simulation. Chapter 30 describes *Federated Simulation for System of Systems Engineering* and provides a "look over the fence" to make simulation engineers aware of engineering principles developed in these domains that can support tasks. It provides a framework to deal with the growing complexity in three steps. The first step defines system of systems engineering and provides an overview of engineering and modeling and simulation approaches prescribed for that context. The second step highlights engineering principles that are particularly relevant in a system of systems context. The final step prescribes a systems engineering methodology for federation development comparable to the approaches described in Chapter 13, but written from the systems engineering perspective. Several practical application examples are given to educate the simulation engineer on this broader and more holistic viewpoint.

Chapter 31 follows this topic by providing an introduction to *The Role of Architecture Frameworks in Simulation Models.* The simulation engineer has already been introduced to architecture frameworks in Chapter 12, but the viewpoint presented in this chapter is an extension thereof. Defense systems have to be described by means of the DoD Architecture Framework. This chapter shows possible mutual support of simulation and architecture efforts, such as enriching DoDAF artifacts to allow for executable architectures. The main focus of the chapter is showing how to extend the DoDAF to include the human view components identified in recent NATO efforts.

The final chapter of Part IV, Chapter 32, describes *Multinational Computer Assisted Exercises* from a practitioner's viewpoint. NATO is organizing multiple M&S efforts in support of training, planning, and support of operations. It established several M&S centers to enable such endeavors. This chapter will give an overview of related activities with focus on the particular challenges of multinational environments. In particular for simulation engineers that have to support multinational endeavors this chapter may be a good starting point for further research.

Annexes

Two annexes provide some start-up information for new simulation engineers in tabular form.

Annex 1 enumerates *Organizations of Interest for Professionals.* Using government, industry, and academia as categories, some selected organizations are enumerated in alphabetical order. Web resources are provided to continue the research. This list can neither be complete nor exclusive and should be extended by the simulation engineer to document personal networks.

Annex 2 gives some *Examples for Combat Models and Distributed Simulations.* This annex describes several currently used simulation solutions and gives references for additional information. Examples include Joint Semi-Automated Forces (JSAF), Joint Conflict and Tactical Simulation (JCATS), Virtual Reality Forces (VR-Forces), and others that were provided by the international community in the course of writing and editing this book.

SOME CRITICAL REMARKS

Before ending the first overview chapter of this book, some critical remarks may be in order. All references supporting the topics described so far will be given in the following chapter, but for these concluding remarks some references are needed, as the ideas are not dealt with in detail in the following chapters.

No other domain-applied M&S is as successful as the military domain. Worldwide training and testing events, multi-million dollar budgets for supporting M&S activities, and the recognition by the US Congress as a "National

Critical Technology" in its House Resolution 487 in July 2007 are just some examples that have no comparison in other domains. Not even the sky seems to be the limit, as space operations and cyber operations are already included in the military scenarios. But despite all the supporting standards and guidelines to be presented in this book we must ask the question: How good are we in combat modeling and distributed simulation really?

In their 1991 White Paper on the State of Military Combat Modeling, Davis and Blumenthal identify several problems with the use of M&S within the US Department of Defense. Actually, they made the statement that the use of combat modeling and simulations within the US DoD at this point in time was fatally flawed, so flawed that it could not be corrected by anything less than structural changes in management and concept. The lack of a vigorous military science as the foundation for combat models and their validity resulted in (1) dissonance of approaches in the community; (2) use of inadequate theories, methods, solutions, and models; and (3) ultimately chaos in combat modeling that became worse by unsolved challenges in the software engineering realm. A particular problem was that nobody was really in charge. That resulted in a multitude of not harmonized approaches resulting in incompatible solutions. Twenty years later, history is in danger of repeating itself. Strong and scientifically competent central organizations are gradually replaced with local organizations and steering committees. Whether they will succeed in the challenge of successful alignment and harmonization remains to be seen, but some critical remarks in the community already warn that we may sink back into "dark ages" (Hollenbach, 2009). With the knowledge provided by this book I hope to raise the awareness of a strong and holistic management approach to combat modeling and distributed simulation, as strong engineering principles are necessary, but not sufficient.

The second word of warning regards the technical maturity of our approaches. The methods, tools, and guidelines compiled in this book are mainly derived from physical technical models. They assume that all models are derived from the same real-world systems on the same battlefield following the same laws of physics. Discrepancies in models can be reduced by increasing the resolution and by aligning the data and harmonizing the processes to the same level of aggregation. This viewpoint is rooted in positivism as the driving philosophical foundation, as it is appropriate for the majority of physical-technical models that follow the Newtonian worldview of physics. Unfortunately, this worldview also underlies the current interoperability standards, tools, and guidelines. It would be naïve to apply standards that were developed for physical-technical models based on a common theory representing the positivistic worldview to integrate socio-psychological models derived from competing theories representing interpretivism and expect valid results. Before jumping into constructing a federation the simulation engineer must ensure that this is the right thing to do to address the sponsor's problem. In some cases, the best way forward may be to live with contradicting models. It is highly unlikely that we will be able to address all problems with one common approach based on a common theory resulting in a consistent federation. It is much more likely that the multi-simulation approach based on multi-resolution, multi-stage, and multi-models envisioned by Yilmaz and colleagues (2007) needs to be exploited to support the analysis of these multi-faceted challenges we are faced with as a community. Some additional aspects, in particular to better address the aspects of uncertainty and unpredictability—simulation systems can never be "magic mirrors" that allow the user to look into a certain future—are addressed in this book, but other problems may never be solvable.

However, some recent criticism on the use of modeling and simulation, as documented by Gregor (2010), may create the danger to throw the baby out with the bathing water. Gregor documents the viewpoint by some experts that the use of mathematical methods for predicting and measuring effects shows a trend toward using metrics to assess the essentially unquantifiable aspects of warfare that reinforces the unrealistic view that warfare is a science rather than an art and a science. As the US Joint Publication 1 states: "War is a complex, human undertaking that does not respond to deterministic rules!" As will be shown in this book, there are plenty of tools either that can be used in an orchestrated tool set to address these challenges together with defense modeling and simulation approaches or that can be integrated into the simulation systems to provide the necessary functionality to the simulation solution. While war is complex and does not respond to deterministic rules, the engineering principles of combat modeling and distributed simulation provide the means to reduce the complexity and manage uncertainty in support of the military decision maker and will enable better decisions for the management as well as on the battlefield.

As is well known in the engineering management community, three things are needed to ensure success of an approach: (1) *technical maturity* of the proposed solution—if it does not work properly it will not be applied more than once; (2) *supporting management processes*—if the manager does not support the use, it is not going to last long; and (3) an *educated workforce*—if nobody has been taught how to use it, it cannot be applied beyond the scope of a limited group of experts. With this book I hope to contribute to a better education of the workforce by providing a compendium that is state of the art to the simulation engineer in education and hopefully for practitioners in the field as well.

I would like to thank everyone who contributed to this book and hope that each reader finds in the following chapters at least the core of what he is looking for. I hope for feedback to continuously improve this book and keep it up to date to serve the community and the soldiers they support.

REFERENCES

Davis PK and Blumenthal D (1991). *The Base of Sand Problem: A White Paper on the State of Military Combat Modeling*. RAND Corporation, Santa Monica, CA.

Gregor WJ (2010). Military planning systems and stability operations. *Prism* **1**, 99–114.

Hollenbach JW (2009). Inconsistency, neglect, and confusion; a historical review of DoD Distributed Simulation Architecture Policies. In: *Proceedings of the Spring Simulation Interoperability Workshop*, Spring, San Diego, CA. www.sisostds.org.

United States Department of Defense (2009) *Joint Publication 1: Doctrine for the Armed Forces of the United States*. March 20, Washington, DC.

Yilmaz L, Ören T, Lim A and Bowen S (2007). Requirements and design principles for multi-simulation with multiresolution, multistage multimodels. In: *Proceedings of the 2007 Winter Simulation Conference*, edited by SG Henderson, B Biller, M-H Hsieh, J Shortle, JD Tew and RR Barton. IEEE Press, Piscataway, NJ, pp. 823–832.

Part I

Foundations

Chapter 2

Applicable Codes of Ethics

Andreas Tolk

WHY TEACH ABOUT THE CODES OF ETHICS

Every scientist, researcher, and engineer will agree that following ethical guidelines when doing a job is important, but why deal with this topic as the very first one in a book on engineering principles of combat modeling and distributed simulation? The answer is easy: because it may become more important than all technical topics as soon as results are applied! Combat modeling deals with warfare: analysis of alternatives, training of soldiers, testing of weapon systems, and more. Following rigid ethical guidelines is crucial, as mistakes or even slight variations outside the validity domain of applied models can have deadly consequences for soldiers and persons in their operational domain.

Let us look at two examples of what can go wrong.

1. In April 2000, during a Marine Corps training mission in Arizona, the pilot of an MV-22 airplane dropped his speed to about 40 knots and experienced "Vortex Ring State (VRS)," a rotor stall that results in a loss of lift. Attempts to recover worsened the situation and the aircraft crashed, killing everybody on board. The pilot had 100 hours in the Osprey simulator and nearly 3800 hours of total flight time. However, none of his training or experience involved coping with a vortex ring. In January 2001, the General Accounting Office, in a presentation to the V-22 Blue Ribbon Panel, attested that the flight simulator used to train the soldiers in handling this aircraft did not replicate the VRS loss of a controlled flight regime.

Engineering Principles of Combat Modeling and Distributed Simulation, First Edition.
Edited by Andreas Tolk.
© 2012 John Wiley & Sons, Inc. Published 2012 by John Wiley & Sons, Inc.

2. In March 2003, during Operation Iraqi Freedom, Patriot missiles shot down two allied aircraft and targeted another. The pilot and co-pilot aboard the British Tornado GR4 aircraft died when the aircraft was shot down by a US Patriot missile. Another Patriot missile may have downed a US Navy F/A-18 C Hornet which was flying a mission over Central Iraq. Both cases were investigated and one cause of these failures stemmed from using an invalid simulation to stimulate the Patriot's fire control system during its testing.

In both examples, the simulation systems used were not bad systems. The problem in both examples was that they were used in an invalid context. The training simulator for the experimental aircraft took many things into account and was a great support to educate soldiers, but under extreme circumstances the simulated behavior was wrong. What went well in the simulation led to the deadly crash when applying the training in the real system. The same was true for the testing simulation. The circumstances observed when the UK Tornado was shot down and the US Hornet was engaged were not part of the underlying design. The Patriot crew relied on the validity of the test … and they shot down their own pilots. It is the responsibility of the engineer to know his customers, their environments and constraints, and every detail the system he design will be used in and contribute to. This is his ethical responsibility: to understand the user, the processes, and constraints, and using this knowledge to build the best technical solution possible.

I do not want to imply that the designers of the simulations described in the examples above were aware of these traps, but if they were, not communicating the validity constraints with the customer and user of the simulation would have been a significant breach of ethical behavior. But even if they were not aware of the trap, would it not be their professional duty to make sure that no such traps exist in their solution? Did they seek advice of colleagues with more background knowledge? Did they make sure that they were the best engineers for the job?

The lessons learned can only be to be as open as possible about assumptions, constraints, and applicability limits, and to keep current regarding developments, paradigms, and solutions. If the designer or developer of a combat model or the developer of a potentially distributed implementing simulation is aware of a shortcoming, it needs to be conveyed to customers, or the silence may result in the loss of lives. Combat models are applied in preparation for or conducting warfare. The ethical standards for the engineer must be high to keep the risks for the user low. Concealments of shortcomings or the willful misrepresentations of capabilities, like claiming to fulfill certain standards although that is not the fact, have no place in a professional community.

IEEE, MORS, AND SCS

Many organizations of engineers, analysts, and simulationists being highly skilled professionals decided to follow a common code of ethics in order to make sure that their profession is recognized and valued. In this chapter, the focus is on three codes of particular interest to engineers of combat modeling and distributed simulation, namely

- the Institute of Electrical and Electronics Engineers (IEEE) Code of Ethics;
- the Military Operations Research Society (MORS) Goals and Code of Ethics; and
- the Code of Professional Ethics for Simulationists, as recognized by several simulation organizations, such as the Society for Modeling and Simulation (SCS), the Simulation Interoperability Standards Organization (SISO), and the North Atlantic Treaty Organization (NATO) Research and Technology Organization (RTO) Modeling and Simulation Group (MSG).

The selection of these three codes was conducted purposefully. As combat modeling and distributed simulation engineering falls under engineering (and many will look in particular to computer engineering and software engineering as the most relevant sub-disciplines), the general IEEE code is applicable. As combat modeling is one of the important means of military operations research (questions of how modeling and simulation and operations research are interwoven are in a later chapter), the findings of MORS are of interest too. Finally, the distributed simulations must be developed and maintained by professional simulationists. The codes overlap in their domain, but all three of them look at the question of ethical behavior through a slightly different facet, all of them important to the domain of combat modeling and distributed simulation.

IEEE Code of Ethics

Table 2.1 shows the IEEE Code of Ethics, as approved by the IEEE Board of Directors in February 2006. While the code has been modified several times over the years, it has retained the fundamental principles detailed in the Code first adopted by the American Institute of Electrical Engineers in 1912, namely that the first role of an engineer is to serve his/her community in the best way possible by applying the best available solutions. This requires not only being honest; it also requires staying up to date with technological and methodological developments.

The IEEE Code of Ethics is documented in the IEEE Policies, Section 7 —Professional Activities (Part A—IEEE Policies). Changes to the IEEE Code of Ethics are made only after the proposed changes have been published with a request for comment and all IEEE major boards have had the opportunity to discuss proposed changes before the Board of Directors makes a decision. During the Board meeting, an affirmative vote of two-thirds of the members is required to pass the recommended change.

Ethical conflicts can arise in myriad situations, in particular when engineering goals and business interests are in conflict. It happens quite often that certain features that are generally desired by the customer but not often observed in supported tasks are not implemented by the contractor to reduce development costs. The example of the highly improbable event of VRS could have been a

Table 2.1   IEEE Code of Ethics

We, the members of the IEEE, in recognition of the importance of our technologies in affecting the quality of life throughout the world, and in accepting a personal obligation to our profession, its members and the communities we serve, do hereby commit ourselves to the highest ethical and professional conduct and agree:

1. to accept responsibility in making decisions consistent with the safety, health, and welfare of the public, and to disclose promptly factors that might endanger the public or the environment;
2. to avoid real or perceived conflicts of interest whenever possible, and to disclose them to affected parties when they do exist;
3. to be honest and realistic in stating claims or estimates based on available data;
4. to reject bribery in all its forms;
5. to improve the understanding of technology; its appropriate application, and potential consequences;
6. to maintain and improve our technical competence and to undertake technological tasks for others only if qualified by training or experience, or after full disclosure of pertinent limitations;
7. to seek, accept, and offer honest criticism of technical work, to acknowledge and correct errors, and to credit properly the contributions of others;
8. to treat fairly all persons regardless of such factors as race, religion, gender, disability, age, or national origin;
9. to avoid injuring others, their property, reputation, or employment by false or malicious action;
10. to assist colleagues and co-workers in their professional development and to support them in following this code of ethics.

candidate for such a decision (which I do hope not to have been the case). Other domains of engineering interest are validation and verification and testing, all activities that are often cut by a business decision to save money. The combat modeling and distributed simulation engineer must decide if and when such decisions violate his or her ethics by unnecessarily endangering the users of the product. Another example is when the proposed solution of the competition is clearly superior and the competing team is better educated, but the business decision requires defending the solution of the employing company and attacking the other solution as insufficient in order to win the contract. Although the ethical foundations are clear, the decision in the business environment will not be easy under such constraints.

MORS Goals and Code of Ethics

The MORS has served the Department of Defense analytic community for over 40 years. It comprises all aspects of operations research, including the use of modeling and simulation, where appropriate. The vision of MORS—as stated on the website of the society—is to "become the recognized leader in advancing the national security analytic community through the advancement and application of the interdisciplinary field of Operations Research to national security issues, being responsive to our constituents, enabling collaboration and development opportunities, and expanding our membership and disciplines, while maintaining our profession's heritage." Only recently, MORS integrated more topics on home-land security to its agenda. The society also intensified international collaboration and advancements in the context of allied operations.

Similar to IEEE, MORS focuses on the traditionally high ethics of the society and its heritage. In the society information, the MORS Code of Ethics goes hand in hand with the society goals, and both enumerations of objectives are mutually supportive, as shown in Table 2.2.

As the focus of MORS is conducting analyses in support of decisions, the adjectives *objective, truthful, accurate*, and *accountable* build important pillars of this code. The code also emphasizes the need for respectful collaboration, giving due credit, and refraining from criticism.

For many scientists the code becomes challenging when new insights or paradigm shifts within the analytic community significantly change the importance of their work. This includes new application domains requesting the development and application of new methods or the integration of new subject matter experts.

As an example, combat modeling is bound traditionally by the laws of Newtonian physics. This book follows this paradigm in the first part and copes with the challenge to model movement, observation, sensing, the use of weapons, etc. As long as all participants follow these rules of physics, results are comparable and solutions can be used across the domain. However, new application domains, like human, social, cultural, and behavioral modeling, require the integration of social sciences, which is far from being as homogeneous in predictions and causality as is Newtonian physics. Accepting the need for this new diversity and the need to experiment with sometimes contradicting theories is challenging for established experts, but necessary to meet the needs of the user of the research.

Code of Professional Ethics
for Simulationists

The discussion on the need for a professional code of ethics for simulation experts started in parallel with the rise of modeling and simulation as a tool supporting decision makers and later on as a tool for training and education. One of the first articles on this subject was presented during the Winter Simulation Conference 1983 by John McLeod. He identified in McLeod (1983) the need to better address the extent to which information derived from simulation-based studies is valid and applicable to the decision of interest. The derived problem categories involve the selection of data, the modeler's interpretation of the dynamics of the system

Table 2.2 MORS Goals and Code of Ethics

The Society endeavors to:

- Understand and encourage responsiveness to the needs of the user of military operations research.
- Provide opportunities for professional interchange.
- Educate members on new techniques and approaches to analysis.
- Provide peer critique of analyses.
- Inform and advise decision makers on the potential use of military operations research.
- Encourage conduct consistent with high professional and ethical standards.
- Recognize outstanding contributions to military operations research.
- Assist in the accession and development of career analysts.
- Strive for a membership which is representative of the military operations research community.
- Preserve the heritage of military operations research.
- Preserve the role of MORS as a leader in the analytical community.
- Encourage the use of operations research in support of current military operations.

The Society promotes the following Code of Ethics for its members:

Military and National Security OR Professionals must aspire to be:

- Honest, open and trustworthy in all their relationships.
- Reliable and consistent in the conduct of assignments and responsibilities, always doing what is right rather than expedient.
- Objective, constructive and responsive in all work performed.
- Truthful, complete and accurate in what they say and write.
- Accountable for what they do and choose not to do.
- Respectful of the work of others, giving due credit and refraining from criticism of them unless warranted.
- Free from affiliation with others or with activities that would compromise them, their employers, or the Society.

---

modeled, and the analyst's interpretation of the results. Enriched by many following discussions, the Society for Modeling and Simulation (SCS) sponsored a task group to recommend a Code of Professional Ethics for Simulationists focusing in particular on the needs for modeling and simulation experts. The result was presented during the Summer Computer Simulation Conference 2002 by Ören et al. (2002). The proposed code covers five areas: personal development and the profession, professional competence, trustworthiness, property rights and due credit, and compliance with the code. Table 2.3 presents the five areas.

This Code of Professional Ethics for Simulationists was translated into French and Turkish. Translations into Chinese, Spanish, and Catalan are

Table 2.3 Code of Professional Ethics for Simulationists

1. *Personal development and the profession*

As a simulationist I will:

1.1 Acquire and maintain professional competence and attitude.
1.2 Treat fairly employees, clients, users, colleagues, and employers.
1.3 Encourage and support new entrants to the profession.
1.4 Support fellow practitioners and members of other professions who are engaged in modeling and simulation.
1.5 Assist colleagues to achieve reliable results.
1.6 Promote the reliable and credible use of modeling and simulation.
1.7 Promote the modeling and simulation profession, e.g. advance public knowledge and appreciation of modeling and simulation and clarify and counter false or misleading statements.

2. *Professional competence*

As a simulationist I will:

2.1 Assure product and/or service quality by the use of proper methodologies and technologies.
2.2 Seek, utilize, and provide critical professional review.
2.3 Recommend and stipulate proper and achievable goals for any project.
2.4 Document simulation studies and/or systems comprehensibly and accurately to authorized parties.
2.5 Provide full disclosure of system design assumptions and known limitations and problems to authorized parties.
2.6 Be explicit and unequivocal about the conditions of applicability of specific models and associated simulation results.
2.7 Caution against acceptance of modeling and simulation results when there is insufficient evidence of thorough validation and verification.
2.8 Assure thorough and unbiased interpretations and evaluations of the results of modeling and simulation studies.

3. *Trustworthiness*

As a simulationist I will:

3.1 Be honest about any circumstances that might lead to conflict of interest.
3.2 Honor contracts, agreements, and assigned responsibilities and accountabilities.
3.3 Help develop an organizational environment that is supportive of ethical behavior.
3.4 Support studies which will not harm humans (current and future generations) as well as the environment.

4. *Property rights and due credit*

As a simulationist I will:

4.1 Give full acknowledgement to the contributions of others.
4.2 Give proper credit for intellectual property.
4.3 Honor property rights including copyrights and patents.
4.4 Honor privacy rights of individuals and organizations as well as confidentiality of the relevant data and knowledge.

(*continued*)

Table 2.3 (*Continued*)

---

5. *Compliance with the code*

As a simulationist I will:

5.1 Adhere to this code and encourage other simulationists to adhere to it.

5.2 Treat violations of this code as inconsistent with being a simulationist.

5.3 Seek advice from professional colleagues when faced with an ethical dilemma in modeling and simulation activities.

5.4 Advise any professional society which supports this code of desirable updates.

---

being prepared. The code was adopted by several modeling and simulation organizations, among them the Society for Modeling and Simulation (SCS), the McLeod Institute of Simulation Sciences (MISS) and the McLeod Modeling and Simulation Network (M&SNet), the Simulation Interoperability Standard Organization (SISO), and the NATO Modeling and Simulation Group (NMSG).

Each engineer working in the domain of combat modeling and distributed simulation is bound by these codes and should not only be aware of them, but he/she should follow them in his/her daily routine and activities. No matter which methodology is applied, no matter what the military application domain is, the conducting engineer must make sure that he or she is not only technically, but also ethically, the best engineer for the job.

REFERENCES

McLeod J (1983). Professional ethics and simulation. In: *Proceedings of the 15th Winter Simulation Conference (WSC '83)*, edited by S Roberts. IEEE Press, Piscataway, NJ, Vol. 1, pp. 371–374.

Ören TI, Elzas MS, Smit I and Birta LG (2002). A code of professional ethics for simulationists. *Proceedings of the 2002 Summer Computer Simulation Conference*. Society for Computer Simulation, San Diego, CA.

Chapter 3

The NATO Code of Best Practice for Command and Control Assessment

Andreas Tolk

CONTEXT AND OVERVIEW

The North Atlantic Treaty Organization (NATO) Code of Best Practice (COBP) for Command and Control (C2) Assessment was produced to facilitate high quality assessment in the area of C2. The COBP offers broad guidance on the assessment of C2 for the purposes of supporting a wide variety of decision makers and C2 researchers. It presents a variety of operations and operational research methods related to combat modeling that can be applied and orchestrated in support of analysis and evaluation of C2 related research questions. As such, it is a best practice guide on how to apply various means of operations research within a combat modeling related study.

Furthermore, the COBP is the product of international collaboration that has drawn together the operational and analytical experience of leading military and civilian defense experts from across the NATO nations. It represents the common understanding on how to conduct good C2 research within a coalition.

In summary, the COBP enhances the understanding of best practice and outlines a structured process for the conduct of operational analysis for C2. It shows how to structure a study that utilizes combat modeling and distributed simulation. This chapter focuses on the necessary details in the context of combat modeling and distributed simulation for engineers. The COBP itself can be obtained

Engineering Principles of Combat Modeling and Distributed Simulation, First Edition.
Edited by Andreas Tolk.
© 2012 John Wiley & Sons, Inc. Published 2012 by John Wiley & Sons, Inc.

at no cost from the website of the US Department of Defense Command and Control Research Program (last visited in June 2011: http://www.dodccrp.org). Three different versions are available for download, namely

- the NATO Code of Best Practice (NATO, 2002a);
- the NATO COBP Analyst's Summary Guide (NATO, 2002b); and
- the NATO COBP Decisionmaker's Guide (NATO, 2002c).

The first version is the complete final report of the study group of international experts. However, as neither analysts and researchers nor high level decision makers like to read through books, the expert group furthermore provides two short versions of only 20 pages. The short versions highlight the main topics of interest for these two categories: the Analyst's Summary Guide in support of a Quick Guide (Getting Started) for researchers that have to plan, execute, and evaluate a C2 study and the Decisionmaker's Guide for project managers and engineers conducting the supervisor role in such projects. Both short guides are intended to provide entry points for where to look for details and do not replace the necessity of reading through the complete report.

The NATO Research and Technology Organization

To better appreciate the COBP, I will start this chapter with an overview of NATO's Research and Technology Organization (RTO). The RTO comprises the Research and Technology Board (RTB), which provides the leadership by the board members, and the Research and Technology Agency (RTA), which provides the administrative and organizational support with a permanent staff. Under the RTO, seven technical panels and groups were established to provide management of special tasks in seven main categories. These seven panels and groups with their visions are the following.

- *System Analysis and Studies* (SAS) conducts studies and analyses of an operational and technological nature and promotes the exchange and development of methods and tools for operations analysis (OA) as applied to defense problems.
- *Human Factors and Medicine* (HFM) provides the science and technology base for optimizing health, human protection, well-being and performance of the human in operational environments with a consideration of affordability.
- *Information Systems Technology* (IST) identifies and reviews areas of research of common interest and recommends the establishment of activities in these areas.
- *Applied Vehicle Technology* (AVT) improves the performance, affordability, and safety of vehicles.

- *Systems Concepts and Integration* (SCI) advances knowledge concerning advanced systems, concepts, integration, engineering techniques and technologies across the spectrum of platforms and operating environments to assure cost-effective mission area capabilities.
- *Sensors and Electronics Technology* (SET) advances technology in electronics and passive/active sensors and enhances sensor capabilities through multi-sensor integration/fusion in order to improve the operating capability and contribute to fulfilling strategic military results.
- *NATO Modeling and Simulation Group* (NMSG) promotes cooperation among NATO and its partners to maximize the effective utilization of modeling and simulation (M&S).

In addition to these technical panels and groups, the Information Management Committee (IMC) was recently established. The mission of the IMC is to provide advice and expertise in the area of applied information management to the RTO as a direct support element of the RTA. It also provides support on information policy and management matters to the benefit of NATO and the nations. IMC's activities cover the entire lifecycle of information, including the acquisition, processing, retrieval, exchange and distribution of information. However, in contrast to the technical panels and groups, no technical teams are set up to conduct research.

Each technical panel or group can set up technical teams of recognized experts that collaborate on an agreed statement of work to contribute to solve certain problems in the domain of the panel and conduct research. The NATO COBP was written by such an international expert group established under the lead of the SAS panel: the technical team SAS-026. The team had members supporting the activities from Canada, Denmark, France, Germany, Norway, Spain, The Netherlands, Turkey, the UK, and the United States.

The Context of the NATO COBP

An earlier version of the COBP was published in 1999. The focus of this first version was traditional warfare as defined in NATO's Article 5 of the Charter of NATO signed on April 4, 1949 (NATO, 1949). This article reads as follows:

> The Parties agree that an armed attack against one or more of them in Europe or North America shall be considered an attack against them all and consequently they agree that, if such an armed attack occurs, each of them, in exercise of the right of individual or collective self-defence recognised by Article 51 of the Charter of the United Nations, will assist the Party or Parties so attacked by taking forthwith, individually and in concert with the other Parties, such action as it deems necessary, including the use of armed force, to restore and maintain the security of the North Atlantic area.
>
> Any such armed attack and all measures taken as a result thereof shall immediately be reported to the Security Council. Such measures shall be terminated

when the Security Council has taken the measures necessary to restore and maintain international peace and security.

In other words, the first version of the COBP dealt mainly with scenarios as perceived during the days of the Cold War: massive main battle tank duels supported by air force operations, well understood opponent structures, military doctrine defining the actions on the battlefield, etc.

With the fall of the Berlin Wall and the Warsaw Pact in the last decade of the 20th century a new era of military operations began, often referred to as Operations Other Than War (OOTW). In order to fulfill the promise to offer broad guidance on the assessment of C2 issues for the purpose of supporting a wide variety of decision makers and the conduct of C2 research in this new era as well, a revision was needed.

The characteristics of this new era are the multilateral dynamics including interactions with new categories of players, such as non-governmental organizations (NGO), private volunteer organizations (PVO), international organizations (IO), international corporations, transnational, sub-national, criminal, and other organizations. In addition, they involve action–reaction dynamics characterized by the impact of interacting soft elements such as culture, morale, doctrine, training, and experience. As such, overall uncertainties in the military planning and decision support processes increased and required engineering methods in support of and supported by operations research studies.

Rudyard Kipling summarized the six interrogatives that have to be addressed in each study: "I keep six honest serving men (they taught me all I knew); their names are What and Why and When and How and Where and Who." The COBP recommends a structured approach to conduct C2 studies to find answers to these questions.

Structure of the NATO COBP

The chapters of the COBP follow the recommended phases of an operational study using the means of operations research including combat modeling and potentially distributed simulation, which are

- the initialization phase;
- the preparation phase; and
- the execution, evaluation, and assessment phase.

Starting with the sponsor's problem, the initialization phase provides best practices for the problem formulation and the solution strategy.

For industry-oriented readers, this phase is comparable with writing a proposal. In the problem domain phase, you define your understanding of the problem to be solved, including constraints, possible options, etc. In the solution strategy part, you describe your planned research efforts, possible methods and tools, and deliverables including a schedule and other management means.

For more academically-oriented readers, this first phase is comparable to preparing a PhD proposal. You define the problem as the gap in the body of knowledge that needs to be closed in order to provide answers to the sponsor's questions. In your proposal defense, you identify your research methods, eventually your experiments, your metrics for success, and your research plan including scheduling of research efforts, which is equivalent to capturing the solution strategy.

The initialization phase actually touches all topics detailed in later phases by addressing the six interrogatives of essence to the study, looking at applicable theory and related methods and tools, availability of data, and where possible risks are, summarizing everything in a study plan and study management plan and a list of deliverables.

The preparation phase comprises the main phases of the foundational research that is conducted to answer the six interrogatives and necessary metrics. Human and organizational issues address the entities, activities, and relations of interest. They are placed in scenarios to address the dynamic components in the context of environmental and other relevant constraints. In order to produce the reproducible and explainable results we need to define measures of merit. At the end of the preparation phase we therefore know who is doing what, where, and when and how and why actions are conducted. We also know what scenarios are needed to capture all these aspects, and what and how we take measures in order to produce the results.

When we use the comparison with a PhD thesis again we see that the preparation phase defines everything that needs to be done before the experiments can be conducted. This is also true in industry: the preparation phase comprises everything that is necessary to understand and prepare after the contract is received and before the work is conducted. It details everything promised in the solution strategy and makes it applicable to the engineers who have to perform the work. The result is the blueprint of the experiments to be conducted following the solution strategy answering the sponsor's problem.

The execution, evaluation, and assessment phase comprises the selection of appropriate methods and tools, obtaining the data based on engineering methods, and assessing the risk of recommendable solutions. It is often perceived to be the main part of the study but, as already can be seen, if the preparation part is conducted correctly, this phase is pure execution and requires very limited additional academic work.

One of the main contributions of this phase—as emphasized in the COBP and often not done correctly in practice—is the selection of appropriate tools and the orchestrated execution thereof. An often observed mistake is that the sponsor already defines the tool or tools to be used to solve the problem. This is a significant violation of the recommended best practices: which tools can be used to address the research question can only be decided based on the results of the research conducted in the preparation phase. Conducting the preparation phase already with the mindset of having to use certain tools unnecessarily limits the options available to the team.

All sub-phases within the main phases as well as the main phases themselves are connected via feedback loops. When within the course of the preparation phase certain ideas of the solution strategy prove to be infeasible, the strategy should be updated accordingly. Also, if new emerging methods and tools are discovered in the course of in-depth research, they should be integrated into the study plan in agreement with the constraints of the study. Restructuring of a study plan based on new research findings is not a mistake. If insufficient research in the initialization phase leads to an oversight the later corrections can cause unnecessary costs, but even then the Code of Ethics requires discussing these findings with the customer and correcting shortcomings to conduct the best operational study possible.

It is best practice to use peer reviews in every phase: before major milestones in a study are reached, a peer review with subject matter experts who were not a part of the assessment team of the evaluation phase should provide feedback, ensuring the best results producible for the customer.

Finally, the results of all phases are used to produce the deliverables or products. They do not only address the sponsor's problem—although this is the main reason for them to be produced—but they also contribute to the body of knowledge and contribute to the foundational research reusable in future studies in related domains (Sinclair and Tolk, 2002) as well as to the knowledge base applicable within the community. Figure 3.1 shows the overview as used in the COBP (NATO, 2002a).

In the next sections, these phases and sub-phases will be addressed in more detail, allowing the application of the best practices recommended by NATO as one of the guiding engineering principles for combat modeling and distributed simulation.

THE INITIALIZATION PHASE

As stated in the first section of this chapter, the initialization phase provides the big picture about what the problem of the sponsor is, and which engineering methods and principles can be applied to conduct research to provide answers.

Problem Formulation

Vince Roske, researcher at the Institute for Defense Analysis (IDA) is attributed with the quote: "First find out what the question is—then find out what the real question is!" Similarly, Roy E. Rice recommended in the symposia of the Military Operations Research Society (MORS): "Define the Problem again ... because you didn't do it right the first time!" Both quotes are anecdotal, but make a good point. The practitioners of the expert group stressed in the COBP the importance for a very careful and rigorous problem analysis that should identify the context of the study and aspects of related issues.

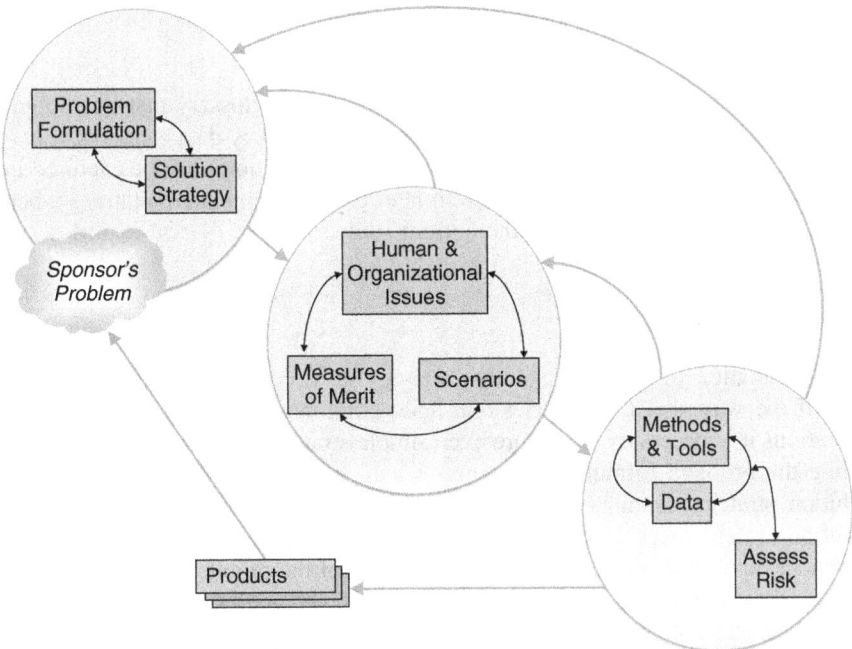

Figure 3.1   Overview of NATO COBP phases.

The context of the study includes the geopolitical context, political, social, historical, economic, geographic and technological environments, actors, threats, aims and objectives of the study, generic C2 issues, relevant previous studies (which includes literature research in related domains), and stakeholders and their organization affiliation. The currently emerging topic of human, social, cultural, and behavioral (HSCB) modeling, which will be dealt with later in this book, supports this activity.

The aspects of the problem, each of which has to be addressed by the assessment team in this first phase, include the issues to be addressed, assumptions, high level measures of merit, controllable and uncontrollable independent variables, constraints on the variables, time constraints on deliverables, and type of decision(s) to be supported (like providing a single report, or providing decision support for a chain of decisions to be made over time).

The following principles were formulated to guide through the problem formulation process.

- Do it before developing concepts for analysis or model selection.
- Understand the decisions to be supported and stakeholder viewpoints.
- Carefully review previous work; do a literature review.
- Include analysis and synthesis.
- Be broad and iterative.

- Practical constraints are modifiers, not drivers.
- Address risks to the study explicitly.

One critical delivery of every study will be a data glossary that unambiguously defines all terms used in all phases. Consequently, key data elements, metadata, information and terms used in the problem formulation must be captured in this phase to avoid miscommunication in later phases or misinterpretations when future researchers want to reuse the current study.

Solution Strategy

The formulation of a formal solution strategy is a recommended best practice even if the way ahead seems clear. In agile environments like today's military operations it is necessary to capture even simple strategies to communicate them. While the problem formulation captures *what* problems need to be solved, the solution strategy captures *how* the problem will be solved. To this end, the solution strategy takes the output from the problem formulations, builds concepts, attributes, and relations to represent the elements of the problem formulation and embeds them in scenarios representing the context, and addresses metrics defining what to measure and how to evaluate the results. This is an iterative process within the team and often includes the sponsor as well. New terms, relations, attributes, etc. need to be captured in the data glossary.

The results of both sub-phases contribute to two artifacts that should be part of the deliverables: the study plan that describes what has to be solved and how it will be solved and the study management plan to guide the direction, management and coordination of the assessment team in the form of a work breakdown structure and a time-phased execution plan that includes necessary resources. The detail of associated supporting plans depends on the individual study, but providing the following artifacts is good practice: analysis plan, deployment and modeling and simulation plan, data collection and engineering plan, configuration management plans, risk and uncertainty management plan, quality assurance plans, security plan, validation and verification plans, and the plan of deliverables. Not all plans are always needed, but using this checklist in the beginning ensures that nothing is overseen.

As stated earlier, the initialization phase is pivotal for the success of a story. If the planning is done correctly, many mistakes can be avoided that will result in costly rollbacks later. Thinking about this phase as establishing a research contract is helpful to guide the engineer to decide what needs to be taken care of.

THE PREPARATION PHASE

The preparation phase is the core of the research. In this phase, the human and organizational issues are refined and conceptualized, the scenarios that will be used to conduct the evaluation are identified, and the measures of merit that will be used to determine the success or failure of evaluated options are defined. This phase refines the six interrogatives who, what, where, when, why, and how in sufficient detail to execute experiments in the following phase.

Human and Organizational Issues

As described in the introduction to this chapter, the significant increase of possible partners and opponents in today's operation in comparison with traditional scenarios as used during the Cold War was one of the drivers behind the revision of the first COBP. The section on Human and Organizational Factors is among the sections that underwent significant change compared with the original code to account for findings of the rapidly evolving research on human and organizational behavior. It is expected that this section will be subject to additional changes in the near future based on new findings in the HSCB domain.

C2 is distinguished largely by the human dimension. Key differences between C2 analyses and traditional military operational analysis applications include the need to deal with distributed military teams and organizations under stress and their decision making behavior as well as the behavior of and interaction with non-military organizations, political groupings, and amorphous groups such as crowds and refugees. Thus, the formulation of the problem and the development of certain strategies cannot be completed without explicit consideration of both human and organizational issues.

Among the human factors of interest are human behavior related to performance degradation, like those produced by stress and fatigue, decision making behavior in the light of the cognitive complexity of the issues and the capacities of commanders and other decision makers, and command style. The organizational factors of interest to the COBP deal with relationships between groups of individuals including connectivity, roles, and organizational structures.

Among the key considerations when structuring the problem is whether individual decision making and behavior of individuals or groups is important to the C2 processes under analysis. Human performance affects behavior and vice versa. Human performance depends on psycho-physiological variables such as stress, fatigue, and alertness, and on ergonomic and external factors limiting performance and behavioral freedom. Individual and group behavior is the result of social interaction among individuals and groups, including the interactions by military commanders and their troops, and the underlying psycho-physiological factors as well as the cultural, educational, and religious background of individuals. HSCB focuses explicitly on these aspects. Any time human performance and behavior are at issue, parameters and models will be needed to reflect those issues. For example, systems that involve human activity, such as watch or command centers, need to be studied in ways that reflect differences in C2 performance that can be traced to human performance and behavior issues, as well as those arising from experience or training, coalition differences (e.g. language, national doctrine and command style), or differences in doctrine and practice between services or branches.

Increasingly, analysts also have to deal with issues where individual decisions are important. This is especially true for operations in which even tactical level decisions by a lower level military commander may have strategic implications because of media presence. The "strategic corporal" in charge of a command post during a riot who has to decide whether to open fire into an increasingly aggressive crowd although a CNN cameraman stands nearby is an example. This represents a major challenge because the variety of human behaviors involved makes modeling decision behavior very difficult and even more difficult to validate. In contrast to simple and contingent decisions that are made on the basis of physical and well established rules, modeling complex decisions relies mostly on "human-in-the-loop" techniques.

Considered to be of significance in actual operations, command style represents a considerable challenge to modeling because it is an elusive and multi-dimensional concept. Differences in command style may be reflected by appropriate attributes such as background and training of commanders, their decision and order style, risk tolerance, and operational experience.

The linkage between human and organizational issues is particularly direct and close. Properly done, organizational design reflects the interaction between the tasks to be completed, the people available to perform them, and the systems or tools that support those people. Organization is a serious subject in military analyses. Military organizational issues are likely to be driven by a fairly small and finite list of principles. Analysts asked to work on C2 issues can use the known list of factors as a checklist for organizational differences to determine whether they need to build organizational matters into their research designs, including the issue of informal relationships that may have evolved in order to overcome organizational deficits and thus streamline day-to-day operations. The principal differences of military organizations are related to structure, function, and capacity. Functional differences include the distribution of responsibility.

The main new issue addressed by the COBP is that organizational factors are closely interrelated. Thus, by changing one factor, other factors may be changing as well, as the system is complex. Review of organizational issues should be treated in a two-step process, guided by a hypothesis testing logic. When assessing new C2 systems, analysts will often need to search for potential role gaps or role overlaps. Either of these would be dysfunctional in military operations over time. Changes in information structures also have considerable potential for creating problems of this type, particularly in a coalition, inter-agency, and other operation and information distillation—such as collaboration—as well as alternating roles and on results in role gaps or role overlaps.

Adequate treatment and modeling of human and organizational behavior demands a truly interdisciplinary approach involving, in addition to technical analysts grounded in the physical sciences and mathematics, experts grounded in the social and human sciences, especially psychology.

As Davis pointed out in Tolk et al. (2010), modeling population-centric activities requires serious use of social science. Fortunately, the social science literature has a great deal to offer. However, the literature is fragmented along boundaries between academic disciplines, between basic and applied research, and between qualitative and quantitative research. The uncertainties and disagreements are profound, on subject-area facts and even on the nature of evidence and the appropriateness of different methodologies. It is therefore pivotal to capture definitions and assumptions in detail to support the credibility of the study and the applicability of recommendations and reuse of study artifacts.

Scenarios

While the human and organizational issues address the concepts, their attributes, and their relations, scenarios put these concepts into a dynamic context. Based on the general scenario framework developed by the NATO Panel 7 Ad Hoc Working Group, the COBP defines a scenario as the description of the area, the environment, means, objectives, and events related to a conflict or a crisis during a specified time frame suited for satisfactory study objectives and the problem analysis directives. The scenario provides the context for and bounds the arena of the analysis. It is used by the analyst to focus the analysis on central issues.

While traditional military operations are characterized by a good mutual understanding of means, weapon systems, doctrine, tactics, and other elements of such scenarios, new operations are defined by uncertainty, disagreement, and a multitude of options. Instead of defining scenarios the analyst may instead use *vignettes*. The COBP defines vignettes as scenarios that are not yet approved, or as smaller scenarios that are excursions from the main scenario. In the context of this book, this definition is extended to mini-scenarios or scenario components that are typical for the questions under discourse. Examples for such vignettes can be setting up a control post on a street under threat of terrorist acts like road bombs or manning an armed observation post at the corner of a marketplace where a crowd is on the edge of rioting.

Each concept identified in the section on human and organizational issues needs to be part of at least one scenario or vignette. If an issue has been identified in the problem formulation but does not become part of a scenario, it cannot be evaluated, as the evaluation happens within the scenarios setting the necessary context. It is therefore good practice to use checklists to ensure that all issues are addressed accordingly.

Measures of Merit

While scenarios define elements, relations, and dynamics to be evaluated within the study related to the questions of the customer, measures of merit enable the evaluation of interim states, the final state, and particular events of interest to deliver the data needed to answer the questions of the customer. They provide the metrics enabling repeatable and understandable evaluation within the study.

Furthermore, it should be pointed out that the ultimate purpose of C2 systems analysis is not to assess the C2 process or C2 performance, but to assess how and

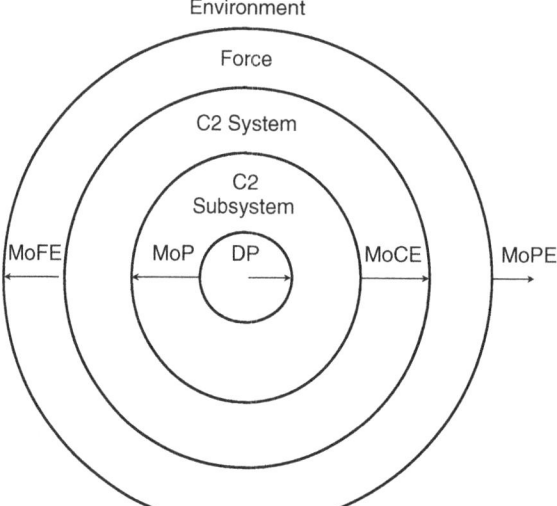

Figure 3.2   Hierarchy of measures of merit.

to what degree C2 affects the accomplishment of military missions it controls and the political objectives that drive the military missions. Consequently, a metrics hierarchy is needed. This idea is not new. In 1994 the NATO Report on the "Impact of C3I on the Battlefield" recommended that a hierarchy of measures be established as an important step in understanding overall system effectiveness. C2 systems should be analyzed at different levels of detail. The NATO COBP modified this hierarchy with the objective of characterizing the contribution of military actions to broader policy societal outcomes. The following five levels of measures of merit have been adopted (see Figure 3.2):

- Measures of Policy Effectiveness (MoPE) which focus on policy societal outcomes ("Is suppressing the riot in the marketplace stabilizing the situation?");
- Measures of Force Effectiveness (MoFE) which focus on how a force performs its mission or the degree to which it meets its objectives ("Does my C2 system enable me to bring forces coordinated to the four corners of the marketplace to suppress the riot?");
- Measures of C2 Effectiveness (MoCE) which focus on the impact of C2 systems within the operational context ("Can I reach all forces supporting me in time with the communications means I have?");
- Measures of Performance (MoP) which focus on internal system structure, characteristics and behavior; performance measures of a system may be reduced to measures based on time, accuracy, capacity or a combination that may be interdependent ("How far do I reach with my tactical radio, and how many messages can I send?");

- Dimensional Parameters (DP) that are the properties or characteristics inherent in the physical C2 systems ("What is the baud rate of my radio?").

As with all other relevant terms, the metrics and measures of merit need to be defined in the data glossary for the study, preferably including their mathematical definition as well. As a general rule, however, measures of performance provide metrics for system performance (usually based on engineering specifications of the systems) while measures of effectiveness provide metrics for the overall operational effectiveness (usually based on specifications of the operational experts).

Within this book, several metrics will be introduced and applied in later chapters to measure success for combat modeling in the traditional sense. Typical metrics are movement and attrition related, such as kill ratios (enemy forces destroyed compared with own forces destroyed over time), attack or defense success probabilities, damage inflicted, terrain and assets occupied or defended, etc. While such metrics are appropriate for combat, OOTW requires other measures. The COBP gives examples to measure success using normality indicators, as enumerated in Table 3.1.

The COBP also gives several examples for successfully applied measures of merit hierarchies and supporting tools. The analyst guide recommends the following process to support defining appropriate metrics:

- Establish evaluation environment
- Define evaluation goals
- State context, assumptions, constraints
- Define domain—MoPE, MoFE, MoCE, MoP, DP
- Identify particular measures
- Specify measures
- Establish scenario or stimulus → Scenario
- Establish data collection means → Data
- Pilot test, revise measures and procedures
- Conduct the tests, debrief, and analyze

Table 3.1 Normality Indicators

Criterion	Examples
Political	Elections, political participation
Economic	Unemployment, interest rates, market baskets
Social	Number of students in schools, number of refugees
Technological	Telephone system availability, Internet availability
Legal	Judicial system functioning
Environmental	Roads, water supply, energy supply, critical infrastructure
Cultural	Sports events, concerts, cultural life

For the engineer of combat modeling and distributed simulation it is important not only to understand these principles, but to understand the responsibility of ensuring that the necessary concepts, attributes, and relations are correctly modeled, the dynamics are represented in the scenario and supported by the model, and the dimensional parameters are accessible during the execution of a distributed simulation event, as will be addressed in the next section.

THE EXECUTION, EVALUATION, AND ASSESSMENT PHASE

While up to this point everything was theoretical research, in this phase the rubber finally meets the road: models and tools are selected, data are obtained, experiments are conducted, results are evaluated, and risk is assessed. The NATO COBP groups the two sub-phases methods and tools and data together, followed by assessing the risk. However, in the context of this book it makes sense to address all three sub-phases and also include the products (although artifacts are already collected that belong to the products in the very first activity).

Selecting Methods and Tools

In this section we consider the best quantitative or qualitative methods and tools for assessing C2 processes, performance, and effectiveness. We cover methods and tools used for analysis, training or operations, each of which has different requirements. In principle, any analytic method or tool has potential application for C2 assessments. It is the task of the engineer to select the best available ones for the objectives defined in the problem formulation, to support the solution strategy, and to fulfill the constraints and requirements identified in detail in the preparation phase. The tools are grouped into four principal categories:

- *Data Collection/Generation Tools* used to collect or generate data for subsequent analysis;
- *Data Organization/Relationship Tools* used to organize data or to establish relationships between data;
- *"Solving" Tools* used to mathematically derive solutions to problems; and
- *Support Tools* used to organize, store, and explore typically large sets of empirical data.

Examples for data collection/generation tools are data obtained from analyzing real-world operations lessons learned, after action reviews and reports, or historical analysis of operations. It is also possible to elicit information from subject matter experts, particularly when other options are not available. Specifically, in the operational analysis community, simulations are perceived as data generation tools, as are applications of war games. Game theory can be applied to generate data as well, in particular when it is used in massive mixed strategy applications (game theory is dealt with in a later chapter of this book). Finally, brainstorming is an option to collect or generate data as well.

Tools supporting organizing data and identifying relationships are causal mapping, using experts of the domain, and statistical tools, such as multi-criteria decision analysis, regression analysis, or factor analysis. It is also possible to apply neural nets to identify hidden connections. In addition, the use of other systemic approaches, like the use of systems dynamics, is identified to support this phase in the NATO COBP. New generations of tools of growing interest to the C2 and the analyst community that are not mentioned in the COBP are ontology tools. Ontologies take the ideas behind data modeling to a new level by allowing automatic reasoning over concepts, attributes, and relations and their consistency. More and more academic communities no longer capture their core terms in simple data glossaries but in ontologies. The C2 community is evaluating options in this direction as well (Winters and Tolk, 2009).

Solving tools are those tools that help extract answers to the questions of the customer from obtained and refined data. Obviously, mathematical means play a major role in this category: mathematical analysis, linear programming, goal programming, or integer programming, as specified in standard operations research textbooks, still build the backbone of solving problems. However, as problems are getting so complex that closed solutions without oversimplifying the problem are hardly possible, heuristic searching and optimization techniques such as genetic algorithms but also simulation-based heuristic optimization belong in this category.

Finally, support can be provided by data analysis tools, in particular when being supported by visualization tools. Databases and data repositories allow efficient storage, access, and support evaluation methods. Geographical information systems (GIS) are increasingly needed in support of analysis as well, but also simple tools like checklists, spreadsheets, and other planning tools contribute.

Which tools belong in which category can be discussed in several instances; however, the need to get the data supporting the analysis, preparing the data for the evaluation, solving the problem using the refined data, and presenting the results are typical steps needed to be supported by each tool set.

It is very unlikely to identify one model that supports all required categories and functions. It is therefore *good practice to evaluate and select an orchestrated set of methods and tools*. You should apply models effectively in a complementary way during all phases of the analysis where required and link methods and tools from problem formulation through problem solution. Similarly, it is recommended to link simple tools for scanning the scenario space and complex models for examination of specific points and link a hierarchy of models when required. We shall learn about applicable standards in a later chapter in this book and understand the limits regarding the interoperability and composability of tools and methods.

Obtaining the Data

One fact often neglected when selecting tools or even when developing models is that every datum modeled for the simulation system or applied within the tool has to be obtained from somewhere. The quality of the system is not only driven by the quality of the model itself, but equally by the quality of the data. The best model is useless if no data can be obtained to drive it. Data therefore play a central role. While the first version only implicitly dealt with the issue, the current COBP devotes an entire chapter to best practices on data.

With the increasing availability of data in repositories, metadata—which are data about the data—are becoming as important as data. The main use of metadata is the description of data to allow applicable data to be identified, in case alternatives are available selecting the best data sources for the research question to answer or the task to fulfill, and providing details on assumptions and constraints for the use of these data.

As we often use the same terms to describe different things—or we use different terms to describe the same thing—syntactic consistence of data types does not imply semantic consistence of data content. Metadata, common namespaces, and the use of reference ontologies can help to overcome this problem, but the community is still in its infancy regarding semantic alignment of data, which is the reason that having a study data glossary with definitions at hand to support the analysis for reusability of information is important.

In order to obtain all necessary data the assessment team needs to know, and should capture, these needs in a data engineering plan:

- What data are needed (semantics)?
- In which structures are the data needed (syntax)?
- Who owns the data?
- Are there any security issues?
- What are the costs to buy/collect/generate the data?

If data are not available and cannot be obtained, it is good practice to use the knowledge of subject matter experts to generate the needed data. However, if empirical data become available at a later phase of the study, it is good practice to replace expert-generated data. Consequently, the metadata should capture the origin of the data. But this is not the only datum needing to be documented. The NATO COBP identifies the following categories for metadata:

- sources of data, like official sources (such as authoritative databases, legacy studies, exercises, or data collected in real operations), open sources (such as data found on the Internet or data derived from public domain literature research), or generated as "best guess" or based on experience in related domains by subject matter experts;
- version and age of the data (making sure that no out-of-date information is used in the study);
- reliability (does the assessment team trust the source?) and accuracy (are the data accurate enough for the application? are they fuzzy or vague?) of the data source;

- which transformation was applied to the data, have they been aggregated from higher resolution sources or disaggregated from lower resolution data sources (and what algorithms were used to do this), are they statistical derivations, have they been interpolated or otherwise mathematically smoothed, or are they based on some form of interpretation (by whom)?

All this information becomes essential when the credibility of the study results is evaluated. It also becomes essential when new data become available and the assessment team needs to trace which assumptions will be affected and may have to be reevaluated based on this new finding. The best practices recommended for metadata support traceability and credibility.

Because the data needed will seldom be available in a form and format ideal for the assessment, data engineering is often needed to gather, organize, and transform the available data. Data engineering was recognized and described in the COBP and specified in more detail in Tolk (2003). When following the steps of data engineering, the team discovers the format and location of data through a data administration process, discovers and maps similar data elements through a data management process, asserts the need for model extension and gap elimination through a data alignment process, and finally resolves resolution issues through a data transformation process.

- *Data administration* is the process of managing the information exchange needs that exist within a group of systems, including the documentation of the source, the format, context of validity, and fidelity and credibility of the data. Data administration is therefore part of the overall information management process.
- *Data management* is the planning, organizing and managing of data by defining and using rules, methods, tools and respective resources to identify, clarify, define and standardize the meaning of data and of their relations.
- *Data alignment* ensures that the data to be exchanged exist in the participating systems as an information entity or that the necessary information can be derived from the data available, e.g. using the means of aggregation or disaggregation.
- *Data transformation* is the technical process—often implemented by respective algorithms within the gateways and interfaces—of aggregation and/or disaggregation of the information entities of the embedding systems to match the information exchange requirements including the adjustment of the data formats as needed.

If the right structures and metadata are provided, a lot of this work can effectively be supported by machines and many processes can even be semi-automated (Tolk and Diallo, 2008). A later chapter in this book will highlight the recent developments in this engineering method in more detail.

Finally, the COBP recognizes that there is an urgent need to agree on standards for data, metadata, and data management between the different relevant communities, which are the operational analyst, the C2 system developers, and combat modeling and simulation experts. The reason is that *data being used today by analysts will be the data needed tomorrow by systems engineers, decision makers, and commanders*. Early alignment of standardization with the C2 systems community in accordance with modeling and simulation experts based on the research recommendations of operation analysts will not only save costs, it will also support better decision making, as the data required for systematic answers as conducted by operational experts are known, well defined, and provided and accessible by the supporting infrastructure.

Assessing the Risk

How good is the study, how reliable are the results? Every researcher will be confronted with these questions. Furthermore, the NATO COBP recognizes that there are risks associated with the decision maker's situation that are an inherent part of the analysis, and there are risks related to the conduct of the analysis itself. Assessing the risk phase deals with both of these sets of risks, focusing upon reducing uncertainty and other contributors to risk as well as the mitigation of their effects.

As mentioned earlier in the chapter, the new reality of military operations is characterized by an increase of uncertainty in all components discussed so far. So, one of the first things to address is the question: what is uncertainty?

The NATO COBP defines uncertainty as *the inability to determine a variable value or system state (or nature) or to predict its future evolution*. As stated before, uncertainty is a fact (that is certain): real-world data will be uncertain, incomplete, contradictory, and vague, and this uncertainty can never be reduced completely—it can only be managed. This management is essential and the researcher needs to make the customer aware of it by, for example, visualizing uncertainty. Some sources of these uncertainties are study assumptions, like uncertainties in scenarios and the model input (we may use the wrong scenario or vignettes to answer the questions, or some of our data are not adequate), modeling assumptions, like having uncertainties in the model itself (appropriateness of chosen equations and algorithms), or structural uncertainty, and also model sensitivity producing uncertainties in the outcome. In particular in complex models with many model components that are highly interconnected via a multitude of nonlinear interfaces, small errors can produces tremendous mistakes. Models can even expose chaotic behavior.

While uncertainty can only be managed, the real objective of studies is reducing risk. The COPB defines risk as the possibility of suffering harm or loss. The NATO COBP—when applied and understood by all team participants, including the customer and user of the study—helps to minimize in particular those risks that result from applying study recommendations in the wrong context or without considering the constraints sufficiently.

A key element is to make study recommendation users aware of the robustness of a recommended solution. If nothing else, sensitivity analyses of recommended solutions are mandatory. Even better is the application of risk-based analysis. Solving problems using single expected values leads to fragile solutions, which do not allow decision makers to deal with inherent uncertainty and risk. A risk-based approach can overcome some major pitfalls by focusing on the multiplicity of possible outcomes, open up the possibility of richer solutions, and recommend portfolios of action instead of single solutions. In other word, the focus becomes robustness versus narrow optimality.

In summary, the explicit treatment of risk and uncertainty is best practice in all studies, especially C2 assessment. Even when study resources are limited, it is best practice to do sensitivity analyses and to take a risk-based approach. The use of checklists is recommended to ensure a rigorous treatment. This is not only best practice; it also follows the principles of the Code of Ethics discussed in an earlier chapter of this book.

Products

Finally, the customer expects answers to questions at the end of the study. That is what has been defined in the deliverables. However, there is more that is worthwhile surviving the delivery of the final report than just the answers to the problem of the sponsor.

Generally, products are used within and between studies for

- managing the program (study plan and study execution plan);
- communicating within the team (interim results);
- communicating with the customer (interim briefings, interview results); and
- conserving the knowledge gained during the study (knowledge repositories).

The following deliverables should be part of every product list:

- the study plan, including in particular descriptions of the problem formulation (what) and the solution strategy (how);
- periodic status/progress reports, including briefing to sponsor and stakeholders, conference reports, workshop summaries, and feedback from peer reviewers; and
- interim reports and final report with documentation on scope and assumptions, applied research methods and approaches (including tools that were applied and data that were used), and findings and results.

It is furthermore highly recommended to deliver the study management plan to ensure better reuse of results and interim results, but also methods for the academic as well as for the administrative work. During this chapter, several of these artifacts were explicitly dealt with. Here is an enumeration of useful plans:

- Study glossary

- Analysis plan
- Tool deployment and M&S plan
- Data collection/engineering plan
- Configuration management plan
- Study risk register
- Quality assurance plan
- Security plan
- Review plan
- Plan of deliverables

The advantage of reusing the study glossary should be emphasized. In other chapters of this book, the idea of using a common set of terms with accepted definitions will come up again and again. Current practice is that the glossary is reinvented over and over. If you are a soldier, think about how easy it would be to communicate your needs with the researcher if terms were commonly understood? If you work as a program manager, think about how much cost you can save by using the same terms in several studies: you not only do not have to write the glossary, you also do not have to educate the assessment team members.

To summarize this section on products: whenever you have to decide what should be delivered, remember that the purpose of study products is threefold—to communicate results to sponsors and stakeholders, to provide a lasting record of what went into the planning, and to establish credibility within the technical community.

This chapter summarizes the NATO COBP from the perspective of an engineer for combat modeling and distributed simulation and provides some updates. Many additional viewpoints and nuances are captured in the full version of the COBP. The reader is strongly encouraged to utilize the original in addition to this chapter to support future efforts as a professional combat modeling and distributed simulation engineer.

REFERENCES

NATO (1949). *The North Atlantic Treaty*, Washington, DC, April 4, 1949.

NATO (1994). *Final Report of the NATO Research Study Group AC/243 Panel 7 Ad Hoc Working Group (AHWG) on the Impact of C3I on the Battlefield*, Report AC/243(Panel 7)TR/4, Brussels, Belgium.

NATO (2002a). *NATO Code of Best Practice*, CCRP Publication Series, Washington, DC.

NATO (2002b). *NATO Code of Best Practice Analyst's Guide*, CCRP Publication Series, Washington, DC.

NATO (2002c). *NATO Code of Best Practice Decisionmaker's Guide*, CCRP Publication Series, Washington, DC.

Sinclair MR and Tolk A (2002). Building up a common data infrastructure. *Proceedings of the NATO Studies, Analysis and Simulation (SAS) Symposium on Analyses of the Military Effectiveness of Future C2 Concepts and Systems*, Section B1, AC/323(SAS-039)TP/32, The Hague.

Tolk A (2003). Common data administration, data management, and data alignment as a necessary requirement for coupling C4ISR systems and M&S. *Info Security* **12**, 164–174.

Tolk A, Davis PK, Huiskamp W, Schaub H, Klein GL and Wall JA (2010). Challenges of human, social, cultural, and behavioral (HSCB) modeling. In: *Proceedings of the 2010 Winter Simulation Conference, Baltimore, MD*, edited by B Johansson, S Jain, J Montoya-Torres, J Hugan and E Yucesan. IEEE Press, Piscataway, NJ, pp. 912–924.

Tolk A and Diallo SY (2008). Model-based data engineering for web services. In: *Evolution of the Web in Artificial Intelligence Environment*, edited by R Nayak, N Ichalkaranje and LC Jain. Springer Berlin, Germany, SCI 130, pp. 137–161.

Winters LS and Tolk A (2009). C2 domain ontology within our lifetime. In: *Proceedings of ICCRTS 2009: International Command and Control Research and Technology Symposium*, June 15–17, CCRP, Washington, DC.

Chapter 4

Terms and Application Domains

Andreas Tolk

INTRODUCTION TO DOMAINS AND TERMS OF MILITARY SIMULATION

After the last chapter set the context on how the systems designed and developed by combat modeling and distributed simulation engineers may be used, this last introductory chapter establishes the vocabulary used in this book. The main terms will be defined and the relations between them will be presented, so that the reader should be able to read the following sections without access to the US Department of Defense *Dictionary of Military and Associated Terms* (2010) amended by a glossary of modeling and simulation terms. All terms and concepts will be detailed in the following chapter, but this chapter will give a general overview, similar to the efforts of Page and Smith (1998), who have been among the pioneers of these efforts.

A good introduction to the general engineering principles of modeling and simulation was compiled by Sokolowski and Banks (2010). This chapter cannot replace reading a detailed introduction into modeling and simulation, such as provided by Fishwick (1995) or Law and Kelton (2006), but it will place several of the basic principles into a new context.

But why is this chapter necessary? Like many other communities, the military simulation community developed its own terms and taxonomy, a kind of language of its own. What makes this environment particularly challenging for newcomers is the fact that not only does the military domain have its own language, but the

Engineering Principles of Combat Modeling and Distributed Simulation, First Edition.
Edited by Andreas Tolk.
© 2012 John Wiley & Sons, Inc. Published 2012 by John Wiley & Sons, Inc.

modeling and simulation community has its own language, and the intersection of the two developed a third. Furthermore, soldiers as well as engineers obviously love abbreviations, and the child of this connection loves them as well. It seems to be a highly interesting occupation for military simulationists to come up with a meaningful interpretation of every possible combination of up to five letters of the alphabet (and sometimes—as the example of C2 for command and control in the last chapter shows—even allows the inclusion of numbers). This chapter is designed to help the newcomer over these initial hurdles and get comfortable in this "ARE"—sorry, this "acronym-rich environment."

In addition, a word of caution is in order: the administration of military modeling and simulation is continuously changing within the governing bodies, in the examples used in this book: the US Department of Defense (DoD) and the NATO Research and Technology Organization (RTO). With such administrative changes, terms are often changed as well. Although we shall focus on foundational concepts in this chapter and this book, some terms may change over time, in particular when they imply the allocation of money and responsibility for certain programs or efforts.

SOME BASIC GENERAL DEFINITIONS

Within this first section, we will learn about some basic general definitions that may have a slightly different meaning when used in the context of military simulation. Where appropriate, we will combine two or three terms for comparison to show their commonalities and differences. Where appropriate, official sources will be used to support definitions, but the more important aspect we are trying to accomplish in this chapter is to get a better understanding about the concepts and related challenges, some of them being addressed in the following chapters of this book.

Although many papers and discussions with subject matter experts shaped the context of this chapter, the four most often used references are

- the US *DoD Modeling and Simulation Glossary* (US Department of Defense, 1998) (which is currently undergoing a revision to bring all terms up-to-date);
- the *Training with Simulations: A Handbook for Commanders and Trainers* of the National Simulation Center (2000);
- the *Military Operations Research Analyst's Handbook* of the Military Operations Research Society (Youngren, 1994); and
- the *NATO Modelling and Simulation Master Plan* (NATO, 1996).

What Page and Smith observed in 1998 is still true: even a well-educated simulation engineer will be surprised by the level of difficulty needed to communicate without the understanding of the military simulation viewpoints summarized in this chapter.

Operations Research and Modeling and Simulation

Operations research (OR) is widely understood as the use of quantitative techniques in the domain of decision making in management, government, industry, and the military. It is largely overlapping with systems analysis, i.e. applied methodologies helping decision makers to make better decisions faster. To this end, it includes, among other things, optimization techniques, dynamic programming, data analysis including statistics, decision theory, simulation and planning theory. In principle, these means fall into one of the following two categories: *scientific solutions* and *engineering heuristics*.

Scientific solutions are proven solutions under given constraints, using scientific method and calculus. It should be pointed out that if an exact mathematical solution to a problem exists, this solution should be used. However, most problems observed in "the real world" are too complicated and complex to fit into the constraints of scientific solutions. Nonetheless, solutions are needed, and they apply the ideas of engineering heuristics. The term heuristics is of Greek origin (*heuriskein*) and means "to find a way." As such, heuristics can be understood as a collection of applicable or executable experience: if a certain method worked well in the past, there is a high likelihood that it works in comparable situations in the future as well. Simplified: if an exact solution is known, the scientists have to find it; if they do not succeed, the engineers have to make it work anyhow!

Modeling and simulation as a discipline (we will look at both terms in the next section in more detail) is the use of all kind of models—as they will be defined later in this chapter—to produce a basis for managerial or technical decisions. They are applied to obtain, display and evaluate operationally relevant data in agile contexts by executing models using operational data exploiting the full potential of modeling and simulation and producing numerical insight into the behavior of complex systems. As such, modeling and simulation has several interesting connections to OR:

- OR can and should be applied within modeling and simulation systems whenever possible. OR allows for building better modeling and simulation systems: in order to find the best way using a network of streets and paths, the principles of dynamic programming can be applied.

- Modeling and simulation can be applied as a method within OR. In particular, when a closed solution cannot be formulated, but it is possible to define entities and their relations and their behavior over time or in given contexts and embed them into a situated environment, simulations can be used as an engineering heuristic within OR, as defined before.

- OR methods have to be applied to support internal decisions for modeled entities. If several alternatives exist, and the modeled entity has to choose between them without interaction with a human, algorithms have to be used that model the decision process.

- OR methods have to be applied to evaluate the data obtained by conducting simulation runs. As specified in the NATO Code of Best Practice, the data needed for the measures of merit need to be provided, but they also need to be evaluated, which happens using OR methods.

Interestingly, these categories can be mixed: an OR expert can use a simulation system to support a study, while this simulation system extensively uses OR methods to optimize internal allocations of resources. It is also possible to use a simulation within a simulation in support of making a decision: the various courses of action between which the modeled entity has to choose are internally simulated and evaluated and the alternative with the highest expected value is selected. Alternative to simulations, game theory can be applied, or within game theory, simulations can be applied to compute the expected pay-out value for possible alternatives.

Models and Simulations

While we looked at modeling and simulation as a discipline in the last section, we will look at models and simulation as two—or even three—different concepts here. In many publications, the terms model and simulation are used interchangeably. However, this leads to unnecessary confusion and will be avoided.

Within the US DoD, as in National Simulation Center (2000), a model is defined as a representation of some or all of the properties of a device, system, or object. There are three basic classes of models: physical, mathematical, and procedural.

- A *physical model* is a physical representation of a real-world object as it relates to symbolic models in the form of simulators. Physical models consist of objects such as scaled down versions of airfoils and ship contours for use in wind tunnels and construction projects such as new buildings. The more properties represented by a model, the more complex the model becomes, creating a tradeoff between completeness and complexity.

- A *mathematical model* is a representation composed of procedures or algorithms and mathematical equations. These models consist of a series of mathematical equations or relationships that can be discretely solved. Usually the models employ techniques of numerical approximation to solve complex mathematical functions for which specific values cannot be derived.

- A *procedural model* is an expression of dynamic relationships of a situation expressed by mathematical and logical processes. These models are commonly referred to as simulations.

Again, the borders between these definitions are not always clear. A procedural model is based on algorithms and follows the constraints of mathematics: every executable function within a simulation is an instant of a computable function, which is a domain of mathematics. However, the idea behind the definitions can be traced back to the previous section: if you can define a closed mathematical solution, you build a mathematical model; if you write a simulation to solve your problem, then you build this simulation based on a procedural model. A physical mock-up is an example for a physical model.

However, mixed forms are more likely than not: you can set a mock-up of a tank into a dome that displays static models of physical representations of other real-world objects as three-dimensional digital images (physical). Some of these images are actually driven by a simulation (procedural), while the positions of others are computed based on linear programming algorithms (mathematical).

The definition of a *simulation* given in the Joint Publications (US Department of Defense, 1998) is that it "represents activities and interactions over time. A simulation may be fully automated (i.e., it executes without human intervention), or it may be interactive or interruptible (i.e., the user may intervene during execution). A simulation is an operating representation of selected features of real-world or hypothetical events and processes. It is conducted in accordance with known or assumed procedures and data, and with the aid of methods and equipment ranging from the simplest to the most sophisticated." The *Modeling and Simulation Glossary* (US Department of Defense, 1998) simply defines a simulation as a "method for implementing a model over time."

We will look into the details of simulations, simulators, and stimulators as well as live, virtual, and constructive simulations later in this chapter. In any case, however, simulation is connected with implementing and executing. If this is done on a digital computer, we use the term *computer simulation*. The main focus of this book is computer simulation, but the reader should be aware that computer simulations are only a fraction of the simulation domain.

Law and Kelton (2006) use Figure 4.1 to categorize the approaches on studying a system, resulting in the same categories used above but showing the dependences better than derivable from the definitions alone.

When building such orchestrated sets of complementary tools, as recommended by the NATO Code of Best Practice and applicable to most modeling and simulation tasks as well, it must be considered that modeling and simulation comprise two disciplines: modeling and simulation. While modeling is the process of abstracting, theorizing, and capturing the resulting concepts and relations in a model, simulation is the process of specifying, implementing, and executing this model. *Modeling resides on the abstraction level, whereas simulation resides on the implementation level.* Models are conceptualizations resulting from the purposeful abstraction and simplification of a task-specific perception of reality. Simulation systems are model-based implementations. This definition will help us later to better understand the challenges when building systems that are built up by reusing existing simulation systems to provide functionalities that implement the task driven capabilities. These ideas will be defined more in the section on system engineering support in this chapter.

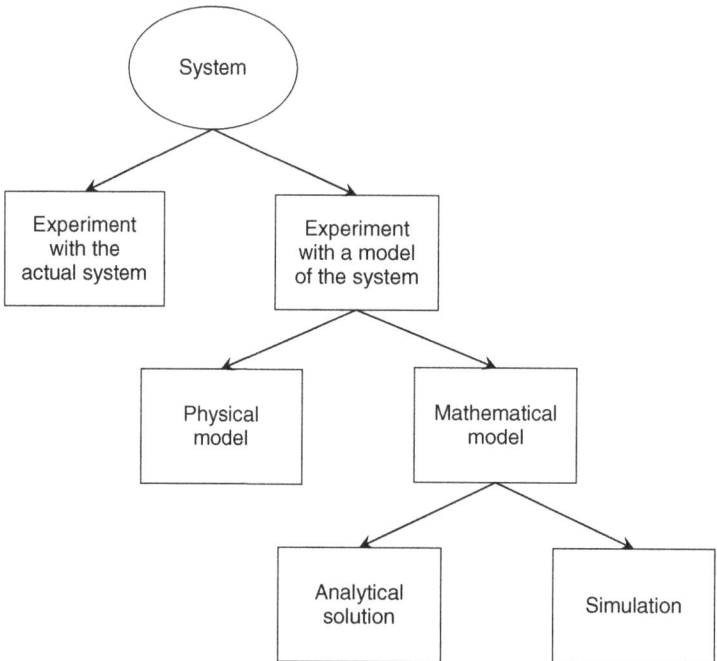

Figure 4.1   Ways to study a system (Law and Kelton, 2006).

Simulations, Simulators, and Stimulators

We will deal with application domains within military simulation later in this chapter. However, depending on which application domain you are in, different terms may be used to refer to how simulation systems are used.

Simulation is the general term applied in the military domain. Whenever simulation systems are applied to support a task with an application domain, we are simulating.

While many engineers know the IEEE definition of a *simulator* as a device, computer program, or system that performs simulation, in the context of military simulation simulators are devices that provide stimuli and feedback for a trainee or a group of trainees. Typical examples are driving simulators or battle simulators where a crew member or a crew is participating with their simulator in a virtual operation, often embedded into a virtual environment that displays other components and participants of the operation.

A *stimulator* is a simulation system that explicitly provides stimuli for a target system in predefined structure via predefined interfaces. This type is often used in testing applications, e.g. providing radar information to a new display to test a new radar screen, but also to conduct stress tests for a new command and control system by producing new input data in the form of messages at a very high rate to ensure that the system functions correctly under such extreme circumstances.

Live, Virtual, and Constructive Simulations

A related set of terms describes live, virtual, and constructive simulation. The easiest way to understand the three concepts is to look at people, systems, and the operation. If real people use the real systems to participate in a simulated operation, then we are talking about *live* simulations. If real people use simulated systems or simulators to participate in a simulated operation, then we are talking about *virtual* simulations. If simulated people use simulated systems to participate in a simulated operation, then we are talking about *constructive* simulations. Table 4.1 summarizes these concepts and terms.

It is possible to think about live simulations as maneuvers, but traditional maneuvers can be augmented with many simulation devices that still qualify the participants to be real people using real systems. Examples are the use of computers, global position devices, and laser technologies that allow the outcome of duel engagements to be computed without involving real ammunition. Also, the effect of artillery support, close air support, landmines, and other means can be simulated and communicated to the real systems to make the exercise very realistic.

Furthermore, mixed forms are supported by mixing live, virtual, and constructive simulation technologies. Using stimulators for the sensor systems, a real system can fight shoulder to shoulder with virtual or constructive partners, which someone can actually perceive via stimulated sensors. Emerging technology even allows displaying such information as overlays in optical devices. Figure 4.2 shows a superimposed virtual target in the real optic of a live weapon system using the DEVIL demonstrator presented by Reimer and Walter (1996). Since these early demonstrations, the underlying visualization technology was improved and driven toward perfection.

Model Hierarchies

Military simulation is applied on all levels of military operations. Simulations are applied on the strategic level, which is the highest level of operations interested in theatres of war, like *Operation Desert Storm*, where decisions have to be made if a country shall go to war or continue in other activities, as much as down to the technical level of individual platforms, where engineering models are used to capture the physical properties of materials like armor, aerodynamic consequences of new equipment, or the effects of several electronic circuits on

Table 4.1   Live, Virtual, and Constructive Simulations

People	Systems	Operation	*Simulation*
Real	Real	Simulated	*Live*
Real	Simulated	Simulated	*Virtual*
Simulated	Simulated	Simulated	*Constructive*

Figure 4.2 Superimposed virtual target in a live simulation.

each other (such as in avionics systems). These different models on different levels build a model hierarchy as depicted in Figure 4.3.

- On the highest military level, the *strategic* level, *theatre/campaign* models are used. They are used to conduct experiments on the structure and the design of the Armed Forces. Decisions on the composition of services and the allocation of strategic capabilities are supported here and need to be supported by models. Often, these models have a low resolution, which means that many contributing entities are aggregated into a single modeled object.

- On the *operational* level, *missions and battles* are simulated to support the analyses of applicability of doctrine, the planning of missions, and the deployment and potential modernization of forces. Questions answered on this level are, for example, "Should an infantry attack always be preceded by a massive air strike?" or "Is it more efficient to use artillery or combat helicopters to escort a tank operation in a given area against infantry-heavy opponents?"

- The *tactical* level breaks battles and combat operations normally down to the *engagement* or duel level. Tactical improvements and optimization of engagements are evaluated here. As discussed earlier in this book, the new military operations are often decided on this level, so that many modeling activities in the recent decade focused on this level.

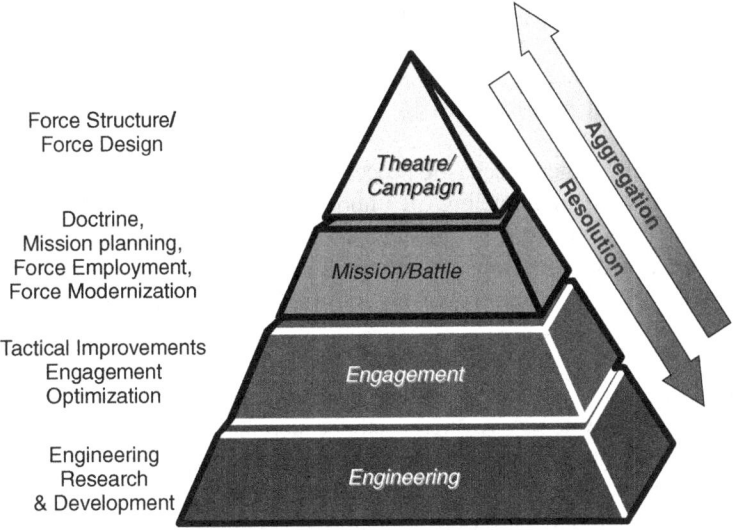

Force Structure/
Force Design

Doctrine,
Mission planning,
Force Employment,
Force Modernization

Tactical Improvements
Engagement
Optimization

Engineering
Research
& Development

Figure 4.3 Model hierarchy of military simulation.

- The *technical* level finally comprises the *engineering* models. The models are physics-based and model the real system components increasingly close to reality. These digital models can be as predictive as physical tests with a prototype. But as digital models are often cheaper than building a prototype, and as engineering models allow thousands of variations to be evaluated instead of being limited to a set of selected experiments with a handful of prototypes, these models are well established for procurement and testing.

Physics-Based, Mathematical, Stochastic, and Logical Models

The differentiation between physics-based, mathematical, stochastic, and logical models often goes hand-in-hand with the model hierarchy definitions, but it is advantageous to describe them as related but individual views on models.

Physics-based models have already been mentioned in relation to engineering models. These models mimic physical behavior in detail: aerodynamics, hydraulics, atmospheric, and other effects are modeled and simulated in highest detail. Motion, momentum, mass, and other physical attributes are captured in the highest degree of freedom necessary to ensure that measurements using the real system are one-to-one reproduced and respectively predicted by the model.

Mathematical models are used to support physics-based models, but they can also be applied to aggregates of system components or even several platforms. While physics-based models address the micro-level of systems, mathematics allows the meso-level and even up to the macro-level to be addressed. Many traditional simulationists represent the view that each model represents a function, and bending this function as close to real observations as possible makes good simulation engineers. Modern simulationists do not always share this view, as they also look at the emerging behavior of systems that results from simple rules on the micro-level, which is in contrast to the dominance of functions on the system level.

As simulations grow larger, stochastic processes and resulting statistics allow *stochastic* models. Instead of modeling all important details, these models use the distribution functions produced by the details. Consequently, stochastic models are more often found at the top of the model hierarchy, as they require a solid foundation allowing their applicability to be validated using a profound set of data.

Finally, *logical* models may capture decisions and resulting sequences of events, but the causality of effects is more important than the physical detail. Many current simulation systems use rule-based decision mechanisms. Whenever these rules are more important than the underlying physics, an appropriate logical model is a good choice. One of the current research topics is how to combine the ideas of emerging simulations based on simple rules on the micro-level and resulting norms that can be observed on the macro-level, representing an emerging logical model instead of an implemented one.

Discrete and Continuous Simulations

The differentiation between discrete and continuous simulation is motivated by technical arguments, not military viewpoints. When objects are simulated, they are characterized by a set of state variables that capture the overall status of the object. Typical variables are location, velocity, but also remaining ammunition, degree of preparedness for a given operation, and more. Within the *discrete* simulation paradigm, these state variables change instantaneously. This is often triggered by an event. If the states change continuously with time, the *continuous* simulation paradigm is utilized. Mixed forms are possible, and are sometimes referred to as heterogeneous simulation methods. As the computers used for the execution are, as a rule, digital computers, solving continuous state changes requires numerical approximation, which always introduces numerical errors.

If continuous simulation is supported, the simulation engineer must ensure that the numerical errors do not falsify the results. It is an unfortunately ill-documented anecdote that the result of several combat simulations changed significantly when NATO upgraded their computer systems. The new computers did solve the same equations using the same numerical approximation methods, but based on the higher accuracy of interim results the battle often took a different direction.

Deterministic and Stochastic Simulations

One of the main challenges in decision making is coping with uncertainties and minimizing the associated risks. This was already emphasized in the chapter on the NATO Code of Best Practice. In order to be able to support this, uncertainties must be represented in the simulation systems, and one of the later chapters evaluates the current state of the art in more detail.

If a simulation does not contain any random parameters and always produces the same output for a given input, it is *deterministic*. If probabilistic components are used not only to represent point estimates but to generate variations in the simulation following the laws of statistics, the simulation becomes *stochastic*.

It is of course unrealistic to assume that any problem operates in a completely deterministic domain. A deterministic environment is a simplification, but it is often necessary to reduce computational effort. Additional methods to account for uncertainty include approaches such as designing with a built-in factor of safety and incorporation of redundant system components. While these approaches avoid the computational burden of stochastic analysis, they do not address the underlying issue of the uncertainty present in the system. Further, they may lead to overly conservative engineering solutions.

It is also a widely distributed incorrect assumption that deterministic simulations are going to provide data "in the most likely range" of possible outcomes. In a study conducted in support of a NATO project, the simulation system KOSMOS—developed as a research project at the University of the Federal Armed Forces of Germany in Munich—was applied to conduct some 70,000 simulation runs (NATO, 1995). KOSMOS primarily modeled movement and attrition for tanks, infantry, artillery, and helicopters. Each process was implemented in a deterministic version and a stochastic version, and it was emphasized within the development that the mean value of each stochastic version clearly maps to the deterministic version. The expectation was that, applying the model to a scenario, the resulting mean value of the measure of performance would map in a similar way. However, Figure 4.4 shows the results of 50 simulation runs of the same input data and their resulting mean value in comparison to the deterministic run. The applied metrics is the force ratio; the scenario was a traditional breakthrough attempt of an attacker against a fortified defender.

In the simulation study for NATO, numerous scenarios were used to produce the data needed for the decision makers, and all of them showed the same result: even with all contributing processes being modeled unbiased, the deterministic run did not match up with the mean value of the stochastic runs. Contrarily, the deterministic value did not carry any more information than any other of the stochastic runs. In other words, if the decision had been carried only by the deterministic run, the wrong advice would have been given. In the example shown, the estimation generated by the deterministic run is far more optimistic that a defender is successful than can be observed as an average when using the stochastic processes.

Entity Level and Aggregate Level Simulation

Differentiating between entity level simulation and aggregate level simulation has principally been covered already. If a simulation models single weapon system

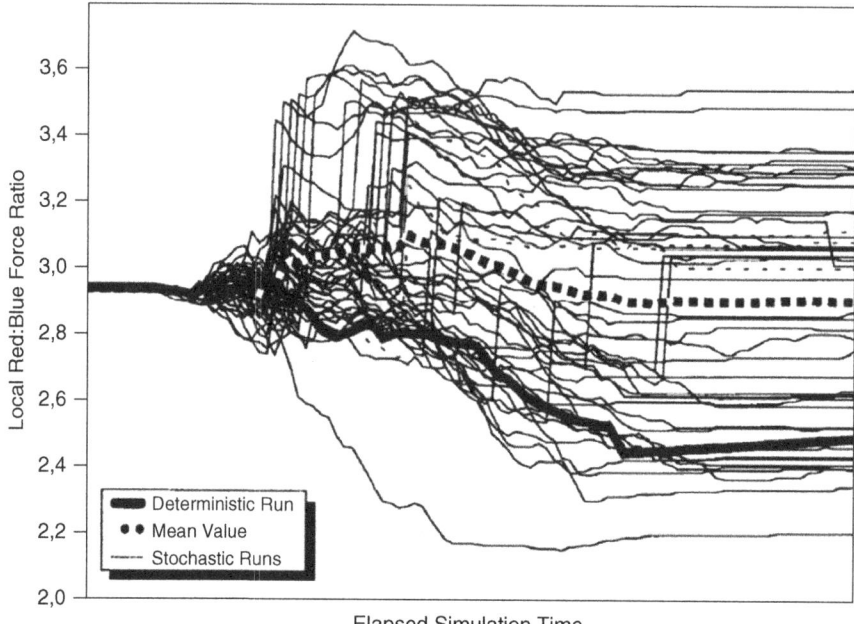

Figure 4.4   Comparison of deterministic and stochastic results.

platforms, like tanks, soldiers, single aircraft or sorties, etc., we are talking about *entity level* simulations. This is the realm of individual duels on the tactical level as well as the level of most simulators. If several of these entities are aggregated into a higher object, like a company with 10 tanks, or an air strike involving 12 aircraft, we talk about *aggregate* level simulation. If all entities with an aggregate are of the same type, like 10 tanks of the same type being in the tank company, we talk about homogeneous aggregates. If this is not the case, like an air strike group comprising eight bombers and four fighter aircraft, we are talking about heterogeneous aggregates.

Modeling Time and Time Advance

Another criterion often used to differentiate between simulation systems is if they are supporting the training or exercise in *real time* or if they are able to run faster than real time or maybe even are slower than real time.

Most simulators have a requirement to run in real time to provide a realistic training experience, in particular in the realm of weapon system simulators. However, on a higher command post level, the time between issuing an order and being able to observe any effects thereof can be long and unproductive for the training audience. In this case, it is an advantage if a system can be executed faster than real time and you only have to wait a couple of minutes instead of several hours to get feedback on your decisions. Also, when a commander wants to use a simulation in support of a decision process and wants to execute simulations of alternative courses of actions, the simulation should execute much faster than real time to allow alternatives to be compared as well as their sensitivity. The other extreme is very computationally intensive detailed calculations of physics-based engineering models.

Simulators that do not execute in real time need an internal representation of time, which is the *logical time* in the simulation system. If the underlying paradigm of a model is a single-server queue, the simulation necessarily is *sequential*, as all events and state changes must be modeled as a sequence in the queue. If several events can be simulated simultaneously, this introduces *parallel* simulation. While scheduling is the main challenge in sequential simulation, synchronization becomes a new challenge in parallel simulation. Whenever several simulation systems are integrated into a federation, we are faced with parallel execution. The synchronization of these federated systems in accordance with their internal times is one of the main challenges for the simulation engineer.

In general, simulationists differentiate between three time advance mechanisms.

- *Real time systems* use a wall clock time to synchronize their events. They occur in the same time as the real event would occur.

- *Event driven systems* use an internal event queue comprising events with time stamps to drive their logical time. The events in the queue are chronologically ordered by the time stamps. Each time a new event is taken from the queue and computed, the logical time is set to the time stamp of this event.

- *Time step driven systems* also use an event queue, but they use fixed time steps to collect all events that happened in this period from the last time step. The advantage is that this approach allows an easy mix of discrete events with continuous processes, as the time step can be used for the discretization of the continuous process as well as for the calculation of event effects.

In practice, mixed forms can be observed as well, like systems that follow the time step driven paradigm for main events but allow this list to be superimposed with special events. An example is an infantry simulation in which an air strike occurs. While the main activities of the infantry occur in time steps, the event of the air strike is of special interest and is computed as a special event outside the general schema.

Stand-Alone Simulations and Federations of Simulations

We already mentioned that an orchestrated set of tools is good practice. A special case is to combine the functions provided by several simulation systems into a new system of systems that provides the union of these functions. This requires a deep understanding of interoperability and composability challenges and possible solutions (see the chapter on this topic in Sokolowski and Banks, 2010).

Many simulation systems are developed as a *stand-alone* tool to deal with a specific viewpoint to understand a problem better or to evaluate a possible solution. To compose the modeling and simulation applications with other applications is often not considered at all, or as an afterthought even when the system has been successfully applied. The military simulation domain had a pioneer role in building *federations* of simulation systems.

We will deal with the resulting standards and solutions in more detail in this chapter as well as in this book, as distributed simulation is based on the federation principle.

Multi-Resolution Modeling: Resolution, Scope, and Structure

Several chapters will address the problem of multi-resolution modeling (MRM). MRM is always an issue when two simulation systems use different resolutions, scope, or structure to represent the same real-world object.

Resolution answers the question of how detailed something is modeled in the simulation system. The more detail is added, the higher the resolution. *Scope* answers the question about what is represented in the simulation system. What has been recognized as important in the viewpoint of one simulation may have been seen to be neglectable in another simulation system. *Structure* describes how observed details are grouped into concepts. The same attributes can be used to describe different concepts in two simulation models, resulting in different structures that are used to categorize the observations of the real system. Figure 4.5 depicts these three terms. In real systems, they are often combined and hard to recognize.

Whenever two systems differ in the way they scope their view, the detail used, or the way they categorize their observations, this creates a challenge to be solved by the simulation engineer.

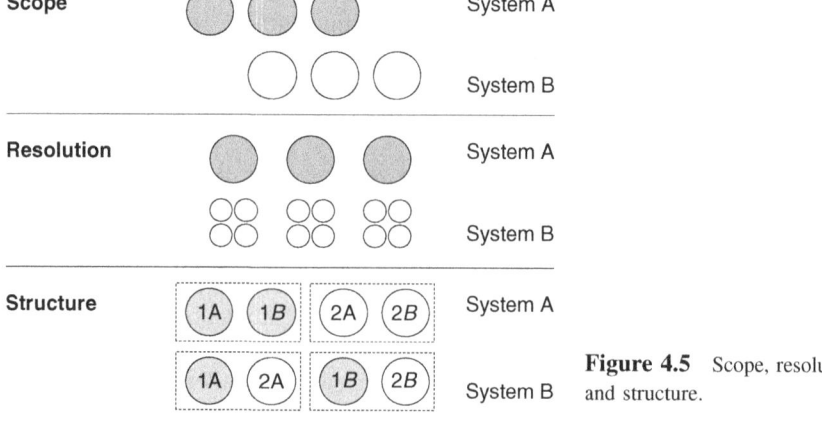

Figure 4.5   Scope, resolution, and structure.

Authoritative Data Sources and Common Simulation Services

When several simulation systems are integrated into a federation, one of the challenges is that all systems start with the same initialization data. While the scope, structure, and resolution may be different, the common data that the simulation data are derived from should be the same. In particular common terrain, including houses, streets, trees, bushes, and elevation should be derived for the simulation systems from the same source.

Another domain that should be addressed in a common way is the initial structure of entities—the order of battle (OOB)—and the current equipment—the table of equipment (TOE). The OOB defines the organizational structure, such as who is logistically responsible to support whom, as well as operation-specific information, e.g. if two units are exchanged between superior commands to be able to better fulfill a given task. The TOE not only describes how much personnel and material is currently within the weapon systems and the unit, it can also specify how much of personnel and material should be there.

Several military organizations provide official data. These data are provided via *authoritative data sources*. These are recognized as official data production sources with a designated mission statement or source or product to publish reliable and accurate data for subsequent use by customers. An authoritative data source may be the functional combination of multiple, separate data sources.

In particular in international federations, there are several challenges to overcome regarding use of such data, many of them being more administrative than technical, but the simulation engineer most be aware of these challenges, which will be dealt with in the advanced section of this book.

Finally, not all these data remain the same over the execution of a federation. Recently, *services* to update the global influence factors are increasingly used to ensure that changes within weather and terrain are consistently communicated. It should be pointed out that it is still the responsibility of each and every simulation to convert this general information into simulation-specific data, which in general includes meeting the already discussed multi-resolution challenges. In recent research projects, the applicability to common services of military simulation systems, command and control systems, and geographical information systems has been a topic of interest.

Fidelity, Resolution, and Credibility of Simulations

A widespread misconception is the idea that fidelity, resolution, and credibility are correlating concepts. This is generally not the case; more detail does not result in better simulation systems.

- The *fidelity* of a simulation is the accuracy of the representation compared with the real-world system represented. A simulation is said to have fidelity if it accurately corresponds to or represents the item or experience it was created to emulate. How realistically does the simulation react?

- The *resolution* of a model or a simulation is the degree of detail and precision used in the representation of real-world aspects in a model or simulation. Resolution means the fineness of detail that can be represented or distinguished in an image. How much detail do I observe?

- *Credibility* is the level of trust of the user of the model. This level can vary and is user dependent as well as application domain dependent. Credibility is the quality or power of inspiring belief, or the capacity for belief. Do I trust the simulation results?

You can have a highly detailed game that reacts very differently from a real system. As an example, just imagine how few people would like first shooter games that require a high level of detailed knowledge before you can actually start to fire a gun. On the other side, an aggregated system with low resolution can produce very accurate results and is trusted by decision makers, like the models used for missile defense during the Cold War show. Actually, all combinations between the three concepts are possible. It is the role of the engineer to ensure that decision makers do not fall into the trap of trusting a nice interface more than valuable academic research (as powerfully exemplified by Roman (2005) in his paper "Garbage in, Hollywood out!").

Validation, Verification, and Accreditation

The engineering method applied to ensure that the right tool is used to solve a problem is known as validation and verification (V&V). For military simulation—in particular in the United States—accreditation is added to these two steps. In other countries, the formal process of accreditation is often replaced by informal procedures evaluating the acceptability of a solution. In any case, validation, verification, and accreditation (VV&A) is the process of determining if a simulation is correct and usable to solve a given problem.

In summary, *validation* answers the question "Is the right thing modeled or coded?", *verification* answers the question "Is the thing modeled or coded right?", and *accreditation* answers the question "Is the resulting simulation approved for a particular use?" In some countries and organizations, the term accreditation is replaced with *acceptance*, as they have not established a formal administrative process.

We will deal with VV&A related questions in more detail in the following chapters of this book.

APPLICATION DOMAINS

The topic of application domains was already mentioned a couple of times during the previous discussions. In this section, the main ideas will be summarized to build a common access point.

There are many ways that application domains have been categorized. For some time, the US DoD used three categories.

- The first category looks at future needs for military equipment, but also new doctrine and missions regarding performance and associated costs. This category was labeled "research, development, and acquisition."

- The second category was related, but left more space for experimentation regarding new concepts and the evaluation of alternatives on how to provide certain capabilities. While the first group focused on weapon system needs, this category focused on doctrinal questions and was called "advanced concepts and requirements."

- The last category was the biggest one in military simulation and was called "training, education, and military operations." Whenever soldiers needed to be trained or educated, simulators and simulation systems could be used to provide a safe environment to practice new tasks and learn new abilities.

The motivation behind this categorization was more driven by associated budgets than by the academic structure of supporting simulation systems. The result is a multitude of simulation systems that provide comparable capabilities but that were developed under different contractual constraints.

To avoid this unwanted redundancy, the new management structure for simulation applications and development is captured in Figure 4.6, often referred to as "modeling and simulation surfboards." This approach recognizes six application domains, namely Acquisition (Acquisition, Technology, and Logistics: AT&L), Analysis (Program, Analysis, and Evaluation, and the Joint Staff: PA&E & JS), Planning (Joint Staff and Policy), Testing (Director, Operational Test and Evaluation and Acquisition, Technology, and Logistics: DOT&E & AT&L), Training (Personnel and Readiness: P&R), and Experimentation (Joint Forces Command: JFCOM).

Furthermore, modeling and simulation is categorized into three supporting activities, namely provision of common and cross-cutting *tools*, common and cross-cutting *data*, and common and cross-cutting *services* (which includes simulation systems). A good overview, as well as detailed motivation for this viewpoint and resulting policies, is given by Allen (2009).

NATO recognizes comparable application domains in the *NATO Modeling and Simulation Master Plan* (1998), namely

- *Defense planning*, which includes force planning, armaments planning, infrastructure planning, communication and information systems planning, determination of logistics needed to support required forces, stockpile planning, nuclear planning, and civil emergency planning. Of particular interest here is to understand that NATO is only responsible for part of its operations, as forces and resources are provided by the participating nations that also have to integrate their models.

- *Training*, which refers to the education of individuals and small functional groups at NATO Headquarters, Combined Joint Task Force (CJTF) and Component Commander, as well as Partnership for Peace (PfP), Western European Union (WEU), Non-governmental Organizations (NGO), and

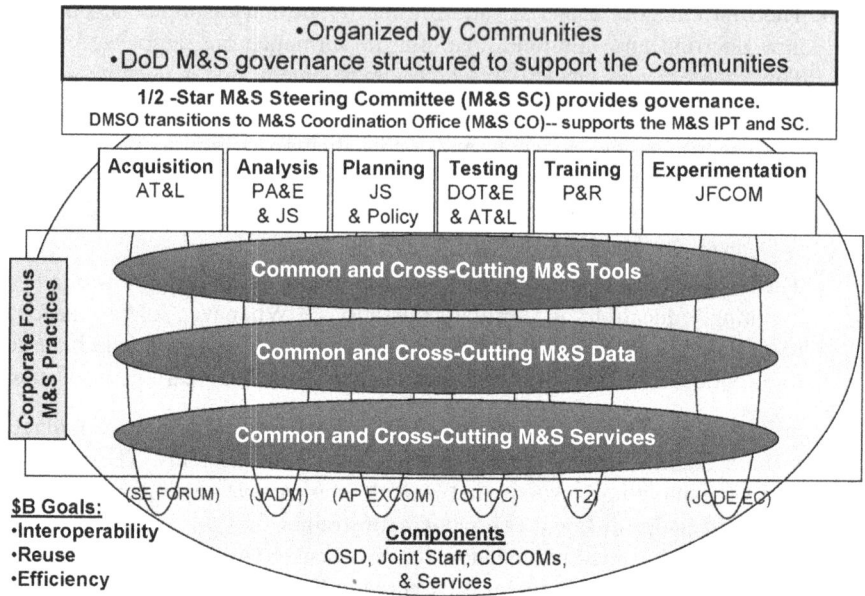

Figure 4.6 Modeling and simulation coordination.

International Organization (IO) staffs, in the conduct of their assigned tasks, is the second domain. Again, the integration of national resources for training is essential.

- *Exercises* are formalized gatherings of a particular command element in which individuals who have undergone training are given the opportunity to interact with a wider community, and the performance of the staff as a whole is measured. The focus here is groups and units, not individuals.

- *Support to Operations* supports military activity involving the conduct of real-world operations that include high levels of combat, continuous, methodical search, and peace support operations. Operational staffs at each level of command need estimates of the impact of various courses of action taken by any participant (friendly, neutral or hostile) before a decision is made.

In summary, military simulation systems are used to determine by analysis if there is a capability gap and what the system should look like that can fill the gap. Procurement refines the requirement for the system and uses these requirements to test the system technically. The military community then tests the system operationally as well and uses simulations to support exercises and simulators to support training to teach users how to use the system. In real operations, simulations deliver decision support to the operations. Experimentation looks into future needs and necessary improvements.

Selected Military Terms

The simulation expert who is new to the domain of military simulation will be confronted with a magnitude of new terms, such as those enumerated in the Joint Publication 1-02 (US Department of Defense, 2010). This section will give a very short overview of terms and concepts that will be encountered.

Command and Control Systems

Command and control is an important topic of many studies and exercises. Within NATO and the US DoD, the abbreviations used are not always completely aligned. Table 4.2 shows the abbreviations and definitions for the US DoD and NATO.

Military Units

Military units are as a rule hierarchically organized. The Armed Services provide the means to conduct military operations.

The *Army* represents the land forces. It comprises at least infantry soldiers, battle tanks, and artillery systems plus many additional support systems. The *Air Force* represents the air component with airplanes such as fighters, bombers, or transporters. In many cases, missiles are under the control of the Air Force as well. The *Navy* conducts sea operations on the surface as well as with submarines under water. In addition to these three standard services, many nations have Gendarmeries or paramilitary police forces. Furthermore, many variations can be observed. In the United States, *Marines* and *Coast Guard* are recognized as services of their own. In other countries, military police and medical and health services are recognized as services. We will focus in this book on the three standard services army, air force, and navy, as a rule.

Traditionally and from the way operations are conducted, the army shows strong hierarchies in all aspects of organization as well as war fighting. The

Table 4.2 US and NATO Abbreviations

Abbreviation	Meaning
C2	Command and Control (US, NATO)
C3	Command, Control, and Communication (US)
	Command, Control, and Consultation (NATO)
C4	Command, Control, Communication, and Computers (US)
C4I	Command, Control, Communication, Computers, and Intelligence (US)
C4ISR	Command, Control, Communication, Computers, Intelligence, Surveillance, and Reconnaissance (US)
CIS	Communication and Information System (NATO)
C2IS	Command and Control Information System (US)

smallest unit of infantry soldiers is the *squad* or the *patrol*. Several groups, often three to five, build a *platoon*. Platoons build up into a *company* (in artillery units also called *battery*). Companies, normally three to five, build a *battalion*. Companies and battalions can form *regiments*. Up to this level, the structures are, in general, homogeneous, which means the soldiers belong to the same branch of the service, like tank soldier, artillery soldier, etc. Above this level, the branches are combined to allow fighting in an operation using all options of mutual support, which means that infantry, tanks, and artillery build a *brigade*. Brigades form *divisions*, and divisions form *corps* and *armies*.

The air force shows a structure that is looser. The individual aircraft is much more important than a single soldier, as the assigned firepower and mobility is higher. While land forces occupy ground, air forces are designed to destroy opponent forces and provide fast transportation of personnel and material. The smallest recognized unit is the *flight* which is between two and five aircraft. The basic unit in many aircraft is the *squadron*, comprising three to five flights. The squadrons form *groups* and *wings*. The wings have a distinct mission with a specific scope and are the operational units of the air force. Some air forces support *air divisions*, but in new operations with the new efficient and flat organizational structures, the wing is often the highest air force element.

The navy structures its units around their type and number of vessels. A single vessel is often referred to as a *task element*. A small number of vessels, usually of the same or similar types, are referred to as a *task unit* or *flotilla*. If they include an important war ship, they are also called *naval squadrons*. A *task group* is designed to fight together and is made up out of a collection of complementary vessels (comparable with the idea of the brigade in the army). A large number of vessels of all types build a *task force* or *battle group*. The vessels in one ocean under common command are referred to as a *fleet*. There are many mixed forms and sub-structures used in reality.

In *joint operations*, the services fight together. Many nations are building joint task forces on lower levels to support synergies and more efficient operations with fewer forces. If several nations fight together in an alliance, this is often referred to as *combined operations*.

Orders and Messages

Military units communicate in a structured form using orders and messages. An increasing number of these orders is produced in machine readable form to support the use within command and control information systems. Some are even coded in binary for higher efficiency, such as messages related to a missile attack sent from an observer to the missile defense unit. As such, there are free text messages, formatted text messages, such as the Allied Data Publication Number 3 specified NATO Message Text Systems or the US Message Text Format (USMTF), and binary messages, such as the Tactical Data Link Messages

(TADL). Knowing these messages is important for modeling communications on higher military levels as well as for engineering models to evaluate the necessary throughput of protocols and infrastructures.

Another aspect of interest is the content of such messages. The military message systems specify on all levels which events have to be reported and how orders are given. Many of these orders are standardized to allow easier communication. For example, a new mission is given to a unit using an operational order (OPORD). It is often preceded by a warning order with first directives, so that the unit can get ready (WARNO). If only parts of the order are changed, the affected fragment can be modified and communicated in a fragmentary order (FRAGO). However, such a FRAGO only makes sense if the OPORD is received as well. The structure of these orders supports in general the six interrogatives—Who, What, Where, When, Why, and How—and give additional constraints for administration, logistics, command and control structures, etc. To what degree orders are specified depends on the doctrinal foundation of the country, in particular how much freedom for decision each subordinate commander is given.

The content is of particular interest when cognitive and decision making aspects have to be modeled as well. It is important for example if decisions are made based on the real situation as currently modeled—also known as "ground truth"—or based on a perception thereof. If a perception is used, report and intelligence message content will be used to update this perception. It is therefore essential to understand what communications are about.

Finally, both—content and form—are essential when information is provided for or obtained from operational command and control information systems. Use cases for these data exchanges are training of soldiers who use their command and control systems to display the simulated situation or using simulation to create test data for command and control systems on the one side, and to generate orders for a simulated unit using the command and control system on the other side.

Military Symbols

Whenever a military situation is displayed on a map, a set of standardized symbols is used. The military is doing this on handwritten maps as well as on large flat panel displays using digital maps and symbols. The visualization of military simulation systems should use the same symbols easily recognizable by the military users, and the simulation engineer needs to know them to understand inputs and recommendations received from the military user. There is more than one standard, but within NATO and the United States the same symbols are understood. In the United States, the Military Standard MIL-STD-2525B defines the symbols also recognized by the NATO Allied Procedural Publication 6B (APP-6A). The interested reader is referred to numerous openly accessible publications for detailed information.

System Engineering Support: Architecture Frameworks

Last but not least, the reader needs to be aware of the architecture frameworks that are used in the military domain. Architecture frameworks provide the foundation for developing and representing formal architecture descriptions that allow standardized artifacts and communicating architectures to be exchanged across branches, services, and nations. Within NATO, the *NATO Architecture Framework* (NAF) is used. Within the United States, the *Department of Defense Architecture Framework* (DoDAF) is applied. The UK uses the *Ministry of Defense Architecture Framework* (MoDAF). Canada developed the *Department of National Defence and the Canadian Forces Architecture Framework* (DNDAF).

All these architecture frameworks provide different views and viewpoints to define all aspects of an architecture, such as what needs to accomplished within the architecture and who is doing it (operational view), what systems and services provide the functionality to conduct these ideas (systems view), and which technical standards support the necessary collaboration of these systems (technical view). They also require a common glossary that comprises definitions for each term that is used within at least one of the architecture artifacts. Some architecture frameworks allow portfolio management and budgets to be addressed as well; others focus more on technical specification and details. Some are extensible to represent new concepts and relations, although the uses of these new information elements are outside the scope of standardized approaches.

Recently, many common domains have been identified for a combination of military modeling and simulation and architecture frameworks. One of these domains is the domain of executable architectures. These executables derived from the framework are by themselves simulations, and they should be able to utilize the resources already provided by military simulations. Another domain is the use of military simulation to experiment on a new system and, as soon as the simulated system provides the capabilities the user desires, being able to produce the specifications of this system as required by the architecture framework used. Finally, architecture frameworks can be used to describe and document simulation systems in the same way as they describe operational system structures.

Some of these cases will be discussed in the advanced section of this book in more detail.

Simulation Standards

As shown so far, the armed forces were among the first pioneers of modeling and simulation applications, in particular for use in training and exercises, and also for supporting procurement, testing, and analysis. The history of military simulation is captured in another chapter in this book in more detail. For this short overview of standardization efforts it is sufficient to know that military simulation was often at the forefront of new application categories. No other application domain outside of military simulation did work on the worldwide distribution of heterogeneously developed simulation systems as early as the military did. The result of these efforts is that the current modeling and simulation standards have been shaped by requirements and constraints of military simulation.

The IEEE Standard 1278 on Distributed Interactive Simulation (DIS) was developed as a large community effort in support of simulators and entity level simulation systems. The standard is defined for real-time systems that exchange data packages, so called protocol data units (PDU), that describe preconceived and standardized events, such as firing at another simulator, communicating with radio signals, collision with objects, and more.

On the higher level, where several systems are aggregated into units—and several lower units are aggregated into higher units—another activity looked at possible standard: the Aggregate Level Simulation Protocol (ALSP). Although it never became a standard, many ideas that made it into a later standard were developed here. ALSP was designed for simulation systems that executed faster than real time. It provided a confederation of simulation systems of different services to provide higher command posts with training and exercise simulation systems.

The IEEE Standard 1516 defines the High Level Architecture (HLA) for modeling and simulation. It defines rules for simulation systems—so called federates—as well as the federation of these federates, the interface between federates and the runtime infrastructure (RTI) providing a set of services, and the object model template (OMT) being used to define the data to exchange between the systems. HLA is the recognized standard for NATO as well as for many organizations. Although defined primarily for the military, HLA was applied in other domains as well, such as aerospace operations, transportation, and manufacturing.

In parallel to HLA, the Test and Training Enabling Architecture (TENA) was developed to support the test organization, in particular the distributed mix of live systems and simulation systems on test ranges.

All these efforts will be dealt with in more detail in other chapters of this book.

SUMMARY ON TERMS AND DOMAINS

Among all the chapters in this book, this chapter is most likely the one that needs continuous updates. New terms are coined frequently not only to unambiguously refer to a topic of common interest, but also to create and enable new funding and support opportunities. Often, this leads to a confusion of homonyms and synonyms that seem to exclude the reuse of earlier work on related domains, as researchers have difficulty finding applicable literature. The challenge to realistically model human beings within military simulation applications, for example, has been referred to as computer generated forces (CGF), human behavior representation (HBR) methods, behavior representation in modeling and simulations (BRIMS), and recently human, social, cultural, and behavioral (HSCB) modeling. The resulting multitude of terms created by simulation engineers in addition to the originally existing multitude within the domain—as different theories address related concepts using different terms as well—creates a very challenging environment for the combat modeling and distributed simulation engineer that can only be facilitated by continuous discussions with all stakeholders and the agreement on common terms as recommended in the NATO Code of Best Practice.

REFERENCES

Allen T (2009). US Department of Defense modeling and simulation: new approaches and initiatives. *Info Security* **23**, 32–48.

Fishwick PA (1995). *Simulation Model Design and Execution: Building Digital Worlds*. Prentice Hall, Upper Saddle River, NJ.

Law AM and Kelton DM (2006). *Simulation Modeling and Analysis*, 4th edn. McGraw-Hill Higher Education, New York, NY.

National Simulation Center (2000). *Training with Simulations: A Handbook for Commanders and Trainers*. Fort Leavenworth, KS.

NATO (1995). *Stable Defence Final Report*. Report AC/243(Panel 7)TR/5, Brussels, Belgium.

NATO (1998). *NATO Modelling and Simulation Master Plan, Version 1.0*. Report AC/323(SGMS) D/2, Brussels, Belgium.

Page EH and Smith R (1998). Introduction to military training simulation: a guide for discrete event simulationists. In: *Proceedings of the 1998 Winter Simulation Conference, Washington, DC*, edited by DJ Medeiros, E Watson, J Carson and M Manivannan. IEEE Press, Piscataway, NJ.

Reimer J and Walter E (1996). Connection of live simulation and virtual simulation. *Proceedings 18th Interservice/Industry Training Systems and Education Conference (I/ITSEC)*. Orlando, Florida.

Roman PA (2005). Garbage in, Hollywood out! *Proceedings of SimtecT*. Sydney, Australia, May 9–12.

Sokolowski JA and Banks CM (eds) (2010). *Modeling and Simulation Fundamentals: Theoretical Underpinnings and Practical Domains*. John Wiley and Sons, Hoboken, NJ.

US Department of Defense (1998). *DoD Modeling and Simulation Glossary*. DoDD 5000.59M, Undersecretary of Defense for Acquisition Technology, Washington, DC.

US Department of Defense (2010). *Dictionary of Military and Associated Terms*. Joint Publication JP 1-02, 8 November 2010. http://www.dtic.mil/doctrine/dod_dictionary (last accessed May 2011).

Youngren MA (ed.) (1994). *Military Operations Research Analyst's Handbook*; *Volume I: Terrain, Unit Movement, and Environment*; *Volume II: Search, Detection, and Tracking, Conventional Weapon Effects*. Military Operations Research Society (MORS), Alexandria, VA.

Chapter 5

Scenario Elements

Andreas Tolk

WHAT NEEDS TO BE MODELED?

From all the information provided so far it should be clear that there is no—and never will be—a one-size-fits-all solution. The question "What needs to be modeled?" can only be answered in the context of the task to be supported, such as:

- Do we support an analysis task to answer a research question?
- Do we provide training for individual soldiers in simulators, or for a command post using computer-assisted exercises to drive their command and control infrastructure?
- Do we want to support a test and have to stimulate a system under test? In other words: do we want to place the system under test into a virtual environment that generates realistic input data for the test?
- Do we support the warfighter by using the simulation system as a decision support system within real-world operations?

As already specified in the context of the NATO Code of Best Practice for Command and Control Assessment in an earlier chapter, we need to address the six interrogatives in a systematic way. In this chapter, we will give an overview of scenario elements that may be addressed during these efforts. While standard literature, like Youngren (1994) and Bracken et al. (1995), provide detailed algorithms and ideas once the elements are identified, the scenario elements to be defined are highly dependent on the task, as we have to understand down to the modeling details,

- What is the mission (the big picture)?

Engineering Principles of Combat Modeling and Distributed Simulation, First Edition.
Edited by Andreas Tolk.
© 2012 John Wiley & Sons, Inc. Published 2012 by John Wiley & Sons, Inc.

- What are the required capabilities to conduct the mission successfully?
- What relations are necessary in order to support an orchestrated set of functions that provide the required capabilities?
- What systems can provide these functions, and do they support the necessary interfaces to other systems to allow for the required orchestration (i.e. being part of the required communications infrastructure as well as having the capability to provide and utilize required messages and formats)?
- What are my time constraints?

Generally speaking, similar to our earlier evaluations a scenario includes descriptions of the following:

- the geospatial definition of the scenario;
- a description of the environment, including terrain and weather, relevant to the scenario;
- the definition of the mission and the means required to achieve the mission of the scenario;
- the objectives that define the mission of the scenario on all relevant levels, which results in measures of merit on various levels, providing an objective hierarchy as known from systems engineering processes; and
- the events that will take place during the time the scenario is intended to span.

The challenge the simulation engineer is faced with is that on one side everything that is needed to support must be provided in order for the task to be complete, but on the other side only what is needed should be provided to support the task to avoid too much and unnecessary complexity. We can only simulate and analyze what we model, but every additional piece of detail added increases the complexity. The following old nursery rhyme about "The Nail that Lost the War" exemplifies the dilemma:

For want of a nail the shoe was lost.

For want of a shoe the horse was lost.

For want of a horse the rider was lost.

For want of the rider the message was lost.

For want of a message the battle was lost.

For want of a battle the kingdom was lost.

And all for the want of a horseshoe nail.

A principle that will be discussed in more detail in the course of this book is the need to align and harmonize what we model (represented concepts), the internal rules driving the behavior (decision logic), and the applied measures of merit defining the success (evaluation logic). Figure 5.1 shows this harmonization triangle visualizing the generally important *harmonization and alignment principle!*

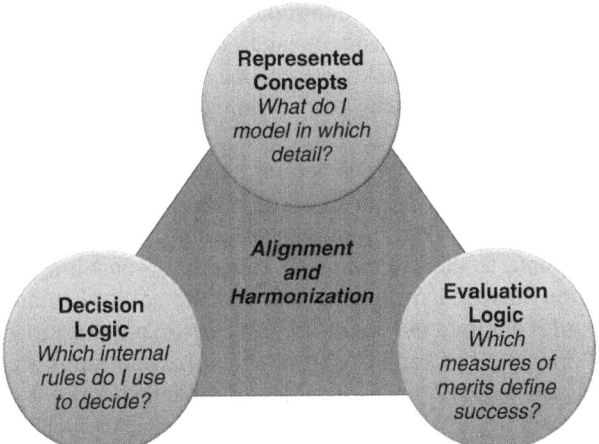

Figure 5.1 Harmonization and alignment principle.

The detail and accessibility of the represented concepts must be aligned with the measures of merit used for the internal decision logic as well as those for the external evaluation logic. If the represented concepts are not rich enough to provide the data points needed for the applied metrics, the respective logic cannot be applied. Furthermore, only if algorithms make use of an additional detail used to model a concept does it make sense to add this detail to the model. Otherwise, the simulation engineer would only do "smoke and mirrors," i.e. deceiving the customer. In addition, if something is modeled it needs to use internal rules and be evaluated by external rules. If, for example, the enemy uses howitzers but the destroying of howitzers is not counted for success, this can either introduce confusing decision rules or result in strange simulation results. Furthermore, the internal decision logic and the external evaluation logic must be harmonized; otherwise the internal decision logic will follow different objectives than are expected by the external evaluation logic. If, for example, the internal logic mainly engages infantry but the success is measured by the number of tanks being destroyed, the results of the simulation can easily become counterintuitive.

When considering the elements of a scenario, it is helpful to remember that the scenario itself is a model. We define a model as a purposeful abstraction of reality, and so it is with a scenario. Each element can be described to an almost non-ending degree of detail and specifics, bringing in more and more items to be considered, but as this is done, the overall complexity of the model increases exponentially. The interaction of the elements is a case of nonlinear dynamics, where the scenario is a highly meshed and netted system. This becomes a case where selections of "What is important?" must be decided, "What can be relevantly influenced?" must be included, and "What can be neglected?" should be excluded, all in the light of the task to be supported. Following these three principles can help a scenario developer keep the complexity to a minimum, while still being able to satisfy the goals of the scenario.

The goal of the remainder of this chapter is to introduce examples for typical scenario elements, to highlight different domains of land operations, air operations, and sea operations, and to lay the groundwork for the following chapters that go into greater detail in related topics, such as movement, effects, perception, and communication.

The *harmonization and alignment principle* is helpful to remember when we consider the scenario, its elements, and its role in all of these related topics. When something is modeled, it needs to be driven by internal decisions. If something is modeled, it needs to contribute to the evaluation of success. If something is decided internally, the decision must be harmonized with the external evaluation. In particular, when internal rules are used that make use of heuristic optimization, it is essential to apply this principle before: artificial intelligence algorithms are applied on what is modeled, not on what it was intended to model!

In order to apply this principle, of course the question of what to include in the scenario is key, but in answering that key question the measures of merit applied must be considered as well. Important and guiding questions can be:

- What do I know about the underlying entities, processes, and relations?
- Where do I get the necessary data?
- How do I manage aggregation and de-aggregation of information?
- Are there solutions to similar problems available (including lessons learned)?

Another metaphor that helps us to understand what is needed to support the task is that of a board game: the situating environment is the board, the different figures used on the board are the scenario elements or represented concepts, the rules how to move the figures on the board represent the internal decision logic, and the rules on who wins the game represent the external evaluation logic.

ENVIRONMENT, TERRAIN, AND WEATHER

All scenario elements play in a situating environment. As mentioned in the last chapter, this environment was often modeled as being static, but modern military simulation systems allow for more agile environments. Terrain characteristics can change with the weather (slippery roads when it rains) or due to activities of the scenario elements (like bombs that destroy streets), and highly sophisticated weather models can influence the activities of the simulated entities and units (including sunrise and sunset influencing infrared sensors, or precipitation influencing the quality of radio communications).

In land warfare, the elevation of the terrain, the surface type, and the coverage is going to influence the line of sight. If two opposing tanks are separated by bushes, they normally cannot see each other (although the soldiers may hear the other tank, if this is modeled). However, when the bushes lose their leaves in the wintertime, it may be possible to see through the bushes that constitute a visual barrier in other seasons. In particular when federating two simulation systems it becomes essential to make sure that the environment is represented in both systems in a way that avoids giving a systematic advantage to one simulation system.

In air force scenarios, clouds, wind, and temperature can play important roles. Depending on the applied sensors, layers of air with different temperatures can be as challenging to aircraft as bushes and trees are for land forces. These atmospheric characteristics can be observed in underwater scenarios for submarines as well: temperature, currents, and salinity have similar effects on underwater warfare.

Littoral operations often require a high detail of modeling the elevation, surface type, and current in order to allow the modeling of landing on the shore and building beachhead missions. Whether tanks can drive the final yards from the landing craft to the shore or if the underwater surface of the ground is too slippery or steep can influence the success of such operations.

Of general interest can also be the question of how night and day are modeled. Some simulation systems simply provide data sets for influenced activities, like how well a system can shoot under night conditions, or how fast an aircraft can be reloaded at night and day. However, there are some critical issues with such an approach, as many military operations are purposefully conducted during dawn or dusk, using the sunrise and sunset to cover operations and make it harder for the opponent to sense their own systems. Furthermore, moonlight can play an important role.

In summary, the environment controls the vision and affects the target–background contrast. It determines how forces can move and hide. Attack routes as well as defense positions will often be governed by the environment as much as by underlying doctrine, and it will often determine location and timing of engagement opportunities. Whenever such characteristics of the environment are evaluated, the simulation engineer should be guided by the questions: "How are these characteristics going to influence my main activities of shooting, looking, moving, and communicating, and is this relevant for the task I have to support?" Whenever two simulation systems represent the environment significantly differently, the simulation engineer needs to apply the same questions in order to decide if a federation in support of the common task makes sense at all, or if significant and often expensive modifications in the simulation systems are needed in order to support such a federated approach meaningfully.

LAND COMPONENTS

Land-based operations, such as conducted by the army, are characterized by the distribution and range of the weapon systems, sensors, and communication means, or aggregations thereof. When selecting the scenario elements, quite often a standardized table of equipment (TOE) can be used to decide how much material will be assigned to each platform, such as different types of ammunition, fuel, etc. Also, the number and type of weapon systems per unit are defined. Depending on the modeled viewpoint, soldiers can simply be modeled as weapon systems or as individuals that are the objective of medical services, cognition, and other possible topics of study. The command and control structure and organization of forces is captured in the order of battle (OOB). Very often, the structure of units in peace time is modified for battles and units are exchanged to increase the success of operations. The OOB is often completed by the electronic order of battle (EOB) that defines frequencies and all communication relevant activities, including jamming operations against opposing forces.

Land forces, in general, use tactical lines to coordinate their activities. Such tactical lines are the forward edge of a unit, the borders to the left and right neighbors, the rear area of a unit, the areas of interest and of responsibility for each unit, and many more. Operations like marching, attacking, defending, delaying, withdrawing, receiving, disengaging, and more are coordinated by such tactical lines. Very often, this line map correlates to characteristics in the terrain, like a line of bushes or trees, streets, telephone lines, etc. The harmonization of operations is one of the major challenges in land operations, and borders between units are always a vulnerable point. Also, if one unit moves slower or faster than planned, the result is an open flank, which is also more vulnerable than the front of a unit.

As discussed before, the weapon, sensor, and communication systems are of particular interest for land forces. Therefore, we need parameters that capture their important characteristics regarding mobility and vulnerability. How fast can a system move in different kinds of terrain? How likely is it that the system is damaged in an attack? Other factors are transport capacity, like usability for mounted infantry or transportation of logistic components and material, command and control functions, and radar and other emitter characteristics. If command and control is modeled explicitly, destroying the leader of a tank unit is an effective way to decrease the combat value. If radars or radios are emitting, they are easier to detect than if the systems are powered off. Consequently, the modeling of decoys as a counter-measure can be of interest as well. What is needed is defined by the task to be provided.

Other scenario element characteristics of interest are those that are used to describe the attrition: a direct fire weapon system needs a line of sight established to shoot at an opponent. Tanks, individual soldiers' weapons, and smaller guided missiles fall into this category. The alternative is indirect fire, like artillery systems provide. Howitzers can fire at opponents that are far away without having to see them. However, quite often they are guided by forward observers that observe the artillery attack and give corrections via radio back to the firing unit. If such observers are modeled they are often primary targets, as without them the accuracy of the artillery is significantly decreased. The use of smart ammunition that is self-guiding in the final phase of an approach reduces the need for such risky jobs, but smart ammunition is expensive. Finally, mine warfare needs to be taken into account as well.

Air defense and missiles are land-based, but are often integrated into the air warfare scenario elements as they deal with ground-to-air as well as air-to-ground fire. Multiple Launch Rocket Systems (MLRS) are sometimes modeled as artillery systems; sometimes they build their own category. In particular, when nuclear warheads are modeled, the units with this capability are modeled separately due to their high impact and political significance.

Among the typical weapon systems generally used when land systems are modeled are the following:

- Infantry is made up of soldiers, sometimes modeled as squads, with hand-held firearms, like rifles, machine guns, or even anti-tank missiles. They may be protected by body armor but, in general, are soft targets that should avoid direct fire without protection.

- Infantry transportation is provided by off-road capable vehicles, like Jeeps or High Mobility Multipurpose Wheeled Vehicles. They are normally not armored and have only light weapons, like mounted machine guns.

- Armored Fighting Vehicles, sometimes referred to as Infantry Fighting Vehicles or Mechanized Infantry Combat Vehicles, are light tanks designed to carry infantry into battle and provide fire support for them. They carry several soldiers that may be able to engage in battles while being in the tank and have light to medium weapon systems for direct fire.

- Main Battle Tanks carry the main part of direct fire battles. They have strong armor and heavy weapons.

- Mortars are high-angle-of-fire weapons that fire ammunition at a high angle so that it falls onto the enemy. Mortars come in several sizes, from small mortars that can be used and carried by infantries to bigger mortars that are part of the artillery.

- Main artillery systems are howitzers and rocket launchers. Howitzers can be towed or self-propelled. As a rule, howitzers fire ammunition while rocket launchers, often MLRS, launch self-propelled rockets, but there are exceptions for modern howitzers.

- Army aviation focuses most often on helicopters, often referred to as rotary wing aircraft, but also uses fixed wing aircraft. These are used mainly for transportation and air based fire support.

The aggregated units used in land-based scenarios, where several weapon systems are composed into a unit or even several units are composed into a higher unit as described in a recent chapter, fall usually into one of the following categories:

- combat or maneuver units (tanks, infantry, armored and infantry fighting vehicles, etc.)

- fire support units (artillery, missile units, close air support, etc.)

- combat engineers (obstacles and mine warfare, active and passive)

- air defense (escorting air defense, air defense facilities, radar, etc.)

- aviation (helicopters, fixed wing aircraft, unmanned air vehicles, etc.)

- command and control (headquarters)
- intelligence and reconnaissance (sensors and facilities)
- communications and networks (infrastructure and systems)
- logistics and supply (transportation and facilities)
- maintenance and medical (in-field support and facilities)
- nuclear, biological, and chemical (NBC) warfare
- others

In new scenarios, a multitude of new options emerged for opponents, starting with former military material sold or stolen with the breakdown of the Warsaw Pact and including current threats with insurgent forces that mount weapons on civilian pickup trucks. Of particular interest are also roadside bombs and other improvised explosive devices (IED). Some sources attribute more than 50% of the casualties in the recent war in Iraq to IEDs; therefore it is critical to increase training and awareness of this danger.

AIR COMPONENTS

Due to the structure of air forces the main scenario elements of air-based activities are far more focused on the individual aircraft. In particular, when focusing on the action on the battlefield, we often model "sorties" representing the appearance of an aircraft in the relevant scenario. The reason is that aircraft can show up over the battlefield, engage in the battle, then fly back to get reloaded, and reappear with new weapons to reengage again. For the scenario it does not matter if the same aircraft appears twice or if two different aircraft of the same type appear in sequence: what is important is that a certain type of aircraft for a given number of engagements is active within the scenario.

Beside the types of aircraft—fighters for air-to-air combat, bombers for air-to-ground combat, transportation aircraft, and many special aircraft for airborne operations, command and control, intelligence, surveillance, and reconnaissance, ballistic missile defense, and more—the kind of weapons used and sensors applied are of particular interest. The same aircraft can fulfill different roles depending on the equipment, but often the amount of equipment that can be carried is limited, so that an aircraft can only serve one role at the time. The European aircraft Tornado, used by the air force of the UK, Germany, Italy, and Saudi Arabia, is a multirole fighter that can engage opponents with guns and missiles and drop bombs, or carry special equipment for surveillance, laser markers for smart ammunition, and more ... but not at the same time. To change roles, time on the ground in facilities with the respective equipment is needed, so this may need to be modeled.

Airports with their runways and shelters are of interest as well. Not every aircraft can land on every runway. Certain minimal length restrictions and conditions have to be ensured to allow jet aircraft to land and start again. In some scenarios, the slip factor of a runway can be as crucial for an operation as discussed for land components.

Air defense facilities are of particular interest as well. Whenever air defense is active, aircraft have to operate with special care. It is therefore a standing objective of the air force to suppress the enemy air defense (SEAD) as early and effectively as possible. The same is true for opposing fighters. Many air operations on a strategic level are therefore starting with operations to ensure air superiority, which means destroying the air defense and fighter capabilities of the opponent with all means allowed under the rules of engagement.

What the tactical lines are for the army, the methods of air space control (ACM) do for air operations: providing means for coordination. There are positive identification means as well as procedural means. The positive identification means are based on identifying an aircraft as friend or foe, which is done by observation, sensors, or special devices, like Identification Friend and Foe (IFF) emitters. Procedural means are defined in an air space control plan. This plan defines special air zones and routes to coordinate the use of air space in operations. Every user of air space must be aware of this plan. It would be a bad idea, for example, to have your helicopters conducting a medical evacuation flying through the air space that is currently used by the artillery to deliver a major attack against the opponent.

Defining coordinating altitudes, Low-Level Transit Routes (LLTR), Minimum-Risk Routes (MRR), Restricted Operations Areas (ROA), or High-Density Airspace Control Zones (HIDACS) are used to coordinate in particular joint operations. Base Defense Zones (BDZ), Weapon Engagement Zones (WEZ), Weapons Free Zones (WFZ), and Air Defense Identification Zones (AIDZ) are furthermore used to coordinate air defense operations better. It is standing procedure, for example, to have BDZ around air bases to protect them against opposing aircraft. Additional zones and routes can be of interest as well, depending on the task to be supported.

Of interest are theatre ballistic missiles (TBM). These are missiles that can engage opponents at very long distances (theatre-wide). TBMs usually ascend so high that they even leave the lower layers of the atmosphere and may travel parts in space before they descend to their impact point, which can include the release of multiple war heads in the final stage. Accordingly, there are three main phases to defend against such a missile, shown in Figure 5.2.

- In the boost or ascending phase (1), the thrust is enormous. If in this phase the hull of the missile is weakened, the missile will probably break apart. The air force experiments with airborne lasers (ABL), for example, to weaken the hull of enemy missiles in their ascending phase. Alternatively, new interceptors are designed to engage missiles in the ascending phase already. Engaging a missile in the boost or ascending phase requires that these missile defense devices are close to the launch point, however, which is not always possible.

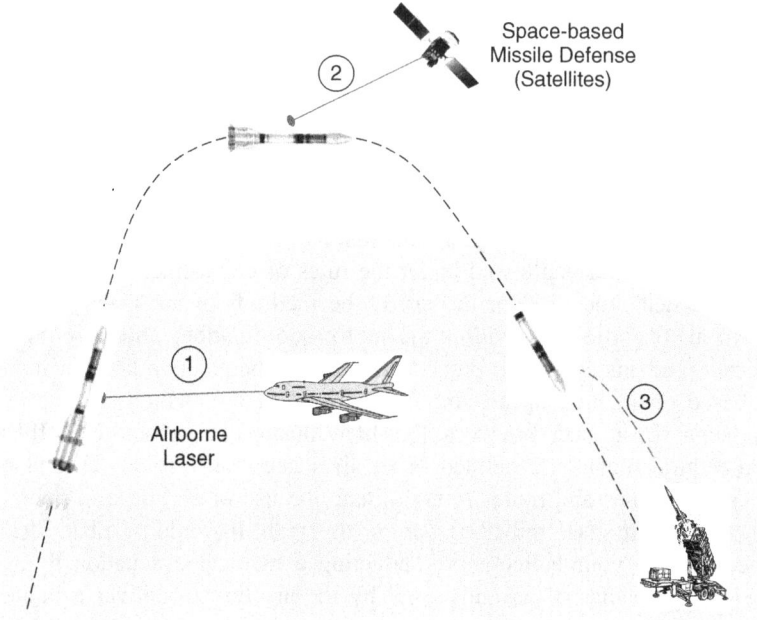

Figure 5.2   Phases of theatre ballistic missile attacks.

- While in the stratosphere (2), the mid-course phase, the missile is vulnerable against space-based or satellite-based anti-missile attacks. However, the distance between satellite and missile is of critical importance, so that this kind of defense is normally limited to self-defense of countries.

- The most common defense against TBMs on the battlefield happens in the descending approaching phase (3), often also referred to as terminal phase. Here, the defense is usually conducted by air defense missiles. One of the major disadvantages of this approach is that hazardous debris may negatively affect the target even if the defense is successful, in particular when NBC warheads are used for the attack.

Some missile defense models actually use four phases and distinguish between initial boost phase and the following ascending phase. While ABLs are only effective while the missile is still in the boost phase, other means—like interceptor missiles—can be used before the missile enters the stratosphere. Looking at available segments of the integrated ballistic missile defense system, that covers all phases from launch to last seconds before the potential impact, its components are also referred to as:

- boost segment components;
- ascent segment components;
- midcourse segment components; and
- terminal segment components.

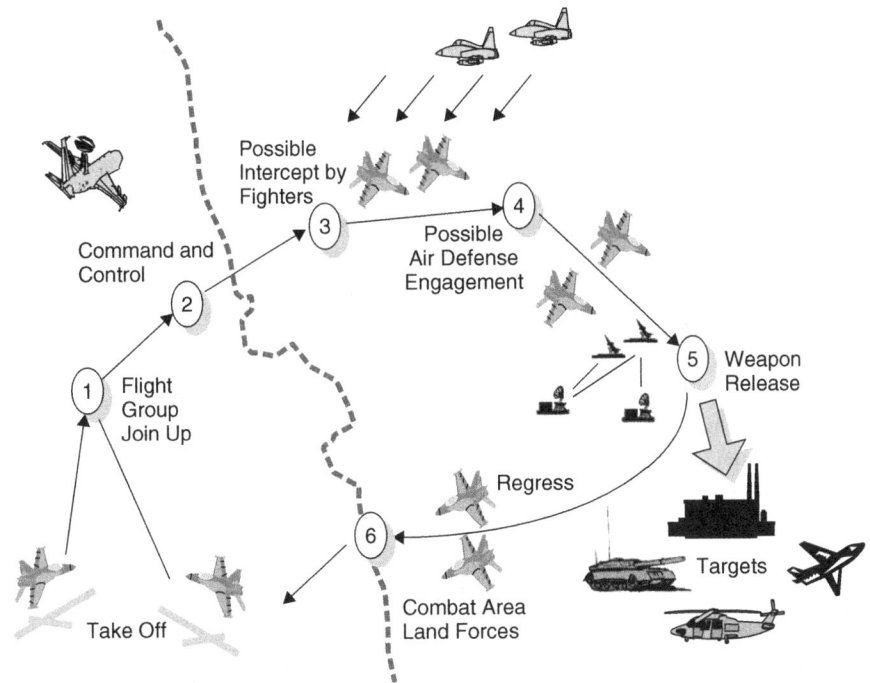

Figure 5.3 Phases of an air operation.

Figure 5.3 is derived from public domain material provided by Dr. Alexander (George Mason University) and shows typical phases of an air force operation flown against some facilities in the rear of the opponent, behind the line of combat activities of the ground forces.

- The participating aircraft take off from their airports. As they may come from different airports, the first thing is to join the flight group (1) that stays together.
- The flight group receives final briefings regarding new developments and latest reconnaissance information via command and control (2).
- During the flight, there is always a possibility of interception by opposing fighters (3). In this case, the flight group needs fighter aircraft as well; otherwise bombers are helpless against such an air-to-air attack.
- When coming closer to the target area and descending for the attack, the engagement of enemy air defense is likely (4). This can be surface-to-air missiles (SAM) or other forms of air defense.
- Finally, the weapons are released against the assigned targets (5), which can be a facility or weapon systems.
- The most critical moment in the regressing phase is when the aircraft are reentering their own terrain (6), as it is essential that the own ground forces

are aware that their own aircraft are coming back from an operation and that these are not hostile aircraft that may fly an attack against them.

Air operations have been a major part in all recent military operations. Quite often, massive missile and air attacks precede the use of land forces. Once air superiority is gained by suppressing the enemy air defense and disenabling opposing fighter aircraft, the air force can support land operations and become a significant force multiplier. However, air forces cannot occupy enemy terrain, so land forces are continuing to remain the most contributing factor for this task.

NAVAL COMPONENTS

The naval forces are playing many roles and consequently may provide many scenario elements of interest that need to be modeled. Naval forces conduct surface operations, underwater operations, and littoral operations. They can provide massive fire power by naval artillery, including missiles, as well as air power by naval air forces. They engage in sea mine warfare against surface and underwater vessels, actively as well as passively.

The weapon systems of the navy are manifold. There are many vessel types in the various naval forces, such as:

- Aircraft carriers that are deployable air bases on the sea. Their primary task is to deploy and recover naval air power, but they are also used to launch missiles.

- Battle cruisers and battle ships provide the artillery firepower and missile launching capability of naval force.

- Frigates and corvettes are normally used to protect battle ships and aircraft carriers, in particular against opposing submarines. They are also often used to protect merchant vessels against pirates or support convoys by providing basic protection. Special submarine hunters focus exclusively on battles against enemy submarines.

- Destroyers and cruisers fulfill a similar role to frigates and corvettes, but their main weapon system is the torpedo.

- Special vessels for sea mine warfare are mine sweepers and mine hunters. Some naval forces also use specially designed mine layers.

- Tenders provide logistics and maintenance for the navy and its related systems.

- Submarines are used for underwater warfare. They come in many roles and shapes, ranging from nuclear submarines that can remain under water for weeks to tactical small submarines for operations in littoral areas or shallow waters, like the Baltic Sea.

- Hospital ships provide medical supply. As the Geneva Convention, which set the standards in international law for humanitarian treatment of the victims of war, forbids attacking Red Cross organizations, they need to be organizationally and physically separated from war fighting scenario elements.

What radar is for the air force sonar is for the navy. There are many models of variant fidelity that are in use to simulate sonar to provide training and analysis. As the ocean and sea have an atmosphere as well, the computation of the sound within the water, including the modeling of ray-path bending, propagation loss, and acoustic noise from other sources, as well as the computation of the reflection of the sound of the sea bottom as well as of various objects, including the target being searched, can be very complex.

JOINT OPERATIONS AND COMMAND AND CONTROL

Modeling joint operations is more than just the sum of scenario elements of all services. A joint operation requires the alignment and harmonization of the planning and decision rules of all participating services and activities. If, for example, land forces are modeled without interaction with other service simulations, the use of main streets and interstates to move fast and efficiently is out of the question. If such a simulation is now coupled with an air force simulation system, major roads, main streets, and interstates are often used as orientation to help pilots find their way to their targets. As the background–target ratio on streets is normally pretty high, there is a significantly high likelihood that land forces marching on roads are not only detected but also engaged. What was a good rule for the stand-alone operation becomes a bad one in the joint context.

Another problem is that participating service simulations must have a minimal understanding of what the other services are doing, and this needs to be simulated. If, for example, a naval battleship simulation has no idea about where and how the land forces are engaged in combat, this simulator cannot provide the naval artillery support that could be decisive in reality. It gets even worse when the weapon systems of one simulation system can engage the weapon systems in the other, but not vice versa. If, for example, the aircraft of the air simulation system can drop bombs on ships, but these ships have no model to provide air defense against an attacker, this unfair fight would lead to very unrealistic results.

Not only within land, air, and navy operations, but particularly in joint operations, the simulation of command and control becomes increasingly important. Headquarters with their capabilities and equipment and their ability to exchange information with other headquarters to provide a distributed planning and execution process are important to winning military operations; therefore these elements need to be modeled in modern military simulation systems in sufficient detail as well. This includes details of the provided infrastructure, interfaces and provided protocols, what kind of messages in which formats can be produced and understood, how much time is needed by which personnel to conduct certain tasks, etc.

Of increasing importance and so far not often efficiently supported by models and simulations are cyber attacks and cyber warfare. Everything from the introduction of false information to the use of computer viruses or massive denial-of-service attacks can be of incomparable importance for successful command and control and therefore for the successful operation. Most of the current solutions are classified based on the sensitivity of the information included in the models, including vulnerabilities that should not be exposed to potential enemies.

SUMMARY

What elements need to be included in a scenario is driven by the task to be supported. The same is true for the resolution of the modeled concepts. However, the harmonization and alignment principle must be applied in any case: if something is modeled, the modeled detail must support the metrics used for internal decision logic as well as for the external evaluation logic. Also, whatever is modeled must be captured through internal decisions as well as external evaluations, at least indirectly. As the external evaluation logic represents the measures of merit used to decide if the operation is executed successfully or unsuccessfully, it needs to be harmonized with the internal decision logic. In other words, the objectives of the simulated concepts must be the same as the objective evaluated at the end of the experiment. Kass (2006) gives several examples for what can go wrong otherwise when he defines the requirement for a good experiment that includes the use of military models and simulation: (1) the ability to use the new capability, i.e. the representing scenario element must be modeled with appropriate internal decision logic; (2) the ability to detect change, i.e. the metrics applied to measure the success must include the parameters affected by the new capability; (3) the ability to isolate the reasons for change, i.e. ensuring that the complexity is not so high that causality of events is no longer traceable; and (4) the ability to relate results to actual operations, i.e. that fidelity and credibility of the experiments are given. These four principles are applicable to all military application domains for modeling and simulation.

REFERENCES

Bracken J, Kress M and Rosenthal RE (eds) (1995). *Warfare Modeling*. Military Operations Research Society (MORS), Alexandria, VA.

Kass RA (2006). *The Logic of Warfighting Experiments, DoD Command and Control Research Program*. CCRP Press, Washington, DC.

Youngren MA (ed.) (1994). *Military Operations Research Analyst's Handbook*; *Volume I: Terrain, Unit Movement, and Environment*; *Volume II: Search, Detection, and Tracking, Conventional Weapon Effects*. Military Operations Research Society (MORS), Alexandria, VA.

Part II

Combat Modeling

Chapter 6

Modeling the Environment

Andreas Tolk

WHAT IS THE ENVIRONMENT?

In Chapter 5 we introduced the notion of a *situated environment* for the scenario elements. The term "situated" originally means "having a place or location." For modeling and simulation, the term needs to be broadened a bit, as the situated environment provides the spatial–temporal constraints for all participating actors as represented by the scenario elements. While Parent and colleagues (2006) show that this is not an exclusive challenge for simulation, in particular in distributed simulations common approaches are required to ensure that the common environment for all participants really is common.

Terrain

The first component of the environment we are dealing with in this book is terrain. Terrain is defined by many possible characteristics that may have to be taken into account when modeling the environment (again, this depends on the task to be supported).

One of the first decisions will be if a two-dimensional or a three-dimensional representation of the terrain is needed, and if the same representation will be used for the execution of the model and for the visualization of the results or not. In particular in game-like virtual environments, as discussed in a later chapter in more detail, a lot of emphasis is placed on realistic visualizations: it needs to look real. The request for detail for a potentially photo-realistic picture is in general higher than that required for the execution of a model, which leads us back to

Engineering Principles of Combat Modeling and Distributed Simulation, First Edition.
Edited by Andreas Tolk.
© 2012 John Wiley & Sons, Inc. Published 2012 by John Wiley & Sons, Inc.

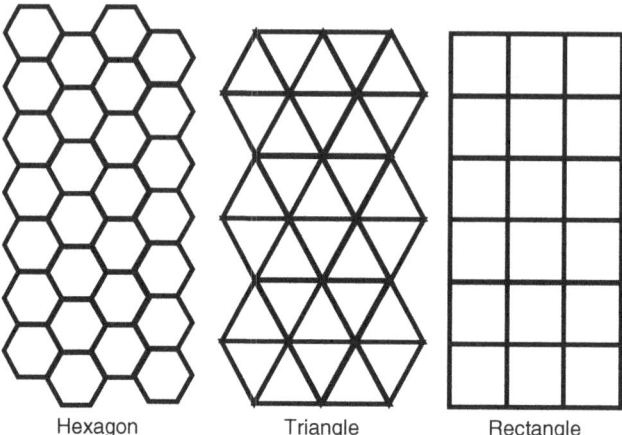

Hexagon Triangle Rectangle

Figure 6.1 Gridforms for explicit terrain models.

the problem that everything that is represented needs to be submitted to the harmonization and alignment principle to avoid strange effects. It is good practice to use the same terrain model for internal reasoning as well as representation, as this not only ensures the appropriate look but also ensures the correct look and feel.

When modeling terrain, elevation, surface type, and covering have already been identified as important characteristics. The terrain itself is often seen as the "game board" on which the scenario elements are moved around, resulting in so-called explicit terrain models, which store the characteristics, in contrast to implicit terrain models, where the characteristics are used to provide aggregates using this information. We will deal with implicit models when we describe the algorithms for shooting, moving, looking, and communicating in more detail in a later chapter.

In particular, we often distinguish between explicit grid models that use a grid to store the required terrain characteristics or explicit patch models. Patch models break the battlefield up into areas where the characteristics do not change. In grids, the cells all have the same size and form, allowing advantage to be taken through geometric characteristics of the used form, as will be discussed next. In patches, potentially large portions of homogeneous terrain can be treated as just one patch, reducing storage space as well as computation time when implementing them.

When building grids, the Euclidean plane can be covered by three kinds of regular polygons without gaps or overlapping areas: the *parallelogram*, the *triangle*, and the *hexagon*. Figure 6.1 shows them with the rectangle/square as the most often used form of a parallelogram or "rectangular parallelepiped."

The three different grid polygons have different advantages.

- Rectangles are very intuitive for most people used to working with maps, in particular as this approach establishes an Euclidean coordinate system.
- Hexagons were the main polygon of choice for a long time, as they have the most efficient perimeter to area ratio of the three possible shapes.

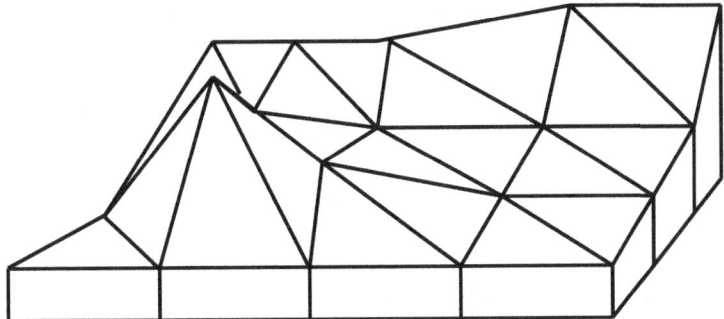

Figure 6.2 Explicit surface terrain model.

Furthermore, the distance between the center of each hexagon to any of the centers of its six neighboring hexagons is equal, which allows the distance between two hexagons to be computed based on the hexagon number (Luczak and Rosenfeld, 1976), which is based on simple arithmetic and is computationally less costly than doing distance computation in an Euclidean coordinate system.

• When using grid models, there are usually instantaneous changes when moving from one grid to the next. The triangle is the best form to transform a grid into an explicit surface terrain model. Instead of measuring the elevation for the grid element in the center, the elevation is measured at the corners, and only the triangles result in planes connecting the three corners. Figure 6.2 shows an explicit surface model made up of triangles.

With the development in computer systems in recent decades, many of the advantages of hexagons and triangles are no longer perceived to be necessary to drive modeling decisions. Nonetheless, they are still necessary to be known, as several legacy systems make use of them.

Of particular interest becomes the use of data that are also obtained in support of operational systems, such as geographic information system (GIS) data that will be dealt with in more detail in the following sections. GIS data are delivered in various formats. Some formats are specified by GIS related government organizations, others evolve from successful software approaches. Typical categories for these data are *raster data* that specify single data points with associated characteristics, and *vector data* that define patterns and shapes of terrain sections with common characteristics. One of the main characteristics of interest is the elevation of the terrain. This elevation is very often stored in the form of a elevation points within a grid. Digital terrain elevation data (DTED) are one of the most often used in military simulation systems, although many applications require a higher resolution than provided by this early GIS standard. Furthermore, non-military organizations use their own data formats, so that improved alignment of such data is required.

While elevation data are important, the remarks made so far show that there are additional factors of interest, in particular what things are on this terrain.

Houses, streets, and other facilities of the represented infrastructure need to be modeled using GIS data to visualize them as well as to represent them within the simulation systems. The need for such data in support of calculating movements and effects will be discussed in the upcoming chapters. The level of agreement on these data is even lower than for elevation data, which often leads to individual point solutions with little potential for reuse and the need for bringing subject matter experts into the team to address the resulting challenges.

Of particular interest are coordinate systems. They should align well with what the trainee or analyst normally uses anyhow, e.g. latitude–longitude coordinates, or variations of the Universal Transverse Mercator (UTM) method as often applied in the US Army. Depending on the approach taken within the terrain modeling process, different assumptions, simplifications, and abstractions lead to very different implementations: some take the curvature of the earth into account, others abstract this into a plan (which brings the challenge of projection systems into the equation); some use fixed reference points, others compute coordinates relative to a mobile reference weapon system; some use Cartesian coordinates, others polar coordinates, etc. The resulting spatial reference systems need to be converted. If this is not done correctly, the results can be very disturbing when visualized. Some tools support the visualization of such potential correlation inconsistencies among different forms of the same terrain area.

Streets, Rivers, and Bridges

In particular when simulating movement, streets, rivers, and bridges become very important. While the explicit terrain models define the general attributes of the terrain, in particular for land forces streets, rivers, and bridges add something special to these general terrain features. While a tank normally has problems maneuvering through a densely wooded area, a driver can easily follow a street through this terrain. While tanks and soldiers cannot easily cross a river, a bridge helps them to cross at given points.

When modeling streets, rivers, and bridges, the characteristic attributes used for the models are important for the modeling of related processes. For example, not all bridges are stable enough to allow the crossing of a whole tank battalion, some streets are too narrow to allow big trucks—such as are used for maintenance—to drive on them through cities, and the width and depth of a river as well as the speed of the current can be used to define whether boats can use the river as well as whether the river becomes an obstacle for land forces.

It is important and quite often a challenge to align streets, rivers, and bridges, in particular when different data sources are used: a street should lead over a river using a bridge, so making sure that the polygon of the street leads over the polygon of the river at the point where the feature of a bridge is positioned is a requirement that needs to be ensured by the simulation engineer.

Some models align streets and rivers with the edges of the grid, but most modern simulation models place them as polygonal overlays on top of the terrain. When using artificial terrain instead of real terrain, i.e. when making up your own terrain in which a virtual operation will be conducted, additional tools can be helpful to ensure that rivers do not have to flow uphill! Some of these tools for alignment are discussed later in this chapter.

Buildings

In particular with new operations focusing on police-like activities in urban terrain the need for good models of buildings grows. The number of levels is as important as the interior structure for high resolution models. The modeled features need to be aligned between different models, as otherwise very strange effects can be observed, such as infantry soldiers that hover several meters above the roof of a building, as the visualization shows a two-storey building but the simulation system uses a four-storey building. Windows, doors, and even small details like curtains and interior illumination can have a significant influence for the detection of soldiers and house-to-house or house-to-street duels. The alignment and harmonization principle will guide the simulation engineer in what needs to be modeled in support of other processes and metrics, and what should be simplified or ignored.

Weather

In addition to the common factor terrain, all participating simulations should also use the same weather in the general sense: atmospheric weather, meso and micro atmospheres in cities and or street blocks, oceanic weather, etc.

As before, the main focus will be on how these different aspects of weather are going to influence the processes of the simulated entities, like shooting, moving, looking, and communicating. Some properties are invisible but affect the operations significantly. Examples are layers of different air temperature and resulting different pressure systems. Other properties have very different influence factors on different categories: depending on which sensors are available, clouds can significantly influence the availability of air force entities while they are often of minor importance for ground operations, if it is not raining, which may influence the trafficability of the terrain. Snow and ice are of special interest, as they not only change slip factors, they also change the background–target ratio mentioned in the last chapter. Thunderstorms can make communications impossible, strong winds will influence the accuracy of air-to-ground operations, etc.

The three big environment categories currently identified by the military simulation community are ocean, atmosphere, and space. The approach taken to ensure the consistency of required weather information is to use central weather services that distribute the same information to all participants. However, the responsibility of mapping this common information to a simulation's internal representation belongs to the individual simulation engineers. The use of a common weather service is therefore necessary but not sufficient to ensure consistency of weather effects in all participating simulation systems.

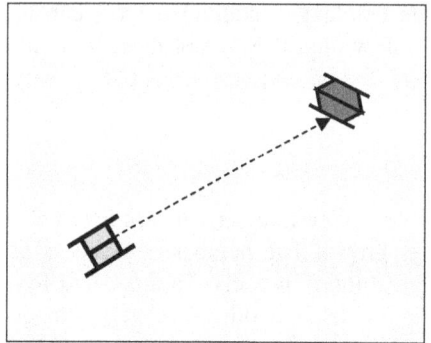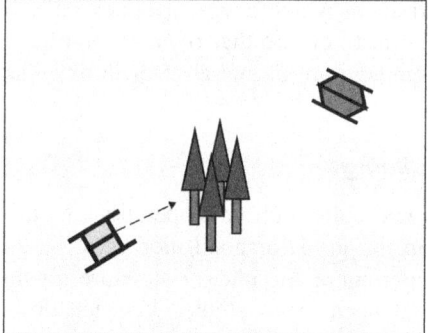

Figure 6.3   Fair fight challenge.

Fair Fight

The main reason to request alignment of data and harmonization of effects of terrain and weather and other components of the environment is to ensure "fair fights" of simulated entities under control of different simulation systems. Figure 6.3 shows the same duel situation of the same tanks in two simulations with different environment representations. In the simulation on the left, only the elevation is used to create the terrain. Both tanks have a line-of-sight connection and can observe and engage each other. In the simulation on the right, in addition to elevation also coverage is modeled. Due to the trees represented in the used data, the tanks cannot see each other and consequently cannot open fire. If in a distributed simulation the tank in the lower left corner is controlled by the left simulation, and the tank in the upper right corner is controlled by the right simulation, the left tank can see and shoot the right tank, and the right tank does not even know where the attack came from. The fair fight challenge is the general quest to avoid such situations. While this example is quite obvious and easy to identify, many real-world relations are far more subtle and complex and hard to trace. Consequently, it is good practice for simulation engineers to avoid such situations and, if they are a necessary part of the solution for a simulation task, ensure traceability based on alignment of data and harmonization of processes and effects.

The use of a common framework to represent all components of the environment with support to mediate general concepts to simulation-system-specific solutions is a great help. In the next two sections, we will have a look at two examples of how this can be done. This does not exclude other frameworks and is not meant as an exclusive support of these two approaches, but they were selected based on their availability and international recognition.

SEDRIS

The Synthetic Environment Data Representation and Interchange Specification (SEDRIS) was developed as a US Government-led program with the maximum participation of industry, the use of international and commercial standards, as well as the support of commercial and government products. All information provided here is specified in more detail on the SEDRIS website (SEDRIS, 2011). In their own words, SEDRIS is an infrastructure technology that enables information technology applications to express, understand, share, and reuse environmental data. SEDRIS technologies provide the means to represent environmental data (terrain, ocean, air, and space) and promote the unambiguous, loss-less and non-proprietary interchange of environmental data. It is internationally recognized and used. Figure 6.4 has been used to introduce the various elements and components that make up the environment by SEDRIS. This view is well aligned with the findings captured in this chapter so far.

SEDRIS began the process of establishing international standards through the combined International Organization for Standardization (ISO) and the International Electrotechnical Commission (IEC) several years ago and successfully standardized several of its recommended approaches. There are four technology components that will guide us for this overview of SEDRIS.

- The *Environmental Data Coding Specification* (EDCS) is a dictionary of terms to be used for specification of classes, attributes, etc. In a collection of several dictionaries, controlled vocabularies embedded in a recommended taxonomy are specified to ensure that only common terms are used to refer to environmental data described in SEDRIS.

- The *Data Representation Model* (DRM) defines an object model to be used for representing environmental data and databases. These objects have to be tagged by terms that are defined in the EDCS.

- The *Spatial Reference Model* (SRM) allows the unified and robust description of coordinate systems including software for the conversion. Objects described using the DRM and labeled using the EDCS are placed into the common environment using coordinates as defined by the SRM.

- The *SEDRIS Transmittal Format* (STF) allows the platform independent storage and transmission of data.

Figure 6.5 exemplifies the relations between these four components and their external access point in the form of application program interfaces (APIs).

The DRM is the common format to describe all elements. It is therefore the embedding middle part of the components. Some applications may be able to directly use the DRM, in particular SEDRIS-based tools; that is why the DRM API is provided. All objects modeled are tagged with a standardized term that needs to be defined within the EDCS. It is possible to add project-specific terms, which need to be defined in EDCS (in addition to their definition in the study-specific glossary as recommended by the COBP in earlier chapters). In order to inject this new term, the EDCS API is provided. The SRM provides the various spatial location information pieces. In order to allow access to the powerful subroutines that convert the most often observed spatial reference systems into each other, not only is the SRM API provided but headers for their access in the

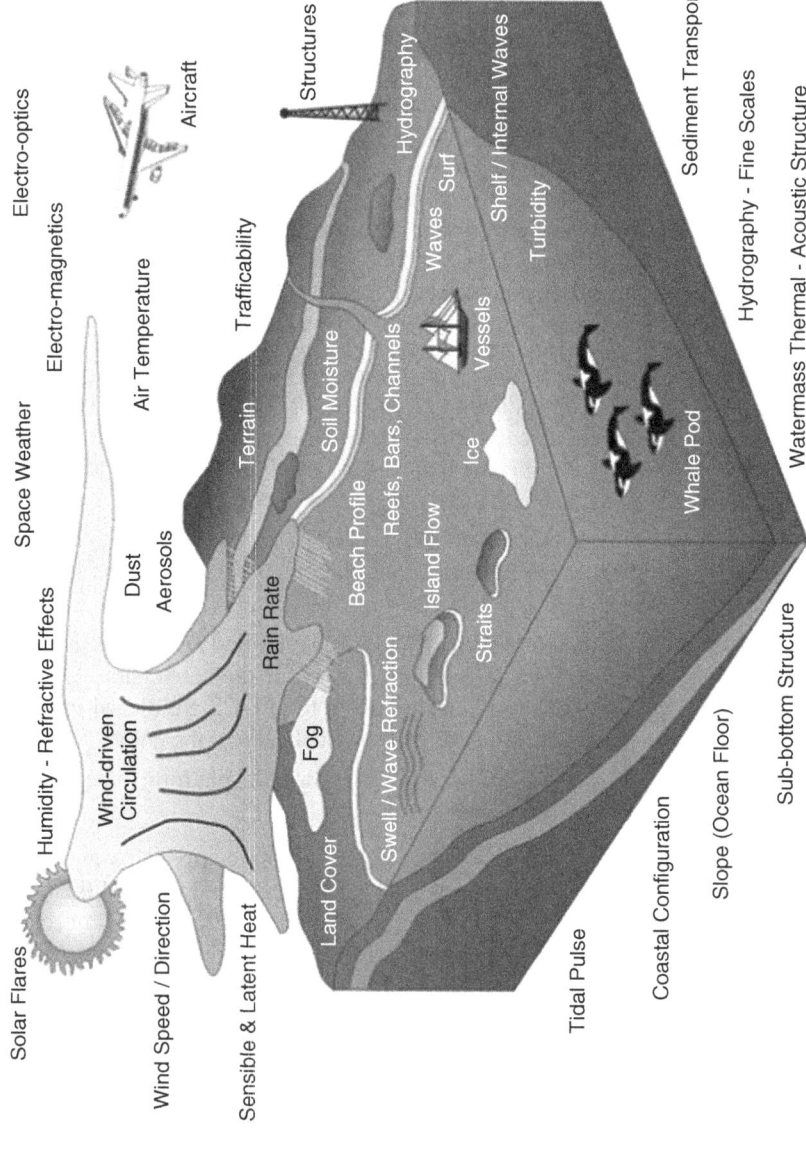

Figure 6.4 Examples for components of the environment (SEDRIS).

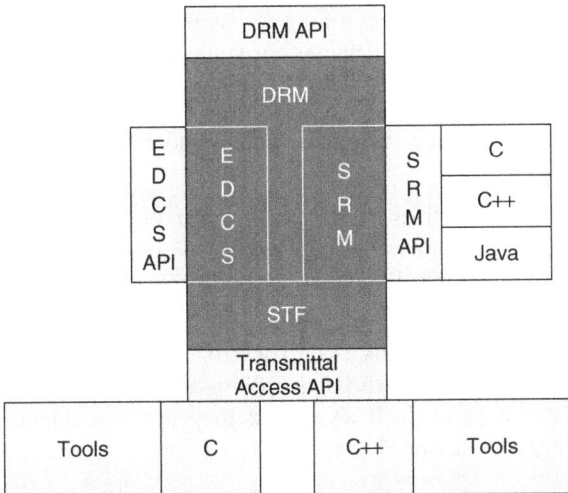

Figure 6.5 Technical components of SEDRIS and APIs.

computer languages C, C++, and Java are provided as well, which allows an easy integration of such conversion routines into programs.

Most systems, however, will use the STF to exchange environmental data. STF defines so-called transmittals, which are consistent data sets based on the DRM in STF. These pieces of data will not compromise the integrity of any data representation into which they will be imported. Readers with a database background may think about them as transactions on databases. STF has therefore a transmittal access API, so that users cannot compromise the consistency with bad access or manipulation calls. This API can be accessed natively or via headers in C and C++. Furthermore, several tools are distributed using these interfaces that visualize the content and allow the comparison of data sets to ensure not only the technical compatibility, which is provided by the STF transmittal, but also the application compatibility of results. For example, a comparison tool can visualize the terrain used in the left simulation of the example given earlier in this chapter, and side by side the terrain used in the right example. Such a comparison would immediately show that the trees are missing in the left simulation, which at least would generate an alarm for the observer that unfair fights may occur.

Environmental Data Coding Specification

EDCS was the first product of the SEDRIS community that was standardized by ISO/IEC 18025:2005(E). EDCS has the objective to identify and standardize the classification and characteristics of environmental objects. It therefore defines not only the terms but also the structure. For example, a "tree" has branches, and a "branch" has "leaves." It also supports various units of measures to allow for conversion between them. In summary, EDCS is structured to provide three questions needed when environmental data are exchanged:

- What is exchanged (terms, labels)?
- What are the characteristics (defining attributes, structure)?
- What units are used to measure those characteristics (unit of measure)?

In order to provide this information, nine dictionaries are standardized within EDCS:

- The *Classification Dictionary* defines all terms used for environmental objects. This is an exhaustive enumeration of what can be represented as an object and communicates as such. Examples are buildings, rivers, trees, among others.

- The *Attribute Dictionary* defines all terms used to describe features or characteristics of environmental objects. Attributes specify an environmental object further, giving the state of such an object; they are not objects themselves. Examples are height, speed, etc.

- The *Attribute Value Characteristic Dictionary* describes possible values of instances that can be used to describe attributes of an environmental object. Examples are real, integer, index, etc.

- The *Attribute Enumerant Dictionary* describes possible enumerations that can be used to specify attributes. They are sets of terms that can be used as enumerants and are grouped into enumerants usable for the same enumeration. Examples are (light, medium, heavy), (red, white, blue, ...), etc.

- The *Unit Dictionary* defines the unit of measure that can be used to complete a real valued parameter to ensure the right interpretation. Examples for specifying a distance are meters, yards, inches, etc. As a rule, different units of measure measuring the same unit should be convertible into each other, and usually be a simple factor, e.g. 1 inch = 2.54 centimeters, 1 yard = 0.9144 meters, and so on.

- The *Unit Equivalence Class Dictionary* defines equivalence classes for units that can measure the same kind. Examples are mass, length, volume, speed, among others. The conversion between units of the same equivalence class may become a little bit more complicated than simply multiplying the values with a scalar, but in principle they should be convertible into each other as well.

- The *Unit Scale Dictionary* defines scale factors for different units of the same equivalence class. Scale factors are used as a multiplicative constant to avoid either excessively large or small attribute values, similar to using the scientific notation on calculators. Examples are KILO, MILLI, MICRO, NANO, etc.

- The *Organizational Schema Dictionary* has been added as a help to SEDRIS users to more easily identify which environmental classifications and attributes usually belong together. Other domains describe these ideas as propertied concepts: which concepts—or environment classifications as defined in the Classification Dictionary—are characterized by a set of given properties or attributes as defined in the Attribute Dictionary.

- The *Group Dictionary* generalizes this idea by defining sets of environment objects that define an environmental group. The easiest way to understand this idea is to think of this as a part list of smaller objects defining together a bigger object. While the environment objects defined in the Organizational Schema Dictionary are minimal and defined by their attributes, the environmental objects defined in the Group Dictionary comprise smaller objects. A village, for example, comprises houses, streets, trees, and more objects.

The EDCS is used stand-alone as a "thing-level" semantic, as the EDCS defines in detail what environmental objects can exist, how they are composed (if they are not atomic), what the characteristic attributes are, what values the attributes can have, how they are measured, and what equivalence classes between different measurements exist. When used in the context of the DRM, EDCS is applied to specify classes and properties modeled within DRM using classification and attribute codes of the EDCS.

There are several ways the EDCS dictionaries can be accessed. The SEDRIS homepage offers downloads in the form of standard websites in hypertext format, as a Microsoft Excel® Workbook, as header files for the programming language C, and as an EDCS Query Tool.

Data Representation Model

With the DRM, SEDRIS provides a common representation model allowing the users to specify their own environmental data unambiguously as well as to understand the environmental data of other users clearly. To support this, the DRM specifies a set of classes including concepts and properties to model things, formal relationships between these things, and applicable constraints applicable to related things. To do this, DRM utilizes a subset of the Unified Modeling Language (UML) as shown in the SEDRIS notation reference sheet depicted in Figure 6.6.

An environmental collection of data, as supported by the DRM, contains data sets that represent the features of a common synthetic environment for simulation applications. The data sets must support not only the visualization of these environmental objects, which is the presentation of the environmental scene for all participants, but also the alignment of the underlying modeling activities, whilst allowing the users to communicate the environment effects on the simulation-application-specific actions. The DRM therefore separates the semantics—what is modeled—from the description—how it is represented. This approach introduces data primitives that are the shared minimal components understood and shared by all applications that use SEDRIS.

The DRM contains over 300 classes that can be used to model environmental objects and share them between applications. Going into the details lies beyond

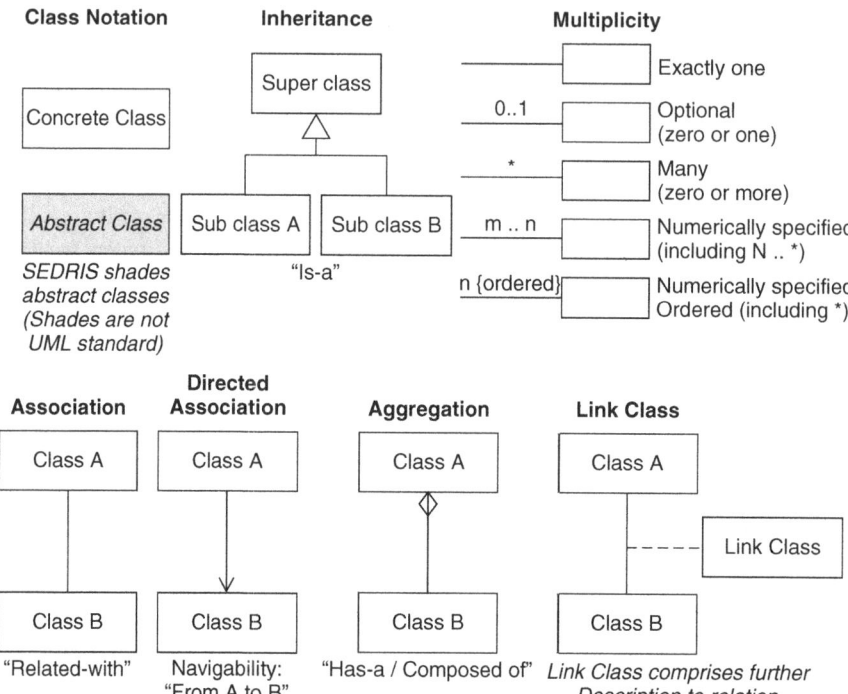

Figure 6.6   UML notation used in DRM.

the scope of this book, so we will focus on the high level categories of these classes.

The first category comprises the *Primitives* addressed before. To these primitives belong:

- point
- line
- polygon
- image
- sound
- point feature
- linear feature
- areal feature

The second category comprises the *Metadata*. As discussed in an earlier chapter, these are data about data that describe an object better and give context information that is not directly derivable from the structure of the environmental object or the tags used to describe it. These metadata comprise:

- keyword
- description

The third category holds the *Modifiers* that are needed to model the environments' effects on the participants/users' action as defined before. They hold classes for:

- property value
- color
- classification data
- image mapping function

Finally, the fourth and final top level category comprises *Organizers and Containers*. These classes are used to group the environmental object models in a similar way to the organizational schema and group dictionaries group the environmental object descriptions. They hold classes for:

- transmittal root
- library
- environment root
- feature hierarchy
- geometry hierarchy

The four categories of DRM and the nine dictionaries of EDCS allow the representation of what is modeled using the standardized and controlled vocabulary of EDCS and how it is modeled using the categories defined by the DRM.

Spatial Reference Model

The Spatial Reference Model (SRM) unambiguously places the modeled object in the common synthetic environment. It provides a unified approach for the representation of spatial location information as well as the use thereof. The SRM has been designed to provide complete and precise treatment of spatial location information and to precisely define the relationship between various spatial reference frames.

Spatial data describe geometric properties such as position, direction, and distance. The SRM makes sure that different ways to do this can be communicated between the systems. To describe the functionality of the SRM, the best approach is to describe the typical questions that are answered by respective functions:

- Where am I in the synthetic environment (location)?
- Where are the other objects I am interested in?
- With whom can I interact, and who can interact with me?
  - Location
  - Range
  - Direction (azimuth, elevation angle, etc.)

○ Geometrical interactivity (line of sight, etc.)

It should be pointed out that conversion between different spatial reference frames does not automatically preserve all information. Only if the used spatial reference frames are equivalent can all information be mapped accordingly. The necessity for SRM was born out of the observation that different communities use different models, some using the earth as a reference model (ERM), others being more general and allowing other reference models as well (ORM). But even when ERM is used, there are many ways to derive coordinate systems. In addition, the used spatial reference frames may be relative to a moving system, such as an aircraft or a planet, and often several of these spatial reference frames have to be supported at the same time. An example is a group of tank simulators that share a common synthetic environment but every simulator displays the objects relative to its simulated tank.

To address these issues of transforming spatial reference frames into each other, SRM uses the concept of an *object space*, which is the real—or abstract—universe that contains spatial objects. This is a three-dimensional Euclidean vector space with well defined reference points and orientation of the dimensions. The spatial objects are placed within this object space and can be real physical objects as well as abstract objects. The object space can be generalized in the *position space*, which builds the logical extension of the object space along all dimensions; this now allows the location and orientation of other objects to be described in the position space defined by the object space. Objects are embedded into the position space. Therefore, if the location and orientation of an object is described, it is always necessary to refer to the position space that is used to do this, which is known as spatial binding. The resulting construct combines location, orientation, and position space and is referred to as the Object Reference Model Template (ORMT) that builds the backbone of the SRM. In summary, the *Spatial Reference Frame* (SRF) serves to locate coordinates in a multidimensional vector space, the ORM is a geometric description or model of the reference object embedded in the SRF, and the *coordinate system* is spatially bound allowing the location and orientation of objects with respect to the origin of that frame to be specified. Geocentric models use the ERM and object centric models use the ORM as their reference object.

The alignment of different spatial reference frames is hereby transformed in loss-free mediation between vector spaces. If two vector spaces are equivalent, objects can be referenced in both spatial reference frames without loss of information. If they are not equivalent, the common syntax of the ORMT allows the problem domains to be identified. In detail, *coordinate conversion* is the process of determining the equivalent spatial location of a point in an SRF which is based on the same object reference model but a different coordinate system, and *coordinate transformation* is the process of determining the equivalent spatial location of a point in an SRF which is based on the same coordinate system but which uses a different object reference model. Figure 6.7 shows two three-dimensional

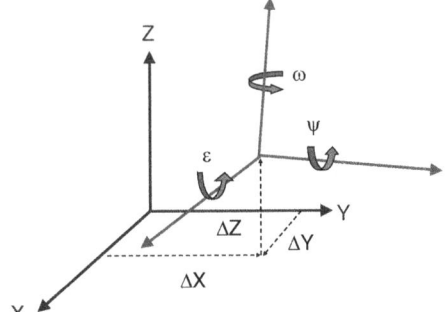

Figure 6.7 Mapping of coordinate systems.

Euclidean spatial reference frames that can be mapped to each other without loss of information.

In this three-dimensional space mapping, there are nine possible differences that have to be taken into consideration: the difference of the center in all three directions (ΔX, ΔY, and ΔZ), the rotations around the three dimensions (ε, ψ, and ω), and the scale changes between the three dimensions. All operations are well defined in the vector space.

One very convenient feature is that this approach allows the standard-based conversion of equivalent coordinate systems that are used in different organizations. It also allows the systematic enrichment of simplified coordinate systems, like map sections that do not take the curvature of the earth into account with geocentric views, or the use of different projection algorithms. Several coordinate transformation functions are provided in SRM-based libraries.

SEDRIS Transmittal Format

The focus of this chapter is modeling of the environment. While it is helpful to understand the concepts of distinguishing between semantics (what environment objects are modeled), as described in the EDCS, syntax (how the environment objects are modeled), as described in the DRM, and a spatial reference framework to describe location and orientation, as described in the SRM, there is no need to go into the details of the STF. For this chapter it is sufficient to understand that the platform independent format for storage and exchange of data is necessary to ensure the usability of these ideas in computer-assisted environments.

ENVIRONMENT SERVICES

While the section on SEDRIS provides a good overview on how to model and name environmental objects to make them exchangeable between different systems, another approach—that already points towards the ideas of distributed simulation in the upcoming chapters—is to provide these common scenario elements needed for all participating simulation systems from a central place in the form of a service. In other words, common components like terrain, weather, ocean, space, and more are not collected from all systems, but one service provides this information as data to all participants. This means that if it rains, the service provides all simulation with the location of the rain, the intensity of the rain, and so forth. If the terrain changes, the updates are distributed to all participating systems so that they can change the way their systems move, etc.

The Ocean, Atmosphere, and Space Environmental Services (OASES) is a first example (Trott, 2005). OASES was designed to be the authoritative provider of data describing the scenario-wide environmental objects and provides these descriptions via services to participating simulation systems. Instead of having simulation-system-specific models for weather, terrain, or space, OASES provides this information via standardized means to all participating simulation systems based on authoritative sources. While the simulation systems can focus on simulating their tactical scenario elements, OASES focuses on providing weather, space, and so forth. The work on OASES was kicked off as a research effort under the Defense Advanced Research Project Agency (DARPA). The development was orchestrated by the US Defense Modeling and Simulation Office (DMSO) and aligned contributions from the Defense Threat Reduction Agency (DTRA), the Office of Naval Research (ONR), the Naval Research Laboratory (NRL), and the Naval Warfare Development Center (NWDC). The product was placed under configuration control of the US Joint Forces Command to support computer simulation supported exercises of all services.

The synthetic environments created by OASES are based on authoritative, validated numerical models. OASES uses the same models that are also used by the meteorological/oceanographic (METOC) community in support of real-world applications. In addition, OASES provides tools for converting authoritative model outputs to a data format recognized by all participating systems. OASES also provides editors with the ability to tailor the synthetic environment to support all participating systems and cuts unutilized information out. This can be done in the initialization phase as well as during the execution. OASES produces information about temperature, air pressure, precipitation, and more in several possible formats via the IEEE Standard 1516 for distributed simulation systems from the server to the users.

This approach is also increasingly used in international efforts. Neugebauer and colleagues (2009) describe an architecture for a distributed integrated testbed that utilizes the Geo-referenced Environment, Object and Infrastructure Service (GOIS). This service is used for initialization of all participating systems with environmental objects as well as for the dynamic distribution of changes. In the given example, Neugebauer and colleagues describe how GOIS was used to manage buildings and their damage state and to provide critical information to all participating systems during runtime. Furthermore, the idea of services was even more supported, as common effects (communication and weapon effects) were also computed by services to ensure alignment of events, and GOIS was aligned with these services as well. If the weapon service, for example, determined that an artillery attack destroyed a house, the weapon effect server informed GOIS, and GOIS distributed the new data to display the house as destroyed in all participating systems.

SUMMARY ON MODELING THE ENVIRONMENT

Within this section, we learned about how to model the environment. The environment is everything in which the scenario elements are situated. It comprises terrain, weather, and even space. With SEDRIS, a systematic approach was introduced on how to name environmental objects, how to model them for sharing and visualization, and how to address their spatial–temporal relationships. These ideas are increasingly used in services that provide a common synthetic environment for exercises.

The decision on how to model the environment influences the modeling of processes like shooting, moving, and communicating significantly, and we will revisit parts of this chapter when these topics are dealt with. As before, the harmonization and alignment principle will govern the simulation engineer's choices. It is of no value to distribute weather to participating simulation systems if they do not represent the influence of weather accordingly. Actually, this may even introduce new sources for unfair fight issues. The same is true for terrain, and so forth. In models that include a human decision maker, visualization can start to play a similar important role, and also can become a pitfall: if the visualization for the sake of looking more realistic includes features that are not modeled, the human decision maker may base his selection on the arbitrarily chosen visualized features that are actually not modeled, creating a bias that is hard to find.

This chapter only gives a first introduction to the principles of modeling the environment as they are needed to understand the following models for shooting, moving, and communicating. The reader is refereed to chapter 22 that addresses more details of interest to the simulation engineer, in particular when being tasked to utilize geographical information system data for his solutions.

REFERENCES

Luczak E and Rosenfeld A (1976). Distance on a hexagonal grid. *IEEE Trans Comp* **25**, 532–533.

Neugebauer E, Nitsch D and Henne O (2009). Architecture for a distributed integrated test bed. *Proceedings of the Annual NATO Modeling and Simulation Conference*, RTO-MP-MSG-069, Brussels, Belgium. NATO Report.

Parent C, Spaccapietra S and Zimányi E (2006). *Conceptual Modeling for Traditional and Spatio-Temporal Applications: The MADS Approach*. Springer, Berlin.

SEDRIS (2011). http://www.sedris.org (last accessed June 2011).

Trott KC (2005). *Simulation Development for Dynamic Situation Awareness and Prediction*. Final Technical Report AFRL-IF-RS-TR-2005-333, Northrop Grumman Mission Systems, Rome, NY.

Chapter 7

Modeling Movement

Andreas Tolk

HIGH RESOLUTION MODELS AND AGGREGATE MODELS

In the previous chapter we focused on general principles. This chapter is the first that looks into the details of the core processes of combat modeling: shooting, moving, looking, and communicating. As mentioned before, when dealing with these core processes, we will often have to distinguish between models that use high resolution models to represent the virtual battlefield and those that use aggregates. That is why we have to look at the differences between these two categories of models.

As a rule, *high resolution models* model individual weapon systems (like tanks, planes, and soldiers) and their activities (like driving specific routes during an attack, giving mutual fire support to hold the opponents down, or directing supporting fire from the air force or the artillery onto the target) on the battlefield. In these high resolution models, detailed interactions of individual combatants are modeled. Each modeled entity, meaning each individual system on the battlefield, has an individual state vector describing the unique situation and own perception. For each system we know the location, the orientation, the speed, the amount of resources on board (such as ammunition, fuel, etc.), what the system knows about its environment (including its own forces, neutral forces, and opposing forces), and so on. Furthermore, all interactions are generally modeled one-on-one: duals, sensor activities, and other activities are modeled as individual processes, and generally every single process is computed individually. This also allows for each process to be modeled stochastically with individual probability and density

Engineering Principles of Combat Modeling and Distributed Simulation, First Edition.
Edited by Andreas Tolk.
© 2012 John Wiley & Sons, Inc. Published 2012 by John Wiley & Sons, Inc.

functions. In summary, we have a highly detailed understanding for what is going on in each individual system.

This approach is not only computationally very expensive, it can also become counter-productive when applied on higher echelons: the soldiers responsible for hundreds of tanks and soldiers and other individual systems that make up a battalion are not able to track all this individual activity and status information—they would not be able to see "the forest for the trees." They are interested in aggregated information as provided by aggregated models.

As the name suggests, *aggregated models* deal with aggregated interactions of groups of combatants. Within the domain of combat modeling, aggregation is often defined as the ability to group entities while preserving the effects of entity behavior and interaction while grouped (US Department of Defense, 1998). Generally, aggregation is the composition of several lower level elements into a higher aggregate. Although each modeled entity still has its individual state vector describing the unique situation and perception, the entity is no longer referring to individual systems and soldiers but is describing a group of combatants sharing this state vector, like a unit of tanks, a logistic group, and so on. Several examples are given in the chapter on scenario elements. As the modeled entity represents a group, the interactions are generally modeled many-on-many as well. Consequently, processes are computed based on group assumptions (as will be discussed in the chapter dealing with the four core processes). It should be pointed out that aggregation always results in information loss, e.g. the average amount of ammunition in all tanks and the standard deviation represents less information than having all individual data for each tank.

These aggregated processes can be modeled deterministically as well as stochastically. When they are modeled stochastically the probability and density functions should be derived from the underlying individual processes' probability and density functions. It is often assumed that random effects will cancel each other out on higher aggregate levels so that deterministic models are sufficient. The examples given earlier in this book show that this is not the case: the need for stochastic modeling exists on all levels of aggregation. The interested scholar is referred to the essay of Tom Lucas (2000) who gives several excellent examples.

GENERAL REMARKS ON MOVEMENT MODELING

Obviously, movement modeling is tightly coupled with environment coupling. Naval forces need waterways to move their vessels, aircraft cannot fly through mountains, and terrain influences the movement of land forces.

The military operations research community often refers to one of the following movement model categories in their studies: explicit grid models, explicit patch models, implicit mobility models, and network models. These models build the foundation for high resolution as well as aggregated models, so they are important for both approaches.

Explicit grid models use a grid to store the required terrain characteristics for the movement model. This grid can be in the form of a square or a hexagon as defined in Chapter 6. When movement is modeled, the modeled entity–individual or group—uses the information stored in the grid cells they are moving through to determine the appropriate movement. A tank, for example, can move much faster in open terrain than in a wooded area. In particular, in areas that expose the same characteristics in many neighboring grid cells, *explicit patch models* are useful. Patches break up the battlefield into areas where the characteristics do not change. In principle, patch and grid models are equivalent, but as the movement procedures must be reevaluated for every grid cell that is passed, a patch model is computationally the better solution. However, the characteristics of terrain that influence the movement of different weapon system or group types can differ, so that type-specific patches may be needed. An example is main battle tanks versus armored fighting vehicles: many swampy areas that cannot be crossed by main battle tanks due to their weight can still be passed by fighting vehicles, so the same terrain results in different usable patches for these two types. If a third and a fourth type is introduced, such as infantry groups or logistics and maintenance trucks, additional type-specific patches may be needed. Consequently, whether patches or grids are the best solution for a given problem needs to be determined by the simulation engineer.

Implicit mobility models quasi outsource the computation of mobility factors or multipliers used to estimate the mobility based on a base mobility for the overall battlefield. They were often used when computing power was still an issue, as in the combat models used during the Cold War era to determine movement constraints in Middle Europe. Implicit models are only rarely used nowadays but should be understood by engineers who may have to migrate legacy solutions based on this category to new approaches.

The last category is often used today. *Network models* use physical nodes and connecting paths to model mobility. They are often used to better model roads, bridges, etc. The advantage is that networks can be defined on top of every other category, so that mixed approaches can be supported. For example, a land unit can march on streets (network model) until it reaches the rear of the combat area, and then it can leave the street and move through the terrain to the engagement area (grid model). Similarly, air forces use networks to model their Low-Level Transit Routes (LLTR) and Minimum-Risk Routes (MRR) that are described in an earlier chapter. When an airplane uses these routes, the entry point is either part of the order or is computed by the modeled entity. Just as the army uses streets, the air force uses these routes. Many combat models create their own networks in the form of *polygons*. When an entity receives an order to move to a given point, the algorithms compute the best path (which can change, as we shall see later in this chapter), and store it in the form of a polygon. The entity follows this polygon until other events force it to react or it reaches the point of destination.

EXAMPLE FOR GRID AND PATCH PARAMETERS: THE ARMY MOBILITY MODEL

To better understand the parameters of terrain cells—which can be grid cells or patches—a concrete example is given in this section. One of the most sophisticated models was developed by the US Army in the 1970s (Turnage and Smith, 1983). The Army Material Command (AMC) invested in the development of two models to support modeling of movement: the AMC-71 Mobility Model followed the Army Mobility Model (AMM) in 1975. The long-term effort targeted to combine all mobility, trafficability, and military environmental knowledge in commonly accessible form. The first model of 1971 modeled patches with 13 parameters describing each patch. Using the rapidly improving technology for quantification of terrain areas, the AMM increased this model to deal with 22 mathematically independent parameters. Both models contributed to significant revision and updated map-making methods to use computers extensively, from development of single-factor terrain unit maps through the production of comprehensive mobility maps, in particular in preparation for the use of computer models, but it took until the 1990s before a close connection was established, as the models of 1970 were not advanced enough to take advantage of the mobility information.

The AMM uses the parameters as stored for the current patches a vehicle is in and the parameters describing the characteristics of the vehicle to calculate the maximum speed the vehicle can attain on this specified terrain. It takes the environmental factors and the vehicle parameters to compute the mobility characteristics. To the environmental factors belong:

- surface type
- surface strength
- surface roughness
- slope
- season
- precipitation form (rain, snow)
- precipitation amount
- obstacle geometry
- obstacle spacing
- vegetation size
- vegetation density and
- visibility characteristics

Surface type, strength, roughness, and slope build the foundation for the mobility that is modified by the other parameters. Different seasons, in particular the associated precipitation form (rain, snow) and their amount, will influence the mobility by making the ground more stable or destabilizing the ground. The same hill that can readily be passed under dry conditions can easily turn into a mudslide with heavy rain. What looks like an easy terrain first may turn out to be nearly impassable when paying attention to the details: the wide open areas in northern Germany look very favorable for battle tanks until the small rivers and creeks are taken into account—their steep slopes become dangerous traps for tanks that cannot cross them without combat engineering support. In addition to the initial conditions, the terrain can be modified by manmade obstacles as well as by natural obstacles. The focus of AMM is real terrain fortifications, such as barbed wire and anti-tank obstacles. But nature itself can become an obstacle as well: a forest easily crossed in the winter is filled with thorns, brushwood, and bushes in the summer.

The effects of these terrain features are different for the various weapon system types. While tracked vehicles can cross smaller trenches and various obstacles, vehicles with wheels are, as a rule, easier to hinder in their movement. However, if the terrain is swampy, heavy tanks can easily get stuck where lighter armed fighting vehicles can still drive through. This information could be stored in terrain patches as well, but the AMM decided to model them in a second input parameter block, the vehicle parameters. These include:

- vehicle weight
- vehicle geometry (in particular ground contact geometry)
- vehicle power characteristics
- dynamic reaction to obstacle impact
- vehicle braking characteristics
- front end strength
- dynamic reaction to rough terrain
- driver's tolerance to longitudinal shock, etc.

As stated earlier, the result of the calculation for combat models is the maximum technical speed of the vehicle of a given type in the current terrain patch. This speed is the upper bound for moving around in the patch, but normally vehicles will move slower, as the movement of all vehicles needs to be orchestrated to maximize their effect on contributing to the mission. If, for example, an attack is driven against an opponent in a fortified position, a mix of infantry fighting vehicles and tanks is needed. The tanks mainly provide fire power and protection, while infantry is needed to take the positions. To allow for an orchestrated operation of tanks and infantry fighting vehicles, they should drive together. For the resulting maximum speed, the minimum of all possible maximum speed needs to be used to avoid the slowest vehicle being left behind. In addition, several other factors may influence the tactical speed, which always needs to be lower than the maximal technical speed.

In high resolution models, one of the challenges is the orchestration of the different vehicle types that belong to a tactical formation. Common practice is to compute the individual technical speeds and use the minimum as the technical speed for the formation. In most cases this makes sense; however, in some cases it may be useful to make an exception: if a formation has to escape a surprise ambush or needs to retreat as fast as possible to avoid a massive artillery strike, the tactical constraints of the formation are no longer binding. In such a case, each weapon platform will drive as fast as possible and—if still possible—regroup or reorganize once free from such extraordinary danger.

In aggregate models, finding the average set of parameters for heterogeneous aggregates can be very challenging, and using the minimum of possible combinations is only one option. There is no general rule that can be applied; the simulation engineer has to apply personal and learned knowledge to tailor the set of possible solutions to meet the constraints given by the research question best.

Another thing to take into consideration is how long the computed speed is valid regarding the time step. If the patch is big or the time step is small enough, it may be sufficient to compute the speed once for each time step, but generally more computation is required. This discretization (i.e. using time steps) of a continuous process—namely driving through terrain—can otherwise lead to strange effects. In Figure 7.1, stripes of very easy terrain and very rough terrain are mixed. Each stripe is three units wide. In the easy terrain, the vehicle can drive at six units per time step; in the rough terrain, the vehicle needs four time steps per unit. To drive from point A to point B, the vehicle needs to cross two rough terrain stripes and two easy terrain stripes ($\frac{1}{2} + 1 + \frac{1}{2}$). For the two rough terrain stripes, $2 \times 3 \times 4 = 24$ time steps are needed. For the two easy terrain stripes, only one time step is needed. The tank should arrive at point B 25 time steps after leaving point A.

However, if the implementation does not take change of terrain into account and just defines the possible speed at point A (six units per time step) to compute the next point on the polygon (which would be the interim point ab, which is six units away from point A), the tank would arrive already after just two time steps at point B. The simulation engineer must make sure that such mistakes do not happen.

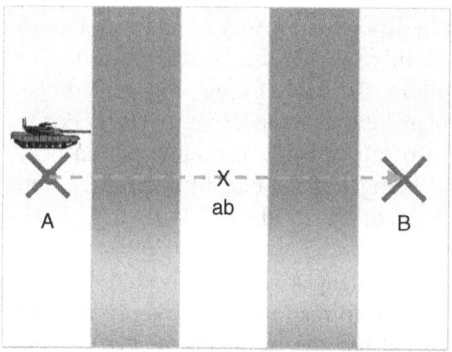

Figure 7.1 Driving through mixed terrain.

POINT MOVEMENT AND AUTOMATED ROUTE PLANNING

An alternative to computing velocity and technical speed is the computation of hindrance points for grids and/or grid borderlines and the use of movement points per simulated entity to compute how far the entity can move. This approach avoids the trap described at the end of the last section as, for each movement, the entity has to use some of its movement points. Several of the examples used in this section are motivated by the work published in Strickland (2011) in more detail. In Figure 7.2, a tank has to move from left to right through three different terrain types:

- easy terrain is supporting tank movement and has a point value of one in this example
- medium terrain is still supporting, but requires three points to pass in this scenario
- rough terrain can be crossed, but requires five points in this example.

Furthermore, there is a river flowing from the lower left to the upper right corner. To cross the river without combat engineering support requires additional points; in this example three points are required. Each entity gets per time step a certain number of moving points. To cross all three grid cells plus the river, the tank needs $5 + 3 + 3 + 1 = 12$ movement points.

The hindrance points per grid cell can be computed for every modeled entity type. If a tank unit is supported by combat engineers, the terrain factors may be lower than for a tank unit without such support. Also, certain ways through the grid can be excluded by giving them hindrance values that are greater than all movement points of participating entity types.

In a later chapter we will look at sensing the environment in more detail. However, we already know from earlier chapters that entities normally do not know everything about the battle field; in particular they do not know everything about the opponent and its weapon systems. Every time step, the perceived situation may change, which may require changing plans, including plans for movement. In support of this, a tactical hindrance value can be added to the

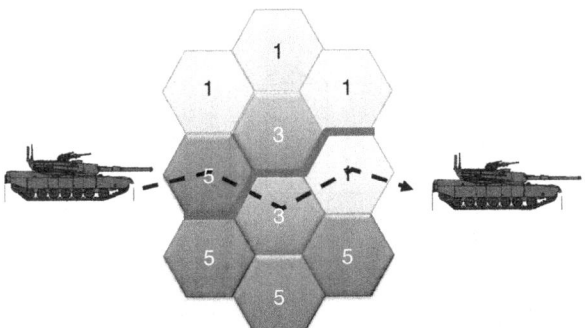

Figure 7.2 Movement points and hindrance values.

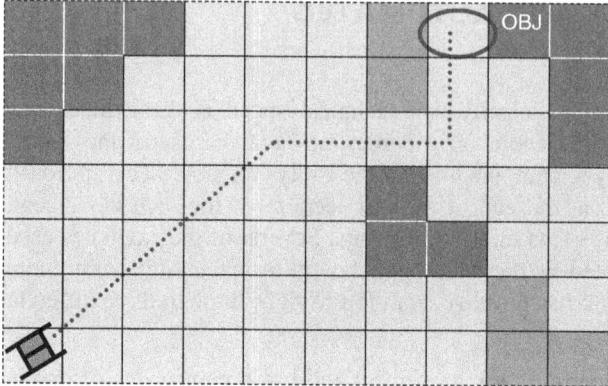

Figure 7.3 Initial plan.

terrain hindrance value. Figures 7.3–7.5 exemplify a dynamic change of plans based on new insights.

Figure 7.3 shows three types of terrain: open fields which are good for rapid movement, wooded areas which hinder tank movement, and urban areas in the lower right and the upper corner that allow tank movement but that normally are avoided by tanks as tanks are vulnerable in urban terrain. Therefore, the best path requiring the least movement points—or minimizing the overall hindrance values—avoids terrain that is not open. The resulting polygon shows the plan for the movement. The tank follows this path until it suddenly hits a minefield that blocks its way, shown in Figure 7.4.

Passing the minefield on the left has two disadvantages: the wooded area hinders movement, and minefields are often guarded by infantry and infantry loves the cover and camouflage they find in wooded terrain. Therefore, computing a new path with the fewest hindrance points is needed. Again, the tank follows

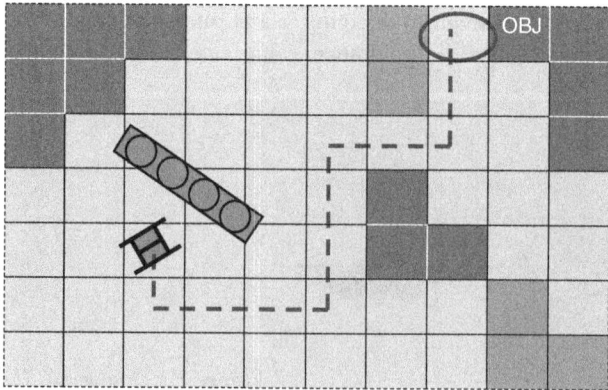

Figure 7.4 First new insight: a minefield.

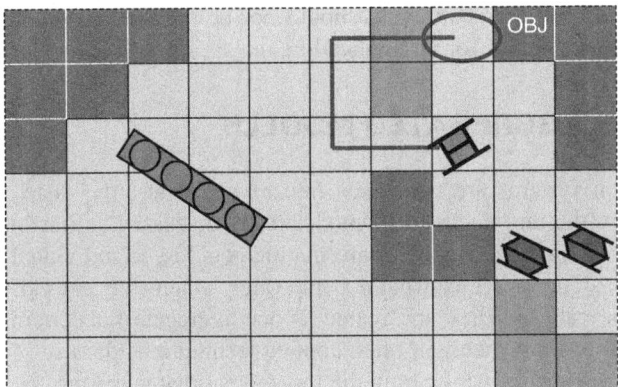

Figure 7.5 New insight: opposing forces.

this new path until it passes the straight between the upper city and the small forest where it discovers two of the opposition's fighting vehicles (Figure 7.5).

Even the short distance through open terrain towards the objective is too dangerous, as two opposing forces have enough opportunities to kill the tank. Therefore, the new movement retreats and crosses the urban terrain, as potential combat in the city is better than ensured combat against fighting vehicles in the open terrain. All this tactical knowledge needs to be encoded into the tactical values that have to be added to the terrain values.

This approach can be applied to individual weapon systems as well as to aggregated units. It allows the application of operations research methods to dynamically compute the optimal path. The movement represents a utility function that needs to be minimized. The A* Search algorithm developed for minimal path computation in the field of artificial intelligence is often applied in this context (Hart et al., 1968).

As not every change of perception may lead to a new plan, different characteristics of leaders can be modeled as well. If a leader is risk avoiding, the safest path will always be the first option. If a leader is willing to take a risk in order to reach a higher objective, the original plan may be maintained, even if this involves some danger. If, in the given example, the tank is trying to evacuate its own soldiers under opposing fire who are waiting for the tanks to arrive as soon as possible, a risk tolerant leader may decide to bull-through the open terrain risking being fired on and potentially getting killed by the two fighting vehicles in order to be at the objective early, while the risk avoiding leader retreats and re-approaches the objectives through the urban terrain.

Another effect that influences the core processes in general and the movement process in particular is the "dirty battlefield." This term is used to describe the effect that with ongoing battles the battlefield changes dramatically, as burning and killed vehicles change the environment. These effects have to be taken into account when hindrance values are computed. As visibility is one factor that influences technical speed, the dirty battlefield should not be overlooked when choosing the right movement model for an application.

DISTRIBUTIONS IN AGGREGATED MODELS

Whenever several weapon systems are composed into an aggregate, the distribution of these systems in the unit needs to be modeled. As discussed earlier in this book, we can build real aggregates where information gets lost as individual systems are merged into units, or we can build composites, where the individual systems remain as they are but they are treated as one aggregate for certain processes. When it comes to movement, an often applied technique is the use of schemas representing formations in which only the movement characteristics for one reference system are computed explicitly.

Small Unit Formations

Small units, made up of only a few individual systems, are often modeled in typical tactical formations. The three most common formations are:

- *The column:* the systems are lined behind each other forming a long column. This formation is often used when a small unit moves along streets or paths. It is also called file.
- *The line:* the systems move shoulder to shoulder. This formation is often used when approaching an opponent, as the systems can support each other with fire.
- *The wedge:* this formation combines the advantages of column and line by forming a V-shaped wedge. Systems can still support each other with fire, but not all systems are equally close to the opponent. The wedge as well as the inverted wedge—or Vee—is often used by small units, in particular infantry.

Figure 7.6 shows the formations. One system is highlighted as the reference system. The reason is that movement modelers often only compute the detailed movement characteristic for the reference system and let the other systems "follow within the schema of the formation." The underlying assumption is that the

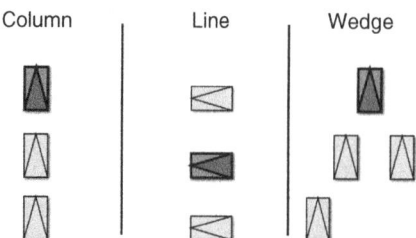

Figure 7.6   Column, line, and wedge.

terrain for all systems of the small unit is sufficiently similar. The simulation engineer must check if this assumption is justified to avoid strange things happening, such as for example tanks swimming in a river as the reference system drives on a shore road.

The naval forces and to some degree the air force use similar patterns for their operations. However, the most often used formation in air force combat operations only comprises two identical airplanes flying one "sortie."

This idea of schemas using distributions can be generalized for aggregated units as well.

Distribution in Aggregated Units

Larger units, like companies and battalions, can use tactical—or tactical/operational—schemas comparable to the simple formations. Figure 7.7 shows several examples often used in particular for land warfare models. The easiest form of aggregation is not even captured in the depicted schema: we assume no distribution at all but treat the aggregated unit as a mass point. All weapon systems are merged into this one mass point, which makes movement easy: only the point representing the unit has to be taken into consideration.

The left three forms are also quite common and quite simple. The area normally occupied by the unit is modeled as a circle, a square, or a rectangle. The rectangle can be used to show marching units as well as defending or attacking units. In many applications, the systems are assumed to be equally distributed within these schemas. Some models, however, place all their systems at the border.

The two schemas on the right hand side are closer to tactical schemas representing the cluster of systems in operations. The left of these two schemas can

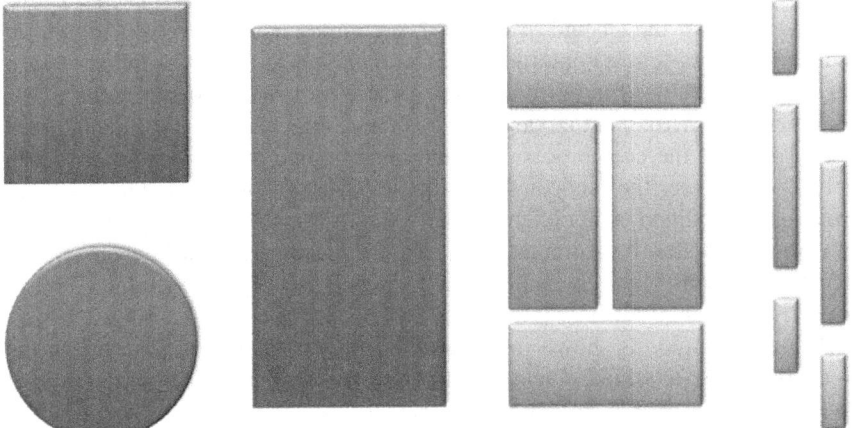

Figure 7.7   Schemas for land warfare aggregates.

be used to model the front, the two flanks, and the rear of a land unit. Quite often, two-thirds of the weapon systems are in the front while the remaining third stays in the rear as reserve and protects the flanks. The right of the schemas can represent a larger unit marching on two parallel roads, each of them using a vanguard and a rearguard with the main forces in between. It should be pointed out that units can change between these schemas. A unit can be in the assembly area getting ready for an operation. For this phase, modeling the unit as equally distributed in a circle may be appropriate. Next they march using two parallel roads from the assembly area to the defense sector, using the just introduced schema for marching. Finally, they move into the defense section, using the defense schema. Transformation rules guide these processes.

The enumerations given in this section are neither complete nor exclusive. There are innumerable ways to model the movement and the distribution of systems within an aggregate. They reach from simply mass point models to complex fluid models that take order movement plus attrition plus entropic dispersion into account to compute the density of systems within a unit.

Applied Tactical and Terrain Modified Schemas

We have already discussed some counterintuitive situations that can occur when schemas are applied to a unit without consistency checks. The example is a small unit driving in a wedge formation with the reference system on the road next to a river, leading to one of the tanks swimming in formation through the river. The general way to cope with this sort of challenge is to use the tactical schema as the rule, but to modify it to reflect environmental constraints. In the example, either the wedge has to move, or the schema needs to be modified as supported by the terrain.

In general, *applied tactical schemas* reflect that, in military forces, the current tactical status often implies the formation, as discussed for high resolution as well as for aggregate models in the earlier sections. Transformation rules can be applied to support the change from one schema into another, like moving out of the assembly area (circle) onto the road for marching (column). Applied tactical schemas are often used in uniform terrain where no strange effects occur.

When the environment model leads to strange effects in the movement model (or any other of the core process models), *terrain modified schemas* need to be applied. As before, the tactical status implies a basic formation, but the basic formation is modified to avoid strange effects. This requires that adaptation and transformation rules be formulated so that they can be applied when strange effects occur. This requires explicit terrain models, such as streets, hills, terrain cover, rivers, and so forth.

In our example, as shown in Figure 7.8, the strange effect is a tank in the river. The terrain modified schema implies that tanks cannot drive on terrain that is not suitable for them (river, swamp, urban terrain with too narrow streets, etc.). The transformation rule moves the tank towards the center of the formation until the strange effect no longer occurs.

Tactical Schema

Terrain
Modified Schema

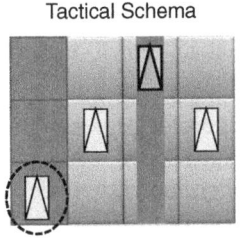
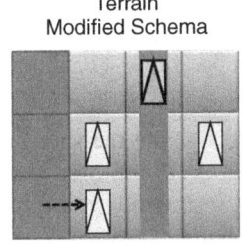

Figure 7.8 Tactical and terrain modified schema.

It is necessary to ensure that this new schema is applied for all core processes; otherwise other strange effects can occur (similar to the fair fight issues discussed earlier in the book). If different schemas are used to present the same simulated entity differently in different processes, we call this *polymorphism* (from the Greek word for "many bodies"). In particular when dealing with building federations of several simulation systems that use different combat models to represent the core processes, this becomes a challenge for the simulation engineer.

ADDITIONAL MOVEMENT MODIFICATIONS

In summary, in high resolution models we model the movement of each individual system. In aggregate models, we can use detailed modeling of one reference system and then let the rest of the unit follow within the geometric shape. This shape is based on tactics and modified to fit the current terrain. Average velocities within terrain patches can be used for movement models for units, or just the current movement characteristic of the reference system. Explicit network models including simple polygons to follow can be used as well. In any case, there are technical upper limits (like the maximum speed that a system can drive in a given terrain) as well as tactical upper limits (like the maximum speed of the slowest system type within a unit that will stay together within an operation). As a rule, the simulated entities will move with the ordered speed that is governed by these upper limits.

Beside terrain and obstacles, firefights in particular have been shown to significantly influence movement. This is not surprising: imagine marching on a road and suddenly being under enemy fire! Not many will continue as if nothing had happened. Military operations research identified two factors now often used to model the effect of fire on movements.

- Casualties influence movement. The more casualties there are, the slower the modeled unit becomes. The reasons are many and go from simply shock and having to deal with this new situation of being under attack to physical blocks by its own systems that cannot move any more, "dirty battlefield" effects of the fight reducing the visibility, and so forth.
- The casualty rate, which means the number of additional casualties that are observed over a time step, is another factor that slows movement down.

The reasons are similar to casualties, but the intensity of an attack—including getting under artillery fire or being attacked by opposing air forces in close air support operations—can slow an attack down or stop it completely, even if not all systems are destroyed.

There are too many models to compute the speed reduction factor to handle in this chapter. Youngren (1994) gives several examples on such speed reduction models. He also gives examples of historically relevant implicit mobility models for theatres of the Cold War era, but such models are no longer in operational use.

Some simulation systems furthermore use models of "panic modes." When a catastrophic event occurs, such as a massive attack or air strike, the tactical rules may no longer be applicable and every system is simply moving as fast away from the catastrophe as it can. The upper tactical velocity or tactical schemas are no longer applicable.

Current research based on agent-directed simulation—as described in Chapter 27—looks to algorithms for doing route selection by a group moving through a terrain based on flocking models. Most interestingly, the small unit formations are often observed and can be derived from the underlying tactical consideration.

In summary, what tactical schema to modify—or how to coordinate individual systems to support unit behavior—as well as how to compute the movement are among the main influencers of how a model behaves. It is surprising that the literature focuses either on physical models for movement, which are tactically overruled in standard cases anyway, or on attrition. However, the simulation engineer needs to understand how this core process is modeled in detail in order to avoid strange effects, in particular when simulation systems are federated to work together in support of a common research question or training task. Polymorph models for movement are a frequent reason for strange behavior and wrong results, if the simulation engineer is not aware of this challenge.

REFERENCES

Hart PE, Nilsson NJ and Raphael B (1968). A formal basis for the heuristic determination of minimum cost paths. *IEEE Trans Syst Sci Cyber* **4**, 100–107.

Lucas TW (2000). The stochastic versus deterministic argument for combat simulations: tales of when the average won't do. *Mil Oper Res* **5**, 9–28.

Strickland JS (2011). *Mathematical Modeling of Warfare and Combat Phenomenon*. Lulu, Inc., Colorado Springs CO.

Turnage GW and Smith JL (1983). *Adaptation and Condensation of the Army Mobility Model for Cross-Country Mobility Mapping*. Technical Report GL-83-12. Waterways Experiment Station (WES), Vicksburg.

United States Department of Defense (1998). *DoD Modeling and Simulation Glossary*. DoDD 5000.59M, Undersecretary of Defense for Acquisition Technology, Washington, DC.

Youngren MA (ed.) (1994). *Military Operations Research Analyst's Handbook; Volume I: Terrain, Unit Movement, and Environment; Volume II: Search, Detection, and Tracking, Conventional Weapon Effects*. Military Operations Research Society (MORS), Alexandria, VA.

Chapter 8

Modeling Sensing

Andreas Tolk

GROUND TRUTH AND PERCEPTION

In this chapter, we will deal with the core process of sensing or looking. In many combat models, the explicit modeling of perceiving is relatively new. Until recently, many combat models used *ground truth* for their planning purposes. Ground truth is the term describing the simulated reality within a combat model: the state vectors describing the simulated entities are known completely to the simulated entities. This is true for high resolution as well as aggregated models. If a system is moving on the battlefield, when ground truth is used the other systems not only know where the system is, they also know in which direction it is going and at what speed it is moving. What is modeled, simulated, and therefore known within the system—the ground truth of the simulation—is known to everybody. Similar to a chess player who knows exactly where his as well as his opponent's figures are and what they are capable of, every entity has full awareness of everything that happens—and can happen—in the simulation.

The term ground truth is not exclusive for land forces. Air forces and navies may also use "ground truth" for their combat models and simulation; it simply refers to the fact that no perception based on sensor results created to show what a player knows, but that every player knows all positions and activities as they are represented in the model.

This approach is not very realistic. In real combat, we do not know everything about the opposing forces and often not even enough about our own forces. Every headquarter continuously works on improving the *situational awareness*. What do we know about the situation on the battlefield? What do we know about our own troops, such as location, amount of ammunition and fuel on the system,

Engineering Principles of Combat Modeling and Distributed Simulation, First Edition.
Edited by Andreas Tolk.
© 2012 John Wiley & Sons, Inc. Published 2012 by John Wiley & Sons, Inc.

readiness for operations, etc.? What do we know about the enemy? Where are the enemy forces? Where are they going? What are the plans for operations? The answers to these questions are normally captured in a perceived situation that is stored in the command and control system and that is displayed and shared in common operational pictures and maps.

The resulting *perception* reflects what a certain entity assumes to be reality. The perception can look very much like ground truth regarding the structure, but the values assigned to the structure may not be accurate: locations may be a little bit off, velocities are unknown, etc. To build a perception, two things are needed: either the entity has sensors that can observe the battlefield and accordingly produce estimates for the values looked for, or it gets reports from other units and entities that comprise these values. The simulation engineer must ensure that the alignment and harmonization principle is observed: if a property is important to guide a decision, it needs not only to be modeled on the system that is characterized by the property but also to make it into the perception, and sensors are needed that can observe the modeled reality and feed observations on this property into the perception. Furthermore the engineer must ensure that decisions are based on an appropriate perception, not on the ground truth.

In this chapter, we will look at various sensors and how they contribute to a perceived situation. We will look at various line-of-sight algorithms and high resolution and aggregate sensor and perception models, including an introduction to radar and sonar models that are of particular interest in many intelligence, surveillance, and reconnaissance operations.

SENSORS AND SENSOR RESULTS

A sensor is a device that observes the battlefield and produces reports of what it observes. In a combat simulation, sensors can be best understood as filters on the ground truth that produce a mapping of the ground truth to the current observation that can be used to contribute to the perceived situation.

For example, one special sensor can estimate the location of vehicles on the battlefield very accurately. However, the location is the only thing that can be observed. The sensor does not produce information about what type of vehicle is observed, what the velocity is, if it is a hostile, a neutral, or an own vehicle, etc. It only observes the location of vehicles. Another sensor observes movement in given areas. The accuracies of the individual movements are not very high, but it gives a good idea in which parts of the battlefield many movements occur, and where systems are relatively stationary. The intelligence officer gets all these observations and has to take them into account when building the perceived situation: what is known about what is going on on the battlefield? The various information sources must be merged as well as with already known information. In the small example, the intelligence officer knows from the planning where troops should be. The first sensor can be used to update their location and also knows that all other locations are either neutral or hostile. Information is added about movement-intensive areas on top of this to get an insight if troops make the planned progress and what hostile troops may do.

Often, the process of merging the information of different sensors regarding one observed entity is supported by technical processes. This is the domain of *multi-sensor data fusion* (Hall and Llinas, 1997). As Hall and Llinas point out, techniques for multi-sensor data fusion are drawn from a wide range of areas including artificial intelligence, pattern recognition, statistical estimation, and other areas. Whether these complex processes have to be supported by the combat model depends on the research question: if the efficiency of new intelligence procedures and technical support thereof are evaluated, all details of building a perception—including the high resolution of technical sensor fusion details—are needed. If a group of intelligence officers is being trained, the outcome of the technical processes is still important, but the details of how these technical processes work and how they produce the data provided to the officers is not of interest. If the target is to train combat troops, producing a simple map with a perceived situation without revealing any details on how it gets produced may be sufficient.

In any case, the combat model is likely to support a number of sensor types that need to be used appropriately to support the research task efficiently.

Types of Sensors

Sensors can be grouped into categories regarding how they observe the battlefield. A common categorization is what *sense* they are using. Accordingly, sensor types can be categorized into:

- *Acoustic sensors.* These are microphones (air waves) or hydrophone (liquids) that capture signals for processing sound characteristics, such as pitch, timbre, harmonics, loudness, and rhythm. Some new sensors are so accurate that they can listen whether there is a heartbeat in the neighboring room within a house. More conventional sensors can distinguish motor and engine sounds of different vehicles. Also, many vehicles need some sort of motor power to support their combat readiness: a tank cannot move the turret without motor power, and a running engine makes sounds that can be heard, no matter how quietly it runs or how well it is muted.

- *Chemical sensors.* This category is often referred to as artificial noses or tongues, as they are used to identify certain chemical—and therefore potentially also biological—substances in a given area. They are of particular interest in biological and chemical warfare, but they can be used for other sensing purposes as well. For example, they can detect exhaust gases in certain wooded areas, which supports the assumption that some motorized troops took cover in this area.

- *Electromagnetic sensors.* These sensors observe changes in electrical and magnetic fields. Radar systems belong to this group, but also special sensors of the navy that observe if the magnetic field changes due to the presence of vessels. Such sensors can also be used to observe and control movement of own troops, such as using electromagnetic loops to count vehicles leaving the assembly area to know how many systems are already on the move.

- *Thermal sensors.* These are sensors that scan the environment for heat sources. They are also known as infrared sensors. They work particularly well when the difference between entity heat and background heat is significant. Many weapons use heat sensors to track airplanes or land-based systems using the heat of the engines as the target indicator.

- *Optical sensors.* Although technically these could be counted under the electromagnetic sensors, optical sensors build a group of their own, as human sensing is strongly connected with this category. Binoculars, telescopes, but also laser-based range finders and similar sensors using the visible spectrum of light belong here.

The borders between these categories are sometimes fluid. Infrared sensors are sometimes grouped under electromagnetic sensors, sometimes as quasi-optical sensors. Another distinction of interest to the simulation engineer is whether a sensor requires line of sight to be able to make an observation, or if this is not the case. We shall deal with several line-of-sight algorithms later in this chapter. As the name implies, most optical sensors require line of sight, but that is not necessarily the case. It is therefore important to have good technical advice—in form of, say, a subject matter expert—at hand when modeling sensors.

Another aspect common to most sensors is the signal–background noise ratio: if the exposed property is very different from the background property, detection is easy; if they are identical, it is not possible. A warm target before a cold background can easily be detected with a thermal sensor; a shining light can easily be seen in the darkness. Just as camouflage helps to visually merge into the environment, similar techniques can be applied—and need to be modeled—for all sensor categories.

In order for a sensor to detect a target, three requirements have generally to be fulfilled.

- The sensor has to be able to detect a certain property or a combination of properties (like an infrared spectrum).

- The target exposes at least one of the observable properties (like giving out heat in the detectable infrared spectrum).

- The background does not expose the same observable property or at least is significantly different (the environment is colder than the target).

Not fulfilling one requirement leads to not detecting the target. Not taking all into account leads to insufficient sensor models. Sensor characteristics and target characteristics have to match.

Sensor characteristics are dependent on the type or category of the sensor, but generally include the range of the sensor, or a sensor footprint, defining the area in which the sensor can detect. Targets have to be described using detectable features, such as the electromagnetic signature, velocity, mass, etc.

Steps of Creating a Perception

When a simulation engineer has to decide which sensor functions need to be modeled in support of a research question, different sensing principles and contributions need to be understood. The following list is neither complete not exclusive. However, it summarizes the most often used terms and principles the author observed during related tasks.

Search, also referred to as surveillance, involves examining a volume of space for possible targets of interest. This is normally done by directing radar energy in a pattern of beams that cover the search volume.

- *Searching* and *looking* are two closely related but different activities. Searching is directed at a special characteristic. A soldier can search for hostile tanks or approaching infantry. If the soldier is simply looking, the area of responsibility is being observed more generally, without focus on special characteristics. The difference is that searching has normally a higher probability of success for entities exposing the special characteristics but a lower probability for entities not exposing the special characteristics compared with the process of looking.

- *Detection* is determining that an entity is present in the search volume. This is normally done by setting a received-signal threshold that excludes most noise and other interfering signals. Signals that exceed this threshold are the detected targets.

- *Tracking* is determining the course and speed of a target from a series of measurements. Once a sensor detects an entity and then follows it, this is called tracking. Brooks et al. (2003) gives a good overview on this topic. Tracking is of high relevance for missile defense, where the idea of *queuing* also plays a role: a sensor who takes over the tracking from another sensor that is currently aware of the entity of interest receives the information needed to take over the responsibility of tracking the entity.

- *Classifying, recognizing*, and *identifying* are also terms with similar meaning. Anderson (2004) gives good examples of how to differentiate between them. Classifying is the process of determining that an observation leads to categorizing the observed entity, e.g. I know the object is a tank, not a truck. Recognizing is establishing that an object is a certain type of entity I am aware of, like a special battle tank with outstanding characteristics. Identifying is categorizing a special individual entity as such, like the company and platoon the tank belongs to.

- *Target acquisition* is the goal of sensing in many combat situations. In the core process of shooting we will look at this closer. However, in order to acquire a meaningful target, a system must detect the entity, must classify it as a potential target (if a soldier has no tank breaking weapons, shooting at a tank is a stupid thing to do), and must be able to engage it (if the target is out of range, there is no reason to engage).

There are many "side effects" on a battlefield that can distract from sensing the right things. The following are just some of many more examples. The degree of preparation of the target, in particular the degree of camouflage, can influence the detection and classification of a target. Decoys can detract from the real systems. Engagement in firefights gives positions away easily; that is why armed forces normally change their position immediately after a duel, as the old position is known to everyone who witnessed the duel. Artillery suppression may not result in casualties but is very likely to heavily distract soldiers under fire from sensing. The already mentioned dirty battlefield effects and other dynamics of the environment add challenges to finding targets. Not every tank that looks like being destroyed is completely useless for the enemy (if nothing else, it can give cover to infantry approaching under the cover of smoke and dust). Similarly, some useless tanks do not expose their state to the outside: a tank with completely destroyed electronics cannot do anything anymore, but it looks intact from outside.

Overall, modeling sensing and creating a perception from the observation is a very challenging topic and requires a lot of subject matter expertise, way beyond what can be dealt with in a book like this. The reader is encouraged to continue his or her research when being tasked with related questions.

LINE-OF-SIGHT ALGORITHMS

Line-of-sight between two modeled entities within their situated modeled environment means that you can establish a straight line between these two elements, and the line does not cut through any other objects between them, no matter if these are other simulated entities or part of the simulated environment. In other words, you can see one simulated entity from the other and vice versa. Many sensors require that the line of sight between the sensor and the object to be observed exists. For example, if you use binoculars, you cannot see soldiers hiding behind trees: the trees are in the way, and the line-of-sight between you and the soldiers "cuts through the trees." A laser used as a range finder only travels until the lights hits the first object. In addition, many weapons—although being guided—require that the shooter can observe the target in order to guide a weapon to the desired spot, such as anti-tank missiles. As a result, line-of-sight computations are very often required, not only for the sensing part. Consequently, a lot of work has gone into efficient algorithms to compute whether the line-of-sight between two objects exists.

Basic Line-of-Sight Algorithm

Line-of-sight algorithms determine if a straight line between observer and target can be established. The basic idea is to generate a profile of the terrain and use

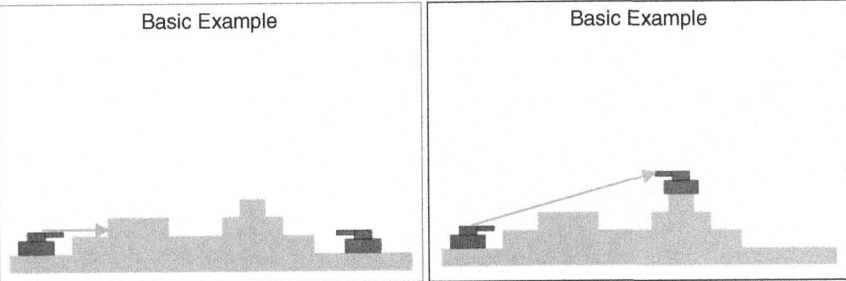

Figure 8.1   Examples for line of sight.

interpolation between pairs of elevation points. Depending on the level of detail, these elevations have to be adjusted by additional ground cover, trees, buildings, etc. The slope of the connection between the observer and the target is compared with all critical elevation points in between. If the slope between observer and target is higher than between the observer and all critical elevation points, the line-of-sight exists. Figure 8.1 shows this in an example: in the left figure, the observer cannot see the target because the line-of-sight intersects the terrain; in the right picture, the target can be seen.

The assumptions behind this simple basic algorithm are that the curvature of the earth can be neglected, that the level of detail is sufficient to support the interpolation, and that a reasonable interpolation schema can be generated. The interpolation schema, which must be based on the resolution chosen to model the environment, is also the limitation to the accuracy of this approach.

Modifications and Extensions of the Basic Algorithm

Many modifications have been developed to make the line-of-sight computations more realistic. Youngren (1994) lists several algorithms in detail.

The first improvement is adding the height of the observer as well as of the target to make the computation more realistic. The higher an observer is above the ground, the less likely it is that terrain features are in the way. However, the observing system gets more exposed as well, so it can become an easy target itself. Another improvement is adding the earth's curvature, which is particularly necessary for sensor systems with a long range. Also, more detail has been added for the terrain as well as for the atmospheric anomalies. Some models take clouds, dust, fog, and different layers of temperature into account, including dirty battlefield effects. The disadvantage of these more and more detailed approaches is that they require a lot of computing power. Therefore, many implicit models have been developed to compute inter-visibility segments (which terrain cells can have line of sight with a given cell). Other approaches define the probability for the line of sight. Analytically derived data were validated and enriched by live fire exercises, such as the Chinese Eye exercises on company and battalion levels and the Kings Ride exercises on platoon and company levels.

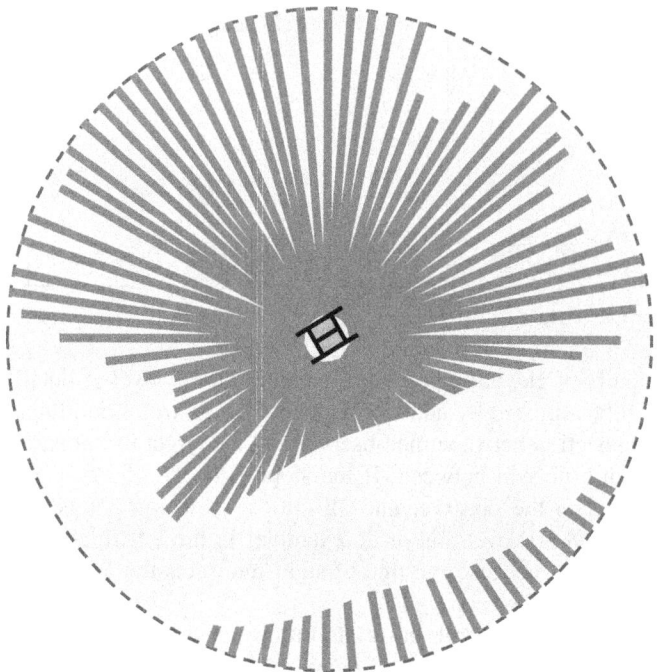

Figure 8.2 Using line of sight to produce visibility maps.

Many combat simulations provide visibility charts for the user showing what a selected system can actually observe (or better, in which part of its surroundings line of sight exists). To compute these graphs, the applicable line-of-sight algorithm is applied in segments around the system until the maximum range is reached or a terrain obstacle is in the way. Figure 8.2 shows such a graph. Of interest are the invisibility islands: these are blind spots for the system where opponents can hide without being detected. Conversely, such blind spots computed for the systems of the opponents can be used for safe approach paths of own systems. The resulting visibility maps depend on the height of the observer as well as on the height of the target, and all other influence components used for line-of-sight computations.

HIGH RESOLUTION SENSING

High resolution sensing algorithms can be applied to both models using weapon systems and models using aggregates.

Requirements

In this requirements section, the focus will be on topics and considerations given to modeling the sensing and perception of a single entity in a high resolution model, in particular its ability to perceive, the direction of an observation, and the frequency with which the observation takes place. This will lead to the two categories of high resolution sensing models: glimpse models and scanning models.

We already dealt, in general, with the ability to sense requirement by listing the three requirements for a detection to occur. High resolution sensor models normally take into account that systems have different capabilities when being in a fortified position, on the march, in a stationary or moving battle situation, and so on. They also have to model the dynamics of the battlefield: infrared sensors are significantly influenced by burning tanks or hot explosions. In order to take this into account, all these effects need to be modeled—and, following the alignment and harmonization principle, what is modeled also needs to be taken into account. These challenges increase when different models are composed when building a federation, but this will be the topic of a later chapter.

Most senses are directional in nature—this is true of hearing, sight and others. As very few combat entities remain perfectly still—whether we model an individual soldier or the ability of a crew to see out of a vehicle–the directions of the senses are moving. When defining the directions of sensing the resolution of time and the environment must be harmonized with the model of the entity. If time steps are very short then the directional cone of seeing should be narrow. If they are longer, the cones can get bigger, or more cones can be used to model the focus of an observation better.

There are two basic procedures for searching and detecting an entity. Continuous sensing uses a detection rate function to estimate the detection time and to create an event that models that at this time the detection occurs. The alternative is the glimpse model that is a time step model. The observer scans the area the target is in several times and with every glimpse the chance to detect the target arises.

Continuous Sensing Model

In continuous sensing models, the time of the detection is computed for an entity as soon as it enters the sensor area. Two assumptions are necessary.

- The success rate for a detection to occur is continuous. If it is easier to detect a target in certain areas, the rate for this area should be modeled individually.
- The detection can be modeled as a Poisson distribution over the expected detection time T. This time needs to be modeled analytically or can be observed in field experiments.

The resulting formula for the expected detection time when using the assumptions is $p(t) = 1 - \exp(-t/T)$. The simulation engineer needs to ensure that at the time the detection occurs all requirements are still fulfilled.

A particular challenge occurs when systems enter and leave the sensor area several times. A naïve approach is to simply draw a new detection time every time the system reenters, but this approach is not only computationally expensive, it can also lead to strange effects in the detection process. Although the detection is using the Poisson distribution, the average detection time is T. If the system leaves shortly before this time is over the sensor area and reenters to get a new number, the system gets "a new chance" to approach without being detected. In the deterministic version of this approach, the detection occurs for sure exactly after the time T. The simulated systems can "cheat" using this information by leaving the footprint of the sensor shortly before they stay in the observed area for this time.

Therefore, some algorithms "remember" if a system has been subjected to the process and store the detection time for this system. When the detection occurs, the algorithm checks if the system is still in the sensor area, and if it is, the detection occurs, regardless of whether the system left the area after the original event was computed. The challenge with this approach is that it may favor the sensor unfairly: if a system takes cover in the sensor footprint area, the search process may really have to be started new when it continues to pursue its objective. Finding the balance between avoiding system artificialities and providing realistic features is one of the tasks of the simulation engineer.

In summary, this approach defines one detection event when the observed entity enters the sensor area. The average detection time needs to be derived analytically or needs to be derived from field experiments.

Glimpse Sensing Model

The glimpse sensing model is the counterpart to the continuous sensing model. It models a series of short glimpses, such as occur with a radar system or when a human observer scans the observation area systematically from the left border to the right border. The observer directs attention to a small and narrow sector. If the target is in this small sector, it may be detected before moving on to pay attention to the next sector.

With every glimpse, a system in the sector may be detected with the sensor–target specific likelihood d. Once a system is detected, it is normally tracked (and often the scanning process stops, depending on the specifics of the modeled observer). After n scans, the probability for a system to be detected is therefore $P_n = 1 - (1 - d)^n$. As for every scan the likelihood of detecting the target is d, the estimated number of scans N until a detection occurs becomes $E(N) = 1/d$.

Scan frequency and detection probability per scan are normally not independent: the longer I look into a sector, the more likely I will detect a system. It is therefore critical for the simulation engineer to understand if and how scan rate and detection probability are modeled. Otherwise, strange effects may be observed. For example, if the detection probability is fixed and the scan rate is an input parameter, the detection rate can be increased or decreased arbitrarily if not controlled by the simulation engineer.

A special challenge occurs if two models that model the same duel between the same systems need to be aligned for a federation, but one model uses the continuous modeling of sensing and the other model uses the glimpse model. The simulation engineer must then come up with meaningful parameters for both models that deliver comparable and similar results.

AGGREGATE SENSING

As high resolution modeling of sensing can be used for both models, so can aggregate models of sensing be used on the weapon platform level as well as on aggregate unit level.

Area-Specific Filtered Perceptions

The easiest way to model perception is not to model it at all: instead of creating a perception, the ground truth is used and unit-specific filters are applied.

The easiest way is to immediately use the values provided by ground truth, assuming perfect information within the system. Many combat models still use this assumption for the own forces assuming that one side knows where the own troops are and in what state they are in. However, friendly fire—own troops shooting at own troops because they perceived them to be enemies—is reported in nearly all operational lessons learned.

The next step is to define observation areas. In these observation areas—which may be computed using the footprints of the available sensors or aggregated into a standardized area based on general sensor assumptions—perfect detection or probability based detection can be modeled. If detection occurs, the data being stored in the ground truth become accessible; otherwise the systems remain unknown to the observer.

The filters can be modified to allow access to only certain attributes or attribute aggregates based on the sensor footprint, the observed entity, and the time in the footprint. This can be used to model the various stages of the perception process from knowing that "something is out there" over the detection that "it is an armored fighting vehicle" to full identification.

Some models apply mixed forms, such as defining an area of certain detection in the main observation direction of the troops (often the areas of responsibility of the assigned units) extended by an area of possible detection (often the areas of interest of the units).

Modeling Reports

So far, the models of perception were mainly driven by the likelihood of detecting a system with its own sensors. In a real operation, the perception of a unit is not only driven by own sensors but also by reports. These reports can be everything from spot reports from small units or entities that report enemy contacts up to huge situation reports generated by the headquarters.

These reports contain information known to other units about the situation. These observations can confirm own information, update own information, or insert new information into the perception of the receiving unit.

Such reports can take the reliability of sources into account when it needs to be decided if the own information or the reported information is more accurate. For example, the own information can be based on a spot report from some forward solider under heavy fire. If the unit receives a report based on the aerial pictures from an unmanned surveillance air vehicle, these data are likely to be more accurate. However, if the data are also based on soldier reports and already several minutes old, the own data may reflect the current situation better.

Some models add errors to the ground truth data based on the time occurring between the last detection/observation and the current simulated time. For example, the perceived location is close to the real location shortly after the detection, but can be modeled to change in the last observed direction with the last observed speed until a new detection report occurs. This algorithm can become as complicated as the modeler wants it to be, as terrain influence and doctrinal constraints can be used to improve the perception process.

Uncertainty and Perception

Of particular interest for modeling sensing and perception is the used of *a priori* knowledge when evaluating the obtained information. Bayes's theorem allows such use of *a priori* probabilities and conditional probabilities of observations to support hypotheses on the implications of observations. The motivation to apply Bayes's theorem in this context can be given as follows.

- We know the *a priori* probability of observing one hypothesis. We know the number of items of a given type.

- We know the conditional probability of items within an interpretation of an observation given that a hypothesis is true.

- We want to know the most likely hypothesis or interpretation given the actual observation.

For example, we know from intelligence resources that the opposing side attacks with 30 units. In our example, there are 10 tank units and 20 mechanized infantry units. Each tank unit comprises 40 tanks and 10 armored fighting vehicles. Each infantry unit is made up out of 10 tanks and 50 armored fighting vehicles. The observation reports that tanks are approaching. What kind of unit has to be expected with what probability?

A priori the likelihood of observing a tank unit is 33.3%, and the likelihood for a mechanized infantry unit is 66.6%. Both of them comprise tanks as well as armored fighting vehicles, but the mix is different: 80% of the weapon systems in a tank unit are tanks, but only 17% of the mechanized infantry unit is made

Table 8.1 *A priori* Probabilities

Probability	Symbol	Value
A priori probability to observe a tank unit	$P(\text{TU})$	$10/30 = 0.33$
A priori probability to observe a mechanized infantry unit	$P(\text{MIU})$	$10/20 = 0.66$
Probability to observe a tank when facing a tank unit	$P(\text{Tk}|\text{TU})$	$40/50 = 0.80$
Probability to observe an armored fighting vehicle when facing a tank unit	$P(\text{AFV}|\text{TU})$	$10/50 = 0.20$
Probability to observe a tank when facing a mechanized infantry unit	$P(\text{Tk}|\text{MIU})$	$10/60 = 0.13$
Probability to observe an armored fighting vehicle when facing a mechanized infantry unit	$P(\text{AFV}|\text{MIU})$	$50/60 = 0.87$
A priori probability to observe a tank	$P(\text{Tk})$	$600/1700 = 0.35$
A priori probability to observe an armored fighting vehicle	$P(\text{AFV})$	$1100/1700 = 0.65$

up of tanks. In total, 600 tanks and 1100 armored fighting vehicles are on the battlefield. To apply Bayes's theorem, these numbers have to be captured as conditional probabilities. Table 8.1 enumerates them.

Bayes's theorem is usually expressed as:

$$P(H|O) = P(H)\frac{P(O|H)}{P(O)}$$

with $P(H|O)$ the conditional probability of hypothesis H when observation O is made, $P(H)$ the *a priori* probability of hypothesis H, $P(O)$ the *a priori* probability to make the observation O, and $P(O/H)$ the conditional probability of observation O when hypothesis H is true.

Applying Bayes's theorem to the perception question of the above example, the answer to the question "How likely is it that we are faced with a tank unit given that we observe a tank?" computes to:

$$P(\text{TU}|\text{Tk}) = P(\text{TU})\frac{P(\text{Tk}|\text{TU})}{P(\text{Tk})} = 0.33\frac{0.80}{0.35} = 0.75$$

In other words, by applying Bayes's theorem we can now be 75% certain that we are facing a tank unit, which is much better than the 33% we had before. The general Bayes theorem is not limited to only two alternatives. The general formula for n observations and m hypotheses is:

$$P(H_i/O_1 \wedge \ldots \wedge O_n) = \frac{P(H_i) \cdot P(O_1/H_i) \cdot \ldots \cdot P(O_n/H_i)}{\sum_{k=1}^{m} P(H_k) \cdot P(O_1/H_k) \cdot \ldots \cdot P(O_n/H_k)}$$

In order to apply this theorem, we have to assume we know all alternatives (closed world assumptions), that the alternatives do not overlap (single fault assumptions), that observations are independent, and that the observations are only dependent on the alternatives.

There are many additional ways to express related concepts of uncertainty, incompleteness, similarity, vagueness, and more. Certainty factors, interval propagation and applications of fuzzy logic are just a few examples of successful applications of mathematical concepts to cope with sensor and perception issues. Some additional concepts are dealt with in an advanced chapter later in this book. A good overview of concepts and their computational representation is given by Kruse et al. (1991).

Radar and Sonar Models

Radar and sonar systems play an important role in sensing and contributing to building a perception of the current situation. Both are used to search, detect, and track targets as well as guiding weapons into the target area. However, while radar uses the electromagnetic spectrum through the atmosphere sonar systems use acoustic waves traveling through water. A detailed overview of modeling possibilities goes beyond the scope of this chapter. The interested reader is referred to Barton (2004) or Curry (2004) for radar performances and to Ainslie (2010) or Urick (1996) for sonar modeling details.

Radars are used in many application domains, from air defense and missile defense to ground mapping tasks. They can be land-based, airborne, sea-based, and even be located in space on satellites. Radar produces an electromagnetic wave that is reflected by objects. Detection is based on evaluation of the reflections. A typical differentiation factor is the waveform used: pulse radars resolve using spatial differences (reflection time, location of the object), Doppler radars resolve using Doppler effects (frequency modulation, speed of the object), and frequency modulated continuous wave (FMCW) radars do both (location and speed; handheld FMCW radars are often used by friendly neighborhood policemen to make you aware of speed limits).

The radar equation—there are actually many radar equations, but this is the one that is used to start with in most textbooks—is used to compute the signal to noise ratio that still allows for detection. The noise represents the background as described in the signal–background contrast discussion earlier in this chapter: if many waves similar to the reflected wave exist in the background, it will be hard to detect the reflection! The radar equation is as follows:

$$\frac{S}{N} = \frac{P_p \tau G_T \sigma A_R}{(4\pi)^2 R^4 k T_S L}$$

where P_p is the peak transmitted power (W), τ is the transmitted pulse width (s), G_T is the transmit antenna gain (factor), σ is the target radar cross-section

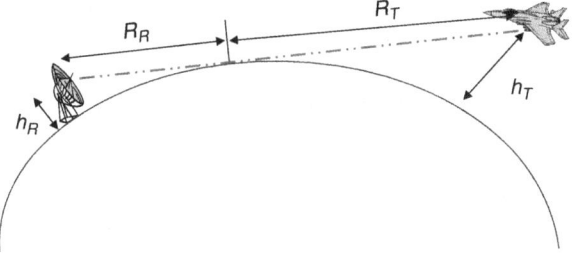

Figure 8.3 Computing the maximum radar detection range.

(m^2), A_R is the receive antenna effective aperture (m^2), R is the antenna–target range, K is Boltzmann's constant $(1.38 \times 10^{-23}$ J/K$)$, T_S is the radar system noise temperature (K), and L are the radar system losses (factor).

A naïve interpretation of this simple model is that detection occurs when the reflected signal S is bigger than the existing background noise N. In other words, if S/N is greater than 0.5, a detection can occur. In probabilistic models, this ratio can be used as the expected value for a detection to occur.

In the line-of-sight section, we already looked into the necessity of taking the curvature of the earth into account for some systems. This is particularly necessary for radar systems that can detect reflected waves over several miles. As both systems, the radar systems as well as the potential target, are often air borne high above the ground, the line-of-sight computation needs to be adjusted accordingly. Furthermore, radar equations take the effective not the actual earth radius into account when computing possible detection ranges, as radar waves can bend to a certain degree. Figure 8.3 shows radar, radar height, target, target height, and the resulting maximal distances that still allow for an effective line of sight between radar and target.

The maximum detection range R_{max} is the sum of the maximum range for the radar system R_R and the maximum range for the target system R_T. These ranges take the radar height above ground h_R, the target height above ground h_T, and the effective earth radius r_E into account:

$$R_R = (h_R^2 + 2r_E h_R)^{1/2}$$
$$R_T = (h_T^2 + 2r_E h_T)^{1/2}$$
$$R_{max} = R_R + R_T$$

These formulae deliver a simple and optimistic estimate for the maximal detection range. When looking at a realistic detection range, many more factors are important, such as mountains, trees, fog, cloud, different temperatures in atmosphere layers, etc. Several of these have been examined in the chapter on modeling the environment.

Sonar systems have similar characteristics to radars. They are also used to determine the position and speed of objects based on the reflection of waves, but sound waves in water have some additional characteristics that have to be taken into account and need to be modeled. Of particular interest is the transmission loss of sound waves when travelling through water. The interested reader is referred to Ainslie's (2010) formulae for details, but for the context of this book it is sufficient to understand that the likelihood of detecting a reflection is increased by the strength of the signal and the target strength (how well sound is reflected), and it is decreased by transmission loss and the noise level.

In contrast to radars, that always are active, sonar systems can be passive as well. However, a passive sonar system can also be described as a microphone in the water that listens to sound, very similar to hydrophones. These devices can be used to listen to special sounds created by mechanical engines, turbines, or simply large objects like submarines gliding through the water.

Space-based sensing introduces a new set of challenges that are outside the scope of this chapter. Oberg (1999) gives an interesting overview of some of the current possibilities.

Obviously, modeling sensors and sensing requires highly detailed technical models of the equipment represented, as well as highly involved modeling decisions concerning rates and directions of scans, methods for determining detection, knowledge of the types of entities involved, etc. Strickland (2011) provides an interesting summary of applicable mathematical models of target acquisition and detection on various resolution levels, similar to those presented in this chapter.

Modeling sensors and sensing need to be well aligned and harmonized with models of the environment as well as of entities and core processes. It makes no sense to use a highly sophisticated model for an active sonar in a homogeneous bunch of still water with no objects beside hostile submarines; or to use highly specialized high energy radar equations for long range detection together with a flat earth model. The simulation engineer must ensure that these principles are understood and followed within the team responsible for the simulation.

INTELLIGENCE, SURVEILLANCE, AND RECONNAISSANCE

Modeling sensing and perception is closely related to the demands of modeling intelligence, surveillance, and reconnaissance (ISR) functions. *Intelligence* deals with managing information and knowledge about an adversary and the operational environment in general. In the context of this chapter, the intelligence group provides the *a priori* information of the composition of hostile units and the numbers to be expected to enable the evaluation of sensor information. *Surveillance* describes the systematic observation of aerospace, surface, or subsurface areas in the search for information. *Reconnaissance* is responsible for planning and conducting missions to obtain additionally needed information. All three concepts are closely related.

When modeling ISR, it is often necessary to clearly distinguish between modeling the ISR physics and the ISR processes. ISR physics focuses on the explicit modeling of the physical process or the statistical approximations to answer questions such as:

- How many sensors can acquire a target?
- How many targets can be detected?
- How many targets are detected?
- What target types are detected?
- What size and unit types does the ISR unit conclude?

ISR processes focus on the explicit modeling of the processes or the statistical approximations of the ISR cells within the headquarters. They answer questions such as:

- How many sensors should be used in support of ISR tasks?
- How should the sensors be distributed to cover the battlefield?
- What rules can be applied to predict size and unit types based on the sensor results?
- What rules should be applied for data fusion?
- What frequency plans are needed to avoid the jamming of own resources?

When modeling sensing in detail, the necessity of coordination efforts of ISR experts quickly becomes obvious. The active signals of one group rapidly become the background noise of another group. The higher the signals of active sonar systems, the more noise they produce for other partners. Jamming of an opponent can also disturb allied partners and their efforts to detect new targets.

The result of modeling sensing affects all other core processes. Without knowing what is going on in the battle sphere planning becomes superfluous: entities will not know where to move or what to shoot at, and communications is meaningless. The importance of good ISR in real operations must be reflected by good models in combat modeling. The diversity of such models is a challenge for distributed simulation, but there is no alternative: the simulation engineer must understand and align the different approaches to optimally support the operational tasks, from training to support of real-world operations.

REFERENCES

Ainslie M (2010). *Principles of Sonar Performance Modelling*. Springer Praxis Books, Springer Verlag, Berlin, Germany.

Anderson SJ (2004). Target classification, recognition and identification with HF radar. *Proceedings of the RTO SET Symposium on Target Identification and Recognition Using RF Systems*. NATO Report RTO-MP-SET-080, Paper 25.

Barton DK (2004). *Radar System Analysis and Modeling*. Artech House Publishers, London.

Brooks RR, Ramanathan P and Sayeed AM (2003). Distributed target classification and tracking in sensor networks. *Proc IEEE* **91**, 1163–1171.

Curry GR (2004). *Radar System Performance Modeling* (2nd edn). Artech House Publishers, London.

Hall DL and Llinas J (1997). An introduction to multisensor data fusion. *Proc IEEE* **85**, 6–23.

Kruse R, Schwecke E and Heinsohn J (1991). *Uncertainty and Vagueness in Knowledge Based Systems*. Springer Verlag, Berlin.

Oberg J (1999). Spying for dummies. *IEEE Spectrum* **36**, 62–69.

Strickland JS (2011). *Mathematical Modeling of Warfare and Combat Phenomenon*. Lulu, Inc., Colorado Springs CO.

Urick RJ (1996). *Principles of Underwater Sound* (3rd edn). McGraw Hill Publishers, New York, NY.

Youngren MA (ed.) (1994). *Military Operations Research Analyst's Handbook*. Vol. I: *Terrain, Unit Movement, and Environment*; Vol. II: *Search, Detection, and Tracking, Conventional Weapon Effects*. Military Operations Research Society (MORS), Alexandria, VA.

Chapter 9

Modeling Effects

Andreas Tolk

WHAT ARE EFFECTS ON THE BATTLEFIELD?

Until not too long ago, combat modeling was focused on shooting and attrition when it came to modeling effects. In these models, within the trio of the Big Three "shoot, look, and move," looking was often reduced to target acquisition, and moving was reduced to bringing the weapon systems into more favorable positions to shoot.

This chapter will focus on modeling duels and engagements as well, but as already described in earlier chapters, many additional effects became of interest under the terms "operations other than war" and "human, social, cultural, and behavioral modeling." Generally, every action that leads to a change in the parameters of the forces has an effect. When the commander gives orders, the *commander's intent* — or command's intent — describes the desired end-state of an operation by specifying the plan's goals and objectives (US Department of Defense, 1993). Every action should drive the parameters used in the metrics to measure success regarding this intent into the desired direction, so every action that changes the parameters towards the desired end-state has an effect. This allows the use of optimization algorithms to drive decisions, as long as the end-state is captured in computable form representing the metrics appropriately (Tolk, 2010). Consequently, every decision — including the decision not to decide but to wait — has an effect that either increases or decreases the success of the ongoing operation.

The importance of choosing the right metrics based on the measures of merit has been highlighted already in the chapter discussing the NATO Code of Best Practice. Many combat models still support the attrition oriented metrics that were

Engineering Principles of Combat Modeling and Distributed Simulation, First Edition.
Edited by Andreas Tolk.
© 2012 John Wiley & Sons, Inc. Published 2012 by John Wiley & Sons, Inc.

applied very successfully in Cold War scenarios, in particular force ratios and force ratio changes. Force ratio is the comparison of the own forces versus the opponents' forces. Force ratio change is the comparison of changes in the force ratio: is the own side more efficient in killing opposing forces than vice versa? To compute this metric, the parameters were often collected in killer-victim tables that showed not only how many systems were destroyed but also who destroyed those systems. This allowed how effective systems were to be analyzed in detail: the more enemies were destroyed, the better a system was.

However, this direct computation of system values only taking direct effects of destroying systems into account is insufficient for analyses, as the effect of enabling—or disabling—supporting systems is not taken into account. A tank company can only shoot at enemies if they have enough ammunition and fuel, so even if the supporting logistic systems do not kill a single enemy their contribution is enormous. Another classic example is measuring the effect of air defense systems on ships protecting allied conveys in the Mediterranean Sea during World War II. German Stukas (from the German term *Sturzkampfflugzeug* or dive bomber) successfully bombarded allied conveys in the beginning phases. The allies equipped the escorting ships with air defense systems, but when evaluating the results several weeks later it showed that they were only able to shoot down a handful of German airplanes. Before the decision to use the air defense systems somewhere else, however, deeper analysis showed that this was an insufficient metric: since the air defense systems were used, the efficiency of German attacks decreased significantly, as the Stukas could no longer engage the ships as efficiently as before as they had to stay out of range of the allied air defense. While no opponent was killed by the systems, they avoided the engagement of own defenseless systems.

These examples show the need for careful choices of the right metrics to avoid bad decisions. They also show the need for good experiment design to evaluate the influence of all possible factors.

Within the next sections, we will look at various models for direct fire, indirect fire, and some smart weapon systems. We will look at models applicable on the weapon platform level as well as aggregate models. We will deal with variations of the famous Lanchester models of warfare (Taylor, 1983). Finally, we will look at the difference of intention-based modeling and effect-based modeling.

DIRECT FIRE, INDIRECT FIRE, AND SMART WEAPONS

We already looked at the differences of direct fire weapons, indirect fire weapons, and smart weapons in the chapter on scenario elements. By now, we have the tools and models to add more detail to this categorization.

Direct fire weapons are tanks that need line of sight for engagement. They engage by shooting toward the opponent in a straight line. Direct fire weapon systems can be infantry weapons such as rifles, but anti-tank missiles and similar systems that can be guided to a moving target belong in this category. The difference between these two categories is that rifles follow the line-of-sight as established in the moment of shooting while guided missiles can change their course slightly following the guidance of the shooter who keeps the line of sight established between the shooter's location and the target. Many systems, in particular helicopters, use periscopes for this phase of the guided flight. This allows them to keep the target in sight without exposing the system to the opponent. The various line-of-sight algorithms discussed in the sensing chapter are applicable here as well. When airplanes are used to provide close air support they engage in direct fire duels as well.

Indirect fire systems engage the opponent without requiring line of sight between the shooting system and the target; in other words, the ammunition does not fly following the line-of-sight but follows ballistic curves. Infantry uses small howitzers, artillery systems provide fire support for combat units, navy artillery contributes to many successes by providing their fire power, missiles can be fired land based or sea based, and so on. All these indirect fire systems are fired as a rule not at a target but into a target area. The lethal weapon is directed into this area, and when systems are in this area they become targets of the fire attack.

Smart weapons combine both aspects. In the first phase, they are brought into the target area, often fired by artillery or carried by missiles. Once they are over the target area, the second phase uses advanced technology to guide the smart weapon into a target. Special sensors can be used in combination with video, pattern recognition, and others. As such, they combine sensing and shooting. Some systems, like cruise missiles, even add the component of moving following topological structures of the terrain or programmed paths to the challenges the modeler faces.

ENTITY LEVEL ENGAGEMENT AND ATTRITION MODELS

In this section, we will look at the entity level engagement where normally two or more weapon systems fight against each other. The duels are modeled as individual events between two systems.

Game-Based Point Systems

In particular when the simulation is supported by gaming approaches, as described in more detail in a later chapter in this book, point systems are often used. Each weapon system has a certain number of points representing the health status for particular components or the overall system. Every action may change this status: the higher the effect, the higher is the loss of points resulting from this action on the side of the target. In the section on movement, we already looked into the option to have a certain number of movement points that have to be used in order to move through terrain. The same approach can be used to model whether a system is still healthy enough to participate in activities, or if the health level is below certain thresholds.

Each engagement with an opponent normally results in a decrease of health values. If the value falls under a certain threshold, some selected actions may no longer be conducted. The health of the system or its components can be mapped to certain damage classes as defined later in this section.

Often, systems can regain their health by waiting a certain time while they increase their heath at a predefined health rate. Alternatively, some action from another entity may be needed: if paramedics are present, certain wounds can be treated and healed; if special mechanics are with the weapon system, certain damages can be repaired, etc.

The change of points is normally dependent on three factors: the event (what happens, such as a duel), the target (infantry ammunition affects soldiers differently from how it affects tanks), and the effecter (if you are shot at by an infantry soldier you suffer different consequences than when being shot at by a machine gun).

These point-based calculations are often combined with other modeling techniques, like the probability driven approaches described in the following section: only if the effecter hits the target and the hit results in a certain damage is this damage subtracted from the point value of the target. Thresholds and breakpoints are special point based models and will be dealt with later in the chapter. Furthermore, we will also deal with point systems based on extensive operational analyses later in this chapter, the combat power or firepower scoring systems.

Hit, Kill, and Kill-when-Hit Probabilities

Many books on combat modeling focus extensively on this topic. Others use it at least as an introduction to other forms of attrition centric models. A good source for in-depth reading on related topics—besides the already several times referenced collection of Youngren (1994)—are the recent books of Washburn and Kress (2009) and Strickland (2011) on combat modeling. Many ideas presented in this chapter are influenced by the books of these colleagues. A classic in this domain is Morse and Kimball's (1950) *Methods of Operations Research*.

The idea behind this modeling approach on entity level is based on statistics. Three main categories are used to categorize different approaches:

- Shooting with or without feedback: the shooter can observe what the effect of engagement is on the target or feedback is not received (due to the lack of sensors, other activities on the battlefield, etc.)
- Single shot or multiple shots: the shooter fires one time or multiple times—maybe even several shots in high cadence, as with a machine gun
- Single target or multiple targets.

Without limiting the applicability to heavier direct fire systems, such as infantry fighting vehicles or tanks, or airplanes, we use an infantry soldier in our example. After the target is detected and the system is selected to be engaged, the soldier first aims at the perceived target. As we already know, this perception must not be identical with the real target: it is possible that the target uses camouflage that creates a wrong perception, or they may even use decoys, etc. This process creates the aim point of the soldier, the point on the perception of the target, but the soldier may aim incorrectly. Furthermore, due to several processes that range from mistakes made by the soldier, such as snatching of the weapon when pulling the trigger, to the dispersion of the weapon or weather influence on the projectile, the impact point may differ, resulting from the ballistic error. If these errors are independent from round to round, they are called *dispersion*; if they are systematic errors, they are called *bias*. Washburn and Kress (2009) use the following definition: "*Dispersion errors* are firing errors that are independent and identically distributed among multiple shots. Such errors are sometimes also called ballistic errors. ... *Bias errors* are errors common to all shots. Such errors are sometimes also called systematic errors or aiming errors."

Figure 9.1 shows how the three typical errors result in potentially significant differences between the doctrinal aim point (where the soldier should shoot), the perceived aim point (the area where the soldier thinks to shoot toward due to target error), the real aim point (where the soldier aims due to aiming error), and the impact point (where the bullet hits due to ballistic error).

Figure 9.1 can be interpreted as follows. The doctrinal aim point to engage the tank is the weak spot between the turret and the main body (point 1). Wrong perception (point 2) and poor aim (point 3) plus ballistic error lead to the impact point (point 4).

Each of these processes can be modeled using independent random experiments, or several of these processes can be combined. In order to use the idea in a combat model, the fitting data need to be obtained in field experiments or in the evaluation of real-world engagements, which can be quite challenging.

Once such errors are estimated, they can be used to compute the probability of hitting a target with a single shot using statistics. To explain the main idea, let us assume that the error is only distributed in one dimension. We know the location of the target including the extension in the dimension of the error (in our example the length). We also know where the aim point is. Figure 9.2 shows the resulting situation.

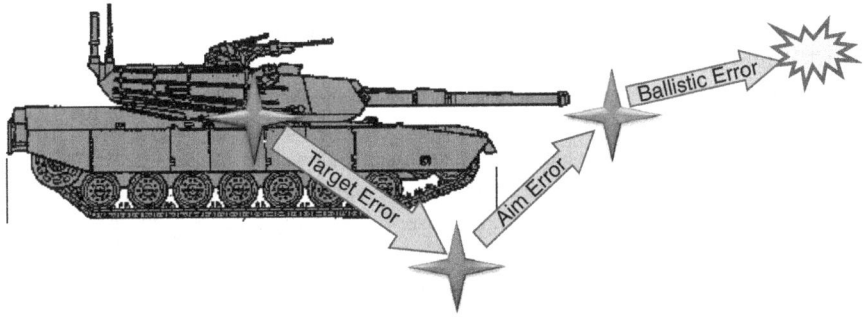

Figure 9.1   Finding the impact point.

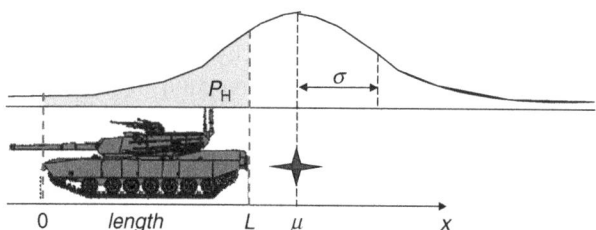

Figure 9.2 Computing the hit probability.

To compute the hit probability for this example, we assume the law of large numbers for these kinds of errors. As we assumed distribution in only one dimension, we can therefore compute the probability of hitting the tank using the following two formulae:

$$F(x) = \frac{1}{\sqrt{2\pi}\sigma} \int_0^L \exp^{-\frac{1}{2}(\frac{x-\mu}{\sigma})^2} dx$$

for the distribution of the error, and

$$P_H = Z\left(\frac{L-\mu}{\sigma}\right) - Z\left(\frac{-\mu}{\sigma}\right)$$

for the resulting hit probability. The value of Z in the second formula is the value of the normal distribution. Tables can be found in every book on statistics.

As a rule, the error is distributed in at least two dimensions (you can aim wrong in the left–right as well as the up–down directions). Often, the third dimension—shooting too short or too long—has to be modeled as well. The resulting bivariate and multivariate extensions of the formulae can be looked up in Washburn and Kress (2009) or Youngren (1994), but the idea remains the same.

A related concept of interest often used to characterize the quality of a weapon is the circular error probable (CEP), which is a radius around the mean point of impact that contains 50% of the shots. In the one-dimension example, this would become the linear error probable (LEP), which is an interval around the mean point of impact that contains half of the shots. In general, the smaller the CEP the better—or the more accurate—is the weapon. Again, formulae can be looked up in the detailed references. It may be of interest that the error measurements for global positioning systems make use of the same idea when they define their accuracy based on "position error" within the CEP.

If the shooter receives no feedback on the damage caused, the shots remain independent. If feedback is received, by observation or by partners, the hit rate can be improved. If the bias, for example, is recognized by an experienced shooter, this can be taken into account and the shooter's aim can be improved.

Another aspect often needed in combat models is the time required for these processes. As discussed in an earlier chapter, a duel does not only consist of the shooting. First, an effecter needs to observe the area and discover a target. Next, a decision has to be made to engage the target and—in the case of multiple targets—a victim must be selected. The shooter has to adjust and prepare weapons and select the weapon of choice. Next, the target must be hit and—in the case of feedback—it must be decided if the hit fulfilled the purpose or if a new fire process is needed. If a second fire fight occurs, this one is normally shorter than the first engagement, as many processes are already fulfilled (including the adjustment of the weapon, which may be replaced with minor readjustments to correct errors).

When feedback is provided, the effecter can choose between different combinations of the three core processes shoot, look, and move. As with the engagement the effecter potentially gives the shooter's exact position away as all other opponents often see where the fire comes from (it is hard to camouflage a firing main battle tank–see Figure 9.3 to get an idea about the visual effect, which is supported by a massive sound effect as well), land weapon systems often change their position immediately after a duel. It is also common practice to shoot twice if a high hit probability is needed but the time between the first fire contact and the opportunity to observe is too long (this is often the case in missile defense in particular). In the combat modeling literature, these practices are often referred to as shoot-look-shoot (S-L-S) or shoot-shoot-look (S-S-L). Other combinations are used as well, but not as commonly as these two.

When computing the kill probability of several independent shots in one round, beginners often overlook the basic rules of stochastics. If the probability for each shot in the round is 30%, for example, the probability of hitting (P_h) the target with at least one shot in this round computes to $P_h(3) = 1.0 - \prod_{i=1}^{3}(1.0 - 0.3) = 0.657$.

Damage Classes

Not every hit results in damage. What damage occurs can be modeled as a follow on to computing the hit probability. Very often, the likelihood of damage is expressed as a conditional probability under the assumption that the shooter actually hits the target. The probability of hitting the target (P_h) can be computed using the equations found earlier in this chapter; the conditional probability of

National Geographic –Tanks: Machines of War

Figure 9.3 US Battle Tank Abrams firing the main gun.

creating damage when hitting the target can be derived from field experiments or controlled experiments on a shooting farm.

The probability of creating damage is also often highly dependent on the angle at which the ammunition hits the target, in particular when the target is armored. Many land systems have heavy armor in the front and the front flanks, as these are the parts of the weapon system that most often were exposed to opponents' fire in traditional battles.

The Russian attack helicopter MI-24 Hind was so well armored in the front area that they did not even have to fear light air defense guns with a caliber up to 20 mm. However, it was very vulnerable against fire and other objects from above, a fact that the Afghan fighters in the Russian invasion of Afghanistan were often able to use when they downed Hind helicopters in mountain valleys by throwing massive rocks onto them from above. Similar weak points in the armor of battle tanks are often used in urban battles: Molotov cocktails—improvised incendiary weapons—thrown on top of the hull close to the turret can create serious damage when the burning substances leak into the gear ring of the turret.

The traditional objective of military engagement is to destroy the opponent. Many combat models only model one damage class: to kill the weapon system. The probability of killing a system is referenced as P_k. Due to its high importance, several probabilities deal with this damage class. As described earlier, the conditional kill probability when the system is hit is referred to as $P_{k|h}$. The probability of killing a system with a single shot is called the single shot kill probability and referred to as P_{ssk}.

However, when a system is hit, the resulting damage is not always a kill that destroys the target. Referring to the main core processes described in this book as well, the following damage classes are often used:

- *movement kill:* the system can no longer move on the battlefield, but all other processes are still possible;
- *firepower kill:* the system can no longer shoot at others, but can do all the rest;
- *communication kill:* the system can neither send nor receive messages, but the other core functions are still ok;
- *catastrophic kill:* the system is completely destroyed.

Combinations of movement, firepower, and communication kills are possible. Some models also use sensor kills. Many models apply the point system as discussed at the beginning of this chapter to compute a degree of availability.

The general application of these damage classes to weapon systems of all forces needs to be well discussed: an aircraft with a movement kill does not make much sense if it is not sitting on the ground and can be repaired. However, some high resolution models defined such classes for human soldiers as well to model the effect of arm or leg wounds in contrast to lethal injuries.

The interested reader is referred to many additional detailed evaluations of options of entity level engagements described in Youngren (1994) and Washburn and Kress (2009). The latter is also an excellent source for even more detailed journal contributions to related fields that are outside the scope of this book.

Strickland (2011) furthermore introduces detailed physical models of attrition for passive targets with detailed mappings to expected damage and more. Similar to Washburn and Kress (2009) he shows the mathematical connections between such elementary experiments and the derivation of attrition coefficients as described in the next section of this chapter.

AGGREGATE LEVEL ENGAGEMENT AND ATTRITION MODELS

So far, the focus has been on entity level models that looked at the effects of fire on single or at least individually modeled weapon systems. In this part of the chapter, we will look at aggregate level engagement and attrition models. As before, the main questions will be (a) how many enemy units can be destroyed, and (b) how many casualties are to be expected within the own forces?

Although several alternatives are offered in the professional literature, the predominant approach of modeling engagement and attrition on higher levels is based on the ideas formulated by Frederick W. Lanchester (1916) approximately 100 years ago. He looked at units as force collections that mainly decrease the number of opponents within duels while simultaneously being decimated by them as well, both based on attrition coefficients depending on the duel situation. This view results in differential equations describing the battle and the number of forces to be expected on both sides over time. It should be pointed out that, although the resulting models are generally referred to as Lanchester models of warfare (Taylor, 1983), a Russian scientist discovered similar models independently. Helmbold (1993) reports that M. Osipov published his series of articles on these observations in the Tzarist Russian Journal Military Collection. A second name that deserves recognition is Lt J.V. Chase. His 1905 publication in American Naval Policy on the "Force on Force Effectiveness Model for Battle Lines" may be the first mathematical model of warfare on this engagement level (Fiske, 1916). Without taking credit from any one of them, we will use the term Lanchester model in this book.

The driving question behind all these efforts were two questions: (a) how many forces are needed to ensure a successful operation, and (b) is there a main difference between modern weapons and older weapons resulting in different principles? The American Civil War with its unbelievably high numbers of casualties on both sides due to fighting with significantly improved weapons and the continuous development of new weapon systems opened very new opportunities. Airplanes and battle tanks are the two most dominant developments, but new mines for land and sea, improvements in the artillery, and better rifles and howitzers contributed as well.

The resulting equations are models of engagement and attrition that can be easily implemented and are based on solid theory, although many objections do exist. But despite all criticism, they are still in use in many combat models.

Classic Lanchester Models

Lanchester and Osipov both used ordinary differential equations to better understand how the numbers of forces on two opposing sides change over time when they fight against each other. Lanchester was in particular interested if his hypothesis would hold that in modern warfare the concentration of forces has advantages compared with ancient warfare. He postulated in his observations that in duels each system fought against opposing systems with a certain quality that could be expressed with an attrition coefficient, which are now often called Lanchester coefficients. The number of destroyed opposing systems would then depend on the number of participating systems on both sides and this coefficient. While the resulting functions would remain the same, the coefficients would be specific.

Lanchester first looked at ancient battles that were defined but short range weapons and the lack of fast movement and command and control. Man-to-man sword fights or the battles of the Greek phalanx are examples. In these battles, the number of duels is the important factor. The more one-on-one duels that could be fought, the more casualties could be inflicted on both sides.

Using blue forces and red forces as names for the opposing sides, this leads to the observations that the more blue forces are available, the more duels can be fought, and the more red forces are available, the more duels can be fought too. Therefore, the number of blue losses will depend on the number of participating blue forces, participating red forces, and the attrition factor of red forces against blue forces. The number of red losses will also depend on the number of participating blue forces, participating red forces, but they would use the attrition factor of blue forces against red forces. In this simple model, no new forces arrive, destroyed forces remain destroyed, and each effect is lethal and final. As blue force losses and red force losses happen in parallel, the resulting ordinary differential equation for each time in a battle is

$$dB = l_r B(t) R(t) \quad dR = l_b R(t) B(t)$$

with $B(t)$ the number of blue forces at time t, $R(t)$ the number of red forces at time t, dB the number of blue casualties, dR the number of red casualties, l_b the attrition coefficient of blue forces against red forces, and l_r the attrition coefficient of red forces against blue forces.

When implementing this approach, very often difference equations are used to approximate the differential equation: for each time step, the number of casualties is computed as the product of the coefficient and both force numbers at the beginning of this time step. This results in slightly higher estimates for casualties. Some models with longer time steps therefore use Euler or Runge–Kutta methods to improve the approximation.

The underlying ordinary differential equation for the above model can also be formulated as follows:

$$l_r[R_0 - R(t)] = l_b[B_0 - B(t)]$$

with R_0 the initial red forces at the start of the battle (with $t = 0$) and B_0 the initial blue forces at the start of the battle.

This explains why this model is also called the *linear Lanchester law*. The model can be used to predict the outcome of the battle if fought until a termination criterion is reached. It also shows that quality of fight (as expressed in the coefficient) can be compensated by numbers. If blue, for example, fights twice as good as red, red needs twice as many systems to make the fight even again. The formal implications are that:

$$\frac{dR}{dB} = \frac{l_b}{l_r} = \text{const and} \quad \begin{array}{c} l_r R_0 > l_b B_0 \\ l_r R_0 < l_b B_0 \end{array}$$

define the outcome of a battle.

In modern warfare, weapons can reach farther and do not lead to something which is *de facto* a collection of one-on-one battles. Using command and control means, the soldiers can engage their targets in an orchestrated manner. For such directed and coordinated fire, the number of kills will only depend on the number of firing systems. The resulting formulae to compute casualties for the two sides change to:

$$dB = l_r R(t) \quad dR = l_b B(t)$$

It should be pointed out the Lanchester coefficients have another dimension in this equation than they have in the linear model. However, the equation can be reformulated as before resulting in the following underlying differential equation:

$$l_b[R_0^2 - R^2(t)] = l_r[B_0^2 - B^2(t)]$$

These equations describe or model modern warfare using the *square Lanchester law*. Here, the product of the attrition coefficient and the square of the initial number of systems is the battle determining value. Accordingly,

$$l_r R_0^2 > l_b B_0^2$$
$$l_r R_0^2 < l_b B_0^2$$

determines which side wins.

In order to compensate efficiency, a much higher number is needed. The well known rule of experience that an attacker should have three times the number of fortified defenders implies that the defender should have an efficiency that is nine times higher. Interestingly enough, comparing the P_{ssk} tables of attacking systems against fortified defenders with the P_{ssk} tables of systems that defend against an attacker out of a fortified position shows that the P_{ssk} differ by a factor of approximately 10, which supports these ideas.

It also supports Lanchester's hypotheses that the concentration of forces is more important in modern warfare (where the influence is quadratic) than it has been in ancient battles. However, the many assumptions that go into this model have been the reason for many discussions regarding the appropriateness of such efforts. Some of the assumptions the simulation engineer has to be aware of are

- coordinated and directed fire on both sides
- attrition coefficients are constant over the battle
- no double engagement, which assumes perfect command and control and communications between all systems
- enough targets for all shooters
- both sides are able to aim and concentrate fire at selected targets
- optimal effect assessment on both sides
- fire shifts immediately to new and valid target

Fowler (1995) evaluates the assumptions for both Lanchester laws in more detail and derives the implications for the outcome of the battle, the length of the battle, and more.

One of the hardest arguments against using the Lanchester approach in combat models is the assumption that the coefficients are constants. Several approaches are published in the literature that recommend the computation of battle-specific coefficients using the following observations.

- For the square law, the interpretation of the blue coefficient is the number of red forces successfully engaged in one time step. Using the detailed models discussed earlier, this coefficient can be computed out of the P_{ssk} and the tactical cadence (how often can a system shoot in one time step). When computing these values in the detailed case, the assumptions for the applicability of the square can be taken into account and used to calibrate the value accordingly. Instead of using constant coefficients they are derived based on more detailed sub-models for each phase of a battle.

- For the linear law, the interpretation of the blue coefficient is the fraction that each firing system contributes to the loss of one red system. Similar observations can be made here leading to the analytic definition of coefficients, if needed.

It should be mentioned that the linear law is often applied in combat models as well, but not to model one-on-one duels. The model fits well for artillery attacks: the more blue shells are fired into the target area, the more red systems in this target area will be affected; the more systems are in the area, the more of them are under fire; and the better the blue systems shoots or the shell explodes, the more red systems are affected. The number of red casualties is therefore dependent on the number of firing blue systems, the number of red systems in the target area, and the blue attrition coefficient.

Extended Lanchester Models

The original two Lanchester laws have been modified by many researchers in order to improve their applicability to more specific cases, in particular to be able to use them for modeling purposes as discussed in this book. Taylor (1983) published an impressive collection of Lanchester models and their application. In this section, we will focus on the main extensions that serve as a basis to understand other derivates as well.

Some changes in the number of blue forces are not caused by red activities but by events within the blue forces or they reflect the structure of the blue forces. Examples are soldiers leaving the forces and deserting, casualties based on infections, etc. In such cases, the number of blue casualties only depends on the number of blue forces and a blue attrition coefficient. The interpretation of the coefficient is very much dependent on the modeled effect, but it can be the overall infection rate or similar factors that take the whole blue force number into account. The resulting equations show why these models are said to follow a *logarithmic Lanchester law*:

$$dB = l_r B(t) \quad dR = l_b R(t) \text{ with } l_b \ln \left(\frac{B_0}{B(t)} \right) = l_r \ln \left(\frac{R_0}{R(t)} \right)$$

In many cases, these models have to be mixed to support the analysis of special situations: when, for example, a guerrilla force (red) ambushes a conventional force (blue), the red forces can use directed and coordinated fire against the blue force. However, the blue forces only know the direction where the ambush is coming from. They simply have to fire back into this direction. The more blue forces shoot back, the higher is the likelihood of hitting one of the guerrillas. Also, the more red forces participate in the ambush, the higher is the likelihood that one of them gets hit. In other words, while the attrition of the blue forces can be described using the square law, the red casualties can be described using the linear law. The resulting model for guerrilla warfare is often called the *mixed Lanchester law*, as it results from mixing the square and the linear models. The resulting formulae are:

$$dB = l_r R(t) \quad dR = l_b B(t) R(t)$$

with

$$2l_r [R_0 - R(t)] = l_b [B_0^2 - B^2(t)]$$

Operational analysts continued to look for a more general Lanchester law that could be used to combine all of these forms in a general formula that would result in the linear, square, logarithmic, or mixed forms thereof as special cases.

Taylor (1983) generalized Helmbold's ideas and used the Weiss parameter w (with $0 \le w \le 1$) to formulate a general Lanchester law as follows:

$$dB = l_r B^w(t) R^{1-w}(t) \quad dR = l_b R^w(t) B^{1-w}(t)$$

Depending on the value for the Weiss parameter, the known Lanchester laws can be derived: when choosing $w = 0$, the square law follows; when choosing $w = 1$, the logarithmic law is the result; and for $w = 1/2$, a derivate of the square law can be modeled.

Hartley (2001) targeted an even closer general approximation that included the mixed options as well and avoided the artificialities of the Weiss parameter. His general formulae to model blue and red casualties are:

$$\mathrm{d}B = l_r^C B^D(t) R^E(t) \quad \mathrm{d}R = l_b^F B^G(t) R^H(t)$$

Choosing the appropriate values for the general exponents C, D, E, F, G, and H, all Lanchester laws and their mixed forms can be derived from this approach. All these extensions and generalizations suffer from the same challenge so far not generally met by research results: how to better define Lanchester coefficients that are not fixed but reflect the modeled situation individually.

Heterogeneous and Stochastic Lanchester Models

Two other big challenges remain: heterogeneity and probability. This section introduces some efforts that address the challenges. These are just examples as a general solution is not established, but the documented heuristics have been successfully applied.

So far, the forces are assumed to be homogeneous! The force of each side is computed as one value, namely $B(t)$ or $R(t)$, respectively. However, the higher we climb in the aggregation hierarchy, the more likely it is that we have mixed weapon systems in the respective units. While weapon systems can be homogeneous on lower levels in the military hierarchy, from battalion up a mix of weapons is the rule not the exception. An example of mixed weapon systems on lower levels is the use of infantry and armored fighting vehicles in ground operations for mutual support and protection, as discussed earlier in this book. In summary, homogeneous battles are more likely to be the exception than the rule.

The following approach was recommended and implemented in the German experimental system KOSMOS (Hofmann et al., 1993). The recommended approach derives a serious of parallel homogeneous battles based on geometrical and tactical constraints. Figure 9.4 shows how to model the fraction of a unit that can engage with another unit.

As a first filter, only those units can engage that overlap with the main circle (shown as (1) in the figure). For those units with overlapping circles, a more detailed analysis is conducted. For each segment—front, flanks, and rear—a circle segment (shown as (3)) is constructed as follows: using the diagonals of the rectangle that represents the units, two circle segments are constructed (shown as (2a) and (2b) in the figure) that define the center of the circle segment for the segments. The intersection of these resulting circle segments with opposing circle segments represent the fraction of systems in the segment that can engage in a battle.

Obviously, if a unit is represented as a circle anyhow, the construction of circle segments is not necessary. However, many models represent land units as rectangles, as discussed earlier.

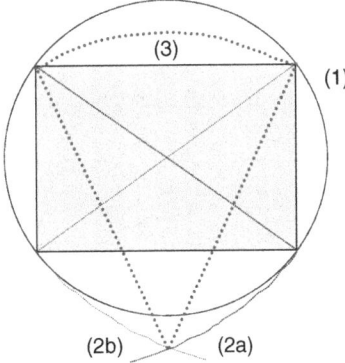

Figure 9.4 Geometrical modeling of engagements.

Using this geometrically motivated active fraction, the amount of potentially active systems for each unit is computed. The next step is to define the allocated fire based on tactical constraints. A system will not fire at an enemy that cannot be affected with the current weapons. Firing at a tank with a rifle does not have much of a success rate, so infantry with rifles is likely to only engage other infantry. However, tanks can be used effectively against tanks as well as against infantry fighting vehicles, so a decision based on tactical constraints needs to be made, which can be modeled as the standard tactical engagement of one weapon system type against other opposing weapon system types.

Figure 9.5 shows an example in which three blue units are engaged with three red units. The non-active fractions are black; the systems in these fractions do not participate in any fights. For each group of weapon systems in each active fraction, the tactical engagement defines which type of enemy is engaged. Red does the same. The resulting formula that can be used to compute the casualties for a weapon system of type i on the red side looks far more complicated than it is: $dR_i = \Sigma_{j=1}^{n}[c_j a_{ji} l_{ji} B_j(t)]$.

The number of casualties a red unit suffers for a weapon system of type i is modeled individually for all battles against the sum of blue weapon systems of type j over all battles. The active fraction c_j is derived from the geometry. The tactical engagement a_{ji} is often normalized regarding the available targets and is derived from tactical constraints. The Lanchester coefficient l_{ji} is known from the square law.

Other approaches to deal with heterogeneity of the blue and red forces may use different heuristics to break the main battle down into individual homogeneous battles, but all follow the same principle to define mini-battles. The same idea also holds true when more than two sides are involved.

The second challenge to be addressed in this section is stochastic Lanchester modeling. The interested reader is referred to Washburn and Kress (2009) for more details and analytic background. So far, all our models are deterministic, but they may result in non-integer numbers. The underlying question is: if the result in a battle determines that 0.472 blue tanks of four participating systems are killed, how should this be interpreted?

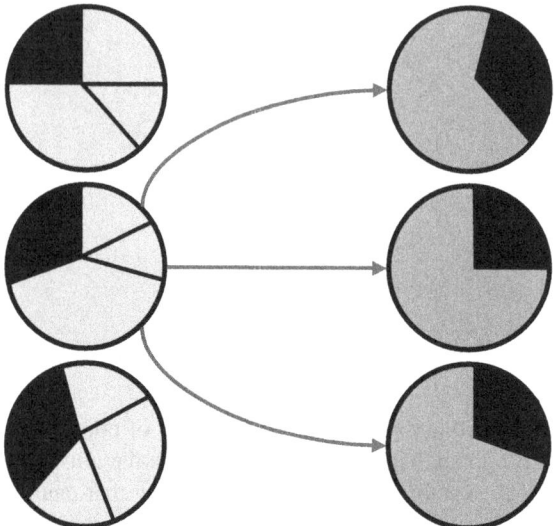

Figure 9.5   Fire allocation and active fractions.

Stochastic Lanchester models treat the results as estimated values of an underlying random experiment. The above mentioned example could be interpreted that each of the four systems is killed with the probability of 0.472/4, which is approximately 12% kill probability. Instead of subtracting 0.472 from the blue force, in stochastic Lanchester models a random experiment results in a kill, or it does not. The result is always an integer, which makes the interpretation easier. The predominant interpretation of such models is that of continuous time Markov chain models. A detailed explanation lies out of the scope of this book. However, the simulation engineer may have to conduct additional literature research when required to support a project that requires more background knowledge. In this case, the references given in this chapter form a good starting point. In addition, chapter 18 deals with the topic of Lanchester equations in some more detail as well.

Reserves and Logistics and Maintenance

Two very important factors increase the forces engaged in battles that have not been taken into account so far: first, fresh forces can be introduced into a battle that were kept as reserves before; second, logistics and maintenance services repair and refresh forces in battle and reintroduce them into battle.

In real combat situations, the use of reserves is critical. The two main questions a military decision maker has to answer are *where* reserves should be engaged and *when* reserves should be engaged. Once these decisions are made, the reserve is integrated into the already fighting force, increasing the forces of the supported side at the time the reserve arrives at the destination. The sooner the reserve arrives, the better for the supported unit in battle. However, the earlier the reserve forces are used, the less operational freedom remains with the decision maker. Therefore, military decision makers wait until decisive moments. When defending, the reserve is normally introduced at the place where an offensive breakthrough might occur. On the offensive side, the same is true, but an attacker will engage reserves to enforce a breakthrough. Once the decision is made, both sides will try to delay the arrival of the opposing reserve force, e.g. by mine warfare, close air support attacks, or by artillery attacks. The simulation engineer may have to orchestrate a complex interplay of movement modeling and effects modeling to ensure the realistic representation of reserve engagements.

Logistics and maintenance are very important for military operations. While early combat models emphasized the combat units, modern combat models take an overall approach to take combat support into account as well. Cutting the logistic lines can be more efficient to deny success to an opponent then engaging his combat troops. Why risk own casualties in direct confrontation when making sure that tanks will be running out of fuel and ammunition with only a small fraction of this risk? Many insurgency operations and other asymmetric warfare activities make use of such options. Therefore, models are needed to train, evaluate, and support related activities.

Furthermore, depending on the damage suffered in a duel, many weapon systems may be recovered and repaired: a tank that drove on a landmine that destroyed its track can get a new track and be reintroduced to the battle; destroyed radios and antennas can be replaced, etc. In particular when using a point system—as introduced at the beginning of this chapter—logistics and maintenance can easily be modeled to increase the points, while hostile effects decrease the points.

Logistics and maintenance can be modeled based on individual processes in high resolution models, or in aggregated form using recovery rates, damage class distributions, main time between failure and repair, etc. The simulation engineer needs to be able to understand and support all these different options.

Breakpoints

Another factor that needs to be taken into account is when a battle will be terminated. Simply continuing fighting until the "last man standing" point is reached is rarely observed and therefore very unrealistic. More often, units stop to function as orchestrated actors much earlier. Therefore, breakpoints are used to model that below certain thresholds units are no longer functioning.

A challenge for the simulation engineer is to understand what happens with units that are no longer functional. Can they disengage and retreat in order to reorganize at a later time? Can several broken units be merged to form a new unit that can be used against the enemy? Historical data provide evidence that these are not unlikely options, which means that models should take this into account. If nothing else, still functioning weapon systems or logistic material, like food and ammunition, can be used. To what degree these can be used by all participating forces is a decision of the modelers to reflect what is most likely.

For example, NATO troops used 7.62×51 mm ammunition while the Warsaw Pact troops used 7.62×39 mm ammunition. Therefore, troops of the Warsaw Pact were able to use the ammunition of NATO rifles in several of their systems, as they used the same caliber with slightly shorter hulls. The same fact, however, precluded NATO troops using Warsaw Pact ammunition.

Even more challenging is modeling the breakpoints for units. They are as important as the attrition coefficients when determining who will win a battle. A low breakpoint is often used to model a high morale for the unit. Some approaches even introduced the notion of nationality factors, as historical data showed that the Japanese and German troops had significantly lower breakpoints than their Italian allies in World War II. To what degree these data still hold today, however, is highly debatable. However, it is without debate that highly motivated forces that are willing to fight "until the last man falls" even with bad weapons can still defeat highly efficient and numerically superior enemies who are faltering with the first casualties. Whether this is the exception to the rule is a decision of the combat modelers. The simulation engineer has to understand the model used.

Technically, two approaches that are often used in mutual support are used. The first approach models a breakpoint depending on the initial force of a unit. For example, a tank unit may be modeled to remain functional as long as at least 35% of the initial systems are still available. If the force strength falls below 35%, the unit no longer participates in the battle. What happens with the remaining systems needs to be decided and documented to avoid strange results. These systems are not really destroyed by anyone, as only the first 65% of casualties happened in battle. On the other hand, they can no longer be used, so they should be counted as losses, if the model does not support the explicit disengagement and potential regrouping as discussed before.

The second approach is to model the force ratio. Even if a unit still has its full fire power, if it is hopelessly outnumbered by opposing forces it may surrender as well. As a side note, if troops surrender, they may bind opposing troops as well, as prisoners of war require resources, personnel and material that otherwise could be used at the front.

While the first breakpoint rule is known as the *absolute breakpoint rule*, the second is the *proportional breakpoint rule*. In real operations, both can lead to the end of a battle. The first one captures the unit internal consistency—the unit is no longer functional due to internal attrition effects; the second rule observes the contextual consistency—a unit is in itself still functional, but the context does not allow for success any more and results in a breakpoint. Figure 9.6 visualizes absolute and proportional breakpoint rules. In both cases, blue wins when red reaches a breakpoint (lower right corner), and red wins when blue reaches a breakpoint (upper left corner).

When both rules are combined, this figure shows who will succeed in the battle depending on where the force trajectory defined by blue forces at a given time $B(t)$ and red forces at the same time $R(t)$ will end. Bringing in reserves, as discussed before, moves the force ratio in favor of the receiving side. Figure 9.7 shows an example.

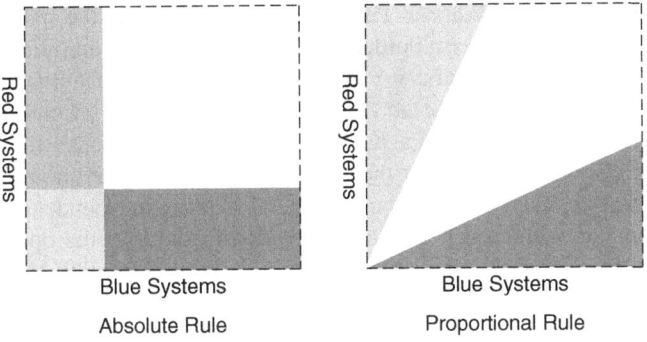

Figure 9.6   Absolute and proportional breakpoint rules.

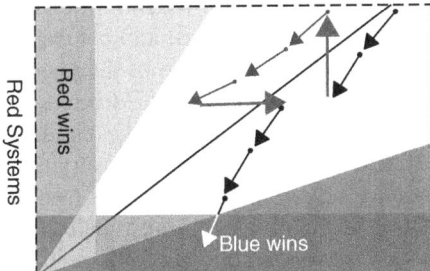

Figure 9.7   Breakpoints with reserves.

The trajectory begins in favor of the blue side. After two time steps, red engages its reserve, leading to a jump in favor of the red side. However, after three more time steps, blue engages its reserve and brings the trajectory back. After three more time steps, blue wins as red reaches the breakpoint (actually both breakpoints, as the figure shows).

ALTERNATIVES TO LANCHESTER

Due to the many constraints, the approach to use in particular homogeneous Lanchester models is often discussed. Several attempts to verify Lanchester models of warfare were as unsuccessful as trying to falsify them. Hartley (2001) and Washburn and Kress (2009) give several examples.

Two opponents of the models proposed earlier in this chapter who came up with alternative approaches were Joshua Epstein (1985) and Trevor Dupuy (1990). Both of their approaches are described in detail in Fowler (1995). In the context of this book, we just look at the underlying ideas and assumptions.

Epstein's Model of Ground Warfare

One of Epstein's major criticisms was that Lanchester had only very limited support of fire and movement and areas of military responsibility. In the 1980s,

NATO was still engaged with the Warsaw Pact in the Cold War. At the main defense line across the inner German border that separated West and East Germany as well as NATO and the Warsaw Pact, nine corps sectors of military responsibility were established to coordinate defense efforts in NATO's Central Region. Figure 9.8 shows the schema. Each corps consisted of brigades that carried the main fight. They moved within the area of responsibility. Although mutual support was possible, each corps was more or less fighting independently within its sector. Of course, withdrawals needed to be coordinated to avoid open flanks of neighboring forces, but the idea was to have nine sectors fighting side by side to push a potential attack back.

Epstein's model was significantly shaped by the doctrine of these days. He recommended looking at land forces as elements within a 'piston' that moved back and forth. These 'pistons' move side by side covering the entire theatre-wide battlefield, just like the nine corps sectors covering NATO's Central Region. Movement of units is calculated within the tracks of the pistons, and combat is calculated between the opposing forces accumulated on each side of the piston, like the Forward Line of the Battle Area (FEBA) between the armed forces in Europe of NATO and the Warsaw Pact. The closer a unit was to the dividing line between blue and red, the more vulnerable it became to opposing activities. In support, mimicking the order of battle areas within corps (with front troops, reserves, and rear area), the pistons were divided in sub-areas. Units could now be moved explicitly towards the FEBA to better support the battle as well as moving away from the FEBA to withdraw and avoid too many casualties. If all units withdraw from the FEBA, the pressure of the opponent moved the piston

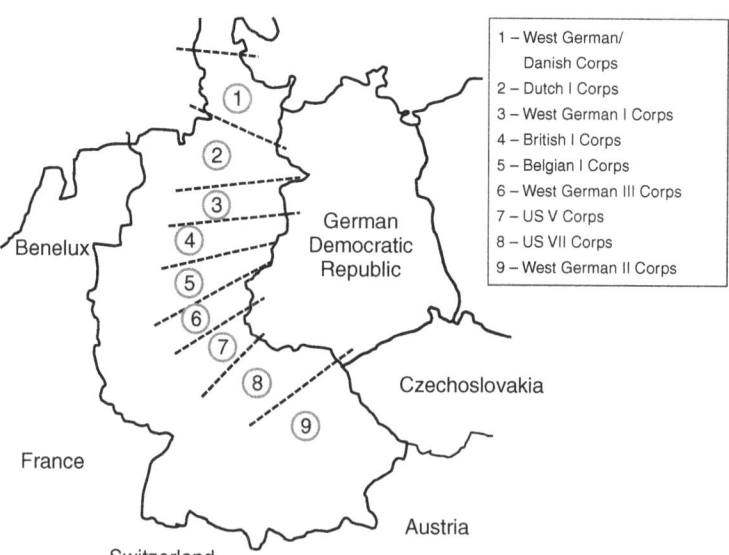

Figure 9.8   Corps sectors of military responsibility in NATO's Central Region during the Cold War.

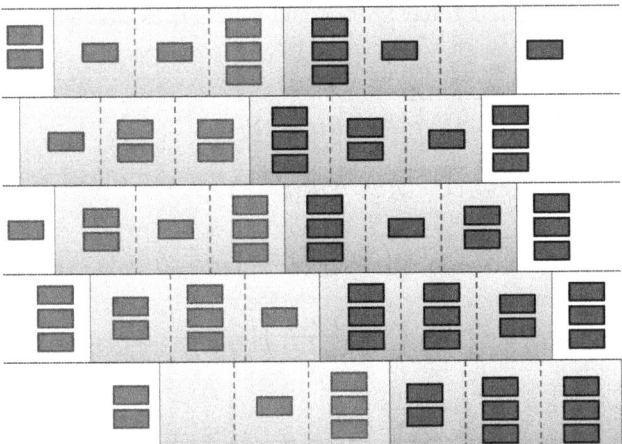

Figure 9.9 Epstein's 'piston' model for land forces.

to the advantage of the attacker, or vice versa. Figure 9.9 shows the resulting schema underlying Epstein's model.

In the chapter about movement, we already addressed that movement is likely to slow down in case of attacks, in particular when the number of casualties or the casualty rate exceeds a certain unit-specific threshold. Epstein observed that similarly in most realistic situations the defender will withdraw from an attack when the attrition rates exceed his acceptable threshold: the defender will not stand in defense but will be pressed backwards. This will disengage troops and slow the attrition rate down. Once the attrition rate is low enough, the defender will continue to defend again.

Epstein's model did therefore explicitly represent the attrition and the ground position. This allowed operational ideas to be used like trading space for time (giving terrain close to the front up in order to have more time to prepare for defense in rear positions), or trading space for forces (avoiding being destroyed by withdrawing and regrouping in the rear). First, it computed the attrition on both sides using attrition factors that were not constant but depended on the ratio between the last ground-lethality attrition rate and the average ground-to-ground casualty exchange ratio. In other words, the momentum of success was enforced; if an operation was successful it tended to remain successful if the opponent did not introduce new forces to counteract. The attrition influenced how fast a defender had to withdraw as well as how fast an attacker could prosecute the defender. Epstein also used the same attrition rate for both sides based on the intensity and the momentum of the battle within the sector.

Epstein needed three input values to compute the expected losses on both sides in each sector as well as the withdrawal and prosecution rates. After each time step, additional coordination efforts between the sectors could be modeled. If the defense in one sector broke together, some support could be provided or the withdrawal rate could be adjusted, but the sections remained constant in the initial model. Assuming that red forces attack and blue forces defend, the resulting formulae are:

Number of blue forces at time t:	$B(t) = B(t-1) - \frac{\alpha(t-1)}{\rho} R(t-1)$
Number of red forces at time t:	$R(t) = R(t-1) - \alpha(t-1)R(t-1)$
Blue withdrawing rate:	$W(t) = W(t-1) + \left(\frac{W_{\max} - W(t-1)}{1 - a_{\text{red}}}\right)[\alpha_{\text{blue}}(t-1) - a_{\text{red}}]$
Red prosecuting rate:	$P(t) = P(t-1) - \left(\frac{a_{\text{blue}} - P(t-1)}{a_{\text{blue}}}\right)[\alpha_{\text{red}}(t-1) - a_{\text{blue}}]$
Attrition rate:	$\alpha(t) = P(t)\left(1 - \frac{W(t)}{W_{\max}}\right)$

with the following interpretation: $\alpha_{\text{blue}}(t)$, defender's total ground-lethality attrition rate, $\alpha_{\text{red}}(t)$, attacker's total ground-lethality attrition rate, a_{blue}, constant for blue threshold attrition rate, a_{red}, constant for red threshold attrition rate, and ρ, average ground-to-ground casualty-exchange ratio.

As with Lanchester models, there are several recommendations on improving Epstein's model as well, in particular allowing for enclosing operations for broken-through opposing forces and adding more flexibility to the sector modeling, such as narrowing the sector for corps that get into trouble and more mutual support between neighbors, as was planned for in NATO operations in the Central Region. The model was successfully implemented in operational combat models, in particular the THUNDER model.

Combat Power Scores

Colonel Trevor Dupuy's (1990) work is an example for calculating combat power—also referred to as fire power scores—for engaging forces and using this combat power to model the engagement result. The idea is very similar to the point models introduced at the beginning of the chapter, but the computation resulting in the combat power can become very complex. Simply counting the number of weapon systems is the first approach.

The next improvement is to define weapon-system-specific weight factors that capture the value of a special weapons system. A main battle tank is probably worth more than a simple jeep. A lot of effort went into defining such combat values of weapon systems, based on evaluations of real operations as well as of simulations.

The next improvement is the differentiation between categories of operations that are supported by the weapon system. A tank in an urban environment has a different value than a tank in open terrain.

One way to compute the value of a system in an operation is the degree to which it contributes to the success of an operation. The potential–anti-potential method has been successfully applied to apply eigenvectors based on the number of destroyed systems to compute an operation-specific value. The idea is that a system that works well against a successful system counts more than one that only destroys unsuccessful systems. The problem is the definition of success, which is often reduced to fire power and destroying first-order systems that themselves destroy opponents. However, a system that successfully destroys the logistic lines and hence stops the attack of an enemy will get no credit, as only fuel tanks and ammunition carriers were destroyed that had no operational value as they do not kill opposing systems.

To take these higher effects into account, sensitivity analysis of results based on the number and structure of participating systems have been conducted. The value of a system is then measured as the degree of change of the overall success factor to the operation. The problem is that this approach may not take correlating effects within system mixes into account. So far, no generally accepted scoring method has been developed. The simulation engineer needs to understand the details behind the heuristics that are applied.

Dupuy's (1990) quantified judgment model is no exception. He recommended computing the power of each side using three contributing terms:

- the *force strength* as a function of the aggregate operational-lethality indexes—similar to the potential–anti-potential values—and weapon effect factors for participating weapons,
- *force effects variables* to take the specific constraints of the operation into account, such as environmental, operational, and behavioral constraints that support or limit the value of weapons, and
- adjustment factors for *combat effectiveness for troop quality* which he defines in detail as the ratio that is a function of mission factors, spatial effectiveness factors, and casualty effectiveness factors.

The general approach using such combat power formulae is to compute values for both opposing units and aggregates and to build the ratio. If the ratio falls into certain intervals, this triggers actions and events for the model. The breakpoint examples discussed earlier in this chapter are actually good examples of this approach.

Intention-Based Modeling versus Effect-Based Modeling

In Tolk et al. (2008), an agent-based alternative to the predominant approach to modeling the intended effect was proposed. In the paper, we observe that traditional combat models reduce the units for modeling to the intended role and the necessary minimal properties to save computational effort, just as discussed in this chapter.

For example, an artillery unit is often modeled as a number of howitzer systems equally distributed in a circle, if the dislocation area is modeled at all.

The intended role for such a unit is to fire with indirect fire weapons at an enemy. If other capabilities are needed, such as that an artillery unit has many soldiers and therefore can be assigned nearly every task an infantry unit can do, this is not possible.

While this example is pretty obvious, a closer look is needed to see what this means for effects. When looking at the models described in this chapter we see that the intent is clearly to hit and damage the opponent. We compute probabilities to hit their systems, the conditional probability of various damage classes when we hit them, resulting attrition coefficients, and more. We therefore model our intention and if we actually fulfill the intent or not pretty well.

But what happens if we miss the opposing tank but instead hit our own tank who is fighting close by? Our models proposed so far do not take this possibility into account. Furthermore, each action can trigger several follow-on effects—just revisit "The Nail that Lost the War" in Chapter 5 to see how a chain of events can be triggered by a rather unimportant event—which are also not modeled so far. The interested reader is referred to Smith (2002).

To cope with these challenges, Tolk et al. (2008) recommend modeling all entities and effects as agents that are initiated with the intention as captured, but that interact with each other so that unintended effects can happen as well. Figure 9.10 shows attrition as an example: the bullet is an influencing agent that decides if the intended target is really hit or if the encountered agent is hit instead. What the damage of the unintended hit will be is computed by the two participating agents. If the encountered agent is, for example, a fuel tank, the unintended effect caused by the effect of hitting the wrong system can be dark and hot clouds and smoke on the battlefield that allows a high value target to escape that otherwise could have been captured.

In general, the recommended implementation—that was actually implemented as documented in Bowen (2008)—relies on three major agent types: environmental agents, combat system agents, and effect agents.

- *Environmental agents* represent trees, walls, and smoke within the situated environment. They are inanimate and serve the role of impediments to the

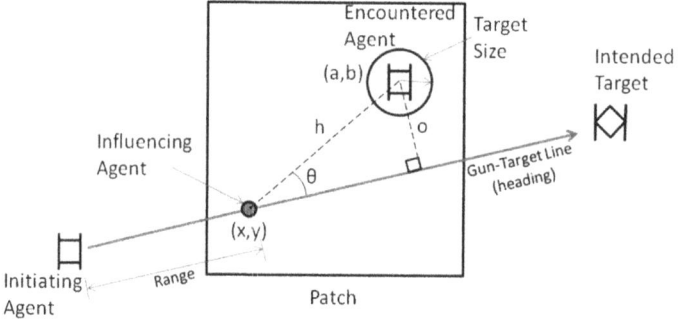

Figure 9.10 Agent-directed attrition modeling.

combat tasks of move, sense, and influence in order to create an adaptable environment.

- The situated environment is populated with *combat system agents* representing tanks, scouts, commanders, and signal trucks. They have the capability to move, sense the environment, communicate, shoot at other combat systems, and make decisions. The agents follow simple commands according to a set of rules that govern how they move and interact with the environment and each other.

- All of the interactions between agents were enabled through the use of *effect agents* representing sensing, shooting, and communicating. Sensing serves the role of enabling perception. The implemented model relies solely on direct line of sight visual sensing, but it can be extended easily.

In the implemented prototype, only influencing agents modeling the actual projectile in flight allowing it to evaluate each contact encountered in turn and communication agents modeling communication, messages, orders, etc. were implemented. Such effect agents were created as needed within the model. When manifested these functions were intended to complete the control and combat decision cycles of which they were a part. If for some reason the intended result did not occur due to missing the target, fratricide, etc., then the network cycle was completed with the effect applied to the unintended target, if the related effects could be observed and were perceived correctly.

Other roles of agents will be dealt with in more detail in a later chapter. As this small episode motivates, many new roles can effectively be modeled using the agent paradigm, and so in particular the human, social, cultural, and behavior modeling community showed interest in this new approach.

REFERENCES

Bowen RJ (2008). *An Agent-based Combat Simulation Framework in Support of Effect-based and Net-centric Evaluation*. PhD Thesis, Old Dominion University, Norfolk.

Dupuy TN (1990). *Attrition: Forecasting Battle Casualties and Equipment Losses in Modern War*. Hero Books, Fairfax, VA.

Epstein JM (1985). *The Calculus of Conventional War: Dynamic Analysis without Lanchester Theory*. Washington, Brookings Institution.

Fiske BA (1916). *The Navy as a Fighting Machine*. Naval Institute Press, Annapolis, MD (Reprint 1988).

Fowler BW (1995). *De Physica Belli—An Introduction to Lanchestrial Attrition Mechanics (Part One). Report DMSTTIAC SOAR 95-01, Defense Modeling*. Simulation and Tactical Technology Information Analysis Center, Huntsville, AL.

Hartley III DS (2001). *Predicting Combat Effects. INFORMS/MAS Topics in Operations Research*. Military Operations Research Society, Alexandria, VA.

Helmbold RL (1993). Osipov: the Russian Lanchester. *Eur J Oper Res* **65**, 278–288.

Hofmann HW, Schnurer R and Tolk A (1993). On the impact of combat support systems on military stability: design and results of simulation experiments with the closed division/corps combat simulation model KOSMOS. In: *Problems of International Stability in a Multipolar World: Issues and Models for Analysis*, edited by RK Huber and R Avenhaus. Nomos-Verlag, Baden-Baden, pp. 181–205.

Lanchester FW (1916). *Aircraft in Warfare: the Dawn of the Fourth Arm*. Constable and Co., London.

Morse PM and Kimball GE (1950). *Methods of Operations Research*. Military Operations Research Heritage Series, Alexandria, VA.

Smith EA (2002). *Effect Based Operations: Applying Network Centric Warfare in Peace, Crisis, and War*. CCRP Press, Washington, DC.

Strickland JS (2011). *Mathematical Modeling of Warfare and Combat Phenomenon*. Lulu, Inc., Colorado Springs CO.

Taylor JG (1983). *Lanchester Models of Warfare*. Operations Research Society of America, Military Applications Section, Arlington, VA.

Tolk A (2010). Towards computational representations of the desired end state. Paper 10F-SIW-056. In: *Proceedings of the Fall Simulation Interoperability Workshop Orlando, FL*. http://www.sisostds.org.

Tolk A, Bowen RJ and Hester PT (2008). Using agent technology to move from intention-based to effect-based models. In: *Proceedings of the Winter Simulation Conference, Miami, FL*. IEEE Press, Piscataway, NJ.

US Department of Defense (1993). *Field Manual 100-5 Operations*. Headquarters Department of the Army, Washington, DC.

Washburn A and Kress M (2009). *Combat Modeling*. Springer, Heidelberg.

Youngren MA (ed.) (1994). *Military Operations Research Analyst's Handbook; Volume I: Terrain, Unit Movement, and Environment; Volume II: Search, Detection, and Tracking, Conventional Weapon Effects*. Military Operations Research Society (MORS), Alexandria, VA.

Chapter 10

Modeling Communications, Command, and Control

Andreas Tolk

COMBAT IN THE INFORMATION AGE

Until recently, many traditional combat models assumed sufficient communication between fighting units. For those models, the view on communication was limited to exchanging tactical orders, supporting the coordination of fire, transmitting spot reports of enemies in the responsibility sector of the fighting units to become new targets, and similar tactical communications on the battlefield. In reality, this communication was based on radio transmission along line-of-sight connections, use of signals, and often the human voice. If these communications failed, detection probability was slightly decreased to take such failures into account. An explicit modeling approach for communications was not perceived to be necessary.

When it came to the support of distributed planning and the integration of command and control systems, the communication of operational orders, order of battle, and other information was needed. Very often, the schemas supported by the command and control community were used. Message text formats (MTFs) were often applied in support of distributed simulation as well. These operational messages also play a pivotal role for the integration of modeling and simulation and operational systems, but we shall take care of this aspect in a later chapter in more detail. In any case, as soon as command and control structures and the supporting infrastructure increased in the interest of the sponsors of combat models, the necessity for explicit communications and command and control modeling arose as well.

Engineering Principles of Combat Modeling and Distributed Simulation, First Edition.
Edited by Andreas Tolk.
© 2012 John Wiley & Sons, Inc. Published 2012 by John Wiley & Sons, Inc.

In this chapter, we shall focus on modeling of communications as well as of command and control and headquarters. In earlier days, all decisions and command and control activities had to be conducted by humans. Only the physical realm of sensing, moving, and shooting was simulated. In modern combat models, more and more of these activities and the headquarters or command posts that execute them are modeled as well. Examples are given in a later chapter on agent-based models for combat modeling. These command posts and headquarters rely heavily on information provided by communication means. The execution and monitoring of their plans—the essence of command and control—needs reliable communication. Consequently, more and more communication modeling is needed.

Today, we are in the information age (Alberts, 2002), and the military underwent several transformations to become the armed forces for the 21st century. Obtaining and distributing the required information between command posts and headquarters became as important as moving the right weapon systems to the right operational sectors. Information operations became a term that described combat for better information on the own side and denying the opponent the opportunity to obtain good information. Communication is no longer limited to radios and signals. In modern headquarters, the TV showing CNN and other world news is as established as a video conference unit, an Internet access point, and other media as they are used in non-military organizations as well. The smart phone and tabloid PC are companions of patrolling soldiers to give them their exact position using internal GPS functionality as well as access to latest information. As such, modeling communications means may range from simple hand signals and shouted commands of an infantry group leader to the satellite communication enabling broadband conferencing straight back to the national command cell.

This vision of modeling decisions and communication has become reality on all levels: if decided, we have the technical means allowing the commander in chief and staff to watch live-feeds from webcams installed on the helmets of Special Forces operation members! Requirements for training to use these new capabilities, as well as for analysis on how important these communication lines are, establish the need to model them. This leads to the need to also model counter-measures, such as jamming of links or denial-of-service attacks on computers, as well as to orchestrate own activities that can interfere with communication needs, such as frequency planning or alignment of jamming operations that can interfere with own or allied systems as well.

Some very similar observations can be made regarding command and control. Until recently, command and control systems with no combat assignment, like higher level headquarters, were represented, if at all, as targets. The special capabilities of soldiers in the headquarters, their tasks, or the participation of a limited number of experts in a variety of teams in support of several operational cells within the headquarters were not taken into account. With the introduction of the agent based paradigm, a lot of progress became possible in modeling this domain, but as a community we are still at the beginning of the full integration of command and control models into our combat models.

Within this chapter, we shall limit the observing to the principles of communications, command and control modeling. This field is currently rapidly evolving and new models are introduced frequently. The simulation engineer needs to be able to select the best available option in support of the needs of current tasks.

The application of the harmonization and alignment principle discussed in Chapter 5 is essential: if something is critical to command, control, and communications, it needs to model consistently over all participating core processes. If an operation has the objective to take out operational communication capability, the systems that provide this capability must be modeled. Further, means to detect and destroy them must also be created, and once they are destroyed the capability should be destroyed as well, which has an effect on other processes on the battlefield.

COMMAND AND CONTROL MODELING

As mentioned in the introduction, many command and control processes were conducted by human players in front of the computer. Command and control did not have to be (or was not) modeled. Each entity received an order and followed this order until it reached an end-state or received a new order. In order to move from an assembly area toward an enemy, all orders had to be given individually and sequentially, like marching, moving in the terrain, firing at an opponent, etc. If such an entity came under fire, the user had to decide what to do. If the user, on the other side, forgot to get gas or ammunition, it was possible that the entity in the middle of the attack stopped driving, because it ran out of gas.

In particular for analysis applications, where hundreds or thousands of runs need to be conducted and evaluated, this kind of command and control was insufficient. The first step was the use of *scripts* for entities. Instead of one order, an entity could work on a series of orders. Once the first order is accomplished, the second order is initiated. This allowed commands like "move until the end of the street, move over the field into the fortified position, begin to defend."

Several command scripts were developed to support advanced concepts, like conditional orders: "march until the end of the street; if no enemy is in sight, move over the field into the fortified positions; if enemies are in sight, move straight into hasty defense." In addition, entities were modeled to exhibit semi-automated behavior. This was a significant step forward regarding making computer generated forces more stable and the resulting simulation systems more realistic. *Sidenote*: When the motorcycle giant Harley Davidson finally introduced the first model with an electric starter in 1965, they called the new series Electra-Glide. Similarly, many combat simulations that used semi-automated behavior models were called SAF for Semi-Automated Forces. Examples still in use are OneSAF of the US Army and JSAF of the US Joint Forces.

The next step in the process of improving the model was enabling the orchestration of several entities using command and control entities, often modeled based on the agent paradigm (McGrath et al., 2000). Instead of human actors these command and control entities received the situation reports of the subordinated combat units. They implement decision algorithms to generate orders for the unit based on the current perceived situation. The perceived situation is composed of the messages received from sensors or from spot reports received from combat entities. As in real headquarters, the planning capability requires resources that may be needed for several competing activities. Modeling priority lists can help to resolve these conflicts.

Explicitly modeling of tangible entities on the battlefield was the next step. These entities could be detected and attacked by enemies, which reduced the capability of the headquarters. It also enabled the electromagnetic profile of the headquarters to be computed that now could be detected by special sensors, etc. Using the same principle for higher headquarters allows today modeling not only of the combat elements, but also of the underlying command and control hierarchy including modeling which tasks are conducted by whom, when, where, and how (use of competing resources).

The German Command and Control Model FIT (Rech, 2002) was designed to utilize established combat models and add command and control capability. All elements were connected with entities modeled as targets in the battlefield, so that explicit casualties affecting the command and control could be modeled and evaluated as well. Each combat entity comprises cells that have a certain amount of resources, which can be personnel, material, or equipment. A cell has to conduct tasks that are conducted by teams. A team is a temporary collection of resources conducting a task. This approach allows one to model that certain tasks take longer if the conducting team does not have the needed expertise, if material runs low, or if the optimal equipment is currently not available. Communication means were part of the equipment. Furthermore, a perceived situation was provided via a standing task within each command agent.

Several combat models use aggregates of headquarters that simply model the time required to conduct the tasks with explicit modeling of subtasks and required resources. They often use a point system, as introduced in the last chapter, to model the command and control capability of the entity: a "healthy" entity with full points is fast, and the more points are lost, the slower the entity becomes until it can no longer support the required tasks. In probabilistic models, it is possible to use the current points versus the optimal points to compute a likelihood that a task can and will be conducted. It is very hard to validate such *ad hoc* approaches, however.

COMMUNICATION MODELING

It should be obvious after reading the last section that, in particular, when modeling command and control entities, the modeling of communication is a key component equal in importance with moving, shooting, and looking. How can we build a perception without getting reports? How can we distribute orders without having the means to reach the combat entities?

The modeling of communication can be either implicit, where all communications between simulated entities are assumed in the behaviors and perceptions of those entities, or explicit, where the act of communicating itself becomes an action undertaken by the simulated entity. As mentioned in the earlier chapter on effects, the communication itself can become a modeled object—or agent—that can interact with other objects, such as creating saturation of frequencies, jamming, or interference of messages.

Consequently, the communication between simulated command elements and simulated commanded elements, the communication between sensing entities, i.e. simulated units and simulated sensors, and the communication of simulated entities to their peers are all necessary elements of modeling and simulation, as the actions and effects they represent are all necessary parts of warfare. The ability of a unit to communicate reports to its command elements and for the commander to communicate tasks to support elements are both crucial, and when a lack of this communication exists, much more than the immediate loss of reporting and task ability is experienced by the elements that have lost the communication. Morale is affected; behavior is affected; and in many cases the extended sense of what is happening in the larger sense of the battle space is lost.

In modern warfare, sensing components are often separated from shooting components. Without proper communication, effectors do not get targets assigned, and decision nodes do not get the information they need to plan, command, and control the execution. All of these things must be modeled if needed to fulfill the underlying task for which the simulation engineer is responsible.

When modeling communications explicitly, an approach that is often chosen is to capture for each information exchange requirement (IER) between two modeled entities:

- necessary communication means (radio, tactical data link, etc.),
- required or usable channels,
- bandwidth required, and
- capacity and time constraints.

Such technically detailed and physics based models furthermore allow for the computation of the maximum distance of communications, which allows for computing the need of repeaters, relay stations, etc. The influence of terrain, such as line of sight, weather, and opposing activities, such as jamming, can and should be part of the model. Furthermore, activities of allies have to be taken into account to compute possible interference as well.

Many models still assume perfect delivery of information but allow for time delays. Other models use a connection probability to determine if a message gets transmitted or not. While these approaches may be sufficient to train soldiers, they seem to be insufficient for the evaluation of new communication infrastructure needs.

In particular for the evaluation of Internet-like infrastructures, network communication models such as the Optimized Network Engineering Tools

(OPNET) models become of interest. Chang (1999) describes how network topology models, data flow models, and control flow models can be used and orchestrated in support of simulating networks and data links.

For an air force, communication between the sorties in flight and their command and control centers is pivotal. As many of their activities are extremely time critical, they rely heavily on data links. To support their high technology weapons, it may become necessary to upload complete new target information data blocks in high detail, literally "on the fly." These data blocks not only contain target coordinates but surrounding environment information, weather updates, updates on the recognized air and land situation, and more. Similar observations can be made for missile defense where the available information about an incoming missile needs to be queued from detecting long-range radars to fire control radars. Particular challenges can arise from multinational collaborations as different nations may support the same message standards but different versions.

Naval communication is a very complex topic as well. It can easily be categorized in internal and external communications, as on-board communication for navy vessels can easily become as complex as army communication for a brigade and higher. The close proximity of electricity, data links, fire components, and many other needs on board a ship with hull-limited space is a particular challenge as complex as avionic systems in airplanes. In particular when a ship gets hit in battle and the battle model needs to determine what exactly got destroyed, this information is highly dependent on the vessel type and even the generation of this type. It is safe to say that no two vessels in the navy are exactly the same; each one is unique when it comes to the distribution of battle decisive components, including communications, command, and control.

COMMAND AND CONTROL MESSAGES

We already touched on various message categories in Chapter 4. As a rule, for every important event category that can happen and be observed on the battlefield a message is defined that allows the occurrence of this event to be communicated. In addition, orders are used to transmit commands, normally starting with the warning order. In general, the message systems standardize the use of equipment as well as the method that has to be used to exchange information—using voice templates or digital data exchange means between participating systems—so that simply based on the fact that a certain message type arrives via a predefined channel necessary activities can immediately be triggered (like getting missile defense systems ready if a report is received that a hostile missile is approaching).

Voice messages are hard to model or to use for integration tasks. Some models integrate voice modules that can generate radio reports or the report delivered via video using avatars (Zyda, 2005), some work has been done to recognize voices and generate data that can be used in a simulation, but in general the use of a voice interface and the modeling of voice based communications in combat models is neglected and—at best—a topic of current research.

Table 10.1   Spot Report as Used in the US Army

LINE 1–DATE AND TIME	(DTG)
LINE 2–UNIT	(unit making report)
LINE 3–SIZE	(size of enemy unit)
LINE 4–ACTIVITY	(enemy activity at DTG of report)
LINE 5–LOCATION	(UTM or six-digit grid coordinate with MGRS grid zone designator) of enemy activity or event observed)
LINE 6–UNIT	(enemy unit)
LINE 7–TIME	(DTG of observation)
LINE 8–EQUIPMENT	(equipment of unit observed)
LINE 9–SENDER'S ASSESSMENT	(specific sender information)
LINE 10–NARRATIVE	(free text for additional information required for clarification of report)
LINE 11–AUTHENTICATION	(report authentication)

Text-based messages are still used. To a certain degree messages based on the Extensible Markup Language (XML) belong to this group as well. The idea is simple: text messages use human readable terms captured in a controlled vocabulary to fill out the structure and form of messages. The terms are separated by special characters that help a parser to read a message and map content to data fields representing the information in machine readable format. Table 10.1 shows the fields of a spot report as used by the US army.

To generate a report, the reporting unit fills out the respective fields with allowed terms. In line 1, the date–time–group (DTG) marks the time of the report; in line 2, the unit identifier of the reporting unit is given, etc. NATO as well as its member nations have enumerations that can be used for some of the other fields, like for the equipment or unit observed. These common terms avoid ambiguity in the information exchange based on using different terms for the same thing or referring to different things using the same term.

Many NATO systems use the Allied Data Publication Number 3 (ADatP-3). This format defines paragraphs made up of sentences. Each sentence ends with "//", and words are separated by "/". Table 10.2 shows an excerpt of an ADatP-3 message.

Each sentence begins with a key term as seen before, like MSGID for message identification, followed by a specified list of terms. Which term is allowed for which spot is standardized in the format. Some terms or groups of terms can be repeated as long as needed. In the example, the forward edge of the battle area (FEBA) is defined by a series of coordination points. At least two coordinates are needed, but more can be used. The example uses four coordinates before the sentence is ended using the "//" marker. The report on units in this example shows how groups of terms are used: a unit is reported using a unit description (UNITDES) followed by the location (UNITLOC) and optional comments

Table 10.2 Example of an ADatP-3 Message

```
EXER/OLIVE DRAB 99//
MSGID/SITREP/AFOP-JT//
REF/A/ORDER/CTG122.4/161500ZJAN1999/0101006//
PERID/172000Z/TO:181800Z/ASOF:171800Z//
MAP/1501/11/1/6-GSFS//
HEADING/ENEMY//
AMPN/LIGHT RESISTANCE, ENEMY CASUALTIES UNKNOWN//
5EUNIT
/DE/CY/ACTTYP/ENUNIT/UNITLOC/TIMPOS
/01/RS/DEPLOY/345 MTR RFL DIV/RIDGELINE CHARLIE/171200Z//
HEADING/OWN SITUATION//
BNDLINE/FEBA/50QRD99109920/50QRD99309940/50QRD99509960/
50QRD99809908
//
5UNIT
/UNITDES            /UNITLOC            /CMNTS
/1BN25INF           /50QRD99109920      /ON LINE 180600Z
/2BN25INF           /50QRD99309940      /ON LINE 180700Z
/4BN25INF           /50QRD99209930      //
AMPN/PERSONNEL ENROUTE TO UNIT//
HEADING/ADMIN AND LOG//
CBCASLTY/10/5/4/0//
GENTEXT/LOGISTICS/POL DUMP AND FUEL LINE IN PLACE,
REQUEST GENESSEE
CLOSE BEACH FOR FUEL TRANSFER AT 181200Z//
GENTEXT/PERSONNEL/4BN, 25INF REPLACEMENTS ENROUTE TO
UNIT//
```

(CMNTS). At least one unit needs to be reported, but the example reports three: 1BN25INF, 2BN25INF, and 4BN25INF (obviously infantry units belonging to the same superior unit, probably three companies belonging to one battalion).

The NATO Headquarters Consultation, Command and Control Systems (NHQC3S) Data Administration (NDA) staff is tasked to establish and maintain a NATO XML Registry that contains all NATO namespaces. A namespace contains all terms and keywords allowed in the domain that uses this namespace. All the keywords and allowed terms of the messages described above form such a namespace. The advantage is that the controlled vocabulary becomes machine readable when organized in a namespace and messages can be generated and parsed by machines out of and into information systems. Many NATO systems support the resulting XML schemas to exchange data with other compliant systems.

Fully digital systems integrate messages into the communication protocols of the underlying systems. Tactical Data Link messages and Link messages mainly used by air forces and missile defense belong to this group. A later chapter deals with these kinds of messages as well as with their presentation in simulation systems in more detail.

Finally, the direct alignment of distributed data using a replication mechanism is also used by modern systems. What information can be exchanged depends on the content of the participating databases plus information specifications. The NATO Data Replication Mechanism (NDRM) for systems supporting the Joint Consultation, Command and Control Information Exchange Data Model (JC3IEDM) allows to be defined in detail which part of the data is shared with whom as triggered by defined events (Kües, 2001)

For command and control modeling and communications modeling challenges, knowing these options and their content is critical: the simulation engineer must know which observations can and will be communicated using which means to capture all aspects important to the warfighter. The means by which the information is received can increase or decrease the reliability of the information or even trigger urgent activities, in particular in time critical environments such as air defense or ballistic missile defense.

Another aspect is that these message types are ready-to-use integration means for command and control: if a simulation system can read and write these messages, they can be used to exchange information between the two systems without having to technically change the systems. We will deal with these ideas in more detail in a later chapter when evaluating the integration of simulation systems into the operational environment.

THE INFORMATION AGE COMBAT MODEL

The Information Age Combat Model (IACM) has been introduced by Cares (2005) to describe combat (or competition) between distributed, networked forces or organizations. The emphasis of this model is the connectivity between all entities, which reflects the need for appropriate communications, command and control. The basic entities of this model are not platforms capable of independent action, but rather nodes that can perform elementary tasks (sense, decide, or influence) and links that connect these nodes. Information flow between the nodes is generally necessary for any useful activity to occur. This focus on *network-centric* rather than *platform-centric* operations is intended to advance the state of the art in combat modeling "by explicitly representing interdependencies, properly representing complex local behaviors and capturing the skewed distribution of networked performance." The IACM employs four types of nodes defined by the following properties:

- *Sensors* receive signals about observable phenomena from other nodes and send them to deciders;
- *Deciders* receive information from sensors and make decisions about the present and future arrangements of other nodes;
- *Influencers* receive directions from deciders and interact with other nodes to affect the state of those nodes;
- *Targets* are nodes that have military value but are not sensors, deciders, or influencers.

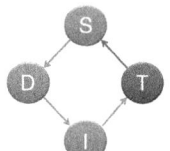

Figure 10.1   Simplest two-sided IACM.

These properties represent the minimum required for each type of node. Other possible characteristics will emerge in the following discussion. Each node belongs to a "side" in the competition, of which there are at least two. We will restrict the present discussion to two sides, conventionally termed BLUE and RED. In principle, any pair of nodes can interact, regardless of side, but some restrictions will be found to occur for both theoretical and practical reasons. It is worth noting that Influencers can act on any type of node, and Sensors can detect any type. The Target type was introduced primarily to reflect the fact that not all military assets fall into one of the other three types. In most situations, however, an Influencer will target an adversary Sensor, Decider, or Influencer.

The basic combat network shown in Figure 10.1 represents the simplest situation in which one side can influence another. The BLUE Sensor (S) detects the RED Target (T) and informs the BLUE Decider (D) of the contact. The Decider then instructs the BLUE influencer (I) to engage the Target. The Influencer initiates effects, such as exerting physical force, psychological or social influence, or other forms of influence on the target. The process may be repeated until the Decider determines that the desired effect has been achieved. It should be noted that the effect assessment requires sensing, which means that this will be conducted in a new circle.

Each of the four links in Figure 10.1 is shown with a different type of line in order to emphasize the fact that the flows across these links may be very different. In particular, some links may represent purely physical interactions, while others may entail both physical processes and information flows. Cares (2005) describes the simplest complete (two-sided) combat network shown in Figure 10.2. The subscripts x and y indicate the side to which a node belongs.

Once the IACM has been defined in terms of a network of nodes and links, the language and tools of graph theory can be used for both description and analysis. A concise description of any graph is provided by the adjacency matrix A, in which the row and column indices represent the nodes, and the matrix elements are either one or zero according to the rule $A_{ij} = 1$ if there exists a link from node i to node j and $A_{ij} = 0$ otherwise. Many properties of a graph or network can be calculated directly from the adjacency matrix, and two are of particular interest here. Since combat power or influence can be exerted only when there exists a connected cycle that includes the node to be influenced, the detection of cycles in the graph is of great importance. One method used in studying the

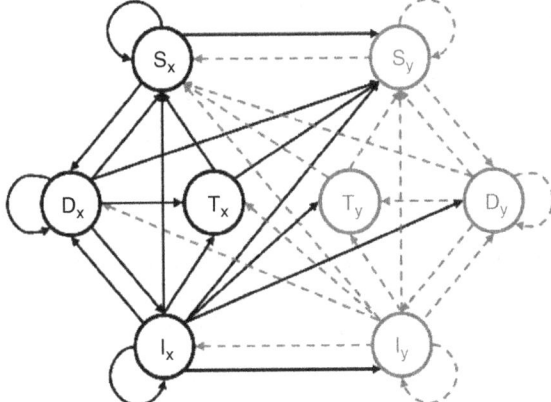

Figure 10.2 Basic information age combat network.

evolution of complex adaptive systems is calculation of the principal eigenvalue of the adjacency matrix.

Extending the original work of Cares (2005), Deller et al. (2009) observe that the existence of a real, positive principal eigenvalue of A_{ij} is guaranteed by the Perron–Frobenius theorem, and this eigenvalue λ_{PFE} (and the corresponding eigenvector) are often referred to as the Perron–Frobenius eigenvalue (eigenvector). It is known that for a graph with no closed cycles $\lambda_{\text{PFE}} = 0$. For a graph with a single cycle of any length, one obtains $\lambda_{\text{PFE}} = 1$. Graphs with more complicated cycle structures have $\lambda_{\text{PFE}} > 1$.

This had led to the proposal (Cares, 2005) that λ_{PFE} be adopted as a measure of the ability of a network to produce feedback effects in general and combat power specifically in the case of the IACM. This is essentially the hypothesis explored in the present work.

An alternative but closely related approach is based on the fact that A^n (the nth power of the adjacency matrix) can be used to obtain the number of distinct paths connecting any pair of nodes. Specifically, $(A^n)_{ij}$, which is the ijth matrix element of A^n, is equal to the number of distinct paths of length n connecting nodes i and j. In particular, $(A^n)_{ii}$ is the number of distinct closed paths from node i back to itself. If node i is an adversary Target T, then $(A^n)_{TT}$ is the number of distinct combat cycles of length n that include T. This represents the number of different ways that T can be engaged by the opposing force. In general, combat cycles must be of length at least 4, and if the links are restricted to the types shown in Figure 10.1 they must be of length exactly 4. In this case, the matrix element $(A^4)_{TT}$ is equal to the number of combat cycles that pass through the Target T and is therefore a potential measure of the combat power that can be brought to bear on it. Deller et al. (2009) hypothesized that λ_{PFE} can be used as a measure of effectiveness for a distributed, networked force or organization.

Approaches like the ICAM promise to improve the modeling of communications, command and control in the future.

NEW COMMAND AND CONTROL CONCEPTS

Hardly any of the other core processes dealt with in this book have increased in importance like communications, command, and control. Unfortunately, most evaluations and reports are classified due to the sensitivity of findings and utilized data. The use of detailed combat models and the federation of high fidelity combat simulations with high fidelity communication simulations are still in their beginnings.

New concepts, like the Global Information Grid (GIG) and the integration of robotic forces open new possibilities (Winters and Tolk, 2005), but also demanding challenges for combat models that conceptualize these ideas. Within the GIG, the planned information infrastructure is highly complex.

- A terrestrial network layer provides the connectivity between weapon systems using radio based data links. Each service provides part of the overall needed functionality.

- A tactical layer based on Link systems supports this ground/sea level network. In particular rotary-wing based components are designed to "bring the infrastructure to where it is needed," such as in infrastructural under-developed or destroyed regions.

- Airborne network layers around Airborne Warning and Control System (AWACS)-like components provide the backbone to connect the tactical and terrestrial components to other operations.

- Global connectivity is supported by the satellite-based space backbone layer.

Alberts (2002) emphasizes that the GIG is not only about technical inter-connection, although connectivity is the enabler, but net-centric warfare is about networking and net-enabling the organizations. Typical hierarchies do not work well in agile and dynamic environments. Command and control structures must change, must become leaner and faster, self-organizing. This is not well captured in the combat models of the current generation and may become one of the main challenges. The Command and Control Research Program (CCRP) conducts annual Command and Control Research and Technology Symposia (CCRTS) coping with related issues, but the use of combat models and distributed simulations is the exception to the rule.

The use of robots on the battlefield, in particular unmanned air vehicles or unmanned underwater vehicles, is less of a challenge for combat modelers, as the capabilities and behavior of such entities are obviously often much better defined than for the man-manned vehicles dealt with so far. Actually, new approaches like the Coalition Battle Management Language (C-BML) can easily use simulation systems as an ideal testbed for communication with robots. C-BML will be dealt with in a later chapter in more detail, but the idea should be mentioned in this chapter anyhow: C-BML provides a common artificial language to communicate command and control relevant information between information systems, such as command and control systems, simulation systems, and robots (Blais et al., 2005). As such, C-BML may replace messages and other means of communication in future scenarios.

New media, in particular the integration of Internet-based information, the effects of social networks on operations, such as Twitter or Facebook, and other new topics, are not yet well enough understood, although we already know that they can definitely shape the perception and potentially the outcome of military operations.

REFERENCES

Alberts DS (2002). *Information Age Transformation: Getting to a 21st Century Military*. Command and Control Research Program (CCRP) Press, Washington, DC.

Blais C, Galvin K and Hieb M (2005). Coalition Battle Management Language (C-BML) Study Group Report 05F-SIW-041. In: *Proceedings of the Fall Simulation Interoperability Workshop, Orlando, FL*. http://www.sisostds.org.

Cares J (2005). *Distributed Networked Operations*. iUniverse, New York, NY.

Chang X (1999). Network simulations with OPNET. In: *Proceedings of the Winter Simulation Conference*, edited by PA Farrington, H Black Nembhard, DT Sturrock and GW Evans. ACM Press, New York, NY, Vol. 1, pp. 307–314.

Deller ST, Bell MI, Bowling SR, Rabadi GA and Tolk A (2009): Applying the information age combat model: quantitative analysis of network centric operations. *Int J Command Control* **3**. http://www.dodccrp.org/files/IC2J_v3n1_06_Deller.pdf.

Kües B (2001). Data management for coalition interoperability. In: *Proceedings of the RTO IST Symposium on Information Management Challenges in Achieving Coalition Interoperability*, NATO Report RTO-MP-064 AC/323(IST-022)TP/11. NATO (NEUILLY-SUR-SEINE CEDEX, FRANCE) Quebec, Canada.

McGrath S, Chacon D and Whitebread K (2000). Intelligent Mobile Agents in the Military Domain, *Proceedings of the Autonomous Agents 2000 Workshop on Agents in Industry*, Barcelona, Spain.

Rech HJ (2002). Command and control assessment using the German simulation system FIT, Report RTO-MP-117. In: *Proceedings of the RTO SAS Symposium on Analysis of the Military Effectiveness of Future C2 Concepts and Systems, The Hague, The Netherlands*. CCRP, Washington, DC.

Winters LS and Tolk A (2005). The integration of modeling and simulation with joint command and control on the global information grid. In: *Proceedings of the Spring Simulation Interoperability Workshop, San Diego, CA*. SISO, Orlando, FL.

Zyda M (2005). From visual simulation to virtual reality to games. *Computer* **39**, 25–32.

Part III

Distributed Simulation

Chapter 11

Challenges of Distributed Simulation

Andreas Tolk

GENERAL DISTRIBUTED SIMULATION CHALLENGES

So far, we have focused on modeling challenges. The way we model entities of combat, their interactions based on the core processes of shoot, look, move, and communicate, their relations, and events and state changes still play a pivotal role for applications, but in the next couple of chapters we will focus on what tasks a simulation engineer will face when executing these models as simulations, in particular when executing them as distributed simulations where independently developed simulation systems are performed on autonomous networked computers supported by information exchange models and protocols that govern the exchange of information between these simulation systems. The tasks of a simulation engineer in this context in general can be summarized as follows:

- selecting the best simulation systems in support of the task,
- composing the simulation systems into a federation,
- exposing the information needed by other simulation systems conforming with the selected interoperability protocol,
- integrating the information provided by other simulation systems via the interoperability protocol into the respective receiving simulation systems,
- avoiding inconsistencies, anomalies, and unfair fight situations,
- addressing additional issues regarding multiple interoperability protocols that are used within the federation,

Engineering Principles of Combat Modeling and Distributed Simulation, First Edition.
Edited by Andreas Tolk.
© 2012 John Wiley & Sons, Inc. Published 2012 by John Wiley & Sons, Inc.

- ensuring that all simulation systems and information exchange models are initialized consistently, and
- ensuring that all information needed can be exchanged via the supported information exchange models and interoperability protocols during execution.

Figure 11.1 shows two simulation systems that need to be federated with selected related challenge domains. In principle, all aspects that are captured in the modeling chapters of this book have to be captured and aligned to ensure consistency in a federation. The two simulation systems in the upper part of the figure have their various concepts of simulated entities, their relations and events, and status changes, and the event queues based on the time advance algorithm implemented. Both simulation systems have to map their view of the world to the infrastructure network, supporting the information exchange model based on the agreed upon communication protocol. We will deal with related standards like the IEEE 1278 and IEEE 1516 in the next chapter.

The main reason for building a federation is the coupling of functionality of contributing systems to provide a new capability. To allow this, the common entities, events, and state changes represented in participating simulation systems must be represented in both systems consistently and synchronized, which means that the challenges of temporal and mapping inconsistencies must be addressed, as discussed later in this chapter. To this end, the infrastructure that supports the interoperability protocol and the information exchange model must support three requirements.

- All information exchange elements must be delivered to the correct simulation systems (effectiveness).

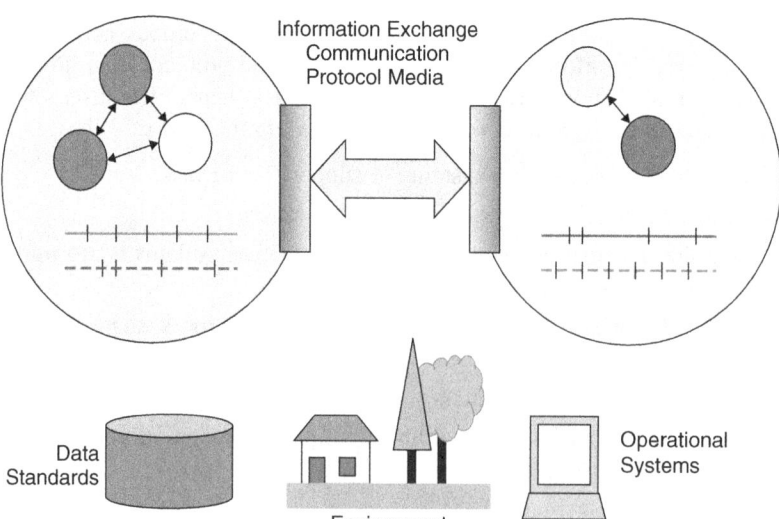

Figure 11.1   Challenge domains.

- Only the required information exchange elements must be delivered to the simulation systems (efficiency).
- The delivery must happen at the right time (correctness and timeliness).

To this end, the infrastructure solution needs to provide several services that support the fulfillment of these requirements. If the infrastructure does not take care of this, either the simulation systems must provide such services, or the management of the development and execution of the federation must ensure the fulfillment with respective rules and regulations.

Every simulation system simulates its entities in a situated environment. The environment representation in the various systems may differ significantly. The fact that different environments can lead to unfair fight situations has already been discussed in the modeling chapters of this book. When building a federation, the simulation engineer must ensure that the environment is represented in such a way that systematic favors or disadvantages are avoided. This is a very difficult task, as the interplay of simulated entities and the situated environment is rarely well documented or supported by information exchange models. Also, only recently has the idea of dynamic updates of the environment been entertained by several simulation systems. Standard solutions have not yet been established.

Common data standards to support the initialization of all systems—and the military scenario definition language (MSDL) will be dealt with as an example later in this book—as well as the exchange of information during the execution need to be taken into consideration as well. Such data standards, which comprise services and object models and ontological means also, are increasingly important when in addition to simulation systems the operational environment needs to be taken into account as well. In two additional chapters the integration into the operational environment as well as the use of geospatial information systems will be addressed in more detail. As already mentioned in the chapter on the Code of Best Practice, authoritative data sources and open sources such as those accessible via the Internet gain in importance and have to be studied as well. The simulation engineer must therefore continuously update knowledge on new developments. The next chapter will give several examples.

Some of the general infrastructure challenges are already addressed in the more general domain of distributed systems. We shall have a short look at them first. Then, we shall look at what we need to address the special challenges for distributed simulations, including the role of conceptual models and metadata to capture them. Next, we look into temporal and mapping inconsistencies and anomalies that need to be avoided, and the general aspect of challenges resulting from multiple scope, resolution, and structure of models resulting in different simulated entities that have to be aligned. Finally, some frameworks to address these interoperability issues will be presented.

DISTRIBUTED SYSTEMS

Distributed simulation is often seen as a sub-domain of distributed systems. Distributed computing systems deal with software that is executed in parallel on multiple autonomous computers. Tanenbaum and Van Steen (2006) give a good introduction on related topics from the theoretic perspective. Ghosh (2007) is more suitable for computer engineers with a focus on practical solutions.

To discuss the resulting challenges from the computer engineering perspective, we use a simple example. Figure 11.2 shows a simple computer network with three computers. One provides a model developed by the US Army for land warfare. Another model was developed by the US Navy. The last one is a US Air Force model. The simulation engineer has the task to support a joint exercise with these three models. Before starting to look into the details of the three models, the computer systems must first be connected.

There are several challenges that have to be addressed when connecting the computer systems, such as answering the following questions.

- *Do the hosting computers provide access to a computer network?*
  As combat models often comprise classified information, access to the computer can be an issue. This is even more important for international exercises or when operational components are part of the networks.

- *Do all computers support compatible versions of network protocols?*
  The ISO/OSI reference model (International Organization for Standardization, 1994) defines what capabilities have to be provided on the basic layers of the reference model. This allows that different providers can support alternative implementations and still be interoperable with

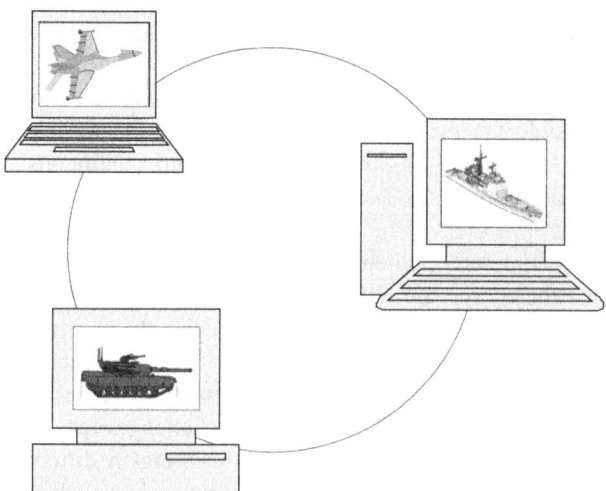

Figure 11.2 Logical view of a computer network supporting a joint exercise.

each other. This further allows that computers of all types and supporting a variety of different operating systems can all participate via the Internet.

- *Do the supporting simulation programs already have a common information exchange standard?*

  Many simulation systems were designed as stand-alone systems and not with the idea of participating in a group. Although in the military simulation domain most simulation systems provide a way to access the data, this is not always the case. In particular when approaching new partners for new simulation tasks, it is possible that the simulation system itself is not designed to support information exchange with other applications. Also, not every information exchange solution works with every simulation system. Often, the programming language used on both implementations can become a hurdle, although more and more infrastructure solutions try to avoid being programming language specific. Another challenge may be that two different information exchange solutions are used that are not compatible with each other, so that brokers and gateways are needed.

- *Is sufficient documentation or simulation system knowledge available?*

  The answer to this question in practice is nearly always "no!" In order to integrate simulation systems, integration expertise on various levels is needed. This starts with what components are needed to successfully execute a simulation and ends with in-depth knowledge of details regarding where and in which format data used in the simulation can be accessed. Some systems use databases; others have all data stored in runtime structures. Some systems use object bases, others XML files, and some of the older legacy systems may still use flat files or comma separated values. The variety is without limits, and rarely has all this information been captured in documentation.

Just providing the network infrastructure is usually a major challenge. Fortunately, many good computer engineering solutions provide the knowledge needed to interconnect systems and ensure that the systems can exchange bits and bytes following protocols that are well known by technicians. Nonetheless, providing the infrastructure for an exercise is still a task that requires sufficient planning and preparation, as also addressed in the NATO exercise chapter later in this book. When these tasks have been successfully taken care of and information exchange between simulation systems is technically possible, the next category of problems can be addressed: "What data need to be exchanged, and when do these data need to be exchanged?"

ENTITIES, EVENTS, AND STATES FOR FEDERATIONS

As already pointed out at several places within this book, in particular in the chapter on the NATO Code of Best Practice for Command and Control Assessment, each proposed solution should always address the problem of the customer.

In other words, the need of an exercise to be supported or an analysis to be conducted must drive the selection and composition of simulation systems. Therefore, the first thing needed to address the question posted at the end of the introduction is a conceptual model of the exercise or analysis that addresses what metrics will be used, what capabilities need to be provided, and which systems with their functionalities can provide these capabilities.

Need for Conceptual Models

What is a conceptual model, and why is it important for distributed simulations?

Balci and Ormsby (2007) define a simulation conceptual model as the model formulated in the mind of the simulation modeler and specified in a variety of communicative forms intended for different users such as managers, analysts, and simulation developers.

Robinson (2008) emphasizes the idea that a conceptual model is a non-software-specific description of the computer simulation model that will be, is or has been developed, describing the objectives, inputs, outputs, content, assumptions and simplifications of the model. He argues that the conceptual model establishes a common viewpoint that is essential to developing the overall model. It specifically addresses understanding the problem, determining the modeling and project objectives, model content, and identifying assumptions and simplifications.

Tolk et al. (2008) emphasizes the need to have artifacts for communication between experts and customers that not only are implementation neutral but are machine-readable as well; which means a conceptual model should be a formal specification of the conceptualization that marks the result of the modeling phase.

In summary, a conceptual model specifies what is needed for an exercise or an analysis task. It defines the blueprint of what needs to be executed within a simulation task. It does this implementation independently and may use various artifacts to do this, but to ensure the support of these processes by tools the artifacts must be rooted in a machine-readable formal approach. That is why ontological means are of particular interest to current research.

Once the blueprint for the exercise is defined, the role of the various simulation systems and their provided functionality can be defined. Tolk et al. (2010) identify the following four general tasks that a simulation engineer has to conduct:

1. Identification of applicable solutions, i.e. finding out which available systems can simulate a required weapon system, the weapon system's effect or activity, and the result of such actions.

2. Selection of the best set of solutions, i.e. based on the customer's problem deciding which combination of the identified solutions is the best.

3. Composing a solution of the selection to support the exercise, i.e. engineering a solution that supports the information exchange required between the different simulation solutions (addressing *what* data need to be exchanged).

4. Orchestrating the execution, i.e. ensuring that the solutions are executed in the right order, that all data are available, that the data are available when they are needed, etc. (addressing *when* data need to be exchanged).

In particular when the simulation systems also provide a conceptual model, a lot of this work can be supported by tools. Professionally engineered conceptual models are necessary that must be part of the model and simulation system documentation. They can be used as metadata when storing information about the model and its simulation in a common repository.

In summary, we can use the conceptual model of the exercise to be supported, that shows what entities are needed, what events occur, and what state changes are expected to support the exercise best, as a blueprint for the expected solution. We can then use the conceptual models of available systems that describe what entities, events, and resulting state changes they model to find a good match for this blueprint. Using the metrics defined by the customer's requirements regarding efficiency and trade-offs we identify and use the best match: this can be the cheapest solution, a solution that favors a set of favorite models, a solution that minimizes the use of simulations not under the control of the customer, and similar ideas that can be used for supporting a rational decision. We then engineer the solution by mapping the systems to the infrastructure. This requires mapping simulation concepts to the information exchange model concepts and vice versa, the synchronization of time, and the orchestration of events to ensure the correct order. The blueprint of the target federation is the yardstick.

This sounds pretty easy, more like a jigsaw puzzle for kindergarten kids, but the truth is that computationally it is a very complex endeavor.

Computational Constraints

The ability to automatically select a set of the best components out of a repository to support an exercise is highly desired by management organizations. Unfortunately, we already know that this problem cannot be solved by computers.

In their work on composability challenges, Page and Opper (1999) showed that the reuse of systems in a new context, such as the reuse of service simulation systems in the context of a joint exercise, introduces new complexity to the problem of selection and composing systems. They extend the work of Overstreet and Nance (1985) who described the condition specification formalism that formally and implementation independently captures the idea of conceptual blueprints (condition) and implementation blueprint (specification) as introduced here. In their work, Overstreet and Nance demonstrate that any condition specification has an equivalent Turing machine specification, which makes sense, as we are talking about simulation systems executable on a computer and hence obeying the laws of computer science. However, this insight has far reaching consequences, as it enables us to utilize the mathematical proofs for Turing machine challenges in the context of the composition of legacy solutions to support a common goal.

Examples of undecidable problems—which means that there cannot exist any algorithm to solve the problem, the best we can do is look for a good heuristic—are questions like "Will the system terminate?", "Are two modeled actions order independent or do I have to orchestrate them?", "Is the specification complete?", "Is the specification minimal?", or "Are two specifications functionally equivalent, in other words, do they deliver the same functionality?"

In their work, Page and Opper (1999) observe that, intuitively, component-oriented design offers a reduction in the complexity of system construction by enabling the designer to reuse appropriate components without having to re-invent them, but they show that this assumption is wrong when applied in the context of condition specification or—as defined in smart grid solutions—conceptualization and implementation. Although determining if a collection of components satisfies a set of requirements becomes feasible under certain assumptions, we still have to solve a potentially computationally intensive problem.

Selecting the right component to fit into an enterprise is a non-trivial task that cannot be generally solved or left to technology. It needs to be done by the simulation engineer based on a heuristic. However, the better the supporting tools and data are, the easier this task becomes. Having machine-readable conceptual models in a common form will help to better solve this in the future.

Metadata

In order to enable the use of a simulation system to support an exercise, the following questions need to be answered and the answers need to be captured in the form of metadata describing the system. The answers build the core of the conceptual model needed to support the ideas presented so far.

- What weapon systems and/or military units are modeled?
- How are the weapon systems/military units related?
- What processes (such as shoot, look, move, communicate) are modeled?
- What effects are modeled (how do systems/units change)?
- What events that affect several systems/units are modeled?

These questions are just a start. Actually, all implementation details of the simulation systems with regard to what is modeled (as discussed in Chapter 5 in the form of an overview) are needed. It should be pointed out that we always have to focus on two meta-categories, the modeling category (what is provided, conceptual level) and the simulation category (how it is simulated, implementation level).

The minimal information needed regarding a simulation component on the modeling level can be summarized as the entity–event–states paradigm, often used for conceptual modeling methods (in particular for those methods that are close to the implementation of the simulation) as well:

- Entities are the simulated things that represent weapon systems or military units that represent no further separatable conceptualizations. They are characterized by attributes.

- Events connect entities, which can be individual entities or groups of entities. When events happen, entities are affected.
- States of entities are equivalent to specified attribute value constellations of these entities. For example, a system may be defined to be in the state of "standing" if the attribute velocity has the value zero.

By defining what entities exist, what events exist, what entities and events are related, and how the state of an entity changes when it is affected by an event simulations can be defined without giving away too many implementation secrets that may by protected by intellectual property laws. However, applying these or similar ideas ensures that the simulation engineer knows the causes and effects of participating simulation systems for all exposed simulated entities, occurring events, and resulting states.

Of particular interest here are the results of related research in computer engineering and systems engineering. Many approaches are currently discussed on how to support these tasks best. Many of them use variations of the Unified Modeling Language (UML) or the Systems Markup Language (SysML). Others focus on applications of the Web Ontology Language (OWL). The simulation engineer should at least be aware of such solutions.

Once the conceptual level is understood, the mapping needed on the implementation level is relatively straightforward. If two simulations use the same concept to represent the same phenomenon, like a date-time-group addressing the date, the time zone, and the time measured in hours, minutes, and seconds in this time zone, the implementations should represent equivalence classes. There should be a lossless translation between the two implementations. One side may model the date using day, month, and year as a composite; the other side may count the days from a certain reference date on. Both views are equally valid and express the same concept. If no lossless translation can be defined, the supported concept is actually different. If both systems define coordinates but one system uses flat-world UTM coordinates and the other system uses geocentric coordinates, only the name 'coordinate' may be the same, but the underlying concepts are different, as we discussed in Chapter 6.

SYNCHRONIZING TIME

As discussed in Chapter 4, the advance of time can be represented quite differently in a simulation system. Some logical systems do not model time at all; some others assume that the simulation progresses in real time. We also introduced event driven systems and time step driven systems. In any case, when several systems are executed on networked but autonomous computers, synchronization becomes an issue and needs to be addressed. The first two categories, no time and real time, are relatively simple, as they can mainly focus on synchronization issues and effective delivery of messages. However, when time is modeled in such a way that parallel executing simulation systems can represent very different times after the same amount of execution time, because they use different ways to advance time and simulate time progress, then maintaining a consistent time between systems can become messy.

In a distributed simulation environment, time management is therefore necessary. A good overview of various challenges and possible solutions has been presented by Fujimoto (1998). He introduced temporal anomalies that need to be avoided. Let us go back to our example shown in Figure 11.2 and assume the following case.

The army is attacked by naval artillery. The air force has an unmanned air vehicle (UAV) in place observing the attack. The army model is a simple straightforward attrition model that focuses on fast casualty modeling. The navy model is a highly complex and sophisticated engineering model that goes down to the last components of the vessel that conducts the artillery attack. The air force model simply observes, which means that this model is waiting to receive messages to display from the partners. The navy models the firing of the missiles and notifies the army (1a). It then updates all the internal states and updates the visualization (1b) notifying all other interested simulation systems. The army receives the fire message (2a) and updates its display immediately (2b). Without time management, the air force UAV may receive the message from the army model showing exploding tanks (3a) before it receives the message from the navy model showing the firing guns (3b). The temporal causal relationship between these events was not preserved: the observer observes the effect before the cause happens, which is an anomaly to be avoided. Figure 11.3 exemplifies this.

The reason for time anomalies can also be message latencies. The effect is the same: things are not received in the logical temporal order. In Figure 11.4, the navy only sends one message out to both other simulations (1), but the latency of the air force model is so long that the army model can receive the

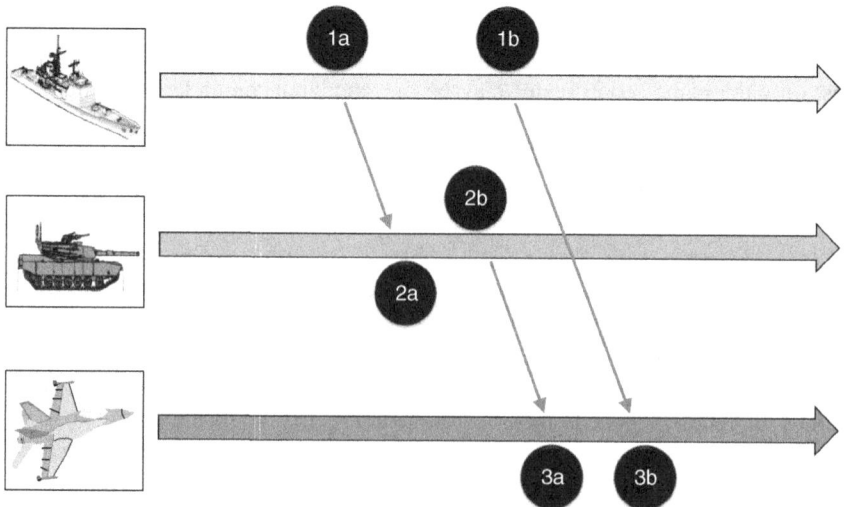

Figure 11.3   Time anomaly based on internal latencies.

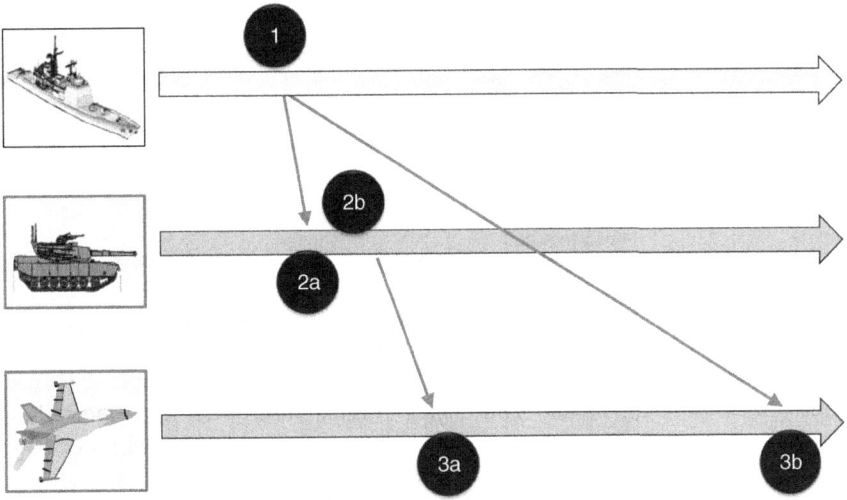

Figure 11.4   Time anomaly due to message latencies.

message (2a), compute the effect, and send the update out (exploding tanks, 2b) which is received by the air force (3a) before the missile message arrives (3b).

In both cases, the time anomaly is a mistake that should not happen and that needs to be taken care of by time management. In essence, time management makes sure that messages are always delivered in the correct order, normally using time stamps with the logical time to be followed by all participating simulation systems.

In parallel with this work, Reynolds et al. (1997) looked into some temporal inconsistencies that can occur when several models and simulations are used to represent the same phenomenon in parallel. This is not a mistake or misalignment of cause and effect; time inconsistency is simply a challenge resulting from the nature of modeling and simulation: two models have a different view of the same phenomenon. Although both models are correct and consistent within themselves, their composition may result in something that may be inconsistent and potentially even wrong. This will be dealt with in the chapter on validation and verification.

Figure 11.5 shows a federation in which the same entity is modeled in two simulation systems. All entities that belong to the left simulation are updated every minute, like army models applied to land force training. All entities on the right side are updated in only 15 minute time steps, like a ground model that computes the targets for an air force exercise. The entity in the middle represents the phenomenon—in this example an army unit—that is a member of both simulation systems. It is updated every minute as a member of the left simulation system and only every 15 minutes as a member of the right.

Reynolds proposed the following definition: "Temporal inconsistency exists when two entities have conflicting or inconsistent representations of the state of a third entity at overlapping simulation times" (Reynolds et al., 1997). There are

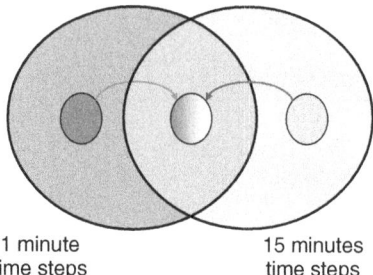

1 minute
time steps

15 minutes
time steps

Figure 11.5   Temporal inconsistency.

two truths regarding the state of the common entity: the truth consistent with the time progression in the left model and the truth consistent with the time progression in the right model. The engineering solution often implemented is to give the control over this entity to only one simulation and simply reflect the changes in the second one, but that may lead to even worse inconsistencies.

If the control is given to the right simulation, the middle entity only gets updated every 15 minutes. If the attrition rates are comparable—which they should be in order to avoid huge obvious discrepancies when observing the evolution of all entities without knowing which system is in control—the state changes in the left simulation will be smaller, approximately only 1/15 of the changes in the right model. If, in the left model, a rule states that an entity changes from attack to hasty defense once 15% attrition occurs and the unit gives up the fight once 25% of the initial strength is reached, it is possible that the entity under control of the right simulation suffers a systematic disadvantage, as the events between the decision points are not dense enough: at the end of the first 15 minutes, the unit may still be slightly above the first breakpoint, which means it continues to attack for another 15 minutes. By the end of this interval it may already have suffered enough additional losses to make the second breakpoint. While all other units that are updated every minute are in hasty defense the 15-minute-controlled unit is destroyed.

The reason for this special case of an unfair fight is that the decision times may not be aligned with the decision thresholds. This kind of problem is very hard to find when it occurs, but it can easily create inconsistent results and hence provide suboptimal training or wrong decision support for the warfighter.

MULTI-RESOLUTION, AGGREGATION, AND DISAGGREGATION

Reynolds et al. (1997) identified more inconsistencies that have their roots in structural challenges, based on the fact that different simulations are likely to be based on models with different scope, resolution, and structure, as discussed in Chapter 4. As in the example with different time steps, in cases where a model differs regarding the entities of the model or the attributes used to model these entities, a common view on an information exchange model needs to be established. In other words, both models have to agree on a common view of entities and properties that describe the situation in a simulation understandable to both systems. This means, however, that either the higher resolution model has to aggregate its view or the lower resolution model has to disaggregate its view. In both cases, mapping inconsistencies can occur.

Reynolds introduced the following definition: "Mapping inconsistency occurs when an entity undergoes a sequence of transitions across levels of resolution resulting in a state it could not have achieved in the simulated time spanned by that sequence" (Reynolds et al., 1997).

To understand this, let us assume that we have two land force simulations. The first simulation is a high resolution model dealing with the behavior of land forces close to rivers. The second simulation is a low resolution model for Lanchester attrition between opposing land forces. Figure 11.6 shows an example of a mapping inconsistency. Blue forces are located close to a river, represented as weapon systems. When under attack, they are aggregated into a unit and engage the opponent. They are pressed toward the river (but cannot cross the river without engineering support). When the battle is over they are disaggregated, but as the unit was pressed towards the river, now one system is on the other side of the river, a resulting state that could not have been achieved otherwise.

Such mapping inconsistencies can result in strange effects. In the example above, a unit can be pressed over a river that otherwise would have been an obstacle. What was supposed to be a deadly trap for the unit may have ultimately saved the weapon systems that magically moved across the river. Again, it is very hard to find such inconsistencies.

In general, multi-resolution modeling is always needed when conceptual and representational differences have to be resolved that are arising from multiple levels of resolution in several models joined for a common objective. Alternative terms are cross-resolution modeling and variable-resolution modeling. Whenever two simulations are federated, it is likely that they differ in scope, resolution, and structure (as well as in time representation). Introducing a common information exchange model to establish a common view only shifts the problem, as now aggregation and disaggregation are needed when mapping to this common information exchange model.

The usefulness of federations that apply several paradigms and resolutions is obvious: the warfighter may want to get a good overview of the general developments on the theatre level but is in particular interested in the details

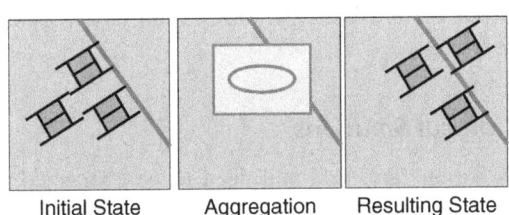

Initial State   Aggregation   Resulting State

Figure 11.6 Mapping inconsistency.

of some selected hot spots, such as key objective areas, urban engagements, special operations, and more. A theatre level simulation is therefore used to get the big picture, and then higher resolution models are used that are interoperably embedded to represent the hot spots. The user may even use more detailed analysis to find out how to improve the situation in these hot spots, leading to three or more levels of detail.

Bowers and Prochnow (2003) document an example in which two models of different resolutions are needed to adequately provide all the capabilities that operational analysis demands. In their paper they describe a mix of forces in an operation the size of Desert Storm or Iraqi Freedom. Analyzing the battle plans from an operational point of view, one is going to have ground forces and air units that will need to interoperate in close cooperation with each other. Operational analysis shows that ground forces and air units will be engaged in theatre-wide warfare, while urban warfare will be supported by air units, but mainly conducted by ground forces. In the example, the situation unfolds in and around the city of Khafji as theater level combat is happening north of and outside the city while tactical combat is also present in the city. The purpose of the study by Bowers and Prochnow (2003) is to create an experimental system that could facilitate multi-resolution modeling and study. The described models of choice were the Joint Theater Level Simulation (JTLS) for theater-wide operations and the Joint Conflict and Tactical Simulation (JCATS) for urban operations. The federation of these models magnifies the challenge of providing continuous support to the warfighter, where continuous in this case means coping with all required capabilities. This work resulted in the Joint Multi-Resolution Model (JMRM) federation that is currently used by NATO for some of their exercises, as discussed in a later chapter. Figure 11.7 shows the general idea using the Battle of 73 Easting as an example for a hot spot during Operation Desert Storm.

A second example is the use of a federation of the aggregate model Eagle and the weapon system level model OneSAF, described by Franceschini et al. (2002). Hester and Tolk (2010) generalized the approach with focus on different modeling paradigms. Additional challenges, alternative viewpoints, and potential solutions are described in Chapter 25 on multi-resolution topics.

Interoperability Challenges

Particularly for global federations it is likely that more than just one standard information exchange infrastructure will be utilized. The findings documented in the Final Report for the Runtime Infrastructure Interoperability Study by Myjak et al. (1999) can be generalized accordingly.

Gateway, Proxy, Broker, and Protocol Solutions

When two different infrastructure solutions are used and need to be connected, four solution categories can be applied: *gateway, proxy, broker*, and *protocol*

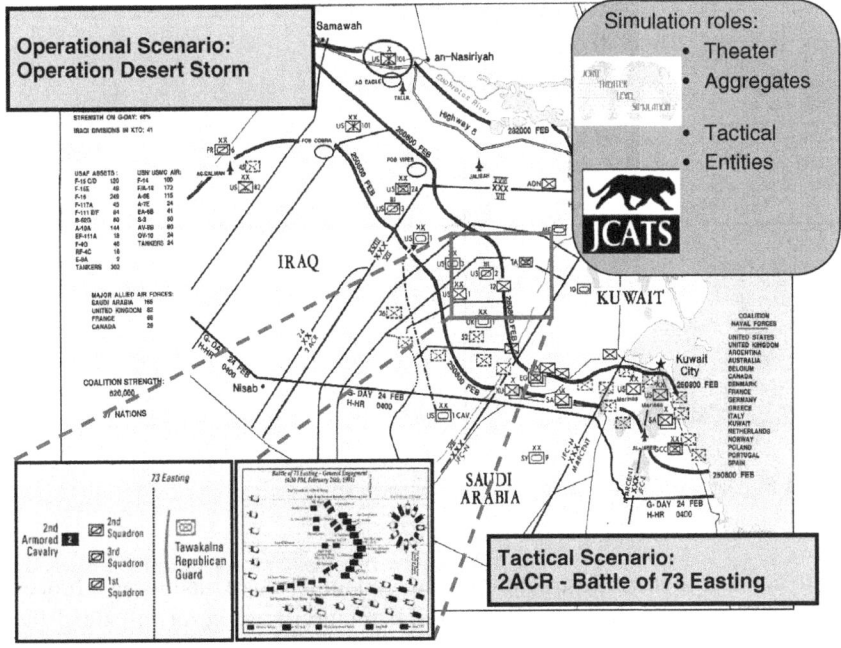

Figure 11.7 JTLS–JCATS multi-resolution modeling federation.

solutions. The report provides the definitions that are simplified and generalized here for better understanding.

- A gateway provides a connection and translation between two simulation systems that are supported by different infrastructure solutions. A gateway focuses on the simulation systems, not the supporting infrastructure.

- A proxy is a simulation system that is connected to two different infrastructure solutions. It comprises the common elements—entities and events—that are shared between the two solutions and uses the interface provided by the infrastructure for simulation systems.

- A broker connects the infrastructures with each other and allows the use of the services of one infrastructure to the other via interface program interfaces. Both infrastructures remain as they are, but the interface provided to other infrastructures is normally richer than the interface provided to simulation systems.

- Protocol solutions extend the functionalities of the infrastructure on the network and protocol level down to binary level interoperability.

Figure 11.8 shows these four categories. It uses four simulation systems: A1, A2, B1, and B2. They are supported by two different interoperability infrastructures I1 and I2. The four options use a gateway to connect two systems, using a

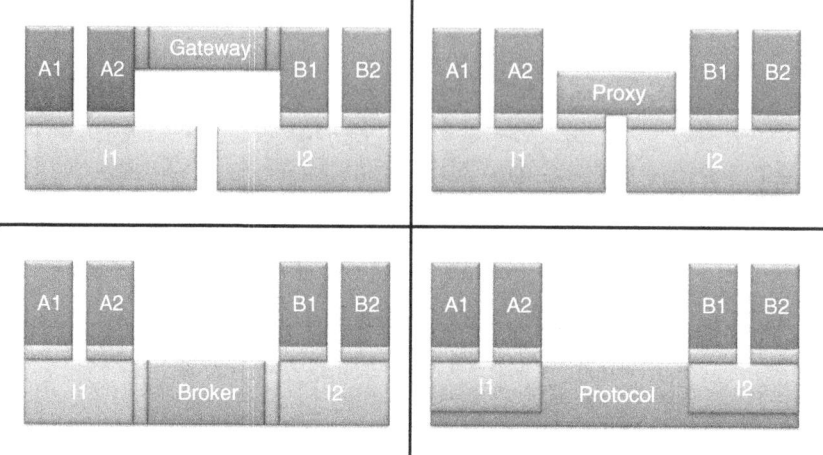

Figure 11.8 Gateway, proxy, broker, and protocol solutions.

proxy to connect to the infrastructure via the system interface, and using a broker to connect to the infrastructure via the infrastructure interface or to extend the protocols on the binary level.

These approaches are not limited to simulation infrastructures, but the same ideas can be used to integrate infrastructures supporting the operational environment as well. The Command and Control to Simulation (C2-Sim) Proxy is a German prototype that connects simulation infrastructures based on IEEE 1516 with command and control infrastructures supporting the Multilateral Interoperability Program (MIP) of NATO; similar approaches were evaluated for the US proxy as well. Both approaches and a common experiment are described by Mayk and Klose (2005). The proxy solution in Figure 11.8 can be interpreted as A1 and A2 being simulation systems coupled via the IEEE 1516 standards, B1 and B2 command and control systems supporting MIP standards, and the proxy connecting both worlds.

Generalizing the Heterogeneous and Homogeneous Approaches

An alternative or generalized view on interoperability and composability challenges was introduced by Myjak et al. (1999). It distinguishes between heterogeneity and homogeneity on the information exchange model level as well as heterogeneity and homogeneity on the infrastructure level. This results in four cases (shown in Figure 11.9):

- homogeneous information exchange models using homogeneous infrastructures (upper left),
- homogeneous information exchange models using heterogeneous infrastructures (upper right),

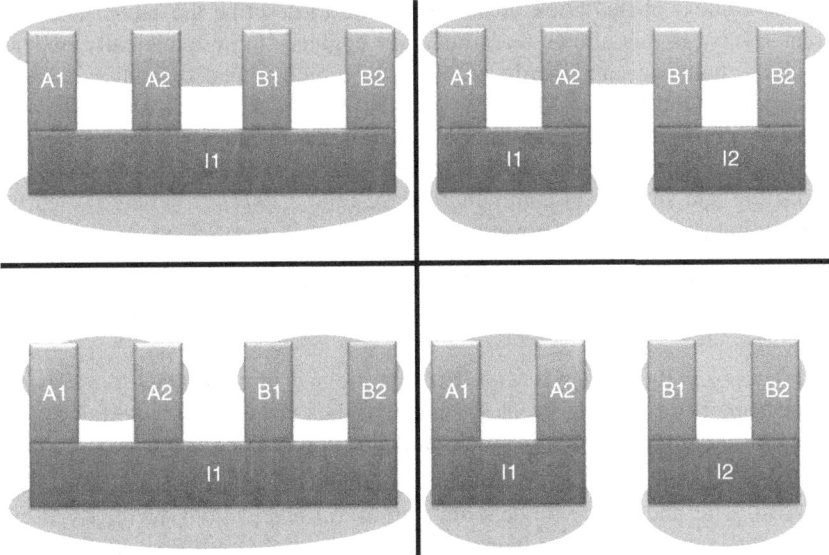

Figure 11.9   Homogeneous and heterogeneous combinations.

- heterogeneous information exchange models using homogeneous infrastructures (lower left), and
- heterogeneous information exchange models using heterogeneous infrastructures (lower right).

The main insight of this generalization is that two different types of problems have to be addressed: the mapping of information content, which represents the modeling level, and the mapping of information structure, which represents the implementation level. Temporal constraints exist on both levels as well, but while, on the modeling level, the conceptualization of time and the synchronization of logical time presentations need to be addressed, the implementation level is concerned with network latency and throughput. All aspects are important, but while the implementation challenges are at least partly addressed by distributed systems, the modeling challenges are unique to the modeling and simulation domain.

Layered Models of Interoperation

This insight of different layers of problems that have to be addressed individually and solved in an aligned way led to the development of the *Levels of Conceptual Interoperability Model* (LCIM). This model and its application to evaluate how well different interoperability approaches satisfy the various levels is described in detail in Tolk (2010) where the LCIM is also compared with some alternative layered models.

In the context of this book, it is worthwhile to describe the model and the different layers, as this allows the simulation engineer to systematically understand which challenges are already addressed by a recommended solution needed to be supported, and which additional alignment is needed and must therefore be provided by other means the simulation engineer has to provide.

In reaction to a first preliminary version of the LCIM being presented to the modeling and simulation community, Page et al. (2004) proposed to clearly distinguish between the three governing concepts of interoperation.

- *Integratability* contends with the physical/technical realms of connections between systems, which include hardware and firmware, protocols, networks, etc.

- *Interoperability* contends with the software and implementation details of interoperations; this includes exchange of data elements via interfaces, the use of middleware, mapping to common information exchange models, etc.

- *Composability* contends with the alignment of issues on the modeling level. The underlying models are purposeful abstractions of reality used for the conceptualization being implemented by the resulting systems.

Successful interoperation of simulation solutions being integrated to build a federation requires integratability of infrastructures, interoperability of the simulation systems, and composability of the underlying combat models. Successful standards for interoperable solutions must address all three categories.

The LCIM was refined based on this recommendation and was used to define seven layers of interoperation addressing clearly distinguishable aspects of interoperation and support engineers in addressing related challenges in a systematic way. The seven layers are depicted in Figure 11.10 and are no interoperability, technical interoperability, syntactic interoperability, semantic interoperability, pragmatic interoperability, dynamic interoperability, and conceptual interoperability. The figure also shows the realms of integratability, interoperability, and composability.

Using Tolk (2010) as a starting point and using examples of the domain dealt with in this book, the layers in the context of combat modeling and distributed simulation can be defined as follows.

- If two systems are stand-alone and are not connected to supporting networks and other infrastructure elements, they obviously cannot exchange anything. There is *no interoperability*.

- The *technical layer* deals with infrastructure and network challenges, enabling systems to exchange carriers of information. This is the domain of integratability. Technical interoperability enables common signals to be exchanged between systems. This layer supports integratability.

- The *syntactic layer* deals with challenges to interpret and structure the information to form symbols within protocols. This layer belongs to the domain of interoperability. Syntactic interoperability enables common symbols that can be exchanged between systems.

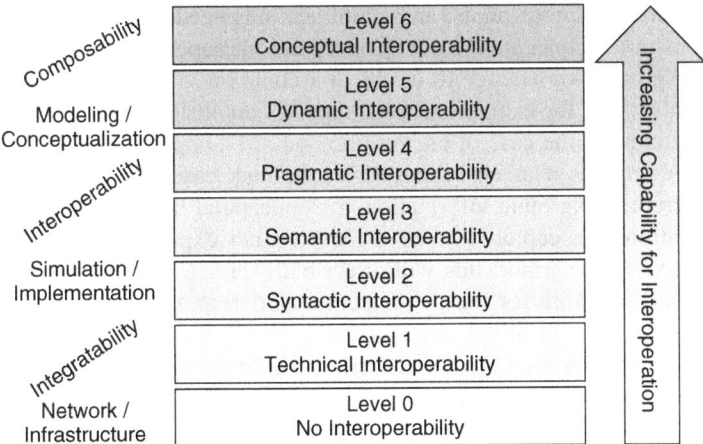

Figure 11.10 The Levels of Conceptual Interoperability Model (LCIM).

- The *semantic layer* provides a "common understanding" of the information exchange by introducing a common terminology. On this level, the pieces of information that can be composed to objects, messages, and other higher structures are identified using common terms to address these structures. Semantic interoperability enables common terms that can be used to tag syntactical structures.

- The *pragmatic layer* recognizes the patterns in which data are organized for the information exchange, which are in particular the inputs and outputs of procedures and methods to be called. This is the context in which data are exchanged as applicable information. These groups are often referred to as business objects. Pragmatic interoperability enables common relations between the tagged structures and relates them to input and output parameters of messages, functions, methods, etc. The information exchange model used earlier in this chapter should capture these tagged structures and relations.

- The *dynamic layer* recognizes various system states, including the possibility for agile and adaptive systems. The same business object exchanged with different systems can trigger very different state changes. It is also possible that the same information sent to the same system at different times can trigger different responses. Dynamic interoperability enables a common functional and mode model. If, for example, the exact same order is sent to an army unit on the march, to another army unit in defense position, and to a third army unit in the assembly area, even if all army units are of the same type and—beside the status they are in—identical regarding their characteristic attributes, the reaction of the three units is likely to be different. If two systems are dynamically interoperable, this knowledge can be taken into account.

• Finally, assumptions, constraints, and simplifications need to be captured. This happens in the *conceptual layer*. Conceptual interoperability requires a lot of background knowledge to establish a common view of the problem to be solved. If, for example, a tank supports an attrition service that computes damage in the case of enemy fire, and also supports a detection service that computes if a sensor can detect the tank based on the signal type and current background information, the conceptual layer makes sure that interrelations are captured, even if they are not explicitly modeled: if the tank gets better armor, this will affect both services. If the damage model uses kill probabilities and the detection model uses detection probabilities, there is no hint in the implementation that kill probabilities and detection probabilities are both influenced by the type of armor. This is conceptual knowledge that needs to be provided somewhere.

The challenges of distributed simulation are not trivial and cannot be solved by computer engineering knowledge alone. Knowledge of distributed systems and computer engineering methods in support thereof is necessary but not sufficient. They only address the realm of integratability and interoperability. To address composability, the modeling aspect needs to be addressed, as this is the unique challenge of modeling and simulation. The following chapters will look at several standards and guidelines that support the simulation engineers in managing these tasks successfully.

SOME LIMITS OF FEDERATIONS

However, before diving into the discussion on methods, standards, and tools that support the development of federated solutions it may be worthwhile to issue a word of caution as well: in particular in new application domains the use of federations may not be the best option. Just because we can federate two systems technically it may not be the best approach to solve the problem of the customer. In particular when starting to understand a problem domain better, the use of many parallel models that all evaluate different facets of the problem domain can be more useful than starting with a federation trying to solve everything.

Davis (2009) calls this a "humble" approach: supporting exploratory analysis under uncertainty and disagreement, and supporting development of strategies that are flexible, adaptive, and robust. On the technical side, it is highly unlikely to be able to address all problems with one common approach. It is much more likely that the multi-simulation approach based on multi-resolution, multi-stage, and multi-models envisioned by Yilmaz et al. (2007) needs to be exploited to support the analysis of these multi-faceted challenges we are faced with as a community.

The simulation engineer should therefore be careful not to bring models together that were never conceptually meant to work together. Many of the methods and tools discussed in the upcoming chapters allow the technical integration of solutions; the conceptual alignment needs to be done by the simulation engineer based on personal knowledge of the modeling side of the integration challenges as we discussed them in Part II on *Combat Modeling* of this book. Technical understanding can solve many interoperability problems, but only conceptual understanding can solve the composability challenges of models that are at least as important as the interoperability challenges of the simulation systems. Chapter 14 on Validation and Verification will address some of these issues, but the current state-of-the-art is still very much focused on the implementation problems.

In his thesis, King (2009) showed that it is possible to execute a federation of conceptually incompatible simulations without technical errors, nonetheless resulting in conceptually questionable recommendations to decision makers. He addressed this as the elusiveness of the conceptual layer in the LCIM. Addressing how to deal with this challenge is subject to current research. Applying formal methods and applying engineering principles to capturing requirements, conceptual models for the federation and the simulation, and other relevant artifacts is necessary, but not sufficient to support the necessary conceptual alignment. However, the simulation engineer who has a basic understanding of the challenges of combat modeling and how important the harmonization and alignment principle introduced in Chapter 5 is for federation development and its limits can contribute to making better decisions in support of the customer.

REFERENCES

Balci O and Ormsby WF (2007). Conceptual modeling for designing large-scale simulations. *J Simulation* **1**, 175–186.

Bowers A and Prochnow DL (2003). Multi-resolution modeling in the JTLS-JCATS Federation. In: *Proceedings of the Fall Simulation Interoperability Workshop, Orlando, FL*. http://www.sisostds.org.

Davis PK (2009). Specifying the content of humble social science models. In: *Proceedings of the Summer Computer Simulation Conference*, Istanbul, Turkey, pp. 452–461. SCS, San Diego.

Franceschini RW, Gerber MB, Schricker SA and Adkins MA (2002). Use of the Eagle/OneSAF testbed to support large scale exercises. In: *Proceedings of the Conference on Computer Generated Forces and Behavior Representation*, Orlando, FL, pp. 411–420. SISO, Orlando, FL.

Fujimoto RM (1998). Time management in the high level architecture. *Simulation*, Part of Taylor & Francis Group, Boca Raton, FL, **71**, 388–400.

Ghosh S (2007). *Distributed Systems—An Algorithmic Approach*. Chapman & Hall/CRC.

Hester PT and Tolk A (2010). Applying methods of the M&S spectrum for complex systems engineering. Emerging Applications of M&S in Industry and Academia (EAIA). In: *Proceedings of the Spring Simulation Multiconference, Orlando, FL*, pp. 17–24. SCS, San Diego, CA.

International Organization for Standardization (1994). *Information Technology—Open Systems Interconnection—Basic Reference Model: The Basic Model*. ISO/IEC 7498-1: 1994(E).

King RD (2009). On the Role of Assertions for Conceptual Modeling as Enablers of Composable Simulation Solutions. Doctoral Dissertation, Frank Batten College of Engineering, Old Dominion University, Norfolk, VA.

Mayk I and Klose D (2005). Technical and operational design, implementation and execution results for SINCE Experiment 1. In: *Proceedings of the International Command and Control, Research and Technology Symposium*, CCRP, Washington, DC.

Myjak MD, Clark D and Lake T (1999). RTI Interoperability Study Group Final Report. 99F-SIW-001. In: *Proceedings of the Fall Simulation Interoperability Workshop, Orlando, FL.* http://www.sisostds.org.

Overstreet CM and Nance RE (1985). A specification language to assist in analysis of discrete event simulation models, *Communications ACM* **28**, 190–201.

Page EH and Opper JM (1999). Observations on the complexity of composable simulation. In: *Proceedings of the Winter Simulation Conference.* ACM, New York, NY, Vol. 1, pp. 553–560.

Page EH, Briggs R and Tufarolo JA (2004). Toward a family of maturity models for the simulation interconnection problem. In: *Proceedings of the Simulation Interoperability Workshop, Arlington, VA.* http://www.sisostds.org.

Reynolds PF, Natrajan A and Srinivasan S (1997). Consistency maintenance in multiresolution simulations. *ACM Trans Modeling Comp Simulation* **7**, 368–392.

Robinson S (2008). Conceptual modeling for simulation part i: definition and requirements. *J Operational Res Soc* **59**, 278–290.

Tanenbaum AS and Van Steen M (2006). *Distributed Systems: Principles and Paradigms* (2nd edn). Prentice Hall, Upper Saddle River, New Jersey.

Tolk A (2010). Interoperability and composability. In: *Modeling and Simulation Fundamentals: Theoretical Underpinnings and Practical Domains*, edited by JA Sokolowski and CM Banks. John Wiley, New York, NY, pp. 403–433.

Tolk A, Diallo SY and Turnitsa CD (2008). Mathematical models towards self-organizing formal federation languages based on conceptual models of information exchange capabilities. In: *Proceedings of the Winter Simulation Conference.* IEEE Computer Society, Washington, DC, pp. 966–974.

Tolk A, Diallo SY, King RD, Turnitsa CD and Padilla JJ (2010). Conceptual modeling for composition of model-based complex systems. In: *Conceptual Modeling for Discrete-Event Simulation*, edited by S Robinson, R Brooks, K Kotiadis and D-J van der Zee. Part of Taylor & Francis Group, Boca Raton, FL. CRC Press, pp. 355–381.

Yilmaz L, Ören T, Lim A and Bowen S (2007). Requirements and design principles for multisimulation with multiresolution, multistage multimodels. In: *Proceedings of the Winter Simulation Conference, Washington, DC.* IEEE Press, Piscataway, NJ, pp. 823–832.

Chapter 12

Standards for Distributed Simulation

Andreas Tolk

WHAT ARE STANDARDS FOR DISTRIBUTED SIMULATION?

As described in the last chapter on the challenges of distributed simulation, the tasks to be conducted by simulation engineering are all driven by the problem of the customer. Every standard that helps

- to capture the objectives of an exercise or another simulation task,
- to derive a conceptual model that can serve as a blueprint to guide the simulation engineer through further decisions,
- to identify potential simulation solutions based on the available documentation,
- to select the best simulation solutions to implement a specific solution for the problem of the customer,
- to compose the solutions into a new system—or a federation—which includes the identification of multiresolution and time challenges,
- to integrate networks and infrastructures (including using proxies, brokers, and protocol solutions),
- to make the simulation systems interoperable (including using gateways) and identify or develop an information exchange model,
- to ensure that the models are composable,

Engineering Principles of Combat Modeling and Distributed Simulation, First Edition.
Edited by Andreas Tolk.
© 2012 John Wiley & Sons, Inc. Published 2012 by John Wiley & Sons, Inc.

- to ensure that data needed for initialization are available or can be obtained,
- and all other tasks and subtasks described so far in this book

… every standard is relevant for distributed simulation. As such, all standards that support distributed systems and computer engineering are relevant. Network standards and standards supporting the Internet are potential candidates.

To give an example, the *Extensible Modeling and Simulation Framework* (XMSF) initiative kicked off by George Mason University in Fairfax, VA, Naval Postgraduate School in Monterey, CA, Old Dominion University in Norfolk, VA, and the Science Applications International Corporation (SAIC) in San Diego, CA, looked at applying World Wide Web standards in support of Internet-based distributed simulation of the future (Brutzman et al., 2002; Blais et al., 2005). A recently conducted peer study showed standards of the semantic web can support distributed simulation better and are expected to be applied more (Strassburger et al., 2008), and recent research shows that this indeed may be the case.

In this chapter, the focus will be on two topics, namely (1) standards officially recognized as modeling and simulation standards and (2) standards that are repeatedly successfully applied in support of distributed simulation in numerous modeling and simulation conferences. Both choices can be neither complete nor exclusive, but they are a compilation that should provide a good starting point for simulation engineers looking for support of their tasks.

Modeling and Simulation Standards

The viewpoint taken in this book to identify modeling and simulation standards is driven by purely practical views: to start looking at modeling and simulation standards, the standards supported by the Simulation Interoperability Standards Organization (SISO) build the initial group.

The reason is simple. SISO states in its vision that it will serve the global community of modeling and simulation professionals, providing an open forum for the collegial exchange of ideas, the examination and advancement of modeling and simulation related technologies and practices, and the development of standards and other products that enable greater modeling and simulation capability, interoperability, credibility, reuse, and cost-effectiveness (SISO, 2010). Furthermore, the Institute of Electrical and Electronics Engineers (IEEE) Computer Society Standards Activities Board voted in November 2003 to unanimously grant the SISO Standards Activities Committee (SAC) status as a recognized IEEE Sponsor Committee. The SISO SAC Chair serves as SISO's primary contact for all IEEE standards activities. In addition, SISO maintains a Liaison Member relationship with Sub-Committee 24 (SC 24) of the Joint Technical Committee 1 (JTC 1) of the International Organization for Standardization (ISO)/International Electrotechnical Commission (IEC). Starting with the standards recognized by SISO as modeling and simulation standards is therefore easily justifiable.

On their website, SISO (2011) enumerates IEEE standards, SISO standards, and ISO standards supporting modeling and simulation, which will be covered at least in the form of an overview here as well. As of June 2011, the following standards were enumerated. As SISO meets at least two times per year in the USA and one time in Europe, this is a list that is constantly updated. The reader is therefore encouraged to check for updates regularly to ensure support from the latest developments.

- IEEE Standards
  - High Level Architecture (will be dealt with in more detail in this chapter as well as in Chapter 19)
  - IEEE Standard 1516—Framework and Rules
  - IEEE Standard 1516.1—Federate Interface Specification
  - IEEE Standard 1516.2—Object Model Template (OMT) Specification
  - IEEE Standard 1516.3—Federation Development and Execution Process (FEDEP) Recommended Practice
  - IEEE Standard 1516.4—Recommended Practice for Verification, Validation, and Accreditation of a Federation—An Overlay to the High Level Architecture Federation Development and Execution Process
  - Distributed Interactive Simulation (will be dealt with in more detail in this chapter)
  - IEEE 1278.1—IEEE Standard for Distributed Interactive Simulation—Application Protocols
  - IEEE 1278.1A—IEEE Standard for Distributed Interactive Simulation—Supplement to Application Protocols—Enumeration and Bit-encoded Values
  - IEEE 1278.2—IEEE Standard for Distributed Interactive Simulation—Communication Services and Profiles
  - IEEE 1278.3—IEEE Standard for Distributed Interactive Simulation Exercise Management and Feedback (EMF)—Recommended Practice
  - IEEE 1278.4—IEEE Standard for Distributed Interactive Simulation—Verification, Validation and Accreditation
- ISO/IEC Standards
  - SEDRIS (see Chapter 6)
  - ISO/IEC 18023-1, SEDRIS—Part 1: Functional specification
  - ISO/IEC 18023-2, SEDRIS—Part 2: Abstract transmittal format
  - ISO/IEC 18023-3, SEDRIS—Part 3: Transmittal format binary encoding
  - ISO/IEC 18024-4, SEDRIS language bindings—Part 4: C
  - ISO/IEC 18025, Environmental Data Coding Specification (EDCS)
  - ISO/IEC 18041-4, EDCS language bindings—Part 4: C
  - ISO/IEC 18026, Spatial Reference Model (SRM)
  - ISO/IEC 18042-4, SRM language bindings—Part 4: C

- SISO Standards
  - ○ Approved Standards
    - – SISO-STD-001-1999: Guidance, Rationale, and Interoperability Modalities for the RPR FOM (GRIM 1.0) (will be dealt with in more detail in this chapter)
    - – SISO-STD-001.1–1999: Real-time Platform Reference Federation Object Model (RPR FOM 1.0) (will be dealt with in more detail in this chapter)
    - – SISO-STD-002-2006: Standard for: Link16 Simulations (see also Chapter 23)
    - – SISO-STD-003-2006: Base Object Model (BOM) Template Specification (see also Chapter 19)
    - – SISO-STD-003.1–2006: Guide for BOM Use and Implementation (see also Chapter 19)
    - – SISO-STD-004-2004: Dynamic Link Compatible HLA API Standard for the HLA Interface Specification Version 1.3
    - – SISO-STD-004.1–2004: Dynamic Link Compatible HLA API Standard for the HLA Interface Specification (IEEE 1516.1 Version)
    - – SISO-STD-005-200X: Link 11 A/B (see also Chapter 23)
    - – SISO-STD-006-2010: Commercial Off-the-Shelf (COTS) Simulation Package Interoperability (CSPI) Reference Models
    - – SISO-STD-007-2008: Military Scenario Definition Language (MSDL) (see also Chapter 24)
    - – SISO-STD-008-2010: Core Manufacturing Simulation Data (CMSD)
  - ○ Product Development Groups (PDGs), Product Support Groups (PSGs), and Standing Support Groups (SSGs)

    PDGs are developing a new standard in a community effort. Based on a product nomination, the group reaches a consensus on the standard and recommends the solution to the SAC. In order to work on regular updates once the standard is accepted, a PSG or SSG may be formed. PSGs meet regularly, SSGs only in case of need.

    - – C-BML—Coalition-Battle Management Language
    - – CMSD—Core Manufacturing Simulation Data
    - – CSPI—Commercial Off-the-Shelf Simulation Package Interoperability
    - – DDCA—Distributed Debrief Control Architecture
    - – DIS—Distributed Interactive Simulation Extension
    - – DSEEP—Distributed Simulation Engineering and Execution Process
    - – DMAO—DSEEP Multi-Architecture Overlay
    - – EPLRS/SADL—Enhanced Position Location Reporting System including Situational Awareness Data Link Simulation Standard
    - – FEAT—Federation Engineering Agreements Template
    - – GM-VV—Generic Methodology for VV&A in the M&S Domain
    - – Link 11 A/B—Link 11 A/B Network Simulation Standard
    - – MSDL—Military Scenario Definition Language

  - RPR FOM—Real-Time Platform Reference Federation Object Model
  - SCM—Simulation Conceptual Modeling
  - SRML—Simulation Reference Markup Language

In addition, SISO also sponsors Study Groups that are made up of experts in the field from academia, industry, and government to evaluate the need and applicability for standardized modeling and simulation solutions for focused topics. These study groups create a final report that often results in the development of a Product Nomination for a PDG. The standardization process including definitions of roles and responsibilities of the PDG and the SAC and all officers is documented in the Balloted Products Development and Support Process (SISO, 2008).

Other Standards in Support of Distributed Simulation

It is much harder to define other standards successfully applied in support of distributed simulation, as there is no organization that tracks such developments. Some professional organizations of interest to simulation engineers are captured in an annex to this book, but there is no central website or point of contact that can be visited for information. The approach taken here was to evaluate the last five years of modeling and simulation workshops and identify other standards that were repeatedly documented as successful in these proceedings by more than one group. The evaluated workshops are

- Winter Simulation Conference (WSC) of the American Statistical Association (ASA), Association for Computing Machinery (ACM), Institute of Electrical and Electronics Engineers (IEEE), Institute for Operations Research and the Management Sciences: Simulation Society (INFORMS), Institute of Industrial Engineers (IIE), National Institute of Standards and Technology (NIST), and SCS
- Principles of Advanced Distributed Simulation (PADS) of ACM, IEEE, and SCS
- Spring and Summer Simulation Multi-conferences of the Society for Modeling and Simulation International (SCS), and
- Spring, Euro, and Fall Simulation Interoperability Workshops (SIW) of SISO

Even with this approach it is likely that this summary only captures a small fraction of standards that can support distributed simulation. As stated before, every standard that supports distributed systems supports by definition distributed simulation as well. New programming languages and maybe even simulation languages are developed, web-enabled development tools become more and more important, and so on. Taking these constraints into account, the application of the following three standard categories was documented repeatedly in the domain of modeling and simulation related research events:

1. using the discrete event system specification (DEVS) and its variants,
2. using modeling languages and architecture languages of the application domain, and
3. applying web standards, in particular semantic web standards, in support of distributed simulation.

Many more topics could have made the list, such as data and process modeling issues, but that would have blown this chapter out of proportion. As it should be self-evident that a simulation engineer has a solid programming background and knows "the usual" programming language well enough that algorithms can be read and understood, it is also a natural necessity to continuously observe the domain of distributed systems and evaluate new ideas regarding their applicability to solve challenges in the domain of distributed simulation.

Discrete Event System Specification Formalism

We understand formalism in this book as a description of something in formal mathematical logical terms so that a machine can read and potentially understand it. Although simulation is not limited to discrete event simulation—we introduced several alternatives in Chapter 4—this modeling paradigm is predominant in current combat models. A formal representation is therefore helpful.

The Discrete Event System Specification (DEVS) formalism was developed by the research group around Bernard Zeigler in support of establishing a theory of modeling and simulation. The latest edition was published by Zeigler et al. (2000). The formalism builds models from the bottom up. The behavior of a system is captured in atomic DEVS components that are defined by the following sets and functions:

- a set of input events X,
- a set of output events Y,
- a set of states S (states the atomic DEVS component can be in),
- the time advance function t_a (that defines how long the component remains in a state),
- the external transition function δ_{ext} (that defines how an input event changes the state of the system),
- the internal transition function δ_{int} (that defines how a state of the system changes internally), and
- the output function λ (that defines how a state of the system generates an output event).

All state transitions are not only governed by the input, they are also governed by the time advance function. Therefore, the atomic DEVS component is well defined by the seven-tuple of the four sets and the three functions: $\langle X, Y, S, t_a, \delta_{ext}, \delta_{int}, \lambda \rangle$. For parallel execution, authors also include an additional

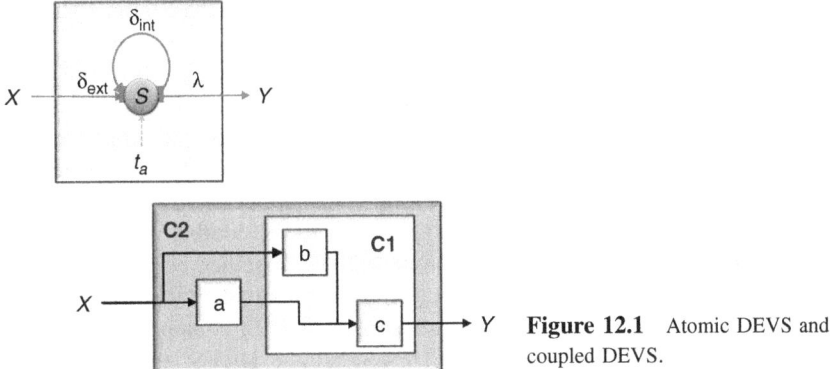

Figure 12.1 Atomic DEVS and coupled DEVS.

confluent transition function that decides the next state in cases of collision between external and internal events or multiple external events. Coupled DEVS use DEVS components and couple them to represent the structure of the system. The lowest layer must be atomic DEVS components, but several layers are allowed. Figure 12.1 shows an atomic DEVS component as well as a coupled DEVS component.

In this example, **a, b**, and **c** are atomic DEVS components; **C1** and **C2** are coupled components. The **C1** couples **b** and **c**. The **C2** couples **a** and **C1**. It also shows that potential conflicts have to be resolved when conflicting events arrive simultaneously. In general, coupled DEVS is therefore well defined by an eight-tuple:

- a set of input events X,
- a set of output events Y,
- a set of names for the subcomponents D (which is used for unambiguous identification of subcomponents),
- a set of subcomponents C_i (which can be atomic DEVS components or coupled DEVS components and which are named by D),
- mapping M_{Eili} of external inputs to internal inputs (all external input events must be mapped to internal input events via ports of the components),
- mapping M_{Ioli} of internal outputs to internal inputs (as a result one internal component can become the input for another internal component),
- mapping M_{IoEo} of internal outputs to external outputs (all results must be produced by an internal component),
- a tie-breaking function that defines which event to select in case of simultaneous events, which is equivalent to the confluent function of the atomic DEVS component.

The behavior is exclusively defined by the state changes and the related time on the atomic DEVS level. If these state changes are not modeled deterministically but stochastically, DEVS becomes stochastic as well. The mappings become intuitively clear when comparing the definitions with the example shown in the figure: the output of **a** is mapped to become the input to **C1**, where it is mapped to become a subset of the input for **c**. The mappings are therefore best understood as the connection instructions between the components. The change of input parameter to output parameters as well as the required time for the execution of this activity is exclusively done by the atomic DEVS components.

The formalism can be used for very practical applications, as shown in many examples by Wainer (2009). The application to the combat modeling task is shown in Chapter 21. Atomic and coupled DEVS together build the classic sequential DEVS approach. With the advancements in computer technology, there have also been numerous extensions, also explained with examples in Wainer (2009), such as parallel DEVS, dynamic DEVS, cellular DEVS, and more.

It is important to distinguish between the DEVS formalism and *DEVS implementations* based on this formalism. Several simulation development frameworks have been proposed based on the DEVS formalism that are united by the common roots, but that does not mean that they are interoperable. Recently, a DEVS standardization effort was launched that tries to establish a community of DEVS practice that works on common standards to support interoperable implementations as well. Examples for several DEVS implementations as well as resulting interoperability challenges are given by Wainer et al. (2010). Most implementations use Java, C++, or C# as programming languages. Examples for DEVS implementations and platforms are

- adevs: A C++ library for constructing discrete event simulations based on the parallel DEVS and dynamic DEVS
- CD++: An environment for developments based on DEVS and cellular DEVS
- CoSMo-Sim: An integrated developer environment for DEVS and parallel DEVS, also supports cellular automata and XML based models
- DEVS++: Open Source Library for C++
- DEVS#: Open Source Library for C#
- DEVSJAVA and DEVS Suite: Development environment for JAVA
- DEVSSOA: Service oriented architectures based on DEVS Modeling Language (DEVSML), and
- DEVSim++: Extension on the basic simulator for DEVS, used in Chapter 21

This list is just a small subset. Many of these environments are open source, in particular those developed by the academic community. In addition, Mittal (2010) developed the DEVS Unified Process which connects DEVS to system architecture frameworks.

The DEVS formalism and DEVS implementations have been extensively dealt with in proceedings by academicians and practitioners in many simulation domains. Many implementations and development environments are open source and therefore often used. The simulation engineers for combat engineering should at least know of these developments. In particular when it comes to the need to federate current combat simulation solutions with new simulation domains needed for new military tasks—such as human, social, cultural, and behavior models as discussed later in this book—it is more than likely that some of them will follow the DEVS formalism and make use of its development environments and implementation. A small group is already looking into DEVS/HLA applications and interoperability challenges, but this research is still in its infancy.

Modeling and Architecture Languages

The more general block of modeling and architecture languages is in particular needed to communicate models and implementations among different stakeholders, such as communications between customers, potential users of the product, and engineers, but also between simulation engineers active in different phases of the lifecycle of a simulation. We shall look at different phases in a following chapter of this book in more detail. Modeling and architecture languages are used to specify descriptions and documentations of systems and architectures based on a set of agreed and aligned artifacts, such as tables, diagrams, or figures. It is highly desirable that these artifacts are based on a common repository containing all described elements only once to ensure consistency in descriptions.

Examples we want to have a look at in this section of the chapter are the Unified Modeling Language (UML), the System Modeling Language (SysML), and the US Department of Defense Architecture Framework (DoDAF). Again, this is only a very limited subset, but it will be used to address the main ideas the simulation engineer has to know in order to apply comparable solutions to facilitate work. In addition, such modeling languages have become often the *lingua franca* between engineers and simply have to belong to the tool set of a simulation engineer supporting combat modeling.

UML and SysML are closely related. They are both governed by the Object Management Group (OMG, 2011) and build the basis for many community standards. They even overlap significantly, as SysML was defined as a subset of UML that was extended to better support system modeling while UML was defined mainly in support of software engineering (although UML was applied for many other approaches as well, such as business modeling). UML artifacts are also used to describe several DoDAF artifacts. The descriptions here cannot replace a more detailed introduction or a tutorial, but should be sufficient to motivate to more deeply deal with these topics.

UML uses different artifacts and diagrams that can be used to provide different points of views on one system. The latest official version of UML is 2.3, but in March 2011 the beta version of 2.4 was released. UML is continuously improved and enriched by a huge variety of users and developers.

The traditional view of systems modeling distinguishes between static views describing the structures and functional views describing the behavior exposed. These views describe what entities are modeled (as classes describing the types

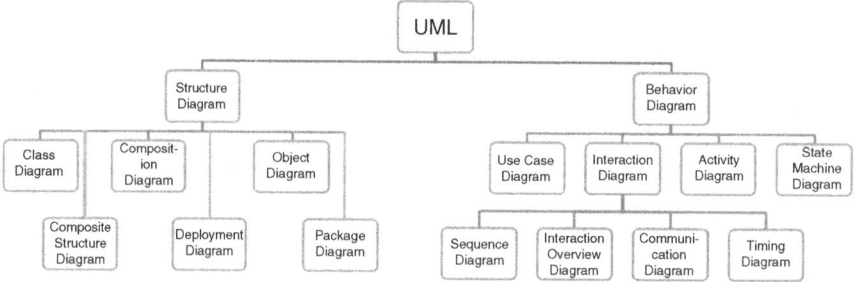

Figure 12.2   UML diagram hierarchy.

and objects describing the instances of things), how entities react when they receive input (as the state changes, which can be broken down if the entity is a composite made up of several smaller entities), and how entities interact with each other (in the form of activities in which more than one entity work together and exchange messages with each other).

UML extends these categories for object oriented software engineering resulting in the diagrams shown in Figure 12.2.

The diagrams of UML are used to explain either the structure or the behavior. UML supports the object oriented paradigm and as such distinguishes between classes that represent type information (properties shared by all things of this type) and objects that represent instance information (properties exposed by the individual instantiations of a thing of a given type).

The structure diagrams deal with modeling the objects representing the entities on the battlefield. Similar to the ideas of DEVS, the atomic entity is an object that has only attributes and methods assigned. Using such objects, composites can be built that combine "atomic" objects and potentially already existing composites. More diagrams are used to add more ideas to facilitate the work of object oriented software engineers, and many of them can be used by simulation engineers as well. Here is an overview of the structure diagrams.

- Class diagrams describe the entity types including their attribute relationships between the types.

- Object diagrams show views of the battlefield with instantiated objects, based on the types, for specific snapshots: how should the battlefield look when the simulation is executed?

- Composite structure diagrams describe the internal structure of composite objects, i.e. objects that have at least one object among their attributes. This allows the simulation engineer to treat the composite as one structure from the outside and describe how the composite works on the inside.

- Component diagrams describe how the system is split up into components, e. g. units, areas of interest, etc. These diagrams also show interdependences and interactions between these components.

- Package diagrams describe logical groupings and their interdependences and define software packages.
- Deployment diagrams describe the allocation of software to hardware components: what software package is executed on which hardware component.

Another diagram type has been added recently that is called a profile diagram. This profile diagram operates at the meta-model level and shows stereotypes: instead of defining a special object, such as tank or AFV, to conduct a certain operation, a stereotype like Direct Fire Weapon System is used to describe the needed actions. Each object that fulfills the requirements of the stereotype can be used in the operation in the respective role.

The most important three diagrams of the functional view of the systems are activity diagrams (how entities interact), state machine diagrams (how entities react), and use case diagrams (how the overall system behaves). These three core diagrams are supported by more detailed diagrams that often focus on software engineering details.

- Use case diagrams describe the interactions of actors with the main functionality of the simulation system in terms. These diagrams are often used to communicate what the simulation system does and why it does it with the users.
- Activity diagrams can be best described as workflows between the entities of the simulation system. They show the overall flow of control, including synchronization between activities, interim results, etc.
- State machine diagrams show how the objects or entities react to certain inputs they receive. While activities focus on the entity external communication state machine diagrams focus on the internal changes.
- Interaction diagrams use four subtypes to better describe the details of interactions within the system:
  - Interaction overview diagrams provide the big picture and build the frame for the other three diagram types to fill in the details.
  - Sequence diagrams visualize causal orders of an interaction and define the lifespan of a temporary object used within such interactions.
  - Timing diagrams focus on timing constraints of sequences to define the temporal orders.
  - Communication diagrams specify the interactions between entities that are part of the modeled sequence. They are very often used as system internal use cases, as they show several aspects of which entities come together how and when.

UML has been applied in a variety of communities and has become something like the *lingua franca* between modelers who specify software intensive systems. As such, the simulation engineer needs to be at least able to read and interpret the artifacts. The main disadvantage of UML, however, is the focus on object oriented software systems. While UML can easily be interpreted to support other systems as well, the software engineering roots often shine through.

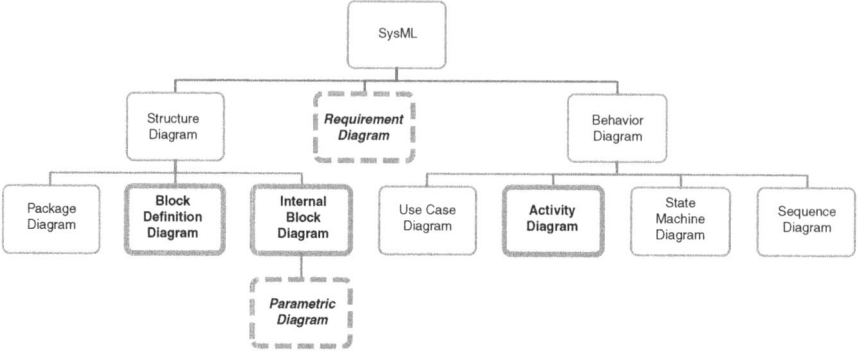

Figure 12.3 SysML diagram hierarchy.

These led to the development of SysML, also under the umbrella of the Object Management Group. SysML is a general modeling language for systems and gets more and more support from systems engineers (Friedenthal et al., 2008). There is a huge overlap between UML and SysML, but the focus is on the system, not on the software. As such, SysML started by stripping all software-specific diagrams from UML, modifying the other in case of need, and adding system-specific new diagrams where needed. The resulting diagram types are shown in Figure 12.3.

The diagrams with the standard frame are identical to the UML diagrams with the same name, those with the bold frame are modified, and those with the dashed-bold frame are new. In particular, the modifications are as follows.

- Requirement diagrams represent a visual modeling of requirements for the system, which are pivotal for systems engineering.
- Parametric diagrams show the relations between parameters for system components on all levels. They also provide the metrics to be used for system performance. While requirements define what is needed, parametrics define how this is measured and evaluated.
- Block definition diagrams are based on the class diagrams, but they use system blocks instead of classes. System block diagrams are well known by systems engineers and easy to map to class representations.
- Internal block diagrams are extending the UML composite structure diagrams by restrictions and extensions that govern system component interactions and interrelations.
- Activity diagrams are extended to allow modeling of continuous flows as often needed in systems. Another extension supports control and control operator modeling. Finally, logical operators are introduced to allow the use of Extended Functional Flow Block Diagrams (EFFBD) as often used in systems engineering.

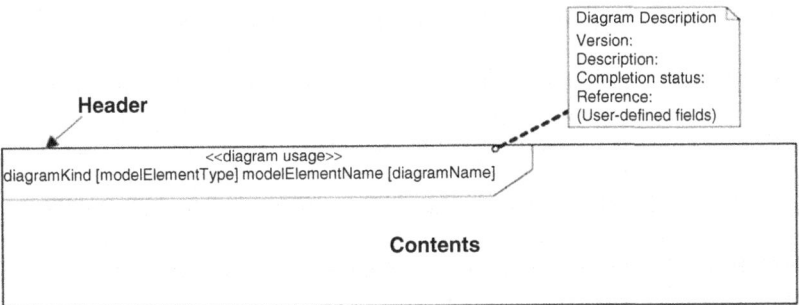

Figure 12.4   System diagram frame in SysML.

Another aspect new in SysML is the use of System Diagram Frames that are part of every artifact. They describe the meta-information needed to better understand what the diagram shows in the context of the whole system. The header specifies the version, what kind of diagram is used (a three-letter abbreviation referring to the types shown in Figure 12.3), what kind of model element is used (block, activity, interaction), if the diagram is complete or has elements elided, etc. Figure 12.4 shows the principle. Every diagram—no matter which content is displayed—always has the same header. In addition, there is also a systematic standardized numbering scheme that allows the engineer to understand with one number on which system level the artifact being looked at is located.

One of the better introductions to SysML has been prepared by Object Management Group for the International Council on Systems Engineering (INCOSE). This tutorial is freely distributed on the OMG SysML website.

The simulation engineer can use SysML specifications for the development of federations, but also many weapon systems and combat support systems are documented in SysML. A feature of particular interest is the traceability of system component functionality to requirements, which can be a significant help when simulation developers have to decide which details should be included in a model. If, for example, a component of a system hosts many of the functionalities driven by high priority requirements it may be wise to represent this component explicitly to allow for better damage evaluations in the light of original requirements.

The last standard supporting a simulation engineer in combat modeling to be explicitly mentioned in this section is DoDAF. A later chapter will deal with this topic in some more detail, but under a slightly different viewpoint.

DoDAF is currently published and applied in different versions. The most current version is DoDAF 2.0, but many organizations are still using DoDAF 1.5. We shall have a look at both versions to show the main difference. As all systems used within the US Department of Defense principally should be documented using the artifacts defined by DoDAF, it is very helpful for the simulation engineer to know the framework. Additional arguments for the use of DoDAF in the context of combat modeling and distributed simulation are compiled by Atkinson (2004).

DoDAF is rooted in earlier systems engineering approaches. The C4ISR Architecture Framework was created in response to the passage of the Clinger-Cohen Act and addressed in the 1995 Deputy Secretary of Defense directive that a DoD wide effort be undertaken to define and develop a better means and process for ensuring that C4ISR capabilities were interoperable and met the needs of the warfighter. The first version was published in June 1996, rapidly followed by version 2.0 in December 1997. The second version was the result of the continued development effort by the C4ISR Architecture Working Group and was mandated for all C4ISR architecture descriptions in a February 1998 memorandum by the Architecture Coordination Council, co-chaired by the Under Secretary of Defense for Acquisition and Technology (USD A&T), the Assistant Secretary of Defense for Command, Control, Communications, and Intelligence (ASD C3I), and the Command, Control, Communications, and Computer Systems Directorate, Joint Staff (J6).

In August 2003, the DoD Architecture Framework Version 1.0 was released. It restructured the C4ISR Framework v2.0 to offer guidance, product descriptions, and supplementary information in two volumes and a desk book. The objective was to broaden the applicability of architecture tenets and practices to all mission areas rather than just the C4ISR community. Therefore, the first version explicitly addressed usage, integrated architectures, DoD and federal policies, value of architectures, architecture measures, DoD decision support processes, development techniques, and analytical techniques.

In April 2007, the DoD Architecture Framework Version 1.5 was released as an evolution of the first version. It reflects and leverages the experience that the DoD components have gained in developing and using architecture descriptions. However, it was designed as a transitional version providing additional guidance on how to reflect net-centric concepts within architecture descriptions and including information on architecture data management and federating architectures through the DoD. The second version also incorporates the Core Architecture Data Model (CADM), a data model that consistently describes all modeled aspects as reflected in various artifacts in support of providing a repository. It also emphasizes the use of UML based artifacts to express the different views needed to describe the system.

Since May 2009, DoDAF Version 2.0 has been available. This version is defined as the overarching, comprehensive framework and conceptual model enabling the development of architectures to facilitate DoD managers at all levels to make key decisions more effectively through organized information sharing across DoD, Joint Capability Areas, Component, and Program boundaries. This new version provides an overarching set of architecture concepts, guidance, best practices, and methods to enable and facilitate architecture development. The focus is the fit-for-purpose concept which allows individual viewpoints to be defined that support special needs. To this end, the artifact focus has to shift from the different diagrams used in various views and that are harmonized based on a common repository to a common data and metadata model that drives different viewpoints, including user defined ones.

DoDAF 1.5 uses four principal categories of views to describe a system with its architecture artifacts.

- The general architecture, or all architecture view, is a very high level view of the system and what it does in the big picture. Consequently, general architecture products—the all views (AVs)—give a holistic view/high level description.

- The operational architecture view is a description of the tasks and activities, operational elements, and information flows required to accomplish or support a military operation. The products, summarized in the operational views (OVs), introduce operational activities to be conducted and performing operational elements and define the required capability (similar to the stereotype ideas described in UML).

- The systems architecture view is a description, including graphics, of systems and their interconnections providing for, or supporting, warfighting functions. The resulting system views (SVs) introduce systems and functions implementing the functionality (similar to the instantiation diagrams of UML –classes, composites, components, etc.).

- The technical architecture view is the minimal set of rules governing the arrangement, interaction, and interdependence of system parts or elements, whose purpose is to ensure that a conformant system satisfies a specified set of requirements. The technical views (TVs) are a set of rules and standards enabling the collaboration of systems and components of the present and the foreseeable future.

Figure 12.5 captures the dependences between operations, systems, and technical views: OVs provide processing and inter-nodal levels of information exchange requirements for the SVs, and the SVs provide the system associations to nodes, activities, need-lines, and requirements. Similarly, the TVs provide basic technology supportability and new capabilities for these levels of information exchange. The SVs define specific capabilities identified to satisfy

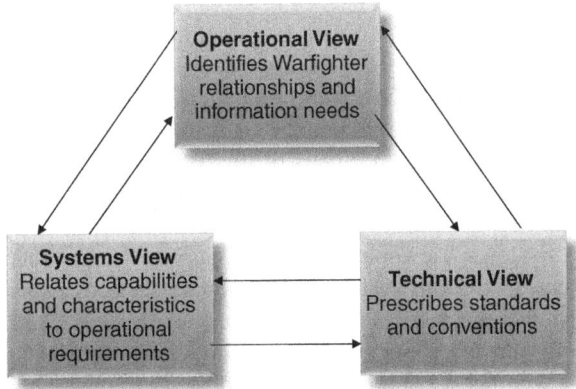

Figure 12.5 Views in DoDAF 1.5.

information exchange levels and other operational requirements, for which the TVs provide technical criteria governing interoperable implementation/ procurement of the selected system capabilities.

It is worthwhile having a closer look at the characteristics of OV products and SV products in some additional detail. The OVs describe the tasks and activities of concern to successfully perform a mission, the participating nodes, and the associated information exchanges. The resulting descriptions are useful for facilitating numerous actions and assessments across DoD. They are generally independent of organization or force structures and driven by doctrine. OVs are generally independent of technology. They can describe activities and information exchange requirements at any level of detail, but are implementation neutral.

In contrast, SVs are used to define the system to be implemented. For a domain, the SV shows how multiple systems link and interoperate and may describe the internal construction and operations of particular systems within the architecture. For the individual system, the SV includes the physical connection, location, and identification of key hardware and software; it may also include data stores, circuits, and networks and may specify system and component performance parameters.

Table 12.1 summarizes the various views that define DoDAF 1.5 artifacts, most of them expressible in UML. The original documents actually have individual views for systems and services that are identical to the diagrams used to describe them. The main difference is that a system provides functionality by well defined components. A service is well defined regarding the functionality as well, but it can be provided by many systems via service oriented architectures from everywhere within the net-centric environment. We therefore only list the systems views here.

DoDAF 2.0 extends the view categories significantly. This version also uses the term viewpoint to emphasize that the view itself should not be the product, but the data that are represented via each viewpoint. The CADM of an earlier version has been replaced by the DoDAF Meta-model (DM2) that supports the architecture content of the DoD core processes by design. While CADM was derived from the various views that needed to be aligned to avoid inconsistencies, DM2 was designed to capture the relationships needed to support qualitative and quantitative analysis. The resulting ability to relate architectural descriptions, based on a common architecture vocabulary as used by decision makers and warfighters as envisioned in the NATO COBP, allows better analysis across architectural descriptions. The price is a more complex data model underlying the viewpoints. As all core processes are in the focus of this version, the scope had to be moved from a single group of systems towards DoD as an enterprise.

As already foreseen in the last update, in DoDAF 2.0 the original systems view has been separated into Systems Viewpoints (SV) and Service Viewpoints (SvcV) to accommodate extension to both systems and software/services engineering practice (Figure 12.6). All the models of data (conceptual, logical, and physical) have been moved into the new Data and Information Viewpoint (DIV) category. The Operational Viewpoints (OV) now describes rules and constraints

Table 12.1   DoDAF 1.5 views (AV, OV, SV, and TV)

AV-1	Overview and summary information
AV-2	Integrated dictionary
OV-1	High-level operational concept graphics
OV-2	Operational node connectivity description
OV-3	Operational information exchange matrix
OV-4	Organizational relationship chart
OV-5	Operational activity model
OV-6 (a, b, c)	Operational rules, state transitions, event–trace description
OV-7	Logical data model
SV-1	System/Services interface description
SV-2	System/Services communications description
SV-3	System/Services matrix (System–System, System–Services, Services–Services)
SV-4 (a, b)	System/Services functionality description
SV-5 (a, b, c)	Operational activity traceability matrix (Systems Function, Systems, Services)
SV-6	System/Services data exchange matrix
SV-7	System/Services parameters matrix
SV-8	System/Services evolution description
SV-9	System/Services technology forecast
SV-10 (a, b, c)	System/Services rules, transition, event–trace description
SV-11	Physical data model (physical schema)
TV-1	Technical standards profile
TV-2	Technical standards forecast

for any function (business, intelligence, warfighting, etc). Technical Views (TV) have been updated to Standards Viewpoints (StdV) to describe business, commercial, and doctrinal standards in addition to technical standards. The All Viewpoints (AV) were not significantly modified. As new viewpoints to support enterprise-wide cross-system analysis and harmonized decisions, the Capability Viewpoints (CV) with focus on capability data in support of capability development and to standardize the capture of capability data were introduced. In addition, Project Viewpoints (PV) focus on the portfolio management information to explicitly tie architecture data to project efforts.

For examples the interested reader is referred to the original DoDAF documents that are freely distributed via the Internet. It is worth pointing out that NATO has the NATO Architecture Framework (NAF) and the UK has the Ministry of Defence Architecture Framework (MoDAF). In order to support a modeling method in support of these main frameworks, Object Management Group defined the Unified Profile for DoDAF/MODAF (UPDM) mainly using UML products as the common denominator. Other nations support similar efforts, such

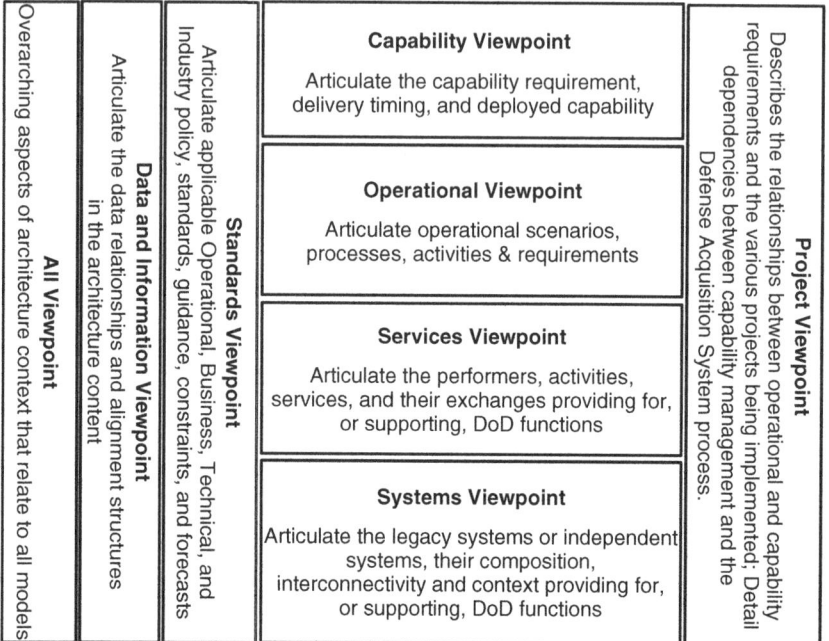

Figure 12.6 Viewpoints in DoDAF 2.0.

as the Department of National Defence and Canadian Forces Architecture Framework (DNDAF) of Canada.

The better the simulation engineer knows these modeling and architecture languages, the easier is communication in the international community as well as within the community of engineers with different professions that support an exercise, or maybe in real operations in which the simulation engineer's expertise is required to ensure success.

Semantic Web Standards

Tim Berners-Lee coined the term "semantic web." The semantic web is understood by the community as a web of data. While the hyperlinked documents of the Internet were merely designed to support humans, the idea of the semantic web is to add machine-readable and machine-interpretable metadata to the websites that allow machines to find data, make connections and relations between websites, etc.

Several methods and technologies are needed to make this vision a reality. The needed standards and tools as currently understood to build the semantic web stack are shown in Figure 12.7.

The set of standards that make up the lower left block of the stack (and that are written in bold font) allowed movement towards a new level of machine support and automatic reasoning in recent years that was never possible before.

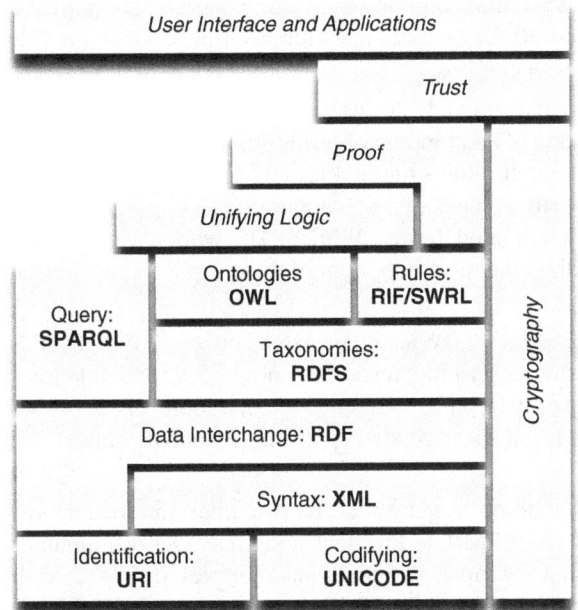

Figure 12.7 Semantic web stack.

In addition, technologies are continuously developed and exchanged within the semantic web community to foster agreement and improvement.

- The basis for general communication on the web is the use of a machine and vendor independent codification allowing the interpretation of signals as symbols. This is supported by the use of *Unicode* as the lowest building block.

- The second foundational leg is the use of *Uniform Resource Identifiers* (URI). Everything on the web becomes a resource and can be addressed as such in a standardized way.

- The *Extensible Markup Language* (XML) provides the elementary syntax for resources, which allows communication within a resource that can be searched for, extracted, etc. It is possible to provide and restrict the structure and content of elements through the use of an XML Schema. The elementary syntax can alternatively be provided by Turtle. Turtle allows descriptions in a compact and natural text form, with abbreviations for common usage patterns and datatypes. Turtle is a team submission for the World Wide Web Consortium (W3C) that is gaining support by practitioners.

- The *Resource Description Framework* (RDF) allows resources and their relations to be structured, similar to using a data model. Most of these frameworks use XML, but that is no longer necessary as alternatives exist. However, whenever data need to be exchanged in a consistent way, which includes the idea of transactions on one or several resources, support of a common RDF is needed. RDF organizes resources as triples, which provide the structure for higher operations.

- While RDF is more a data model, the use of an *RDF Schema* (RDFS) compares to the definition of a taxonomy of terms describing properties and classes of RDF resources. It allows hierarchies and complex relationships to be built using well-defined types of associations connecting well-defined structures tagged with common terms. While RDF works with general triples, RDFS standardizes the most important of these triples as recognized by the community.

- The *Web Ontology Language* (OWL) is an extension of RDFS. It adds more standardized terms describing resources, properties, and relations, such as cardinality, equality, typing, enumerations, and more. OWL allows reasoning over these extensions, if all supporting resources follow the standard accordingly.

- Queries about resources do not require that the resources are documented in RDFS and OWL. The SPARQL Protocol and RDF Query Language (SPARQL) allows search for triples, conjunctions of triples, disjunctions of triples, and optional patterns. As such, SPARQL works with RDF, RDFS, and OWL.

- The *Rule Interchange Format* (RIF) is not yet standardized but only recommended. It allows the communication of constraints within the semantic web stack to ensure consistency between the layers. An alternative for OWL based resources is the Semantic Web Rule Language (SWRL) that utilizes description logic subsets of OWL (the dialect OWL-DL).

The use of OWL should not be decided without proper consideration of the issues of decidability and computational complexity. Not all dialects of OWL are decidable, and even if they are decidable, their application can lead to non-polynomial computing time. Several papers deal with these challenges. The simulation engineer should focus on OWL-DL (which is decidable) and preferably even the profiles defined for OWL 2: OWL 2 EL, OWL 2 QL, and OWL 2 RL. For more information, a good review of available semantic web tools and standards and their computational constraints is given by Hitzler et al. (2009).

Building a unifying logic that supports all alternatives and that becomes the foundation of solid proofs is an ongoing effort. Together with cryptography (which was originally envisioned to be based on XML as well, but this idea is no longer supported generally) the proof layer becomes the basis for trusted information providers that can support the applications. The importance of these ideas and protocols becomes immediately obvious when we go back to the data discussions in the NATO COBP, the need to initialize simulation from trusted sources, the need for common initialization, and more. If the infrastructure that supports the distributed simulation is based on the semantic web stack, many of the tasks of the simulation engineer are taken care of. Vice versa, the simulation engineer can use personal experiences in distributed simulation to enhance the ideas accordingly. Tolk (2006) contributed to enhancing the vision by actively pushing the idea of semantic web based modeling and simulation for the next generation of the web. An interesting perspective on future directions for semantic systems is given by Sowa (2011).

To apply these ideas in practice, the simulation engineer has two immediate options: (a) using the concept to better describe modeling and simulation resources using metadata that are stored in registries, and (b) using web services to access modeling and simulation resources that are available. This connects directly back to the four tasks in support of distributed simulation: to identify applicable solutions, to select the best options, to compose the best solution, and to orchestrate the execution.

The community has already started to work on identifying metadata that can be used to describe modeling and simulation resources, which includes data sources, simulation systems, and more (US DoD M&S CO, 2008). The approach is summarized by Gustavson et al. (2008). The purpose of this specification is to standardize on the set of metadata used to describe resources in Modeling and Simulation Resource Repository (MSRR) nodes and similar applications, and to ensure that the product metadata templates will align with the DoD Discovery Metadata Specification (DDMS) as part of the Global Information Grid (GIG)/Net-Centric Data Strategy. The idea was generalized in support of a Modeling and Simulation Information System (MSIS) for the US DoD. Similar approaches are conducted in other countries as well. Several MSRRs are already in use, and there is potential to merge these different approaches based on a common set of metadata. Current efforts, application of the metadata, tools, and the guiding vision were recently presented by Gustavson et al. (2009).

To access and utilize modeling and simulation services, web services are the method of choice in most cases. In general, web services are services that can be accessed via the Internet based on the specifications of the service published as well. The uses for military combat modeling and distributed simulation applications were described by Tolk et al. (2006), and the general ideas were covered by the already mentioned XMSF projects (Brutzman et al., 2002; Blais et al., 2005).

There are several different categories of web services. The earlier versions were based on a combination of using XML to define the data to be exchanged, the Simple Object Access Protocol (SOAP) to access the service, the Web Service Description Language (WSDL)—which is based on XML—to describe the access characteristics of the service, and Universal Description Discovery and Integration (UDDI), also based on XML, to support publication and discovery. The general idea was that a web service provider described the service characteristics in WSDL and posted them to the UDDI server. Whoever needed to find such a service downloaded the descriptions from the UDDI and looked for applicable solutions based on the WSDL description. If a service was found and accessed, the searcher used XML to provide the needed data and accessed the service via SOAP. The disadvantage was that this procedure, from the logic very close to a remote procedure call (RPC), required a lot of knowledge regarding the service interface, the information exchange requirements, etc.

To overcome these constraints, Representational State Transfer (REST) methods were applied to define RESTful services. The idea of REST is to generalize the interface as much as possible and instead put the needed information into the data exchanged, not the application programming interface. This allows component interactions to be made scalable, deploying them independently, reducing latency, and enforcing security. As REST allows building wrappers around existing systems, such encapsulation of legacy systems supports migration towards this new infrastructure. To allow for these approaches, RESTful services expose all services as resources that are addressed via URI. They are connected via uniform connectors or channels that are used to exchange messages that comprise all information in the form of metadata and data that the receiving service needs to act on the message. This allows every RESTful service to provide a uniform interface that addresses the same fundamentals. To these fundamentals belong the following:

- within the messages, resources are identified by URIs and can be manipulated by the receiver;
- each message includes enough information to describe how to process the message;
- the receiver uses hypermedia to make state transitions; and
- all transactions are handed by the providing server.

RESTful services have been successfully applied in simulation infrastructures. Examples are given by, among others, Al-Zoubi and Wainer (2009).

The simulation engineer should observe the developments in the domain of distributed computing. These tools and standards can support tasks and may even develop to the point that they can replace the specific simulation interoperability standards given in the last section of this chapter. In any case they can support these efforts of distributed simulation significantly.

Of additional and increasing interest is the use of standards supporting the gaming industry. This topic will be dealt with in more detail in the context of Chapter 17 of this book.

MODELING AND SIMULATION INTEROPERABILITY STANDARDS

This last section gives an overview of the two IEEE Modeling and Simulation Interoperability Standards: IEEE 1278 Distributed Interactive Simulation (DIS) and IEEE 1516 High Level Architecture (HLA). A more technology-oriented overview is given by Tolk (2010). These standards are dealt with in more detail under different viewpoints also in other chapters of this book, including the history as well as their role for multiresolution challenges.

The detail given in this section cannot replace a comprehensive introduction. Readers interested in a comprehensive guide to DIS are referred to Neyland (1997). For the HLA, Kuhl et al. (2000) is a good place to start, but there are also very good tutorials available during interoperability focused workshops, such as the Simulation Interoperability Workshops of SISO.

Another effort that needs to be mentioned is the Test and Training Enabling Architecture (TENA). Although not being a *de jure* standard, the ideas supported by TENA are pivotal to interoperable simulation systems. TENA is described in detail in Chapter 20.

IEEE 1278 Distributed Interactive Simulation

In the 1980s, the Defense Advanced Research Project Agency (DARPA) and the US Army initiated the development of a simulation network (SIMNET) that allowed coupling former stand-alone simulators to support better training. In this prototype, SIMNET showed how to combine individual tank simulators of the Combined Arms Tactical Training System (CATT) to enable tank crews to operate side-by-side in a common synthetic battle space. The individual simulators represented weapon systems on this common virtual battlefield that had a well defined set of actions and interactions: tanks could move, observe, shoot at each other, exchange radio communication, etc. Individual activities led to status changes that were communicated via status reports. Interactions were communicated via messages.

If two tanks engaged in a duel, the order of activities and the data to be exchanged between these entities were well defined. A shooter decides to engage a victim. The shooter moves the weapon system—and potentially platform components like turret and cannon—into the best direction, always providing status updates, so that other simulators could update their visualization showing that the tank/turret/cannon is moving. The shooter shoots the victim. These data were sent to everyone as well. All observing systems could visualize the shooting (smoke, flash, etc.). The victim also received information on the ammunition shot, velocity, angle, etc. The victim computed the result of this engagement—like catastrophic kill, movement kill, firepower kill, etc.—and communicated the result. All observers, including the shooter, updated their visualization of the victim (like being on fire, smoking, or no effect beside the impact explosion). Based on his assessment of the effect, the shooter could reengage or continue with a new task. The tasks of who is doing what based on what data were well understood by those simulators embedded as individual independent entities in the common battle space.

As the set of information exchange specifications could be well defined, this resulted in the idea to standardize these messages, which led to the IEEE1278 Distributed Interactive Simulation (DIS) standard: the Protocol Data Units (PDUs) captured syntactically and semantically all possible actions and interactions based on the idea that individual simulators represent individual weapon platforms. Only later, instead of individual platforms also groups and aggregates (like platoons or companies) were accepted as receivers and producers of such PDUs, but these groups were understood as individual entities in the battle space as well.

Following the principles learned in SIMNET, the DIS community defined and standardized PDUs for all sorts of possible events that could happen during such a military training. Whenever a preconceived event happens—such as one tank firing at another, two systems colliding, artillery ammunition being used to shoot into a special area, a report being transmitted using radio, a jammer being used to suppress the use of communication or detection devices, and more—the appropriate PDU is selected from the list of available PDUs and used to transmit the standardized information describing this event. Within a PDU, syntax and semantics are merged into the information exchange specification. Table 12.2 shows the principal structure of a PDU.

As the DIS infrastructure is usually a ring or a bus without any other control besides transporting the PDUs, each simulator always receives all PDUs and has to decide using the **Receiving Entity ID** if a particular PDU concerns him or not. The **PDU Type** defined how the following fields had to be interpreted. Table 12.3 shows what types are standardized.

DIS is primarily used for military simulators that represent weapon systems. The objects that DIS can represent are categorized in IEEE 1278 as platforms, ammunition, life forms, environmental cultural features, supplies, radios, sensors, and others. DIS also supports the notion of an "expendable" object that allows user-specific representations, but this object is not standardized.

The general characteristics of DIS are the absence of any central management; all simulations remain autonomous and are just interconnected by information exchange via PDUs; each simulator has an autonomous perception of the situation; cause–effect responsibilities are distributed for the PDUs to minimize data traffic. There is no time management or data distribution management. The PDUs are transmitted in a ring or on a bus and each simulator uses PDUs that are directed at one of his entities.

Table 12.2   Structure of a PDU

PDU header	Protocol version	8 bit enumeration
	Exercise ID	8 bit unsigned integer
	PDU type	8 bit enumeration
	Protocol family	8 bit enumeration
	Timestamp	32 bit unsigned integer
	Length	16 bit unsigned integer
	Padding	16 bit unsigned integer
Originally entity ID	Site	16 bit unsigned integer
	Application	16 bit unsigned integer
	Entity	16 bit unsigned integer
Receiving entity ID	Site	16 bit unsigned integer
	Application	16 bit unsigned integer
	Entity	16 bit unsigned integer

Table 12.3 DIS PDU types

Value	Description	Value	Description
0	Other	129	Announce object
1	Entity state	130	Delete object
2	Fire	131	Describe application
3	Detonation	132	Describe event
4	Collision	133	Describe object
5	Service request	134	Request event
6	Resupply offer	135	Request object
7	Resupply received	140	Time space position indicator—FI
8	Resupply cancel	141	Appearance—FI
9	Repair complete	142	Articulated parts—FI
10	Repair response	143	Fire—FI
11	Create entity	144	Detonation—FI
12	Remove entity	150	Point object state
13	Start/resume	151	Linear object state
14	Stop/freeze	152	Areal object state
15	Acknowledge	153	Environment
16	Action request	155	Transfer control request
17	Action response	156	Transfer control
18	Data query	157	Transfer control acknowledge
19	Set data	160	Intercom control
20	Data	161	Intercom signal
21	Event report	170	Aggregate
22	Comment		
23	Electromagnetic emission		
24	Designator		
25	Transmitter		
26	Signal		
27	Receiver		

DIS is still successfully used and supported by a large user community. As mentioned earlier in this chapter there are still standardization efforts going on under the umbrella of SISO. Furthermore, the Realtime-Platform-Reference Federation Object Model (RPR-FOM) activities described in the following section migrated many DIS solutions into the new IEEE 1516 world. Also, DIS-HLA Gateways are often applied to integrate the functionality provided by DIS conform simulation systems into HLA federations.

IEEE 1516 High Level Architecture

Although Chapter 19 will deal with the high level architecture in more detail, a short summary will be given to allow for a better presentation of the follow-on ideas in the next section. HLA was developed to unify various distributed simulation approaches within the US DoD and has been adopted by NATO as well. Under the lead of the US Defense Modeling and Simulation Office (DMSO), several prototypes were developed and distributed in several versions. The last version that was submitted to IEEE for standardization was the HLA 1.3 NG. Many US companies are still using this version to this day, as many tools were freely distributed by DMSO. An international standardization group under IEEE improved this version by bringing it up-to-date regarding supported standards and generalizing some of the ideas. For example, the Backus–Naur nomenclature used in 1.3 NG was replaced by XML. Furthermore, hard-coded enumerations were replaced by reconfigurable solutions and configuration tables. The result was the IEEE 1516—2000 HLA, which became the main version implemented in Europe and parts of Asia and Australia. Although most differences between the two versions are of an editorial nature, gateways are needed to connect federations that are based on different versions.

Like every IEEE standard, HLA was reviewed and updated after 10 years. Under the title HLA Evolved, this work was conducted under the lead of SISO. The result was the updated version of the standard: IEEE 1516—2010 HLA. The main difference between this new version and the older ones is that the information exchange model became modular, so that part of the information exchange agreement can be changed during runtime. Furthermore, dynamic link capabilities, extended XML support, increased fault tolerance, and web-based standards were integrated into the concepts supported by HLA-based federations. At the point in time that this book was written, only some early adopters used the 1516—2010, but several supporters and organizations had already announced the decision to go straight to this new standard version when they update their federations.

The objective of defining the HLA was to define a general purpose architecture for distributed computer simulation systems. It defines a federation made up out of federates, which are the simulation systems, and the connection middleware that allows the information exchange between the simulation systems. To this end, three components are defined by the technical parts of the standards:

- the *HLA Rules* describe the general principles defining how the federation and the participating federates work together, i.e. how responsibilities for updates are shared, who does what when, etc.;
- the *Interface Specification* between the connection middleware—which is called Runtime Infrastructure (RTI)—and a federate, which provides an application interface in both directions: what services provided by the RTI the simulation system can call, and what services the RTI will call in order to request something from the simulation system;
- the *Object Model Template* (OMT) that defines the structure of the information exchange between the federates via the RTI.

In order to make sure that (1) all information required is provided to the right federate; (2) only the information required is provided to the right federate;

and (3) the information is provided at the correct time, six management areas are provided for effectiveness, efficiency, and timeliness. The six RTI management areas are defined as follows.

- The purpose of *Federation Management* is to determine the federation. Federates join and leave the federation using the functions defined in this group.

- The purpose of *Declaration Management* is to identify which federate can publish and/or subscribe to which information exchange elements. This defines the type of information that can be shared.

- The purpose of *Object Management* is managing the instances of shareable objects that actually are shared in the federation. Sending and receiving and updating belong in this group.

- The purpose of *Data Distribution Management* is to ensure the efficiency of information exchange. By adding additional filters this group ensures that only data of interest are broadcast.

- The purpose of *Time Management* is the synchronization of federates.

- The purpose of *Ownership Management* is to enable the transfer of responsibility for instances or attributes between federates.

The Object Model Template (OMT) defines what information can be exchanged. In principle, there are two categories of information that can be exchanged, which are persistent objects and transient interactions. From the structure, both categories are very similar: objects have attributes, interactions have parameters. Both use types that have to be defined in the OMT tables; both expose characteristics used by the RTI services. Both are defined in tree-based object structures providing single inheritance from classes to sub-classes. The main difference is that interactions are distributed just once while objects are created; they can be updated, they can change ownership, and they can be destroyed. All interactions and objects including parameters and attributes and other definitions build the *Federation Object Model* (FOM).

The information exchange within the federation is done in orchestration of RTI services with the OMT definitions. The information provided in the OMT defines *what information can be exchanged* between the participating federates; the services provided by the RTI define *how the information can be exchanged*.

Of particular interest is also the time management in HLA. As HLA allows the federation of simulations that support different models of time—including those that do not represent time in their model at all—temporal anomalies would occur without time management. Within HLA, these anomalies are eliminated by assigning the logical simulation time of an event to the information exchange element—which can be an interaction or the creation, update or destruction of an object—as a time stamp. The RTI uses these time stamps to deliver events in time stamp order. Besides providing time stamp ordered events, a federate can also request to progress in time in case it is idle too long. This avoids deadlock situations where no messages are in the system that can be delivered to drive the federation forward. Fujimoto (1998) describes details.

Of additional interest to the simulation engineer can also be the Management Object Model (MOM), which is part of the standard of the OMT. However, while simulation data build the FOM, management data and extensions thereof belong in the MOM. Some examples of management data needed to be known by the RTI to enable their services are:

- How many federates joined the federation?
- What time management measures do the participating federates use?
  - What instances of simulated entities are registered?
  - In which regions are they registered?
  - Are there any subscribers for these instances?
- Did any late arrival of additional federates occur?
  - What is the minimal logical time that can be assigned?
  - What instances are already registered and updated?
  - What are the latest values of these attributes?

The MOM is the general facility that manages an HLA federation. The core data are standardized, but they can be extended to support user required additional functionality. The standard functionality is provided by the RTI, but extensions can be used by "management federates" as well.

Finally, additional federation agreements are needed to ensure coping with advanced distributed simulation challenges, like defining who is responsible to conduct operations that are in the realm of more than one federate, or how to resolve multiresolution mappings within a federation. These federation agreements are often provided in the form of websites for the developers and engineers and are a pivotal part of making a federation execution work smoothly. The following activities are often captured and specified for all participating federates in these agreements:

- orchestration of publication and subscription responsibilities, including change of ownership of entities,
- data management and data distribution,
- time management,
- synchronization points, including save/restore strategies,
- initialization and start-up procedures,
- use of supporting databases, in particular authoritative data sources, and
- security procedures.

The FEAT mentioned under the SISO standard activities earlier in this chapter attempts to capture some of these agreements in a common format as a template.

RPR-FOM and GRIM

When HLA was introduced to the combat modeling and simulation community, most distributed simulation systems supported DIS in a real-time environment. Many systems had just been modified to support working not only as stand-alone simulators but as part of a common virtual battle space. In particular the community of virtual training simulators saw no real need to conduct another potentially expensive conversion to support a new simulation interoperability protocol.

In order to facilitate the integration of these simulators without expensive conversions of the systems, two approaches were used: the use of DIS-HLA Gateways as discussed by Steel (2000), and the development of a technical standard in the form of the Realtime-Platform-Reference Federated Object Model (RPR-FOM) and the supporting Guidance, Rationale, and Interoperability Modalities for the RPR-FOM (GRIM). We shall focus on the second approach in this section.

The RPR-FOM and GRIM were developed under the leadership of SISO with the goal to implement the DIS PDU structures within HLA objects and attributes and interaction and parameters as well as to provide an intelligent translation of the concepts used in DIS to an HLA environment. While the RPR-FOM defines the information exchange means needed by DIS simulation systems, the GRIM documents the guidelines on how to use them most efficiently. As the DIS standards evolve continuously, so do the RPR-FOM and the GRIM as well. As mentioned earlier in this chapter, RPR-FOM and GRIM are SISO standards in version 1.0, but versions 2.0 and 3.0 supporting new developments and agreements in DIS are being worked on by standardization groups already.

The main idea is that the RPR-FOM maps PDUs into HLA object and interaction classes. In general, individual PDU fields are mapped into corresponding class attributes or parameters. An individual PDU may be mapped across one or more HLA object or interaction class. This change in structure is designed to take advantage of the RTI services declaration management and object management. The RTI is able to send any subset of a class attributes or parameters. This capability can be used to limit network traffic in two ways, namely reducing the transmission of unchanged data, and providing delivery only to federates which have expressed interest.

The resulting object structure is shown in Figure 12.8. The structure was designed to maximize the use of declaration and object management. To fully support management and filtering, federates will publish all objects at the leaf nodes of the RPR FOM. If a RPR FOM leaf node is subclassed, then the federate may publish objects at the newly created leaf nodes. A leaf node is defined as the lowest level available in the object class hierarchy table, i.e. there are no additional subclasses. In contrast, class subscription should be used at the highest level—farthest from the leaf nodes—that supports all of the attributes and DM filtering required by the receiving federate.

RPR-FOM Version 1.0 supported the IEEE 1278.1–1995 PDU definitions and functionality. It was build on OMT as defined by HLA 1.3 NG. Many

Class 1	Class 2	Class 3	Class 4
BaseEntity	PhysicalEntity	Platform	Aircraft
			AmphibiousVehicle
			GroundVehicle
			Spacecraft
			SurfaceVessel
			SubmersibleVessel
			MultiDomainPlatform
		Lifeform	Human
			NonHuman
		Sensor	
		Radio	
		Munition	
		CulturalFeature	
		Expendables	
		Supplies	
	EnvironmentalEntity		
EmbeddedSystem	Designator		
	EmitterSystem		
	RadioReceiver		
	RadioTransmitter		
EmitterBeam	RadarBeam		
	JammerBeam		

Figure 12.8   RPR-FOM classes.

other reference FOM proposals are rooted in this effort. RPR-FOM Version 2.0 incorporates the IEEE 1278.1a-1998 functionality and updates the OMT to IEEE 1516–2000. However, the standard was not officially finalized. In parallel, the group is already working on Version 3.0 which includes aggregated entities as well as additional enumerations and functionality.

Each RPR-FOM is accompanied by a GRIM that captures common guidelines on how to best use the RPR-FOM in orchestration with RTI services to maximize the use and interoperability. The GRIM addresses many aspects that have to be captured in the federation agreements as well.

Applying the Levels of Conceptual Interoperability Model to Select and Align Simulation Standards

In Chapter 11, we introduced the Levels of Conceptual Interoperability Model (LCIM) developed to support gaining a better understanding of the theoretical underpinnings of interoperation between two federate simulation systems. As discussed in Tolk (2006), this approach has been proven useful to better understand what needs to be addressed by modeling and simulation interoperability standards as well. Using the definitions for integratability, interoperability, and composability provided by Page, Briggs, and Tufarolo (2004), we can use the LCIM to describe what simulation standard can contribute to support a common infrastructure, interoperable implementations, and composable models. In order to achieve successful interoperation of solutions, all three aspects are necessary. Successful standards for interoperable solutions must therefore address all three categories, and it must be done in a holistic way. The LCIM defines seven layers of interoperability to address distinguishable issues. In order to accomplish a working solution, all levels need to be addressed in an orchestrated way by selected standards, very similar to the layered cake developed for the semantic web, as discussed earlier in this chapter.

Using the definitions for the seven levels of the LCIM, they can be applied to the selection and alignment of simulation standards as follows:

- If simulation systems are stand-alone applications with no interconnection, there is obviously *no interoperability*. No standards are needed.

- The *technical layer* deals with infrastructure and network challenges, enabling systems to exchange carriers of information. This is the domain of integratability; a communication protocol to exchange signals exists. All technical standards that support the exchange of bits and bytes in a common infrastructure are applicable, from Internet-based connections to the integration of radio-based digital communication devices or data collection devices in test facilities.

- The *syntactic layer* deals with challenges to interpret and structure the information to form symbols within protocols. This layer belongs to the domain of interoperability; common symbols—like the use of Unicode or ANSI code—are identified. This is where the layer cake for the Semantic web begins.

- The *semantic layer* provides a common understanding of the information exchange. On this level, the pieces of information that can be composed to objects, messages, and other higher structures are identified. This level also support interoperability: it introduces common terms to tag structures that represent tags that are used to name functions, variables, and constants. Standards have to allow to address entities and their properties as well as relations to other entities.

- The *pragmatic layer* recognizes the patterns in which data are organized for the information exchange, which are in particular the inputs and outputs of procedures and methods to be called. This is the context in which data are exchanged as applicable information. These groups are often referred to as (business) objects. This is the highest level still supporting the realm interoperability; the relations between functions and there input and output parameters are captured. The standards have to recognize and define which semantic entities are group together under which operational contexts to be used as input parameters or are produced as output parameters. The context may include the intent of this information exchange as well.

- The *dynamic layer* recognizes various system states, including the possibility for agile and adaptive systems. The same business object exchanged with different systems can trigger very different reactions. It is also possible that the same information sent to the same system at different times can trigger different responses. This level is the first level in the realm of composability, as it requires the alignment of assumptions and constraints.

- Finally, general assumptions, constraints, and simplifications need to be captured. This happens in the *conceptual layer*. As recognized in the description of the LCIM in the last chapter, this layer is elusive and needs to be addressed explicitly in metadata.

Tolk (2010) shows how to use the LCIM in descriptive as well as prescriptive roles when evaluating the support of proposed solutions in support of modeling and simulation interoperability. The LCIM can therefore be used to guide the need for metadata that allow better support by intelligent software, as described by Tolk and Diallo (2010). In any case, it can help to guide the simulation engineer when addressing the various challenges of selecting the right standards to best support his federation tasks of identifying applicable solutions, select the best solution subset for his task, compose these solution into a supporting system, and orchestrate their execution to solve the customers problem.

REFERENCES

Al-Zoubi K and Wainer G (2009). Performing distributed simulation with RESTful web-services approach. In: *Proceedings of the Winter Simulation Conference, Austin, TX*, IEEE Press, Pistcataway, NJ, pp. 1323–1334.

Atkinson K (2004). Modeling and simulation foundation for capabilities based planning. In: *Proceedings Spring Simulation Interoperability Workshop*. http://www.sisostds.org.

Blais CL, Brutzman D, Drake D, Moen DM, Morse KL, Pullen JM and Tolk A (2005). *Extensible Modeling and Simulation Framework (XMSF) 2004 Project Summary Report. Final Report NPS-MV-05-002*. Naval Postgraduate School, Monterey, CA.

Brutzman D, Zyda M, Pullen JM and Morse KL (2002). *Extensible Modeling and Simulation Framework (XMSF) Challenges for Web-Based Modeling and Simulation. Workshop Report*. Naval Postgraduate School, Monterey, CA.

Fujimoto RM (1998). Time management in the high level architecture. *Simulation* **71**, 388–400.

Friedenthal S, Moore A and Steiner R (2008). *A Practical Guide to SysML:, The Systems Modeling Language*. Morgan Kaufmann, Burlington, MA.

Gustavson P, Nikolai A and Scrudder R (2008). Maximizing discovery of M&S resources—an inside look at the M&S COI discovery metadata specification. In: *Proceedings of the Fall Simulation Interoperability Workshop*, Orlando, FL. Accessed at www.sisostds.org.

Gustavson P, Nikolai A, Scrudder R, Blais CL and Daehler-Wilking R (2009). Discovery and reuse of modeling and simulation assets. *Modeling and Simulation Information Analysis Center (MSIAC) Journal* **4**, 11–20.

Hitzler P, Krötzsch M and Rudolph S (2009). *Foundations of Semantic Web Technologies*. Chapman & Hall/CRC, Boca Raton, FL.

Institute of Electrical and Electronics Engineers. *IEEE 1278 Standard for Distributed Interactive Simulation*. IEEE Publication, Washington, DC.

Institute of Electrical and Electronics Engineers. *IEEE 1516 Standard for Modeling and Simulation High Level Architecture*. IEEE Publication, Washington, DC.

Kuhl F, Dahmann J and Weatherly R (2000) *Creating Computer Simulation Systems: An Introduction to the High Level Architecture*. Prentice Hall PTR Upper Saddle River, NJ.

Mittal S (2010). Agile net-centric systems using DEVS unified process. In: *Intelligence-based Systems Engineering*, edited by Andreas Tolk and Lakhmi Jain. ISRL 10, Springer, Berlin, pp. 159–199.

Neyland DL (1997). *Virtual Combat: A Guide to Distributed Interactive Simulation*. Stackpole Books, Mechanicsburg, PA.

Object Management Group (OMG) (2011). http://www.omg.org/ (last accessed June 2011).

Page EH, Briggs R and Tufarolo JA (2004). Toward a family of maturity models for the simulation interconnection problem. In: *Proceedings of the Simulation Interoperability Workshop, Arlington, VA*. http://www.sisostds.org.

SISO (2008). *Balloted Products Development and Support Process. Administrative Document SISO-ADM-003-2008*. SISO, Orlando, FL.

SISO (2010). *The SISO Vision. Administrative Document SISO-ADM-004-2010*. SISO, Orlando, FL.

SISO (2011). http://www.sisostds.org (last accessed June 2011).

Sowa JF (2011). Future directions for semantic systems. In: *Intelligence-based Systems Engineering*, edited by A Tolk and LC Jain. ISRL 10, Springer-Verlag, Berlin.

Steel J (2000). The use of DIS and HLA for real-time, virtual simulation—a discussion. In: *Proceedings of the RTO NMSG Conference on "The Second NATO Modelling and Simulation Conference"*, NATO Report RTO MP-071, Shrivenham, UK.

Strassburger S, Schulze T and Fujimoto R (2008). Future trends in distributed simulation and distributed virtual environments: results of a peer study. In: *Proceedings of the Winter Simulation Conference, Miami, FL*. IEEE Press, Piscataway, NJ, pp. 777–785.

Tolk A (2006). What comes after the semantic web, PADS implications for the dynamic web. In: *Proceedings of the 20th ACM/IEEE/SCS Workshop on Principles of Advanced and Distributed Simulation, Singapore*. IEEE CS Press, Orlando, FL, pp. 55–62.

Tolk A (2010). Interoperability and composability. In: *Modeling and Simulation Fundamentals—Theoretical Underpinnings and Practical Domains*, edited by JA Sokolowski and CM Banks. John Wiley, New York, NY, pp. 403–433.

Tolk A, Diallo SY, Turnitsa CD and Winters LS (2006). Composable M&S web services for net-centric applications. *J Defense Model Simulation* **3**, 27–44.

Tolk A and Diallo SY (2010). Using a formal approach to simulation interoperability to specify languages for ambassador agents. In: *Proceedings of Winter Simulation Conference*. IEEE Press, Piscataway, NJ, pp. 359–370.

US Department of Defense (2007). *DoD Architecture Framework Version 1.5. Volume I: Definitions and Guidelines; Volume II: Product Descriptions; Volume III: Architecture Data Description*. US DoD Chief Information Officer, Washington, DC.

US Department of Defense (2009). *DoD Architecture Framework Version 2.0. Volume I: Introduction, Overview, and Concepts: Manager's Guide; Volume II: Architectural Data and Models: Architect's Guide; Volume III: DoDAF Meta-model: Physical Exchange Specification Developer's Guide*. US DoD Chief Information Officer, Washington, DC.

US Department of Defense Modeling and Simulation Coordination Office (US DoD M&S CO) (2008). *Modeling and Simulation (M&S) Community of Interest (COI) Discovery Metadata Specification (MSC-DMS) Version 1.1*. Alexandria, VA.

Wainer GA (2009). *Discrete-Event Modeling and Simulation: A Practitioner's Approach*. CRC Taylor & Francis, Boca Raton, FL.

Wainer GA, Al-Zoubi K, Mittal S, Risco Martín JL, Sarjoughian H and Zeigler BP (2010). Standardizing DEVS simulation middleware. In: *Discrete-Event Modeling and Simulation: Theory and Applications*, edited by G Wainer and P Mosterman. CRC Press, Taylor and Francis, Boca Raton, FL, pp. 459–493.

Zeigler BP, Praehofer H and Kim TG (2000). *Theory of Modeling and Simulation* (2nd edn). Academic Press, San Diego, CA.

Chapter 13

Modeling and Simulation Development and Preparation Processes

Andreas Tolk

MODELING AND SIMULATION PROCESSES FOR PROBLEM SOLVING

The last two chapters showed that current simulation interoperability standards—also likely to be the case for future standards—do not seamlessly cover all challenge areas of advanced distributed simulation. There will always be details that have to be taken into account by the simulation engineer through organizational means and implemented project specifics based on respective decisions. Processes help to guide the simulation engineer ensuring that all challenge fields are addressed. We already dealt with the NATO Code of Best Practice (COBP) described in Chapter 3. This is already a process in place that can and should help the simulation engineer.

In this chapter, we shall look at two more simulation-specific processes, namely the general *Problem Solving Process* as defined in the Verification, Validation, and Accreditation (VV&A) Recommended Practices Guide (RPG), published by the US Modeling and Simulation Coordination Office (2006), and guidelines defining *Federation Development Processes*. Many of these processes are copyrighted by IEEE, so that we cannot use the original standard documents. However, where possible, openly accessible papers are referenced supporting more information. In addition, the Simulation Interoperability

Engineering Principles of Combat Modeling and Distributed Simulation, First Edition.
Edited by Andreas Tolk.
© 2012 John Wiley & Sons, Inc. Published 2012 by John Wiley & Sons, Inc.

Standards Organization (SISO) provides access to the core documentation in the process of standard developments and evaluations.

In addition to the topics dealt with in the following sections, Chapter 30 documents results of recent research on system engineering principles and system of systems engineering in support of federation development and the integration of simulation solutions.

THE PROBLEM SOLVING PROCESS

The US Modeling and Simulation Coordination Office (MSCO) succeeded the US Defense Modeling and Simulation Office (DMSO) as the administrative organization that coordinates the modeling and simulation-(M&S) related activities within the US Department of Defense. DMSO supported the development of the VV&A RPG that was published in 2006 and is still frequently used. As the document is freely available, many other organizations are using the provided recommended practices as well. Of particular interest in this chapter is the general problem solving process, shown in Figure 13.1. This general process defines the processes related to the application of M&S based solutions that are applicable to contribute to solving the problem of the customer.

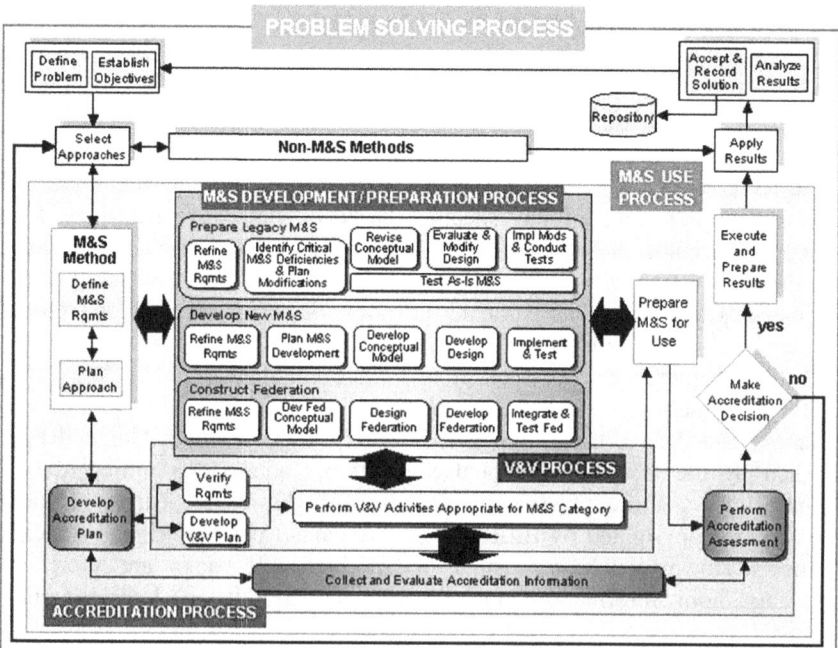

Figure 13.1   Problem solving process (VV&A RPG).

As with the NATO COBP, everything starts with a problem that needs to be solved. In every case, the problem solving process starts with the definition of the problem (what problem needs to be solved) and establishing the objectives (how the problem will be solved). These two steps go hand-in-hand with problem definition and the solution strategy, as we learned in Chapter 3. They build the basis for deciding and selecting possible approaches. As with the NATO COBP, the use of simulation is only one option. However, if simulation is part of the solution, the process defined here requires much more detail.

To select the appropriate M&S method, the M&S requirements have to first be derived from the established objectives. Ideally, each M&S requirement can be traced back to original requirements, if possible using machine readable repositories supporting this task. The simulation engineer uses these requirements to define the best M&S approach. These decisions do not only build the basis for the *M&S development/preparation process*, they also become the foundation for the *validation, verification, and accreditation processes* that will be dealt with in a later chapter in detail. There are three process categories that are possible, which are not mutually exclusive. Actually, mixed forms applying all three tracks are more the rule than the exception, but the processes become clearer when they are separated:

- Whenever an existing M&S system can satisfy the requirements, the *prepare legacy M&S* track becomes important. No new development is needed, only slight modifications of an existing simulation to be applicable to solve the problem.
- When the analysis shows gaps in the required M&S capability, new solutions are needed and the *develop new M&S* track becomes relevant. This means that a new model needs to be developed and the simulation implemented and tested. This new simulation system is then applied to solve the problem.
- As already a significant variety of simulation systems is available of which all may be able to contribute to parts of the solution, the track *construct federation* will often belong to the choices. The federation development processes described in the following section all specify this track in more detail. This means that the required capability is provided by at least two independent simulation systems. To solve the problem, these two—or more—simulation systems need to be federated and executed together.

The different steps are captured in Figure 13.1. No matter which of the three tracks is selected, in every case a track starts with the activity *refining the M&S requirements*. This is motivated technically as well as contractually: only what is needed will be reused, implemented, or federated; and only what is needed will be paid for. Several quality ensuring activities and standards explicitly require capturing the requirements at predefined project phases for these reasons.

The second element all three tracks have in common is establishing a *conceptual model* that reflects how the M&S solution fulfills the M&S requirements.

Although we have just dealt with the concepts in Chapter 11, it is worth revisiting some of the ideas. In this section, we rely on recent work by Robinson (2008). Although the M&S community does not unite on these definitions, literature reviews underlying Robinson's choice of definitions suggest significant adherence within the M&S community. In Robinson's view conceptual modeling for simulation boils down to a process of abstraction in which essentials of a real or proposed system that address all M&S requirements are captured. Conceptual modeling is then defined as "a non-software specific description of the computer simulation model (that will be, is, or has been developed), describing the objectives, inputs, outputs, content, assumptions, and simplifications of the model" (Robinson, 2008). In all observed examples, conceptual modeling is understood as an iterative process of selecting importing elements, attributes, and behaviors combined with abstraction and simplification thereof, resulting in a structure that can be shared and communicated within a simulation team and the customer. The work of Pace (2000) needs to be mentioned in this context as well as it gives a good overview of additional possible views on conceptual model definitions. One of the more restricted views on conceptual modeling emphasizes the primary function of the conceptual model as the mechanism by which user requirements are transformed into detailed simulation specifications that fully satisfy the requirements. In any case, each of the tracks requires either the update or a new development of the conceptual model that becomes the foundation for design decisions as well as validation and verification processes.

Finally, all three tracks require testing at the end of the process. The VV&S RPG gives more details and examples for all tracks. The borders between them can become fluent. Sometimes the redesign of a legacy system grows from minor to major changes and even new developments. In particular in new service oriented approaches, a new development of a service for a particular simulation system can be generalized into a service that can be federated into several other simulation systems as well, and so on.

The general problem solving process described in the VV&A RPG is a good example on how to use cascading definitions to better communicate within a heterogeneous group with different focus points. In this example, it was used to better communicate what the simulation engineer has to do when conducting M&S development and preparation processes. In the original document, the simulation engineer's activities are then connected to the verification and validation processes that are building the foundation for the accreditation. The same approach can be applied to specify in detail how the activity *select approaches* is conducted. It is likely that some sort of metrics are applied that can—and should—be reused when deciding on the metrics that address the customer's problem when conducting the experiments. This kind of capturing the tasks and results using engineering methods as used to produce Figure 13.1 help not only to communicate better in the simulation team, they can also be used to identify double efforts and possible inconsistencies in the overall process.

The process discussed here is very focused on military simulation, as to be expected in a book about combat modeling. However, Balci and Ormsby (2007) generalized these ideas into an M&S lifecycle model that may also be of interest to the reader, in particular when discussing with simulation developers who are not familiar with the often very special vocabulary of combat models and simulation.

FEDERATION DEVELOPMENT PROCESSES

Several reasons are often used to justify a preference of federation development over developing new M&S systems or modifying existing M&S systems, namely

- federations allow the *reuse of existing solutions*,
- federations support *modularity*, and therefore help to *reduce complexity*,
- federations allow *costs to be reduced*, and
- federations can be used to *compose cross-domain solutions*.

To what degree these assumptions are generally justified is a topic of several academic discussions, but they are generally accepted by the community of practitioners.

All simulation interoperability solutions have more or less formalized processes that support the successful application of the standards. In this section, we shall deal with two examples that are important for practitioners as well.

The IEEE 1516.3–2003 Federation Development and Execution Process (FEDEP)

The Federation Development and Execution Process (FEDEP) was developed as a guideline and recommended practice standard in support of the High Level Architecture (HLA). Although the IEEE 1516.3–2003 standard is still valid, it has been superseded by the IEEE 1739–2010, which we will describe later in the chapter. However, the FEDEP is still used and referenced. The IEEE 1516.3–2003 is also the basis of overlay activities for verification and validation, as will be shown in the next chapter. Although it can be assumed that in the future most overlays will be updated to reflect new terms and ideas, the FEDEP is still the foundation of many of the activities and should be known to the simulation engineer.

Like the other products related to the HLA, the FEDEP went through different stages until the guidelines were standardized by IEEE. In several phases, the M&S community under the lead of SISO developed FEDEP versions that increased in detail and aligned terms to be consistent with other related standards. The FEDEP looked explicitly at the support of the IEEE 1516 High Level Architecture, but the experience within US DoD as well as NATO showed that it is unlikely that all simulation systems that are integrated in support of answering the customer's questions and solving the problems will support this standard. It showed that it is more likely to be heterogeneous federations that support a variety of different standards and require several gateways and/or proxies as defined in Chapter 11. As many of the steps described in the FEDEP are general engineering method steps that are independent of HLA, i.e. they are valid and useful no matter which interoperability standard profile is applied, the M&S community decided to generalize the FEDEP and make it applicable to all forms of distributed simulation engineering and execution, which resulted in IEEE 1730–2010 described later.

The FEDEP was developed as a generic, commonsense systems engineering methodology that could be tailored to meet the varying needs of disparate user applications. It purposefully focuses exclusively on federation activities and excludes federation development or modification. As such, the clear application track within the problem solving process described earlier is *construct federation*.

The FEDEP is only on the top level view comparable with the problem solving process, where it defines steps. Each step is defined by several activities, and for each activity detailed input, outcome, and tasks lists with detailed descriptions of each are provided. Some activities may be complex enough to require an additional layer of sub-activities. The graphical notation introduced for the FEDEP and used in most other approaches also shows the step/activity/sub-activity as a box with the input(s) as incoming arrows and the outcome(s) as outgoing arrows. Each income is either provided from the outside in support of the project or is produced as an outcome in another activity. Each outcome is either used as an input for another activity or contributes immediately to the results of the project. Inputs do not simply appear nor do outputs disappear.

Although each federation development project has individual challenges, there are sufficient similarities on a higher abstraction level. On the highest level, the FEDEP identifies the following seven steps:

- Step 1: Define federation objectives
- Step 2: Perform conceptual analysis
- Step 3: Design federation
- Step 4: Develop federation
- Step 5: Plan, integrate, and test federation
- Step 6: Execute federation and prepare outputs
- Step 7: Analyze data and evaluate results

In the first step, similar to the NATO COBP recommendations, the sponsor's problem drives the initial planning. All team members such as the sponsor, operational users and other stakeholders, and simulation engineers for the development and integration team define the federation objectives. These objectives are documented to capture the agreed contract within the team and build the foundation for the following activities.

In the second step, the operational requirements of the federation are evaluated to derive technical requirements, concepts of operations, and enabling scenarios. This step can be compared with the solution strategy of the NATO COBP, as the general design and conceptual constraints are agreed on. The result is a conceptual blueprint of what needs to be modeled in the form of required orchestrated capabilities.

Based on the operational and technical requirements, the third step creates the design. Using the metaphor of the blueprint, simulation engineers identify applicable solutions and select the best ones. Each capability needs to be supported by at least one of the existing solutions. Each information exchange needed to connect and orchestrate capabilities needs to be supported within the supporting system or via interfaces between the solutions. When all required functionalities are allocated, a plan for the development can be agreed on that becomes the basis for the next step.

The fourth step focuses on the federation development. As a rule, this step focuses on adapting the participating simulation systems—the federates—to fulfill the information exchange requirements. The *federate object model* (FOM) that supports these information exchange requirements is defined and the federates are modified so that they support the FOM, which is the main outcome. In addition, the parameters that need to support the metrics of success should be obtainable, either via the FOM or other means. In some cases, limited new development is required if new functionality is needed to implement an important required capability, but developing completely new federates is not in the scope of the FEDEP.

The fifth step is what simulation engineers normally like most: planning, integrating, and testing the federation. Particular focus in this step lies on the interoperability requirements: do the federates provide the needed data at the right time in the right format? Are the federates providing the functionality in the correct order? Are all time anomalies taken care of? All necessary tests need to be planned and documented; agreements need to be captured, and so on. Of particular interest are the federation agreements mentioned in the last chapter: Who publishes and subscribes to what objects? Does the time management work as planned? Are all security needs addressed? Finally, the initialization with preferably authoritative data in standardized format also needs to be assured and tested. At the end of these steps, the federation should be ready to be applied as required by the customer in the first place.

This execution happens in the sixth step: the exercise is conducted, the operational analysis is supported, a test of new equipment is conducted, and so forth. During the execution, all data required for the measures of merit are obtained as well, either directly supported by the simulation federation or by other means as documented in the data collection plan—as recommended in the NATO COBP. An execution can be a one-time or a recurring event. The recommendations of Kass (2006) that we already examined in Chapter 5 are very valuable for the planning process of this step.

In the seventh and final step, the obtained data are analyzed and the results are evaluated. This includes the preparation of deliverables, documenting lessons-learned, capturing recommendations for future improvements of federates, etc. It is good practice to revisit the federation objectives to make sure that all activities have been conducted, all data collected, and questions answered as originally agreed. In earlier versions of the FEDEP, steps six and seven were combined and seen as one step at the end of the chain of steps.

Every experienced simulation engineer recognizes immediately that these seven consecutive steps will often have to be revisited due to corrective actions needed, and not only when iterative development is done. For example, when working on the scenario, additional operational requirements may be discovered. It is also very likely that the original operational ideas may have to be adjusted due to the available functionality of federates and other constraints. Authoritative data may not be available as needed, or security constraints may be in the way of exchanging data as technically desirable.

Figure 13.2 shows the top level view of the seven steps of the FEDEP. The solid line shows the planned consecutive execution and the dotted line represents corrective actions and other feedback. As to be expected, feedback is not limited to the preceding step but can reach from every step all the way back.

As mentioned before, each step is documented in additional detail by a set of necessary activities that transform the inputs into outputs. On the highest level, the following things belong to the inputs:

- program objectives of the sponsor (what needs to be provided by the federation?)
- available solutions, in particular federates, FOMs developed for similar federations, lessons learned, etc.

The high level outputs produced by the FEDEP are the deliverables of the federations, including reusable components, the FOM, lessons learned, and more. Figure 13.3 shows the high level inputs and outputs of the top level steps, as captured in IEEE 1516.3–2003.

The list of external inputs and required outcomes are not complete. They reflect just the generally to be expected type of documents, deliverables, and products that can be expected to be observed in this environment. However, the lists and other checklists provided by the standard can help to jump-start projects by producing a core of required inputs and a list of desirable outcomes that can then be tailored to reflect specific project requirements.

Each step is further broken down into activities. To ensure traceability of inputs and outcomes, the arrows are numbered with the activities in which they were produced or where they are needed. Only those external inputs and outcomes

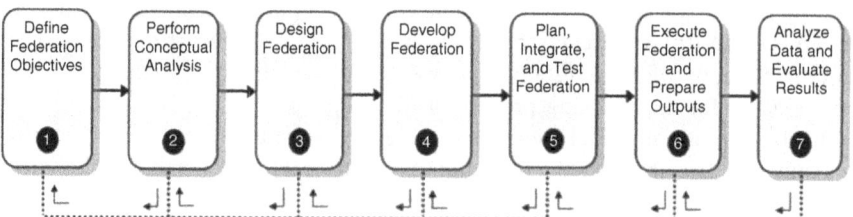

Figure 13.2   Top level view of the FEDEP.

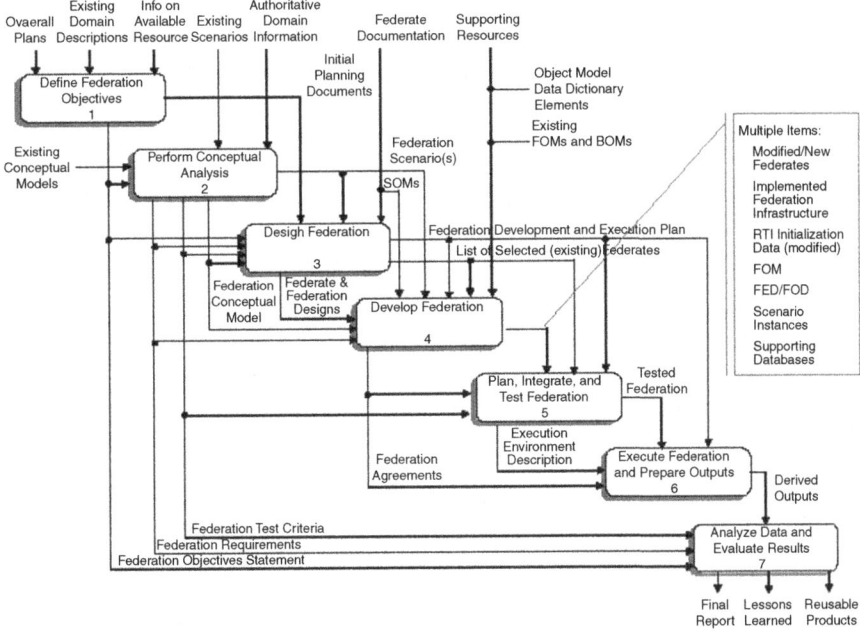

Figure 13.3   Detailed input and outcome flow on the top level of the FEDEP.

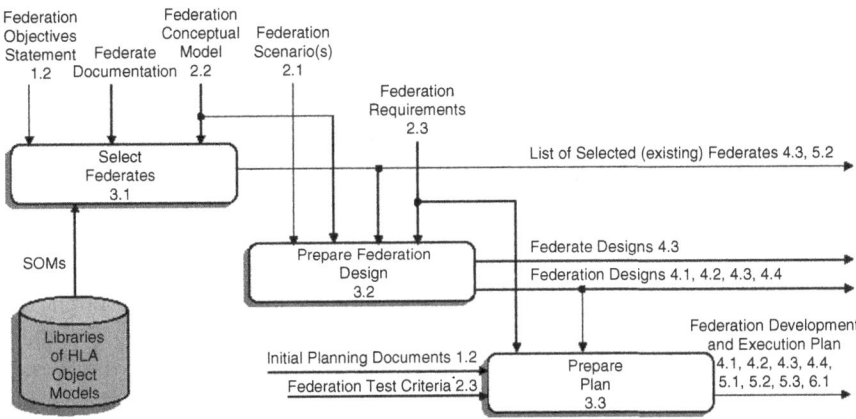

Figure 13.4   Activities for step 3 of the FEDEP—design federation.

that already are captured on the top level for the overall FEDEP do not have a number. Figure 13.4 shows how the IEEE 1516.3–2003 breaks down the step *design federation* into three activities.

In order to explain the principles of the graphical representation, here are some examples of how to read Figure 13.4. For more details, the reader is referred to the IEEE standard.

- The second activity of step 1 (step 1.2) produces the *federation objectives statement* that is part of the inputs for the first activity of step 3. Other inputs are the *federate documentation*, which was provided from outside of the project, and the *federation conceptual model*, which was produced in activity 2 of step 2.

- Activity 3.1 produces a *list of selected (existing) federates* that becomes input to several follow-on activities. Within step 3, it feeds directly into activity 3.2: *prepare federation design*. In addition, the list is also needed as input in the third activity of step 4 (4.3) and the second activity in step 5 (5.2).

- In order to support the selection of federates, *simulation object models* (SOMs) are used to support the decision. (A SOM describes the information exchange capability of a federate using object model template as the format.)

The FEDEP depicts all activities at least in this detail. In addition, all inputs and outcomes are defined at least as enumerations, and where of particular interest in support of the management processes, additional checklists that can be used in projects are provided as well.

Furthermore, the FEDEP specifies the connection between process and product(s). For every activity input and output are specified, and the development of deliverables via their various contributing parts is traceable. These products, that contain deliverables as well as documentation, play a pivotal role for quality control as well as for the validation and verification activities that will be discussed in the next chapter.

The Euclid RTP 11.13 Synthetic Environment Development and Exploitation Process (SEDEP)

The Western European Armaments Organization (WEAO) is a subsidiary of the Western European Union (WEU) and, at the time this book was written, had 19 members: Austria, Belgium, Czech Republic, Denmark, Finland, France, Germany, Greece, Hungary, Italy, Luxembourg, Netherlands, Norway, Poland, Portugal, Sweden, Spain, Turkey, and the UK. The executive body is the Western European Armaments Group (WEAG). The original idea was to establish a European Armament Agency, but at this time the work is still mainly conducted by panel-like program groups, WEAG Research Cells (WRC). They are categorized into 16 Common European Priority Areas (CEPA) that cover a wide range of technologies considered to be fundamental to satisfy the operational requirements of defense systems. Each CEPA carries several Research Technology Programs (RTPs). The European Cooperation for the Long Term in Defense (Euclid) has been integrated into this structure. Its RTP category 11 deals with "human factors" including technology for training and simulation and is aligned with CEPA 11: Defense Modeling and Simulation Technologies.

Euclid RTP 11.13 was the largest European Defense Research and Technology activity undertaken by the WEAG (more than 17 million euros). The program started in November 2000 and the final demonstration was held in November 2003. Under the title "Realizing the Potential of Networked Simulations in Europe," RTP 11.13 aimed to overcome the obstacles that prevent synthetic environments (SEs) being exploited in Europe by developing a *process* and an integrated set of prototype *tools*, which will reduce the cost and timescale of creating and utilizing SEs for training, mission rehearsal, and simulation based acquisition.

Using FEDEP as a starting point, RTP 11.13 improved the descriptions, added additional processes to improve the underlying systems engineering methodologies (addressing the objective of *developing a process*), and also developed several software tools that directly support the processes and their management (addressing the *prototype tools* mentioned in the objective). The result of the process activities is the Synthetic Environment Development and Exploitation Process (SEDEP) (Ford, 2005); the result of the prototype tools activities is a Federation Composition Tool (Brasse et al., 2003).

One of the main concerns of the European experts was that the FEDEP was predominantly driven by technical needs and viewpoints. The management aspect of steering and controlling the process of federation development was not sufficiently addressed. Furthermore, the driving objective was not emphasized in the FEDEP—*why* the federation was built was overshadowed by *how* the federation was built. Many of the results were used to improve the FEDEP and also did feed into the DSEEP.

The SEDEP identifies eight necessary steps on the top level.

- *Analyze user's need*: Understanding the objectives and requirements and capturing them in a *user needs analysis document* (high level view). Such an effort is not part of the FEDEP.

- *Define federation user requirements*: Evaluating the technical implications of the high level needs and capturing the results in the *user requirements documents* (operational view). This document bridges the operational needs with the technical requirements. The FEDEP starts where this process ends.

- *Define federation system requirements*: Mapping the operational needs and the technical requirements to the engineering level results in the *system requirements documents* (system/technical view).

- *Design federation*: The technical specifications are agreed upon and captured in detailed *design documents* in which design decisions are traceable to the originating requirements.

- *Implement federation*: By implementing the design idea, *federation components* are produced that not only serve as solution providers in one federation, but are stored with the accompanying metadata in a repository.

- *Integrate and test federation*: This step fulfills the users' requirements technically by producing the *federation ready for operations*. All components are coupled and integrated, the interoperability constraints have been tested, the needed data have been obtained, etc.

- *Operate federation*: By running the simulations within the federation, the *federation execution outputs* provide the data that are needed to fulfill the users' requirements operationally as well. However, the data as they are produced by the federation executions still need preparation and additional processing to provide an answer to the original modeling or research question.

- *Perform evaluation*: Using the data in metrics that support the measures of merit for successful operations results in *evaluation results and report documents*. The SEDEP documentation also emphasizes that the operational questions often require taking additional aspects into account that cannot be addressed by the technically possible federations, so that, in this last step, all needed additional phases of solid operational analysis have to be integrated. This view expressed in the SEDEP embeds the simulation work into the broader activity of defense analysis and operational research as one tool among many,

Figure 13.5 shows the SEDEP steps and related products. Each step is broken down into detailing activities with inputs and outcomes as described before for the FEDEP as well. Many of these ideas were introduced into the improvement

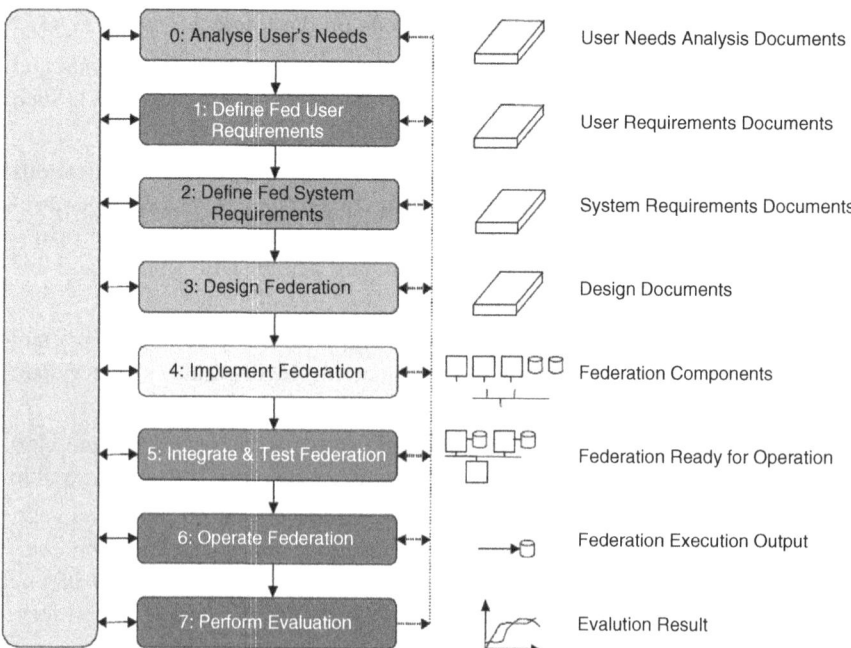

Figure 13.5   Top level view of the SEDEP and products.

process for the FEDEP and the follow-on standardization activity, Distributed Simulation Engineering and Execution Process (DSEEP), in particular by European partners who were members in both groups.

When comparing the top level views of FEDEP and SEDEP, the conceptual integration of the repository that is used to store the artifacts and components is a main difference. This use of a common repository to support the discovery of possible tools, solutions, or other reusable components was one of the driving requirements for the European group. By integrating the necessary metadata and procedures into the standard development process, many approaches that we looked at in earlier chapters, such as the metadata registry studies and the Modeling and Simulation Resource Repository (MSRR) activities, were automatically aligned and their application facilitated. From the management standpoint this was a very good move.

From the simulation engineer standpoint, this integration of a repository looks like additional work at first glance. Such a repository may pay off once it is populated and many components are often reused, but for the start, it means more work. The engineers of Euclid RTP 11.13 initially focused on identifying processes that can be supported by tools. Brasse et al. (2003) give several examples in their paper. The tools should all support the methodology as standardized in the SEDEP and therefore be easily alignable. As the data to be developed are defined for the repository, all tools share a common meta-model for their data from the beginning, similarly to DoD Architecture Framework 2.0 as later decided (as described in Chapter 12). Figure 13.6 shows an example of the federation composition tool.

Another outcome of the Euclid RTP 11.13 worth mentioning is the initiation of a glossary of terms. In order to be able to identify the various categories of the tool descriptions with unambiguous terms, they were defined and became part of the documentation. To these definitions belong

- *Federation*: the top level (HLA) federation
- *Complex federate:* a federate whose internal structure consists of a set of federates that is federated in the same architectural manner as the federation
- *Simple federate*: a federate whose internal structure is considered a black-box
- *Federation bridge*: an architectural runtime element that connects two different federations
- *Federation architecture*: the hierarchical structure of the federation, containing federates and their required capabilities
- *Federation/system specification*: input for the composition process (federation scenario, federation conceptual model, system constraints, evaluation data requirements)
- *Federation agreements*: federation-wide descriptions needed for implementation and execution planning not covered by the FOM.

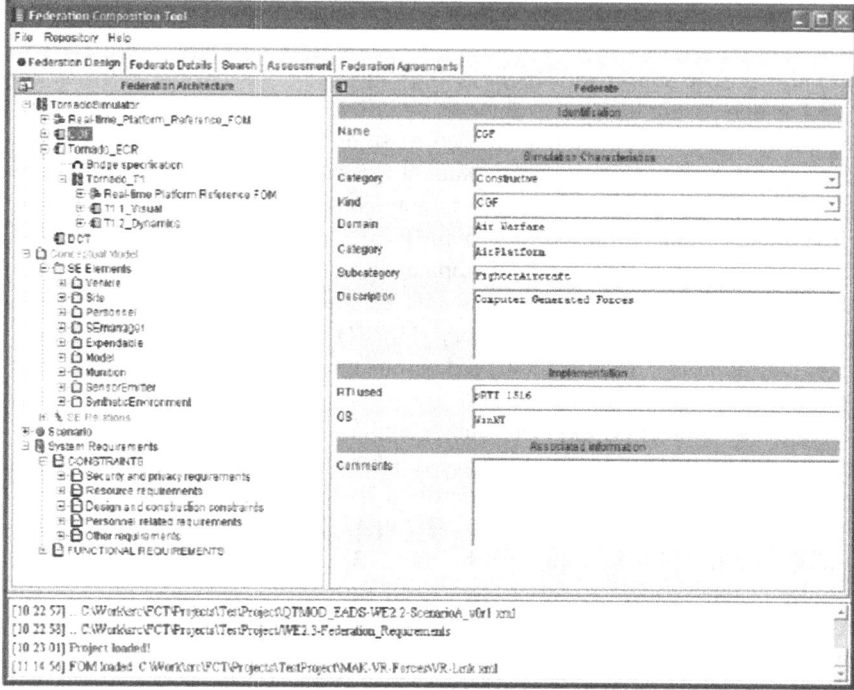

Figure 13.6   Federation composition tool (Brasse et al., 2003).

Many of these definitions made it back into the Simulation Interoperability Standards Organization (SISO) as well as into NATO, where they were used to support tagging entries into the Simulation Resource Library.

The SEDEP is frequently used in Europe for developments. Also, more tools have been developed that support the applications, such as initialization and execution support tools.

The IEEE 1730–2010 Distributed Simulation Engineering and Execution Process (DSEEP)

As mentioned at the beginning of this chapter, the FEDEP is still valid but has been superseded by the IEEE 1730–2010 Distributed Simulation Engineering and Execution Process (DSEEP). Among the first publications presenting the new ideas is Lutz et al. (2008). Two developments led to the need to broaden the ideas as captured in the FEDEP and the SEDEP beyond the scope of HLA federations.

First, the diversity of M&S interoperability solutions was not ended with the introduction of HLA. Alternative protocols, in particular DIS and TENA, continue to be supported by a broad and even growing user community. The "no

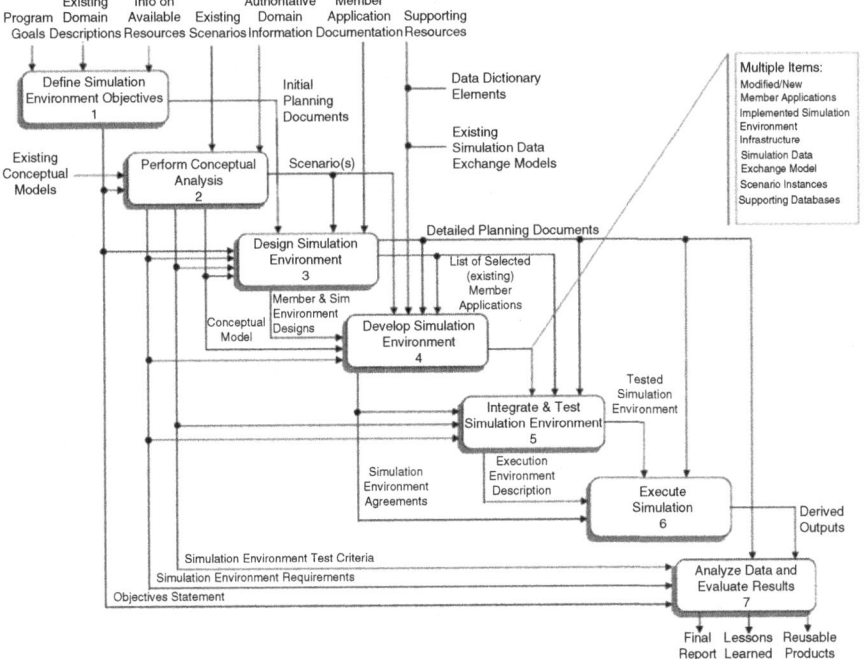

Figure 13.7   Detailed input and outcome flow on the top level of the DSEEP.

play—no pay" policy of 1996, which envisioned that in the future only HLA compatible simulation systems would be funded, did not happen.

Second, the technological advance and the emergence of XML as a common data description language allowed more and more diversified entities from the live, virtual, and constructive simulation community to be integrated into *de facto* common architectures (Loper and Cutts, 2008).

On one side, both solutions described so far focused on federation development in homogeneous HLA environments that did not support such diversity. On the other side, they were both rooted in general systems engineering principles that were derived from general integration methods. The logical idea was to preserve the accomplishment of FEDEP and SEDEP but generalize them to allow applying them to integrating non-HLA components and not limiting the infrastructure to HLA conforming solutions. The result is the DSEEP as shown in the high level top view with inputs and outcomes in Figure 13.7. The resemblance to the IEEE 1516.3 FEDEP becomes immediately obvious.

Without going into the details specified in the IEEE 1730–2010 standard, the activities to be conducted in support of these top level steps were derived from the FEDEP and SEDEP recommendations.

- *Step 1: Define simulation environment objectives.* In this initial step, the team has to identify the needs of the user and sponsor, develop the objectives and document them, and conduct the initial planning.

- *Step 2: Perform conceptual analysis.* In the following step, a scenario is developed allowing the objectives to be evaluated, a conceptual model is developed that captures the needed capabilities, and the resulting simulation requirements are derived and documented.

- *Step 3: Design simulation environment.* In this step, member applications are selected and the supporting simulation environment is designed. A detailed plan of necessary changes and adaptations for the selected members and the supporting infrastructure—or infrastructures—is prepared and documented.

- *Step 4: Develop simulation environment.* The focus of this step is the development of the simulation data exchange model and the establishment of simulation environment agreements: who is responsible for delivering the functionality to implement the required capabilities, in which order, based on which information exchange, etc. Based on these decisions and other related documents produced in earlier steps, the member application designs are implemented and the simulation environment infrastructure is implemented. If more than one infrastructure is needed, the connection of these infrastructures is conducted during these activities as well.

- *Step 5: Integrate and test simulation environment.* Integrating and testing the simulation environment are the main activities in this step. The tests are driven by the operational needs that are captured in a detailed execution plan that is created and documented in this step as well, based on earlier decisions in the process.

- *Step 6: Execute simulation.* Executing the simulation as planned and collecting and preparing the simulation environment outputs for the final steps happen here. Again, the data to be collected are not necessarily limited to simulation outputs but comprise much more, in particular in LVC environments. Also, in heterogeneous infrastructure environments, different collection options have to be synchronized and composed.

- *Step 7: Analyze data and evaluate results.* In this final step, data are analyzed and results are evaluated in order to give feedback in the form of deliverables and documents.

The IEEE 1730–2010 standard specifies recommended tasks for each activity that are all tied together by the inputs and outcomes, as described before. There are two main contributions that will be noticed in comparison with earlier approaches.

First, the terms need to be harmonized. Top level steps, activities, and tasks are used throughout the document showing the simulation engineer immediately which description level he or she is on. The same is true for the consistency of terms used to define the inputs and outcomes. Overall, the modeling approach used to describe the DSEEP became very mature and supportive.

Second, and even more important, these common terms and descriptions of steps, activities, tasks, inputs, and outcomes are now mapped in new sections of the standard to relevant candidate environments. There are three annexes that define normative overlays for the High Level Architecture (Annex A), the Distributed Interactive Simulation (Annex B) and the Test and Training Enabling Architecture (Annex C). An additional effort to define a Life-Virtual-Constructive (LVC) overlay to the DSEEP is currently being undertaken by the Simulation Interoperability Standards Organization.

Each overlay comprises a section that maps the specific terminology and definitions to those used in the DSEEP. For example, DSEEP uses the terms *simulation environment* and *member application* in the core part of the standard document. HLA uses the terms *federation* and *federate* to address these concepts. In DIS, the respective terms are *simulation exercise* and *simulation application*, while TENA uses the terms *logical range* and *range resource application*. These mappings now allow recommendations from these different approaches—HLA, DIS, and TENA—to be compared and applied across all supported alternatives.

The mapping, however, is not limited to terms and lexical definitions. Each annex comprises a global mapping section that compares and maps specific architecture and methodology steps, phases, or operations to the steps, activities, and tasks defined in the DSEEP. While this section is minor for the HLA—as the DSEEP steps are derived from FEDEP, which was designed for HLA—there is some more work to do for DIS and TENA.

The DIS Exercise Management and Feedback Process, e.g. a recommended practice within the group of IEEE 1278 standard documents, only recognizes the following five top level steps:

- plan the exercise and develop requirements;
- design, construct, and test the exercise;
- conduct the exercise;
- conduct exercise review activity; and
- provide results to decision makers.

The second top level step of DIS—design, construct, and test the exercise—needs to be mapped to four top level steps of the DSEEP, as the activities are conducted in the DSEEP steps 2–5. On the other hand, the two distinguished steps of "conduct exercise review activity" and "provide results to decision makers" are both conducted as activities or tasks within the DSEEP step 7. Similar challenges exist on all levels. They also exist for TENA. By bringing HLA, DIS, and TENA experts together, the DSEEP was enriched by additional viewpoints that could contribute and be important for conducting a successful exercise or other simulation-based event, and also all three groups were helped to establish a common perspective that allows better architecture independent dialogue on how to solve implementation-specific challenges.

Knowing the DSEEP and its application is therefore a useful tool for simulation engineers who have to support heterogeneous simulation events. Knowing the DSEEP is as important as knowing the interoperability-standard-specific details of the participating infrastructures and simulation systems.

SCRUM FOR M&S SERVICES

This last section is different from the standardized processes for M&S solution, but as it is rapidly gaining importance in the software engineering community it will at least be mentioned: the Scrum methodology supporting agile project management in particular for software development (Schwaber and Beedle, 2002). Landaeta and Tolk (2010) give examples how these ideas can be applied to M&S challenges.

The idea behind Scrum is actually pretty simple: while most approaches start with the definition of requirements that become the basis for contracts, designs, and other fix point solutions at the beginning of a system process that requires to be managed, Scrum recognizes that requirements cannot be fixed in an agile environment. What was of significant importance at the beginning of a project can decrease in importance over the course of in particular multi-year projects. What is needed to react appropriately in such agile environments is the following: (1) an agile project management process and (2) a mature implementation technology that together allow for reactions during project execution to make necessary adjustments in the prioritization of requirements.

The term *scrum* comes from the word *scrummage*, which describes a way of continuing a rugby game after accidental infringements. As such it describes the restart at a well defined point to continue into a potentially new direction after something unforeseen happens. In the world of software development this translates into continuing with your project in the new direction when the customer suddenly changes requirements. Figure 13.8 displays the idealized scrum process.

The driving idea behind the Scrum process is to find a structure that allows reacting to changing requirements. To this end, the product is not seen as a monolithic block but as a set of components that provide functionality in support of required capabilities. The capabilities are derived from the requirements. If requirements change, required capabilities change as well.

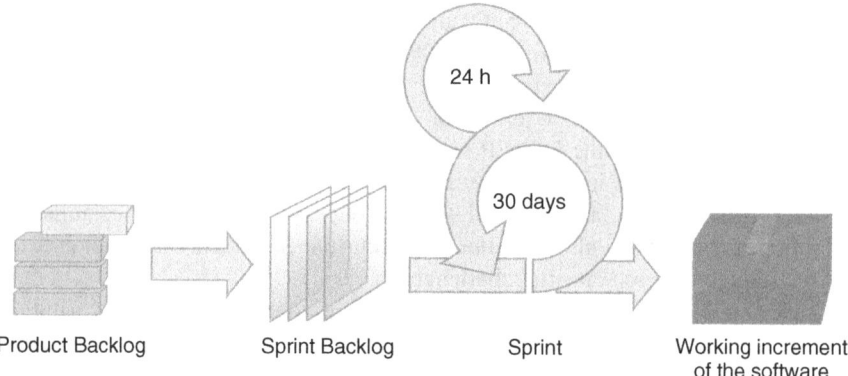

Figure 13.8   Scrum process for agile software development.

The process starts with product backlogs that actually reflect components that host functions to provide capability. These are broad descriptions of potential features that are prioritized based on business values, or, as expressed by Schwaber and Beedle (2002), *what* will be built sorted by importance for the customer. The most important component is selected in accordance with the customer to become a *sprint*. A sprint is the collective term for the work that can be conducted in a week up to a month (30 days) to address one selected particular user need. The product owner helps in selecting the most needed piece of the product backlog together with the team, and the team comes up with a plan for the sprint that captures *how* the capability will be provided with the sprint.

Daily scrums are held at fixed times in fixed places (or telecom numbers with distributed teams). This daily scrum addresses what has been accomplished with a focus on the last day, and what will be attacked in the next work day.

At the end of the sprint itself, a review meeting is conducted with the product owner to show the new functionality and how it fits in the current capabilities. In addition, the team meets to improve internal processes for a sprint retrospective meeting.

The Scrum method defines clear roles for the product owner and the team and has been successfully applied in several organizations. The Scrum Alliance is a not-for-profit professional membership organization created to share the Scrum framework and transform the world of work that supports networking and exchange of lessons learned.

As mentioned before, the successful application of Scrum requires two things, the agile project management process just discussed as well as a mature implementation technology. For M&S in general, and for combat modeling and distributed simulation in particular, the use of composable M&S services (Tolk et al., 2006) builds such a mature technology, in particular with the increasing use of supporting semantic web technology as discussed in Chapter 12 and documented in Diallo et al. (2011).

REFERENCES

Balci O and Ormsby WF (2007). Conceptual modeling for designing large-scale simulations. *J Simulation* 1, 175–186.

Brasse M, Mevassvik OM and Skoglund T (2003). Federation Composition Process and Tools Support in EUCLID RTP 11.13. *Proceedings of the Fall Simulation Interoperability Workshop, Orlando, FL*. Accessed at www.sisostds.org.

Diallo SY, Tolk A, Graff J and Barraco A (2011). Using the levels of conceptual interoperability model and model-based data engineering to develop a modular interoperability framework. In: *Proceedings of the Winter Simulation Conference, Phoenix, AZ*. IEEE Press, Piscataway, NJ.

Euclid RTP 11.13-TT&S SA-WE1.5. *Volume 2: The Synthetic Environment Development and Exploitation Process (SEDEP), Version 1.0, July 2001*, Brussels, Belgium.

Ford K (2005). The Euclid RTP 11.13 Synthetic Environment Development and Exploitation Process (SEDEP). *Virtual Reality* 8, 168–176.

IEEE (2003). *IEEE 1516.3 Recommended Practice for High Level Architecture Federation Development and Execution Process (FEDEP)*. IEEE Publication, Washington, DC.

IEEE (2010). *IEEE 1730 Distributed Simulation Engineering and Execution Process (DSEEP)*. IEEE Publication, Washington, DC.

Kass RA (2006). *The Logic of Warfighting Experiments, DoD Command and Control Research Program*, CCRP Press, Washington, DC.

Landaeta RE and Tolk A (2010). Project management challenges for agile federation development: a paradigm shift. In: *Proceedings of the Fall Simulation Interoperability Workshop, Orlando, FL*. IEEE Press, Piscataway, NJ.

Loper ML and Cutts D (2008). *Live Virtual Constructive Architecture Roadmap (LVCAR) Comparative Analysis of Standards Management*. Final Report, M&S CO Project No. 06OC-TR-001.

Lutz R, Gustavson P and Morse KL (2008). Toward a standard systems engineering process for distributed simulation. In: *Proceedings of the Interservice/Industry Training, Simulation & Education Conference (I/ITSEC), Paper No. 8068*, Orlando, FL.

Pace DK (2000). Conceptual model development for C4ISR simulations. In: *Proceedings of the 5th International Command and Control Research and Technology Symposium*. CCRP Press, Washington, DC.

Robinson S (2008). Conceptual modelling for simulation part i: definition and requirements. *J Operational Res Soc* **59**, 278–290.

Schwaber K and Beedle MA (2002). *Agile Software Development with Scrum*. Prentice Hall, Upper Saddle River, NJ.

Scrum Alliance Website. http://www.scrumalliance.org.

Tolk A, Diallo SY, Turnitsa CD and Winters LS (2006). Composable M&S web services for net-centric applications. *J Def Model Simul* **3**, 27–44.

US Modeling and Simulation Coordination Office (2006). *Verification, Validation, and Accreditation (VV&A) Recommended Practices Guide (RPG)*. Alexandria, VA. http://vva.msco.mil (last accessed June 2011).

Chapter 14

Verification and Validation

Andreas Tolk

OBJECTIVES OF VERIFICATION AND VALIDATION

In earlier chapters of this book, we already addressed the need to be serious about making sure that soldiers and decision makers can rely on combat models and distributed simulation systems provided by simulation engineers. All application domains of military modeling and simulation (M&S) are directly connected with the well-being of humans; wrong decisions and insufficient training will probably result in harm and even the loss of lives. Verification and validation are the processes designed to support the necessary evaluation processes.

Validation and verification processes are conducted to avoid two main error categories regarding the use of M&S.

- The first error category is that *valid simulation results are not accepted.* Balci (1998) refers to this as the modeler's risk that the model is not accepted although applicable and categorizes the error to be less serious. However, after the Columbia accident where the space shuttle and its crew were lost due to the fact that a valid simulation that forecast the destruction of the shuttle during reentry into the atmosphere was considered to be not applicable, this evaluation needed to be revisited: not accepting good advice from the correct model can lead to serious mistakes and consequences.

- The second error category is that *non-valid simulation results are trusted.* This is the model user's risk and the results are generally serious, as the user follows bad advice.

Some texts use a third category: not relevant simulation results are used. This is the authorities risk! In this chapter we follow the interpretation that irrelevant

Engineering Principles of Combat Modeling and Distributed Simulation, First Edition.
Edited by Andreas Tolk.
© 2012 John Wiley & Sons, Inc. Published 2012 by John Wiley & Sons, Inc.

results are not valid for the use case so that this third category is considered to be a special case of the second category.

The chapters on combat modeling have shown that there is a huge variety of models and modeling assumptions to be taken into consideration. *Validation* has the objective to ensure that these models and their assumptions are correct with reference to the real or envisioned systems that will be represented. As stated before, validation ensures that we build the right model!

The chapters on distributed simulation have shown that many implementation details have to be planned, implemented, integrated, and tested. All these activities are transformations that may introduce unintended behavior or structural changes of the implemented systems. *Verification* has the objective to ensure that these transformations are correct and result in a simulation system or federation that implements the valid model correctly. As stated before, verification ensures that we build the simulation system or the federation correctly!

Validation and verification (V&V) should not be confused with quality control and measures of the underlying software engineering processes. These measures need to be taken to ensure that the resulting software is stable, that it correctly accesses peripheral devices, that it interacts correctly with other software components of the operating system or other applications, that the program does not enter undefined states or infinite loops, that all valid user inputs are accepted, that initialization data are read and output data are produced, and all the other functions that are expected from computer programs in general. These characteristics are important, but they are assumed to be addressed in the software testing and integration phases. V&V focuses on the special needs of models and simulation.

In particular in the defense domain, the process of *accreditation* becomes important as well, leading to the acronym VV&A. Accreditation is a formal administrative process using V&V results. While there is a strong emphasis on accreditation in particular in the US DoD, the term *acceptance* is preferred in many other countries, if the process is addressed at all. Some publications distinguish between accreditation and *certification*, in particular when data are the focus of interest, but more recent studies no longer distinguish between formal processes for M&S and formal processes for data. The idea behind all these activities is that an independent organization officially determines that a special model or simulation in light of the conducted V&V is perceived to be suitable for the intended purpose.

Shannon (1975) was among the first dealing with validation of simulations as a serious topic. The work of Balci (1997, 2001) and Sargent (2000, 2001) belong to the foundations of today's view on V&V. To support common processes and documentation, the US Modeling and Simulation Coordination Office (2006) provides the VV&A RPG addressed in the earlier chapters, as this document ties the M&S development processes with the V&V processes and the accreditation processes together. Under the chair of SISO, the community also standardized a VV&A Overlay to the FEDEP which is currently updated for the DSEEP, as they were discussed in the last chapter. NATO is involved in this standardization effort as well.

In this chapter, we will look at various aspects of VV&A, not limited to the US DoD view, but this view has been at least influencing alternative approaches. This chapter cannot replace the requirement to become familiar with V&V/VV&A rules and regulations that are legally binding within projects that the simulation engineer will support, but it should contribute to build a solid foundation of what can be done and has been done within the security to avoid fatal mistakes as discussed at the beginning of Chapter 2.

ACADEMIC FOUNDATIONS

Many researchers contributed to a better understanding of what it means for a model to be valid and what is needed to verify simulation implementations. We will give an overview of some selected approaches, methods, and viewpoints before dealing in more detail with the guidelines of government and industry.

Selected Approaches

One of the most often referenced representations of V&V processes was published by Sargent (2000). Figure 14.1 exemplifies the processes.

Sargent (2000) starts with the problem entity, which is the referent of a real or assumed system. Within the process of analyzing the customers' problem and understanding what characteristics of the problem entity are important to address the concerns of the customers, the simulation engineer conceptualizes the needed entities, behaviors, relations, and other elements needed to model the

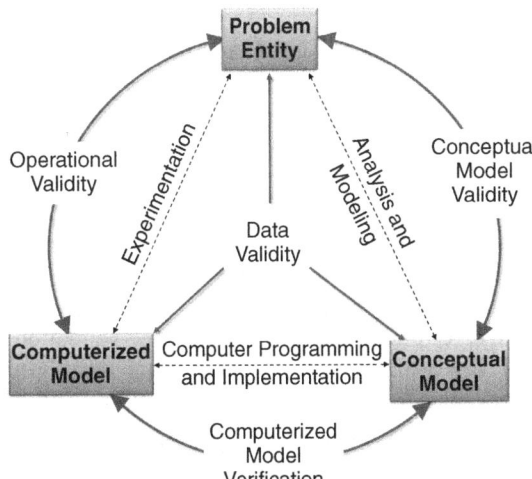

Figure 14.1 Sargent's (2000) model for V&V.

conceptual model. This conceptual model needs to be documented and validated. This is done by comparing the conceptual model with the problem entity: are all important characteristics captured, are all relations taken into account, and so on. This process is the conceptual model validation.

In the next process, the conceptual model gets implemented as a computerized model. This is accomplished by the processes discussed in the last chapter: computer programming, implementation, integration, etc. The resulting simulation system needs to represent the conceptual model appropriately, which means that all implemented entities, behaviors, and relations need to act as prescribed by the conceptual model. This process of ensuring that the conceptual model was correctly transformed into a simulation system—or a computerized model—is called the computerized model verification.

Finally, once the model is implemented, it can be executed and observed to see if the implemented behaviors and interactions actually reflect the expected or—even more important for real systems—the observed behavior and interactions of the problem entity. This is the process of evaluating the operational validity.

For all phases of these V&V processes and activities, the supporting data have to be valid. If authoritative data sources are used, transformations must be correct, the accuracy needs to be appropriate, and other challenges must be dealt with, as also addressed in the data section of the NATO COBP in Chapter 3.

The view expressed in this basic model is that the characteristics of the real systems are exposed and easily observable. Furthermore, it assumes that the purpose of the system is well understood and also reflected in the simulation. These two assumptions allow the evaluation of whether the implementation is *correct* and *suitable*. Correct means that in comparison with the real system the model and implementation thereof are complete and consistent. Suitable means that the model and implementation provide the required capability for the purpose with the necessary fidelity and accuracy. These ideal constraints are not often observed in reality.

Sargent (2001) extended the principles and applied system theory insights to V&V for simulation systems. Hester and Tolk (2010) extended his contributions by generalizing the application of system theory further in order to allow for the application of systems engineering principles to support V&V. Figure 14.2 couples the idea of systems theories with real-world experimentations and virtual-world experiments. It does not assume that all data are easily accessible but replaces the observable system with a system theory that is rooted in experimentation. The figure exemplifies this significant challenge for conducting V&V for M&S: the question regarding the availability of real-world problem entities and—even if available—how to retrieve valid data and conduct model validation.

As documented in Hester and Tolk (2010), systems engineering inherently involves the inclusion of models when systems theory is applied; instead of dealing with the system as it is, it is described using differential equations, states and state changes, Petri nets, etc. The application of complex simulation systems is a logical next step. Instead of solving a simple differential equation, typical

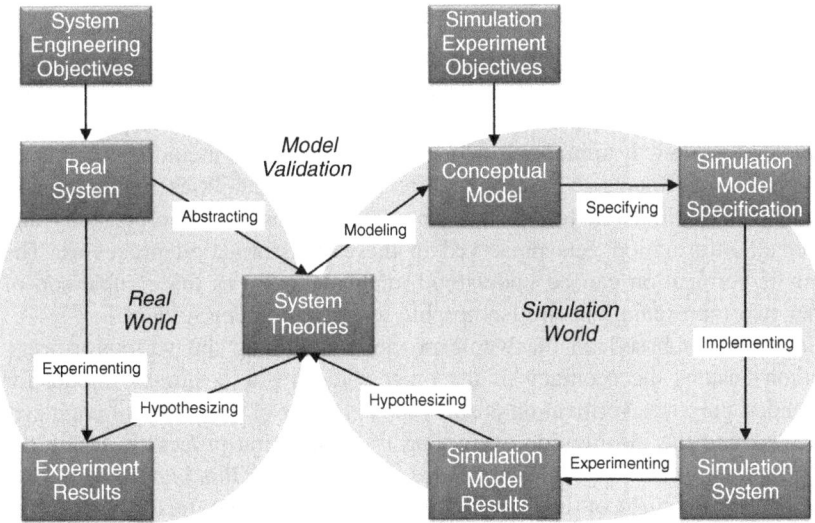

Figure 14.2   Real system and simulation.

cases are simulated in detail to explore the solution space defined by all possible solutions that fit the model.

However, replacing the system with a system theory results in the effect that neither correctness nor suitability with regard to the real system are possible. Instead, V&V can only aim at establishing *credibility*. As stated by Brade in his observations:

> *The credibility of a model is based on the perceived suitability and the perceived correctness of all intermediate products created during model development. The correctness and suitability of simulation results require correctness and suitability of the model and its embedded data, but also suitable and correct runtime input data and use or operation of the model. Verification and validation aim to increase the credibility of models and simulation results by providing evidence and indication of correctness and suitability. (Brade, 2004, p. 13)*

Like every other model the system theory, or the context in which a system is evaluated, is driven by the purpose. The intended use therefore plays a pivotal role in what the user and the simulation engineer consider important and what is neglectable. Suitability is not an independent characteristic; it can only be answered in the context of the question "suitable for what intended uses?"

Validation under these constraints can best be defined as the process of determining the degree to which a model and its implementing simulation are an accurate representation of the real-world problem entity in the context of the intended use. Validation deals with behavioral or representational accuracy of representing the system or system theory. Verification is then the process of determining that a model and the resulting simulation implementation accurately represent the conceptual description and specifications, which can require several models that gradually add implementation detail until the result is conceptual models as described in the last chapter of this book. To support this view, verification deals with transformational accuracy between all steps of the systems engineering process. This is not limited to the transformation between the conceptual model and implementation, it applies also to all interim models, including tracing of operational requirements to system requirements to capabilities to implementing functions. Verification focuses on whether all steps are done correctly and whether all information gets preserved in these transformation processes. The process of verification can be understood mathematically as the comparison of whether two representations are isomorphic under the modeling question.

Credibility is based on the whole process integrating the various aspects. Validation ensures the accuracy of the process leading to a suitable model for the intended purpose. Verification ensures the accuracy of the software engineering process and the information preserving transformation processes. Using the correct data and scenario elements ensures data accuracy; this is done with data validation in the context of the intended use. All efforts ensure the usability of the product. In summary, credibility means doing the right thing (validation) using a correct implementation (verification) using the appropriate data and scenario for the intended use.

To ensure the availability of required information, Brade (2004) recommends the following artifacts to support the general processes of V&V.

- The V&V process starts with sponsor's needs. In the step of *problem definition* these needs are transformed from natural language in the dialect of the sponsor into a structured problem description that addresses all information needed by the simulation engineer.

- The structured problem description is used in the next step together with system observations or the application of system theory approaches as inputs for the *system analysis*. This results in a conceptual model that captures entities, behavior, and relations using artifacts that support communication within the team as well as with the sponsor.

- Choosing the best modeling method, e.g. such as described in Hester and Tolk (2010), the ideas captured in this conceptual model are transformed by the *formalization* step into a formal model. It is worth mentioning that while Brade and many European colleagues distinguish between these two model steps, Balci (1997) and many US colleagues recommend using the formal model as the conceptual model, as it captures all information needed for the implementation while the "conceptual model" as defined by Brade (2004) is only a preliminary product.

- Using appropriate solution techniques, program languages, and development environments, the formal model is transformed via *implementation* into the executable model: the simulation system. It has been discussed what represents the executable model, as again various interpretations from algorithms in pseudo-code to the binary executable can be used here.

- The last step is the *experimentation* that transforms input data into simulation results that by themselves need to be validated. Which of these data will be derived from which data sources, and whether these sources are accredited or need to be authoritative data sources, needs to be specified by the sponsor. Respective agreements need to become part of the early definitions.

Figure 14.3 shows Brade's (2004) model development and execution process as well as the deliverables and artifacts connected with these recommended steps. How to use these processes and artifacts in support of V&V is shown in Figure 14.4.

The project starts with the project initialization driven by the sponsor's needs. Task 1 is the problem definition and is conducted in two steps: to actually produce the structured problem description (1.1) and than to verify that this transformation covers—or sufficiently addresses—the sponsor's needs (1.2). In task 2, the system analysis produces the conceptual model (2.1), but part of this task is also to check the consistency—and transformational accuracy—between the structured problem description and the conceptual model, but also between the sponsor's needs and the conceptual model. This schema is repeated for five tasks, ensuring that no inconsistencies or transformational inaccuracies occur in the process. By following these recommendations, the process ensures internal consistency and

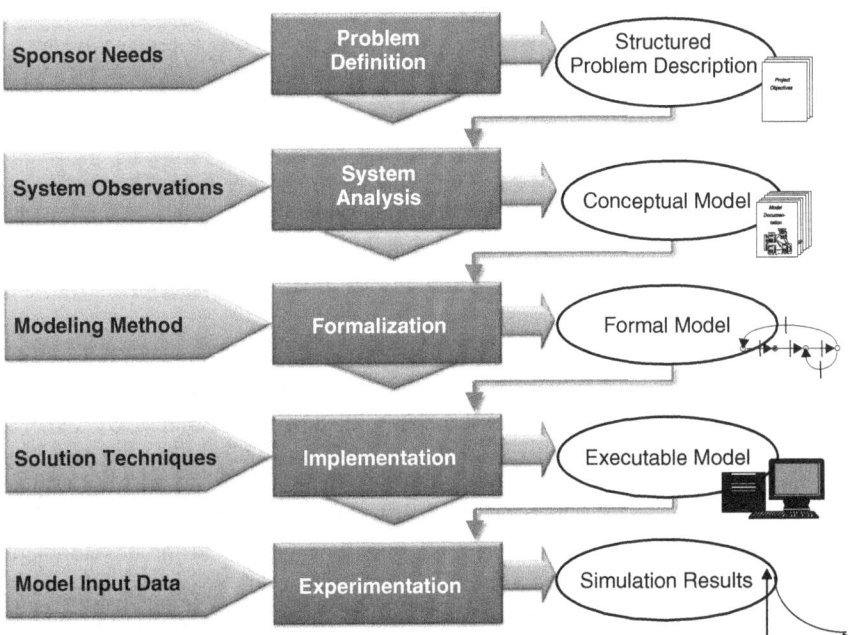

Figure 14.3   Model development and execution.

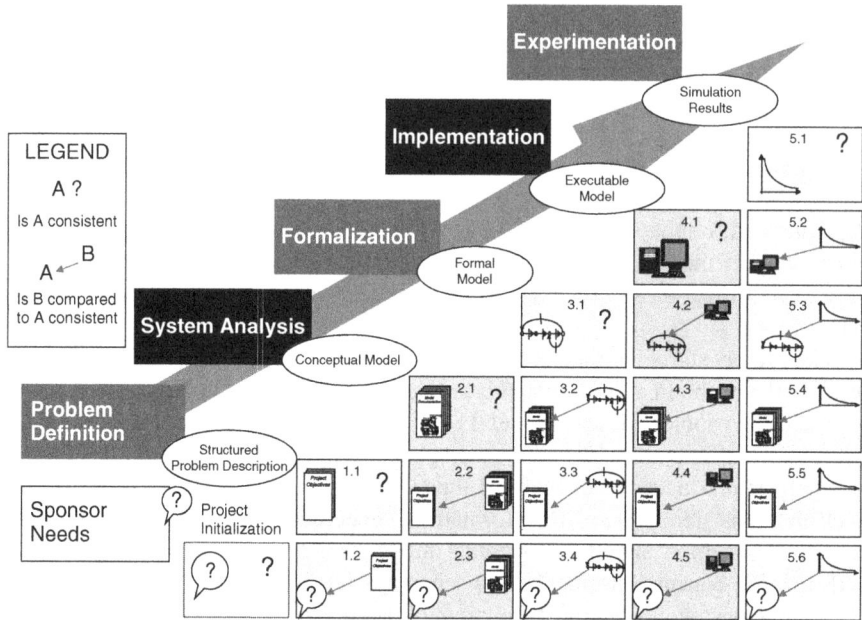

Figure 14.4 Recommended triangle of validation.

completeness as well as the pair-wise consistency and traceability of requirements and decisions throughout the system.

Principles of V&V

All academic approaches identify a set of principles that have to be taken into consideration to ensure successful V&V. While the government processes described later in this chapter do recognize these principles and support them as well, it is worthwhile to enumerate them as principles in this section to emphasize their importance.

- *V&V must be an integrated activity with model selection, development, and integration activities.* It makes no sense to do V&V after the fact. Interaction of subject matter experts with developers and V&V experts in parallel is necessary to avoid design mistakes before they happen. For instance, it ensures that the rationale for development, scoping, and other assumptions are understood and the appropriate documentation is captured. This documentation of rationale is critically important if the development team needs to choose among multiple, perhaps more or less equally valid, choices of implementation, make hard choices owing to scope, and if there are vocal subject matter experts with differing opinions in the field of study. If V&V is done for the finished product, it is too late to address critical issues and the rationale is often lost. As it becomes too expensive to modify the system again, V&V suddenly finds itself in the role of finding "excuses why to use the software despite some errors," which cannot be best practice.

- *The intended purpose needs to be specified precisely.* Too often, the intended purpose is captured in the form of a paragraph of explanations that do not provide the necessary rigor. The intended purpose must explicitly consider and capture specific instances of use and must define metrics that allow one to judge if the system is useful or not for those instances. This requires narrowing the intended purpose to the specific analytic, training, or other objectives and required scope to be met by the model and simulation. It is good practice to apply systems engineering principles, such as described in Buede (2009): the requirements are organized in an objective hierarchy. Each leaf of the resulting tree is specified regarding what acceptable outcomes are based on various use cases. Each use case represents a certain constellation, and all use cases together represent the purpose. Balci (2001) shows how to apply this principle for simulation V&V. The new semantic web standards allow going even further. As envisioned in Tolk et al. (2011), the use of ontological approaches even allows one to reason over requirements to ensure their consistency and to reason over system functionalities to evaluate if a system fulfills the requirements. In any case, "battlefield prose" describing the purpose of a system is insufficient and needs to be avoided.

- *Sufficient knowledge about the system to be simulated is necessary.* It should go without saying that without enough knowledge about the problem entity there is no way to build a good model. However, as discussed in Reynolds (2008), there are two main application categories for simulations. On one side, simulations provide solutions to support training and optimization by representing systems as close to reality as possible. On the other side, simulations are also used to gain understanding about complex systems that are not well understood. Applying V&V to the second category is academically not justifiable. The new domain of human, social, cultural, and behavior (HSCB) modeling often uses models to better understand how social and cultural aspects effect operations, but we have no theory that supports modeling socio-psychological effects as we have for physics based models. But even some technical systems do not expose all needed attributes, relations, and behaviors. There are huge limitations to having the necessary system knowledge available.

- *Sufficient V&V tools and V&V experts are available.* We will have a look at V&V tools in the next section and the various V&V roles in the section on the recommended V&V practices. However, as in every systems engineering environment the success is influenced by three constraints: (a) the support by the management, which includes the presence of *management processes* to guide and support the engineering tasks; (b) the availability of *mature technology*; and (c) an *educated workforce* that understands the objectives and knows how to use the tools. Despite all publications and other efforts, the need for V&V is not yet deeply embedded in the community.

The more formal methods are applied in this process, the better is V&V supported. Examples are using the requirements diagrams and the parametrics diagrams defined in SysML – as discussed in the previous chapter – to capture what needs to be achieved (requirement) and how we measure the achievements (parametrics). Formal and holistic approaches are therefore preferable to independent processes. Sound engineering principles support better V&V, and the simulation engineer shall support respective solutions.

Given the constraints discussed so far it is obvious that we will only be able under very special and unrealistic circumstances to prove sustainability and correctness of systems. However, this cannot be an excuse not to do as much as possible. The methods discussed in the next section can help the simulation engineer to better fulfill these tasks.

Methods Supporting V&V

Balci (1998) categorized the methods that a simulation engineer can use to support tasks as follows:

- *Informal testing methods* are approaches that rely heavily on human intuition and subjective evaluation without rigorous mathematical formalism. Subject matter experts mainly conduct these tests based on their experience with comparable solutions that can be used as a reference.

- *Static testing methods* are approaches that conduct their assessment on the basis of characteristics of the model design and code without execution thereof. The experts evaluate the "blueprints" and static models, such as architecture descriptions or artifacts of frameworks, as discussed in Chapter 12.

- *Dynamic testing methods* are approaches that assess by executing the simulation system and evaluating its results, including comparison with other models or observations in experiments conducted in the real world.

- *Formal testing methods* conduct V&V based on rigorous mathematical proofs of correctness.

Table 14.1 presents the work of Balci (1998). The categorization has been proven to be stable for more than a decade. Even new methods based on the semantic web, like reasoning over OWL expressions to prove consistency and evaluate pair-wise consistency, are captured implicitly under the formal testing methods, as OWL utilizes mathematical logic, predominantly first order logic, to express statements about requirements, entities, relations, behaviors, etc.

Table 14.1 V&V methods

Informal Testing Methods	Static Testing Methods	Dynamic Testing Methods	Formal Testing Methods
• Audit	• Cause–effect graphing	• Acceptance testing	• Induction
• Desk checking	• Control analysis	• Alpha testing	• Inductive assertions
• Documentation	• Data analysis	• Assertion checking	• Inference
• checking	• Fault/failure analysis	• Beta testing	• Logical deduction
• Face validation	• Interface analysis	• Bottom-up testing	• Lambda calculus
• Inspections	• Semantic analysis	• Comparison testing	• Predicate calculus
• Reviews	• Structural analysis	• Statistical techniques	• Predicate transformation
• Turing test	• Symbolic evaluation	• Structural testing	• Proof of correctness
• Walk-throughs	• Syntax analysis	• Submodel/ correctness module testing	• ...
• ...	• Traceability assessment	• Visualization/ animation	
	• ...	• ...	

The VV&A Recommended Practice Guide (2006) has a section on V&V tools that support these tasks and are based on these methods, but even the current version—last updated 2006—focuses mainly on the support of the software engineering process and does not address the modeling level sufficiently. However, the document shows how these software engineering tools can be applied to support the different phases of the M&S development and integration process, as defined in the last chapter. Therefore, the reference document is a valuable starting point on the search for tools that can support the team.

The topic of coping with uncertainty is still not well covered and may not even be well understood. Chapter 28 gives examples of current research and how it can be applied to support the simulation engineer faced with such challenges.

How Much V&V is Enough: Risk-Based V&V

The examples used to show the need for V&V all showed that, potentially, the life of soldiers and other human beings is at risk. In particular in the UK the need for a risk based approach to V&V has been emphasized. The idea is similar to NASA's failure rating system that defines critical components that—if they do fail during an operation—put the life of the crew and other personnel in danger.

Mugridge (1999) presented the approach during several related NATO discussions. He defines severity categories that define how much can go wrong based on not knowing enough about the correctness and sustainability of the system and taking the risk to use it anyhow. These are *catastrophic, critical, marginal*, or *negligible*. While catastrophes shall be avoided at all costs (or nearly all costs, as man-made catastrophes based on purposeful violation of best practices and legally binding procedures show too often), negligible components are acceptable under many circumstances. He also defines a list of impact domains to provide examples of the severity categories. The results are presented in Table 14.2.

The table builds the core and needs to be extended to reflect the values of the sponsor within affected application domains. Also, it gives examples, not norms. Depending on the sponsor, the severity can be defined quite differently, which becomes easily obvious when looking at the entries for financial loss: small companies may become bankrupt with very much less than $1 million.

It is therefore good practice to use a project-specific extension of this table to capture the critical fields of a simulation project. If a simulator is used to replace a radar system in the training of a radar crew, the table can be used to capture what is important and what are the consequences if this specific aspect is not validated accordingly. Also, the values in the supporting culture shift as well. While it was acceptable to suffer some level of injuries in military exercises only some decades ago, today even minor injuries can lead adverse effects. In particular, when simulators are involved the tolerance limit is often well below that for training with operational systems. It is therefore good practice to adjust concrete values to new cultural and social realities.

The underlying question for the efforts described above is "How much V&V is enough?" In particular as we know that only in trivial cases can we prove that a system is correct and suitable, the method described above at least helps to answer this question with "As much V&V is needed as to ensure that no catastrophes happen and to avoid critical impacts!" In complex environments, such as V&V for combat modeling and distributed simulation, this question unfortunately only shifts the problem to another level without answering it, as we still have to understand the systems, their interactions in reality and their implementations well enough, which due to the nature of the problem is not always possible.

In order to support the simulation engineer as much as possible, government agencies and professional societies worked on recommended practices. The next three sections deal with examples.

VERIFICATION, VALIDATION, AND ACCREDITATION (VV&A) RECOMMENDED PRACTICE GUIDE (RPG)

In particular the military services have recognized the need to ensure that simulation systems they apply to procure and test military equipment, train soldiers how to use their equipment efficiently and safely, and to optimize doctrines and strategy are fit for this purpose. Consequently, many different VV&A directives and guidelines exist in the military domain, and they often differ between organizations by emphasizing different viewpoints. The US Modeling and Simulation Coordination Office (US MSCO) compiled the Verification, Validation,

Table 13.2 Examples for Severity Categories in [Assessment] ( )

	Catastrophic	Critical	Marginal	Negligible
Personal safety	Death	Severe injury	Minor injury	Less than minor injury
Occupational illness	Severe and broad scale	Severe or broad scale	Minor and small scale	Minor or small scale
System damage	Loss of system	Major system damage	Minor system damage	Less than minor system damage
Environmental impact	Severe environmental damage	Major environmental damage	Minor environmental damage	Trivial environmental damage
Operator workload	Operator cannot continue to operate system	Severe reduction in the ability to operate system	Major reduction in the ability to operate system	Minor reduction in the ability to operate system
Financial loss	More than $1m	$250k to $1m	$10k to $250k	Less than $10k
Security breach	Top secret or higher	Secret	Confidential	Restricted or less
Reliability	Total loss of functional capability	Severe reduction in functional capability	Significant reduction in functional capability	Slight reduction in functional capability
Project schedule	Slip impacts on overall capability	Slip impacts on other projects	Slip results in major internal schedule reorganization	Schedules republished
Mission impact	Mission loss (operational)	Severe mission degradation (operational)	Slight mission degradation (operational); mission loss (training)	Mission delayed (operational); mission degraded (training)
Criminal liability	Custodial sentence imposed	Large fine imposed ($5k plus)	Small fine imposed (up to 5k)	Conditional discharge etc.
Civil liability	Multiple, large civil suits ($10k plus)	Single, large civil suit ($10k plus)	Multiple, small civil suits (up to $10k)	Single, small civil suit (up to $10k)
Maintenance burden	Project servicing schedules severely adversely affected	Unscheduled maintenance predictions severely adversely affected	Project servicing schedules slightly adversely affected	Unscheduled maintenance predictions slightly adversely affected
Political impact	Government falls	High political official resigns	Congress debates, national press aware	Parliamentary questions, local press aware
Delivered system performance	Design does not meet requirement in critical areas—leading to a failure to accept system	Design does not meet requirement in non-critical areas—leading to major modification program	Impact on operating procedures—leading to small modification program	Some trivial deficiencies—leading to minor corrective measures

and Accreditation (VV&A) Recommended Practice Guide (RPG) to facilitate the application of those directives and guidelines and to promote the effective application of VV&A. The compilation takes several roles and viewpoints into account and presents documents to deal with them. The RPG is a living web-based document that has been influenced by contributions from V&V experts all over the world and is not limited to the US DoD.

Structure of the VV&A RPG

The structure of the VV&A RPG documents reflects the idea that there are three main categories for M&S development and preparation within the general problem solving process as introduced in Chapter 13: (1) reuse of legacy simulations; (2) development of new simulations; and (3) federating simulations to provide a new set of capability. As the VV&A Overlay to the FEDEP that we shall deal with later in the chapter explicitly addresses the third case, the VV&A RPG focuses on the first two cases.

The following topics are covered in the version as available at the time this book was written. When new topics emerge, either new documents are added to the MSCO VV&A website or the documents are updated and specially marked as updates.

- *Key Concepts:* This document is a general description of VV&A concepts applicable to all categories.
- *Core Documents:* Legacy Simulation Applications: This document collection focuses on VV&A challenges when simulation systems are reused. As this reuse is normally based on a new intended use, the challenge of how much of former V&V documents can be reused needs to be addressed. The provided documents are:
  - Legacy Overview
  - User's Role in the VV&A of Legacy Simulations
  - V&V Agent's Role in the VV&A of Legacy Simulations
  - Accreditation Agent's Role in the VV&A of Legacy Simulations
- *Core Documents: New Development*: This collection defines the roles of users, developers, program managers, V&V agents, and accreditation agents when new simulation systems are developed. The provided documents are:
  - User's Role in the VV&A of New Simulations
  - Developer's Role in the VV&A of New Simulations
  - M&S PM's Role in the VV&A of New Simulations
  - V&V Agent's Role in the VV&A of New Simulations
  - Accreditation Agent's Role in the VV&A of New Simulations
- *Special Topics:* This collection of documents addresses fields of general interest that require more in-depth evaluation. These are more guidelines

and ideas for consideration, not legally binding directives. The topics of the comprised documents are:

- ○ Conceptual Model Development and Validation
- ○ Data V&V for New Simulations
- ○ Data V&V for Legacy Simulations
- ○ Fidelity
- ○ Human Behavior Representation Validation
- ○ Paradigms for M&S Development
- ○ Requirements
- ○ Risk and its Impact on VV&A
- ○ Subject Matter Experts and VV&A
- ○ Validation
- ○ Foundations for V&V of the Natural Environment in a Simulation
- ○ Measures
- *Reference Documents:* This last collection comprises papers on special topics that often influence V&V and potentially can facilitate the work of V&V and accreditation agents. The documents in this collection are:
  - ○ Human Behavior Representation (HBR) Literature Review
  - ○ M&S Data Concepts and Terms
  - ○ A Practitioner's Perspective on Simulation Validation
  - ○ T&E/V&V Checklist
  - ○ T&E and V&V Integration
  - ○ V&V Techniques
  - ○ V&V Tools

A critical remark about this otherwise very support worthy collection: it is surprising that only very few of the academic foundations are referenced or taken explicitly into consideration. Furthermore, the focus lies on tools listings that support the verification of software, and as such the simulation aspects are more emphasized than the modeling aspect, where many hard problems for V&V are situated.

On the positive side, the VV&A RPG is a collection of examples and guidelines that help practitioners understand and communicate the need of various VV&A processes, steps, and products and fulfill the roles captured in this guideline.

VV&A Processes and Products

It cannot be the objective of this book to reprint the VV&A RPG. Instead, the interested reader should study the documents for additional interest. What will be done instead is to give a summary of the main ideas to enable an easy start for simulation engineers in the context of VV&A for military systems.

Military V&V has the same objectives as described before. The processes are conducted to determine if a simulation (or a federation) is suitable and correct in the light of the intended purpose. The focus of the processes described in the VV&A RPG is in particular on traceability of evaluation processes and their results in the form of products that can be used later to document the work that has been done. This is particularly important in support of the different development, V&V, and accreditation processes that need to be conducted.

The requirement for accreditation is unique to the US DoD. Accreditation is a certification by a single individual—usually a senior officer or high level civil servant—who has responsibility for the specific program or system that is utilizing the results of the model or simulation. Obviously, this person depends on the recommendations of his/her staff that performed the V&V process, but US DoD policy holds that senior officer personally responsible for accuracy of the VV&A process. This ensures that appropriate command attention is given to the VV&A process to guarantee its correctness. It also ensures that the processes and their results are properly documented in case something goes wrong with the program and the individual needs to show that all directives and guidelines were followed.

Figure 14.5 shows again the overview of the general problem solving process already known from Chapter 13, but this time we focus not only on the core processes of M&S development and preparation, but on V&V and accreditation activities.

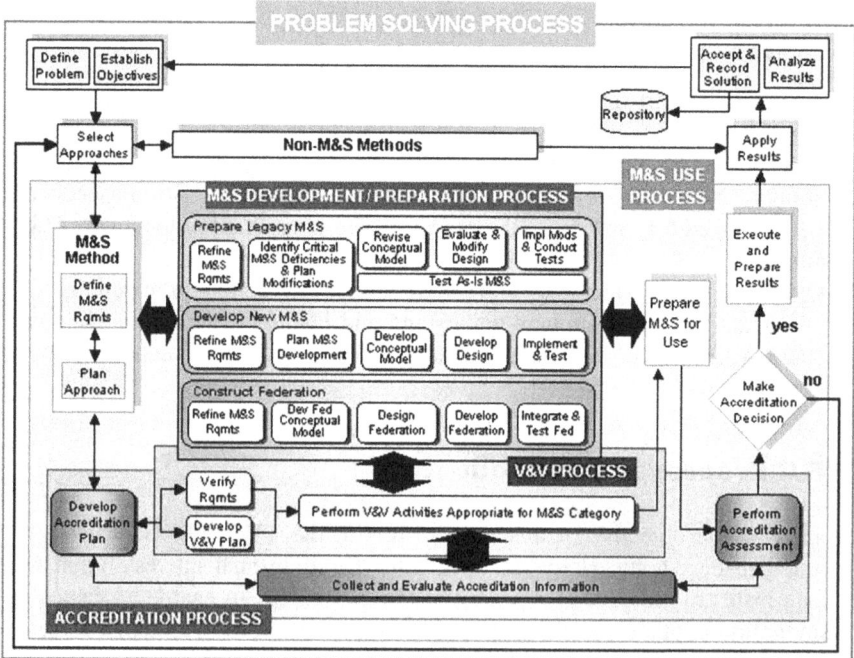

Figure 14.5 Problem solving process (VV&A RPG).

All three processes—the M&S Development/Preparation Process, the V&V Process, and the Accreditation Process—are rooted in the planned approach and the accompanying M&S requirements. Independent from the category chosen to provide the required M&S functionality, three things need to happen.

- The V&V team needs to verify the requirements and their refinements as developed by the developing team in the first step of the development process. *V&V requirements* need to be identified or derived in collaboration with the development team, as the V&V experts define what is needed to support the process, and the developer must ensure that the system exposes the required attributes in the test phase.

- Based on the refined and agreed requirements, a *V&V plan* is established that defines in detail how the V&V requirements will be fulfilled, which functional and operational tests can be utilized, etc. The V&V plan becomes a schema that defines the tests and results. Each test results later in a V&V report that describes the results in detail.

- Based on the V&V plan, the accreditation team establishes an *accreditation plan.* This plan defines the required V&V reports and the expected results in order to be able to accredit the simulation. It is good practice to collect the V&V plan and the V&V reports together with the accreditation plan in a repository to compile accreditation packages in case of need. The accreditation decision itself is also captured in a report that should become part of this collection.

Figure 14.6 shows what needs to be validated and verified using a high level view (and we shall revisit these ideas when we discuss the VV&A Overlay later).

The resemblance to Sargent's (2000) model is clear: requirements are verified, as they are transformed from the sponsor's need. The conceptual model is validated to fulfill the needs. The design is verified to be correctly derived from the conceptual model. The implementation is verified as it is derived from the design. The results are validated, as they have to be the same as observed with the real system. During the whole time, the accreditation team collects and evaluates accreditation information in the form of V&V results and reports to be able to conduct the accreditation assessment at the end of the process. These processes are part of all three categories to provide M&S solutions.

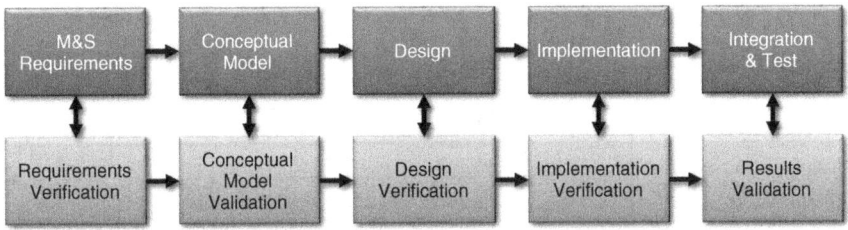

Figure 14.6 High level view on V&V activities.

Like the NATO COBP, the VV&A RPG addresses *data V&V* in extra sections and documents. Although data V&V should be interwoven with other V&V processes, because of the large number of data categories used in a simulation and the amount of time needed to obtain the data, even when a data engineering plan is in place, data V&V has a very unique nature and deserves special attention. The V&V plan should address, among other data-centric questions, the following challenges:

- How to deal with different data set categories?
- How to deal with data sets obtained in phases?
- How to identify the needed expertise required for different data sets?
- How reliable former data certifications are regarded?
- How to deal with data sets that are mixed from authoritative and other sources?

The VV&A RPG provides detailed process structures for V&V processes for preparing legacy simulations as well as for developing new simulations. Figure 14.7 shows the processes for the preparation of legacy simulation systems. The left corner of the figure shows various roles of experts that conduct, support, or are responsible for these processes and activities. We shall look at some selective definitions in the next section.

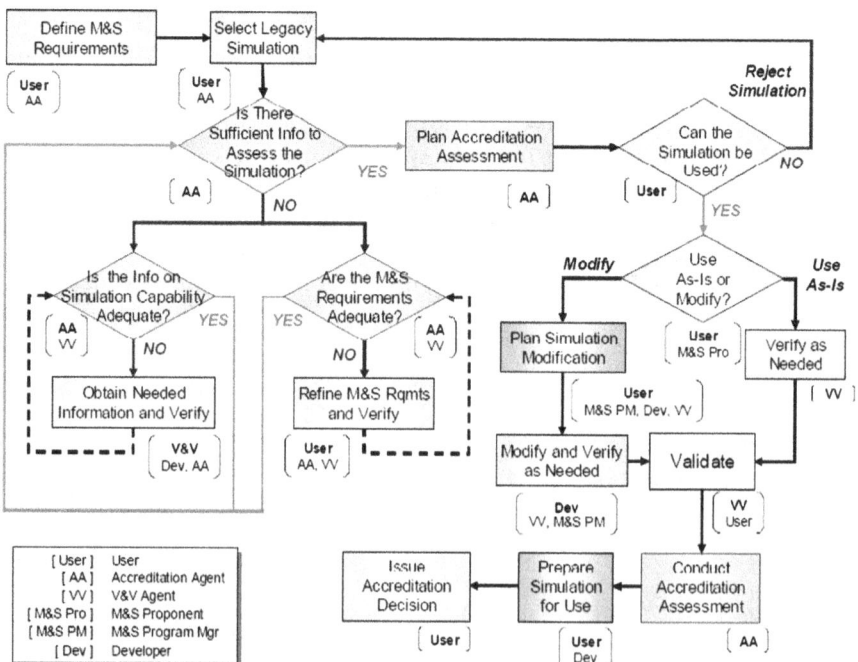

Figure 14.7 VV&A processes for legacy simulations.

VV&A Roles and Responsibilities

Six roles are defined to explain the various responsibilities within the V&V processes and the parallel accreditation process. As already discussed, V&V should be as objective as possible, optimally conducted by a set of people who are not liaised with the developers. The reason is that affiliations that are too close may blur the objectivity and ability to conduct an independent assessment. Every expert knows that after having worked on a particular topic often a fresh pair of eyes is needed to see what the team no longer perceives, as they are too focused on their solutions and underlying assumptions. That is why the NATO COBP highly recommends peer reviews by independent experts.

On the other hand, the three processes of development, V&V, and accreditation have to be connected to deliver optimal results. An error or misperception should be discovered as early as possible to minimize costs and risk connected with it. This requires, quite often, a cultural change that rewards such collaboration efforts and necessary team building integrating all experts.

To support both requirements, V&V roles are introduced. In particular in small teams some persons can play more than one role, but the independence of development and V&V needs to be maintained as well as the independence of V&V and resulting accreditation decision. The Reagan quotation "Trust but verify" plays a central role here. The five roles that are used throughout the VV&A RPG are the following.

1. *M&S User* is the term used throughout the RPG to represent the organization, group, or person responsible for the overall application. The user needs to solve a problem or make a decision and wants to use simulation to do so. The user defines the requirements, establishes the criteria by which simulation fitness will be assessed, determines what method or methods to use, makes the accreditation decision, and ultimately accepts the results. The user also has been referred to as the sponsor of the study or the customer at other places in this book.

2. *M&S Program Manager* (M&S PM) is the term used to define the role responsible for planning and managing resources for simulation development, directing the overall simulation effort, and overseeing configuration management and maintenance of the simulation. In legacy simulation reuse when a major modification effort is involved, the user may designate an M&S PM to plan and manage the modification effort. These are usually senior decision makers or high ranked officers that are supported by a staff.

3. *M&S Developer* is the term used to define the role responsible for actually constructing or modifying the simulation, preparing the data for use in the simulation, and providing technical expertise regarding simulation capabilities as needed by the other roles. The simulation engineer is normally in this role, but not exclusively.

4. *Verification and Validation Agent* (V&V Agent) is the term used to define the role responsible for providing evidence of the simulation's fitness for the intended use by ensuring that all the V&V tasks are properly carried out. This is another role the simulation engineer may have to fill as well—but not at the same time of the development.

5. *Accreditation Agent* is the term used to define the role responsible for conducting the accreditation assessment. The accreditation agent provides guidance to the V&V agent to ensure that all the necessary evidence regarding simulation fitness for use is obtained in the form of V&V reports, the evidence is collected and assessed, and the results are provided to the accreditation authority.

These are the five main roles, but they are often completed with additional ones. In Figure 14.7, for example, the *M&S Proponent* is mentioned. This role is responsible for managing the simulation throughout its lifecycle, establishing and ensuring configuration control of the official version of the simulation, maintaining and enhancing its capabilities, managing its usage, and protecting it from damage and misuse. The M&S Proponent owns the simulation that several M&S users use to solve their problems.

Very often, *subject matter experts* (SMEs) are needed to support special tasks. They may be needed to gain insight into the system to be modeled as well as the simulation that is used. Whenever special knowledge is required, an SME can help to provide this knowledge, but it remains the responsibility of M&S developers, V&V agents, and accreditation agents to integrate the knowledge in support of their processes.

Similar to bringing the M&S Proponent into the picture when several users use the same simulation system, it may be advantageous to explicitly bring the *Accreditation Agency* in as an explicit role, in particular when accreditation decisions mutually influence each other, as they may compete for resources, etc.

The VV&A RPG furthermore defines six participation categories or responsibilities of role personnel for VV&A tasks:

1. *Perform*: actually conducts the task and "does the work"
2. *Assist*: supports the task by conducting tests, collection data, etc.
3. *Lead*: leads the task as a chair of the performers, which may include performing the task itself as well
4. *Monitor*: oversees the task ensuring it is done correctly
5. *Review*: reviews result and recommends improvements
6. *Approve*: determines the satisfactory completion

Table 14.3 gives an overview of which roles participate with which responsibility in which VV&A processes. The table is just a subset of typical roles and responsibilities associated with M&S VV&A. It is good practice to create a project-specific table that captures roles and responsibilities in the project plan as part of the work breakdown structure, including planning for resources

Table 14.3 Typical Roles and Responsibilities Associated with M&S VV&A

Activity	Role M&S User	M&S PM	M&S Developer	V&V Agent	Accreditation Agent	SME
Define requirements	Lead Approve	Monitor	Assist	Review	Review	Assist
Define metrics	Lead Approve	Monitor	Assist	Assist	Assist	Assist
Define MOP/MOE	Assist Approve	Monitor	Assist	Assist	Lead	Assist
Plan M&S development or modification	Lead Assist Approve	Lead Assist	Assist	Assist		
Develop V&V plan	Review	Assist Approve	Review	Lead	Assist	
Develop accreditation plan	Review Approve	Assist		Assist	Lead	
Verify M&S requirements	Lead Approve	Monitor	Assist	Lead	Assist	Assist
Develop conceptual model	Assist Approve	Monitor	Lead			Assist

(continued)

283

Table 14.3 (*Continued*)

Activity	M&S User	M&S PM	M&S Developer	V&V Agent	Accreditation Agent	SME
				Role		
Validate conceptual model	Assist Approve	Monitor	Assist	Lead		Assist
Develop design		Monitor Approve	Perform			
Verify design	Approve	Monitor	Assist	Lead		Assist
Implement design		Monitor Approve	Perform			
Data V&V	Approve	Monitor	Assist	Lead		Perform
Verify implementation	Approve	Monitor	Assist	Lead		Assist
Test implementation	Approve	Monitor	Lead	Assist		Assist
Validate results	Assist Approve	Monitor	Assist	Lead		Assist
Prepare V&V report			Assist	Perform		
Accreditation of V&V efforts	Monitor	Monitor			Perform	Assist
Prepare accreditation report					Perform	Assist

and budget. All roles, participation types, and processes are detailed in the guidelines with examples.

Particularly with smaller projects, the responsibilities of the M&S PM, the M&S User, and the M&S Proponent may often be undertaken by one person. Very often, the SME responsibilities will be carried out within the team, which must be considered when determining the credibility of assumptions and constraints (you would not trust a medical simulation system giving advice for open heart surgery that was programmed—without the help of a surgeon—by a highly skilled Java programmer using medical textbooks, would you?).

As stated in Chapter 2, combat models and distributed simulations need to be developed as professionally as possible, which includes proper ethical norms and standards. The VV&A RPG cannot replace a code of ethics but can provide the know-how and checklists of tasks recommended to be performed to ensure that credibility, suitability, and correctness of a simulation system are given and justified.

The Codes of Ethics, the NATO Code of Best Practice for C2 Assessments, and the VV&A RPG should therefore be known, understood, and valued by every professional simulation engineer.

IEEE STD 1516.4-2007: RECOMMENDED PRACTICE FOR VV&A OF A FEDERATION

The technical parts of the IEEE 1516 standard were described in some detail in Chapter 12; the FEDEP and DSEEP as recommended practice documents describing how to develop and execute a federation based on this standard were given in Chapter 13. The IEEE 1516.4-2007 is another recommended practice standard describing an overlay to the FEDEP that specifies the VV&A activities in parallel to the development. At the time this book was written the standard was not yet transformed to support the DSEEP, but as FEDEP and DSEEP are sufficiently similar such a transition is to be expected shortly.

On one side, IEEE 1516.4-2007 can be seen as the "missing part" of the VV&A RPG that addresses modification and reuse of a legacy system and the development of new systems but does not specify the details of federation development. On the other side, the efforts in compiling these standards are much broader, as consensus was not only reached with US DoD, but expert groups of SISO as well as a NATO task group (NATO, 2008) contributed to the document which is an official standard (while the VV&A RPG is a non-standardized guideline). Therefore, it is justifiable to deal with IEEE 1516.4-2007 in additional detail. A good overview is given by Masys (2006).

The general idea of IEEE 1516.4-2007 is to provide an overlay to the FEDEP that uses the definitions and descriptions of the steps, tasks, and activities that are already standardized for the development of a federation and adds the required phases and activities for verification, validation, and accreditation in the same detail. Figure 14.8 shows the seven steps of the FEDEP accompanied by the

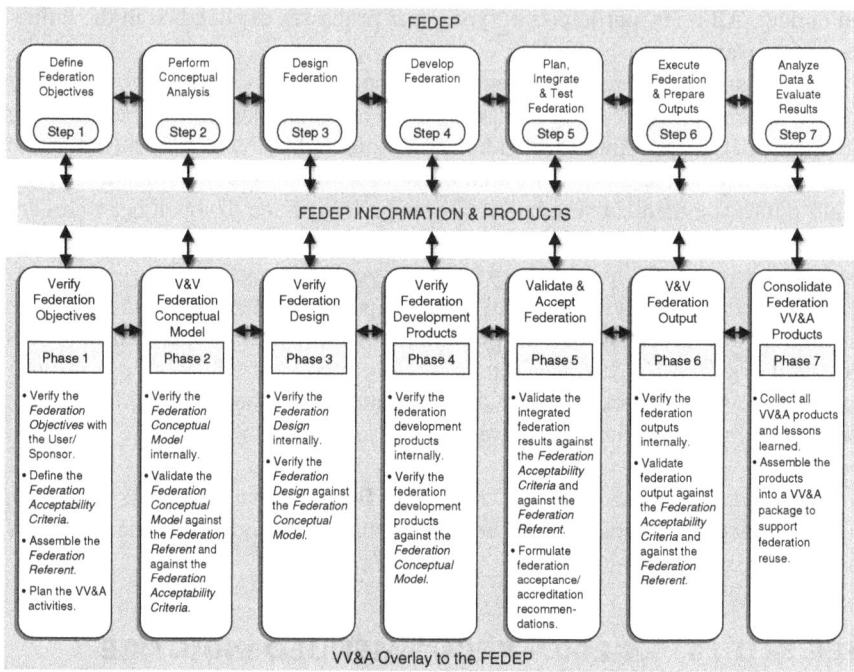

Figure 14.8 Top level view of the VV&A overlay to the FEDEP.

seven top level steps of recognized and standardized VV&A processes. The FEDEP is shown on top; the VV&A processes are shown below. The coupling happens via the products and artifacts as they are defined and standardized in the FEDEP standard (IEEE 1516.3-2003) described in Chapter 13.

The principles are well aligned with the terms introduced so far in this chapter, but some aspects are generalized in comparison with the VV&A RPG. An example is the conceptual model which is validated in the RPG but is internally verified as well as validated against federation referents—often observable real-world problem entities—based on acceptability criteria; in other words, the level of detail and integration with other necessary tasks, such as definition of acceptability criteria in early phases, has been improved.

The standard also defines the same roles and responsibilities as recommended in the RPG and comprises a table showing who should do what for the identified phases (top level) and related activities. The defined activities do not only conduct VV&A, but support and contribute many of the products as well. The community agreed to standardize the following activities, which do not exclude additional activities to be defined and captured in the plans for specific projects.

- Phase 1—*Verify federation objectives* comprises the activities support identifying user/sponsor needs, plans accreditation activities, supports developing federation objectives, contributes to verifying federation objectives, assembles federation referent, defines federation acceptability criteria, and plans V&V activities.

- Phase 2—*The V&V federation conceptual model* comprises the activities support developing federation scenarios, contributes to verifying federation scenarios, supports developing federation conceptual model, contributes to verifying federation conceptual model, validates federation conceptual model, supports developing federation requirements, and contributes to verifying federation requirements.

- Phase 3—*Verify federation design* comprises the activities support selecting federates, supports preparing federation design, contributes to verifying federation design, and supports preparing the federation development and execution plan.

- Phase 4—*Verify federation development products* comprises the activities that support developing the FOM, contributes to verifying the FOM, supports establishing federation agreements, contributes to verifying federation agreements, supports implementing federate designs, supports implementing federation infrastructure, contributes to verifying federation infrastructure, and verifies and validates federation data sets.

- Phase 5—*Validate and accept federation* comprises the activities support planning federation execution, supports integrating federation, contributes to verifying integrated federation, supports testing federation, validates integrated federation results, and performs acceptance assessment.

- Phase 6—*Verify and validate federation output* comprises the activities support executing federation, contributes to verifying raw execution and derived output, and validates federation output.

- Phase 7—*Consolidate federation VV&A products* comprises the activities support analyzing data and prepares federation VV&A products for reuse.

All the activities including how they are interrelated by their input and outcomes are defined in the standard. In addition, assumptions are enumerated and recommendations for tailoring and extensions are given. Although the current vocabulary is focused on the IEEE 1516 High Level Architecture standard, using the HLA Overlay of the DSEEP allows already a generalization of theses ideas to make them applicable to other federation alternatives, as discussed in Chapter 12.

ADDITIONAL VV&A EFFORTS

Both examples so far were supported by international experts but are often seen as very US-centric. To get a better understanding of what alternative and supportive efforts are going on, Masys from Canada, Emmerik from The Netherlands, and Bouc from France (2006) did a comparison of additional VV&A related activities. Their paper is still a good start to get an overview. They address both topics of the last two sections with special focus on the VV&A Overlay. In addition, they also present the European effort Referential for VV&A (REVVA) and other related activities that are currently still being evaluated under the Chair of SISO, the General Model for VV&A (GM VV&A).

REVVA

The kick-off for the European efforts to contribute to ideas on verification and validation may be traced back to a French–UK joint study (AFDRG) on validation of technical-operational simulations. The two-year study concluded in 2002 that the definitions at that time proposed by the US Defense Modeling and Simulation Office (DMSO) were focused too much on technical problems, and that socio-technical aspects are as important for the acceptance of simulation systems. They observed that many difficulties that Europeans had with the US-driven approach had their origin in the concepts underlying VV&A approaches, and that therefore new paradigms are needed to stress that validation is a decision making activity based on socio-technical aspects.

The first REVVA study was conducted under the chair of the West European Armament Group (WEAG) from 2004 to 2005. The French Thales group had the lead; the Netherlands, Denmark, Sweden, and Italy were involved and contributed to this joint project. The objective of REVVA-1, also known as THALES JP 11.20, was to establish the basis of a European VV&A framework. One of the main results was an extensive literature research and evaluation that clearly showed the need for the integration of socio-technical aspects in addition to the physical-technical aspects. In addition, REVVA-1 recommended another viewpoint. Instead of applying processes, the method should be product-oriented. The objective of VV&A must be to produce the relevant V&V products which contribute to a good decision making. It also emphasized that the acceptance of a simulation is a project stakeholder decision, related to the operational objective, which consequently emphasized the purpose of intent even more.

Using these recommendations, the second REVVA study was initiated as the EUROPA 111.104 program and conducted between 2006 and 2007. It was conducted by an international consortium composed of France, the Netherlands, Denmark, Sweden, the UK, and Canada. The objectives of REVVA-2 were (1) to produce a comprehensive VV&A methodology addressing the issues that were raised by the first study, and (2) to deliver four major contributions to be used as drafts for further standardization. These documents include the following (Masys et al., 2006).

- A *user manual* guides users through the VV&A effort and clarifies their responsibilities by explaining how to apply the methodology in practice. It describes the activities to be performed, the products to be produced, the interactions that take place among those involved, the flow of products, and how to tailor the methodology to the specific needs of the M&S project.
- A *reference manual* documents the underlying concepts of the methodology, including the foundations of the chosen terminology, the explanation

of the dependences between activities and products, their meaning for the VV&A endeavor, and the rationale for their creation, tailoring, and execution. The reference manual is referred to whenever a deeper understanding of the methodology is required.

- The *VV&A Recommended Practices Guide* is a document devoted to an audience of M&S and V&V project leaders but without entering into the description of technical details or technical solutions. This document is comparable with the abbreviated *NATO Code of Best Practice—Decision Maker's Guide* described in Chapter 2.
- The *Technical Notes* document is the most detailed technical document of the VV&A methodology. It is used by V&V executioners for a very detailed knowledge of tools and techniques recommended by the methodology and for good practices of their uses.

The resulting recommended activities and products are shown in Figure 14.9. What is new in comparison to the earlier sections is that all stakeholders explicitly agree on Terms of Acceptance (ToA) including acceptance criteria (AC) as well as related metrics. Similarly, the Terms of Verification and Validation (ToVV)

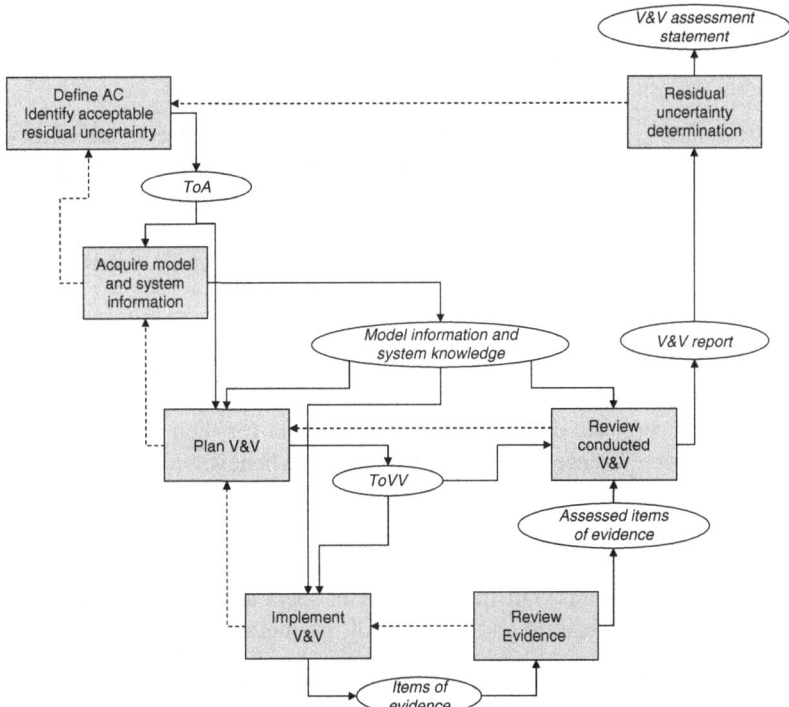

Figure 14.9   REVVA processes and products.

are explicitly captured. These two important documents are the contractual foundation for assessing the suitability and correctness of a simulation or a federation for the operational context and the intended use.

Following the ideas of the SEDEP as described in Chapter 13, all products—ToA, ToVV, V&V report, and V&V assessment statement—are associated with the simulation products in the repository to facilitate the decision making process when the simulation product may be reused.

If, for example, the simulation solution was successfully used for training, a decision maker can evaluate the terms of acceptance and V&V for training in detail. This allows the decision maker to evaluate how much additional work is needed to fulfill acceptance criteria when the solution will be used to generate alternative courses of action in support of a commander in real combat situations.

SISO Product Development Group: Generic Methodology for VV&A in the M&S Domain

As described in Chapter 12, SISO standards are developed by product development groups. One of the current study groups proposes to develop a product for the international community for a generic V&V and acceptance methodology for models, simulation, and data. The product will leverage and harmonize with the contributions from other national and international VV&A initiatives such as those described in this chapter, plus others. The proposed products include the following.

- The *Handbook* guides its users through the V&V and acceptance efforts and clarifies their responsibilities by explaining how to apply the methodology in practice. It describes the activities to perform and the products to produce, the interactions taking place among those involved, the flow of products, and how to tailor the methodology to the specific needs of the M&S project.

- The *Reference Manual* documents the underlying concepts of the methodology, including the foundations of the chosen terminology, the explanation of the dependences between activities and products, their meaning for the V&V and acceptance endeavor, and the rationale for their execution and creation. The reference manual is referred to whenever a deeper understanding of the methodology is required.

- The *Recommended Practices* document provides user-specific guidance with regard to the selection and use of techniques and tools in support of the Handbook. This will include domain-specific case studies thereby illustrating the application and tailoring of the methodology.

In support of these activities, Masys et al. (2006) used the work of their colleague Stephane Chaigneau (Direction Générale de l'Armement, DGA) to establish a vision of how to bring the different activities together. Figure 14.10

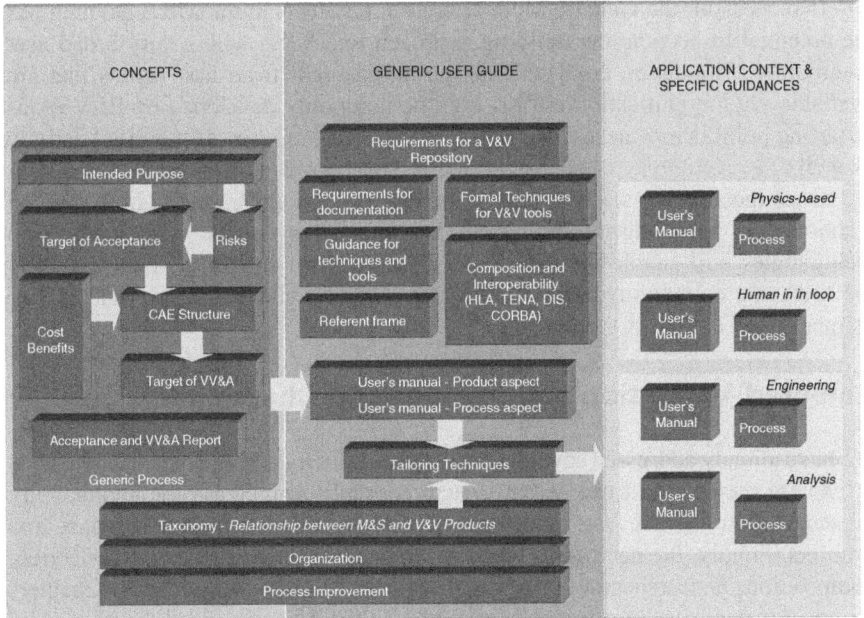

Figure 14.10   Vision for a unified view of international VV&A initiatives.

presents a group of products, the above target products included, as envisioned to provide a solid foundation for a common understanding of VV&A.

Based on a common taxonomy that clearly defines the relationship between M&S products and VV&A products, three parallel but mutually supportive activities can contribute definitions for

- a *generic process* that specifies the concepts of V&V. This process is rooted in academic research and vetted by operational necessities.

- a *generic user guide* that projects the theory-based concept into applicable methods. This user guide becomes a schema that serves as a common core. This core can be tailored and expanded to reflect the special needs of user communities.

- *application-domain-specific user guides and process definitions* that support the application domains and their communities. The four examples in the figure are neither complete nor exclusive; other communities will be supported as well.

Each project can use either the generic process and generic user guide as a starting point, or the already pre-tailored and modified domain-specific recommendations. In any case, the project will adapt these recommended solutions to the specific needs that are characteristic for the operational context and the intended use.

If successful, the Generic Methodology for VV&A in the M&S Domain has the potential to become the unifying approach to VV&A with a very broad user community. Simulation engineers can already benefit from first results that are available via the unification efforts, e.g. the taxonomy developed by REVVA as a starting point is now accessible for other users that are part of the SISO activity as well.

Once the structures and concepts are agreed upon and captured in traditional technologies, migrating the definitions into the richer context of semantic web technologies will enable the use of intelligent system support as envisioned by the academic community and described by—among others—Tolk et al. (2011), as mentioned earlier in the chapter.

Selected Validation and Verification Challenges

We have already addressed several challenges within this chapter, such as the need to capture the intended use of the system formally to increase the applicability of semantic tools. We also addressed the application of formal methods and proofers. Finally, the need for a blueprint to support the identification, selection, composition, and orchestration of valid services, such as described in Chapters 12 and 13, needs to be supported by V&V, as the blueprint needs to be a valid model of the research question to be addressed, only valid components should be selected, and the composition and orchestration needs to be validated and verified as well. It should be pointed out again that the composition of valid components can lead to invalid results, so validation of a federation is essential in any case.

Of particular interest is the domain of agent-directed simulation as defined in more detail in Chapter 27. The traditional system modeling approach defines the system's behavior on the macro level in the form of equations. System dynamics represent this view on how to best model a system. The agent paradigm shifts the definition to the micro level where the interplay between agents is defined. The resulting system behavior emerges from the multitude of entity interactions on the micro level. An example is a tank attack of several units: although the individual tanks follow simple rules on when to shoot and when and where to move in communication with neighboring tanks, looking at the resulting movement of the units you can observe "waves of attack" that emerge from the simple rules without having to be programmed into the system. Chen and colleagues (2009) show how such emergence can be used to engineer systems. The challenge is that traditional V&V approaches are rooted in top-down approaches to system modeling, not for emerging behavior.

How to validate and verify agent-directed simulation systems is topic of current research and most likely will require more discussion. Examples of positions on these topics are given by Windrum and colleagues (2007) and Moss (2008). As agent-directed models are increasingly used in particular to address the human, social, cultural, and behavior modeling challenges described in Chapter 26, these challenges need to be addressed and solutions supporting V&V need to be agreed on soon.

Finally, the question posited at the end of Chapter 11 needs to be addressed systematically: when does it makes sense to federate simulation systems, and when is it preferable to use them in a coordinated effort of multi-simulations to support exploratory analysis of various facets of the problem? In particular in domains where the expert opinions are divergent and the obtainable data are uncertain, incomplete, and contradictory, it is good practice to use several approaches in parallel to get a better understanding of the problem domain as well as of the contribution potentials of several solutions before deciding which approach is the best and what additional solutions can be federated to support this master approach.

As discussed by Tolk (2010), neither current interoperability standards nor V&V techniques support answering these kinds of question sufficiently currently. Simulation engineers keep observing and contributing to this field of ongoing research.

REFERENCES

Balci O (1997). Principles of simulation model validation, verification, and testing. *T Soc Comput Simul I* **14**, 3–12.

Balci O (1998). Verification, validation, and testing. In: *The Handbook of Simulation*, edited by J Banks. John Wiley & Sons, New York, NY, pp. 335–393.

Balci O (2001). A methodology for certification of modeling and simulation applications. *ACM T Model Comp Simul* **11**, 352–377.

Brade D (2004). A generalized process for the verification and validation of models and simulation results. PhD Thesis at the University of the Federal Armed Forces of Germany, Munich.

Buede DM (2009). *The Engineering Design of Systems, Models and Methods* (2nd edition), *Wiley Series in Systems Engineering*. John Wiley and Sons, New York, NY.

Chen C-C, Nagl S and Clack C (2009). Complexity and emergence in engineering systems. In: *Complex Systems in Knowledge-based Environments: Theory, Models and Applications*, edited by A Tolk and L Jain, Studies in Computational Intelligence, Springer, Berlin/Heidelberg, Vol. 168, pp. 99–128.

Hester PT and Tolk A (2010). Applying methods of the M&S spectrum for complex systems engineering. Emerging applications of M&S in Industry and Academia (EAIA). In: *Proceedings of the Spring Multi-Simulation Conference, Proceedings of the Spring Simulation Multiconference*, published by SCS, San Diego, CA, *Orlando, FL*, pp. 17–24.

IEEE (2007). *IEEE Std. 1516.4-2007: Recommended Practice for Verification, Validation, and Accreditation of a Federation—An Overlay to the High Level Architecture Federation Development and Execution Process*. IEEE Publication, Washington, DC.

Masys AJ (2006). Verification, validation and accreditation: an HLA FEDEP overlay. In: *Proceedings of the Canadian Conference on Electrical and Computer Engineering, published by IEEE Ottawa, Canada, Ontario*, pp. 1850–1853.

Masys AJ, van Emmerik ML and Bouc P (2006) Verification, validation and accreditation (VV&A)—leveraging international initiatives. In: *Proceedings of the NATO MSG Conference on Transforming Training and Experimentation through Modelling and Simulation*. NATO Report *RTO-MP-MSG-045*, Paper 10. Neuilly-sur-Seine, France.

Mugridge C (1999). *Verification, Validation and Accreditation of Models and Simulations Used for Test and Evaluation—a Risk/Benefit Based Approach*. Internal Report, Technical Development Group, Defense Evaluation and Research Agency, UK.

Moss S (2008). Alternative approaches to the empirical validation of agent-based models. *J Artif Soc Social Simul* **11**, paper 5; online journal accessible via http://jasss.soc.surrey.ac.uk/11/1/5.html.

NATO (2008). *Verification, Validation, and Accreditation (VV&A) of Federations*. NATO Report RTO-TR-MSG-019, Brussels, Belgium.

Reynolds PF (2008). The role of modeling and simulation. In: *Principles of Modeling and Simulation: A Multidisciplinary Approach*, edited by JA Sokolowski and CM Banks. John Wiley & Sons, New York, NY, pp. 25–43.

REVVA website: http://www.revva.eu.

Sargent RG (2000). Verification, validation, and accreditation of simulation models. In: *Proceedings of the Winter Simulation Conference*. IEEE Computer Press, Piscataway, NJ, pp. 50–59.

Sargent RG (2001). Some approaches and paradigms for verifying and validating simulation models. In: *Proceedings of the Winter Simulation Conference*. IEEE Computer Press, Piscataway, NJ, pp. 106–114.

Shannon RE (1975). *Systems Simulation and the Art of Science*. Prentice Hall, Eaglewood Cliffs, NJ.

Tolk A (2010). M&S body of knowledge: progress report and look ahead. *SCS M&S Magazine*, Vol. 1, No. 4, October 2010; online magazine available via http://www.scs.org/msmagazine (last accessed June 2011).

Tolk A, Adams KM and Keating CB (2011). Towards intelligence-based systems engineering and system of systems engineering. In: *Intelligence-based Systems Engineering*, edited by A Tolk and L Jain, Intelligent Systems Reference Library, Springer, Berlin, Vol. 10, pp. 1–22.

US Modeling and Simulation Coordination Office (2006). *Verification, Validation, and Accreditation (VV&A) Recommended Practices Guide (RPG)*. Alexandria, VA. http://vva.msco.mil (last accessed June 2011).

Windrum P, Fagiolo G and Moneta A (2007). Empirical validation of agent-based models: alternatives and prospects. *J Artif Soc Social Simul* 10, paper 8; online journal accessible via http://jasss.soc.surrey.ac.uk/10/2/8.html (last accessed June 2011).

Chapter 15

Integration of M&S Solutions into the Operational Environment

Andreas Tolk

DEFINING THE OPERATIONAL ENVIRONMENT

In order to understand the research on how to integrate modeling and simulation (M&S) solutions into the operational environment, the first thing to understand better is what the operational environment itself is. Answering this question is particularly difficult, as there are no authorities that address these issues holistically. Furthermore, as we are interested in all topics in the general and international context, we cannot expect to get a common solution that is agreed upon by all potential contributors.

Working Definition

The operational environment can be defined as the infrastructure that supports the military user with all information necessary to optimize the decision process. In this chapter we are in particular interested in soldiers and their supporting infrastructure, in particular the command and control (C2) information technology (IT) infrastructure. Although the environment is not limited to IT—there are many additional things that have to be taken into consideration when addressing the domain of C2, as described in detail in Chapter 3—the simulation engineer

Engineering Principles of Combat Modeling and Distributed Simulation, First Edition.
Edited by Andreas Tolk.
© 2012 John Wiley & Sons, Inc. Published 2012 by John Wiley & Sons, Inc.

will probably focus on integration aspects of simulation services into the C2 IT infrastructure, so that will be the focus and core of this chapter as well.

Our working definition is therefore the following: *The operational environment is the environment in which the military user conducts evaluations and makes decisions. It comprises the C2 IT infrastructure embedded into processes and procedures—including human, organizational, and political components as they are affecting system of systems—that define the C2 system to which the military user contributes.*

In other words, everything that defines the work environment of the military user, in particular but not limited to the IT infrastructure, belongs to the operational environment. Also, whenever M&S supports elements of the operational environment by providing data or even a synthetic environment, or whenever M&S provides services needed within this environment, the question regarding how to support the integration has to be addressed.

As discussed in earlier chapters, there are several standards and common objectives supporting such integration efforts. In particular web technologies have proven to become facilitators for integration, mainly driven by XML, web services, and RESTful services (as described in Chapter 12). Most C2 IT systems now provide XML versions of their data and support their exchange using web services or RESTful services. This does not solve the conceptual challenges of different scopes, different resolution, and different structures, but at least a common exchange mechanism and a common syntax are established based on internationally accepted methods. These standards help us to prepare *how* to exchange information. The focus of the next section is to understand *what* information to exchange and *why*.

Use Cases and Application Examples

Why do we want to integrate simulation services into the operational environment? The answer to this question is relatively simple: because simulation services provide the required functionality that cannot be provided by another service already in the operational environment. Going back to the different use cases and application domains defined in Chapter 4, the following examples motivate the need for the integration of simulation systems and C2 IT infrastructures.

- *Training:* In order to provide good training, one of the most quoted principles is "to train as you fight!" In other words, the training experience will be as realistic as possible, which requires that the operational environment will be integrated into the training environment that provides all information at the right time in the right format to the right operational component that is accessible to the trainee. If a soldier has to use a certain format provided by the C2 IT infrastructure to generate an order, the same steps should be followed and the same programs should be used during training as later during the real operation. If in real operations certain reports come in via certain communication channels, this information will be provided by exactly these channels as well during the training. A simulation system used to train soldiers must therefore integrate the operational environment.

- *Testing:* In order to test new C2 IT systems that will be integrated themselves as new systems into the operational environment these systems should be exposed to inputs that are as realistic as possible in order to test them. It may even be desirable to put some extra stress on the new systems when they are tested to see how they react under extreme circumstances. Nonetheless, the input needs to be generated via the channels and in the format expected in the real-world operation, so that in the ideal case the system can be switched between the virtual test environment based on combat models and the operational environment without additional work.

- *Decision support:* As discussed in Tolk (2009), *decision support systems* are information systems supporting operational—business and organizational—decision making activities of a human decision maker and they help decision makers compile useful information from raw data and documents that are distributed in a potentially heterogeneous IT infrastructure, personal or educational knowledge that can be static or procedural, and business models and strategies to identify and solve problems and make decisions. Adding M&S makes them *decision support simulation systems* that use decision support system means to obtain, display, and evaluate operationally relevant data in agile contexts by executing models using operational data exploiting the full potential of M&S and producing numerical insight into the behavior of complex systems. For this category, the operational environment embeds the simulation services.

Integrating simulation services and operational environments therefore has the potential to allow for better training as the soldiers can use their original C2 equipment to receive operationally relevant information, plan their actions and reactions, orchestrate their efforts with real or potentially also virtual partners, and give orders. It allows for better training as the test data generated by the virtual environment can be calibrated for optimal testing and the test is based on real-world operational interfaces. Finally, by using real-world data to initialize simulation services with a current perception of the situation, simulation services can be used to conduct what-if analysis and can simulate various alternative courses of action, which is supporting making better decisions.

While the use cases for training and testing are relatively well accepted, the application of simulation services as decision support systems is not yet completely accepted everywhere in the community. However, developments in information technology allows the efficient distribution of computing power in the form of loosely coupled services, as discussed in Chapter 12. Consequently, the idea to use an orchestrated set of operational tools—all implemented as services that can be loosely coupled in case of need—to support the decision maker with analysis and evaluation means in an area defined by uncertainty, incompleteness, and vagueness regarding the available information is gaining ground. In order to measure improvement in this domain, in the operational community the value chain of net-centric warfare was introduced; see among others Alberts and Hayes (2003). The idea is close to Ackoff's (1989) ideas captured in his famous paper "From data to wisdom."

- *Data* are factual information. The value chain starts with *data* quality describing the information within the underlying C2 systems.
- *Information* is data placed into context. *Information* quality tracks the completeness, correctness, currency, consistency, and precision of the data items and information statements available.
- *Knowledge* is procedural application of information. *Knowledge* quality deals with procedural knowledge and information embedded in the C2 system such as templates for adversary forces, assumptions about entities such as ranges and weapons, and doctrinal assumptions, often coded as rules.
- Finally, *awareness* quality measures the degree of using the information and knowledge embedded within the C2 system. Awareness is explicitly placed in the cognitive domain.

Following the ideas of Alberts and Hayes (2003), Tolk (2009) proposed the following view: if the thesis is correct that C2 quality is improved by an order of magnitude whenever a new level of quality is reached in this value chain, then simulation services in connection with semantic web technology and artificial intelligence technologies—as described in this book—can support all levels up the awareness quality. Figure 15.1 depicts the quality categories and the supporting technologies.

Data quality is characterized by stand-alone developed systems exchanging data via text messages as used in most C2 systems. Having the same data available at the distributed locations is the first goal to reach.

By the introduction of a common operational picture, data are put into context, which evolve the data into information. The collaborating systems using this common operational picture result in an order of magnitude of improvement of the C2 quality, as decision makers share this common information: "a picture is worth a 1000 words!"

The next step, which is enabled by service oriented web-based infrastructures, is using simulation services for decision support. Simulation systems are the prototype for procedural knowledge, which is the basis for knowledge quality. Instead of just having a picture, an executable simulation system can be used.

Finally, using intelligent software agents to continually observe the battle sphere, to apply simulations to analyze what is going on, to monitor the execution of a plan, and to do all the tasks necessary to make the decision maker aware of what is going on, C2 systems can even support situational awareness, the level in the value chain traditionally limited to pure cognitive methods.

This is the future state-of-the-art that we want to reach, but as the next sections in this chapter will show, we are still far away from implementing

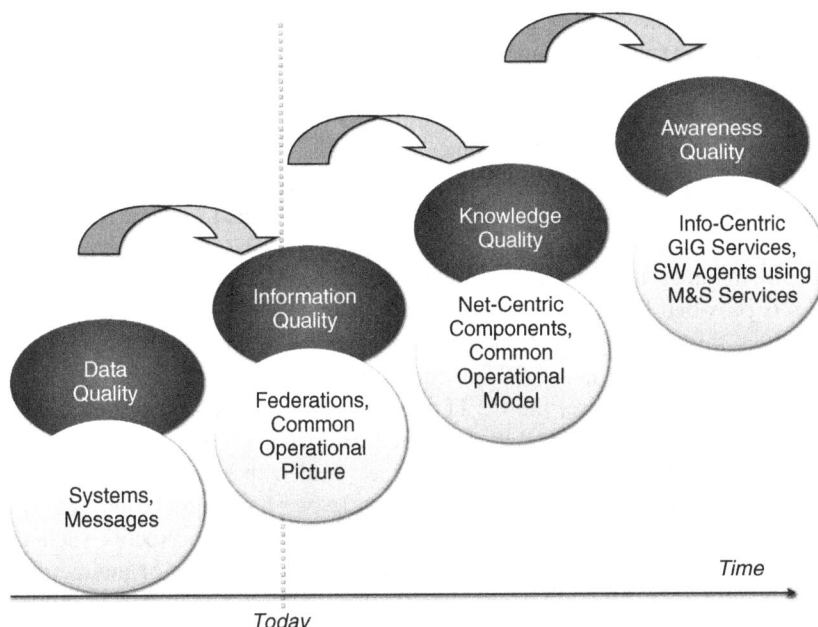

Figure 15.1   Command and control improvements.

this vision. Nonetheless, the vision will help us to better understand the various contributions of existing solutions, as they will be discussed in the remainder of this chapter.

OPERATIONAL C2 ENVIRONMENT

One of the main concerns when bringing operational C2 and M&S together is always security. No matter how well designed the integration methods are, unclassified systems that have not been rigorously tested to support the military security standards are always perceived as a potential risk (and rightfully so). Nobody wants to grant access to vitally important operational data to a system that can endanger them by, for example,

- making these data available to unsecured third parties during exercises;
- changing these data following simulated perceptions which result in wrong data or inconsistent data that may interfere with the functioning of the C2 system; and
- introducing data that create wrong impressions and perceptions for the users.

Some readers may remember the 1983 movie *War Games*. In this movie, a young man gets access to a military central computer through a back door and starts a game on worldwide thermonuclear war. As nobody at the command center is aware if the display of attacking missile is the product of the virtual environment or the real environment, this nearly leads to a third world war. While the movie may not be perceived to be too realistic, the idea is: if our simulations are so close to reality that soldiers can truly realistically be trained with them and emerge into the scenarios than they may lead to confusion with operators who can mix up exercise and reality as well. This must be avoided at all costs.

The current policy is to deny access to components that are not rigorously tested, which excludes a lot of functionality that could be of use, but as it is provided by an unsecure system it cannot be made available. The solution to this problem is provided by secure gateways that are also "gatekeepers." Figure 15.2 demonstrates this idea by showing a secure gateway between an M&S infrastructure on the right side and a C2 infrastructure on the left side.

On the left of the figure, the C2 IT infrastructure is based on the exchange of well defined messages or, and this or is not exclusive, on data replications or database updates of military databases. We shall deal with both options in more detail in the upcoming sections. On the right side, we have simulation systems or federations that are based on simulation interoperability protocols such as discussed in Chapter 12. Information is either updated via events or interactions

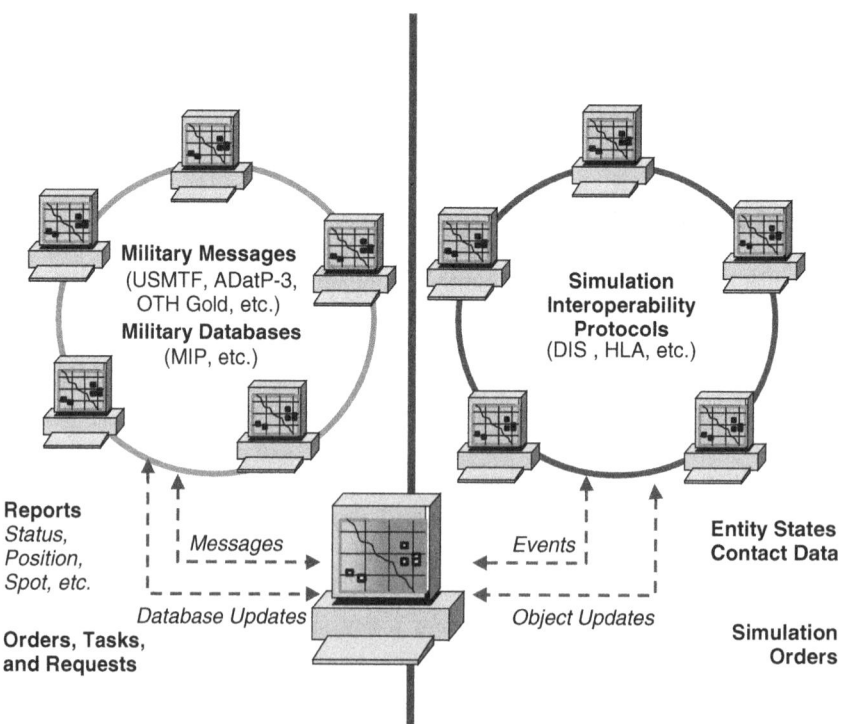

Figure 15.2   State-of-the-art of C2/M&S integration.

using the transient information exchange paradigm, or attributes of persistent objects, such as simulated entities, are updated.

The operational side knows all kinds of reports that are standardized regarding form and content, depending on which message or replication protocol is supported. These reports have to be translated into events and updates on the simulation side. Furthermore, orders are generated, tasks are given, or requests are submitted. This planning information needs to be communicated with the simulation side as well. This communication should be supported in both directions, which means that entity updates and simulation orders should be generated not only based on C2 information but based on entity updates and simulation orders that the C2 side will be notified of by reports and generated orders as well.

These tasks have to be conducted by a system that is home in both worlds: the C2 side as well as the M&S side. The same possibilities to bridge this gap discussed in Chapter 11 (Gateway, Proxy, Broker, and Protocol Solutions) are applicable here as well. Several approaches have been discussed in the literature, such as the Modular Reconfigurable C4I Interface (MRCI) (Hieb et al., 1997), or various C2-Sim proxy servers as presented by Klose et al. (2004). The US Army even established the Simulation-to-C4I Interoperability (SIMCI) Overarching Integrated Product Team (OIPT) (Hieb et al., 2002). For a NATO conference on the topic, Tolk (2003) prepared an overview of approaches. Another good overview of selection criteria and provided functionality for such systems is given by Wittman et al. (2004).

All these approaches have in common that they were rooted in the assumption that the information exchange requirements of the C2 systems were sufficient in support of the information exchange between M&S and C2 systems as well. This will be discussed in more detail in a later section of this chapter. However, before we go into this discussion, we will have a look at the various types and constraints of C2 infrastructures and their information exchange requirements and formats, starting with examples for tactical messages.

Tactical Messages

Whenever information has to be exchanged between two C2 systems, a message is generated between these two systems. As already defined in Chapter 4, these messages can be text-oriented, which makes them human readable, or bit-oriented, which makes them closely tied to communication protocols. They can also be transactions on database systems, or data replication mechanisms. To give the reader new to the field an idea, the following table shows the 162 different text messages that are defined for the US Army as Message Text Formats (Field Manual 101-5-2).

Conceptually, there is no big difference between these approaches. All of them define criteria when the message is sent. A message can be sent at certain times, like a situation report that summarizes at the end of a given period what the overall situation of the reporting unit is. A message can also be triggered

Table 15.1   US Army Message Text Formats

2406 NMC Summary Report [2406NMC]	Cemetery Status [CEMSTAT]
Accident Report/Serious Injury Report [SIR]	Chaplain's Report/Unit Ministry Team Daily Line Report [CHPREP]
2406 NMC Summary Report [2406NMC]	Chemical Downwind Report (CDM) [CDMREP]
Accident Report/Serious Injury Report [SIR]	
Acknowledge Message [AKNLDG]	Civil-Military Operations Status [CIVMILSTAT]
Air Defense Command Message [AIRDEFCOM]	Class IV Bulk Barrier Materials Request [BLKIVREQ]
Air Mission Request Status/Tasking [REQSTATASK]	Close Air Support Summary [CASSUM]
Air Support Request [AIRSUPREQ]	
Airspace Control Means Request [ACMREQ]	Closure Report [CLOSEREP]
	Commander's Situation Report [SITREP]
Airspace Control Order [ACO]	
Ammunition Fire Unit-Ammunition Supply Rate [AFU.ASR]	Computer Attack [COMPATK]
	Crew Manning Report [CREWMNQREP]
Ammunition Fire Unit-Ammunition Status [AFU.AS]	Daily Blood Report [DBLDREP]
Ammunition Fire Unit-Deployment Command [AFU.DC]	Decon Site Report [DECONSTREP]
	Decontamination Request [DECONREQ]
Ammunition Fire Unit-Fire Status [AFU.FS]	Detained Civilian Personnel Report [DETAINCIVREP]
Ammunition Fire Unit-Firing Site Data [AFU.FSD]	
Ammunition Fire Unit-Mission Fired Report [AFU.MFR]	Direct Support Unit Report [DERSPTREP]
Artillery Target Intelligence-Artillery Target Report [ATI.ATRI]	EA (Electronic Attack) Data Message [EADAT]
	Effective Downwind Message [EDM]
Artillery Target Intelligence-Target Criteria [ATI.TCRIT]	Electronic Warfare Frequency Deconfliction Message [EWDECONFLICT]
Asset/Multiple Asset Status Report [ASTSTATREP]	
	Electronic Warfare Mission Summary [EWMSNSUM]
Aviation (Army Rotary Wing) Mission/Support Request [AVIAREQ]	Electronic Warfare Requesting/Tasking Message [EWRTM]
Basic Wind Data Message [BWD]	Enemy/Friendly/Unit Minefield/Obstacle Report [MINOBREP]
Battle Damage Assessment Report [BDAREP]	
Bed Availability and Element Status [BEDAVAIL]	Enemy/Prisoner Of War Report [EPOW]
Bed Designations [BEDDESIG]	Environmental Condition Report [ECR]
Bed Request [BEDREQ]	Explosive Ordnance Disposal Support [EODSPT]
Blood Shipment Report [BLDSHIPREP]	
Bulk Class III Request/Forecast [BLKIIIREQ]	Feeder Report [MIJIFEEDER]
Bulk Petroleum Allocation [POLALOT]	Fire Mission-Beacon Location [FM.BEALOC]
Bulk Petroleum Contingency Report [REPOL]	
Bulk Petroleum Requirements Forecast [POLRQMT]	Fire Mission-Request To Fire [FM.RF]

Table 15.1 (*Continued*)

Fire Planning-Compute a Fire Plan [FP.COMPFP]

Fire Planning-Fire Plan Executive Orders [FP.FPO]

Fire Planning-Fire Plan Target List [FP.FPT]

Fire Planning-Nuclear Schedule [FP.NUCSCD]

Fire Planning-Reserve Fire Unit [FP.RESFU]

Fire Support Element-Commander's Criteria [FSE.CRITER]

Fire Support Element-Friendly Unit Location [FSE.FRD]

Flight Control Information [FLTCONTINFO]

Fragmentary Order [FRAGO]

Friendly Nuclear Strike Warning [Strikwarn] [NUC]

General Administrative Message [GENADMIN]

Handover Message [HANDOVER]

Highway Situation Report [HWYSITREP]

Human Remains Search and Recovery Status Report [REMAINSARSTAT]

Intelligence Report [INTREP]

Intelligence Summary [INTSUM]

Logistics Resupply Request [LOGRESREP]

Logistics Situation Report [LOGSITREP]

Lost Sensitive Item Report [LOSTITEM]

Mail Distribution Scheme Change [MAILDISTCH]

Maintenance Support Request [MAINTSPTREQ]

Maintenance Support Response [MAINTSPTRES]

Major Ammunition Malfunction-Initial Report [AMMOMALFUNCREP]

Meaconing, Intrusion, Jamming, and Interface (MIJI)

Media Contact Report [MEDIACOTREP]

Medical Evacuation Request [MEDEVAC]

Medical Sitrep [MEDSITREP]

Medical Spot Report MEDSPTREP]

Medical Status Report [MEDSTAT]

Message Correction/Cancellation [MSGCORRN]

Meteorological-Computer Message [MET.CM]

Meteorological-Fallout Message [MET.CF]

Meteorological-Target Acquisition Message [MET.TA]

Military Postal Facility Request [POSTREQ]

Modification-Attack Criteria [MOD.ATTACK]

Modification-Exclude Criteria [MOD.XCLUDE] NBC 1/Rota Report [NBC1]

NBC 2/Rota Report [NBC2]

NBC 3/Rota Report [NBC3]

NBC 4 Report [NBC4]

NBC 5 Report [NBC5]

NBC 6/Rota Report [NBC6]

NBC Situation Report [NBCSITREP]

Operation Order [ORDER]

Operation Report [OPREP]

Operations Plan Change [PLANORDCHG]

Operations Summary [OPSUM]

Patrol Report [PATROLREP]

Personnel Status Report [PERSTAT]

Preliminary Technical Report [PRETECHREP]

Psychological Operations Report [PSYOPREP]

Public Affairs Operation Report [PUBAFFOPSREP]

Radar Status Report [RADSTAT]

Rear Area Protection Unit Status [RAPSTAT]

Rear Area Security Activities [RASACT]

Rear Area Security Request [RASREQ]

Reconnaissance Exploitation Report [RECCEXREP]

Reconnaissance Following Report [RECON 4]

Reconnaissance Nickname Report [RECON 1]

Reconnaissance Scheduling Report [RECON 3]

Table 15.1   *(Continued)*

Reconnaissance Track Report [RECON 2]	System-Request For Report [SYS.RFR]
Request Confirmation [REQCONF]	Support-Damage Avoidance Area
Request for Information [RFI]	[SPRT.DAACAT]
Response to Request for Information [RRI]	Surveillance and Reconnaissance Plan
Road Clearance Request [ROADCLRREQ]	Report [SURRECONREP]
Route Report [ROUTEREP]	Survey-Control Point Access Request
Rules of Engagement Authorization	[SURV.TPAC]
[ROEAUTH]	Survey-Control Point Storage
Rules of Engagement Implementation	(Input/Output) Message
[ROEIMPL]	[SURV.SCPST]
SAEDA Report [SAEDAREP]	Survivability Report [SURREP]
Scatterable Minefield Record	System-Reply or Remarks Message
[SCATMINREC]	[SYS.RRM]
Scatterable Minefield Request	Tactical Elint Report [TACELINT]
[SCATMINREQ]	Temporary Burial Site Request
Scatterable Minefield Warning	[TEMPBURIALSITEREQ]
[SCATMINWARN]	Track Management Message
Search And Rescue (SAR) Request	[TRKMAN]
[SARREQ]	Track/Point Report [TRKREP]
Search And Rescue Incident Report	Transportation Support Request
[SARIR]	[TRANSSPTREQ]
Search And Rescue Situation Summary	Transportation Support Response
Report [SARSIT]	[TRANSSPTRES]
Sensitive Items Report [SENITREP]	Unit Situation Report [UNITSITREP]
Severe Weather Waning [SVRWXWARN]	US Medical Status Field Report
Slant Report [SLANTEP]	[USMEDFLDREP]
Sortie Allotment [SORTIEALOT]	War Crime Reportable Incident Report
Spill Report [SPILLREP]	[WCRIR]
Spot Report [SPOTREP]	Warning Message–Air Defense
Stop Jamming Message [STOPJAMMING]	[AIRDEFWARN]
Straggler Status Report [STRAGSTATREP]	Warning Order [WARNORD]
Summary Report Of Nuclear Detonations	Water Supply Point [WTRSUPPT]
[NUDETSUM]	Weather Advisory/Watch
Support-Air Corridor [SPRT.AIRCOR]	[WEATHERWATCH]
Support-Battlefield Geometry	Weather Forecast [WXFCST]
[SPRT.GEOM]	

by internal or external events. An internal event is, for example, that a certain level of material is reached and logistic support needs to provide new supplies. An external event is, for example, that enemy activities have been spotted within the area of responsibility of the unit. Both events normally result in an update in the C2 system of the unit, and this update can trigger sending a message in the form of a text, a bit string, or a database update. The content of the message is also specified, but dependent on the category of the report, as explained in Chapter 10.

Examples for text-oriented message formats are the NATO text message format as defined by the Allied Data Publication No. 3 (ADatP-3), the US Message Text Format (US MTF), or Over the Horizon (OTH) messages. Most of these text oriented messages have been mapped to XML schemas (US Department of Defense, 2003a).

Examples for bit-oriented messages are the Tactical Information Link (TADIL) messages dealt with in detail in Chapter 23, and the Joint Variable Message Format (JVMF). Molitoris (2003) suggested using XML as a common denominator to align text-oriented and bit-oriented messages, and several papers made similar suggestions, but while these ideas were applied for the text-oriented messages, the adoption in the bit-oriented communities is slower.

Translating tactical messages into simulation data is normally not the main challenge when the integration is done on this basis. The other way constitutes a problem, as the simple fact that a certain message is generated may provide additional operational information besides the content provided within the message. For example, the content about an enemy may be the same in a spot report as well as in an enemy situation report. However, while the spot report is sent out immediately when a new enemy is spotted, the enemy situation report summarizes all information collected about hostile forces in a given period of time. One potentially triggers an alarm, the other one is more or less a bookkeeping function. When creating reports on the simulation side such meta-information is significant as well and needs to be taken into account by the simulation engineer.

Common Operating Environment

One approach to make the integration of new IT infrastructure components secure was the definition of a Common Operating Environment. Within the United States, the Defense Information Infrastructure (DII) Common Operating Environment (COE) fulfilled this role (Engert and Surer, 1999). Within NATO, the same ideas were implemented driving the NATO COE.

Beside the security aspect, the COE approach also directly addresses interoperability. Certain capabilities are required in every C2 system, so instead of implementing the providing functions in every C2 system differently and then aligning their different viewpoints, the COE defines various layers around a common core.

- The *COE kernel* is present on every COE compliant workstation. This common core functionality is exclusively developed by the Defense Information Systems Agency (DISA) and is also controlled and deployed by DISA. The kernel comprises the operating system and extensions, the common desktop, software install tools, security extensions, etc.
- On top of the kernel are the *common infrastructure services*. These services may be developed by military services, supporting agencies,

or DISA, but they are exclusively controlled and deployed by DISA. The services provide common functionality across the US Department of Defense C2 communities and emphasize moving data through the network, providing among other things relational database management systems (servers/clients), common network management services, and communications services.

- The infrastructure services are utilized by the *common support applications*. They emphasize interoperability via common view of data by providing functionality common within domains. Examples are the display of geospatial information, common message processing, order of battle enumerations, office automation, correlation/fusion, alerts, and other applications every C2 system uses. These applications are developed by the services and supporting agencies; however, they are controlled and deployed by DISA to ensure security.

- Although not part of the COE, the *mission applications* provide mission-unique functionality. They are developed, controlled, and deployed by the services and agencies as they utilize the DII COE.

Figure 15.3 exemplifies the DII COE with some prototypical services. The figure also shows some of the main C2 systems used in the US context as well

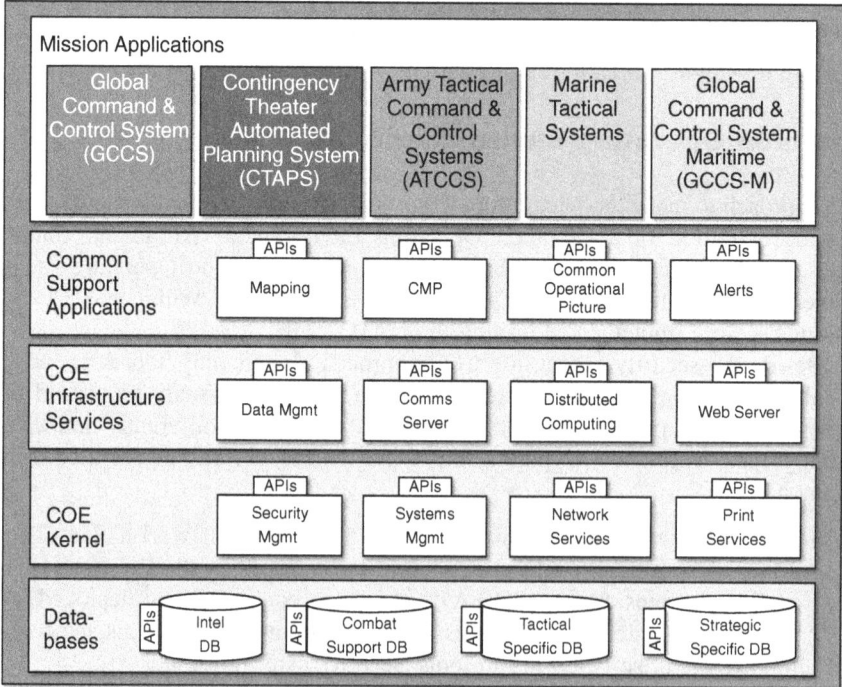

Figure 15.3 Services of the DII COE.

as the databases that are part of the COE structure and that are accessed via the database management systems enumerated above.

To support security and interoperability, the DII COE additionally provided a set of principles that comprise an architecture and approach for building interoperable systems based on these functional categories, an environment for sharing data between applications and systems, an infrastructure for supporting mission-area applications, a collection of reusable software components and data, a rigorous set of requirements for achieving DII compliance, an automated tool set for enforcing COE principles and measuring DII compliance, an approach and methodology for software and data reuse, and a set of Application Program Interfaces for accessing COE components.

In particular the rigorous set of requirements for achieving DII compliance was enforced for all systems that requested access to the C2 infrastructure, which excluded most of the military simulation systems that were developed unaware of such requirements. Several publications focused on using the DII COE, such as Carr and Hieb (2001). Others focused on the NATO COE in the NATO context, although NATO was never as rigorous regarding COE compliance as was the US DoD.

Of particular interest was the work conducted at the Naval Research Laboratory (NRL) (Layman and Dahly, 2002). NRL worked on developing the US Navy's Embedded Simulation Infrastructure (ESI) Program supporting the implementation and management of embedded models and simulations for C2 systems that are compliant with the DII COE. The ESI program objective was to provide a robust set of common reusable DII COE simulation services and object-oriented links between simulations and DII COE services and databases for simulation services. The ESI should satisfy the rigorous requirements for full DII COE compliance but also allow simulation services their flexibility. The result was an integrated development environment for DII COE compliant simulation applications.

Layman and Dahly (2002) document two examples how this development environment has been used. The first example is a Planning Mission Editor that simply allows using the same planning functions that are designed for real-world planning for M&S applications as well. While this seems trivial, many current standardization approaches try to accomplish this goal that was available by design in the NRL ESI approach. The second example is a Weapons of Mass Destruction Analysis Application that simulates Chemical, Biological and Radiological (CBR) contamination clouds in order to support predicting contamination intensity and dispersion over time.

With the rise of service-oriented architectures and the idea of the Global Information Grid, which will be discussed here as well, the importance of the COE approach decreased, but many of the ideas are still applicable and worthy of reevaluation by simulation engineers on the quest for a good solution. Also, many systems based on the COE are still in use, and the layered structure of services is reflected in other approaches as well, as we shall see.

Multilateral Interoperability Program

The Multilateral Interoperability Program (MIP) is conducted by the information system experts of NATO and the supporting nations. All documents and information are provided via their website, although some partitions are password protected.

The objective of MIP is stated on their website as follows: "The aim of the MIP is to achieve international interoperability of Command and Control Information Systems (C2IS) at all levels from corps to battalion, or lowest appropriate level, in order to support multinational, combined and joint operations and the advancement of digitization in the international arena."

As MIP started as a NATO initiative, and as NATO recognizes and emphasizes the sovereignty of their members regarding their national procurement, the objective was never to procure one common system that everyone needs to use, but to identify a set of common guidelines, rules, and specifications that will support interoperability between two MIP compliant systems without interfering with national interests. The focus was therefore on standardizing the information exchange between two MIP systems. How the information exchange was conducted nationally was not touched by this capability. It was only requested that such a national system can exchange information following the MIP specifications in the international context.

These objectives and constraints resulted in the focus points of MIP: (1) the information exchange data model and (2) the exchange mechanism. The information exchange data model is the *Joint Consultation, Command and Control Information Exchange Data Model* (JC3IEDM). Two exchange mechanisms are supported, which are the *Data Replication Mechanism* (DRM) and the *Message Replication Mechanism* (MRM). The MIP website provides documentation for all components.

The focus in this section will be on the JC3IEDM, as this was recognized as a potential interoperability enabler between C2IS and M&S systems by several national contributions in NATO conferences, such as the NATO Modeling and Simulation Group Conference on C3I and M&S Interoperability (NATO, 2003). The ideas were formerly specified into action items for simulation engineers by Tolk (2004).

To understand the special role of the JC3IEDM, it is necessary to understand the history of its development. In 1978, NATO's Long-Term Defense Plan (LTDP) Task Force on Command and Control (C2) recommended that an analysis be undertaken to determine if the future tactical automatic data processing requirements of the nations, including that of interoperability, could be obtained at a significantly reduced cost compared with the approach that had been adopted in the past. In early 1980, the then Deputy Supreme Allied Commander Europe initiated a study to investigate the possibilities of implementing the Task Force's recommendations. This was the birth of the Allied Tactical Command and Control Information Systems (ATCCIS) Permanent Working Group (APWG), which was dealing with the challenge of the future C4I systems of NATO. The ATC-CIS approach comprised more than just another data model. It was designed to be an overall concept for the future C4I systems of the participating nations. One of the most important topics of ATCCIS was that each nation could still build independent systems with their own "view of the world" and respective applications, business rules, implementation details, etc. Thus, ATCCIS defined a common kernel to facilitate common understanding of shared information, and therefore facilitated facing the general challenge to reach interoperability based on various heterogeneous IT solutions.

The technical feasibility was demonstrated several times and ATCCIS-based systems were a reliable part of the annual Joint Warrior Interoperability Demonstrator (JWID) programs. Finally, the ATCCIS data model became a NATO standard with the Allied Data Publication No. 32 (ADatP-32) with the new name Land Command and Control Information Exchange Data Model, which was adopted in 1999. In parallel with this, the Multilateral Interoperability Program (MIP) was established by the project managers of the Army Command and Control Information Systems (C2IS) of Canada, France, Germany, Italy, the UK, and the United States in April 1998 in Calgary, Canada. MIP replaced and enhanced two previous programs: BIP (Battlefield Interoperability Program) and QIP (Quadrilateral Interoperability Program). The aim of these programs was similar to the present MIP but each was active at a different level of command. By 2002, the activities of ATCCIS/LC2IEDM and MIP were very close, expertise was shared, and specifications and technology were almost common. The merger of ATCCIS and MIP was a natural and positive step and this was recognized by the almost immediate publication of a NATO policy that endorsed MIP. LC2IEDM became the data model of MIP, which established the Message Exchange Mechanism (MEM) and the Data Exchange Mechanism (DEM) based on replication mechanism. In 2003 the name was changed to Command and Control Information Exchange Data Model (C2IEDM). Shortly thereafter, the C2IEDM remerged with the NATO Corporate Data Model. This step resulted in the JC3IEDM and the current version. The JC3IEDM is continually improved by its international user group.

Technically, JC3IEDM describes all objects of interest on the battlefield, e.g. organizations, persons, equipment, facilities, geographic features, weather phenomena, and military control measures such as boundaries, using a common and extensible data modeling approach. It is based on information concepts that are modeled as 15 independent entities. Five key information concepts are of fundamental importance in generating the structure of the data model. They are defined in Table 15.2. The distinction of objects and items is essential. The battlefield consists of a large number of objects, each with its own set of characteristics. Objects may be described as a class or type rather than as individually identified items. Actual instances are catered for by use of OBJECT-ITEM. Types are recorded as OBJECT-TYPE. While general attributes are collected on the type side, such as general capabilities and abilities, only the instantiation-specific

Table 15.2   The Five Core Concepts of the JC3IEDM

Concept	Definition
OBJECT-ITEM	An individually identified object that has military significance. Examples are a specific person, a specific item of materiel, a specific geographic feature, a specific coordination measure, or a specific unit
OBJECT-TYPE	An individually identified class of objects that has military significance. Examples are a type of person (e.g. by rank), a type of materiel (e.g. self-propelled howitzer), a type of facility (e.g. airfield), a type of feature (e.g. restricted fire area), or a type of organization (e.g. armored division)
CAPABILITY	The potential ability to do work, perform a function or mission, achieve an objective, or provide a service
LOCATION	A specification of position and geometry with respect to a specified horizontal frame of reference and a vertical distance measured from a specified datum. Examples are point, sequence of points, polygonal line, circle, rectangle, ellipse, fan area, polygonal area, sphere, block of space, and cone. LOCATION specifies both location and dimensionality
ACTION	An activity, or the occurrence of an activity, that may utilize resources and may be focused against an objective. Examples are operation order, operation plan, movement order, movement plan, fire order, fire plan, fire mission, close air support mission, logistics request, event (e.g. incoming unknown aircraft), or incident (e.g. enemy attack)

values are on the item side. Examples are the caliber of the weapon being specified on the type side, but the actual ammunition state and location are on the item side.

Another important issue enabling the JC3IEDM to be extended as required is an extensive use of the categorization mechanism. To show the principle, Table 15.3 shows the category codes applicable to OBJECT_ITEM. In recent versions, the category code UNKNOWN was introduced for an OBJECT-ITEM, which is tracked but has not yet been classified. If new information has to be introduced, this can be done by adding new attribute values (in particular new category and sub-category codes), adding new attributes, adding new tables, or adding new associations.

The resulting top level view for the JC3IEDM data model is captured in Figure 15.4. It shows the five core concepts (enriched by reporting data) and the five object-item and object-type elements resulting from the categories.

Every concept is further defined by an unambiguous set of properties. To implement the categorization mechanism just introduced, JC3IEDM uses category codes and sub-category codes utilizing well defined enumerations. The meaning of each enumeration—often including the source of the definition—is specified

Table 15.3   Category Codes for OBJECT-ITEM

Concept	Definition
FACILITY	An OBJECT-ITEM that is built, installed, or established to serve some particular purpose and is identified by the service it provides rather than by its content (e.g. a refueling point, a field hospital, a command post)
FEATURE	An OBJECT-ITEM that encompasses meteorological, geographic, and control features that are associated with a location to which military significance is attached (e.g. a forest, an area of rain, a river, an area of responsibility)
MATERIEL	An OBJECT-ITEM necessary to equip, maintain, and support military activities without distinction as to its application for administrative or combat purposes (e.g. ships, tanks, self-propelled weapons, aircraft, etc., and related spares, repair parts, and support equipment, but excluding real property, installations, and utilities)
ORGANIZATION	An OBJECT-ITEM that is an administrative or functional structure
PERSON	An OBJECT-ITEM that is a human being to whom military significance is attached

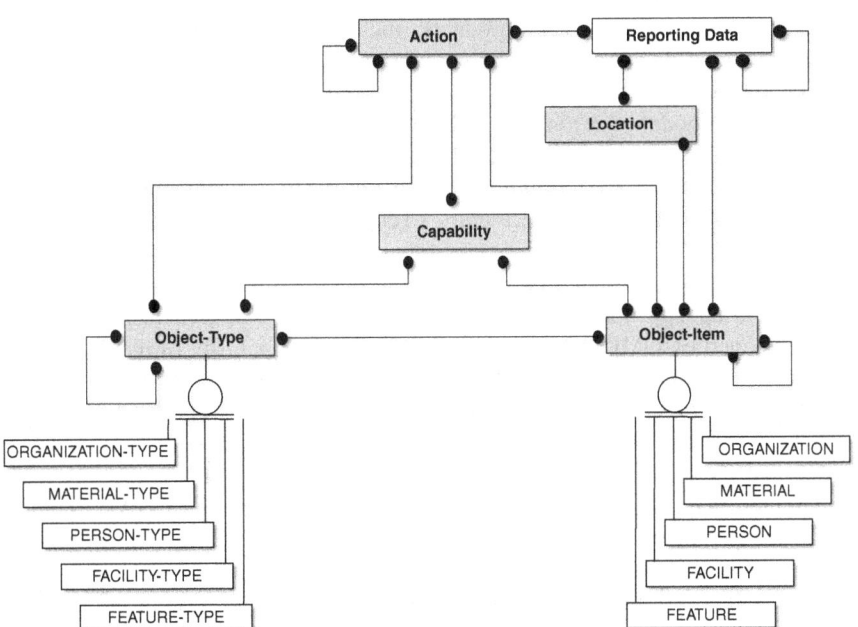

Figure 15.4   JC3IEDM top level view.

as well in the documentation. All JC3IEDM terms are therefore captured in a glossary that reflects the language and definitions of the military user, as captured in the dictionary of military terms discussed in Chapter 4.

As the concepts of JC3IEDM are connected by relations they also build the foundation for the replication mechanism. Some of these relations are mandatory in order to assure that within a semantic association of information spread over various tables all pieces of data are presented. In the language of database engineers, they represent transactions. Every one of these semantic concepts forms a replication domain set. In other words, the set is defined as a group of associated concepts connected by mandatory relations. The concepts are perceived to be fundamental to command and control and must be provided by every participating system. For example, every message has an originator and a time stamp, so every replication set must include this information as well. The replication mechanism defines for all agreed transactions who will receive an update in case data within this group are changed. While the data concept allows one to define automatic triggers that initiate an information exchange, the configurability of the replication mechanism allows one to ensure that only eligible partners receive potentially sensitive data.

The current MIP Baseline 3 was approved in October 2009. The current version of JC3IEDM—Version 3.0.2—has 273 entities. Several efforts are currently being conducted to improve the efficiency of applying these ideas, such as the use of platform independent models as recommended by the Model Driven Architecture (MDA) of the Object Management Group (OMG). Such approaches ensure the consistency of the various products, including the information exchange model, metadata, documentation, business rules, etc., as they all can be derived from a common repository comprising all required information. Current research focuses on the modularization of JC3IEDM to make it faster and easier to adapt as well as on identification of logical layers of the data model.

In this context, another closely related effort under development is of interest as well. The Shared Operational Picture Exchange Services (SOPES) represents an initiative of OMG to develop a set of open standards for generic architectures, interfaces, and technologies that promote interoperability during coalition, partner, or multi-agency operations. By combining industry-proven technical solutions developed by OMG partners with the operational maturity of the JC3IEDM and related MIP ideas, this international effort may contribute to the definition of the next generation of C2 information systems (Chaum and Christman, 2008).

Which of these current research efforts will be supported in the next version of MIP and the JC3IEDM was not yet decided at the time this book was written. Simulation engineers who are interested in the integration of NATO related operational environments should observe the MIP website for the latest updates.

In summary, it should be pointed out that the JC3IEDM is not just another data model. It was designed to support the unambiguous definition of information exchange requirements in the operational domain derived from real information exchange requirements as defined by the user. The contributions of *data modeling experts* as well as *operational experts* and users from more than 20 countries for more than 20 years ensure *technical maturity* and *operational applicability* based on mutual agreement and *multilateral consensus*. This makes the JC3IEDM unique in the technical as well as the operational domain. Every recommended alternative must be measured against these criteria and achievements.

Global Information Grid

The Global Information Grid (GIG) was designed within the last decade to support the new ideas for Joint Command and Control (JC2) (US DoD, 2002). This term was later replaced by the term Net-Enabled Command Capability (NECC). The main idea of NECC is that information is obtainable by the warfighter "wherever the location, whatever the task, and whichever system used." To this end, technically interoperable and conceptually composable services relevant to the full range of application domains had to be brought together in a distributed, heterogeneous, information technology environment. The rigorous constraints of the DII COE approach were not able to support this vision. An alternative structure based on loosely coupled services providing comparable functionality was needed. The resulting structure opened a new door for M&S services to benefit from these new concepts that provided a way to integrate them into the operational environment as loosely coupled service providers.

It needs to be emphasized that NECC was driven by operational needs, not necessarily technical opportunities. The operational need was that due to the new tasks and operations to be conducted by military forces the rapid accessibility of information became very important. The former concept of Task, Process, Exploit, and Disseminate (TPED) regarding data and information was replaced by Task, Post, Process, and Use (TPPU). The reasons for this are twofold. First, even raw, potentially incomplete data can empower many uses in a time constrained environment, and in fact the knowledge gained balances the risk inherent in not waiting for the processed information. Second, data distributors may not be aware of all potential users of their data. The unidentified user would never be reached by the traditional data distributions paradigm of pushing data from the provider to the user. In other words, a culture change in the way we do C2 was needed. C2 organizations had to work together in a network. The underlying C2 infrastructure had to support these new ideas. However, the term net-centric operations (NCO)—or the related NATO term net-enabled capability (NEC)—targeted the need for networking organizations.

The technical backbone chosen to support NCC is the GIG, as defined in US Department of Defense Directive 8100.1 (2002). There have been several additional directives published to specify the details, but this general overview is still valid. It defines the vision for the GIG to be globally interconnected, providing end-to-end sets of information capabilities, associated processes, and personnel for collecting, processing, storing, managing, and disseminating information on demand to warfighters, policy makers, and support personnel. The GIG includes all owned and leased communications and computing systems and services (software, data, security services, and other associated services) necessary to achieve information superiority. As such, it serves not only the US Department of Defense but all contributing other departments and supporting organizations to support efficient planning, execution, and management of defense and defense-related operations.

A stable communication and information exchange infrastructure that is accessible no matter where the user needs the information is necessary, but is also not sufficient. The Joint Tactical Radio Systems (JTRS) with its data link capability plays a pivotal role in the concept known as the tactical Internet. This whole effort is established by several contributing programs. Overall, the tactical Internet is realized as an orchestrated effort bringing together airborne data link devices, radio devices, and strong backbone connections. Many of these backbone connections use fiberglass optical networks, when possible. Since the first vision for the GIG, many changes and updates have been discussed and initiated, such as cyber war defense means, new enabling technologies and more.

Form the management perspective, one of the biggest challenges is data management. In order to support that new and unforeseen users can identify and obtain information that they need for their successful operation, a net-centric data strategy was developed and is continually improved (US Department of Defense, 2003b). This strategy defines the content and the use of metadata that allow providers of data and services to describe the content and the format of their data in standardized meta cards, which are similar to the concepts described in Chapter 12. Potential users look at predefined repositories for this information and, if they find a service that fulfils their need, make contact to establish a contract with the provider. Figure 15.5 exemplifies these ideas.

At the top of the figure are two systems that are connected in the traditional way across their engineered and well designed interfaces. This allows exchanging information between them, because the engineers were aware that these two systems would conduct an operation together and have to exchange well understood information in support of this operation. In new scenarios, however, a new coalition partner can show up, or a new organization needs support, or even an unforeseen collaboration with another service can take place. To support these unforeseen collaborations, the net-centric system (here System A) posts the metadata description of what it can provide in the form of a meta card to the discovery metadata catalog (1) as well the structural description of required and provided data to the structural metadata section (2). If another system needs such services, it first searches in the discovery metadata catalog (3) for a match. If the service the system is looking for is provided, it checks whether it can provide the needed data in the desired format and can utilize the provided information (4). If these constraints can be fulfilled and the other systems are eligible to make a contract with System A, the source system posts the required information to the data section (5) where it is received by the receiving system (6). The focus of the net-centric data strategy is the exchange of information in eligible but often unforeseen contexts. From the concept there is no reason that the other system cannot be a simulation system that requests an update for operationally relevant information in order to provide decision support as discussed at the beginning of

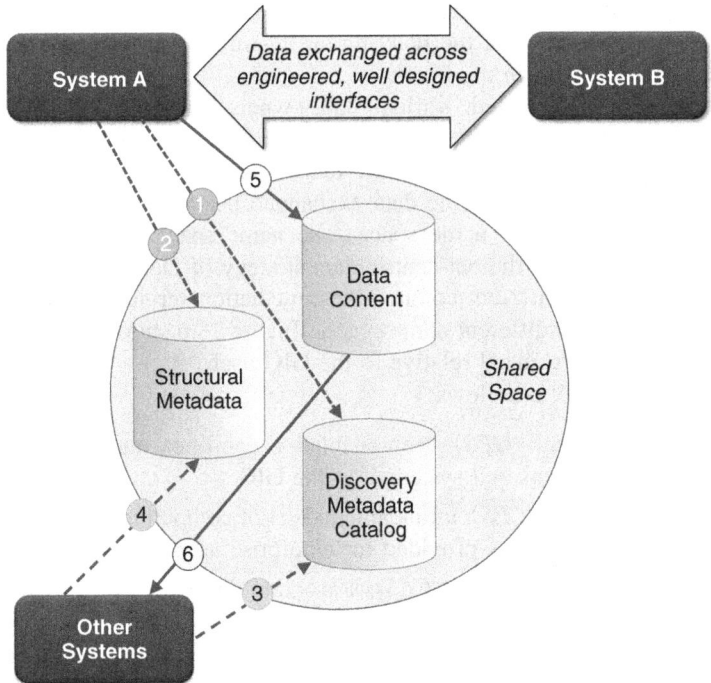

Figure 15.5   Scope of the net-centric data strategy.

this chapter. The net-centric services have to provide the infrastructure to allow for such exchange scenarios.

From the organizational standpoint, the most important idea to make this possible is the distinction between Core Enterprise Services, which are applied in all domains and which are developed and maintained centrally, and services of the different communities of interests (COI). The term COI is used to describe any collaborative group of users who must exchange information in pursuit of their shared goals, interests, missions, or business processes, and who therefore must have shared vocabulary for the information they exchange. While the services are technically identical, the organizational constraints can be described as follows.

- *Core Enterprise Services* (CES) are provided for all participating systems and services. Whenever someone needs the service of data mediation or storage, etc., the same core service must be invoked, no matter which COI the user belongs.

- *Community of Interest Services* are provided, implemented, and maintained by the COI for the COI. Namespaces, unambiguous definitions, etc., are specific to the COI. It is therefore possible that similar services are implemented in the various COIs; however, they will be COI-specific and address specific needs.

There is no technical difference between a CES and a COI service. Organizationally, however, the distinction is the determining factor in who is going to maintain and update the service in the future.

As COI membership may include various data owners and producers (e.g. developers, program managers, subject matter experts, users, etc.) who need to share the same semantic knowledge, one of the main issues of COI services is enabling a common understanding of the data exchanged between the services. This is established by a common name space. The name space management efforts of all COIs are based on the net-centric data strategy of DoD. To enable information sharing between different communities, mediation services are provided to translate between the different name spaces. Figure 15.6 shows how the CES and COI services are organized relative to the GIG user.

The terms can be defined as follows:

- *GIG Enterprise Services (GES)*: web-enabled capabilities and services available to users (humans and systems) on the GIG
- *Core Enterprise Services (CES)*: a fundamental set of computing, networking, and data sharing services provided for enterprise user support
- *Net-Centric Enterprise Services (NCES)*: a program designated to provide CES on the GIG
- *Domain*: a major area of functional responsibility, less than the DoD enterprise scope, comprising persistent requirements and resources, spanning organizations and other COIs
- *Community of Interest (COI)*: a collaborative group of users who exchange information in pursuit of their shared goals, interests, missions, or business processes.

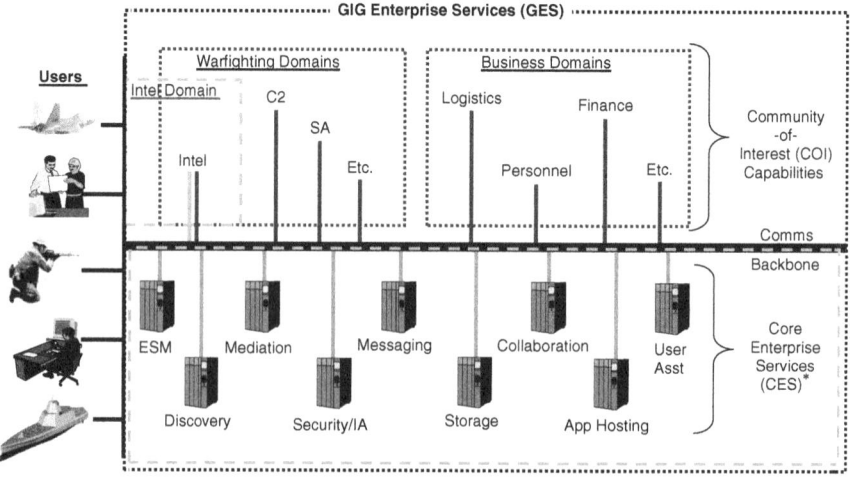

Figure 15.6 Net-centric enterprise services.

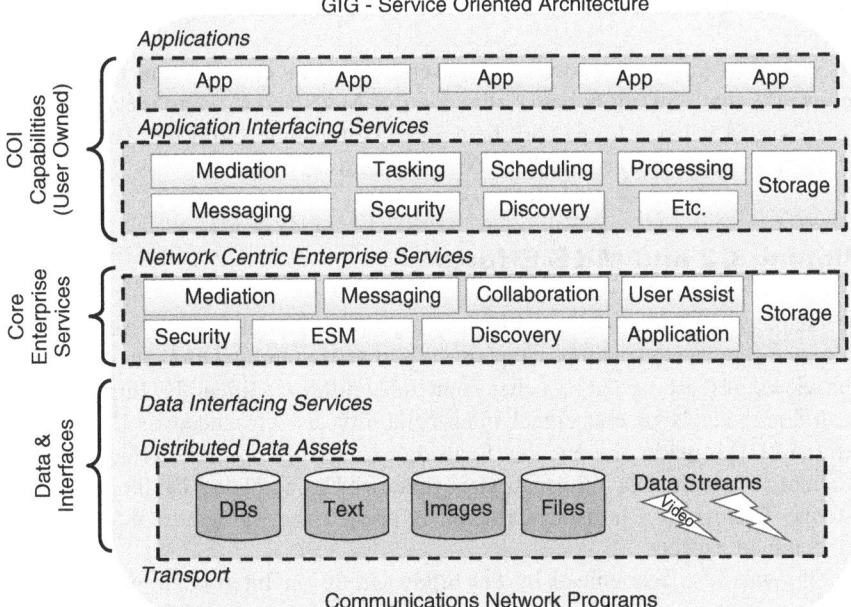

Figure 15.7   Conceptual view on net-centric enterprise services.

The similarity to the categories of the DII COE are by design, although the services no longer need to be provided on approved hardware by approved software components from DISA. Figure 15.7 shows a conceptual schema of the services that makes the relation even more obvious.

Taking into account the description of the GIG just given, there are significant implications for M&S programs. It is good practice that M&S applications and services provided by programs will need to use CES rather than develop their own. M&S programs will also need to participate in COIs to ensure that their data are visible and accessible to GIG users. However, they can also be integrated more easily than ever before into the operational environment as services that can be accessed by the user. The SOPES effort mentioned in the last section is actually trying to merge the ideas from MIP and NCO in a more orchestrated effort (Chaum and Christman, 2008).

All these ideas on loose coupling and net-centric service support are directly supported by semantic web technologies, including name spaces, taxonomies, and ontologies, and other means that support the easier identification, selecting, composition, and orchestration of services (Tolk et al., 2006). For the simulation engineer, this new environment is challenging, but also full of opportunities for new applications and integration ideas.

FRAMEWORKS AND CURRENT RESEARCH

This section of the chapter focuses on several efforts on contributions to common frameworks that will allow better alignment of M&S and C2 activities. We shall also look at a general framework helping to better understand the data sharing and aligning needs between C2 and M&S systems.

Aligning C2 and M&S Efforts

The chapter on simulation systems presented in this book and the sections on C2 systems presented in this chapter show that the architectural and management views of these two approaches sometimes differ significantly. Furthermore, the different levels of conceptual interoperability, as introduced in Chapter 12 with the LCIM, made obvious that focusing exclusively on data exchange is not sufficient to ensure that information is processed as intended by the receiving systems. Therefore, a holistic approach is needed that integrates management and technical aspects.

This was also recognized by the Study Group on Interoperability between C4I Systems and Simulation Systems (SG-C4I) of the Simulation Interoperability Standardization Organization (SISO) that developed a framework to cope with these issues (Timian et al., 2000). Figure 15.8 shows the domains that need to be harmonized to provide shared solutions. This figure was originally introduced

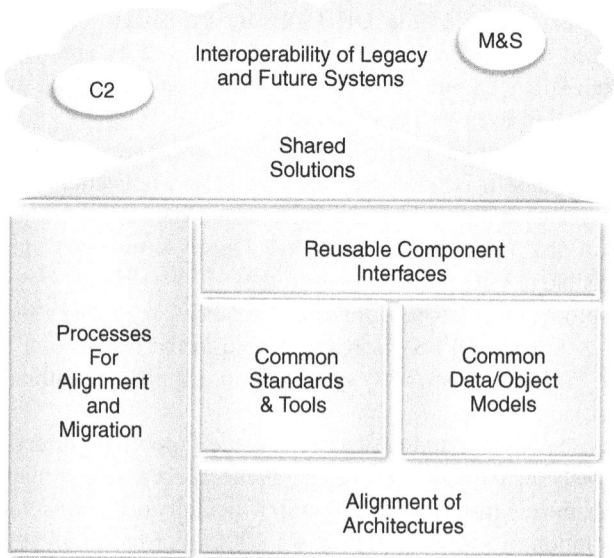

Figure 15.8   The house diagram.

by Michael R. Hieb and Andreas Tolk during the 1999 Fall Simulation Interoperability Workshop in preparation for briefings of SG-C4I and was consecutively published by Hieb and Sprinkle (2000) and successfully applied in the SIMCI OIPT efforts discussed earlier to address the different domains of interoperability.

This *house diagram* (Figure 15.8) blueprints the complexity of interfacing simulation systems and C2 systems. This holistic view emphasizes the interdependence of five major factors involved in the effort to secure shared solutions for C4ISR/M&S interoperability, namely architectures alignment, defining common data models and object models, agreeing on the application of common standards, identification of reusable components and their interfaces, and the alignment of management processes in support of these activities plus enabling the migration of legacy solutions towards the agreed solutions.

Chapter 12 focused on architectural constraints for many M&S systems while the sections of this chapter discuss several of these for C2 systems. These constructs establish the foundations that set the requirements for fundamental interoperability between components of these domains. The architecture alignment needs to be able to resolve differences in viewpoints or fundamental representations of the problem space. Although the ideas of service-oriented architectures help to resolve some issues, the need for alignment remains apparent.

Chapter 29 deals with the application of model-based data engineering to support information exchange requirements in lieu of common data models. The common data and object models as shown in the house diagram actually refer to the definition of equivalence classes that allow that different viewpoints can be mediated into each other. Some additional challenges are captured in Chapter 25. It should be pointed out that current research no longer focuses on establishing common data or object models but on ways to ensure common consistent interpretation of such models. The semantic web technologies discussed in Chapter 12 play a pivotal role here.

Common standards are most effective when they are part of the initial system design, which starts with the common architecture, but continue via design patterns to particular solutions. All standards, including those discussed in Chapter 12 as well as in this chapter, that are applied in M&S and C2 systems need to be aligned and supportive of each other.

Reusable components are sitting on top of the right conceptual pillar. Although interface specifications and reuse are often a hotbed of activity without addressing the underlying requirements, the resulting solutions are often spot solutions that have to be done again and again and in general tend to lack efficiency with regard to costs, time, and flexibility. However, if they are rooted in common information understanding, common standards, and aligned architectures, basic incompatibilities between the systems are taken care of and real reuse becomes possible.

The left side addresses the need to align processes, which are not only technical processes and routines; these include in particular the management processes and education processes. If technical solutions are not supported by the management, they are probably not applied successfully in organizations.

If people do not have the proper education to use the technology, the same shortcoming is observed.

Only when processes and technology are aligned and balanced does the final vision of truly shared solutions between legacy and future M&S and C2 components become feasible.

General Framework

While the house diagram focuses on the interoperability domains, the general framework discussed here focuses on information that needs to be exchanged and managed in systems that comprise C2 and M&S components. The simulation engineer is encouraged to revisit the description of the Levels of Conceptual Interoperability Model in Chapter 12 in parallel with reading this section. The focus on this framework is to identify the main categories of data that are of interest in C2 and M&S components and how to exchange them.

The framework in the presented form was first published in Tolk (2003), but the original idea for such a categorization was first published by Carr and Hieb (1998). They made the case that there should be no significant difference between data representing concepts of operational interest in C2 systems and data representing the same concepts in realistic training simulation systems. Their observation is similar to the NATO COBP summary on the importance of common understanding of data: "As the data being used today by the analysts will be the data needed tomorrow by systems engineers, decision makers, and commanders for their operations, alignment of the standardization processes and the respective toolsets as early as possible with the command and control systems community is good practice" (NATO, 2002, p. 241).

The viewpoint presented here is driven by the perspective of a simulation engineer who has to integrate a simulation system and a C2 system by identifying which data categories are present in both systems and which not, and when they have to be exchanged: as initialization data before the operation begins, or as information exchange data during runtime. The resulting four categories are shown in Figure 15.9.

The four categories are as follows.

1. Simulation- and federation-specific data are simulation-specific artifacts that are needed to run and manage the execution of a simulation. Run, Reset, Update, and other commands belong in this group that may have to be represented on the C2 side to allow smooth execution and collaboration.

2. On the other side, IT-specific coordination data needed for the operational system may need to be mimicked by the simulation side. Hopefully, the closer the architectures will grow together, the more these two categories will merge as well. Currently, some significant effort may be needed to support the smooth execution of both systems.

**Operational
System**

**Decision Support
System**

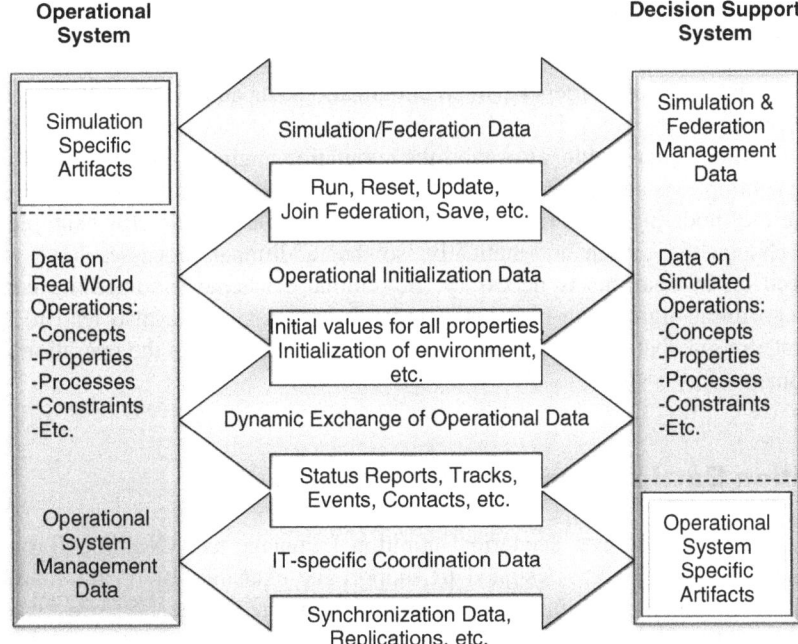

Figure 15.9   Information exchange categories.

The remaining data represent the operational data needed on the C2 side as well as on the M&S side. Idealistically, as both sets of data represent the same tactical or operational entities of the real world, they should be close to each other, but several studies show that this is not necessarily the case. As shown in Chapter 29 in more detail, model based data engineering can support to derive a common model that support the viewpoints represented on both sides.

We now have to motivate why we are distinguishing between data that is used for the initialization and data that is exchanged between participating systems during the execution.

3. Operational initialization data comprise all data used to prepare both systems to support an operation. All initial attribute values of represented entities, all data describing the synthetic environment and all other scenario-relevant data need to be exchanged. The Military Scenario Definition Language (MSDL) described in Chapter 24 and the Army C4ISR and Simulation Initialization System (ACSIS) (Shane and Hallenbeck, 2004) are examples of how this can be accomplished.

4. The dynamic exchange of operational data is the category that has received the most attention so far. On the C2 side, this is done via replication mechanisms and messages on status updates, reports, orders, etc. On the simulation side, protocol data units or objects are used to exchange the information. The Coalition Battle Management Language (C-BML) that

will be dealt with in the next section is a promising candidate to support this category.

There should be no borders between initialization data and dynamic exchange data regarding the content. Everything that is initialized can be changed, or at least it should be changeable. However, the simulation engineer has to be careful that the assumptions and constraints of the simulation components—as we talked about in the modeling part of this book—are met. Some systems, for example, cannot change the terrain automatically, so that additional processes have to be agreed on in case this is necessary. In summary, whenever something can lead to an unfair-fight issue between simulation systems, it can also lead to a major integration challenge when the simulation is integrated into the operational environment.

Coalition Battle Management Language

As just discussed, Military Scenario Definition Language and Coalition Battle Management Language are designed to support the exchange of operationally relevant information. The concept of creating a language for the US Army was presented by Carey et al. (2001). The Coalition Battle Management Language extended the concepts to address coalition operation and was first introduced to the M&S community by Tolk and Hieb (USA), Galvin (UK) and Khimeche (France) (Tolk et al., 2004). This presentation launched myriad activities, including supporting activities by the NATO Modeling and Simulation Group as well as standardization efforts by SISO.

> *Coalition Battle Management Language is defined as the unambiguous language used to command and control forces and equipment conducting military operations and to provide for situational awareness and a shared, common operational picture.*

The scope of C-BML was identified by Carey et al. (2001) and accepted by the broader community in follow-on efforts as C2 related information, in particular tasks and reports that are exchanged between live forces using their C2 systems, simulated forces, and robotics. Figure 15.10 shows the information exchange regarding tasking information exchange supporting command and control as well as reporting information exchange supporting situational awareness.

The driving idea behind such a language is that there should be no difference when specifying a task or delivering a report depending on the receiving system. If an order is given to clear a minefield, in principle this order should be the same no matter if it is sent to the C2 system of a combat engineer unit in the field, or if a simulated unit conducts the operation on the virtual battlefield, or if a group of unmanned engineer robots have to clear the minefield. The same is true for reports that contribute to the situational awareness. It should play no role if the report comes from a live unit, from a simulated unit, or from an unmanned vehicle in the field.

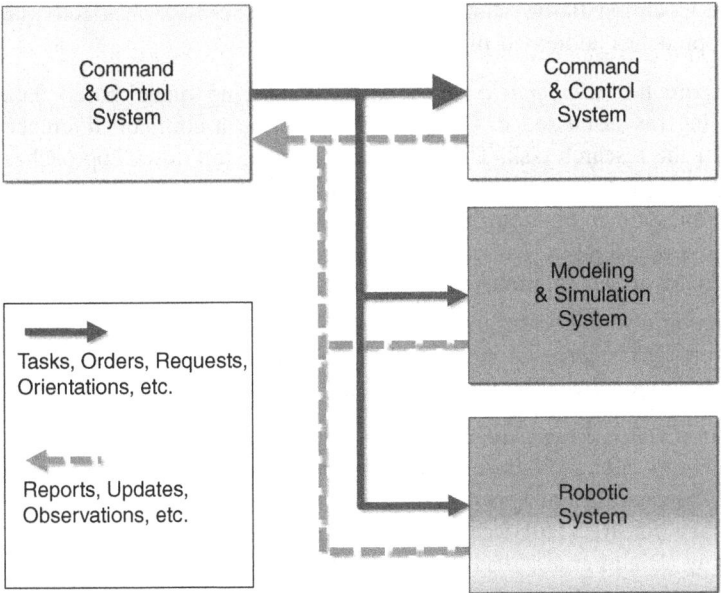

Figure 15.10   C-BML scope.

One of the main challenges for this effort is that we are looking at IT systems that exchange information, so we are talking about an artificial language with all its constraints on computability and complexity. The need for such a language is clearly perceived, but the technical implementations are debated. To be able to address these complex issues, the standardization efforts are conducted in phases. Phase One defines an XML schema that allows the information exchange of standardized C2 relevant data. Phase Two defines a grammar that produces valid sentences. Phase Three defines an ontology. The phases build on each other and gradually increase the expressiveness. Furthermore, the results will be aligned with MSDL in overlapping domains.

There are two approaches under evaluation that ultimately will be joined. One focuses on the use of common standards that will minimize the adaptation effort in particular on the side of the C2 systems (Tolk et al., 2007). The other approach is driven by computer linguistics with a focus on defining a lexical-functional grammar (Schade and Hieb, 2007). How they are related and can be merged was discussed by Kunde et al. (2009). The first standard drafts are currently under evaluation.

Despite some divergence in the school of thought, several prototypes were used to demonstrate the feasibility and usefulness of these ideas. The NATO Modeling and Simulation Group activities on Coalition Battle Management Language (MSG-048, MSG-079) are documented on the NATO website.

In support of the Coalition Warfighting Program, the ideas first published in Tolk et al. (2006) were implemented based on the latest Internet technologies, resulting in the Coalition Battle Management Services. These services combine many of the approaches addressed in this chapter.

- They ensure the *homogeneity of models* by applying model-based data engineering (as discussed in Chapter 29) to derive a common reference model for the research task. This results in merging top-down approaches that are driven by the information exchange need derived from the research question and bottom-up approaches that represent the information exchange capabilities of participating systems, and allows the integration of JC3IEDM driven solutions in a net-centric environment.

- They support also the *heterogeneity of simulations*, as different viewpoints are accepted and addressed by mediating them based on the common reference model conducted by the systems.

The Coalition Battle Management Services are implemented by leveraging the latest technology, such as X-Base for document persistence and XQuery processing, the Atmosphere framework for Hypertext Transmission Protocol based messaging, Jersey for RESTful web servicing, and xLightweb for client-side HTTP processing (Diallo et al., 2011).

CONCLUDING REMARKS

Interoperability across military simulation C2 systems continues to be a significant problem. Simulation is well established in military training. It can also be a valuable asset for planning and mission rehearsal. For efficient support, simulation and C2 systems need to be able to exchange information, plans, and orders effectively. To better support and enrich the C2 capabilities with simulation based capabilities an open standards based framework is needed that establishes operational and technical coherence between these systems. However, how to technically support these objectives is still discussed in the community. A main obstacle remains that simulation quite often is still seen as a non-essential after-the-fact function instead of being treated as a core capability within command and control capabilities.

During a conference addressing current grand challenges for M&S, Tolk and Hieb (2003) presented six theses regarding interoperability of C2 and simulation systems. Although some of the infrastructure constraints and implementation details have changed over the last decade, the theses themselves remain valid and may serve as a guideline for future developments.

- Thesis One: C2 systems and military simulation systems are *complementary techniques* needed for future military operations
- Thesis Two: C2 systems as well as military simulation systems must use the same *commercial standards*
- Thesis Three: C2 systems and military simulation systems must use the same *common architecture*

- Thesis Four: C2 systems and military simulation systems must use data and object models based on the same *common ontology*
- Thesis Five: C2 systems and military simulation systems must use the same *common set of algorithms and methods*
- Thesis Six: A *common overarching concept* for C2 systems and military simulation systems development and evolution is needed

As long as both system families continue to decide on implementation and management issues without taking the other side sufficiently into account, the simulation engineer has to be at least aware of approaches, architectures, standards, and other relevant efforts on both sides. Knowing the architecture framework is as important as knowing the recent developments in semantic web methods. This chapter can only be a first introduction to the ideas, methods, and solutions relevant to fulfill the tasks.

In summary, the job of a simulation engineer for combat modeling and distributed simulation remains as one of the most challenging—but also most interesting—professions in the domain of M&S.

REFERENCES

Ackoff RL (1989). From data to wisdom. *J Appl Sys Anal* **16**, 3–9.

Alberts DS and Hayes RE (2003). *Power to the Edge, Command and Control in the Information Age*. Department of Defense Command and Control Program; Information Age Transformation Series, Washington, DC.

Carey SA, Kleiner MS, Hieb MR and Brown R (2001). Standardizing battle management language—a vital move towards the army transformation. In: *Proceedings of the Fall Simulation Interoperability Workshop, Orlando, FL*. http://www.sisostds.org.

Carr FH and Hieb MR (1998). Issues and requirements for future C4ISR and M&S interoperability. In: *Proceedings of the 7th Conference on Computer Generated Forces and Behavioral Representation, Orlando*, FL, pp. 407–420, IEEE CS Press.

Carr FH and Hieb MR (2001). M&S interoperability within the DII COE: building a technical requirements specification. In: *Proceedings of the Spring Simulation Interoperability Workshop, Orlando, FL*. http://www.sisostds.org.

Chaum E and Christman G (2008). Making stability operations less complex while improving interoperability. In: *Proceedings of the International Command and Control Research and Technology Symposium. Paper 168. Seattle, WA*. CRC Press.

Diallo SY, Tolk A, Graff J and Barraco A (2011). Using the levels of conceptual interoperability model and model-based data engineering to develop a modular interoperability framework. In: *Proceedings of the Winter Simulation Conference, Phoenix, AZ*.

Engert P and Surer J (1999). Introduction to the Defense Information Infrastructure (DII) Common Operating Environment (COE). *CROSSTALK J Def Software Eng* **12**, 4–5.

Hieb MR and Sprinkle R (2000). Simulation infrastructure for the DII COE architecture: the army vision. In: *Proceedings of the Fall Simulation Interoperability Workshop, Orlando, FL*. http://www.sisostds.org.

Hieb MR, Cosby M, Griggs L, McKenzie F, Tiernan T and Zeswitz S (1997). MRCI: transcending barriers between live systems and simulations. In: *Proceedings of the Spring Simulation Interoperability Workshop, Orlando, FL*. http://www.sisostds.org.

Hieb MR, Sudnikovich WP, Sprinkle R, Whitson SR and Kelso T (2002). The SIMCI OIPT: a systematic approach to solving C4I/M&S interoperability. In: *Proceedings of the Fall Simulation Interoperability Workshop, Orlando, FL*. http://www.sisostds.org.

Klose D, Mayk I, Sieber M and Menzler HP (2004). Train as you fight: SINCE—the key enabler. In: *Proceedings of the RTO NMSG Symposium on Modeling and Simulation to Address NATO's New and Existing Military Requirements*. NATO Report RTO-MP-MSG-028, Koblenz, Germany.

Kunde D, Orichel T, Schade U, Tolk A and Hieb MR (2009). Harmonizing BML approaches: grammars and data models for a BML standard. In: *Proceedings of the Spring Simulation Interoperability Workshop, San Diego, CA*. http://www.sisostds.org.

Layman G and Dahly JJ (2002). C4I tactical applications utilizing embedded simulations. In: *Proceedings of the Command and Control Research and Technology Symposium, Anaheim, CA*, Publisher CCRP Press.

Molitoris JJ (2003). Use of COTS XML and web technology for current and future C2 systems. In: *Proceedings of the Military Communications Conference (MILCOM)*. IEEE Press, Piscataway, NJ, Vol.1, pp. 221–226.

Multilateral Interoperability Program website: http://www.mip-site.org/ (last accessed June 2011).

NATO (2002). *NATO Code of Best Practice Analyst's Guide, CCRP Publication Series*. NATO, Washington, DC.

NATO (2003). *Proceedings of the Conference on C3I and M&S Interoperability*, NATO Report RTO-MP-123, Antalya, Turkey.

Schade U and Hieb MR (2007). Improving planning and replanning: using a formal grammar to automate processing of command and control information for decision support. *Int C2J* **1**, 69–90.

Shane R and Hallenbeck P (2004). Managing data for interoperability: the Army C4ISR and Simulation Initialization System (ACSIS). In: *Proceedings of the International Command and Control Research and Technology Symposium*. Paper 087. Copenhagen, Denmark.

Timian DH, Lacetera J, Wertman C, Hieb MR, Tolk A and Brandt K (2000). Report out of the C4I Study Group. In: *Proceedings of the Fall Simulation Interoperability Workshop, Orlando, FL*. http://www.sisostds.org.

Tolk A (2003). Overview of recent findings of the study groups of the simulation interoperability workshop dealing with C3I and M&S interoperability. In: *Proceedings of the Conference on C3I and M&S Interoperability*, NATO Report RTO-MP-123, Antalya, Turkey.

Tolk A (2004). Moving towards a Lingua Franca for M&S and C3I—Developments concerning the C2IEDM. In: *Proceedings of the European Simulation Interoperability Workshop, Edinburgh, Scotland*. http://www.sisostds.org.

Tolk A (2009). Using simulation systems for decision support. In *Handbook of Research on Discrete Event Simulation Environments: Technologies and Applications*, edited by Abu-Taieh and El Sheikh. IGI Global, Hershey, PA, pp. 317–336.

Tolk A and Hieb MR (2003). Building and integrating M&S components into C4ISR systems for supporting future military operations - six theses on interoperability of C4ISR and M&S systems. In: *Proceedings of the Western Multi Conference (WMC'03), International Conference on Virtual Worlds and Simulation (VWSIM'03), Orlando, FL*. SCS.

Tolk A, Galvin K, Hieb MR and Khimeche L (2004). Coalition battle management language. In: *Proceedings of the Fall Simulation Interoperability Workshop, Orlando, FL*. http://www.sisostds.org.

Tolk A, Diallo SY, Turnitsa CD and Winters LS (2006). Composable M&S web services for net-centric applications. *J Def Model Simul* **3**, 27–44.

Tolk A, Diallo SY and Turnitsa CD (2007). A system view on C-BML. In: *Proceedings of the Fall Simulation Interoperability Workshop, Orlando, FL*. http://www.sisostds.org.

US Department of Defense (2002). *DoD Directive 8100.1: Global Information Grid (GIG) Overarching Policy*. Washington, DC.

US Department of Defense (2003a). *Command, Control, Communications, Computers, and Intelligence (C4I) Joint Extensible Markup Language (XML) Message Text Format (MTF) Roadmap (JXMR)*. October 29, Washington, DC.

US Department of Defense (2003b). *Department of Defense Net-Centric Data Strategy*. Department of Defense Chief Information Officer (CIO). Washington, DC.

Wittman RL, Lopez-Couto S and Topor L (2004). C4I Adapter Reuse Experience Report. In: *Proceedings of the Spring Simulation Interoperability Workshop, San Diego, CA*. http://www.sisostds.org.

Part IV

Advanced Topics

Chapter 16

History of Combat Modeling and Distributed Simulation

Margaret L. Loper and Charles Turnitsa

INTRODUCTION

This chapter presents a brief history of both combat modeling and distributed simulation (especially as it has developed in order to support the simulation of combat models). It is presented in two broad sections: the first covers the long history of combat modeling and mock warfare, and the second discusses the developments within the 20th century that have led to the modern state of distributed simulation for combat modeling.

HISTORY OF COMBAT MODELING

Combat modeling has a long and interesting history, much of which influences the modern state of affairs in the community. Before beginning on the history, the subject deserves a brief look at some of the terms that will be used while examining it. The first term considered is "combat." In this chapter, and in much of the practice of combat modeling, the word combat is taken to mean more than just physical violence between two combatants. While there might be a possible need for such modeling, especially if models of the sort will be used to train law enforcement or security professionals, the term is usually confined to mean combat in the context of warfare. In that light, combat is considered to mean physical violence between two or more organizations or nations, undertaken to either forward some goal or protect against encroachment. The second term is

Engineering Principles of Combat Modeling and Distributed Simulation, First Edition.
Edited by Andreas Tolk.
© 2012 John Wiley & Sons, Inc. Published 2012 by John Wiley & Sons, Inc.

"modeling." What is meant, for the purposes of this chapter, by that term is the imitation or representation of the real thing—in this case, combat. The reasons for imitating or representing combat vary widely. This chapter will review some of them.

Combat Games and Gaming

The earliest forms of combat modeling go back to the beginning of civilization, and the beginning of warfare. Civilization first formed about 5000–6000 years ago, in the various major river valleys of Asia and North Africa. In those locations, agriculture had progressed to the point where the construction of cities could begin. Soon, these cities were competing with each other for space and resources, and the beginning of warfare became manifest. Evidence for this development is spread throughout the earliest civilizations (Archer et al., 2002). The early art and recorded writings from Sumeria show aspects of warfare in Mesopotamia going back to 3200 BC. There are fortified citadels and portions of the cities in the Indus valley with remains of both Mohenjo Daro and Harappa also going back as early as 3200 BC. Shortly after, the introduction of games (some still recognizable and playable today) came about. These games were used to teach would-be rulers and nobles the ideas and tenets of combat. While early examples include *Senet* (Nile Valley) and *The Royal Game of Sumer* (Mesopotamia), the game *Wei Hai* (Yangtze Valley) goes back to approximately 3000 BC and survives to this day in the version popularized in 2300 BC as the Japanese game *Go* (Bell, 1979). What these games had in common, including later contributions to the field such as *Chaturanga* (Indus Valley, 500 BC) and eventually *Chess* (Persia, at least 500 AD, perhaps earlier), was that they showed a strategic situation on a game board where approaches to the play of the game were intended to solve that strategic situation (Bell, 1979). This allows military leaders, rulers, and nobles to practice thinking about solving such problems. Even though specific attributes of military units are missing from these early games, the training in strategic thinking is there.

It is thought that the use of games specifically designed to represent particular forms of combat, rather than abstract strategic situations, was developed with the introduction of *Königspiel* in 1664 (by Christopher Weikmann). The chief contribution to the field of simulation based gaming of *Königspiel* is that, while it is superficially chess-like, it does introduce different terrain types on the board, and the units behave appropriately on those terrain spaces. While earlier examples (such as the Roman game *Ludus Latrunculum*, or the "game of mercenaries") had a very clear basis in military settings, perhaps more so than the earlier mentioned abstract military games such as Chess, they still did not seek to represent the characteristics of the units involved, or their ability to engage in combat. From a few centuries prior, in northern Europe, there arose a class of Viking and Celtic "board" or "table" games, of which *Hnefatafl* is perhaps the best remembered title. Although these games had some differentiation in the behavior of some of the pieces (for instance, a king similar to a Chess king, surrounded by warriors, similar to Chess pawns), in general they were similar to the other earlier abstract games (Murray, 1951). The chief contribution from the Viking and Celtic table games was the variety of different starting conditions or "scenarios" that were possible. Each of these led to different strategic problems that needed to be solved through play. Representing different strategic situations is an important contribution to using games to teach military training.

Weikmann's offering, and those that followed, are revolutionary in being the first of the simulations in this sense—that they are attempting to simulate (simply at first, more complex later on) the available actions of the military units represented (Halter, 2006). More than a century later, *Königspiel* was followed up by *War Chess* in 1780 (by C.L. Helwig) and an unnamed British naval tool developed in 1781 (by John Clerk). This last consisted of tabletop model ships that were used to explore tactical situations against enemy formations. While John Clerk's naval tool could certainly be described as a naval simulation, it lacked the ability to codify risk and to allow decisions to be evaluated. Those elements are key to games and game theory. All three of these (Weikmann, Helwig and Clerk) intended to show specific examples of the location and posture of various military units to each other. The commonality is that they were all developed with the purpose of training military professionals.

In 1811, the first true "wargame" was developed by Baron von Reisswitz, the war counselor in Prussia in 1811. His wargame, named *Kriegsspiel*, had a table-top covered in model terrain, representing a miniature version of the battlefield. He used wooden blocks to represent the placement of units. Players would deter-mine their intended actions, report these to a referee, and then await the referee to update the situation on the tabletop. There were rules and charts governing the outcome of actions in the game. To represent randomness in adjudicating the results of combat, a dice toss was introduced as a stochastic factor affecting the outcome of each action. An interesting side note to consider about this develop-ment was that it was during a time when Prussia was occupied and under the rule of a foreign power. There was already a plan within certain elements of the country to rise up against Napoleon's France (the occupying power), but little could be done to muster or train the army, aside from what France allowed under a national guard training regime, so other means were necessary to train officers and come up with war plans—hence, the wargame.

In 1824, the second von Reisswitz in our history adapts his father's *Kriegsspiel* to be played on paper maps (Reisswitz, 1983). This makes the game portable and easily reproducible. Seeing the value in using the tool in training officers and also in exploring possible battle plans, the Prussian Chief of Staff, von Muffling, has the game introduced throughout the Prussian Army.

In 1837, the then Prussian Chief of Staff von Moltke makes the use of wargames and wargaming much more widespread in the Prussian Army. At the same time, the concept of the Staff Ride is introduced. In conjunction with playing out a battle using *Kriegsspiel*, the training audience will take a tour of the battlefield represented in the game, and the timing of events, and results of combat clashes, will allow them to talk through the narrative of the unfolding battle, so that the effects of space, time and geography can be appreciated. The widespread success of Prussia in a variety of European wars in the 1850s, 1860s, and early 1870s is attributed in part to the use of wargaming by the officer cadre, and this leads to broad interest by other national military organizations (Zuber, 2008). The following nations are recorded as having adopted the idea of the wargame in the year listed (Young, 1957):

- Austro-Hungarian Empire (1866)
- UK (1872)
- Italy (1873)
- France (1874)
- Russia (1875)

In 1883, Major Livermore of the US Army Corps of Engineers takes the Prussian *Kriegsspiel* and improves upon its mathematical attrition tables by applying historical data (from engagements in the American Civil War, and also Prussia's wars in 1866 and 1870–71). He urges the use of wargaming within the US Military. General William T. Sherman, then the US Army Chief of Staff, discourages the use of such tools. A quote attributed to Sherman (Caffrey, 2000) has him saying, "I know there exist many good men, who honestly believe that one may, by the aid of modem science, sit in comfort and ease in his office chair, and with little blocks of wood to represent men or even with algebraic symbols, master the great game of war. I think this is an insidious and most dangerous mistake.... You must understand men, without which your past knowledge were vain." The temptation to view Sherman as a man against progress must be resisted. His chief objection was that the games of the time (based on *Kriegsspiel*) all represented military units fighting to the last man. The horrors he witnessed in the American Civil War were a testament to Sherman that men will often quit the fight long before the last man in a unit dies.

At approximately the same time, however, William McCarty Little is successful in introducing wargaming at the US Naval War College (Little, 1912). There, wargaming is widely received, and the first lecture wargame is presented by Little in 1889. That same lecture wargame continues to this day. Some wargaming activity continued within the US Army. In 1887 the first Army–Navy exercise based on a wargame was held, sponsored by Little and Livermore.

Following the turn of the century, there is a period introducing the production of the first civilian wargames. The first civilian naval wargame came from Fred Jane, and featured rules for engagement of battleships and statistics for four different models of British ships. The game was quite popular, but information on additional classes of ships was requested by the audience. Soon a version of statistics for the entire British Navy was available, but fighting British ships against British ships was equally unsatisfactory. Jane followed up by offering information describing German ships (which caused a political uproar), and finally by offering information on all of the world's fighting ships. This led to the development of the annual yearbooks on information concerning all the world's fighting ships, and the formation of the Jane's Defense publications empire—still recognized today as one of the best international sources of information concerning all weapon systems.

The first civilian land wargame, introduced in this period, came from none other than H.G. Wells (the author famous for *The War of the Worlds* among many other science fiction novels). He came up with a set of rules for staging small battles on the living room floor, called "Little Wars." The production of inexpensive and attractive metal toy soldiers at the time made the spread of this game practical.

Leading up to, and during, the World War I the great powers of Europe all engaged in wargaming; however, several situations decreased the impact that such gaming could provide. In Germany a nephew of the original von Moltke had pushed for making the *Kriegsspiel* easier and cheaper to stage, by allowing veteran officers to quickly give adjudication results rather than relying on the time and staff that mathematical calculations would take. This was criticized by the German military historian Delbrück as problematic, as the only officers with practical experience were those from the 1870–1871 war with France, and they were all retiring out of the military.

In Russia, in the early days of the war, two chief generals wargamed a situation that should have provided insight to actual operations that would unfold later, but unfortunately (for the Russians) it did not. In the scenario, two mutually supporting Russian army groups would proceed west to engage an invading German force. In the wargaming exercise, the staff members representing the German force halted one group with a screening force and proceeded to overwhelm the second group, dismantling it in detail. Later on, in the war, the same operation was carried out in real life by the same generals. In the same region as the wargame represented, the two generals were confronted by a German force that behaved as the fictional forces did in the wargame. The results—dismantling of half of the Russian force by the entire German force—played out in real life (Caffrey, 2000). The Russian generals should have known better.

In the period of peace following the war, all nations would begin wargaming possible situations, once international tensions made further war a possibility. Under the arms limitation treaties that ended the Great War, this was an excellent way for nations to practice their war plans, much as Prussia did during the Napoleonic wars with the original 1811 *Kriegsspiel*. Wargaming would continue to play a huge role for all nations involved in World War II.

Following World War II, in the 1950s, the development of wargames continued in a different light. At that time, Charles Roberts, in the United States, developed a board game called *Tactics*. This allowed the movement of pieces on a chess-like grid, on a board that represented a simplified military map. The game, and its further refinement into *Tactics II*, led to the launch of the Avalon Hill Game Company, which would go on to produce dozens of military simulation wargames. These board wargames were quite popular and grew in number and popularity throughout the 1950s, 1960s, 1970s, and 1980s. There are many other wargame companies that came about in this period, like Simulation Publications Incorporated (SPI), publishing many games on an annual basis covering a wide variety of topics. Almost all campaigns of history have been covered at least once, and many fictional (historical "what if" scenarios) and fanciful (including science fiction, and sword and sorcery) campaigns. The table top wargame hobby continues to be strongly supported and followed today; however, in many ways civilian computer wargaming has replaced it somewhat.

Some related civilian activities have also come to rise along with the development of the commercial wargames market. Beginning largely in the 1960s and increasing in pace until today, historical military re-enacting has become a popular hobby. Participants research military history, acquire accurate replicas of uniforms and equipment, form into units to drill, and participate in mock combat.

In addition to re-enactment, the development and rise of role playing games came out of the commercial military game market. Beginning in the 1970s with medieval fantasy games such as *Dungeons and Dragons*, this hobby has grown large and now is represented by many games highlighting many different military and non-military topics from different eras of history, fantasy, or science fiction. In this sort of game, the player participates by choosing (and then adjudicating) the actions of a single character, rather than an entire military unit or army. Role-play is also very popular as a business gaming venture, where the potential decisions and actions are decided on by corporate participants, and the results are gamed out to see possible outcomes. These types of training games may be adjudicated by mathematical models (either paper or computerized), or they may be a lightweight form of *Kriegsspiel* as advocated for by the younger von Moltke, where experts give narrative results and adjudication to the play by participants.

Role-playing has proven to be a useful gaming activity not only for recreational adventure and business simulation, but also for training in military and diplomatic arenas. The Model United Nations organization, available for participation by students at many secondary schools and colleges, is a form of role-play.

Beginning slowly in the 1960s and 1970s and developing with a heightened pace of development in the 1980s, and finally overtaking their tabletop comrades in the 1990s, computer wargames eventually took over much of the military wargaming scene, as well as the civilian wargaming scene. Tabletop and board wargames are still used by both audiences, but it is much more common these days to encounter a computer moderated wargame than the other way around. Curiously, as a side effect of commercial (civilian) board wargames being largely (but not completely) supplanted by computer wargames, there has been a marked rise in and return to tabletop wargames using military miniatures (toy soldiers and models representing military equipment) in the civilian wargaming community.

Simulated Combat and Mock Warfare

Imitating combat, or warfare, is a practice almost as old as warfare itself. As soon as organizations (tribes, cities, states) realized that they could perform better in conflict against their peers by having a class of trained warriors, the process of training began to adopt simulated combat.

This is a very important portion of the history of combat modeling because it eventually culminates in modern training techniques that involve stimulating actual combat systems with modeled combatants and combat events. To see how to get to the modern case, it is useful to retrace history. At each step the individual historical examples will be addressed as to whether they help with the practice of just individual combat or if they assist with organized combat.

As with so many other elements of military science, one can trace at least one beginning of simulated warfare to the Romans. Certainly other, older, civilizations were participating in mock combat and practicing military maneuvers, but the Romans have it well documented. The Roman author Vegetius in his training manual *Epitoma Rei Militaris* (*De Re Militari*) discusses a number of aspects of the Roman Army and its training, including the practice of maneuvers, drill, and weapon training (Clarke, 1767). The date of Vegetius's publication is not known, but it is thought to be sometime in the late 4th century (it mentions the deeds under the Emperor Gratian, who died in 383 AD) (Charles, 2007). The surviving portion of the work is divided into four sections, the first of which is entirely given to the training and exercises of soldiers. With the organization and bureaucracy of the Roman military establishment, there were probably earlier training manuals that discussed training exercises and mock battles.

Vegetius describes an event during the training of Roman soldiers called the Armatura, which involved mock combat using wooden weapons. These weapons weighed roughly twice what the normal weapons weighed, so while they were training in the proper technique of using the weapons (alone or in small groups), there was also physical conditioning going on. As units, Roman trainees also participated in other military training and maneuvers such as infantry marches and practicing the tasks related to building march forts and other field defenses. The manual goes so far as to say that participation and success in the event is so important to the training of the soldier that "their normal daily pay of grain should be withheld if they do not perform well enough." *De Re Militari* is one of the few surviving primary texts concerning the training of the Roman Army, and the one that made the point that the secret of Roman success was via the extensive training and maneuvers that the army performed. As mentioned it is likely that earlier manuals at one time existed, and there is documentation that mock battles certainly occurred in an earlier age. One of the strongest corroborations of this supposition is the famous quote from Josephus. The Jewish historian describes the state of Roman preparedness for combat this way: "... their exercises were as unbloody battles, and their battles as bloody exercises." It is hard to imagine "unbloody battles" as anything other than simulated combat.

In the Middle Ages, as feudalism grew, military science developed initially from the defense of domains and the economy (peasant farmers) through strong-man military techniques (the local lord and his select men) into a military system with well defined weapon systems (the armored knight, castle building, siege craft, longbowmen, schiltrons). The professional soldier class (chivalry, or mounted knights) were in a peculiar situation—they were being paid by their sponsoring agents (lords of land domains, typically) to be professional, full-time soldiers (Oman, 1937).

In order to provide for the training of knights, a rigorous system developed that began with training in pre-adolescent years and continued through the life of the knight. This training involved mock weapon training –with practice weapons or against practice targets—and eventually it involved fighting in mock combats such as jousts (mounted combat with knights facing each other with weakened lances), melees (groups of foot soldiers and knights in teams fighting each other), and other events. Jousts and melees, which are normally thought of as elements of medieval feudal culture, are mentioned here because Vegetius's work survived and was referred to throughout the Middle Ages, and much of what came about later on in military science can be traced back to *De Re Militari* (Rogers, 2003).

The medieval period finally began to give way to military science advances in the 14th and 15th centuries as infantry systems proved capable of beating the chivalry (knights). These systems involved the notable victories of the longbow equipped English armies in certain battles of the Hundred Years War, the use of Scottish "pike" formations (schiltrons) during the wars between England and Scotland, and eventually the adoption of the pike by the Swiss cantons and the armies of the various Italian states. Each of these combat systems, which developed through a series of various revolutions in military affairs (RMAs), involves not only the weapon but also its use in an organized and orchestrated way on the battlefield. As they are a new system, they almost demand from their nature that a certain amount of mock combat takes place in training (Oman, 1937).

In the case of the English longbow, there were certainly archery contests and tournaments that frequently coincided with the chivalry tournaments in England. In the case of the Swiss pike formations (the Keil), there were practice maneuvers undertaken by units. It is considered that the simulated combat and maneuvers that the Swiss pikemen undertook allowed them to be far more successful than ancient world pike formations (made famous by the Greek as the Hoplite era gave way to the Phalanx era, but really developed into a fine military weapon by the Macedonians—Phillip and Alexander) that the weapon system was adopted from. A study of the mock combats that could be taken by a unit of pikemen during training was made by the German historian Hans Delbrück (1990).

The eventual victory of gunpowder as the main weapon of the battlefield led to the formation of professional standing armies funded by the rising nation states of the 17th and 18th centuries. It was realized that the earlier reliance on nobility or special talent or skill could be democratized and leveled by the adoption of a weapon system (musket) and mass training and drilling that made all men equals (Parker, 1996).

As with the Romans in an earlier age, one of the military powers of the period became great (partially) due to his reliance on training and drilling. The man who led that military power was Frederick the Great of Prussia. His amazing victories during the Seven Years War were due (in part) to the ability of his army to rapidly assume position on the battlefield, especially compared with the time it took for his opponents. Much of this expertise was due to training and drilling.

During the Napoleonic Wars, much has been made of the increased effectiveness of British musketry over that of their foes (France and her allies). After their experiences in North America during the American War of Independence, the British army was reformed under a new training regimen with a manual written by Dundas (1788). This manual specified all of the maneuvers and formations for the infantry, as well as goals for training. The British Infantry is "the only military force not to suffer a major reverse at the hands of Napoleonic France" (Haythornthwaite, 1987). Much of the reason for this is due to the fact that British infantry units participated in much more musketry practice with live ammunition—a practice that was anything but standard at the time. The statistics frequently given are 30 rounds a year for a British infantry unit, rather than the 10 rounds per year for other states (Haythornthwaite, 1987).

The age of gunpowder gave way to the age of rifles during the 19th century as changes in the technology improved, and this led to the first mock battles of the modern period. In 1837, von Moltke became Chief of Staff of the Prussian Army and increased the use of tabletop wargames introduced as early as 1811 (covered earlier in this chapter). He also instituted the Staff Ride.

The Staff Ride was a technique where the Chief of Staff would gather all of his senior officers to a field that might be typical of an expected battle. He would lay out, for the officers, the location and posture of the enemy and Prussian troops, and then ask the Generals, one after another, what they would do if in command of the battle. Finally, some comments would be made as to the effectiveness of the plans and what was expected. In a modern examination of the legends of Moltke in the areas of training and planning, Zuber points out that the Staff Ride was quite useful for teaching the officers about maneuver, placement, terrain, and timing of battlefield actions (Zuber, 2008). It is pointed out that understanding of certain core concepts such as logistics are not helped at all by the simulated warfare of the Staff Ride.

This technique, while very good for introducing the ideas of seeing the battlefield and picturing the troops where they would actually be, had some fallbacks—mostly due to the psychology of the trainees. If you are a junior officer and you are asked your opinion, it would be in your best interest to give the same opinion as your senior officer, so as not to risk embarrassment, or worse, sounding better than your senior.

The Staff Ride (or field exercise), however, was improved and followed up through World War I and is still practiced today. Additionally, it was adopted by many countries. It was followed as a technique to use in combination with tabletop wargaming to predict the outcome and effectiveness of proposed plans. The staff would perform a Staff Ride, make predictions concerning a battle on a particular piece of terrain, and then wargame the situation as a tabletop wargame (using a technique based on the *Kriegsspiel*). The combined technique was able to simultaneously train officers in valuable tactical decision making as well as provide an evaluation for planning exercises.

The Staff Ride and tabletop exercise combination continued to be developed in Germany until the preparations for what would become World War I. Schlieffen perfected the plan to invade France, which was eventually followed by Germany upon her entry into the Great War (1914). This plan was wargamed out extensively, and success was accurately predicted. Unfortunately, several of the involved units did not perform as predicted and/or as ordered, so the plan was not successful in real life. This is seen as one of many of the real-life situations where wargaming alone is not sufficient; rather mock combat, maneuvers, and field exercises are also required by the troops who are expected to follow the plans coming from wargames.

As the mobilization of modern rifle armed armies grew the need for morale boosting training, especially with the bayonet, was seen and adopted by almost all modern nations. Mock combat with the bayonet against other live combatants or, much more likely, against dummies and targets was seen to have several benefits. First it taught the specifics of thrusting and retrieving the bayonet itself. Second it was thought to be a great morale booster. The advent of mechanized elements into warfare towards the end of the Great War and the interwar period (1920s–1930s) brought with it training techniques that sought to imitate combat.

We have already seen the Link Trainer (ASME International, 2000), which was an attempt to imitate an airplane, but on the ground there were actual maneuvers and mock combat techniques involving mechanical combat elements. The development of the Link Trainer and its predecessors leading up to the current state of the art in modern computer based simulations is covered in the second half of this chapter.

Before the successful advent of computerized vehicle simulators, one vehicle (cheaper and easier to operate) would be substituted for another when maneuvers were attempted. In this way, for instance, the early training by Germany in Blitzkrieg tactics, employing the rapid and coordinated maneuvers of tanks with other mechanized assets, were accomplished by using (for instance) bicycles and simple trucks in place of other vehicles. The United States and other countries developed weapons testing sites where imitation battles could be staged with both actual vehicles and mock vehicles. This was seen as one of the final tests of such equipment before applying it directly to the order of equipment of actual units. This is unfortunate in that it would be much more useful if such testing (simulated combat) was able to be held much earlier in the development phase of such weapons so that lessons learned from the imitation battles could influence the design process itself.

Naval operations participate in mock combat by undertaking maneuvers in and around areas where hostilities are expected, and also by taking part in gunfire against a variety of different targets, including the surrendered assets of recently defeated nations. In this way, many of the German assets that escaped the scuttling of the fleet at the end of World War I were used by the British Navy as targets in the 1920s to practice gunfire against a "realistic" target. Of course, the ship would be moored, but it was still the right size and elevation.

The introduction of anti-aircraft fire necessitated new training techniques as well, and the US Navy (and other organizations) came up with mock combat techniques to simulate the actual event. These mock combat techniques included not only drones (a more modern invention) but also towed targets and kites.

Finally, modern mechanized warfare (including land, sea, and air combat) has evolved to include simulated training (mock combat) by using computer assisted simulation systems to stimulate modern command and control devices (introducing electronic signals representing imagined enemy units to train against). Much more of this is covered, in great detail, in the second half of this chapter.

The use of laser based weapons and sensors, as used in infantry exercises, is often seen as the Live component of the modern concept of Live, Virtual, and Constructive combat training. The Virtual component is provided by a simulator (such as a combat simulator), and the Constructive component is provided by a traditional, wargame-style, military simulation system.

Such electronic devices as land warfare MILES gear (representing the professional version of Laser Tag!), and simulated foes in the form of the Opposition Force aircraft that fly at training facilities like the Aggressor Squadron at the US Navy's Top Gun school, provide the ultimate in modern mock combat. The actual weapons and platforms are employed, under realistic conditions, against human foes that are simulating an enemy force—but can behave with all the savvy and fickleness of an actual entity.

HISTORY OF DISTRIBUTED SIMULATION

As the technological advances of the 20th century have been applied to combat modeling, there has been a swift march to the state of distributed simulation. These changes have altered the way modern militaries train, plan, and perform many of their other functions. The distributed simulation has augmented, or in some cases replaced, the traditional methods of wargaming and simulating combat that were described in the first half of this chapter.

A Brief History of the First Computer Simulators

Simulator technology began in 1927 with the development of the Link Trainer. It was a significant contribution to the development of aviation as it provided a means to train pilots in realistic conditions without sacrificing their safety. After earning his pilot's license, Ed Link used his experiences from flying and his knowledge from working in his father's piano and organ company to develop a flight trainer. "The earliest trainer sat on a series of organ bellows, which would inflate or deflate to various heights to cause the trainer to bank, climb, and dive" (ASME International, 2000). The trainer allowed Link to reduce the cost of flying lessons by enabling student pilots to learn some flying skills on the ground. The first fully instrumented Link Trainer sold to the US Navy in 1931 for $1500.00, and the Army subsequently accepted delivery of its Link Trainers in 1934. The Link Trainer was credited with saving vast sums of money and time, and importantly the lives of pilots. According to a report to a subcommittee of the US House of Representatives, Link Trainers were estimated to have saved the Army Air Corp at least 524 lives, $129,613,105 and 30,692,263 man-hours in one year (US Congress, Office of Technology Assessment, 1994).

Another innovation came in the mid-1940s when physicists at Los Alamos Laboratory were investigating radiation shielding. Despite having most of the necessary data, the problem could not be solved with analytical calculations. John von Neumann and Stanislaw Ulam suggested using statistical sampling techniques on an electronic digital computer to model the experiment. The name "Monte Carlo" was used to refer to this technique since it involved probability and chance. Monte Carlo methods were central to the simulations required for the Manhattan Project, though they were severely limited by the computational capabilities at the time. It was only after electronic computers were first built in 1945 that Monte Carlo methods began to be studied in depth. In the 1950s these methods were used at Los Alamos for early work relating to the development of the hydrogen bomb and became popular in the fields of physics, physical chemistry, and operations research (Anderson, 1986).

During World War II, the US Navy approached the Massachusetts Institute of Technology (MIT) about the possibility of creating a computer to drive a flight simulator for training bomber crews. They envisioned a system where the computer would continually update a simulated instrument panel based on control inputs from the pilots. Unlike the Link Trainer, they envisioned a system with more realistic aerodynamics that could be adapted to any type of plane. After a demonstration in 1945 of the Electronic Numerical Integrator And Computer (ENIAC), the military decided a digital computer was the solution. Such a machine had the potential to allow the accuracy of the simulation to be improved by adding more code to the computer program, as opposed to adding fixed parts to an analog machine. This began Project Whirlwind, the first computer that would operate in real time and use video displays for output. Its development led directly to the US Air Force's Semi Automatic Ground Environment (SAGE) system, an air defense system that simulated combat from the perspective of more than one combatant. This work was motivated by the 1949 detonation of an atomic device by the Soviet Union. The US Department of Defense (DoD) became serious about improving air defense systems, and in 1950 began transmitting digital radar data from the Microwave Early Warning radar at Hanscom Field to SAGE for processing. By 1953 a scaled down version of SAGE, called the Cape Cod System, allowed radar operators and weapon controllers to react to simulated targets presented to them as real targets would appear in actual conflict (US Congress, Office of Technology Assessment, 1994).

An early example of separate simulations being used together appeared in 1964 with the Base Operations-Maintenance Simulator (BOMS). This simulator modeled the essential characteristics of an Air Force base, specifically a SAC B-52/KC-135 organization. Three separate simulation programs made up BOMS: (1) the data generator combined a flying schedule with random failures (generated by a Monte Carlo process) and resources required for repair; (2) the main program simulated the minute-by-minute activities of the base; and (3) the analysis program summarized the output of the main program and printed reports. One reason the simulation was broken into three separate programs was because the size of the model and the dimension of the experiments were so large they exceeded the available computer memory (Ginsberg and Barbara, 1964).

To this point computer simulation technology focused on providing input to users in a single direction to enable them to learn appropriate responses. There was no ability to interact with another person through the simulation. This changed in 1961 when Steve Russell, a student at MIT, created the first interactive computer game called *Spacewar* (Brand, 1972). *Spacewar* was a two-player game involving warring spaceships firing photon torpedoes. Each player could maneuver a spaceship and score by firing missiles at the opponent. It ran on a Digital PDP-1 mainframe computer and the graphics were ASCII text characters. By the mid-1960s it was very common to find *Spacewar* on research computers in the United States. As computer technology continued to evolve, so did the ability to play against an opponent in a simulation. In 1974 Howard Palmer and Steve Colley developed the first networked, 3D multi-user first person shooter game called *Maze War* (DigiBarn Computer Museum, 2010). Developed at the NASA Ames Research Center, *Maze War* used two Imlac computers connected with serial links. It introduced the concept of online players as "eyeball" avatars chasing each other around in a maze.

Early systems focused on a functional distribution of the simulator, including instructor stations and tasks. One of the earliest examples of linking full up training simulators came in 1978 with the B-52/KC-135 system. The B-52 and KC-135 simulators, some built in old railroad cars, operated together so that the B-52s could include the tanker rendezvous in their long missions. These simulators had an opaque windscreen and used lights to model the appearance of lightning strikes. This program made significant progress in connecting disparate simulators, simulating system problems and weather; however, visual cues to the outside world were still missing.

Technology Evolves

Early systems were limited from several perspectives. Most simulators required identical computer hardware and software in order to operate and create the environment. In this sense they were closed systems and they were not expandable. Also, computer networks were synchronous; the fixed nature of synchronous systems using shared memory and time slicing of data arrivals made adding simulators to these networks impractical. Technically it was a challenge to separate simulator capabilities so that functions could be performed autonomously.

This changed with the evolution of computer and communications technologies. In 1969 the development of UNIX, a portable, multi-tasking and multi-user operating system, brought the ability to handle asynchronous events, such as non-blocking input/output and inter-process communication. That development was followed by expanded computer-to-computer communications with the creation of the Advanced Research Projects Agency Network (ARPANET). The ARPANET was based on packet switching, which enabled a computer to communicate with more than one machine by transmitting packets of data (datagrams) on a network link. The invention of Ethernet technology in 1973 enabled the connectivity of a variety of computer systems within a local geographical area. Thus, simulators no longer needed to be located in the same room, but could be distributed. The following year the Transmission Control Protocol (TCP) and the Internet Protocol (IP) emerged as a means to enable intercomputer communications. This development supported heterogeneous computer communications, which would change simulators from being bounded by their computational resources to being bounded by their ability to send and receive information.

Distributed Simulation Begins

As computing technology advanced, applications for networked simulation emerged. These applications included inter-crew training, live tests and training, and analysis. For inter-crew training, improving tactical proficiency and training large groups was a growing need. In live tests and training, reducing the number of events was desired due to safety, environmental, and cost concerns. In the discipline of analysis, there was a need for individual subsystem evaluation and fine-tuning of tactics and strategies. The emerging application areas and the advances in computing and communication technology led to the first generation of distributed simulation architectures.

Initiated in 1983 by the Defense Advanced Research Projects Agency (DARPA), the SIMulator NETworking (SIMNET) project was the first attempt to exploit the developments in communications technology for simulation. SIMNET emphasized tactical team performance in a battlefield context, including armor, mechanized infantry, helicopters, artillery, communications, and logistics. Combatants could visually see each other "out the window" and communicate with each other over radio channels. This distributed battlefield was based on selective fidelity (only provided features for inter-crew skills training), asynchronous message passing, commercial computer networks, and replicated state information. SIMNET was based on six design principles.

- *Object/event architecture* — the world is modeled as a collection of objects which interact using events
- *Common environment* — the world shares a common understanding of terrain and other cultural features
- *Autonomous simulation nodes* — simulations send events to other simulations and receivers determine if that information is relevant

- *Transmission of ground truth information*—each simulation is responsible for local perception and modeling the effects of events on its objects
- *Transmission of state change information*—simulations transmit only changes in the behavior of the object(s) they represent
- *Dead reckoning algorithms*—simulations extrapolate the current position of a moving object based on its last reported position

The first platoon level system was installed in April 1986 and over 250 networked simulators at 11 sites were transitioned to the US Army in 1990. Two mobile platoon sets (in semi-trailers) were delivered to the Army National Guard in 1991.

Among the many contributions of the SIMNET program was the invention of semi-automated forces (SAF). The purpose of a SAF simulation was to mimic the behavior of different objects on the battlefield, whether vehicles or soldiers. Realizing it was impractical to have large numbers of operators to control both friendly and opposing forces, Colonel James Shiflett created the idea of SAF as a way to put opposing forces on the battlefield. His inspiration for SAF was *Night of the Living Dead*, a movie in which teenagers are attacked by zombies. Colonel Shiflett wanted a simulation that could produce a large number of "dumb" targets (i.e. zombies) to roam the battlefield and provide targets for SIMNET operators (Loper interview with COL James Shiflett, December 2008). Eventually SAF simulations turned into something much more intelligent: they could plan routes, avoid obstacles, stay in proper formations, and detect and engage targets.

Soon after the SIMNET project, DARPA recognized the need to connect aggregate level combat simulations. The Aggregate Level Simulation Protocol (ALSP) extended the benefits of distributed simulation to the force level training community so that different aggregate level simulations could cooperate to provide theater level experiences for battle staff training. In contrast to the SIMNET simulators, these simulations were event stepped and maintaining causality was a primary concern. The ALSP program recognized that various time management schemes and more complex simulated object attribute management requirements were needed. The requirements for ALSP were derived from the SIMNET philosophy and included the following (Weatherly et al., 1991).

- Simulations need to be able to *cooperate over a common network* to form confederations.
- Within a confederation, *temporal causality* must be maintained.
- Simulations should be able to *join and exit a confederation* without major impact on the balance of the participating simulations.
- The system should be network based with *no central controllers* or arbitrators.
- *Interactions do not require knowledge of confederation participants* and should support an object oriented view of interactions.

Life After SIMNET — The Need for Standards

Several efforts to evaluate simulation technology during this time frame supported and encouraged the need to develop and invest in distributed simulation. The Defense Science Board (DSB) Task Force on Computer Applications to Training and Wargaming stated: "Computer-based, simulated scenarios offer the only practical and affordable means to improve the training of joint operational commanders, their staffs, and the commanders and staffs who report to them" (Defense Science Board, 1988). This was followed by the report on *Improving Test and Evaluation Effectiveness*, which found that modeling and simulation (M&S) could be an effective tool in the acquisition process throughout the system's lifecycle, especially if employed at the inception of the system's existence (Defense Science Board, 1989).

Then, in 1991, the potential for distributed simulation for the military was realized in an operational context. The Battle of 73 Easting was a tank battle fought during the Gulf War between the US Army and the Iraqi Republican Guard. Despite being alone, outnumbered and out-gunned, the 2nd Armored Cavalry (ACR) struck a decisive blow destroying Iraqi tanks, personnel carriers, and wheeled vehicles during the battle. The 2nd ACR had trained intensely before the battle both in the field and on SIMNET. Immediately, SIMNET's potential for network training was confirmed.

The following year, the DSB looked at the impact of advanced distributed simulation on readiness, training, and prototyping (Defense Science Board, 1993). It concluded that distributed simulation technology could provide the means to substantially improve training and readiness; create an environment for operational and technical innovation for revolutionary improvements; and transform the acquisition process.

Recognizing the importance of the SIMNET program and concerned that activity related to networked simulation was occurring in isolation, a small conference was held in April 1989 called Interactive Networked Simulation for Training. The group believed that if there were a means to exchange information between companies, distributed simulation technology would advance more rapidly. The group also believed that technology had stabilized enough to begin standardization. The conference soon developed into the Distributed Interactive Simulation (DIS) Workshops.

Through these workshops, networked simulation technology and the consensus of the community were captured in proceedings and standards. The standards initially focused on SIMNET, but quickly evolved to include a broader range of technology areas. In 1996 the DIS Workshops transformed into a more functional organization called the Simulation Interoperability Standards Organization (SISO). An international organization, SISO is dedicated to the promotion of M&S interoperability and reuse for the benefit of a broad range of M&S communities.

One of the significant contributions of the DIS Workshops was the definition of Live, Virtual, and Constructive (LVC) simulations. Live simulation refers to

M&S involving real people operating real systems (e.g. a pilot flying a jet). A virtual simulation is one that involves real people operating simulated systems (e.g. a pilot flying a simulated jet). Constructive simulations are those that involve simulated people operating simulated systems (e.g. a simulated pilot flying a simulated jet). The LVC taxonomy is a commonly used way of classifying models and simulation.

Distributed Simulation Science

Distributed simulation technology is based on the science of distributed systems. A distributed system is a collection of independent computers that appear to the users of the system as a single computer (Tanenbaum, 1995). This definition has two aspects. The first deals with hardware: the machines are autonomous. The second aspect deals with software: the users think of the system as a single computer. This characterization provides a good foundation for distributed simulation technology. The goal of a distributed simulation is to create the illusion in the minds of the users that the entire network of simulations is a single system rather than a collection of distinct machines. Therefore, understanding how to separate the hardware and software design issues is key to developing the technology.

There are numerous challenges associated with building software to support distributed simulation. These include transparency, openness, scalability, performance, fault tolerance, and security. Transparency is specifically important as it refers to hiding the distribution of components, so the system is perceived as "whole" and not a collection of "independent" simulations. Tools are needed to support the construction of distributed simulation software, specifically protocols that support the patterns of communication as well as naming and locating simulation processes.

There are two types of characteristics that distinguish the basic patterns of communication in distributed simulations: communication mechanisms and event synchronization. Communication mechanisms refer to the approach for exchanging data among two or more simulations. This includes message passing, shared memory, remote procedure call, and remote method invocation. With message passing there are several variations of delivery depending on the number of receivers. Data can be sent unicast to individual simulations, broadcast to every simulation, or multicast to a selected subset of simulations. Mechanisms such as publish/subscribe can also be used to define subsets of potential receivers.

Event synchronization refers to the approach for synchronizing the sending and receiving of data among the participants of a distributed simulation. Important properties include time, event ordering, and time synchronization. Each simulation in a distributed simulation is assumed to maintain an understanding of time. That can include an informal relationship or a very strict adherence to a simulation or wall clock. In either case, simulations assign a time stamp to each message they generate. Event ordering refers to the way in which events are delivered to each simulation. There are several choices of event ordering. Receive order delivers events regardless of the message time stamp and its relationship to the global distributed system. Timestamp order delivers events in an order directly related to a global interpretation of time. Time synchronization is related to both time and event ordering in that it is concerned with the global understanding of time in the distributed system. If global time is needed, there are a number of conservative and optimistic synchronization algorithms that can be used to achieve this state.

Communication mechanisms and event synchronization can be implemented in one of two ways: by individual simulations or by an operating system. There are three types of operating systems commonly used in distributed systems. A network operating system is focused on providing local services to remote clients. A distributed operating system focuses on providing transparency to users. Middleware combines the scalability and openness of a network operating system and the transparency and ease of use of a distributed operating system to provide general purpose services. There are a number of tradeoffs with the different approaches such as performance, scalability, and openness. Modern distributed simulation has implemented a range of these approaches.

LIVE, VIRTUAL, CONSTRUCTIVE SIMULATION ARCHITECTURES

The most widely used LVC simulation architectures in the DoD are Distributed Interactive Simulation (DIS), High Level Architecture (HLA), and Test and Training Enabling Architecture (TENA). This second generation of distributed simulation architectures has evolved over the last 20 years using different technology, standards, and funding strategies. The following sections will give a brief description of each architecture and then characterize its approach for communication and event synchronization.

Distributed Interactive Simulation

"The primary mission of DIS is to define an infrastructure for linking simulations of various types at multiple locations to create realistic, complex, virtual 'worlds' for the simulation of highly interactive activities" (DIS Steering Committee, 1994). DIS is based on the fundamental design principles of SIMNET. The goal of DIS is to create a common, consistent simulated world where different types of simulators can interact. Central to achieving this goal are protocol data units (PDUs) which use standard messages exchanged to convey state about entities and events. The PDUs comprise object data related to a common function; e.g. entity state, fire, detonation, and emissions are all frequently used PDUs. The Institute of Electrical and Electronics Engineers (IEEE) approved the first DIS standard in 1993 with 10 PDUs; the most recently published standard has 67 PDUs. The approved IEEE Standards for Distributed Interactive Simulation include

- IEEE 1278.1—Application Protocols
- IEEE 1278.1A—Enumeration and Bit-Encoded Values
- IEEE 1278.2—Communication Services and Profiles
- IEEE 1278.3—Exercise Management and Feedback (EMF)
- IEEE 1278.4—Verification, Validation, and Accreditation

From an implementation perspective, simulation owners either custom develop DIS interfaces or buy commercial products. There is also an open source initiative, Open-DIS, to provide a full implementation of the DIS protocols in C++ and Java (McGregor and Brutzman, 2008). The first DIS demonstration was held at the 1992 Interservice/Industry Training, Simulation and Education Conference in San Antonio, Texas. The demonstration included 20 companies, 25 simulators, and one long haul connection. A minimal set of PDUs (entity state, fire, and detonation) was used, and the interaction among participants was focused mainly on unscripted free-play.

From a distributed system viewpoint, DIS is based on the idea that the network and simulators are integrated, i.e. there is minimal transparency. All communication about simulation entities and their interactions occurs via the PDUs. Reasonably reliable delivery is sufficient; dead reckoning algorithms are robust, so 1%–2% missing datagram (randomly distributed) does not have an adverse impact on performance. As a result, most PDUs are sent using the best effort user datagram protocol (UDP). The network is assumed to provide a certain level of assured services including 300 ms end-to-end latency for "loosely coupled" interactions and 100 ms total latency for "tightly coupled" interactions. Due to the potential for high latency in wide area networks, DIS is best for exercises on local area networks.

Interaction among DIS simulations is peer-to-peer and occurs using a message passing paradigm. Since PDUs are broadcast to everyone on the network, bandwidth and computing resources can be consumed processing data that is not relevant to a specific simulation. A study of multicast communications occurred in the early 1990s with the idea of developing a new protocol for highly interactive applications. Developing a new protocol proved problematic and was abandoned. However, progress was made in understanding how to create multicast groups. One of the most commonly understood approaches to grouping information was called *area of interest* (Macedonia et al., 1995). Multicast was difficult to implement in DIS due to the lack of middleware or a distributed operating system, which could provide transparency to the simulations.

Time in DIS simulations is managed locally. Each simulation advances time at its own pace and clocks are managed locally using a local understanding of time. There is no attempt to manage time globally. Each PDU has a time stamp assigned by the sending simulation and PDUs are delivered to simulations in the order received. Simulations provide ordering locally, based on their understanding of time.

High Level Architecture

The High Level Architecture (HLA) program emerged in the mid-1990s based on several assumptions. The first premise is no one simulation can solve all the DoD functional needs for modeling and simulation. The needs of the users are too diverse. Changing user needs defines the second premise: simply, it is not possible to anticipate how simulations will be used in the future or in which combinations. It is important, therefore, to think in terms of multiple simulations that can be reused in a variety of ways. This means that, as simulations are developed, they must be constructed so that they can be easily brought together with other simulations, to support new and different applications. These assumptions have affected the HLA design in several ways. Clearly, the architecture itself must have modular components with well defined functionality and interfaces. Further, the HLA separated the functionality needed for individual simulations (or federates) from the hardware infrastructure required to support interoperability. The HLA architecture is defined by three components:

- *rules* that simulations must obey to be compliant to the standard
- an *object model template (OMT)* that specifies what information is communicated between simulations and how it is documented
- an *interface specification* document that defines a set of services that simulators use to communicate information

The HLA standards began in 1995 under a government standards process managed by the Architecture Management Group. The DoD adopted the baseline HLA architecture in 1996 and the standards were moved to an open standards process managed by SISO. The IEEE standards for HLA were first approved in 2000 and included

- 1516—Framework and Rules
- 1516.1—Federate Interface Specification
- 1516.2—Object Model Template (OMT) Specification
- 1516.3—Federation Development and Execution Process (FEDEP) Recommended Practice
- 1516.4—Recommended Practice for Verification, Validation, and Accreditation of a Federation—An Overlay to the HLA FEDEP

From a distributed system viewpoint, HLA is based on the idea of separating the functionality of simulations from the infrastructure required for communication among simulations. This separation is accomplished by a distributed operating system called the Runtime Infrastructure (RTI). The RTI provides common services to simulation systems and provides efficient communications to logical groups of federates. Data can be sent using both best effort (UDP) and reliable (TCP) internetwork protocols. An important distinction is that the HLA is not the same as the RTI. The RTI is an implementation of the HLA interface standard, and thus there can be many different RTIs that meet the HLA interface standard. From an implementation perspective, HLA follows a commercial business model. There have been a variety of open source initiatives, but none has produced an HLA compliant RTI.

In contrast to the static DIS PDUs, HLA uses the concept of OMTs to specify the information communicated between simulations. This enables simulation users to customize the types of information communicated among simulations based on the needs of the federation (what DIS called an exercise). When the OMT is used to define the data for a federation, the Federation Object Model (FOM) describes shared information (e.g. objects, interactions) and inter-federate issues (e.g. data encoding schemes). It did not take long, however, for the community to understand the difficulty in developing FOMs. This led to the emergence of reference FOMs (SISO, 2001), a mechanism for representing commonly used information, and Base Object Models (BOMs), a mechanism for representing a single set of object model data (SISO, 1998).

From a communications perspective, HLA learned that broadcasting information to all simulations has serious implications on performance. The HLA defined a publication/subscription paradigm whereby producers of information describe data it can produce and receivers describe data it is interested in receiving. The RTI then matches what is published to what has been subscribed. This approach maximizes network performance by allowing individual simulations to filter data it wants to receive at many different levels.

The HLA does include time management services to support event ordering. Both time stamp order, where messages are delivered to simulations in order of time stamp, and receive order, where messages are delivered to simulations in order received, are supported. Global time advance and event ordering is implemented by means of synchronization algorithms. The HLA interface specification supports the two commonly defined approaches: conservative and optimistic. While HLA provides global time management, use of these services is not required. Simulations can choose to advance time at its own pace, not synchronized with other simulations.

Test and Training Enabling Architecture

The Test and Training Enabling Architecture (TENA) emerged in the late 1990s, after the HLA initiative was under way. The purpose of TENA is to provide the architecture and the software implementation necessary to do three things. First, TENA enables interoperability among range systems, facilities, simulations, and C4ISR systems in a quick, cost-efficient manner. It also fosters reuse for range asset utilization and for future developments. Lastly, TENA provides composability to rapidly assemble, initialize, test, and execute a system from a pool of reusable, interoperable elements.

The principles of the TENA architecture include constrained composition, dynamic runtime characterization, subscription service, controlled information access, and negotiated quality of service. Constrained composition refers to the ability to compose the system for specific intended purposes that may be either transitory or permanent in nature. Constraints apply to use of assets including physical proximity and location, coverage regions, performance capabilities, and subsystem compatibility. Dynamic, runtime characterization is focused on responding to many allowable compositions and permitting rapid reconfigurations. This is accomplished by establishing methods for self-description of data representations prior to or concurrent with data transfer or negotiating representation issues before operation starts. Similar to HLA, the subscription service is an object based approach to data access, which matches producers and consumers of information. Due to the nature of many range assets, controlled information access is particularly important. Levels of access allow users to limit information access to a desired subset of all users. Since some services have significant performance and cost implications (e.g. data streams with large capacity requirements or strict latency tolerance), users can request specialized assets be allocated when needed. The negotiated quality of service protocols relies on the principle of separation of control information from data.

The TENA project uses a government standards process and is managed by the Architecture Management Team (AMT). The AMT controls implementation content and government members of the AMT recommend implementation changes. As such, no open standards have been published for TENA; however, they do follow a formal process for standardizing object data.

From a distributed systems view, TENA separates the functionality of range assets from the infrastructure required to communicate among assets using middleware. The TENA Middleware is a common communication mechanism across all applications, providing a single, universal data exchange solution. Data exchanged among range assets is defined in object models, which can be sent using both best effort (UDP) and reliable (TCP) internetwork protocols. A logical range object model is defined for a given execution, and can include both standard (time, position, orientation, etc.) and user defined objects.

The TENA Middleware combines several communication paradigms, including distributed shared memory, anonymous publish–subscribe, remote method invocations, and native support for data streams (audio, video, telemetry, and tactical data links). Central to TENA is the concept of a Stateful Distributed Object (SDO). This is a combination of a CORBA[1] distributed object with data or state. It is disseminated using a publish–subscribe paradigm, and subscribers can read the SDO as if it were a local object. An SDO may have remotely invocable methods.

Given the nature of real-time range assets, there is no requirement for time management to support event ordering. Messages are delivered to assets in the order they are received. The clock services defined in TENA are to manage time issues for the test facility. This includes synchronization and time setting services, as well as maintaining a global clock for exercises.

---

[1] Common Object Request Broker Architecture.

EPILOGUE

Distributed simulation architectures in use within the DoD today have all been designed to meet the needs of one or more user communities. These architectures continue to evolve and mature based on changing requirements. The existence of multiple architectures allows users to select the methodology that best meets their individual needs. It also provides an incentive for architecture developers and maintainers to competitively keep pace with technology and stay closely engaged with emerging user requirements (Henninger et al., 2008).

One of the challenges in achieving the transparency desired in distributed simulation is that multiple architectures exist. Incompatibilities between DIS, HLA, and TENA require the development of point solutions to effectively integrate the various architectures into a single, unified set of simulation services. Integration is typically achieved through gateway solutions, which can often restrict users to a limited set of capabilities that are common across the architectures. The successful integration of distributed simulations will continue to rely upon the development of simulation standards.

Despite the advances in distributed simulation technology and standards, challenges remain. In a 2008 survey on future trends in distributed simulation, the most promising areas of research for the simulation community were identified as distributed simulation middleware, human–computer interfaces, and the semantic web/interoperability (Strassburger et al., 2008). Within simulation middleware, the greatest needs identified were plug-and-play capability, standardization and interoperability between different standards, semantic connectivity, and ubiquity (accessible anywhere with any device). The results of this survey combined with the findings of the Live, Virtual, Constructive Architecture Roadmap panel (Henninger et al., 2008) define the needs for the next generation of distributed simulation. The DoD has been a driving force in shaping the technology and standards for nearly 30 years, and it will continue to have a major role defining the way forward.

FURTHER READING

Caffrey M Jr (2000). Toward a history-based doctrine for wargaming. *Aerospace Power Journal* vol XIV, No. 3; pp. 33–56.

Smith, R. (2010). The long history of gaming in military training. *Simulation and Gaming*, 40th Anniversary Issue **41**(1), February.

REFERENCES

Anderson HL (1986). Metropolis, Monte Carlo and the MANIAC. *Los Alamos Science* **14**, 96–108.

Archer CI, Ferris JR, Herwig HH and Travers TH (2002). *World History of Warfare*. University of Nebraska Press, Lincoln, Nebraska, pp. 1–61.

ASME International (2000). *The Link Flight Trainer: A Historic Mechanical Engineering Landmark*. Roberson Museum and Science Center Publication, Binghamton, New York.

Bell RC (1979). *Board and Table Games from Many Civilizations* (revised edn). Dover Publications, New York.

Brand S (1972). SPACEWAR: Fanatic life and symbolic death among the computer bums. *Rolling Stone Magazine*, December 7.

Caffrey M (2000). Toward a history-based doctrine for wargaming. *Aerospace Power Journal*, vol XIV, No. 3; pp. 33–56.

Charles MB (2007). Vegetius in Context. Establishing the Date of the Epitoma Rei Militaris. *Historia*, Paper 194.

Clarke J (1767). *The Military Institutions of the Romans*. London (trans. of Vegetius' *De Re Militari*) Original publisher unknown; reprint published in 1944, by Greenwood Press, Westport Connecticut.

Defense Science Board (1988). *Computer Applications to Training and Wargaming*. Final Report, ADA199456.

Defense Science Board (1989). *Improving Test and Evaluation Effectiveness*. Final Report, ADA274809.

Defense Science Board (1993). *Simulation, Readiness and Prototyping*. Final Report, ADA26612.

Delbrück H (1990). *History of the Art of War. Volume III: Medieval Warfare*. University of Nebraska Press (reprint), Lincoln, Nebraska; Trans. J. Renfroe Walter.

DigiBarn Computer Museum (2010). The maze war 30 year retrospective at the DigiBarn: celebrating thirty years of the world's first multiplayer 3D game environment! http://www.digibarn.com/collections/games/xerox-maze-war/index.html#intro (last accessed 1 December 2010).

DIS Steering Committee (1994). *The DIS Vision*. Institute for Simulation and Training, Orlando, FL.

Dundas D (1788). *The Principles of Military Movements chiefly applicable to Infantry*. Printed for T. Cadell, in the Strand, 1788; London.

Ginsberg AS and Barbara AK (1964). *Base Operations-Maintenance Simulator*. RAND Memorandum, RM-4072-PR, September.

Halter E (2006). *From Sun Tzu to Xbox: War and Video Games*. Thunder Mouth Press, New York.

Haythornthwaite PJ (1987). *British Infantry of the Napoleonic Wars*. Arms and Armour Press, London.

Haythornthwaite PJ (1996). *Weapons & Equipment of the Napoleonic Wars*. Arms and Armour Press, London.

Henninger A, Cutts D, Loper M, Lutz R, Richbourg R, Saunders R and Swenson S (2008). *Live Virtual Constructive Architecture Roadmap (LVCAR)*. Final Report. M&S CO Project No. 06OC-TR-001.

Josephus. *The Wars of the Jews or the History of the Destruction of Jerusalem*, Book III, Chap. 5, Sect. 1, 17 A.D.

Little WM (1912). *The Strategic Naval War Game or Chart Maneuver*. US Naval Institute Proceedings 38, December 1912.

Macedonia M, Zyda M, Pratt D, Brutzman D and Barnham P (1995). *Exploiting Reality with Multicast Groups*. IEEE Computer Graphics and Applications, September.

McGregor D, Brutzman D (2008). Open-DIS: an open source implementation of the DIS protocol for C++ and Java. In: *Proceedings of the Fall Simulation Interoperability Workshop, 15–19 September 2008, Orlando, FL*. http://www.sisostds.org.

Murray HJR (1951). *A History of Board-Games Other than Chess*. Oxford University Press, Oxford.

Oman C (1937). *A History of the Art of War in the Sixteenth Century*. Methuen & Co, London.

Oman C (1960). *The Art of War in the Middle Ages: A.D. 378–1515*. Cornell University Press, Ithaca, New York.

Parker G (1996). *The Military Revolution: Military Innovation and the Rise of the West 1500–1800*. Cambridge University Press, Cambridge.

Reisswitz B (1983). *Instructions for the Representation of Military Maneovres with the Kriegsspiel Apparatus*. Berlin, 1824, Bill Leeson, Hemel Hempstead, Hertfordshire.

Rogers CJ (2003). The vegetian 'science of warfare' in the Middle Ages. *J Medieval Military Hist* **1**, 1–19.

SISO (2001). *BOM Study Group Final Report*. SISO-REF-005-2001, 15 May.

SISO (1998). *Reference FOM Final Report*. SISO-REF-001-1998, 9 March.

Strassburger S, Schulze T and Fujimoto R (2008). Future trends in distributed simulation and distributed virtual environments: results of a peer study. In: *Proceedings of the Winter Simulation Conference, 7–10 December 2008, Miami, FL*. IEEE Press, Piscataway, NJ.

Tanenbaum AS (1995). *Distributed Operating Systems*. Prentice Hall, Englewood Cliffs, NJ.

US Congress, Office of Technology Assessment (1994). *Virtual Reality and Technologies for Combat Simulation—Background Paper*. OTA-BP-ISS-136. US Government Printing Office, Washington, DC.

Weatherly R, Seidel D and Weissman J (1991). Aggregate level simulation protocol. In: *Proceedings of the Summer Computer Simulation Conference, July 1991, Baltimore, MA*. Society for Computer Simulation, San Diego, CA.

Young JP (1957). *History and Bibliography of War Gaming*. Department of the Army, Bethesda, MA.

Zuber T (2008). *The Moltke Myth: Prussian War Planning, 1857–1871*. University Press of America, Lanham, MA.

Chapter 17

Serious Games, Virtual Worlds, and Interactive Digital Worlds

Roger D. Smith

The computer gaming industry has created powerful technologies in visualization, computation, networking, and social interaction. These applications were created as a form of entertainment, but their integration into the larger user experience of the Internet and web has brought them to the attention of a number of industries beyond entertainment. This attention and the early successes of several game-based products have given additional impetus to the use of game technologies for serious applications in science, engineering, education, training, and marketing. This chapter will explore the business and social forces driving the adoption of game technologies in multiple industries. We introduce five specific forces that compel industries to adopt game technologies for their core products and services. These five forces are computer hardware costs, game software power, social acceptance, industry successes, and native industry experimentation. Together these influence the degree and rapidity at which game technologies are adopted in a number of industries.

The chapter also identifies five industry sectors which are actively experimenting with serious games and in which many successful applications have been created. These sectors are education and training, engineering design, product and service marketing, scientific exploration and analysis, and business decision making. Examples in each of these sectors will be provided along with discussions of the forces that have made games an effective choice.

Engineering Principles of Combat Modeling and Distributed Simulation, First Edition.
Edited by Andreas Tolk.
© 2012 John Wiley & Sons, Inc. Published 2012 by John Wiley & Sons, Inc.

The conclusion is that games and the technologies that they embody will continue to proliferate through a wide variety of industries, following a similar pattern to that taken by electronics and the Internet.

INTRODUCTION

In *Growing Up Digital* (1997) and *Grown Up Digital* (2008), Don Tapscott explored the differences between the current generation of digital natives and their older siblings and parents. Their tendency to live with a constant form of multitasking while plugged into numerous forms of electronic communication and entertainment was one of the key distinguishing features of the generation. Regardless of whether the effects of this splintering of consciousness are good or bad, they are a real and enduring characteristic of the young people who are growing into young professionals with these technologies. This generation has grown up with the expectation of handling information digitally and working on multiple streams of data simultaneously. They have also grown up with both a physical body and a number of virtual avatars. They socialize, work, and think in both the physical reality and one or more created virtual realities. They carry with them a persona that exists in multiple social networks, virtual worlds, and computer games. Some of these personas are disposable, while others are a persistent and evolving portrait of their identity. To this generation, a digital persona is not a waste of time or a distraction from their physical life; instead it is an electronic extension of who they are and how they function both at work and at play. As they grow into positions of responsibility, these electronic extensions will be applied to the jobs that they take and the companies that they create in every industry. Social networks and virtual worlds are not a flash in the pan for a young generation. They are a fundamental part of a generation that has grown up with them and will be carried with them as they mature and move through society. These forms of communication and entertainment will be as persistent and prevalent as the telephone, radio, television, computer, and cell phone have been to previous generations. In a learning society, these electronic extensions of individuality will be one of the primary means through which current and future generations gather and digest new information.

Many of us who have not grown up as digital natives perceive computer games and social networks as distractions, toys, and competitors for more "real" forms of communication or work. But computer games are one of the most advanced implementations of several leading edge technologies. They are more sophisticated than the word processor, email, or the World Wide Web. A really rich computer game brings together the latest advances in personal computer hardware, 3D graphics rendering, intuitive user interfaces, artificial intelligence, networked communication, and persistent storage for a virtual world. No other application uses this diversity of technology in such an advanced form.

These games began in government and academic research labs just as their predecessors—electronics and the Internet. They have a commercial trajectory similar to the radio, the VCR, and television as described in Alfred Chandler's history of the electronics industry (Chandler, 2001). Just as the radio was able to replace telegraph and letter communications and television was able to replace theater as the primary means of communicating information, the technologies in social networks, computer games, and virtual worlds will replace the telephone, cell phone, and email as the preferred means of communication. Each year these technologies deliver more capabilities in smaller electronic packages that can be carried into more social activities. The Apple iPhone that has become so prevalent in just a few years puts more communication and computation into the hands of a teenager than was possessed by entire armies in World War II or that were built into the original Space Shuttle. The existence and wide availability of this power is a disruptive force wherever it goes. Computer games are not a separate and restricted piece of this disruptive force; they are integrated into it at the core. As we enter the 21st century, our communication and computation devices are changing every aspect of society. Games will enter and change the internal functioning of education, science, engineering, marketing, religion, and dozens of other industry segments.

This chapter will explore five forces that are driving the universal adoption and application of computer technologies. It will also identify five industry segments that are already being impacted by this technology and which provide examples of early changes.

BACKGROUND

In his 1957 Nobel Prize winning research, Robert Solow demonstrated that the advancement of technology has been responsible for 50% of the economic growth of nations during the 20th century. The application of these new technologies created new businesses and transformed established institutions, allowing them to be significantly more productive than they had been prior to the introduction of the technology (Solow, 1957). Industrialists and innovators have come to appreciate the economic advantages of new technology and are now very focused on identifying, capturing, and applying these technologies as rapidly as possible. Those who do not will find themselves falling behind in productivity, cost reduction, and customer satisfaction. Through a process of evolution and renewal, the effect that Solow identified rewards companies and projects that leverage new technology and it replaces those that do not.

In his 1992 doctoral dissertation and his 1997 bestselling book, Clayton Christensen labeled the transforming effect of new technology as a "disruptive innovation." Technologies that are initially too immature to satisfy the needs of established customers contain within them the seeds to grow into the next dominant solution in an industry. Figure 17.1 illustrates this by contrasting the growth curve of the needs of an industry with the growth curve of the capabilities of a new technology. When the technology can grow and improve faster than the changes needed in industry, an initially inferior technology can grow into a market satisfying product. At first it will meet the needs of low-end and niche customers who are

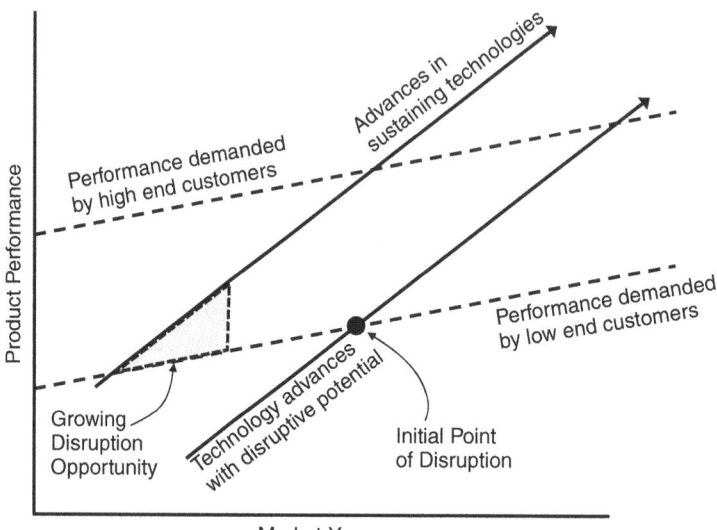

Figure 17.1   Christensen's theory of disruptive innovations explains how new technologies overthrow established businesses by offering better performance at lower prices (adapted from Christensen, 1997).

not in the sweet spot of the established industry. But continued improvement of the technology will eventually move it into the heart of the industry and displace the technology that was previously dominant. But more importantly, companies that hold onto the old technology will also be displaced by those who grasp and apply the new technology. Christensen and his readers were more interested in the fate of established and emerging companies than in the fate of a disembodied technology. His model illuminates the causes of many of the commercial disruptions that occurred during the dot-com boom and that have continued unabated since then.

The technology in computer games is an example of this type of industry disrupting change. In its early form, the computer game was able to satisfy only the most basic needs of education, training, scientific experimentation, engineering design, and architectural visualization. Only customers who had very primitive requirements found game technologies useful. As computer hardware and software applications have become more powerful, they have reached a level of capability that satisfies more mainstream tasks in all of these industries. It is still early in the evolution and adoption process, but the evidence of erosion and displacement are already visible.

In 1942 Charles Schumpeter wrote *Capitalism, Socialism and Democracy* in which he discussed the long-term viability of a capitalistic economy (Schumpeter, 1942). In that book he also described the impact that change has on business and the economy.

We must now recognize the further fact that restrictive practices … as far as they are effective, acquire a new significance in the perennial gale of creative destruction, a significance which they would not have in a stationary state or in a state of slow and balanced growth.

This description and one phrase in particular have become famous among modern economists and business leaders. Schumpeter cast a positive light on "creative destruction," the impact that change, technology, and knowledge have on the current practices of business and their influence on society. An economy that remains static or grows slowly is not in the clutches of the forces of growth. But when significant growth is occurring it will lead to the destruction of existing structures and the devaluation of skills that were previously essential. However, Schumpeter characterized this as a positive force that creates more opportunities than it destroys. It can also be a very painful force because most people and businesses cannot see clearly where the new value is being created and must often suffer the loss of income and jobs before they can enjoy the fruits of renewal and growth.

Schumpeter's ideas were reinforced by the writings of David Wells, who insisted that

Society has practically abandoned—and from the very necessity of the case has got to abandon, unless it proposes to war against progress and civilization—the prohibition of industrial concentrations and combinations.

It seems to be in the nature or natural law that no advanced stage of civilization can be attained, except at the expense of destroying in a greater or less degree the value of the instrumentalities by which all previous attainments have been affected. (Perelman, 1995)

Wells recognized the necessity of abandoning past practices in order to achieve new and more powerful methods of building an economy. He describes the necessity of "industrial concentration"—what we would call mergers and acquisitions today. He realized that the overcapacity that is generated within any successful industry is damaging to that industry and to the economy as a whole. Overcapacity robs the economy of the productivity that could be realized by applying those human, financial, and industrial assets to new and more valuable efforts (McDaniels, 2005). When 50 people are employed in a company that really only requires the labor of 25 to do the work, the company and the economy suffer for it. The company experiences higher costs and lower profits. This leads to customers paying higher prices for their products. If a new technology can enable 25 people to do the same work previously done by 50, then the company and the customer both benefit. The 25 people displaced by this process cannot simply move to a new company where they can pick up their former profession because the displacement process will soon become universal. These displaced workers can only reenter the workforce when they learn to leverage the new technologies that caused their displacement. They must find a way to be productive in the new and changed society, rather than struggling to remain viable with their old beliefs, practices, and mind sets.

The ultimate effect of the growth and displacement caused by new technology is to free up the efforts of 25 people to create additional value someplace else in the society. Through this painful process, society and industry can increase its output many times over. The classic example of this creative destruction is the transformation of agriculture that was brought about by the Industrial Revolution. Today's developed countries can be fed through the efforts of only 3% of its population working in agriculture. But before the Industrial Revolution and the application of machinery on the farm, it required the efforts of over 50% of the population to produce enough food to sustain the country. Today we take for granted the fact that over a period of 100 years this process displaced 47% of the American workforce. This was a very painful change for the children of the farmers and laborers who had grown up with the expectation of living and working on the farm. But this process also freed up the physical and intellectual energy of millions of people to advance all of society and to create the modern conveniences and lifestyle that we enjoy today.

The Information Revolution has had a similar effect on the industrial and agricultural businesses that emerged from the heart of the Industrial Revolution. This transformation is based on a new speed of moving and manipulating information. It has eliminated the need for millions of human administrative positions. Industrial methods used humans to move information and compute new data. Today, all of that is done by electronic networks and computer processors. As a result, there are millions of people whose time and talents have been released to create new industries.

Game technologies are just one of the most recent expressions of this information revolution. They have brought the power of television, computers, and the Internet into the entertainment space of the common household. From that foundation they have expanded into more serious applications in non-entertainment industries (Figure 17.2).

Figure 17.2 illustrates a path of technical innovation and transformation that has created the commercial and serious games industries. The evolution of technology has brought us the television, mainframe computer, personal computer, and Internet. As they became available to large audiences, each of these have been applied to electronic entertainment, first in the form of game consoles, then as PC games, and most recently as games on mobile devices. The figure also reminds us that there are many more forms of entertainment than games, but many of these can be improved through the use of game technologies.

ISSUES, CONTROVERSIES, PROBLEMS

Reinganum (1985) explained that new companies entering a market always invest more in innovation and changes than do the incumbents. This means that monopolies are often short lived because a challenger will soon innovate and spend their way into a leadership position. Games and the technologies that support them have received an unprecedented level of investment over the last

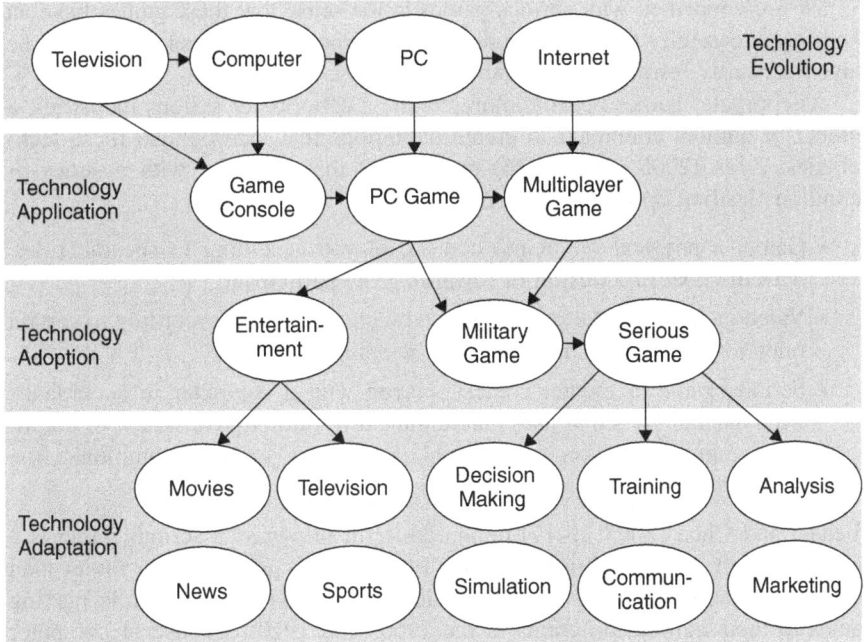

Figure 17.2   Computer games are one expression of a much longer evolution and adaptation of technologies to meet the needs of various industries and users.

Source: Created by Roger D. Smith.

20 years. In a move to capture the growing entertainment market, vendors of computer chips, graphics equipment, human interface devices, personal computers, and entertainment consoles have focused their attention on the needs of this market. Tens of billions of dollars have been invested specifically in game technologies and this investment is now being leveraged by a number of other industries.

Spread of Game Technologies

In his 1970 book, Dr Clark Abt introduced the term "serious games," which has been applied more recently to computer based games. Though Abt's games focused on manual role playing, board games, and card games, the concepts behind using entertainment techniques for more serious purposes was identical to the transformation that is coming from computer game technologies today. Abt described "serious games" in this way:

> *Reduced to its formal essence, a game is an activity among two or more independent decision-makers seeking to achieve their objectives in some limiting context. A more conventional definition would say that a game is a context with rules among adversaries trying to win objectives.*

We are concerned with serious games in the sense that these games have an explicit and carefully thought-out educational purpose and are not intended to be played primarily for amusement (Abt, 1970, p. 9).

As computer games became more common as tools for serious industries, a number of authors attempted to create definitions that incorporated these technologies. Zyda (2005, pp. 25–26) approached the challenge with a series of definitions leading up to one for serious games.

- Game: a physical or mental contest, played according to specific rules, with the goal of amusing or rewarding the participant.

- Video game: a mental contest, played with a computer according to certain rules for amusement, recreation, or winning a stake.

- Serious game: a mental contest, played with a computer in accordance with specific rules that uses entertainment to further government or corporate training, education, health, public policy, and strategic communication objectives.

Michael and Chen (2005, p. 17) define the term this way: a serious game is a game in which education (in its various forms) is the primary goal, rather than entertainment. Just as many different industries applied board and role playing games to their training programs in the 1960s and 1970s, the use of computer games is spreading through industry in the 21st century. Table 17.1 identifies more modern applications that are appearing in numerous industries that see an opportunity to leverage this technology for their own goals. Each of these industries has been encouraged by applications in other serious industries, but has required the creation of unique capabilities to meet their own needs. Together all of these create a supportive evolutionary ecology of game technology applications.

SOLUTIONS AND RECOMMENDATIONS

We suggest that game technologies will continue to expand from one industry to another based on five core forces of the technology and the environment in which it is growing (Figure 17.3).

Hardware Costs

Computer games are designed to take advantage of all of the power available on a consumer-grade computer. Their focus is on reaching the highest number of customers based on the hardware that these customers have available. Therefore, unlike serious industries, game companies do not want to create a product that requires a new hardware purchase. As a result, these technologies are designed to be as efficient as possible, maximizing the amount of work that can be done on a consumer-grade computer. Consumer machines are often an order of magnitude less expensive than a professional workstation, dropping hardware costs from the $20,000–$50,000 range down to the $2000–$5000 range.

Table 17.1 Computer games and their supporting technologies have been adopted by a number of diverse industries

Industry	Game technology impact
Military	Training soldiers and leaders in the tactics and strategies of war. 3D modeling of equipment to illustrate or explore its capabilities
Government	Ethics training for NASA. Project management training for the State of California
Education	Augmenting classroom instruction in nearly every subject, including English, math, physics, and history
Emergency management	Training emergency responders, firefighters, Federal Emergency Management Agency staff, and others to deal with disasters
Architecture	Visually promoting major hotel, casino, and office spaces to potential clients
City and civil planning	Lay out and experimentation with public services in a growing community
Corporate training	Orienting people to company products, facilities, and policies. Also, commercial pilot and safety training
Health care	Educating patients on treatments, rehabilitation, and managing anxieties. A new generation of game-based exercise programs
Politics	Presenting political issues and the consequences of political decisions. Promoting candidates through simulated scenarios
Religion	Interactive versions of sacred texts. Tools to teach religious history
Movies and television	Tools for creating animation and 3D worlds. Alternative forms of storytelling using the virtual world as the set, props, and characters
Scientific visualization and analysis	Rapid display of objects under experimentation and the physical forces acting on them. 3D display of collected and analyzed data
Sports	Recreating live sporting events for review and for prediction of potential outcomes. Design and rehearsal of critical "one time" events like Olympic ceremonies
Exploration	Mission preparation for NASA Mars Lander and other vehicles. Recreate environments around deep sea probes
Law	Illustrating crime scene activities for judge and jury. Analyzing crime scene data

Source: Compiled from Michael and Chen, 2005; Bergeron, 2006; Casti, 1997; Maier and Grobler, 2000.

Software Power

The ability to create a user interface that an average employee or customer can understand and operate is critical to a product's success. For a computer game, the goal is usually for the customer to understand how to use the product without ever reading a manual. Any instruction that is required must be built into the game itself, allowing the customer to learn while they are using it.

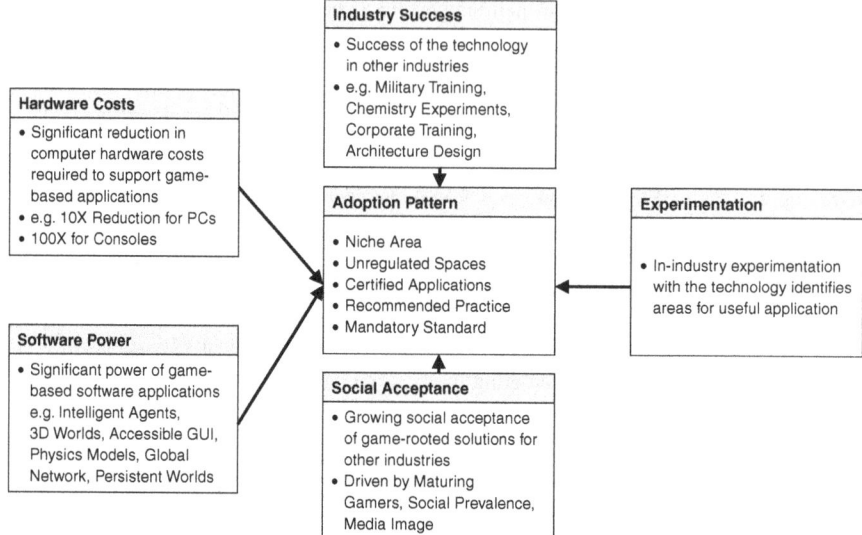

Figure 17.3 A five-force model explains the adoption of game technologies by diverse industries.

Source: Created by Roger D. Smith.

Games also require clever and adaptive artificial intelligence to create game controlled characters that interact with humans in a realistic and engaging manner. Sophisticated artificial intelligence has always required significant computational resources and significant expertise to configure and run the system. Games fit this power into a consumer PC and provide scripting languages that allow a customer to manipulate the behaviors to meet their needs.

The 3D engine that visualizes the virtual world is the most attractive and recognized form of game technology. The visual and auditory stimuli that it produces have the same power of engagement that has been reported for television, movies, and music. It appeals to the mind's core need to experience information in a form similar to the physical world that the brain has evolved to handle.

Finally, the software is no longer constrained to a single user on a single computer. Global networks allow users to connect and interact with similarly equipped people everywhere in the world. The 3D game with visual and voice exchange improves upon every other form of global communication that is currently available.

Social Acceptance

Games have largely overcome the stigma that they are just toys focused on play. The technology has persuaded most critics that these systems can be applied to serious problems. As the children who were raised with these games become the leaders inside companies and government organizations, this level of acceptance has and will continue to increase. When a child plays a game it is considered fun or, at best, educational. But when an adult professional plays a game it becomes a serious learning, communication, and revenue generating experience. Game technology becomes more important and more serious as the people who grew up using it become more important and more serious.

Business and industry are also very much a social experience. All major projects require the communication and cooperation of a large number of people. When game technology is an accepted option to those practicing their profession, it can be applied to larger social challenges and its success will further eliminate the remaining social stigma associated with its entertainment origins.

Most of society has become accustomed to seeing 3D representations in courtrooms, medical facilities, museums, building designs, and military systems. After experiencing the advantages of this type of interface, people are much more willing to accept these technologies in other products and services.

Other Industry Success

Television shows like Modern Marvels, Nova, National Geographic, and a host of programs on the Discovery and History channels have applied 3D visualization and physical modeling to illustrate the behaviors of animals, humans, machinery, and the universe. The clear communications that these game technologies enable motivates other industries to experiment with them as well.

Similarly, the military has incorporated many of these technologies into its training systems. Simulators for tank crews and company commanders all incorporate the 3D game engine, graphical user interface, physical models, artificial intelligence, and global networking that are part of computer games. If the technology can be applied to warfare and teaching soldiers to handle life-and-death situations, then what commercial application could be more serious than that?

Progressive architectural firms used game engines to create 3D models of the inside and outside of construction projects. Future tenants could experience the office or living spaces that they would be purchasing long before these spaces existed in a physical form. The game engine allows more realistic visualization, navigation, and manipulation than the traditional cardboard models.

As each of these examples proliferated through society it stimulated creativity in new areas and provided a foundation of tools and data that new industries could build on.

Experimentation

As managers, programmers, and artists experiment with game technologies within a new industry, they discover useful methods for studying chemical reactions, understanding the stresses on aircraft parts, exploring the effects of the environment and erosion, evaluating the visual appeal of architectural designs, or delivering city services in a growing suburb. Some of these experiments lead to significant improvements over previous methods of doing the job. As with all experimentation and research, new discoveries change the way products are created and services are provided.

Adoption Pattern

At the center of this five-forces model is the adoption pattern of the technologies. Adoption may follow a pattern that is similar to that experienced by the entertainment and military industries. It will begin with a niche area that is closely aligned with at least one of the powerful game technologies. If successful, the method will be applied to applications and activities that are not regulated by outside organizations. These are spaces where local groups can define their own processes and measures of success. From this position, support can grow for the technology across a number of organizations and geographic areas. This can lead to a form of certification for game technologies as an acceptable solution to specific problems. Success at this level may lead to it becoming a recommended practice in which the regulating bodies include it among the proven and preferred approaches to solving a problem. Finally, game technology may become one of the mandatory methods of solving problems across an industry.

The five-forces model provides one view of some of the forces that are driving the adoption of game technologies. Each of the forces is at work in modern societies and, along with demographic changes in society, is driving the acceptance and application of these technologies to a wider set of users.

Disruption of Established Industries

Game technologies are changing competition in many large industry categories. Several of these were used as examples in the previous section. But we want to examine five of them in more detail because of the apparent prevalent use of the technologies and the close alignment of their needs with what game technologies have to offer. These areas are entertainment, training, scientific analysis, decision making, and marketing. There have been a number of technologies that have significantly changed every industry in the last two decades, including computers, the Internet, business information technologies, and cellular communications. The changes wrought by these provide clues to changes that lie ahead in the adoption of game technologies (Figure 17.4).

Entertainment

Computer games began as research experiments with emerging computer technologies in university laboratories. The first is considered to have been "Spacewar!" developed by scientists at the Massachusetts Institute of Technology (MIT) on a Digital Equipment Corporation (DEC) PDP-11 computer in 1961 (Michael and Chen, 2005). That game presented two spaceships that fired small missiles at each other. Each ship was controlled by a joystick input device and displayed

Entertainment	Training	Marketing
Arcade – Casual part-time access to customers. Matched with shopping. **Console** – Specialized in-home access. Matched with television. **Home PC** – Mass consumer access. Matched with IT. **Portable** – Specialized portable access. Matched with radio. **Hybrid** – Exercise equipment, dancing, hunting, **Cellphone** – Mass consumer, portable access. Matched with portable telecom. **Content Spinoffs** – Movies, news, sports, magazines, and web sites.	**Emergency Management** – Immersion in major disasters and collaborative operations **Logistics** – Capture complexity and volume of detailed data. **Medical** – Drill basic skills, standards, and behaviors. **Military** – Reduce combat attrition by eliminating mistakes. **Airlines** – Cost effective flight training, accessible aircraft **Safety** – Industry jobs requiring safety training. Good for scenario-based training	**Advertising** – Lock in audience attention on a message mixed into a game. **Activism** – Deliver a serious message that may be outside of normal human experience and difficult to create empathy. **Politics** – Motivate to take political actions in favor of the point-of-view of the game. **Religion** – Electronic proselytization and retention of the faithful.

Scientific Analysis	Decision Making
Visualization – Explore complex data with the primary human sense. **Graphing and Animation** – Numeric and dynamic representations of systems. **Presentation** – Communicative representations of complex problems. **Immersive Navigation** – Ability to navigate and search data in a three dimensional context. **Networked Research** – Tools for connecting multiple researchers together in an effective manner.	**Reasoning** – Assistance in thinking through a complex problem in a consistent and accurate manner. **Illustration & Demonstration** – Visualize, Communicate ideas to broad variety of people. Show how a decision will play out. **Quantification & Comparison** – Measure the impacts of different decisions. Contrast multiple options and identify the preferred. **Experimentation** – Consider multiple options and variations that are beyond physical experiments and prototypes.

Figure 17.4 At least five major industry segments are actively adopting game technologies for specific functions within their business.
Source: Created by Roger D. Smith.

on a circular monitor. Even in its infancy we can see the similarities between this experiment and the Asteroids arcade game that was released in 1979.

PONG was the first commercial arcade game that was delivered to pinball arcades, pool halls, and bowling alleys around the country in 1972. Developed by Nolan Bushnell and distributed through his newly formed Atari Corp., this game began the transition from analog pinball games to the digital arcades that we see today. At the same time, Magnavox released the Odyssey home gaming system. This also included the game of PONG, which they claimed Bushnell had copied and Magnavox filed a law suit claiming patent infringement (Sheff, 1999; DeMaria and Wilson, 2004).

In 1972, Atari put computers into pinball arcades and Magnavox put them into people's homes. The popularity of the new devices created a rush of imitators and began the growth of game arcades and home entertainment consoles. DeMaria and Wilson (2004) provide an excellent history of the emergence of these arcade games. This opened the door for games that incorporated sports themes like ping pong, tennis, and handball; space themes like Space Invaders, Asteroids, and Defender; and abstract themes like PacMan, Tempest, and Tetris. These games introduced computer graphics, human interfaces, and audio to the general populace and to the entertainment business (Bushnell, 1996).

Home console gaming became a popular leisure activity similar to watching television. It was constantly accessible, required little physical activity, and could be played alone or with other people. This began the ongoing 35 year incursion of gaming into a space previously controlled by the television. As gaming has grown more popular, it has become a serious threat to the advertising revenue that drives television programming (Bushnell, 1996).

Gaming on the PC.   Games on the home PC extend entertainment onto a machine that was previously purchased for "serious" applications like online communications, banking, and education. In moving to this machine, games became instantly accessible to professionals who may not have been inclined to purchase a game-specific console. This move also leveraged the existing expertise that the audiences had with the PC to smoothly transition them into gaming. Gaming was no longer relegated to a specialized device attached to the television. Rather it became a popular function on a general purpose computer, alongside the word processor and web browser.

Portable Handheld Gaming.   The introduction of handheld gaming devices like the Gameboy and the Playstation Portable (PSP) moved gaming from a fixed site activity to one that was portable and accessible in any environment. Gaming can then displace the activities that filled time while traveling or away from a desktop computer. "Down time" and travel time is no longer reserved for reading. Instead, people can always have a game device ready in their pocket.

The newest Gameboy DSi device also contains a wireless communication chip that allows a player to interact with others who are playing Gameboy titles anywhere in the world. This means that multiplayer gaming has also gone mobile.

Cellular Gaming.   The success of cellular telephones has been driven by an increasingly busy and mobile lifestyle and by their ability to sustain networks of relationships via voice conversations and text messaging. As these devices have become more powerful, their growing color screens, computer processors, and network bandwidth have made them attractive devices for gaming applications. Cellular manufacturers and service providers see gaming as a means for selling more advanced devices and services. Just as people upgrade their computers to be able to play the newest games, the cellular providers are hoping that games will cause people to upgrade their cell phones more frequently.

In many ways, cellular gaming is socially similar to portable game devices. But it is also another instance of moving games off a dedicated gaming device like the Gameboy and onto a more general purpose device that is owned by billions of customers.

Hybrid Devices.   Interesting combinations of devices are being created to explore less passive applications for games. The largest category of hybrid devices is in exercise equipment or "exergames" (Michael and Chen, 2005). The first step was when game arcades introduced dancing games like *Dance Dance Revolution* that required a player to stand on a platform and control the game with very basic dance moves. The success of these devices in Japan spread to America and Europe and demonstrated that physical activity in conjunction with gaming was acceptable to the customer base. The incredible popularity of the Wii game console and its exergame peripheral the Wii Fit may stimulate this area in a manner similar to the effect of the Apple iPod on MP3 music and players.

Training

Training is an essential function within all organizations. Distinct from education, training focuses on conveying or mastering specific skills that will be used for specific jobs by specific people. Public and university education prepares people to learn to work in a profession, equipping them with a broad foundation of knowledge. Specific jobs require focused knowledge and skills, often unique to a single position or company. Conveying the necessary knowledge is an expense to the organization in both time and money. The cost of instructors, instructional materials, classrooms, and student time can be significant. Games appear to be able to reduce the costs of training and to improve its effectiveness at the same time.

Bloom (1984) has shown that one-on-one tutoring can improve a student's test scores by as much as two standard deviations. This can effectively move a student from the 50th percentile of a class to the 96th percentile. Though tutoring is extremely effective, it is also very expensive and time consuming. Even if there were no financial costs for a tutor, there are not enough domain experts available to serve all of the students seeking instruction. Because of the cost and limited availability, self-directed study has always been an attractive option to formal classroom or tutored study. Correspondence learning requires that the student learn at his/her own location and often on his/her own time. This extends learning to a much broader set of students and it reduces the cost of delivering the training. Computer-based training (CBT) and web-based training (WBT) have become a large business because of the significant cost savings involved. Computer games offer a richer environment in which to host computer-based training and student collaboration.

Like all learning materials, games attempt to communicate information in a form that is accessible to the student and that motivates the person to learn. A study by The Learning Federation (2006) identified a number of themes that are driving the training business to a game-based solution. These are

- *Accessibility:* Games have lower costs for materials and staffing.
- *Believability:* Games present a 3D environment that is so similar to the real world that lessons fit well into their natural environmental context and are more believable.

- *Engagement:* Games can interact with the student in more ways than simple online tests. These engagements hold the students' attention and increase their learning experience.

- *Accuracy:* Given a richer environment, games can create a more accurate representation of the real situation than can materials like textbooks and 2D images.

- *Variability:* Games can present situations in many different forms and in various contexts to reinforce lessons and to show special cases.

- *Measurability:* Games allow the numeric collection of performance data. These measurements enable scoring and identification of the tasks that the student has not mastered.

- *Repeatability:* Games are infinitely patient. They can take a student through identical or similar scenarios as many times as it takes for them to learn the lessons.

Though the advantages described above do exist, there remains some question as to the effectiveness of game-based training. Like all other forms of training before it, we fear that this method is less effective than real "on the job training," and we do not know how it compares with older forms of training. A recent study by the Federation of American Scientists emphasized that games may develop higher order thinking, but they were careful not to assume that this is true in spite of potential advantages like those listed above (FAS, 2006a).

In another document generated by that same study, the Federation of American Scientists identified some specific learning advantages of games (FAS, 2006b).

- *Contextual bridging:* Students are able to apply lessons from 3D games much more readily to the real world because of the similarity of the two environments.

- *High time-on-task:* Because of the high level of immersion and interactivity, students spend a significantly larger amount of time on tasks in a game environment than with other learning materials.

- *Motivation and goal orientation:* Students are highly motivated to complete tasks successfully. Failure in game situations does not seem to discourage students, but encourages them to try different tactics.

- *Providing learners with cues, hints, and partial solutions:* Game environments can subtly provide students with hints on how to perform tasks. Like a good tutor they can be programmed to recognize when a student is choosing an incorrect path and give hints as to why they should change direction or tactics.

- *Personalization of learning:* Games allow the student to personalize their avatar in the game. This customization builds a stronger relationship between the student and the materials. Given some basic information, the game can also address the student by their preferred name and make references to other personal interests.

- *Infinite patience:* Games do not become irritated or impatient with students who learn at different rates. The game does not insist that people learn within a specific number of iterations or in a specific order. They present an ever-present environment that can be explored.

The Learning Federation has observed that the "expense and related challenges often cause both formal education and corporate training to rely on strategies that ignore the findings of learning research" (Learning Federation, 2006). Like computer-based training before it, games have the potential to bring additional learning techniques into an affordable range for industry, government, and academia. Moving more of the knowledge and skills of human instructors and tutors into an interactive, self-directed environment like a game can potentially extend this limited resource to a much higher number of students and spread the costs of the tools broadly enough to make them affordable.

This approach has been working very well for companies focused on a wide variety of customers. In the training space, successful projects have been created for the following types of customers.

- *Military.* Arguably, the military has been using games for training since 1664 when Christopher Weikenman introduced Kreikspeil to the German Army (Perla, 1990). The American Air Corps purchased a number of flight trainers from Edwin Link in 1930, devices which were also being sold to game arcades. The adoption of computer games is a very natural step for this customer, especially since many of their problems fit well into a 3D virtual world.

- *Emergency management.* There are a number of new game-based trainers for emergency management that present firefighters, medics, policemen, ambulance drivers, and resource managers with situations that require expertise that is not readily gained through daily operations (Hamm, 2006; Musgrove, 2006; Rowland, 2006).

- *Government.* Management of public lands, ecological decisions, public policy consequences, and ethics training have all been incorporated into games for various branches of the government.

- *Education.* Key events in history are often based on geographic location, key people, and machinery, all of which can be compellingly represented in a game. Physics, biology, chemistry, and business are other subjects that have worked well within a game. The communications that occur in massively multiplayer games has been shown to encourage and enhance reading and writing skills (Steinkuehler, 2005).

- *Corporate training.* In the corporate environment games are an economic way to teach ethics, company policy, customer service, safety, and organizational structures.

- *Health care.* Games are entering the health care industry almost as fast as they entered the military. They are being used to educate a wide variety of professionals and even to achieve certification. There have been arguments that games are not sufficiently detailed to train a surgeon, but the real benefit is in using games for the hundreds of functions which do not involve working directly on human patients. Basic skills, standards, and behaviors can be conveyed to a large number of people distributed across many locations. Games are also being used to treat mental disorders like phobias. They allow a person to be in a simulated threatening situation, but with enough detachment and control to avoid triggering a hysterical reaction (Rizzo and Kim, 2005).

- *Airlines.* Like the military, the airlines have always used computer simulation to train pilots. These devices allow the pilots to rehearse potentially lethal situations in a controlled and learning environment. Games are just a miniaturization of the many simulators already used by the airlines.

- *Safety.* Safety training is conducted in most professions. This training is usually limited to classroom discussion, slide presentations, and printed reading materials. Most people never have an opportunity to act out the steps they are expected to follow in a real emergency. Games create an environment in which everyone can walk through a simulated emergency and play a number of different roles.

Scientific Analysis

The 3D engine of a computer game is simply a game-specific version of the 3D scenes that are already created in a number of different scientific disciplines. Computer aided design (CAD) tools allow engineers to create digital models of physical products like automobiles, aircraft, and kitchen appliances. Movie animation tools do the same for computer-generated movie characters like those introduced in Pixar's *Toy Story* and carried on in the many movies that have followed. In games, CAD, and animated movies there are two essential tools for creating the 3D world. Modeling tools are used to design 3D representations of individual physical objects. Then graphics engines combine and animate these models by driving interactions between the objects and their environment.

Data Visualization.   National laboratories employ supercomputers to analyze the behavior of nuclear explosions, geologic movements, weather patterns, astrophysics, and other complex phenomena. These analyses typically begin with a numerical model that represents the movement and interactions of thousands or millions of subcomponents of an item or event. These computations generate extremely large volumes of numerical data that must be analyzed mathematically and statistically. Potentially, all of the knowledge embedded in these data can be extracted numerically. However, in practice, scientists have found that visualization is an extremely powerful tool for understanding the behavior of the data and looking for anomalies in the models that created the data.

Robert Rosner, the Director of Argonne National Laboratory, explained that he was originally skeptical of the real value of visualization in the work of the energy labs. However, in one experiment, scientists at Argonne used a giant, high resolution display to plot every single data point that had been calculated by their models. Prior to that, much smaller plots were created from a thinned or averaged data set. But when every point was visualized, the scientists noticed a pattern in which aberrant behavior appeared at repeating thresholds in their equations. This visualization exposed inaccuracies in the model which were too subtle to be readily identified statistically. Understanding the importance of this, Rosner became a champion of data visualization and now supports its use whenever possible (Rosner, 2006).

Current game engines lack the fidelity to present the high resolution data that come from many scientific models. However, they remain potentially valuable for rapid real-time visualization before more in-depth analyses are conducted.

Graphing and Animation. Animations of solar systems, molecular arrangements, and chemical reactions represent small scale problems for which game engines can be readily applied. These tools present significant value in lowering the price of animation and in extending access to the tools due to their relative ease of use compared with competitors. Commercial quality game engines can be licensed for a few hundred dollars, where more traditional visualization tools cost many thousands of dollars for a similar license. Additionally, game engines are designed with an "application programming interface" that allows programmers to attach new software to the game engine to customize it. This means that the game engines can be connected directly to scientific software that is generating the data to be visualized.

Presentation. Both traditional science experiments and the more recent computational science contain information that is often difficult to grasp or to communicate. 3D visualization engines are a means for communicating this information to audiences at many different levels of scientific sophistication. Rather than creating traditional "science museum" plaster models of human organs, a game engine can create a moving 3D visualization of the inside of the body at work. It can also add the sounds that should be present in such a scene.

These visualizations can be portrayed on desktop computer screens, embedded in head-mounted displays, or projected onto giant walls for viewing and immersion of an entire audience.

Immersive Navigation. Science fiction has presented stories in which military, medical, scientific, and maintenance personnel navigate a real space by immersing themselves into a virtual representation of that same environment. This technique has become reality in many modern systems.

Maintenance personnel for the Boeing 777 aircraft have access to a virtual overlay repair manual. In experiments, a repair person can don see-through display glasses and approach the aircraft. Projected on the glasses is a schematic of the aircraft engine that is superimposed over the real engine. The schematic may include labels to identify all of the parts and may color code the part that is to be worked on. When it comes time to do the operation, the schematic may even show an animated instruction sequence for the best method of performing the operation. Potentially this can allow a repair person to work on a specific problem for which he has not received prior in-class training. In military environments this can allow a skeleton crew to work on a wide variety of equipment by learning specific information just as they need to use it (Boswell, 1998).

Similar displays are being used to position radiation sources around a patient who is about to receive treatment. A 3D overlay augments a technician's native ability to visualize the location of the tumor and the exact angle of alignment for the equipment. The visualization also has the power to identify dangerous situations and to alert the technician to check for these before proceeding with treatment.

Networked Research. When research is carried out in multiple facilities, a means is needed to exchange working data, leverage laboratory resources at each facility, and collaborate among all parties involved. The new generation of Massively Multiplayer Online Games (MMOGs) and virtual worlds are designed to serve these types of needs for the entertainment community. An MMOG creates a shared virtual world that can be entered by a large number of participants from around the globe. The world is persistent because actions taken there or objects deposited need to remain in those locations until acted upon by other participants.

For scientific research, this means that data, or an icon that represents and locates a data set, can be deposited in the virtual world by one player and retrieved, copied, or processed by any number of other players. Currently, Second Life is the environment that is most supportive of this type of operation. It allows participants to load or link graphics, textures, documents, presentations, and a number of digital data forms into the world. These can be viewed and manipulated in the world or downloaded to a local computer for manipulation.

Social structures have also evolved within all MMOGs. Players have created teams, guilds, societies, and alliances that bring them together for a common purpose. These groups hold regular meetings to discuss their missions and to manage the operations of the group. Many of these are run like business operations, even to holding weekly teleconferences with participants to conduct group business and plan future quests (Steinkuehler, 2007).

Second Life is considered a virtual world rather than an MMOG because it does not come equipped with quests built by the creators of the game, but relies more on every participant to create the assets and activities that occur in the world. Because it is so open and flexible, it has become a popular venue for conducting university classes, distributing assignments, and presenting work completed. It can also be used to bring together researchers for collaboration or to partition work out to a number of remote participants.

Decision Making

Leaders in business, government, academia, military, and non-profit organizations are faced with the necessity of making large numbers of decisions every day. In general, people are guided by principles, objectives, and information. Establishing principles and objectives focuses a decision, but evaluating all information available tends to diffuse the process. Also, many decisions involve multiple parties with different objectives and different principles. Therefore, decisions often must be custom designed to partially meet the objectives of each party, rather than optimizing for any one party (Janis, 1989; Jennings and Wattman, 1998).

Historically, decision makers have turned to advisors and reliable sources of information to aid them. More recently, computers have been employed to organize large volumes of information and to present multiple perspectives on a problem. Within the field of artificial intelligence, computers are programmed to provide advice. These programs attempt to codify a body of knowledge and process it according to specified rules. The value of the advice generated is proportional to the degree to which real-world information and considerations can be encoded into a form that can be processed by a computer (Russell and Norvig, 1995). However, limits in technology, time, and funds available to create such systems impact their ability to provide actionable advice. These limitations apply to games and decision making systems derived from them as well.

Reasoning.   Computer games contain simplified artificial intelligence algorithms that enable computer controlled characters to act somewhat intelligently in a very restricted environment. Typically, the artificial intelligence is custom designed for the physical space in which the action is taking place and with prior knowledge of all of the relevant objects that will be involved. Game reasoning is not general purpose or universal. It attempts to mimic intelligence in a limited space, not to be intelligent in a number of different circumstances.

Understanding this limitation indicates that, to be effective, a game may need to be based on a knowledge base of prior situations that were very similar to the one in which a current decision maker is involved. Because game reasoning is so customized, it may not be appropriate for unique new situations for which prior data are not available.

Illustration and Demonstration.   Games can provide a valuable environmental context into which data are injected from other programs. Just as maps are used to provide background and context for reports on the movement of military units on the battlefield, creating a realistic environment in which to portray facts that are relevant to a decision maker can make valuable contributions to understanding those facts.

The animating power of games can also be used to demonstrate the dynamic characteristics of data and to project potential impacts of different decisions. A recorded game session may be shared in the same way that a movie is shared. Setting up a situation and allowing a team of people to experience the same space in which decision makers are working can convey an appreciation for the complexities involved that is not possible with non-dynamic, non-interactive tools.

Quantification and Comparison.   As digital environments, games are able to record numerical results of actions taken. These can be processed objectively and remove much of the human bias that creeps into decision making that includes a number of people with varying agendas. Games like "Mike's Bikes," which is used in a number of business schools, place teams of players into an environment in which they compete against each other to maximize the sale of bicycles and the company share price. These games are formula driven, quantifying every decision and keeping a record of the results of each. In Mike's Bikes the actions of each team have an impact on the results achieved by the others as well, creating an interactive competitive environment. All teams are developing, marketing, and selling products into a single market with a defined level of demand for the product (Reid et al., 2009).

Given the quantification of multiple decision paths, computer systems can be used to compare the results of each and the consequences expected. Games, like business, usually portray a win–lose scenario that results in one or more companies winning a stake while others lose their stake.

Experimentation.   Computer programs are ideal for experimentation with multiple options. When the computer can calculate results very rapidly, it is possible to explore dozens or even hundreds of variations on a decision before identifying the most promising combinations. This digital laboratory is rapid and private. It allows excursion without sharing ideas so widely that they become rumors or competitive intelligence.

Games contribute to this in that they are relatively inexpensive to modify and can be limited to a single computer with access restrictions. They also display results in a visually memorable and comparable form.

Marketing

In earlier sections we made it clear that game environments have a powerful influence on their users. They invite interactive mental engagement that is rare in most other forms of communication. They also encourage audiences to engage with them for long periods of time to repeat and perfect their performance in the environment. These characteristics are very attractive for marketing products, services, and ideas to users. The marketing and advertising industries are exploring the use of games as a form of persuasion. They are also considering the impact of product placement in standard game environments, much the same as product placement in movies.

Advertising.   Some games are created purely as advertisements. Web sites like Postopia.com present children with simple games that prominently feature company products and convey happy, persuasive messages about those products. In this sense they are exactly like television advertisements. However, they improve on traditional advertising because the game engagement can keep a child connected to the product and its message for many minutes or even hours. This is impossible with television and print ads. This type of advertising via game is more closely aligned with event sponsorships, like the Nike logo that appears on football uniforms. Audiences may fix their attention on the game and on the advertisement for hours at a time (Winkler and Buckner, 2006).

Activism.   A few games have been created to alert players to situations like starvation, genocide, discrimination, pollution, and conservation. These games create an accessible and interactive story. Like advertisements, the engagement can capture people's attention for significantly longer periods than traditional narration or printed stories. They convert statistics and data on the situation into realistic, interactive events that place the player in the position of the person at risk. This creates an empathy that cannot be matched in other media. Games like *Darfur is Dying, Food Force*, and *A Force More Powerful* have been used to convey important social problems to audiences around the world.

However, this application of games is not new with computers. Hutchison (1997) describes a board game called *Healthland* that was created by the Red Cross in 1920 to teach children how to reduce the chances of catching diseases. It taught the player to avoid places like "Dirtyville" and the "Mosquito Swamp" while pursuing "Bathtubville" and "Book Land."

Politics.   Games have been used to convey political messages. These may be as simple as illustrating the dynamics of voting in a state like Illinois or exploring the conflict between the Israelis and the Palestinians. In some cases, the game is designed to promote a specific point of view. But in others it is meant to communicate multiple points of view so that a player can appreciate the real difficulty of a situation.

Peacemaker allows a player to act as either the Israelis or the Palestinians in dealing with the various conflicts that arise between the two groups. The lesson is in understanding the impossibility of satisfying the conflicting demands of both parties and the futility of perpetual violence as a form of political action.

Religion.   Several games exist to communicate the core message of a religion and to take advantage of the long-duration contact that people put into games. These also animate historical information that can easily become rote or dull to an audience that has heard it many times over.

Left Behind: The Game takes the message of the antichrist portrayed in the popular series of novels and allows the player to act as one of the primary forces in the game. *Interactive Parables* allows you to play through the Biblical parables and experience them interactively.

FUTURE DIRECTIONS

In this chapter we have explored the many ways that games and game technologies are creeping into different industries. As the breadth of this impact becomes clearer, games begin to look less like toys and more like a 21st century manifestation of advanced computer technologies that are useful for a wide variety of businesses. Games are really an extension of the computer/Internet/web/IT transformation that has been occurring across all industries for the last 20 years. They are unique only in that the technology came from entertainment applications rather than from some other "serious industry." They are similar to movies, film, and radio, which were able to leverage a huge user base to create advances in technology that could be applied in a number of non-entertainment industries.

As game technologies become more prevalent they will face a number of challenges in meeting the needs of demanding customers.

Interoperability

It is unlikely that a single game engine or product will be adopted by all serious industries. However, any group of companies that work together will expect their game tools to interoperate in the same way that their telephones work for teleconferences, their email programs exchange messages and schedule meetings, and their databases shift information from one product to another. Game environments are largely unique and separate from each other. Competitive forces in the entertainment market have kept companies from establishing standards which would allow multiple games to interoperate with each other. The goal has always been for one game company to take customers away from another, not to allow customers to work across multiple competing products.

As these products enter other business industries, there will be a need for the products to be configured to work together. The level of cooperation and interoperability will be one of the selling points to customers who must collaborate with other companies using different game tools.

Reality

It has been said that "virtual reality" has always provided a lot more "virtual" than it has "reality." The point is that the virtual world only appears to be realistic. It looks and often sounds like the world in which we live. However, it is a very shallow representation. There is very little reality in virtual reality. The objects possess few if any physical characteristics which would determine how they can be used in the world. Most objects presented in a game are static decorations that are present for appearances, but cannot be used in any way. Entering a virtual house you will find that water does not come out of the faucets, the curtains cannot be closed, the windows do not break, and the furniture cannot be moved. Almost everything in a game environment is meant to be looked at and accepted as it appears. It is not meant to be interacted with, nor can it interact with other objects in the world.

If game engines and environments are to be used for scientific experimentation they have to be able to incorporate the physical behaviors of the things that are being experimented with. Without this, they are little more than a movie. Virtual world researchers are looking for methods to provide physical and functional characteristics to all of the objects in the world. Up until now "function" or "action" has been reserved for the player's avatar and a few objects that the player was meant to interact with. The remainder of the world was just a digital movie set which created the mood for the interaction, but had no properties of its own.

Extensibility

Serious games will be challenged to grow in both the detail of representation and the breadth of coverage. Some industries will want to represent the fine details of small pieces of machinery. Others will want to represent the entire breadth of the commercial airspace. Adding both breadth and depth to a virtual world is a fundamental challenge. There is no reason to model the physics of water flowing in a river if it is just going to be seen from an aircraft flying over at 5000 feet. However, to the model of plant growth on the banks of the river, this water flow is essential.

To date, most models have focused on providing either breadth of perception or depth of focus, but not both. As people work more and more in the virtual worlds, there will be a need to allow these to coexist under specific conditions, though perhaps not at all times and in all places.

Each of these areas of research points to an exciting future in using games and virtual worlds for serious applications.

CONCLUSION

Game technologies are proving to be very flexible in adapting to industry needs outside of gaming. Entrepreneurs are launching companies to take advantage of this disruption. Because of their very specific focus on the visualization of environments and problems, games will probably never achieve the level of adoption that has come to the Internet or IT services. But they do represent the application of powerful new hardware and software technologies being created in the 21st century. It may be fair to expect that the spread of game technologies is the first disruptive wave of computer technologies in this century, and that it will be followed by others that will be larger.

As I described in a previous paper (Smith, 2006), it is important for companies to continue to evolve and adopt new technologies. Companies that refuse to consider the uses of game technologies may very well hinder their ability to understand and appreciate the waves of technology that will follow. For

---

example, within games there are important lessons to be learned regarding computer graphics, networking, shared data persistence, analysis of large volumes of information, and interoperability among disparate tools. Companies that do not learn these lessons now may be ill equipped to apply them to non-game applications in future waves of disruption.

Recently, games and the technologies embodied in them have proved to be an undeniable force in changing the way many businesses operate. The ability to bring rich visualization and interactivity to every computer desktop adds value to a number of different industries. In this chapter we have explored this value in entertainment, training, scientific analysis, decision making, and marketing. The adoption of games in these areas is reminiscent of the use of the Internet and web in the mid 1990s. Though the future of all applications of the technology is still unclear, there are hundreds of entrepreneurs and researchers pressing the edge of what can be accomplished. We can look forward to a future that has been influenced and completely permeated by these technologies.

REFERENCES

Abt C (1970). *Serious Games*. The Viking Press, New York.

Bergeron B (2006). *Developing Serious Games*. Charles River Media, Boston, MA.

Bloom BS (1984). *Taxonomy of Educational Objectives*. Allyn and Bacon, Boston, MA.

Boswell B (1998). Time to market. *Evolving Enterprise*. http://www.lionhrtpub.com/ee/ee-spring98/boswell.html (last accessed 21 December 2009).

Bushnell N (1996). Relationships between fun and the computer business. *Comm ACM* **39**(8), 31–37.

Casti J (1997). *Would-be Worlds: How Simulation is Changing the Frontiers of Science*. John Wiley & Sons, New York.

Chandler A (2001). *Inventing the Electronic Century: The Epic Story of the Consumer Electronics and Computer Industries*. Free Press, New York.

Christensen C (1992). The innovator's challenge: understanding the influence of market environment on processes of technology development in the rigid disk drive industry. Unpublished Doctoral Dissertation, Harvard Business School, Boston, MA.

Christensen C (1997). *The Innovator's Dilemma: When New Technologies Cause Great Firms to Fail*. Harvard Business School Press, Boston, MA.

DeMaria R and Wilson J (2004). *High Score: The Illustrated History of Electronic Games*. McGraw-Hill Publishing, New York.

FAS (2006a). *R&D Challenges for Games in Learning*. Federation of American Scientists, Washington DC.

FAS (2006b). *Summit on Educational Games: Harnessing the Power of Video Games for Learning*. Federation of American Scientists, Washington DC.

Hamm S (2006). Disaster management 101. *Business Week Magazine*, p. 12.

Hutchison J (1997). The Junior Red Cross goes to Healthland. *Am J Pub Health* **87**, 1816–1823.

Janis IL (1989) *Crucial Decisions: Leadership in Policymaking and Crisis Management*. The Free Press, New York.

Jennings D and Wattman S (1998). *Decision Making: An Integrated Approach*. Pitman Publishing, London.

Learning Federation (2006). *Learning Science and Technology Roadmap*. Federation of American Scientists, Washington DC.

Maier F and Grobler A (2000). What are we talking about?—A Taxonomy of computer simulations to support learning. *Sys Dyn Rev* **16**, 135–148.

McDaniels BA (2005). A contemporary view of Joseph A. Schumpeter's theory of the entrepreneur. *J Econ Iss* **39**, 485–489.

Michael D and Chen S (2005). *Serious Games: Games that Educate, Train, and Inform*. Thompson Publishing, Boston, MA.

Musgrove M (2006). A computer game for real-life crises. W*ashington Post*.

Perelman M (1995). Retrospectives: Schumpeter, David Wells, and creative destruction. *Journal of Economic Perspectives* **9**, 189–197.

Perla P (1990). *The Art of Wargaming: A Guide for Professionals and Hobbyists*. Naval Institute Press, Annapolis, MD.

Reid S, LaBonia L, Liu B, Luoma P, and Asare A. (2009). Capital structure and dividend policy in an intro to business course. *J Instr Tech Financ* **1**(1), 1–5.

Reinganum JF (1985). Innovation and industry evolution. *Q J Econ*, **100**(1), 81–99.

Rizzo AA and Kim G (2005). A SWOT analysis of the field of virtual rehabilitation and therapy. *Presence-Teleop Virt* **14**, 1–28.

Rosner R (2006). Keynote address to the high performance computer users conference, Washington DC. http://www.hpcusersconference.com/agenda.html (last accessed 21 December 2009).

Rowland K (2006). Computer game will train first responders. *Washington Times*.

Russell S and Norvig P (1995). *Artificial Intelligence: A Modern approach*. Prentice Hall, New York, NY.

Schumpeter JA (1942). *Capitalism, Socialism, and Democracy*. Harper & Row Publishers, New York, NY.

Sheff D (1999). *Game Over: Press Start to Continue*. Cyber Active Publishing, Wilton, CT.

Smith R (2006). Technology disruption in the simulation industry. *J Def Model Simul* **3**, 3–10.

Solow R (1957). Technical change and the aggregate production function. *Rev Econ Stat* **39**, 312–320.

Steinkuehler C (2005). The new third place: massively multiplayer online gaming in american youth culture. *Tidskrift J Res Teacher Educ* **3**, 17–32.

Steinkuehler C (2007). Massively multilayer online video gaming as participation in a discourse. *Mind, Culture, Activity* **13**, 38–52.

Tapscott D (1997). *Growing up Digital: The Rise of the Net Generation*. McGrawHill, New York, NY.

Tapscott D (2008). *Grown up Digital: How the Net Generation is Changing Your World*. McGrawHill, New York, NY.

Winkler T and Buckner K (2006). Receptiveness of gamers to embedded brand messages in advergames: Attitudes toward product placement. *J Interactive Advert* **7**, 37–46.

Zyda M (2005). From visual simulation to virtual reality to games. *IEEE Computer* **38**(9), 25–32.

Chapter 18

Mathematical Applications for Combat Modeling

Patrick T. Hester and Andrew Collins

INTRODUCTION

Military strategists face a difficult task when engaged in battle against an unknown adversarial force. They must allocate their resources in the most efficient manner possible so as to maximize their chance of defeating their opponent, while being mindful of their finite supply of soldiers, weapons, ammunition, etc. This task is not straightforward, nor trivial. To assist in this effort, military strategists typically employ computational approaches to help them evaluate potential battle outcomes and positively influence the overall outcome of a particular engagement. Mathematical models provide the theoretical foundation for these computational approaches. These models, while they can be effective, fall into two camps: (1) those that have a basis in Cold War era tactics which focus on large-scale conflicts with massive weaponry, or (2) newer, agent based models which are computationally expensive and, often, unable to be validated (Champagne and Hill, 2007). At the basis of the mathematical underpinnings of these complex computational approaches are two often utilized mathematical models for battle prediction, Lanchester equations and game theoretic models. This chapter begins with a discussion of military planning decision support, including specifics on battle evaluation and prediction. The theoretical foundations of Lanchester and game theoretic models are then discussed. Then, a comprehensive example is provided which showcases how the two techniques can be used together to solve a Colonel

Engineering Principles of Combat Modeling and Distributed Simulation, First Edition.
Edited by Andreas Tolk.
© 2012 John Wiley & Sons, Inc. Published 2012 by John Wiley & Sons, Inc.

Blotto problem. Finally, some conclusions are provided outlining the drawbacks of utilizing these modeling paradigms for combat modeling purposes.

MILITARY PLANNING DECISION SUPPORT

Operations research, or management science, techniques have been applied to solve problems and support decision makers for decades. Specifically, use of these techniques within the domain of military planning has a prolonged history. Anderson, Sweeney, Williams, and Martin offer the following observation:

> *The scientific management revolution of the early 1900s, initiated by Frederic W. Taylor, provided the foundation for the use of quantitative methods in management. But modern management science research is generally considered to have originated during the World War II period, when teams were formed to deal with strategic and tactical problems faced by the military. These teams, which often consisted of people with diverse specialties (e.g. mathematicians, engineers, and behavioral scientists), were joined together to solve a common problem through the utilization of the scientific method. After the war, many of these team members continued their research in the field of management science. (Anderson et al., 2008, p. 2)*

It is apparent that operations research plays a significant role in military planning decision support. Among the many techniques available to an operations researcher, simulation systems provide unique capabilities and advantages, especially compared with live exercises. Gwynne et al. (1999) summarize the advantages as follows.

- Full-scale exercises are never realistic due to the constraints of safety for the participants and lack of realistic emotions on the part of the participants (i.e. effectively reproducing battle emotions and environmental conditions is unrealistic).
- Simulation is cheaper and repeatable and does not interfere with the operational schedule of the organization.
- Data are easier to collect from a simulation than from live exercises.

Additionally, in many cases concerning military planning, live exercises are simply not feasible, due to scale, cost, and danger to participants. However, additional capabilities afforded by simulation do not come without cost. In fact, Tolk outlines a new set of requirements introduced by the use of simulation systems for decision support: all relevant system characteristics need to be modeled adequately; all relevant data for initialization and execution must be obtainable; models, simulations, and data need to be validated and verified; the simulated entities must not require constant human interaction; and the internal decision logic must be aligned with the external evaluation logic, in particular when more than one simulation system is used (Tolk, 2009).

One of the major challenges for the use of simulation as an effective tool in military planning decision support is effective and efficient battle evaluation and prediction of potential battle outcomes. There are two approaches to battle planning that can be undertaken, namely high resolution models and low resolution aggregated models. High resolution models incorporate detailed interactions of individual combatants, each having his or her own intelligence, motivation, characteristics, etc. (see, for example, Zacharias et al., 2008). Each process within this model is individually modeled and detailed probabilistic information can be incorporated for more accurate results. Given the high amount of detail, high resolution models are computationally intense and require significant construction and analysis time. Conversely, aggregated models compute high level views of battle, grouping together combatants and utilizing this grouping to calculate battle outcomes and entity behaviors. Stochastic behaviors can be incorporated into these models by applying probability functions at a group level. The models typically require significantly less computational effort than high resolution models. Hester and Tolk address the advantage of aggregated models:

> *In particular for attrition, aggregated models are preferred for short term analysis and evaluation. To individually track each bullet or even each duel can easily become counterproductive. However, a good theory is needed to deal with aggregations. (Hester and Tolk, 2009, p. 78)*

These aggregated models typically have a theoretical basis in differential equations, allowing for efficient methods like Lanchester equations and game theoretic models to be utilized in battle predictions. These models are not without controversy, however, as many of the underlying assumptions belying initial formulations of the approaches have become antiquated in modern warfare (post Cold War). To rectify these deficiencies, many extensions have been introduced to modernize the mathematical models. The theoretical underpinnings of the two approaches, as well as modern extensions, will be discussed. Examples will be provided to demonstrate their utility. Further, drawbacks of the approaches will be discussed at the conclusion of this chapter.

DETERMINISTIC LANCHESTER SYSTEMS

Lanchester analyzed World War I aircraft engagements and developed a theory to explain why concentration of military effort in these entanglements was advantageous (Lanchester, 1914). This theory was encapsulated as laws, which are low resolution aggregated models for determining a battle outcome between two opposing forces. For the purposes of this discussion, the two forces considered are denoted as the blue force (the force the analyst is planning for, aiding, etc.) and the red force (the force the analyst is trying to defeat). This model results in a system of differential equations as follows:

$$dB/dt = f_1(R, B \ldots) \tag{1}$$

$$dR/dt = f_2(R, B \ldots) \tag{2}$$

where B denotes the blue force and R the red force. This system of differential equations can be solved by substituting unique functions into equations (1) and (2), depending on whether or not the analyst is determining the outcome of an ancient or modern conflict, resulting in Lanchester's linear and square laws, which are each discussed separately in the following sections. Further, these functions can be set to depend on time, separation distance between forces, weapon characteristics, training, experience, or any desired set of variables. The B and R values can represent complex vectors of weapon systems, with f_1 and f_2 summing the results of paired opposing weapon system results (heterogeneous laws), including cases in which some pairs use a linear law and some pairs use a square law. These require simulations to evaluate results.

Linear Law

Lanchester's more simplistic model was his linear law. The linear law was a formulation for ancient warfare in which one-on-one combat was the only mechanism of engagement. Lanchester theorized that adding additional troops to either side in this type of combat did not affect the attrition rate of the forces. The reasoning for this was that, in ancient combat when individuals used weapons such as swords to engage in battle, each combatant could only fight with one other combatant at a time. Thus, the mathematical formulation of Lanchester's law (although not stated directly by Lanchester but able to be inferred from his writing) is determined by setting equations (1) and (2) equal to a constant as follows, referred to as the area fire equations:

$$dB/dt = f_1(R, B \ldots) = -aBR \qquad (3)$$
$$dR/dt = f_2(R, B \ldots) = -bRB \qquad (4)$$

where a and b are the individual fighting values for the red and blue teams, respectively. Given this formulation, equation (3) is divided by equation (4), yielding

$$dB/dR = E \qquad (5)$$

where E is known as the exchange rate of weapon efficiency (i.e. the relative weaponry/skill level between the opposing forces; a value of 1 indicates homogeneity amongst the forces, meaning they are killed at an equal rate, $E < 1$ indicates the red force is killed at a faster rate than the blue force (the blue force is said to be more efficient than the red force), and $E > 1$ indicates the blue force is killed at a faster rate than the red force).

With separation of variables and subsequent integration of equation (5), Lanchester's linear law can now be expressed as

$$B_0 - B_F = E(R_0 - R_F) \qquad (6)$$

where B_0 is the initial number of individuals in the blue force, B_F is the final number of individuals in the blue force, R_0 is initial number of individuals in the red force, and R_F is the final number of individuals in the red force (zero when estimating a fight to total force annihilation).

Table 18.1 Ancient Warfare Example Illustration (Adapted from Taylor, 1983)

B_0	100	150	200	250	300
B_F	0	50	100	150	200
Loss by B	100	100	100	100	100

Taylor provides an excellent illustration for application of Lanchester's linear law (Taylor, 1983). In this example, the exchange rate is one ($E = 1$), and there are 100 red combatants initially ($R_0 = 100$). Further, the size of the blue force begins at 100 and increases in increments of 50, to a maximum of 300. Equation (6) can then be solved in terms of B_F, yielding $B_F = B_0 - [E(R_0 - R_F)] = B_0 - [1 \times (100 - 0)] = B_0 - 100$. Solving for the remaining number of blue combatants when complete annihilation of the red force ($R_F = 0$) is realized is shown in Table 18.1.

The key insight from this law is the understanding that there is no advantage in ancient combat scenarios to concentrating, or grouping, forces. Since individuals can only fight one-on-one, any additional forces are unused. Recognizing that this finding is both intuitively incorrect and empirically unjustified given modern battle scenarios, Lanchester augmented his linear law for more modern combat scenarios by developing his square law, discussed next.

Square Law

Recognizing the limitation of his linear law to only describe one-on-one combat, Lanchester formulated his square law, which could be utilized to describe many-on-many combat. Under the revised formulation, decisions regarding concentration of forces are of paramount importance to a force's ability to defeat its opponent. Given modern warfare capabilities, opposing forces are able to engage multiple individuals in combat, therefore leading to a more complex view of the attrition landscape. This revised attrition model can be conceptualized by setting equations (1) and (2) equal to the following:

$$dB/dt = f_1(R, B \ldots) = -aR \qquad (7)$$
$$dR/dt = f_2(R, B \ldots) = -bB \qquad (8)$$

Dividing equation (7) by equation (8) yields

$$dB/dR = E(R/B) \qquad (9)$$

Separation of variables and subsequent integration of equation (9) yields Lanchester's square law, as follows:

$$B_0^2 - B_F^2 = E(R_0^2 - R_F^2) \qquad (10)$$

Table 18.2   Modern Warfare Example Illustration (Adapted from Taylor, 1983)

B_0	100	150	200	250	300
B_F	0	112	173	229	283
Loss by B	100	38	27	21	17

The immediate consequence of the square law (versus the linear law) is the observation that a force can drastically reduce its losses by committing more individuals to a given battle. To illustrate this point, the example from the previous section can be revisited by first solving equation (10) in terms of B_F, yielding

$$B_F = \sqrt{B_0^2 - E[R_0^2 - R_F^2]} = \sqrt{B_0^2 - 1[100^2 - 0]} = \sqrt{B_0^2 - 10000}$$

Resolving the example from the previous section utilizing this modern relationship yields updated force numbers for the blue force as shown in Table 18.2. It is apparent when comparing Table 18.1 and Table 18.2 that concentration of forces is quite advantageous to a military planner. In fact, the simple addition of one individual to the blue force results in 14 remaining individuals (when comparing 101 versus 100 blue force members). This is a remarkable result as the difference in combating 100 opponents can be drastically different with the simple addition of a single individual.

Additional Parameters

Two further parameters can be derived from the Lanchester equations which provide additional insight for a combat analyst. The first is the parameter that describes whether or not the blue force will win a battle with the red force. For an ancient battle, the blue force will win if $B_0/R_0 > E$. Under modern warfare, the blue force will win if $B_0/R_0 > \sqrt{E}$. Each of these quantities can be derived from the Lanchester models.

The second parameter concerns how many of the blue force will remain after a battle with the red force. This quantity can be derived for the linear law (by substituting $R_F = 0$ into equation (6) and solving for B_F) as

$$B_F = B_0 - R_0/E \tag{11}$$

Further, for modern warfare, B_F can be derived using the square law (by substituting $R_F = 0$ into equation (10) and solving for B_F) as

$$B_F = \sqrt{B_0^2 - E\,R_0^2} \tag{12}$$

These indicators provide simple metrics for battle prediction that can be utilized by an individual when constructing a mathematical battle model. They are simplistic, however, and do not include stochastic behavior (among other limitations which are discussed later in this chapter), an extension which is discussed in the following section.

STOCHASTIC LANCHESTER SYSTEMS

The original formulation of Lanchester's models did not include stochastic behavior. That is, all parameters in the model are taken to be deterministic. This is a limiting assumption of the Lanchester models and does not represent reality, as uncertainty abounds, especially in combat situations. To that end, several extensions which amend Lanchester's model to include stochastic behavior have been developed and are discussed here.

To this point, the deterministic view of battle outcomes was largely concerned with the ratio of the initial number of blue force members to the initial number of red force members necessary to ensure a victory for the blue force. While this is an important characteristic, more important is the probability of the blue force winning any battle with a given number of red force members. This property is more generalized and meaningful to an analyst and would probably matter to a commander who is planning to engage military forces in battle.

Brown (1963) developed an approximation to estimating the probability that B blue force members would neutralize R red force members (denoted as $P_{B,R}$). This approximation is based on Lanchester's square law (equation (10)) and assumes that $E = 1$, i.e. that the forces are equally efficient. $A_{B,R}$ is defined as the probability of a reduction in R's forces if B blue members and R red members are engaged in battle. Brown shows that $P_{B,R}$ can be reduced to the following recurrence relationship (Brown, 1963):

$$P_{B,R} = A_{B,R}P_{B,R-1} - 1 + (1 - A_{B,R})P_{B-1,R} \tag{13}$$

where $A_{B,R} = B/(B + R)$.

This recurrence relationship is mathematically possible due to the total probability theorem as discussed by Papoulis (1965), which states for n events

$$P_{B,R} = P_{B,R|E_1}P_{E_1} + P_{B,R|E_2} + \cdots + P_{B,R|E_n}P_{E_n} \tag{14}$$

The usefulness of equation (14) can be illustrated in terms of a simple battle between B and R individuals. If the quantity of interest for an analyst concerns the probability that B defeats R, then the analyst must consider all scenarios under which this event can occur.

For B to win, there are two possible scenarios:

1. B is completely annihilated in defeating R. That is, both B and R end the battle with zero individuals remaining. This is not accounted for in Brown's work, as it is assumed that attrition of both combating forces cannot occur in a given round.

2. B retains some fraction ($> 0\%$) of its original forces in defeating R. There are B feasible end-states for this scenario.

Each of the above scenarios has a probability associated with it and the summation of these discrete scenarios represents the entire landscape of battle outcomes that leads to victory for B forces. Brown combined the outcomes accounted for in scenario (2) above into an approximation which determines the probability of B defeating R as follows (with at least one B member remaining):

$$P_{B,R} = \sum_{j=1}^{B} \frac{(-1)^{B-j} j^{B+R}}{[(B-j)!(R+j)!]} \tag{15}$$

where i and j are variable indices, and all other variables are as before.

Equation (15) assumes that each force is depleted by one individual per round. That is, two blue and two red members can be reduced to two blue and one red member or one blue and two red members and not to one blue and one red member (in one battle step).

While this approach to calculating $P_{B,R}$ represents an improvement over the deterministic Lanchester-based approach, it does come with several limitations. First, it does not determine the degree to which a battle is won. That is, how many individuals remain in a winning force? This piece of information is crucial for repeated interactions between opposing forces. This is typical of military engagements, especially as reinforcements can strengthen a depleted force (lack of the ability to handle reinforcements is another deficiency of Lanchester's and Brown's work).

An example of engagement with reinforcements involves a battle between blue and red forces in which the blue force wins (thus the red force is depleted to zero individuals), the red force supplies reinforcements, and then the remaining blue force and the second red force battle. The likelihood of the blue force beating both red forces is not simply the product of $P_{B,R}$ for both battle scenarios, as the number of blue combatants remaining for the second battle is not known with certainty; in fact, it is a stochastic quantity. The number of members in the blue force remaining to battle the red force in the second battle ranges from one (assuming blue victory) to the number of individuals in the original blue force. It goes without saying that this quantity drastically influences blue's ability to defeat the red force in the second battle.

For realistic battle modeling, not including replacements is a shortcoming (Taylor, 1983, p. 111). Thus, Brown's approach to calculating $P_{B,R}$ is insufficient. Utilizing equation (15) to calculate $P_{B,R}$ for a battle between two blue force members and two red force members (denoted as a (2,2) battle), $P_{2,2}$ can be calculated as

$$P_{2,2} = \frac{(-1)^{2-1} 1^{2+2}}{[(2-1)!(2+1)!]} + \frac{(-1)^{2-2} 2^{2+2}}{[(2-2)!(2+2)!]} = 0.5$$

The sole insight gained from this result is that two blue combatants have a 0.50 chance of defeating two red members. However, the question still remains, how decidedly has the battle been won by the blue members? In order to evaluate this quantity, the analyst would like to be able to calculate $P_{B=1|R=0}$ and $P_{B=2|R=0}$ (the probability of one blue team member remaining given zero red team members remain and the probability that two blue team members remain given that zero red team members remain, respectively). These quantities are 0.167 and 0.333, respectively (see equation (16) and the ensuing discussion).

To understand the differences in the predictions of the deterministic and stochastic approaches, let us examine a problem in which three blue team members engage in a modern battle with two equally efficient ($E = 1$) red team members. Based on the deterministic prediction, blue will always win and have about 2.24 survivors (see equation (12)). Utilizing stochastic analysis (equation (15)), blue only wins 77.5% of the time. These are drastically different results. The stochastic prediction is more realistic, as red might occasionally get "lucky" and win a battle, even though it is overmatched from a manpower perspective.

Further, Washburn and Kress discuss a state-based, Markov chain approach to calculating any MOE (measure of effectiveness) of interest in a battle environment based on Lanchester attrition models (Washburn and Kress, 2009). This approach avoids computational inefficiencies associated with approaches such as Monte Carlo simulation as it is computationally expedient. Since the parameter of interest for this analysis involves transitions between the number of combatants within the simulation model, Washburn and Kress's model focuses on state transitions occurring as the opposing forces are attrited by one individual. For each state of m blue members and n red members, there are two entering states, $(m + 1, n)$ and $(m, n + 1)$, and two leaving states, $(m - 1, n)$ and $(m, n - 1)$. Using this realization, the following recursion relationship can be established to identify a general $\text{MOE}_{m,n}$ for any given state using its leaving states:

$$\text{MOE}_{m,n} = f_{m,n}\text{MOE}_{m,n-1} + (1 - f_{m,n})\text{MOE}_{m-1,n} \tag{16}$$

where $f_{m,n}$ is the likelihood of transitioning from (m, n) to $(m, n - 1)$. As a result, $1 - f_{m,n}$ represents the likelihood of transitioning from (m, n) to $(m - 1, n)$, by the total probability theorem. For example, if the MOE is $P_{m,n}$, then $f_{m,n}$ is defined as $A_{m,n}$ (as defined in equation (13)).

$\text{MOE}_{m,n}$ is assumed to be known in terminal states, and, in keeping with $P_{m,n}$ as the MOE, $\text{MOE}_{m,0} = 1$ and $\text{MOE}_{0,n} = 0$. The analyst can work backwards to these end-states to determine adjacent $\text{MOE}_{m,n}$ values. After solving equation (16) for the entire battle landscape, $\text{MOE}_{m,n}$ can be determined for any state. $\text{MOE}_{m,n}$ then represents the probability that m blue forces will win a battle with n red forces. This approach is very efficient. Regarding their computational efficiency, Washburn and Kress (2009, pp. 90–91) state:

> The computations are efficient, compared to Monte Carlo simulation or to first employing the Chapman–Kolmogorov equations to find the state probabilities at some time large enough to guarantee battle completion....

Terminal states (and associated MOEs) can be adjusted to obtain other information regarding battle outcomes. To determine the probability that exactly i blue individuals (of m initial) will remain after a battle with n red individuals, the blue force's ith end-state value is set to 1 (and others to zero). Then, $MOE_{m,n}$ is calculated as before.

Hester and Tolk (2009) extended Washburn and Kress's work by proposing an approach for sequential battles utilizing equation (16). For a battle involving red forces engaged in sequential battles, $MOE_{m=B,n=1...R}$ values are copied to $MOE_{0,n}$ values for the second battle (while $MOE_{m,n=0}$ remains as in the first battle). This in effect provides probabilistic outcome data from the first battle as input data for the red force in the second battle. Similarly, for a battle involving blue forces engaged in sequential battles, $MOE_{m=1...B,n=R}$ values are copied to $MOE_{m,0}$ values for the second battle. Note that only one force is retained for the second battle as it is assumed that both forces engage in combat until one is completely annihilated. Thus, only the winners of the original engagement remain to fight. This approach can be used recursively for more than two sequential battles.

This approach is an important extension to Lanchester combat models which represents real-world scenarios. One such example involves a layered defense against one opposing force. Naïve stochastic Lanchester analysis employed by combining multiple sequential battles into a single overarching interaction does not allow for analysis of this battle without introducing major statistical flaws.

A simple illustrative example from Hester and Tolk (2009) will now be considered. Three sequential (1,1) battles take place where the surviving blue combatant continues on to fight in sequential battles. If this were to be modeled as a (1,3) battle, then $P_{B,R} = 0.042$. However, by modeling the interaction as three sequential (1,1) battles, we obtain $P_{B,R} = 0.125$. Thus, the analyst is led to believe, by virtue of insufficient modeling techniques, that their probability of defeating their opponent is lower than in reality. Hester and Tolk concluded that their approach shows that battles cannot be combined and decomposed at will; doing so will lead to drastically different computational results. Further, it is obvious that combining their forces dramatically increases the red members' chance of winning the battle with blue members. Thus, the foundation of Lanchester's square law (that there is an advantage to combining forces) remains valid.

It should be noted that the same probability can be calculated using the total probability theorem to determine the potential outcomes of battle 1 to utilize as inputs to battle 2, and so on. While this is feasible for a small number of combatants, the combinatorial aspects of such an approach make it untenable for large numbers of blue and red combatants. Thus, Hester and Tolk (2009) concluded that their approach is less prone to error and more computationally efficient.

CRITICISMS OF LANCHESTER

Critics of Lanchester models point out several deficiencies in the original formulation of Lanchester's laws that we would be remiss not to mention. Taylor succinctly summarizes them as (Taylor, 1983)

1. Constant attrition rate coefficients
2. No force movement (e.g. no advance or retreat of forces)
3. Homogeneous forces
4. Battle termination not modeled
5. No element of chance
6. Not verified by history
7. No way to predict attrition rate coefficients
8. Tactical decision processes not considered
9. Battlefield intelligence not considered
10. Command, control, and communications not considered
11. Logistics aspects not considered
12. Suppressive effects of weapons not considered
13. Effects of terrain not considered
14. Spatial variations in force capabilities not considered
15. No replacements or withdrawals
16. Symmetric form of attrition
17. Target priority/fire allocation not explicitly considered
18. Target acquisition force level independent in modern warfare model
19. All troops assumed to fire in combat
20. Noncombat losses (e.g. surrenders, desertions) not considered

It should be pointed out that these shortcomings are not necessarily unique to Lanchester models, as many other combat models include these limitations. However, there still remain many individuals opposed to the use of any form of Lanchester models for modern combat battle evaluation. Epstein (1985, p. 4) states that Lanchester's equations "offer a fundamentally implausible representation of combat under all but a very small set of circumstances." Cares (2004) objects to the use of Lanchester models on several grounds, including the nature of modern combat as being by definition linked, and thus unable to be explored in an independent environment. Further, Cares objects to the treatment of forces as single entities rather than distinct combatants.

Further, despite the convenience and appeal to reason of the Lanchester laws, there is real doubt about their accurate representation of reality. Hartley (2001) provides a book-length treatment of the problems of validating linear and square laws against historical data. Hartley and Helmbold (1995) provide a shorter treatment of the same subject. Also see Bracken (1995) for analysis of extensive detailed combat data.

Several extensions to the original formulation help to alleviate many of the objections raised. For example, as discussed, the extension proposed by Hester and Tolk (2009) allowed for a number of discrete "mini-battles," providing for the discretization of a battle landscape into smaller conflicts, answering Cares' objections. Further, the extensions discussed incorporating stochastic behavior further extend the Lanchester model's usefulness. Heterogeneous forces are accounted for in Snow (1948). Large-scale, complex planning models are addressed in Cherry (1975) and Farrell (1975). While these examples represent only a handful of extensions, Taylor (1983) provides a comprehensive discussion of many other extensions to Lanchester models.

Despite these shortcomings, we believe that the novelty and usefulness of Lanchester models lie in part in their simplicity. They are able to very efficiently analyze a combat scenario with large force sizes. When one considers adding extensions such as stochastic behavior and sequential battles, the underlying simplicity of the model disappears and additional complexity is introduced. Thus, an analyst must be careful to balance complexity and efficiency when utilizing Lanchester based attrition models, particularly with post-Lanchester extensions employed. We do not believe that a simpler, more efficient means of analyzing combat is available to today's modeler. Thus, many have adapted Lanchester's original equations to account for modern warfare scenarios and we believe this is both a valid and a useful approach to simple analytical combat modeling, especially when employed in conjunction with the game theoretic environment discussed next.

GAME THEORY INTRODUCTION

Game theory is the study of decision problems involving more than one intelligent agent, or player, and it is used to model how these sophisticated agents interact. The term "game theory" comes from the application of "games of strategy" to economic problems by John von Neumann and Oskar Morgenstern in their seminal book *Theory of Games and Economic Behavior* (1944). There is some controversy about the use of the word "game" especially when game theory is applied to serious problems like war and criminal activity; however, individuals can gain great insight into the essence of these problem areas by applying the paradigm of a game to them. This insight could not necessarily be gained by other training or analytical techniques which tend to assume that unsophisticated behavior is used by the agents being modeled. Examples of games are given throughout this section on game theory to illustrate this insight. The remainder of this section on game theory uses the paradigm of a game which is applied to a variety of different military problems.

J.D. Salinger stated, "Life is a game, boy. Life is a game that one plays according to the rules" in his famous novel *The Catcher in the Rye* (Salinger, 1951, ch. 2). The rules are what distinguish different games from each other and they are the foundational structure of any game. Game theory is about determining the *best* way to play a game for any given set of rules. Defining what is meant by best is non-trivial because any strategy used by a player must take into account the other player's strategy as well. Game theory is unconcerned about what are good strategies for playing the game unless that good strategy happens to be the best strategy; this distinguishes game theory from everyday gaming.

Probably the most famous military training game in history is chess. The exact origins of chess are not completely known but there is evidence that it was used by ancient Indian kings (Shenk, 2007, p. 16):

The game of chess became a school of government and defense; it was consulted in time of war, when military tactics were about to be employed, to study the more or less rapid movements of troops.

No best strategy for chess has yet been found, even though the games' rules are relatively simple; currently, only scaled down versions of the game containing 16 pieces have been solved. The complexity of the best strategy is such that even modern computing power is unable to give an exact best strategy to the game. This failure to *solve* chess does not mean that computer simulations of chess are useless, as many advanced computer programs are more than capable of easily beating any human player. Thus gaming should not be underestimated as a useful, practical tool for training and analysis; it is just focused on ways to beat the current opponent, e.g. the current best chess computer programs, as opposed to focusing on beating any opponent.

Game theory was started when, in the 19th century, Antoine Cournot proposed an idea that economists should look at situations where there are only a few competitors (Cournot, 1838). Economists had, until that point, only looked at markets without competition, called *"Crusoe on his island,"* or markets when there was infinite competition, called *"Multeity of atoms"* (Eatwell et al., 1987). The work by Cournot was virtually ignored until John von Neumann and Oskar Morgenstern (1944) wrote their ground-breaking book during World War II. This book became the bedrock of modern game theory. Seven years later, John Nash (1951) developed his Nash equilibrium concept which allowed game theory to become a useful technique within the modern day analysis community. This development won John Nash (1951), along with John Harsanyi and Reinhard Selten (1988), a Nobel Prize in 1994.

Game theory has been developed and adapted further since the 1950s via tens of thousands of academic publications. Most of the current research into game theory is conducted by micro-economists but the subject is not confined to that subject as there have also been several successful applications of the technique in areas as diverse as computer science (Dash et al., 2003) and evolutionary biology (Smith, 1974).

Classic applications in the defense realm range from looking at international power struggles in Mesquita and Lalman's *War and Reason* (1994) to whether a victor should be magnanimous after the conflict in Bram's *Theory of Moves* (1993). Game theory has also been applied to the nuclear arms control negotiations that occurred between the United States and the USSR during the Cold War (Brams, 1993). It is interesting to note that John von Neumann was also a principal member of the Manhattan project, as well as the father of game theory. In recent times, game theory has been applied to the analysis of terrorist organizations (Collins et al., 2003).

This section focuses on two distinct military applications of game theory: search games and Colonel Blotto games. A description and simple example of these games is described to give a flavor of their application. Several points of theory must first be introduced to give the two examples some context; these theoretical points include how the games are represented and how they are solved.

GAME REPRESENTATION

Game theory attempts to mathematically capture behavior in strategic situations, or games, in which an individual's success in making choices depends on the choices of others. There are two standard forms with which game theory attempts to display the overall game under consideration: *normal* and *extensive*. Normal form games use a payoff matrix as seen in Figure 18.1. Extensive form games use a game tree as seen in Figure 18.3 later.

The rows of a payoff matrix represent the possible actions available to the blue player (blue) and the columns represent the possible actions available to the red player (red). The two values in the cells of the matrix represent that reward or payoff that the players receive. The first number is the payoff to blue and the second number is the payoff to red. In the payoff matrix of Figure 18.1 if blue plays action "Ballet" and red also plays action "Ballet" then the payoff to blue is 4 and the payoff to red is 1.

Displaying numbers without context has little meaning, as with most mathematical techniques, so some explanation is required for the game displayed in Figure 18.1. This game has the politically incorrect[1] title of "battle of the sexes"

Figure 18.1   Battle of the sexes in normal form.

---

[1]Game theory has a history of political-incorrectness in its example games. One such example is the "street walker" game which was updated to the "marriage" game but even this change has caused problems within the community. We leave it to the reader to do their own investigations into the nature of these two games.

but the players will remain genderless in this explanation. The players are planning an evening out and there are two choices for them: attending the ballet or attending a soccer match. Blue would prefer to go to a soccer match, red would prefer to go to the ballet, and both prefer going somewhere with the other player to going somewhere on their own. These preferences are arbitrarily translated into payoff values. Payoff for attending an event on your own, say 0, is less than the payoff for attending a non-preferred event with the other player, say 1; both are less than the payoff for attending a preferred event with the other player, say 4. What the values 0, 1, and 4 actually mean is not important within game theory as it is only important that the correct preferences are captured. The payoff matrix given in Figure 18.1 can then be constructed using these values.

The battle of the sexes is just one of many classical example games used by game theorists. Other examples include the "prisoner's dilemma" game, the "matching pennies" game and "ultimatum" game. All of these examples and many more can be found in Binmore's *Fun and Games* (1991).

Games where one person wins while the other loses are called *zero-sum* games, i.e. the payoffs of the players are the exact opposite for all action combinations. Examples of zero-sum games include most recreational games where there is only one winner, and gambling games. If a game is not a zero-sum then it is called non-zero-sum. Non-zero-sum games allow for the possibility of coordination between players. The battle of the sexes game is an example of a non-zero-sum game.

The normal form is usually used to represent games where the action choices of the players are revealed simultaneously. Simultaneous action selections can occur in a number of ways. The players could reveal their action selection as simultaneously as possible as in the game "rock-paper-scissors," or the players could write their actions down which are then submitted to an independent party as in a sealed-bid auction.[2] The labeling of the players has no meaning in these types of games, i.e. blue and red could also be P1 and P2.

Extensive form is a tree-like representation of a game and it is the primary, and most logical, way to represent sequential games: games where the players take turns to play their actions. An example of an extensive form game is shown in Figure 18.3 which will be discussed later.

When both players know everything about the current state of the game and the history of moves that the other player has made, this is called a game of perfect information; otherwise, the game is called one of imperfect information. An example of a perfect information game is chess. An example of an imperfect information game is poker because the players do not know what cards the other players have in their hands. If you have a sequential game of imperfect information were the second player does not have any information about the first player's move, then this is equivalent to a simultaneous game; however, if the second player gains any information about the first player's move, even if this is

---

[2] Auction theory is considered a sub-field of game theory.

just a smile from the first player after they have made their move, then the game is not equivalent to a simultaneous game.

When the players know the payoffs and strategies of the other players then this game is one of complete information; otherwise the game is one of incomplete information. Not knowing an opponent's payoff can have a drastic effect on the game; for example, if you were playing a game of pool against a "shark" it would be useful to know this information as the shark's payoff is to get you to make increasingly larger and larger bets as opposed to trying to win every game against you.

Only games with two players are considered throughout this section. Games can have more than two players and this is considered under n-person game theory. N-person game theory is studied under a branch of game theory called cooperative game theory. Cooperative game theory takes into account that a game of more than two players can result in coalitions being formed between some players to receive a better than individual play payoff. N-person game theory is usually modeled using tuples instead of extensive or normal form. Games of only two players can also require cooperative interaction of the players and these games are called coordination games. The battle of the sexes game shown in Figure 18.1 is an example of a coordination game.

Modern game theory can be split into three basic types which have been discussed above: zero-sum games, non-zero-sum games and cooperative games. A new type of game has arisen in recent years, called *soft games*, which are related to soft operations research (Howard, 2001; Bryant, 2007). Though they have military applications, soft games are not discussed in this section.

SOLUTION METHODS

Game theory is only concerned with the actual solution of a game under certain criteria. Thus turning to game theory for help with your chess game would be almost pointless because (1) chess has not been solved, and (2) the solution to chess is likely to be so complex that no single human is likely to understand it. Understandable solutions to simple games like "tic-tac-toe" and "nim" (Binmore, 1991) do exist, but this is more due to the simplicity of the games than anything else. It is important to realize that unlike other analytical methods (where finding the maximum or minimum of some value is the sole goal), there is no "one size fits all" solution method to a game. For example, possible solution methods include maximin (von Neumann and Morgenstern, 1944), dominance, and Nash equilibrium (Nash, 1951). All the solution methods make certain assumptions about the player's playing behavior.

The maximin solution method assumes that a player looks at his or her worst possible outcome from any given action and chooses the action that has the least worst outcome. This is the solution mechanism for the paranoid player where the player assumes that the other players are out to get him/her. This solution mechanism makes sense for zero-sum games because an opponent benefits from minimizing the other player's payoff. The solution method does not make sense in other types of games, especially coordination games.

Some games can be solved by dominance. An action is dominant over another action if the player always receives a higher payoff no matter what their opponents choose to do. That is, no matter what the outcome of the game, the player would have been better off playing the dominant action instead of the action it dominates. A simpler version of the game can be constructed by repeatedly removing the dominated actions of the players. If this removal of dominated actions leaves only one action available for each player than a solution to the game has been found, where each player would play their remaining action. Not all games can be solved this way; for instance, the battle of the sexes does not contain any dominated actions.

The other main solution method for game theory is the Nash equilibrium which is the most accepted approach within the field. The Nash equilibrium is concerned with the stability of the strategies chosen by the players. A player's strategy is how they choose their action for a game. If a player chooses to go with a single action then this is called a *pure strategy*. If a player chooses to randomly select an action from several different actions then this is called a *mixed strategy*. When mixed strategies are used by any of the players, we consider the player's payoff to be the expected payoff.

Given a particular set of strategies for the players, it is a Nash equilibrium if the following statement is true for all players: *Each player does not benefit from changing their current strategy, given the current strategy of the other players*. This does not mean that the players get the maximum payoff available to them within the game but that they gain the highest payoff available under the constraint of the other players' strategies. Mathematically, the Nash equilibrium can be represented as a best response function to an opponent's strategy.

In many games, such as the prisoner's dilemma, both players could do better with a different set of strategies than the Nash equilibrium. Both players could agree to undertake their strategies to achieve this higher reward. However, as this new set of strategies is not a Nash equilibrium, at least one player could benefit more by changing their strategy; thus this new set of strategies is unstable as it relies on trust.

There can be more than one Nash equilibrium in a game, which can be seen in the battle of the sexes. If both players choose "soccer" as their action, then this is a Nash equilibrium as neither player would benefit from changing their strategy as this would result in the outcome being soccer–ballet or ballet–soccer which would give both players a payoff of zero. If both players choose "ballet" as their actions then this is also a Nash equilibrium. Multiple Nash equilibriums can occur in games because games tend to be nonlinear by nature.

In many games the payoffs achieved through some Nash equilibriums are better for all players than other Nash equilibriums. When this occurs, it is called Pareto dominance. Deciding which Nash equilibrium is the appropriate solution to a game is a non-trivial task and has been the focus of much research (Harsanyi and Selten, 1988; Herrings et al., 2003).

Mixed strategy

The concept of a mixed strategy can be difficult to interrupt, especially in one-off games. For example, consider advising someone that their Nash equilibrium policy is to play one action 99% of the time and another only 1% of the time. If they were only going to play the game once, you might expect them to just play the first action without bothering to randomize their choice between the two actions; hence they would be playing a pure strategy and not the mixed strategy suggested. This could result in them not gaining the best response benefit that the Nash equilibrium offers (i.e. their opponent is likely to realize that they will only play the pure strategy and will change their strategy accordingly). This concept is an unsolved dilemma that faces game theorists (Binmore, 1990).

The battle of the sexes game has a third Nash equilibrium which is mixed. The strategies for this Nash equilibrium are as follows. Blue randomly chooses ballet 80% of the time and soccer 20% of the time. Similarly red chooses ballet 20% of the time and soccer 80% of the time. This results in an expected payoff of 0.8 for both players. This value is of lower value for both players than found in the previous two Nash equilibriums, where the minimum payoff the players received was one.

If all the possible actions have a positive probability of occurring, then the policy is called a *totally mixed strategy*. A totally mixed Nash equilibrium strategy can have good stability properties, and sometimes game theorists insist that the players only use totally mixed strategies (this version of a game is called the *perturbed* game). Perturbed games are behind the trembling hand perfect equilibrium, which was part of the Nobel Prize winning work of John Harsanyi and Reinhard Selten (1988).

Nash equilibriums are used as the solution method throughout the remainder of this chapter. The following two sections discuss some particular applications of game theory.

SEARCH GAMES

Search games revolve around one player looking for the other player or players. This could be as innocent as the game of "hide and seek" to the more daunting search for submarines during the Cold War. The blue player's strategies are usually concerned with where to hide and the red player's strategies are where to search. As search games involve trying to work out where an opponent will be, they can be used for other applications, for example what is the best mine layout to protect a naval base from submarine attack (Woodward, 2003). It is not necessarily true that search games are about avoidance, and some search games, like rendezvous search games, are dedicated to which strategies are best for the players to find each other. Rendezvous search games are used to figure out the best search patterns for rescue squads that are looking for someone who is lost (Thomas and Hulme, 1997).

In certain games, mixed strategies make more sense than pure strategies and this is especially true for search games. If a player who was being hunted were to use a pure strategy for their evasion plan then the hunter would know exactly where they were. Thus, the hunted player would need to randomize where they were to maximize the number of places that the hunter had to look.

Search Game Example

A simple search game example is shown in Figure 18.2 which revolves around an escaped convict trying to flee to a safe haven. The blue player has escaped from jail and is attempting to flee across the border. There are two ways that he/she can get there, either through rural paths or via urban cities. The red player is the marshal pursuing the blue player. The red player only has limited resources, so he or she will only be able to search the rural paths or the urban ones. It is easier to spot the escapee in a rural setting than an urban one. We model success as probabilities of capture. The probability of capture within a rural area is 80% and within an urban area it is 60%. Let us arbitrarily assume the payoff for the escapee escaping is 1 and the payoff for being captured is −1. If we make this a zero-sum game, then the pursuer's payoffs will be the exact opposite.

If the escapee takes the opposite route to the one being searched by the pursuers, then he/she will escape; hence the payoff will be one. If, however, the escapee takes a route that is being searched, there is a chance that he/she will be recaptured. We use the expected payoff value, or average payoff, to express the payoff in these circumstances. For instance, if both the escapee and pursuer choose urban as their action, then there is a 60% chance that the escapee will receive −1 as the payoff and a 40% chance he/she will receive 1 as the payoff; this results in an expected payoff of −0.2. The expected payoff is also calculated for rural–rural action selection.

Calculating the Nash equilibrium is quite cumbersome and the calculations have been omitted here. The Nash equilibrium results for this game are unsurprisingly a mixed strategy. It results in the blue player choosing the rural route

RED

		Rural	Urban
BLUE	Rural	−0.6, 0.6	1, −1
	Urban	1, −1	−0.2, 0.2

Figure 18.2 Search game example in normal form.

3/7 of the time and the red player also choosing the rural route 3/7 of the time. These strategies result in an expected payoff of 11/35, or 0.314, for the blue player. It is left to the reader to verify that these strategies are in fact a Nash equilibrium.

A different Nash equilibrium would have been obtained if different payoff values were used by the players or if more actions were allowed by the players, e.g. if the red player was able to split the search across the domains. As with all modeling methods, the game and its solutions give insights into the problem and not necessarily an exact solution.

Another problem with this game is assumptions that it makes about the players—not only about their payoffs, but also about their rationale. As quoted in a paper on rendezvous search (Thomas and Hulme, 1997, p. 45): "You cannot assume anything about how someone who is lost will behave. They are probably idiots to have got lost in the first place." Idiots are not considered within game theory; in fact, the perfect player is assumed to be playing. This perfectly rational and infinitely intelligent player has been labeled "Homo Economicus" or "Economic Man" (Persky, 1995).

COLONEL BLOTTO

One of the oldest applications of game theory to the military domain is Colonel Blotto games. Colonel Blotto games look at the allocation of troops and resources when fighting across multiple areas of operation. The game is based around the simplest assumption that having more troops means that you are more likely to "win" or control a particular area of operation. The actions that are available for the players are how they allocate their troops to the various areas of operations.

Figure 18.3 gives an example of a Colonel Blotto game. In this example, the blue player has to defend a front guard post and a rear guard post. Blue player has two soldiers to achieve this task and has a choice of three actions: deploy both soldiers to the rear guard post, split the soldiers between the two guard

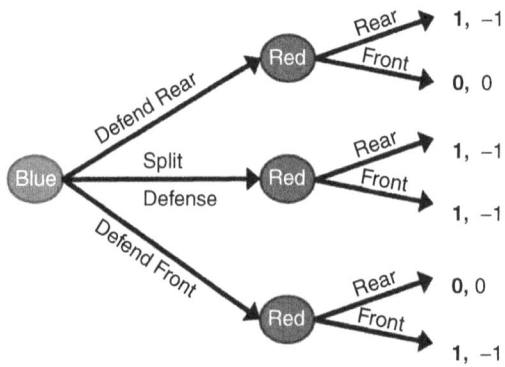

Figure 18.3   Colonel Blotto example in extensive form.

posts, or deploy both to the front. The red player is trying to attack the blue's base with one soldier and can attack either the rear or front post. As the blue player's action is observed by the red player before he/she makes their decision, this is a sequential game and is represented in the extensive form.

Determining the payoffs for this game can be difficult as highlighted in the previous example. It really depends on what the result for a successful or unsuccessful attack by the red player means. If the front and rear guard posts represent entry points to a defended area, then losing either would have dire consequences for the blue player. We shall assume that both guard posts are equally strategic and losing or keeping either is equally weighted. Thus the payoffs for each of the guard posts are determined as follows:

- If the blue player has more soldiers allocated to a guard post than the red player, then that guard post payoff is one.
- If the blue player has fewer soldiers allocated to a guard post than the red player, then that guard post payoff is minus one.
- If the blue player has the same number of soldiers allocated to a guard post as the red player, then that guard post payoff is zero. This includes the case when neither player sends troops to a guard post.

This game is assumed to be zero-sum, so the payoffs for the red player are the opposite of what the blue player gets. The payoffs for each action pair can be determined using this set of simple rules and are shown in Figure 18.3.

The Nash equilibrium for this game can be resolved using backward induction. For instance if the blue player chooses to defend only the front guard post then the red player will attack the rear guard post resulting in draw or zero–zero outcome for each player. The Nash equilibrium for this game is for the blue player to split the forces across both guard posts which results in him/her winning one battle and drawing the other. The red player can respond however he/she likes, but the results will remain the same. Notice that the blue player is not using a mixed strategy here because randomizing has little benefit as the red player will respond to what actual action the blue player ultimately chooses.

There are many variations of the Colonel Blotto game. The opposing sides can have symmetric or asymmetric forces. When each side has the same number of forces to deploy this is symmetric; otherwise it is asymmetric. More complex versions of the game include using different force types like aircraft and tanks, as well as the use of decoys by the defending player.

The original paper on Colonel Blotto games was written by Émile Borel, known for his contributions to measure theory, which acts as the foundation for all modern probability theory (Borel, 1953). Though originally written in 1921, it was translated from French in the 1950s, which was the same time that game theory was becoming popular in the English speaking world due to von Neumann and Morgenstern's work. It is no consequence that game theory's popularity coincided with the start of the Cold War.

GAME THEORY APPLICATIONS AND ISSUES

The previous two examples of search games and Colonel Blotto games highlight some of the diverse areas to which game theory can be applied. One obvious application area that has not been discussed here is negotiation. Once game theory was used extensively for military negotiation research, see Aumann (1995) for an example, but the technique has since fallen out of favor (Goodwin, 2005). The reason that game theory has fallen out of favor is due to the assumptions about the players' rationality and intelligence. Modern negotiation theory is based around the psychology of the participants and not around how a perfectly rational and intelligent player should play the "game." Thus negotiating takes advantage of the weakness that human psychology highlights and these weaknesses are difficult to incorporate in a game.

There are several problems with the application of game theory and many of these problems have already been highlighted in the above examples; a brief summary is given here.

- *Payoff determination*: It can be difficult to determine what the payoffs for a game should be for each player's actions. It might seem that in some cases the payoffs are obvious, i.e. if the game is about winning money, then the money won or lost should be the payoff. However, this has been shown in examples like the ultimatum game not to be the case. As game theory's usefulness is about giving insight, and it is not an exact predictive solution, then the use of approximate payoffs is not a "show-stopper" for its application. In many games the outcome remains the same even with slight changes to the payoff.

- *Determining the Nash equilibrium*: Games can be very easy to construct but can be difficult to solve. This problem can be seen in the game of chess. Chess is a relatively simple game that has been around for centuries, yet no solution to the game has been found, even though a solution has been shown to exist.

- *Mixed strategies*: When the solution to a game is a mixed strategy this implies that players should randomly choose between different actions. This might make some sense if the game is repeatedly played, but for one-off games it becomes problematic. Imagine that the mixed strategy says "play action A 99% of the time and action B 1% of the time." What would you do in this situation? It can be hard to explain this need to randomize to the player. It is also important that your opponent knows that you are going to randomly choose your strategy, but how is this done? Should you phone your opponent and tell him/her you are rolling a die?

- *Rational*: Game theory assumes that the players are perfectly rational and infinitely intelligent. This assumption might be fine when highly skilled game players are being modeled with game theory, but is not necessarily the case for everyday people in their everyday lives.

It is important to remember that any modeling technique will have its own weakness and that there is no perfect way to model complex situations. What game theory gives the modeler is an insight into the multi-player situation under consideration which can be more valuable than some exact solution given by other modeling techniques. This insight that game theory offers is the driving force behind much application for game theory to the "real world." There have been several attempts to bring game theory to the commercial software world. Linguistic geometry (Stilman, 2000) uses game theory as part of combat simulation software; the approach is novel because it attempts to incorporate some level of intelligence by the simulated agents as opposed to the scripted behavior that is normally adopted by most modern strategic combat models. The Gambit software suite is freeware that is available to solve normal-form games; for more information see McKelvey et al. (2007).

There are several other mathematical techniques which are related to game theory but only consider the one-player case; these include Markov decision processes and decision trees (Goodwin and Wright, 1991). There are several books on game theory that are recommended for the interested reader. For an entertaining further introduction, see *Fun and Games* by Binmore (1991), and for a text related to operations research see *Games, Theory and Application* by Thomas (2003).

COLONEL BLOTTO AND LANCHESTER EQUATIONS

As mentioned previously, one major drawback in game theory is in the naïve generation of payoffs for games such as battle scenarios. In an effort to resolve this issue, we propose utilizing the battle prediction of equation (15) (per Lanchester's stochastic square model) to develop the payoffs shown in Table 18.3, which are used to determine the Nash equilibriums in the game theoretic formulation of the Colonel Blotto game.

The game that Table 18.3 represents is one where the players have to allocate up to four soldiers to attack/defend a single guard post. In this version of the Colonel Blotto game, the red player does not know blue player's allocation so the game is considered to have simultaneous action selection. The payoffs for this game are the players' chance of winning the guard post, as calculated using the Lanchester stochastic square model. Unsurprisingly, the Nash equilibrium of this game is when both players allocate the maximum number of soldiers to the guard post.

Now let us consider the case when there are two guard posts to be attacked/defended. The blue player will allocate B1 to the front guard post and B2 to the rear guard post. Similarly, the red player will allocate R1 and R2. Once force distributions are chosen via the Lanchester-assisted game theoretic model, an overall battle success probability can be calculated. Calculation of battle success depends on the military planner's view of success. If a win in any engaged battle is considered a success (e.g. forces are split to gain access to a particular target, thereby ensuring *any* success is a triumph for the planning force), then the battle

Table 18.3   Battle Outcome Prediction as Payoffs in a Game

# of RED soldiers

		0	1	2	3	4
# of BLUE soldiers	0	0,0	0,1	0,1	0,1	0,1
	1	1,0	0.5, 0.5	0.167, 0.833	0.042, 0.958	0.008, 0.992
	2	1,0	0.833, 0.167	0.5, 0.5	0.225, 0.775	0.081, 0.919
	3	1,0	0.958, 0.042	0.775, 0.225	0.5, 0.5	0.26, 0.74
	4	1,0	0.992, 0.008	0.919, 0.081	0.74, 0.26	0.5, 0.5

can be conceptualized as a series system reliability problem (Halder and Mahadevan, 2000, pp. 240–243) for two separate battles as

$$P_{(\text{win either})} = P_{B1,R1} \cup P_{B2,R2} = P_{B1,R1} + P_{B2,R2} - P_{B1,R1}P_{B2,R2} \qquad (17)$$

where $P_{B1,R1}$ and $P_{B2,R2}$ are defined by the payoffs given in Table 18.3. The allusion to series system reliability lies in the conceptualization of the red player as the system to be evaluated and the failure of *any* one of its attacks to be understood as a failure of the system. An underlying assumption is that the individual battles are independent events. This is a reasonable assumption as separating one's force is done in order to divide and conquer an opposing force by splitting their defense into two or more independent altercations.

Conversely, certain battle scenarios require that *all* engaged battles must be victorious in order for success (e.g. forces fight until one of the two sides is completely annihilated); then the battle can be conceptualized as a parallel system reliability problem (Halder and Mahadevan, 2000, pp. 243–245) for two separate battles as

$$P_{(\text{win both})} = P_{B1,R1} \cap P_{B2,R2} = P_{B1,R1}P_{B2,R2} \qquad (18)$$

where all variables are as before. The allusion to parallel system reliability again lies in the conceptualization of the red player as the system to be evaluated and the failure of *all* of its defenses to be understood as a failure of the system. Note that the assumption of independent events remains.

Results for analysis of the underlying four-person battle discussed in the Colonel Blotto section of the chapter and associated force division are shown in Table 18.4. In this case, the feasible scenarios for this battle are shown (with

Table 18.4   Battle Outcome Prediction as Payoffs in a Game

# of RED soldiers at front post

		0	1	2	3	4
# of BLUE soldiers at front post	0	0.5 (0)	0.74 (0)	0.919 (0)	0.992 (0)	1 (0)
	1	1 (0.26)	0.75 (0.25)	0.813 (0.129)	0.96 (0.04)	1 (0.008)
	2	1 (0.081)	0.871 (0.188)	0.75 (0.25)	0.871 (0.188)	1 (0.081)
	3	1 (0.008)	0.96 (0.04)	0.813 (0.129)	0.75 (0.25)	1 (0.26)
	4	1 (0)	0.992 (0)	0.919 (0)	0.74 (0)	0.5 (0)

redundant scenarios eliminated), with the probability of the blue team winning in both a series and parallel (payoffs given in brackets) scenario also provided. Only one player's payoff need be presented in the table as the game is zero-sum. The actions represent the number of soldiers allocated to the front guard post with the remaining being allocated to the rear guard post.

The Nash equilibrium solution to this game is not obvious, unlike the single guard post version. The Gambit software package (McKelvey et al., 2007) was used to find the Nash equilibrium of these games. In the game where only one guard post had to be defended by blue, a mixed strategy was found. Their percentage selections of the five actions for the blue player were (27%, 16%, 13%, 16%, 27%) and for the red player (5%, 36%, 18%, 36%, 5%). This results in an expected payoff for the blue player of 0.86. To gain some understanding about what this payoff means, if the blue player had chosen just to randomly allocate all the soldiers to either of the guard posts, then his or her payoff would have been 0.75.

The mixed strategy for the game when blue must win both guard posts results in a payoff of 0.13. Again this game was solved by Gambit and the strategies are (0, 49%, 2%, 49%, 0) for the blue player and (35%, 0, 30%, 0, 35%) for the red player. The reason for these strategy allocations is not immediately obvious and further investigation will be required.

We believe this novel approach is useful compared with the traditional Colonel Blotto approach, whereby payoffs are not analytically derived. The approach affords the battle planner the ability to understand battle results probabilistically, and it can be extended to more complicated battle scenarios (e.g. more combatants, more than two battles) by way of the mathematics discussed in the Lanchester section of this chapter, combined with the system reliability calculations introduced in this section and discussed in detail in Halder and Mahadevan (2000).

CONCLUSIONS

Military strategists are able to use Lanchester equations and game theory as tools to enable their decision making when planning for a battle against adversarial forces. Using Lanchester equations provides insights into the possible outcomes from a particular conflict, especially the expected attrition levels. Game theory gives insights to what strategies might be employed by an adversarial force and what is the best response to these strategies. A key strength of both of the analytical techniques is their simplicity, which helps any decision maker gain clear insights into the problem being investigated by presenting the solution in a concise manner.

There is a temptation by modelers to increase the complexity of their models and simulations due to cheaply available computing power. As we saw in the example combining Lanchester equations and game theory given in the previous section, extra complexity within the model produces extra complexity within its solutions. This extra complexity of the solution is too often ignored by only presenting the simple statistics of the results to the decision makers. We would argue that unless the richness of results from a complex model is going to be explored for further utility, then the modeler should utilize simple models to ensure they remain fully aware of what the results imply and gain the insight from that knowledge. Complexity can bring another problem during the modeling stage, as it is easy to construct games that cannot currently be solved—like chess.

As with all modeling methodologies, it is important to remember why you are using a particular technique. Both Lanchester models and game theory represent useful methodologies that can be imbedded into a larger simulation to solve the particular problems that they are designed for, i.e. attrition determination and strategy choice. It is in this embedded usage that we envision the true utility of these modeling paradigms.

REFERENCES

Anderson DR, Sweeney DJ, Williams TA and Martin K (2008). *An Introduction to Management Science: Quantitative Approaches to Decision Making* (12th edition). Thomson Higher Education, Mason, OH.

Aumann RJ (1995). *Repeated Games with Incomplete Information*. MIT Press, Cambridge, MA.

Binmore K (1990). *Essays on the Foundations of Game Theory*. Basil Blackwell, Oxford.

Binmore K (1991). *Fun and Games: A Text on Game Theory*. D.C. Heath, Lexington.

Borel E (1953). The theory of play and integral equations with skew symmetric kernels. *Econometrica* **21**, 97–100.

Bracken J (1995). Lanchester models of the Ardennes campaign. *Nav Res Log* **42**, 559–577.

Brams SJ (1993). *Theory of Moves*. Cambridge University Press, Cambridge.

Brown RH (1963). Theory of combat: the probability of winning. *Oper Res* **11**, 418–425.

Bryant JW (2007). Drama theory: dispelling the myths. *J Oper Res Soc* **58**, 602–613.

Cares JR (2004). *An Information Age Combat Model*. Technical Report, Alidade Incorporated, produced for the Director, Net Assessment, Office of the Secretary of Defense under Contract TPD-01-C-0023.

Champagne LE and Hill RR (2007). Agent-model validation based on historical data. *Proceedings of the Winter Simulation Conference*. IEEE Computer Society, Washington, DC, pp. 1223–1231.

Cherry WP (1975). The Role of differential models of combat in fire support analyses, Appendix 4 in Fire Support Requirements Methodology Study Phase II. In: *Proceedings of the Fire Support Methodology Workshop*, edited by RM Thackeray. Ketron, Inc., Arlington, Virginia.

Collins A, Pullum F and Kenyon L (2003). *Applications of Game Theory in Defence Project: Year One Report*. Dstl/CR07880. Defence Science and Technology Laboratories, Ministry of Defence, UK.

Cournot AA (1838). *Recherches sur les principes mathématiques de la théorie des richesses* [*Researches into the Mathematical Principles of the Theory of Wealth*]. Hachette, Paris.

Dash RK, Jennings NR and Parkes DC (2003). Computational-mechanism design: A call to arms. *IEEE Intell Syst* **18**, 40–47.

Eatwell J, Milgate M and Newman P (eds) (1987). *The New Palgrave: Game Theory*. Macmillan Press, London.

Epstein JM (1985). *The Calculus of Conventional War: Dynamic Analysis Without Lanchester Theory*. Brookings Institution Press, Washington, DC.

Farrell R (1975). VECTOR 1 and BATTLE: Two versions of a high-resolution ground and air theater campaign model. In: *Military Strategy and Tactics*, edited by R Huber, L Jones and E Reine. Plenum Press, New York, NY, pp. 233–241.

Goodwin D (2005). *The Military and Negotiation: The Role of the Soldier-Diplomat*. Frankcass, New York, NY.

Goodwin P and Wright G (1991). *Decision Analysis for Management Judgment* (3rd edition). Wiley, New York, NY.

Gwynne S, Galea ER, Owen M, Lawrence PJ and Filippidid L (1999). A review of the methodologies used in the computer simulation of evacuation from the built environment. *Build Environ* **34**, 741–749.

Haldar A and Mahadevan S (2000). *Probability, Reliability, and Statistical Methods in Engineering Design*. Wiley, New York, NY.

Harsanyi JC and Selten R (1988). *A General Theory of Equilibrium Selection in Games*. MIT Press, Cambridge, MA.

Hartley DS (2001). *Predicting Combat Effects (Topics in Operations Research)*. INFORMS, Hanover, MD.

Hartley DS and Helmbold RL (1995). Validating Lanchester's square law and other attrition models. *Nav Res Log* **42**, 609–633.

Herings PJ, Mauleon A and Vannetelbosch VJ (2003). Fuzzy play, matching devices and coordination failures. *Int J Game Theory* **32**, 519–513.

Hester PT and Tolk A (2009). Using Lanchester equations for sequential battle prediction enabling better military decision support. *Int J Intelligent Defence Support Systems* **2**, 76–90.

Howard N (2001). *Drama Without Tears*. Dramatec, Alexandria, VA.

Lanchester FW (1914). Aircraft in warfare: the dawn of the fourth arm—no. v, the principle of concentration. *Engineering* **98**, 422–423.

Maynard Smith J (1974). The theory of games and the evolution of animal conflicts. *J Theor Biol* **47**, 209–221.

McKelvey RD, McLennan AM and Turocy TL (2007). Gambit: software tools for game theory—version 0.2007.01.30. http://econweb.tamu.edu/gambit.

Mesquita BB and Lalman D (1994). *War and Reason: Domestic and International Imperatives*. Yale University Press, New Haven, CT.

Nash J (1951). Non-cooperative games. *Ann Math* **54**, 286–295.

von Neumann J and Morgenstern O (1944). *Theory of Games and Economic Behaviour*. Princeton University Press, Princeton, NJ.

Papoulis A (1965). *Probability, Random Variables, and Stochastic Processes*. McGraw-Hill, New York, NY.

Persky J (1995). Retrospectives: the ethology of homo economicus. *J Econ Perspect* **9**, 221–223.

Salinger JD (1951). *The Catcher in the Rye*. Reissue. Little, Brown and Company, New York, NY.

Shenk D (2007). *The Immortal Game: A History of Chess*. Doubleday, New York, NY.

Snow R (1948). *Contributions to Lanchester Attrition Theory*. Report RA-15078. RAND Corporation, Santa Monica, CA.

Stilman B (2000). *Linguistic Geometry: From Search to Construction. Operations Research/Computer Science Interfaces Series*. Springer, Norwell, MA.

Taylor JG (1983). *Lanchester Models of Warfare, Volume I*. Operations Research Society of America, Arlington, VA.

Thomas LC (2003). *Games, Theory and Applications*. Dover Publications, Mineola, NY.

Thomas LC and Hulme PB (1997). Searching for targets who want to be found. *J Oper Res Soc* **48**, 44–50.

Tolk A (2009). Using simulation systems for decision support. In: *Handbook of Research on Discrete Event Simulation Environments: Technologies and Applications*, edited by Abu-Taieh and El Sheikh. IGI Global, Hershey, PA, pp. 317–336.

Washburn A and Kress M (2009). *Combat Modeling*. Springer, New York, NY.

Woodward ID (2003). Discretization of the continuous ambush game. *Nav Res Log* **50**, 515–529.

Zacharias GL, MacMillan J and Van Hemel SB (eds) (2008). *Behavioral Modeling and Simulation: From Individuals to Societies*. National Academies Press, Washington, DC.

Chapter 19

Combat Modeling with the High Level Architecture and Base Object Models

Mikel D. Petty and Paul Gustavson

INTRODUCTION

Modeling and simulation (M&S) has arguably been used more extensively by the US Department of Defense (DoD) than any other organization. Taken in total, the application and use of M&S has been hugely successful. M&S in general is often used in situations where exercising or experimenting with the real-world subject of the simulation would be too difficult, too expensive, or too dangerous, and military applications in particular include some of the most extreme examples of difficult, expensive, and dangerous situations. Consequently, M&S is used widely and frequently in the DoD to support training, acquisition, analysis, experimentation, engineering, and test and evaluation (Castro et al., 2002). The nearly ubiquitous adoption of M&S technologies throughout the DoD provides incontrovertible evidence of its efficacy and cost-effectiveness.

As would be expected, M&S is often used by the DoD to model combat. This can include the M&S of military and civilian platforms, weapons, known or perceived threat capabilities, and military doctrine. And it may also

Engineering Principles of Combat Modeling and Distributed Simulation, First Edition.
Edited by Andreas Tolk.
© 2012 John Wiley & Sons, Inc. Published 2012 by John Wiley & Sons, Inc.

include the representation and behavior of the underlying tactical system components (weapon systems, radar systems, sensors, and tactical interfaces), known or emergent threat capabilities, environmental conditions, terrain data, and other components that are needed to effectively represent warfare within a simulated battlespace. Combat has been modeled at every scale and resolution (Banks, 2010), and using almost every modeling paradigm and architecture available. Among those architectures, combat modeling has been a primary and motivating application of distributed simulation (Hofer and Loper, 1995), a category of M&S architectures wherein large simulation systems are assembled from a set of independent simulation nodes communicating via a network.

Several different distributed simulation standards and technologies have been developed and applied to combat modeling; of particular interest in this chapter are two related items: the High Level Architecture (HLA), which is a distributed simulation architecture and interoperability protocol standard, and Base Object Models (BOMs), which is a standard for developing reusable and composable components of models.

HLA is a general purpose distributed simulation architecture suitable for a wide range of M&S applications. HLA intentionally does not contain any features or characteristics specific to any application domain, including combat modeling. It is nevertheless well suited to implementing combat models[1] of various types, and has been used to do so many times. Likewise, the BOM standard offers a general purpose modeling architecture for defining components to be represented within a live, virtual, or constructive simulation environment. BOMs are well suited for characterizing combat models including the anticipated behavior of interacting systems, individuals, and other entities. Additionally BOMs lend themselves to encouraging the composability of combat models, resulting in a framework for the assembly of a system (i.e. simulation) or system of systems (i.e. distributed simulation environment).

The essential characteristics of HLA and BOMs, and how they can be used for combat modeling, are the subjects of this chapter. The chapter has four main sections. Following this introduction, a brief tutorial on distributed simulation and HLA is presented. Next, sample applications of combat modeling using HLA are surveyed at varying levels of detail. Then, the features of the BOM standard and how those features can be used for combat models are examined. Finally, discussion of some additional technical features and capabilities that help facilitate interoperability and reuse is presented.

---

[1]To clarify the terminology used in this chapter, when used as a noun "model" is generally defined as a representation of something else (Petty, 2010); in this chapter it will often mean a computer program instantiating a mathematical or logical representation of combat or of a combat system. In contrast, "simulation" as a noun is the execution of a model over time. The common use of "simulation" (noun) as a synonym for "model" (noun) is intentionally avoided. Finally, "simulation" as an adjective describes any computer system, computer program, or model used in the conduct of running simulations. For additional detail and related definitions, see Petty (2010).

DISTRIBUTED SIMULATION AND THE HIGH LEVEL ARCHITECTURE

This section begins by introducing the general concept of distributed simulation, of which HLA is a notable example. It then quickly reviews three distributed simulation architectures and standards that preceded HLA and informed its development. It details the design intent, concepts, and technical capabilities of HLA. Finally, the different versions and standards for HLA are identified and described.

Distributed Simulation

In the US DoD, military M&S standard efforts and investment has been dominated by the development, use, and maintenance of several well known distributed simulation standards. A succession of distributed simulation standards, each differing from its predecessors in important ways but nevertheless overlapping with them to varying degrees in both application and technical characteristics, have been designed, implemented, tested, and promulgated. All of this work has come at the cost of considerable investment of financial and technical labor resources (Henninger et al., 2008). These standards consequently form an interesting and important case study in military M&S standards that both illustrates issues in standards development and governance and serves as the basis for certain findings to follow later.

As defined earlier, distributed simulation standards, also known as interoperability protocols, are intended to allow independently executing models to interoperate, typically via a network, so as to collaboratively simulate a common scenario or environment (Wainer and Al-Zoubi, 2010). Because the interoperating applications may be developed at different times by different persons or organizations, standardization of the interoperability protocol is a prerequisite for such systems to operate successfully (Tolk, 2010). In general, distributed simulation standards can include definitions of the formats of the messages to be exchanged at runtime between the linked models, the data items contained in those messages, and the logical actions and sequences to be performed when models interact via those messages. The details vary with each standard but the concepts are substantially consistent.

To set the context for the individual distributed simulation standard reviews to follow, Figure 19.1 illustrates the historical relationships and content of these standards. In the figure, the arrows indicate ideas and experience flowing from one standard to the benefit of the next, but they do not necessarily imply that one standard is replacing or subsuming another. In fact, three of the five standards shown in the figure (DIS, HLA, and TENA) remain in active use.

In distributed simulation, large simulation systems are assembled from a set of independently executing models running on computational nodes communicating via a network. Crewed simulators of the type described earlier and semi-automated forces systems may be nodes linked in a distributed simulation

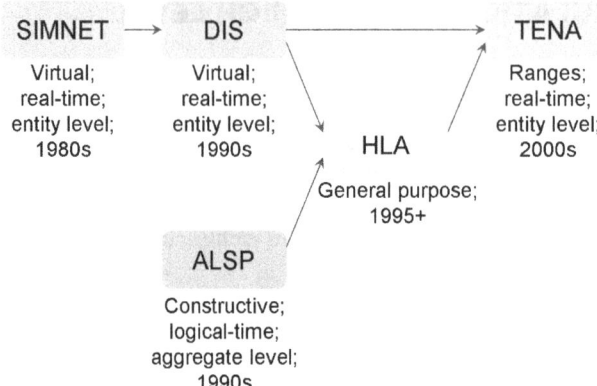

Figure 19.1 Historical relationships between distributed simulation standards.

system. Executing within a distributed simulation architecture adds implementation complexity to the individual models because they must use an interoperability standard, but it has benefits, including *scalability* (larger scenarios can be accommodated by adding more computational nodes to the network), *specialization* (models optimized for a specific purpose can be linked to produce a complete simulation system), and *geographic distribution* (the computational nodes need not all be at the same physical location).

In a distributed simulation system the networked models report the attributes (e.g. location) and actions (e.g. firing a weapon) of interest regarding the entities they are simulating by exchanging network messages. The interoperability standard defines the format of the messages, the conditions under which specific messages should be sent, and the proper processing for a received message. Several standard distributed simulation interoperability standards have been developed over the last 30 years; Simulator Networking (SIMNET) and Aggregate Level Simulation Protocol (ALSP) are no longer in service, but Distributed Interactive Simulation (DIS) (IEEE, 1995), High Level Architecture (HLA) (Dahmann et al., 1998a), and Test and Training Enabling Architecture (TENA) (Powell, 2010) are currently in widespread use.

Predecessors to HLA: SIMNET, DIS, and ALSP

SIMNET was the first distributed simulation interoperability standard developed by the US DoD and the first fully functional distributed simulation protocol of any kind (Goldiez, 1995). Despite the costly and underpowered computer equipment of the era, SIMNET was in many ways a "game changer" because of its ground-breaking ideas (Crosby, 1995). Fundamental technical concepts and approaches for implementing distributed simulation developed first for SIMNET can be found in all of its distributed simulation successors.

SIMNET (the term can refer to either the interoperability standard, the overall distributed architecture, or simulation systems built using it) was designed to model mounted combat in real time in a virtual environment and was intended to train team tactics. Its prime design consideration was to allow multiple vehicle crews to participate and interact with each other in the same virtual battle. Multiple individual vehicle simulators, each consisting of a crew compartment with physical vehicle controls supported by two computers, one to handle vehicle dynamics modeling and network interface and the other to generate visual images of the battlefield within the physical simulator, were linked via a computer network. Figure 19.2(a) on the left shows a SIMNET facility with multiple M1A1 simulators and Figure 19.2(b) on the right shows an example of how the virtual battlefield entities might have appeared to SIMNET users on the image generators of that time.

At runtime the simulators' computers exchanged messages via the network. Different message types had different purposes: some reported the current location, velocities, and appearance of the entities each simulator was controlling, whereas others mediated inter-entity interactions such as direct fire or collisions. The types, format, and content of the messages were defined in the SIMNET standard. Notably, the standard specified not only what each of the messages contained, but when to send each message; for example, a set of "dead reckoning" algorithms and rules controlled how often a simulator should send out a message reporting the new location of the vehicle it was simulating as it moved in the virtual environment.

In contrast to SIMNET, the progenitor of the military distributed simulation interoperability protocols, and DIS, which was SIMNET's conceptual successor, ALSP was not designed for virtual, real-time, entity level simulation. ALSP was instead meant to link constructive combat models, where representation was more often at the level of military units (e.g. battalions), not individual combat entities (e.g. tanks), and the advance of time within the simulation had no relationship with time advance in the real world. Examples of the models linked with ALSP include the US Army's Corps Battle Simulation (CBS), the US Air Force's Air

Figure 19.2   (a) SIMNET facility and (b) virtual battlefield (US Army).

Warfare Simulation (AWSIM), and the US Navy's Research, Evaluation, and Systems Analysis (RESA) model (Fischer, 1994).

ALSP's technical design tenets included some carried forward from SIM-NET, including communication between computational nodes via network message exchange and the assumption that each model controls its own objects in the simulated world. However, in response to the constructive nature of models it was designed to link, ALSP added additional features not seen in SIMNET, including logical time management (explicit collaborative control of simulation time as distinct from real-world time) and architecture independence (technical facilities enabling simulation applications[2] developed by different organizations using different languages and different model architectures to interoperate). A common software module, the ALSP Common Module, was part of each model and mediated the application's runtime interoperation (Wilson and Weatherley, 1994).

Like SIMNET, DIS was designed for virtual, real-time, entity level simulation. Developed from SIMNET beginning in the early 1990s, DIS exploited lessons learned from the earlier standard. Those lessons learned included both technical and governance related issues; in contrast to the closed SIMNET standard, DIS was planned from the outset to be a non-proprietary open standard, accessible to the entire community (Hofer, 1995). The intent was that simulation applications developed by multiple organizations would be able, as a consequence of compliance with the standard, to successfully interoperate.

True to its SIMNET predecessor, DIS was designed to link entity level models of mounted combat in real time in a virtual environment and was likewise intended to train team tactics. Multiple individual vehicle, aircraft, and infantry simulators, of a wider variety than those available with SIMNET, were linked via a computer network (most often an Ethernet local area network) and exchanged network messages. Those messages, which in DIS were termed Protocol Data Units (PDUs), were formatted and sent in accordance with byte-by-byte definitions given in the DIS standard. The messages were broadcast on the network using either the UDP/IP or TCP/IP protocols. Different message types reported the state (location, status, movement) of simulated entities, mediated interactions between entities, and managed or controlled simulation execution. The standard specified the data items to be passed, the format of the individual data items, the grouping of the data items into messages (the PDUs), the conditions under which a simulation application should send a specific PDU type, the processing a simulation application should perform upon receiving a specific PDU type, and key

---

[2]A *simulation application* is one of the executing programs connected in a distributed simulation system. The simulation applications are often *models*, which represent some real-world aspect of the battlefield scenario, such as a tank or a missile. When a model is specifically what is meant, that term will be used. However, there are also non-model applications that may be part of a distributed simulation system; examples include applications that record network messages for later playback and analysis and applications that monitor the execution status of the connected application. When any application is meant, model or non-model, the term simulation application will be used.

Figure 19.3   (a) CCTT facility and (b) virtual battlefield (US Army).

algorithms to be shared among the models (Loper, 1995). The latter included, for example, dead reckoning algorithms, which served to reduce network traffic by reducing the number of PDUs sent to update entity locations as the entities move (Lin, 1995). Ultimately a total of 57 PDU types organized into 11 functional categories called "families" were defined within the standard.

The US Army's Close Combat Tactical Trainer (CCTT) was a major training system implemented using the DIS standard. Analogous to the SIMNET images shown earlier, Figure 19.3(a) on the left shows a CCTT facility with multiple M1A1 simulators and Figure 19.3(b) on the right shows an example of how the virtual battlefield entities might have appeared to CCTT users.

HLA Design Intent

Like its predecessors SIMNET, ALSP, and DIS and its contemporary TENA, HLA is an architecture for constructing distributed simulation systems. It facilitates interoperability among different types of models and simulation applications and promotes reuse of simulation software modules (Dahmann et al., 1998a; Kuhl et al., 1999). Unlike those other architectures, HLA is not intended for a specific type of model, such as constructive logical-time models for ALSP (Wilson, 1994) and virtual real-time models for SIMNET and DIS (Hofer, 1995), or a specific class of applications, such as test and training range applications for TENA (Powell, 2010). By design, HLA is instead intended to provide a general purpose distributed simulation architecture suitable for any type of model and any class of application including training, acquisition, analysis, experimentation, engineering, and test and evaluation. HLA can support virtual, constructive, and live models and has inherent capabilities for both real-time and logical-time execution (Kuhl et al., 1999).

HLA Definitions and Concepts

HLA, both as an architecture and as an IEEE standard, is defined by three documents:

1. The HLA Rules, which "define the principles of HLA in terms of responsibilities federates (simulations, supporting utilities, and interfaces to live systems) and federations (sets of federates working together) must uphold" (IEEE, 2010a). The Rules lay out the essential interoperability expectations and capabilities required of federates and federations.

2. The Interface Specification, which "defines the interface between federates ... and the underlying software services that support interfederate communication in a distributed simulation" (IEEE, 2010b). The Interface Specification is a precise specification of the interoperability related actions that a simulation application may perform, or be asked to perform, during a simulation execution.

3. The Object Model Template, which "defines the format and syntax for recording information in [HLA] object models, to include objects, attributes, interactions, and parameters" (IEEE, 2010c). The Object Model Template does not define specific simulation data; rather it is a format and notation for specifying simulation data in terms of a hierarchy of object classes, and their attributes, and interactions between objects of those classes, and their parameters.

In HLA terminology, a set of collaborating simulation applications is called a *federation*, each of the collaborating simulation applications is a *federate*, and a simulation run is called a *federation execution*.

Data to be exchanged between federates is defined using a two-part class hierarchy in what are called *object models*. In the first part of an object model, a single-inheritance hierarchy of *object classes* is defined. Each class in the hierarchy may have associated data fields termed *attributes*, whose types and sizes are specified in the object model. Instances of the object classes may be instantiated by the federates during a federation execution, and once instantiated, a federate may update the values of an object instance's attributes during the execution. The attribute value updates may be transmitted to the other federates. In this way a federate may create a battlefield object, such as a helicopter, by instantiating an object of a suitable class, and then simulate that battlefield object's actions over time by updating the values of the corresponding object model object's attributes; for example, by changing the x, y, and z location coordinates of a helicopter as it moves. Attribute value updates may be transmitted to other federates in a manner to be described shortly.

In the second part of an object model, a single-inheritance hierarchy of *interactions* is defined. Each class in the hierarchy may have associated data fields termed *parameters*, whose types and sizes are again specified in the object model. Instances of the interactions may be instantiated by the federates during a federation execution. Unlike instances of object classes, which normally represent persistent battlefield objects such as tanks or aircraft, instances of interaction classes typically represent instantaneous interactions between battlefield objects, such as direct fire attacks or radio message transmissions.[3] As with object attribute updates, interactions and their parameters may be transmitted to other federates. The object model for an overall federation, which specifies the object and inter-action classes that will be in use in a federation, is termed a *federation object model* (or FOM). The object model for a particular federate, which specifies the object and interaction classes that a federate can send and receive, is termed a *simulation object model* (or SOM).[4] The FOM for a federation is often a sub-set of the union of the SOMs for that federation's federates. Both SOMs and FOMs are defined according to format and notation defined in the Object Model Template.[5]

Federates that adhere to the Rules can exchange data defined using the Object Model Template by invoking the services defined in the Interface Specification. The HLA Runtime Infrastructure (RTI) is a software implementation of the Inter-face Specification. The RTI actually provides the services defined in the Interface Specification, including services to start and stop a simulation execution, to send simulation data between interoperating federates, to control the amount and rout-ing of data that are passed, and to coordinate the passage of simulated time among the federates. Federates perform those functions by invoking the appropriate RTI services. The RTI may also invoke services that federates must provide to it, such as receiving simulation data; those services are likewise defined in the Interface Specification.

Figure 19.4 shows a logical view of an HLA federation. As the figure suggests, multiple federates exchange data with each other during a federation execution. That data exchange is mediated and supported by the RTI, using the services already mentioned. However, the actual technical architecture of an HLA is bit more complex, at least in part to avoid the latencies associated with a naïve implementation of the logical architecture as shown, wherein a sending federate would pass data to the RTI and then the RTI would pass those data to one or more receiving federates. Such a scheme would require two network sends for each data transmission.

---

[3]Of course, direct fire attacks and radio transmissions actually do have a non-zero or non-instantaneous duration, while the fired round flies toward the target or the radio waves propagate. Nevertheless, at the granularity and resolution of typical HLA combat models, such events usually can be and are treated as instantaneous. However, there is nothing inherent in HLA that forces this approach. Whether battlefield objects such as missiles, which once fired have an existence separate from the firing object that is much shorter than a persistent battlefield object like a tank and much longer than an effectively instantaneous interaction like a direct fire attack, are treated as objects or interactions is up to the designer of the federates and the federations involved.

[4]A more precise name for the object model of a federate would be *federate object model*, of course. However, the resulting acronym (FOM) would conflict with the acronym for a federation object model.

[5]The Object Model Template defines additional information that an object model may contain beyond the already mentioned object classes, attributes, interaction classes, and parameters. These details, which are beyond the scope of this chapter, can be found in the Object Model Template (IEEE, 2010c).

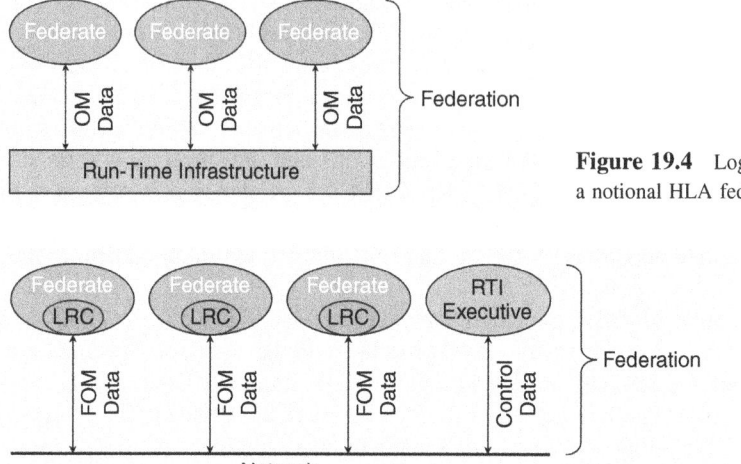

Figure 19.4 Logical view of a notional HLA federation.

Figure 19.5 Technical view of a notional HLA federation.

The actual technical architecture for a notional HLA federation is shown in Figure 19.5. The invocations of HLA services by the federates on the RTI or by the RTI on the federates are possible because the software implementation of each federate in an HLA federation necessarily includes a portion of the RTI software known as the Local RTI Component (LRC). When a federate invokes a service, it does so by calling the LRC; likewise, when the RTI invokes a service on the federate, it is the LRC that calls the federate. Additionally, there are some services that must be provided by a central RTI component; that component is labeled the RTI Executive in the figure.

HLA differs from SIMNET and DIS in a number of important ways. One interesting technical difference is the separation of functionality. For example, in SIMNET and DIS the standards define both data transport (e.g. DIS PDU broadcast via TCP/IP) and data content (e.g. specific fields in a PDU). In contrast, HLA separates those concerns from each other. Data transport is handled within the RTI and so frees the developers from those details, whereas data content is defined by the federation developers using the Object Model Template.

HLA Capabilities and Services

As has already been mentioned, the RTI provides a set of services that the federates use to exchange data and perform other simulation related functions during a federation execution. Those services are grouped into several categories or groups. The categories, and the functionality they provide, are as follows.

   1. *Federation Management.* Create, control, destroy federation executions; join and resign from federation executions; pause, resume, checkpoint, and restart federation executions.

2. *Declaration Management.* Announce capability to send or receive object and interaction data.

3. *Object Management.* Create and delete objects; send and receive object attribute updates; send and receive interactions.

4. *Ownership Management.* Transfer ownership (right to update) object attributes between federates.

5. *Time Management.* Control and synchronize the advance of logical (simulated) time during federation executions.

6. *Data Distribution Management.* Select and filter data to be received by federates during a federation execution.

7. *Support.* Provide infrastructure status information to federates.

Not every federate will use every service, or even every service group. For example, real-time models will often not use Time Management services, and many federates do not support the transfer of object attributes and so do not use Ownership Management. Only those services relevant and necessary to the federate's purpose will be used.

HLA Versions and Standards

HLA development was initiated in 1995. During its early development the standard was governed by an architecture management group (Dahmann et al., 1998b). The group met regularly in open meetings to present the evolving standard's technical details and to receive and discuss suggestions and proposals for changes and additions to the standard.[6] The suggestions were taken into consideration by the architecture management group, which retained control of the final decisions. The standard was given its first large-scale interoperability tests in a set of demonstration "proto-federations," from which a number of technical lessons were learned (Harkrider and Petty, 1996a, 1996b). The culmination of this phase of development was the designation of HLA as the DoD standard architecture for distributed simulation in 1996 and the release of the DoD 1.3 standard for HLA in 1998. Interestingly, the mandated compliance with the HLA standards was met, in many cases, not by actually converting those applications to HLA but instead by connecting non-HLA simulation applications to HLA federations using runtime protocol translators, usually known as "gateways" (e.g. see Wood and Petty (1999) or Drake et al. (2010)).

Following that, the governance and evolution of HLA entered its second phase. A committee was organized within the Simulation Interoperability Standards Organization (SISO) to transform the DoD 1.3 HLA standard into an IEEE

---

[6]Those meetings, many of which the first author attended, were often unexpectedly contentious. Disagreements that were inexplicably and fascinatingly intense arose over seemingly minor technical details of the architecture.

standard. That committee completed its work and produced an officially approved IEEE standard for HLA, designated IEEE 1516–2000, in 2000. While the IEEE 1516–2000 and DoD 1.3 versions of the HLA standard were quite similar, they were not identical (DMSO, 2004). For example, changes in the data distribution management services resulted in different ways of using those services (Morse and Petty, 2004), but did not change their underlying capabilities (Petty, 2002; Petty and Morse, 2004). In the third phase of development, a new version of the HLA standard was carefully developed that addressed issues and opportunities identified in extensive usage experience from 2000 onward (e.g. the Joint Experimental Federation in the Millennium Challenge 2002 exercise (Ceranowicz et al., 2002)). That third version, officially known as IEEE 1516–2010 and colloquially as "HLA Evolved," was officially approved as an IEEE standard in 2010 and adds features that include, for example, extended XML support for object models and standardized time representations.

EXAMPLES OF COMBAT MODELING APPLICATIONS OF HLA

This section describes three combat modeling applications of HLA. Each is an example of a different category of HLA application or system (federates, special purpose federations, and persistent federations), selected from among many examples available in each category. The first, OneSAF, is a constructive entity level combat model that is a federate in the HLA context. The second, the Joint Experimental Federation, was a large special purpose HLA federation that was developed to support a specific, single experiment. The third, MATREX, is a persistent federation, i.e. a federation that once developed is used repeatedly and enhanced over time (Dahmann and Lutz, 1998).

Federate: OneSAF

Virtual environment training simulations often include simulated entities (such as tanks, aircraft, or individual humans), which are generated and controlled by computer software systems rather than individual humans for each entity. These systems, which are known as semi-automated forces (SAF) because the software is monitored and controlled by a human operator, play an important role in virtual environment simulations.

One Semi-Automated Forces (OneSAF, sometimes known as the OneSAF Objective System or OOS) is the US Army's newest constructive battlefield simulation and SAF system. It is the result of an extensive development effort, including extended preparatory experimentation with models and implementation techniques using an enhanced version of an earlier SAF system. OneSAF is intended to replace a number of legacy entity based simulations to serve a range of applications including analysis of alternatives, doctrine development, system design, logistics analysis, team and individual training, and mission rehearsal,

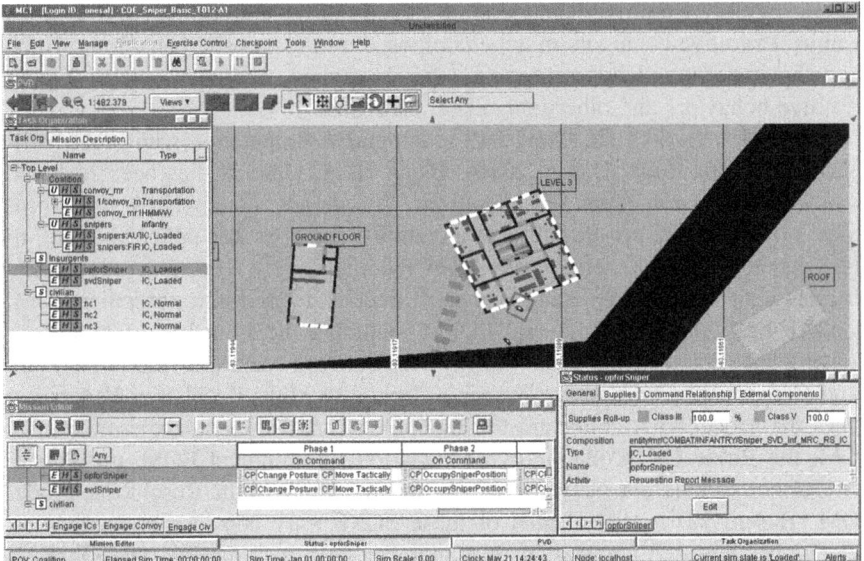

Figure 19.6   OneSAF operator interface screen (US Army).

and to be interoperable in live, virtual, and constructive simulation environments (Parsons, 2007).

OneSAF's capabilities incorporate the best features of previous SAF systems (Henderson and Rodriquez, 2002) and include advanced features such as aspects of the contemporary operating environments (Parsons et al., 2005), multi-resolution terrain databases with high resolution buildings, and command and control systems interoperability (Parsons, 2007). An example of the OneSAF operator's interface screen is shown in Figure 19.6 (Parsons, 2007); it shows individual soldiers in an urban environment under the control of the software. OneSAF has been developed using modern software engineering practices and has a product line architecture that allows the software components of OneSAF to be reusable in different configurations for different applications (Courtemanche and Wittman, 2002).

The behavior generation mechanism in OneSAF combines tested concepts that have appeared multiple times in various forms in SAF research (primitive and composite behaviors) with the latest modeling approaches (agent based modeling) (Henderson and Rodriquez, 2002). As with many earlier SAF systems, behavior generation in OneSAF is behavior emulation rather than cognitive modeling (Petty, 2009).[7] The basic elements of behavior representation in One-SAF are *primitive behaviors*, which implement units' "doctrinal functionality"

---

[7]*Behavior emulation* refers to generating usefully realistic behavior without regard to whether the behavior generation mechanism is in any way similar to human cognition; *cognitive modeling*, on the other hand, seeks to actually model human cognitive processes.

(Parsons, 2007). Primitive behaviors are directly executable by OneSAF entities or units. Primitive behaviors may be composed into composite behaviors, which are behaviors formed by combining other behaviors, and may comprise both primitive behaviors and other composite behaviors.

OneSAF uses both HLA and DIS to support interoperation with other models (Lewis et al., 2008; Hasan, 2010). Using HLA, OneSAF has been included as a federate in a number of federations. As already noted, HLA itself provides infrastructure support for interoperability but does not define the data to be exchanged among federates; any HLA federate in any HLA federation interoperates with the other federates using the data as defined in a particular FOM. OneSAF has worked with several FOMs, including the Joint Land Component Constructive Training Capability Entity Resolution Federation FOM, the Joint Land Component Constructive Training Capability Multi Resolution Federation FOM, the Modeling Architecture for Technology and Research Experimentation FOM, the Battle Lab Collaborative Simulation Environment FOM, the Persistent Partner Simulation Network FOM, and the Joint Multi Resolution Model FOM (Hasan, 2010). The details of these FOMs vary considerably with their federations' intended applications and member federates; the fact that OneSAF has been adapted to use such a range of FOMs suggests the flexibility of its HLA interface. Within OneSAF's software architecture, HLA interoperability is encapsulated in a specific module, and converter classes are provided to convert object attributes and interaction parameters in OneSAF's internal object model to and from the specific external FOM being used in the federation (Lewis et al., 2008). Converter classes are provided in OneSAF for some widely used FOMs; in other cases converters must be developed to connect OneSAF to a new FOM, or to add new objects, attributes, interactions, or parameters to an existing FOM (Lewis et al., 2008).

Special Purpose Federation: Joint Experimental Federation

The Millennium Challenge 2002 (MC02) experiment was organized by the US Joint Forces command to test organizational and doctrinal ideas related the concept of a Standing Joint Force Headquarters, which is a joint force headquarters that has been organized, trained, and equipped prior to a specific requirement for a joint military operation and maintained in a state of readiness so as to be immediately available in response to a crisis or immediate need.[8] MC02 was

---

[8] As is generally well known, during the MC02 experiment the Red force commander used unconventional tactics early in the scenario that resulted in heavy and unanticipated Blue losses. In order to proceed with the experiment, the controllers were obliged to restore the lost Blue forces and place restrictions on what actions the commanders would be allowed to take (Borger, 2002; Arquilla, 2010). While those events are certainly interesting, they are not the focus here; this chapter's interest is in how HLA was used to support the experiment.

quite large, possibly the largest such exercise ever, involving over 13,000 personnel during the actual experiment and costing approximately $250 million to plan, implement, prepare, and conduct (Borger, 2002).

The MC02 exercise included both computer simulation models and live players; the Joint Experimental Federation (JEF) was a large special purpose one-time HLA federation developed to support MC02 by simulating the experimental battlefield scenario so as to stimulate the command and control operations and command decision making of the headquarters (Ceranowicz et al., 2002).[9] A total of 44 different models were integrated into the JEF; all existed prior to developing the JEF, i.e. none were developed specifically for MC02. They ranged from general purpose constructive models capable of simulating a wide range of battlefield entities and actions (e.g. Joint Semi-Automated Forces, or JSAF, for naval and maritime entities and Joint Conflict and Tactical Simulation, or JCATS, for land entities) to special models focused on one particular type of entity or action (e.g. National Wargaming System, or NWARS, which modeled national sensors).

Of the 44 JEF models, 20 used HLA for interoperation before being integrated into JEF, 22 used DIS, and two used some other mechanism (or none). The work of integrating the models into the JEF could be broadly categorized into three types of tasks: (1) converting the non-HLA models to use HLA; (2) improving the RTI and network infrastructure to support the large entity counts required for the MC02 scenario; and (3) reconciling differences in modeling methods and assumptions between the models. Regarding the first type of integration tasks (HLA conversion), the DIS models were ported to HLA either by modification to use HLA natively or through the use of HLA-DIS gateways. For some, a gateway was used initially so as to begin testing the model's integration with the other JEF federates as soon as possible, and then later the model was ported to native HLA. The JEF FOM was based on the Real-time Platform Reference FOM (RPR FOM), an HLA FOM that specifies as closely as possible the same entities, attributes, interactions, and parameters defined in the DIS protocol (SISO, 1999). Because the RPR FOM is based on DIS, the DIS models to be integrated into the JEF were already using the objects and interactions within the JEF FOM, reducing the integration effort. Moreover, many of the HLA federates to be integrated into the JEF were also using versions or extensions of the RPR FOM.

Regarding the second type of tasks (RTI improvements), even within the JEF HLA infrastructure (especially the RTI) technical issues had to be resolved to meet the demands of the MC02 scenario. The one-time nature of the experiment dictated that the federation be able to continue executing even if individual federates were to crash. To address this, the RTI was modified to provide connectionless operation without a central RTI executive that supported only best effort

---

[9]Except where other citations are given, much of this section is based on (Ceranowicz et al., 2002).

transmission.[10] The potential for lost updates and interactions in this mode was considered tolerable given the size of the scenario. Certain RTI queries that could trigger large spikes of network traffic were disallowed. Relaxed dead reckoning update thresholds and lowered heartbeat frequencies reduced the volume of entity attribute updates. The basic question of time management, i.e. whether the federation would execute in logical time or real time, was resolved in favor of real time. Real time was chosen to avoid the overhead of the RTI service calls required to coordinate the advance of logical time, to simplify interoperation with live systems that operate in real time, and to avoid federation-wide pauses that could be caused by a single federate failing to acknowledge time advance requests. Finally, federation-specific customizations to HLA's interest management and data filtering mechanisms (Declaration Management and Data Distribution Management) were developed to further reduce network traffic. The latter involved designating different HLA Data Distribution Management regions for different classes of objects (e.g. Blue air entities) or interactions (e.g. detonations) so that JEF federates could subscribe to just those objects and interactions that interested them. Specific multicast addresses were assigned to each of the regions, with the result that a substantial portion of the data filtering could be done in the federates' host computers' network hardware. In addition to these federation-wide agreements on how the Data Distribution Management regions would be defined and used, there were JEF-specific modifications and optimizations to the RTI software used for the exercise.

Regarding the third type of integration tasks (reconciling modeling differences), even once the JEF federates were communicating reliably via HLA, there were still discrepancies in the modeling methods and assumptions, which required resolution. While these differences were not, strictly speaking, specific to HLA, they were still interesting and several examples will be briefly mentioned. Discrepancies between federates on the continued existence in the virtual world of entities that had been destroyed caused problems; entities controlled by one federate that assumed destroyed entities were removed from the battlefield continued to expend time and ammunition attacking already destroyed entities that were retained in the battlefield after their destruction by another federate. Because of the large and complex scenario, the number of different entity type and munitions type pairs was quite large, and not all those combinations could be tested for validity in advance. Finally, the many federates involved used differing terrain formats, fidelities, and levels of detail, introducing into the JEF a class of issues and problems collectively referred to as terrain correlation (Schiavone et al., 1995). Effort was expended to produce correlated terrain by starting from a single source terrain data set.

---

[10]In a computer network connectionless transmission means that each message carries its own destination information in a header, eliminating the need for a pre-established connection or channel. Connectionless mode reduces network overhead per message but admits the risk of message loss during transmission.

Persistent Federation: MATREX

The US Army's Modeling Architecture for Technology, Research, and Experimentation (MATREX) is a modeling and simulation system that consists of a large set of combat related models and simulation tools interoperable via an established architecture (RDECOM, 2008). From an HLA perspective, MATREX is a persistent federation, i.e. an HLA federation that after initial development is reused repeatedly, perhaps with modifications and enhancements, over a long period of time. Like many persistent federations, MATREX is not specific to any particular application; rather, it is intended to be adaptable to a range of applications as is or through extension.

The MATREX federation includes approximately 30 federates. An important MATREX federate is OneSAF (described earlier), which provides broad capabilities to simulate a range of combat entity types and interactions. Also included are a set of more specialized federates that model specific types of entities, interactions, and phenomena in more detail to provide for higher fidelity or higher resolution simulation as needed. Among them are the Aviation Mobility Server, the Countermine Server, the Missile Server, the Weather Server, the Vehicle Dynamics Mobility Server, the Chemical Biological Simulation Suite (O'Connor et al., 2007), the Comprehensive Munitions and Sensor Simulation, the Logistics Server, and the Vehicle Level Human Performance Model; while the details of these individual models must be omitted, their names alone are sufficient to convey the range of specialized capabilities present in the MATREX federation. There is a set of five MATREX federates, collectively known as the C3 Grid, that provide services related to modeling command, control, and communications actions and effects. In addition to federates that provide modeling capabilities, MATREX includes a set of non-model tools that support testing and debugging, execution control, and results logging and analysis.

The MATREX FOM was initially based on the RPR FOM (described earlier), but has been extended and enhanced with objects and interactions of interest to the MATREX user community. The MATREX architecture includes an interface module common across the various federates, known as the ProtoCore, through which MATREX federates communicate at runtime. The ProtoCore includes an application programming interface and software produced by code generation that abstracts the details of the interoperability protocol and the object model (Metevier et al., 2010). The intent of this structure is to insulate the individual federates from changes to the interoperability protocol and its support software, e.g. different versions of the RTI, as well as differences between object models. The MATREX version of the HLA RTI has a variety of parameters that allow users to configure the HLA services, such as Data Distribution Management, as best suits their application (Lewis and Vagiakos, 2009); proper configuration of the MATREX RTI for Data Distribution Management requires some care (Lewis et al., 2010).

Additional examples

The three previous examples (OneSAF, the Joint Experimental Federation, and MATREX) are examples of the primary categories of HLA-based combat models (federate, special purpose federation, and persistent federation, respectively).[11] A brief list of selected additional examples of combat related HLA federates and federations follows. This list is quite far from exhaustive, and the examples were chosen to illustrate the range of possibilities for combat related applications of HLA.

1. *Platform Proto-Federation (PPF)*. One of the initial prototype federations used to test the emerging HLA standard in 1996, the PPF modeled conventional modern combat at the entity level, executed in real time, and included a virtual M1 tank simulator with human crew (Harkrider and Petty, 1996a, 1996b).

2. *Combat Trauma Patient Simulation (CTPS)*. This federation modeled the end-to-end process of medical treatment for combat casualties, from wound infliction to completion of medical care. CTPS included an instrumented mannequin-type human patient simulator to allow hands-on wound treatment training for medical caregivers (Petty and Windyga, 1999).

3. *MV-22 Operational Evaluation*. HLA was used to connect a high fidelity virtual flight simulator of an MV-22 aircraft, a constructive model of the scenario terrain and hostile entities, such as surface-to-air missile sites, an actual MV-22 aircraft located in a shielded hangar, and emitters in the same hangar stimulating the sensors of the aircraft (Huntt et al., 2000). The federation was used to perform operational evaluation of the MV-22.

4. *EnviroFed*. This federation focused on the representation of the natural environment in combat simulations. EnviroFed federates either produced or made use of high fidelity environmental features or capabilities, such as dynamic terrain and weather (Lutz and Richbourg, 2001).

5. *Probability of Raid Annihilation* (P_{RA}). This analysis federation modeled a single ship with its combat systems attempting to defend itself against varying threat raids, e.g. a set of incoming missiles. The PRA federation calculated the probability of the raid's annihilation using a combination of live testing and high fidelity modeling (Grigsby and Blake, 2001).

6. *NATO/Partnership for Peace Interoperability and Reuse Study (NIREUS)*. This federation, which consisted of engineering-level federates, modeled vertical take-off and landing air vehicles, initially unmanned air vehicles, landing on ships (Reading et al., 2002).

---

[11]The categories referred to are architectural types, not modeling paradigms.

7. *Army Constructive Training Federation—Multi-Resolution Modeling (ACTF-MRM)*. One of several multi-resolution combat models linked via HLA, the ACTF-MRM federation linked the unit level Corps Battle Simulation with the entity level Joint Conflict and Tactical Simulation and was designed to drive command post training exercises for commanders and their staffs (Zabek et al., 2004).

COMBAT MODELING USING THE BASE OBJECT MODEL (BOM) STANDARD

The Base Object Model (BOM) was standardized by SISO in 2006. It provides a simulation standard that allows combat model developers and simulation engineers to create modular conceptual models and composable object models, which can be used as the basis for a simulation or simulation environment (SISO, 2006). The BOM concept is based on the assumption that components of models, simulations, and federations can either be decomposed or newly developed, and then reused as building blocks in the development of a new simulation or a federation.

The BOM is unique because it provides a means to capture and represent aspects of a conceptual model, which is used to provide a clear understanding of "what is to be represented, the assumptions limiting those representations, and other capabilities needed to satisfy users' requirements" (IEEE, 2003). The BOM is also unique because it allows those conceptual models to be mapped to one or more class definitions, which may be used by a software design or distributed simulation architecture such as HLA. Additionally, one of the areas where the BOM standard is showing applicability is in support of human behavior modeling (Turnitsa et al., 2010).

Using BOMs

The BOM standard is used for such things as defining system functionality, representing mission threads (such as combat effects), and defining useful class structures that may be pertinent to a variety of programming languages, or simulation architectures such as HLA or TENA.

An example of how a BOM can be used to reflect mission thread functionality is reflected in the sequence diagram in Figure 19.7. In this example, part of the behavior of a close air support mission is documented. Specifically what is represented is the execution of a directed engagement. This represents a common pattern used for combat modeling. Across the top of our sequence diagram several conceptual entities have been identified. These include an Air Operations Center, Air Mission Control Function, the Aircraft Interceptor, and Target. Each of these entities can be further characterized in the BOM, and then mapped to specific object model classes that can support the conceptual entity if they are known. Potential model mappings might include the following:

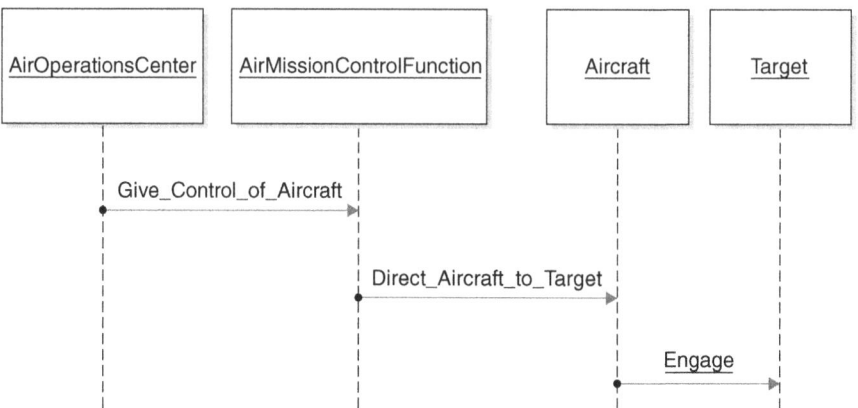

Figure 19.7 Close air support mission thread example.

1. Air Mission Control Function entity may map to an AWACS object model used for the simulation exercise,

2. Air Operations Center entity may map to a CAOC or ASOC object model,

3. Aircraft entity may map to an F18 or F22 object model, and

4. Target entity may map to a MIG object model.

Also reflected in the example sequence diagram are the pattern actions that take place among the conceptual entities. The actions identified include the following:

1. give control of aircraft

2. direct aircraft to target

3. engage

Within a BOM these actions can be characterized two ways:

1. either by a *conceptual event*, which defines a directed message or trigger that takes place within the simulation environment, or

2. by another *conceptual model pattern* defining more details of how that action is represented.

Like conceptual entities, conceptual events reflected within the BOM can map to specific object model classes (attribute updates) and/or interaction classes. The mappings from conceptual entities and events to supporting object model classes provide a useful mechanism when choosing the underlying simulations and systems that are needed to support an exercise.

If a pattern action is not documented by a conceptual event but by another conceptual model pattern, then it points to another BOM. This highlights the support that exists for creating layers and hierarchies of BOMs. For example,

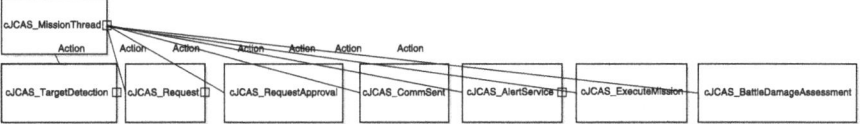

Figure 19.8   Dependence diagram for the close air support mission thread example.

Figure 19.8 depicts a dependency diagram that illustrates a conceptual model pattern hierarchy for the Joint Close Air Support (JCAS) mission thread.

The top block is defined by a conceptual model pattern of interplay representing the JCAS mission thread, but the actions identified in that pattern point to other conceptual model patterns defined in other BOMs. They include the following:

1. target detection
2. request support
3. request approval
4. communication sent
5. alert strike
6. execution mission (which is the direct strike combat model pattern identified in Figure 19.7)
7. battle damage assessment

The benefit of breaking these aspects of our mission thread into separate more detailed conceptual models is that it allows such conceptual models to be used for multiple purposes. For instance a pattern action of a mission thread identifying multiple conceptual models can be experimented with by swapping out one underlying conceptual model with another. Additionally new mission threads can be quickly constructed and experimented with early in the development process. The mapping to the object and interaction class models provides a context for knowing how to compose and build a simulation environment reflecting the desired mission thread functional.

A BOM mapping can be made from the conceptual model of a BOM to the class definitions of either other BOMs or HLA FOMs or even TENA Logical Range Object Model (LROM) components. The key in the mapping is the correlation with the class definitions found in the BOM, FOM, or TENA LROM. The difference is that the FOM characterizes HLA-specific implementation details, and the TENA LROM characterizes TENA implementation details; both are aspects that are not intended to be captured by a BOM.

Comparing BOMs with HLA FOMs

BOMs are often compared and confused with the capabilities offered by HLA's FOM. While there are some similarities the differences are stark. Unlike a BOM,

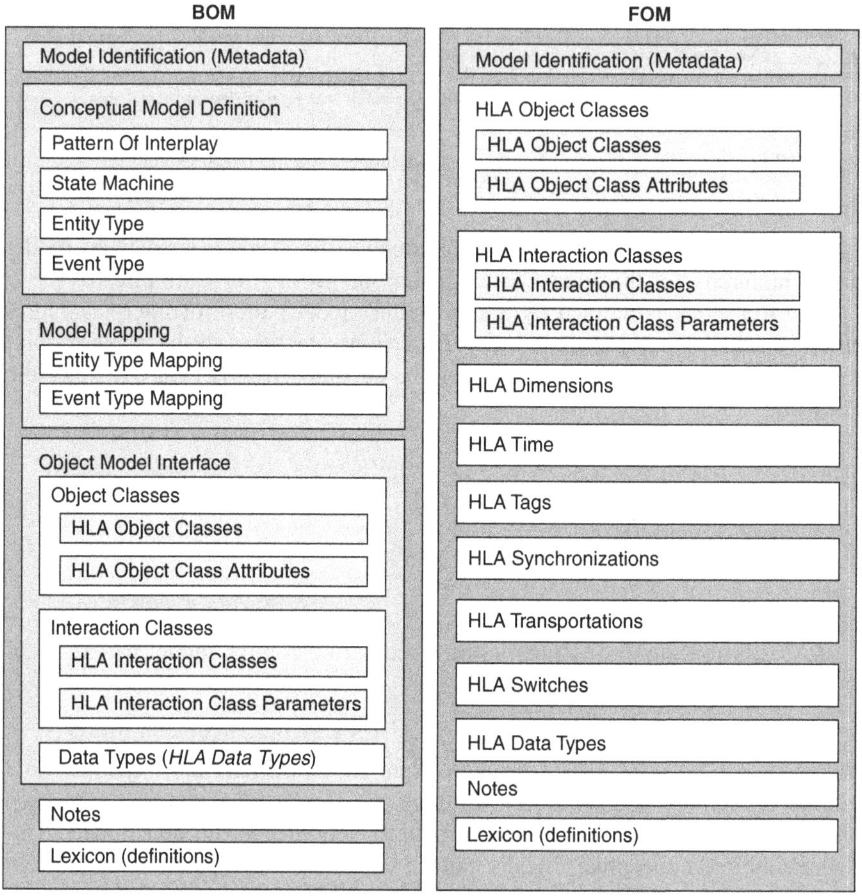

BOM

Model Identification (Metadata)	
Conceptual Model Definition	
Pattern Of Interplay	
State Machine	
Entity Type	
Event Type	
Model Mapping	
Entity Type Mapping	
Event Type Mapping	
Object Model Interface	
Object Classes	
HLA Object Classes	
HLA Object Class Attributes	
Interaction Classes	
HLA Interaction Classes	
HLA Interaction Class Parameters	
Data Types (*HLA Data Types*)	
Notes	
Lexicon (definitions)	

FOM

Model Identification (Metadata)
HLA Object Classes
HLA Object Classes
HLA Object Class Attributes
HLA Interaction Classes
HLA Interaction Classes
HLA Interaction Class Parameters
HLA Dimensions
HLA Time
HLA Tags
HLA Synchronizations
HLA Transportations
HLA Switches
HLA Data Types
Notes
Lexicon (definitions)

Figure 19.9 Comparison of BOM and FOM structures.

a FOM lacks the ability to document a conceptual model. To be fair to HLA, conceptual modeling is not within the scope of an HLA data exchange as intended in the Object Model Template (OMT).

However, because of the conceptual model support, BOMs can be used in the early phases of development. They provide a mechanism to capture and document conceptual models, which enables the intended capability of a combat system or model to be understood early on. It enables sponsors, architects and developers to properly focus on the analysis and design phase before moving to the development and integration phase, which is the focus of FOMs. A comparison of the BOM to the structure of an HLA FOM is illustrated in Figure 19.9.

The common shared capabilities between the FOM and BOM include the following:

1. *Model identification.* Used to provide a general description of the model also known as metadata

2. *Class definition.* Includes the object classes and attributes, and interaction classes and parameters to be represented

3. *Data types.* Used to characterize attributes and parameters

4. *Notes.* Used to provide comments

The BOM also provides the following architecture independent characteristics not supported by HLA:

1. *Conceptual model.* Provides a mechanism to document the conceptual entities, events, behavior patterns that take place among one or more entities (Fowler, 1996), and state machines that describe the behavior of an entity.

2. *Model mapping.* Allows mappings of conceptual models with the specific class definitions defined in the same or other models.

The FOM, on the other hand, provides the following unique implementation focused characteristics.

1. *DDM information.* Provides a description of data to be used for value based filtering.

2. *Transportation types.* Identifies how the specified attributes and interactions will be updated.

3. *Update rates.* Specifies the frequency of attribute updates.

4. *Synchronization points.* Used to mark when to perform certain actions such as scenario start.

5. *User defined tags.* Provide a means to provide application-specific data.

6. *Time representation.* Identifies how simulated, logical time will march for an exercise, including the time interval.

7. *Switches.* Controls the behavior of the RTI; the RTI provides the network interface for an HLA federate.

The predominate benefit of the BOM standard is that it has provided a viable mechanism for capturing the conceptual models that are to be represented in a simulation environment, and it can be used to support a variety of architecture standards such as HLA or TENA.

ADDITIONAL DEVELOPMENTS AND CAPABILITIES

There have been some additional intriguing capabilities that facilitate the advancement and application of combat modeling. These include the following: (1) HLA evolved FOM modules, (2) Runtime Infrastructure (RTI) capabilities, and (3) metadata discovery.

HLA Evolved FOM Modules

The BOM standard described earlier was the first M&S standard to focus on the aspect of defining and using modular object models to support composability. A similar but implementation focused capability has emerged with an update to the HLA standard known as HLA Evolved. HLA is used to support the interoperability of constructive and virtual simulations and also supports the integration of live systems by providing a peer-to-peer message-passing paradigm. The HLA has always focused on ensuring that the data model to be exchanged at runtime has been agreed upon by members of the federation prior to the exercise. This user-defined data exchange model is known as the Federation Object Model (FOM) (IEEE, 2010c).

While FOMs have been part of the HLA architecture since it was formally introduced in the 1990s, the HLA Evolved standard allows FOMs to be divided into smaller, reusable components called FOM modules (Möller et al., 2007b). This newer and attractive aspect of the architecture allows FOMs to be constructed not just prior to runtime but also during runtime using these FOM modules.

Comparing FOM Modules with BOMs

The questions that are invariably asked when discussing BOMs and Modular FOMs include the following.

1. How do BOMs and Modular FOMs relate to each other?
2. How are they different from one another?
3. How can they both be used to support one another?

These common questions are often asked because there are some striking similarities between the two standards. In fact, Modular FOMs and BOMs are based loosely on the same premise, which is to define a "piece-part" of a model that characterizes the structure of an entity or event that is to be modeled and/or exchanged within a simulation execution. Both are well suited for representing combat models, both use similar constructs to define an object model, and both use XML as the mechanism to document the model information. However, the similarities end there.

Modular FOMs and BOMs are intended for two different purposes. Modular FOMs are intended solely for supporting an HLA environment for representing what will be exchanged during an HLA federation execution. A FOM module is used solely for building up a FOM either before runtime or dynamically during the runtime of an HLA federation. BOMs, on the other hand, are not focused on the interoperability architecture, but on characterizing what is to be represented by one or more simulations or systems independently of the platform or architecture implementation. In this way, a BOM can be used for supporting HLA, or TENA, or any other distributed architecture interoperability standard or programming language.

Both BOMs and FOM modules used in conjunction provide an even more effective mechanism to support combat modeling. BOMs can be used in the early phases of development. They provide a mechanism to capture and document conceptual models (i.e. BOMs), which enables the intended capability of a combat system or model to be understood early on. It enables architects and developers to properly focus on the analysis and design phase before moving to the development and integration phase, which is the focus of FOMs. The BOM, in this capacity, serves as a mechanism for building either a full FOM or, better yet, individual FOM modules.

FOM modules have a key play in the middle and later phases of the development and execution process. Consider that FOM modules are unique to a specific architecture (HLA) and that the characteristics of HLA are not vital until the specific development and execution phase of the process. Furthermore, FOM modules can provide an aspect of reuse not evident until the execution and run-time of a federation. Consider that a FOM can be orchestrated at runtime, rather than just design time, simply by submitting new FOM module parts to federation participants during the execution.

Using FOM Modules

A FOM module can be used to add new object and interaction classes, data types, update rates, synchronization points. The way that a FOM module works is as follows. Suppose a new capability needs to be added to an exercise that is already in play. For example, a new air platform needs to be modeled and represented in the environment, but yet the current executing system FOM does not define the air platform. With a FOM module, however, the new classification of the air platform can be potentially added to the environment during runtime. There are a few caveats though. A general outline of how this process works is described below.

1. *Compare classes.* When the FOM module is submitted to be inserted into the system, a comparison is made of its classes against the current system FOM. (Go to Step 2.)

2. *Determine duplicates Part 1.* A check is made based on the results to determine what duplicates (equivalent-like classes) already exist in the current system FOM.

   (a) If no duplicate class is found based on like names and the FOM module can be attached to an ancestor class (i.e. Platform) then success. (Go to Step 3.)

   (b) If a duplicate of that FOM module is already part of the current system FOM, then further checks are needed to provide feedback. (Go to Step 4.)

3. *Merge FOM module.* The FOM module does not already exist and it can be added to the system FOM.

4. *Check internals of class structure.* Every sub-element of class from the FOM module is compared against matching classes in the system FOM. (Go to Step 5.)

5. *Determine duplicates Part 2.* A check is made to determine what duplicate class attributes and data types already exist in the "system."

   (a) If some but not all sub-elements are equivalent then share this feedback. (Go to Step 6.)

   (b) If all the sub-elements exist then share this feedback. (Go to Step 7.)

   (c) If no sub-elements are equivalent, then stop.

6. *Issue warning message—not equivalent.* A warning message is provided to the user that a class to be inserted already exists but is not equivalent and that the FOM module needs to be modified before it is integrated. Stop.

7. *Issue warning message—object model already exists.* A warning message is returned indicating that the FOM module already exists. It cannot be used. Stop.

The recommended way to reuse a FOM module is to use the BOM first to document the capabilities intended and then map that information to a supporting FOM module. FOM modules will be more likely to be reused if they are understood and the conceptual model aspect of the BOM provides such insight. For example, a BOM that documents the behavior of an aircraft engagement and reflects the conceptual entity representing the interceptor (the aircraft entity) can be mapped to specific object models that have been defined, which characterize the interceptor. These object models may be a FOM module used for a federate or federation used to fulfill the role of the aircraft. Furthermore it is easier to find FOM modules of common functionality.

Runtime Infrastructure Capability Extensions

The HLA RTI provides a unique set of services that drive the distribution of FOM based data for a simulation environment. The RTI's job is to deliver FOM data from a publishing simulation application to one or more subscribing simulation applications. The enhancements of the HLA standard have brought some additional RTI capabilities that also help facilitate combat modeling (Möller et al., 2008). They include the following:

1. Fault tolerance support
2. Smart update rates
3. Dynamic link compatibility
4. Web services

Fault Tolerance Support

Within an HLA federation execution, during runtime, a fault may occur that prevents the federation from continuing (Möller et al., 2005). When this occurs, one or more simulation applications that are connected to the exercise may be disconnected from the session. However, the underlying RTI component of the HLA infrastructure is designed to post a special message identified as the "HLA Report Federate Lost" interaction. If the dropped simulation application was properly time synched by a functioning RTI with a valid known time stamp, then the RTI can perform a proper resignation of the simulation application. This is accomplished under what is called an Automatic Resign Directive. When this is issued, then the participants in the federation can be restarted with the other active systems. This functionality is illustrated in Figure 19.10.

It is important to note again that this capability is provided by the latest version of the HLA specification. The fault detection capability depends on the underlying RTI. If the RTI conforms to an up-to-date HLA interface implementation, then fault tolerance should be supported.

Smart Update Rates

The frequency that a federate requires an update and reception of information is not necessarily the same across a given federation. Rates may vary. The various federates that are participating in an HLA federation may not all require information at the same update rate. For example, within a distributed simulation exercise, a Navy Surface Ship combatant may need to monitor air track information at a higher rate than ground based sensors used to support Marine or Army ground combatants. The smart update rate reduction capability provided by HLA Evolved makes it possible for multiple federates to subscribe to the same information, but with varying update rates (Möller and Karlsson, 2005b). This allows the exercise to be optimized for the needs of the simulation applications and helps reduce overall network traffic.

Once all subscriptions are in regarding an information component defined by a FOM, the requested update rates for the information can be analyzed by the RTI. The fastest update rate identified dictates the update rate required by the publishing simulation application. Knowing the highest update rate allows the simulation to perform only at that level, even if it was capable of performing updates faster. Knowing this value is a key factor for the RTI in helping reduce overall network traffic.

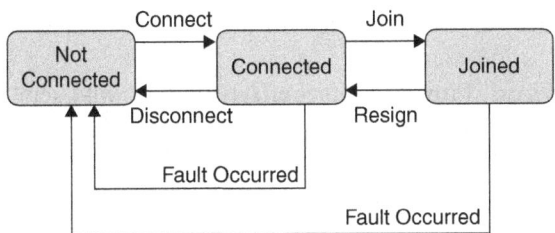

Figure 19.10 Federate restart functionality state diagram. Note: change occured to occurred in the figure, occurs twice.

Dynamic Link Compatibility

Currently, in order to properly exchange information and interoperate with other federates over a network all federates in a federation need to use the same vendor RTI implementation. The RTI, although compliant with the interface calls defined by the IEEE specification, communicates with the other RTI nodes on the network using its own implementation framework. Thus, mixing RTI vendor components introduces potential risk to federation developments.

Even though the interface from simulation application to RTI is common, making a simulation application work with different RTI implementations is a key problem. Preparing to participate in an exercise requires knowing what vendor RTI implementation will be used. Refactoring a simulation application to work with the selected RTI implementation for the federation is a concern for many. If the simulation application was built initially to work with RTI from Vendor A, it may not easily connect on its own local host with the RTI from Vendor B. Recompilation and relinking is typically required.

However, advancements within the SISO community have produced an API identified as the Dynamic Link Compatibility (DLC) API, and a follow-up identified as the Evolved DLC (EDLC) API, that help simplify linking a simulation application to new RTI implementations (SISO, 2004). Both the DLC and EDLC APIs make it possible to dynamically switch a simulation application from one RTI to another RTI without any recompilation support. This is considered a key feature for those who need to support and switch to different RTI implementations as dictated by each federation.

Web Services

Web services are a common mechanism used in the enterprise marketplace to share and exchange information among distributed system applications. A remote method call from an application is made to an outside server or another remote application with a request for information. The connection to a web service is loosely coupled. It is provided via a web service interface similar to an Application Programmer's Interface (API). This API is often documented using a Web Service Description Language (WSDL) document. The WSDL reveals a set of remote services (i.e. functions) that can be invoked over a LAN or WAN, either securely or insecurely.

Web services offer a unique adjunct capability useful for supporting a simulation exercise and combat modeling. Some examples include the following:

1. *Data terrain web services.* Provide terrain information to participating systems when a simulated (or live represented) entity or entities enter a new battlespace region.

2. *Environmental web services.* Provide weather effects such as wind, temperature, and conditions (including historic recorded information).

3. *Aggregation web services.* Provide a mechanism for entity level simulations to request aggregate solutions that can help reduce network load (Sisson et al., 2006).

4. *Repository based web services.* Used to find and share assets of interest during the design and development stage and to share feedback after an exercise.

This list provides examples of some of the web services that are available or emerging that can be used to support the design time or runtime aspects of an exercise. An additional web service that can be added to this list for supporting M&S is that of the RTI. An RTI WSDL API is available to support the implementation of an RTI enabled web service. This means that the functionality typically provided by an RTI that is tightly coupled to a simulation application is now accessible using a loosely coupled mechanism through web service calls.

The RTI WSDL reflects the equivalent functionality to the C++ and Java APIs that are provided with the HLA Evolved standard (Möller et al., 2007a). A simulation application can use the RTI web services to connect to a vendor's RTI across local and wide area networks, as opposed to linking on the host box. Performance will not be as high as a locally executing RTI supported using C++ or Java, but it does offer unique opportunities to enable other devices to operate as simulation application. For example, it is quite possible to use the web service to turn a smartphone or tablet device into a simulation application executing in a distributed simulation environment. Furthermore, it is possible to use https-based encryption and authentication to secure and harden such mobile applications. Considering that a number of applications within the smartphone market place utilize web services to access and display information to the mobile user, leveraging these devices in the future to support combat modeling is quite possible.

Metadata Discovery

Both modular object models and RTI capabilities discussed previously only provide value if BOMs, FOM modules and the RTI web services can be discovered and reused. This is also true of the simulations, tools, and data models used to support a simulation exercise. This aspect of discovery is an important facet that should not be ignored. It is a need highlighted by the DoD Net-Centric Data Strategy.

DoD Net-Centric Data Strategy

The DoD Net-Centric Data Strategy (dated May 9, 2003) defines goals and approaches for users and systems to discover and access a wide range of data assets throughout the DoD Enterprise. The desire and need is to support the aid and discovery of M&S assets for these communities and services as directed by the DoD Net-Centric Data Strategy. Discovery in this context is defined as "the ability to locate data assets through a consistent and flexible search" (DoD, 2007).

M&S COI Discovery Metadata Specification (MSC-DMS)

The MSC-DMS is the intended standard for the DoD M&S community to catalog M&S resources as metacards (Gustavson et al., 2010). Metacards provide a consistent labeling convention of resources including simulations, federations, adjunct tools, data models, interface models, and related documents. The MSC-DMS, which is available through the M&SCO website and the DoD Metadata Registry, is based on the DoD Discovery Metadata Specification (DoD, 2007) but tailored specifically for the M&S community.

The opportunity exists to ensure that the FOM modules and BOMs that are developed, as well as the simulations, web based RTI implementations, and tools used to support an exercise are cataloged using a common consistent mechanism such as the MSC-DMS labeling convention.

M&S Catalog

The DoD M&S Catalog is one of the first DoD repository projects to apply a search engine against resources that are commonly tagged with the same type of metadata. Figure 19.11 provides an illustration of how the M&S Catalog works. The Catalog provides entrée into several M&S based repositories. Users of the M&S Catalog and MSC-DMS include the Army, Navy, Air Force, and DoD MSRRs, and the joint analysis community among others. A user interested in a resource can learn of availability from multiple repositories with a single query.

The underlying mechanism to catalog the resources for the M&S Catalog is the MSC-DMS. Sources representing the services and communities provide their descriptions of resources to the M&S Catalog using the MSC-DMS. The Catalog provides a user interface capability for a consumer to search and look for resources of interest based on what has been captured in the set of MSC-DMS based resource metacards. Additionally, the M&S Catalog is intended to provide a set of web services that allow tools and other repositories to submit MSC-DMS based resource metacards.

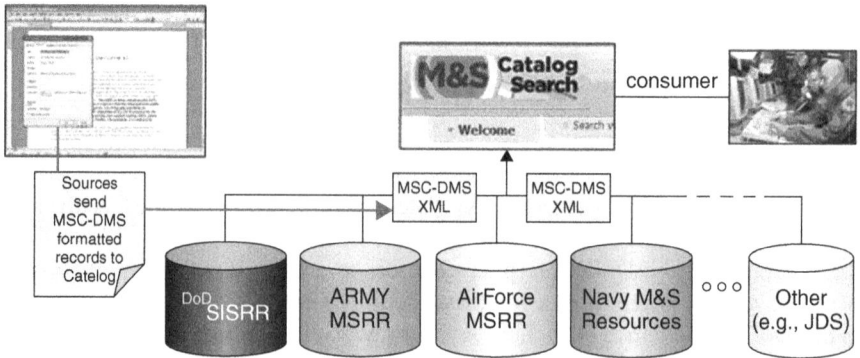

Figure 19.11 M&S Catalog structure.

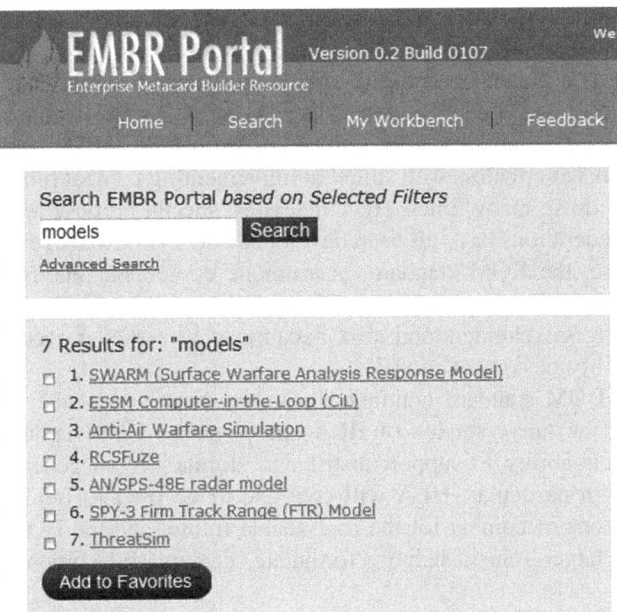

Figure 19.12 EMBR portal home page.

Enterprise Metacard Builder Resource

Another emerging web-based system is the EMBR Portal. The EMBR Portal is an online service that allows users to search and build metacards that describe M&S resource assets. The EMBR portal includes an online tool that enables users to author and modify metacards with requiring XML expertise. It allows collaboration with stakeholders and team members and the ability to store and submit metacards to the M&S Catalog and other resources (Riggs et al., 2011). Consumers can search, evaluate, and provide feedback on M&S assets. The home page for the EMBR Portal is reflected in Figure 19.12.

Future functionality of the Portal will facilitate the transfer of metadata directly to M&S repositories and the DoD M&S Catalog. Both the M&S Catalog and EMBR Portal support the DoD Net-Centric Data Strategy, and help facilitate discovery for supporting combat modeling.

SUMMARY

Almost every modeling paradigm and architecture available has been used by the DoD to model combat. Prominent among them is distributed simulation: no fewer than five distinct simulation architecture standards have been developed and used to help support combat modeling. These include SIMNET, ALSP, DIS,

TENA, HLA, interoperability protocol standards, and the BOM, which is a model component standard.

One of these, HLA, is a general purpose distributed simulation architecture suitable for a wide range of M&S applications; it does not contain any features or characteristics specific to combat modeling, differing from DIS, ALSP, and TENA in this respect. It is nevertheless well suited to implementing combat models and has been used to do so many times. HLA federates, special purpose federations, and persistent federations have all been developed and used extensively.

Another one of these, the BOM standard, is a unique component interoperability simulation standard that encourages the sharing and communication of conceptual designs among stakeholders and also, because of its modular focus, facilitates the composability of combat models.

Both HLA and the BOM standard continue to evolve. New technical features and capabilities in the latest version of HLA and proposed BOM update significantly enhance their ability to support distributed simulations in general and combat simulations in particular. HLA will continue to be used to implement distributed simulations of combat for the foreseeable future, and the BOM is anticipated to play a larger role in helping formulate, design, and compose simulation environments.

REFERENCES

Arquilla J (2010). The new rules of war. *Foreign Policy* **178**, 60–67.

Banks CM (2010). Introduction to modeling and simulation. In: *Modeling and Simulation Fundamentals: Theoretical Underpinnings and Practical Domains*, edited by JA Sokolowski and CM Banks. John Wiley & Sons, Hoboken, NJ, pp. 1–24.

Borger J (2002). Wake-up call. *The Guardian*, September 6, section G2, p. 2.

Castro PE et al. (2002). *Modeling and Simulation in Manufacturing and Defense Acquisition: Pathways to Success*. National Research Council, National Academy Press, Washington DC.

Ceranowicz A, Torpey M, Helfinstine B and Evans J (2002). Reflections on building the joint experimental federation. In: *Proceedings of the 2002 Interservice/Industry Training, Simulation, and Education Conference, Orlando FL, December 2–5*, pp. 1349–1359.

Courtemanche AJ and Wittman RL (2002). OneSAF: a product line approach for a next-generation CGF. In: *Proceedings of the Eleventh Conference on Computer Generated Forces and Behavioral Representation, Orlando FL, May 7–9 2002*, pp. 349–361.

Crosby LN (1995). SIMNET: an insider's perspective. In: *Distributed Interactive Simulation Systems for Simulation and Training in the Aerospace Environment*, edited by TL Clarke. SPIE Optical Engineering Press, Bellingham WA, pp. 59–72.

Dahmann JS and Lutz R (1998). Persistent federations. In: *Proceedings of the Spring 1998 Simulation Interoperability Workshop, Orlando FL, March 9–13*. http://www.sisostds.org

Dahmann JS, Kuhl F and Weatherly R (1998a). Standards for simulation: as simple as possible but not simpler: the high level architecture for simulation. *Simulation* **71**, 378–387.

Dahmann JS, Fujimoto RM and Weatherly RM (1998b). The DoD high level architecture: an update. In: *Proceedings of the 1998 Winter Simulation Conference, December 13–16, Washington DC*. IEEE Press, Piscataway, NJ, pp. 797–804.

DMSO (2004). *Transition of the DoD High Level Architecture to IEEE Standard 1516*. Technical Report, October 21.

DoD (2007). *Department of Defense Discovery Metadata Specification (DDMS)*. Version 1.4.1, August 10.

Drake DL, Lutz RR, Lessmann K, Cutts DE and O'Connor MJ (2010). LVC common gateways and bridges. In: *Proceedings of the 2010 Interservice/Industry Training, Simulation, and Education Conference, Orlando FL, November 29–December 2 2010*, pp. 2865–2875.

Fisher MC (1994). Aggregate Level Simulation Protocol (ALSP) managing confederation development. In: *Proceedings of the 1994 Winter Simulation Conference, Orlando FL, December 11–14*. IEEE Press, Piscataway, NJ, pp. 775–780.

Fowler M (1996). *Analysis Patterns: Reusable Object Models*. Addison-Wesley, Reading, MA.

Goldiez BF (1995). History of networked simulations. In: *Distributed Interactive Simulation Systems for Simulation and Training in the Aerospace Environment*, edited by TL Clarke. SPIE Optical Engineering Press, Bellingham, WA, pp. 39–58.

Grigsby S and Blake DW (2001). Assessing uncertainties in the probability of raid annihilation. In: *Proceedings of the Fall 2001 Simulation Interoperability Workshop, Orlando FL, September 9–14*. http://www.sisostds.org.

Gustavson P, Dumanoir P, Blais C and Daehler-Wilking R (2010). Discovery and reuse of modeling and simulation assets. In: *Proceedings of the Spring 2010 Simulation Interoperability Workshop, Orlando FL, April 12–16*. http://www.sisostds.org.

Harkrider SM and Petty MD (1996a). Results of the HLA platform proto-federation experiment. In: *Proceedings of the 15th DIS Workshop on Standards for the Interoperability of Defense Simulations*, Orlando FL, September 16–20, pp. 441–450.

Harkrider SM and Petty MD (1996b). High level architecture and the platform proto-federation. In: *Proceedings of the 18th Interservice/Industry Training Systems and Education Conference, Orlando FL, December 3–6*.

Hasan O (2010). OneSAF interoperability overview. Presentation, 2010 OneSAF Users Conference, Orlando FL, April 19–23, http://www.onesaf.net (last accessed 7 May 2011).

Henderson C and Rodriquez A (2002). Modeling in OneSAF. In: *Proceedings of the Eleventh Conference on Computer Generated Forces and Behavioral Representation, Orlando FL, May 7–9*, pp. 337–347.

Henninger AE, Cutts D, Loper ML, Lutz R, Richbourg R, Saunders R and Swenson S (2008). *Live Virtual Constructive Architecture Roadmap (LVCAR) Final Report*, Institute for Defense Analyses. http://www.msco.mil/MSCO%20Online%20Library.html (last accessed 20 September 2010).

Hofer RC and Loper ML (1995). DIS today. *Proc IEEE* **83**, 1124–1137.

Huntt L, Markowich A and Michelletti L (2000). Modeling and simulation augments V-22 operational testing. In: *Proceedings of the 2000 Interservice/Industry Training, Simulation and Education Conference, November 27–30, Orlando FL*, pp. 945–953.

IEEE (1995). *IEEE Standard for Distributed Interactive Simulation—Application Protocols*. IEEE Std 1278.1-1995, September 21.

IEEE (2010a). *IEEE Standard for Modeling and Simulation (M&S) High Level Architecture (HLA)—Framework and Rules*. IEEE Std 1516™-2010, August 18.

IEEE (2010b). *IEEE Standard for Modeling and Simulation (M&S) High Level Architecture (HLA)—Federate Interface Specification*. IEEE Std 1516.1™-2010, August 18.

IEEE (2010c). *IEEE Standard for Modeling and Simulation (M&S) High Level Architecture (HLA)—Object Model Template (OMT) Specification*. IEEE Std 1516.2™-2010, August 18.

IEEE (2003). *IEEE Recommended Practice for High Level Architecture (HLA) Federation Development and Execution Process*. IEEE Std 1516.3™-2003, August 23.

Kuhl F, Weatherly R and Dahmann J (1999). *Creating Computer Simulation Systems*. Prentice Hall, Englewood Cliffs, NJ.

Lewis J and Vagiakos D (2009). Combating network load in high entity count federations. In: *Proceedings of the 2009 Interservice/Industry Training, Simulation, and Education Conference, Orlando FL, November 30–December 3*, pp. 1351–1358.

Lewis J, Kemmler KE and Do K (2008). The hitchhiker's guide to developing OneSAF HLA interfaces. In: *Proceedings of the 2008 Interservice/Industry Training, Simulation, and Education Conference, Orlando FL, December 1–4*, pp. 1118–1124.

Lewis J, Do K and Vagiakos D (2010). DDM explained: lessons for data distribution management developers and strategists. In: *Proceedings of the 2010 Interservice/Industry Training, Simulation, and Education Conference, Orlando FL, November 29–December 2*, pp. 2911–2920.

Lin K (1995). Dead reckoning and distributed interactive simulation. In: *Distributed Interactive Simulation Systems for Simulation and Training in the Aerospace Environment*, edited by TL Clarke. SPIE Optical Engineering Press, Bellingham, WA, pp. 16–36.

Loper ML (1995). Introduction to distributed interactive simulation. In: *Distributed Interactive Simulation Systems for Simulation and Training in the Aerospace Environment*, edited by TL Clarke. SPIE Optical Engineering Press, Bellingham, WA, pp. 3–15.

Lutz R and Richbourg R (2001). The enviroment federation: simulated environments impact tactical operations. In: *Proceedings of the Spring 2001 Simulation Interoperability Workshop, Orlando FL, March 25–30*. http://www.sisostds.org.

Metevier C, Gaughan G, Gallant S, Truong K and Smith G (2010). A path forward to protocol independent distributed M&S. In: *Proceedings of the 2010 Interservice/Industry Training, Simulation, and Education Conference, Orlando FL, November 29–December 2*, pp. 2764–2775.

Möller B and Karlsson M (2005). Developing well-balanced federations using the HLA evolved smart update rate reduction. In: *Proceedings of Fall 2005 Simulation Interoperability Workshop, Orlando FL, September 18–23*. http://www.sisostds.org.

Möller B, Löfstrand B and Karlsson M (2005). Developing fault tolerant federations using HLA evolved. In: *Proceedings of the 2005 European Simulation Interoperability Workshop, Toulouse France, June 27–29*. http://www.sisostds.org.

Möller B, Dahlin C and Karlsson M (2007a). Developing web centric federates and federations using the HLA evolved web services API. In: *Proceedings of the Spring 2007 Simulation Interoperability Workshop, Norfolk VA, March 25–30*. http://www.sisostds.org.

Möller B, Gustavson P, Lutz B and Löfstrand B (2007b). Making your BOMs and FOM modules play together. In: *Proceedings of the Fall 2007 Simulation Interoperability Workshop, Orlando FL, September 16–21*. http://www.sisostds.org.

Möller B, Morse KL, Lightner M, Little R and Lutz B (2008). HLA evolved—a summary of major technical improvements. In: *Proceedings of the Fall 2008 Simulation Interoperability Workshop, Orlando FL, September 15–19*. http://www.sisostds.org.

Morse KL and Petty MD (2004). High level architecture data distribution management migration from DoD 1.3 to IEEE 1516. *Concurrency Comp: Pract Experience* 16, 1527–1543.

O'Connor MJ, Fann JD and Jones DL (2007). CB Defense Modeling & Simulation (M&S) suite. In: *Proceedings of the 2007 Huntsville Simulation Conference, Huntsville AL, October 30–November 1*.

Parsons D, Surdu J and Jordan B (2005). OneSAF: A next generation simulation modeling the contemporary operating environment. In: *Proceedings of the 2005 European Simulation Interoperability Workshop, Toulouse France, June 27–30*. http://www.sisostds.org.

Parsons D (2007). One semi-automated forces (OncSAF). In: *Proceedings of the 2007 Department of Defense Modeling and Simulation Conference, Hampton VA, May 7–11*. http://www.onesaf.net (last accessed 4 January 2011).

Petty MD (2002). Comparing high level architecture data distribution management specifications 1.3 and 1516. *Simulat Pract Theory* **9**, 95–119.

Petty MD (2009). Behavior generation in semi-automated forces. In: *The PSI Handbook of Virtual Environment Training and Education: Developments for the Military and Beyond; Volume 2:*

VE Components and Training Technologies, edited by D Nicholson, D Schmorrow and J Cohn. Praeger Security International, Westport, CT, pp. 189–204.

Petty MD (2010). Verification, validation, and accreditation. In: *Modeling and Simulation Fundamentals: Theoretical Underpinnings and Practical Domains*, edited by JA Sokolowski and CM Banks. John Wiley & Sons, Hoboken, NJ, pp. 325–372.

Petty MD and Morse KL (2004). The computational complexity of the high level architecture data distribution management matching and connecting processes. *Simulat Modelling Pract Theory* **12**, 217–237.

Petty MD and Windyga PS (1999). A high level architecture-based medical simulation. *Simulation* **73**, 279–285.

Powell ET (2010). The benefits of TENA for distributed testing and training. In: *Proceedings of the 2010 Interservice/Industry Training, Simulation, and Education Conference, Orlando FL, November 29–December 2*, pp. 1157–1206.

Reading R, Örnfelt M and Duncan JM (2002). Results and lessons learned from a multinational HLA federation development supporting simulation based acquisition. In: *Proceedings of the Spring 2002 Simulation Interoperability Workshop, Orlando FL, March 10–15*. http://www.sisostds.org.

Riggs WC, Morse KL, Brunton R, Gustavson P, Rutherford H, Chase T and Belcher J (2011). Emerging solutions for LVC asset reuse. In: *Proceedings of the Spring 2011 Simulation Interoperability Workshop, Boston MA, April 4–8*. http://www.sisostds.org.

RDECOM (2008). MATREX simulation architecture. In: *Proceedings of the 2008 Department of Defense Modeling and Simulation Conference, Orlando FL, March 10–14*. http://www.matrex.rdecom.army.mil (last accessed 7 May 2011).

Schiavone GA, Nelson RS and Hardis KC (1995). Interoperability issues for terrain databases in distributed interactive simulation. In: *Distributed Interactive Simulation Systems for Simulation and Training in the Aerospace Environment, SPIE Critical Reviews of Optical Science and Technology*, edited by TL Clarke. SPIE Press, Bellingham, WA, Vol. CR58, pp. 281–298.

SISO (1999). *Real-Time Platform Reference Federation Object Model (RPR FOM 1.0)*. SISO-STD-001.1-1999, http://www.sisostds.org (last accessed 7 May 2011).

SISO (2004). *Dynamic Link Compatible HLA API Standard for the HLA Interface Specification (IEEE 1516.1 Version)*. SISO-STD-004.1-2004, http://www.sisostds.org (last accessed 7 May 2011).

SISO (2006). *Base Object Model (BOM) Template Specification*. SISO-STD-003-2006, http://www.sisostds.org (last accessed 7 May 2011).

Sisson B, Gustavson P and Crosson K (2006). Adding aggregate services to the mix: an SOA implementation use case. In: *Proceedings of the Spring 2006 Simulation Interoperability Workshop, Huntsville AL*. http://www.sisostds.org.

Tolk A (2010). Interoperability and composability. In: *Modeling and Simulation Fundamentals: Theoretical Underpinnings and Practical Domains*, edited by JA Sokolowski and CM Banks. John Wiley & Sons, Hoboken, NJ, pp. 403–433.

Turnista C, Gustavson P and Blais C (2010). Exploring multi-resolution human behavior modeling using base object models. In: *Proceedings of the Fall 2010 Simulation Interoperability Workshop, Orlando FL, September 20–24*. http://www.sisostds.org.

Wainer GA and Al-Zoubi K (2010). An introduction to distributed simulation. In: *Modeling and Simulation Fundamentals: Theoretical Underpinnings and Practical Domains*, edited by JA Sokolowski and CM Banks. John Wiley & Sons, Hoboken, NJ, pp. 373–402.

Wilson AL and Weatherly RM (1994). The aggregate level simulation protocol: an evolving system. In: *Proceedings of the 1994 Winter Simulation Conference, Orlando FL, December 11–14*. IEEE Press, Piscataway, NJ, pp. 781–787.

Wood DD and Petty MD (1999). HLA gateway 1999. In: *Proceedings of the Spring 1999 Simulation Interoperability Workshop, Orlando FL, March 14–19*. pp. 302–307. http://www.sisostds.org.

Zabek A, Henry H, Prochnow D and Wright M (2004). The army constructive training federation—multi-resolution modeling: the next generation of land component commander training at the unit of employment echelon of command. In: *Proceedings of the Fall 2004 Simulation Interoperability Workshop*, *Orlando FL, September 19–24*. http://www.sisostds.org.

Chapter 20

The Test and Training Enabling Architecture (TENA)

Edward T. Powell and J. Russell Noseworthy

INTRODUCTION

This chapter explores the Test and Training Enabling Architecture (TENA), designed to bring interoperability to America's test and training ranges and their customers. TENA is designed to promote integrated testing and simulation based acquisition through the use of a large-scale, distributed, real-time synthetic environment, which integrates testing, training, simulation, and high performance computing technologies, distributed across many facilities, using a common architecture. While highly capable, America's ranges have many "stovepipe" systems, built with different suites of sensors, networks, protocols, hardware, and software. This situation must be changed if the military is to make the most efficient use of its current and future range resources. Future testing and training requires the integration of systems from multiple ranges with hardware-in-the-loop facilities and advanced simulations. As weapon systems are becoming more sophisticated and complex, our ability to test and train with them is becoming more difficult. Using TENA, real military assets can interact with each other and with simulated weapons and forces, on what are called "logical ranges," no matter where these forces actually exist in the world.

Department of Defense (DoD) testing and training activities are increasingly composed of live, virtual, and constructive (LVC) distributed simulations

Engineering Principles of Combat Modeling and Distributed Simulation, First Edition.
Edited by Andreas Tolk.
© 2012 John Wiley & Sons, Inc. Published 2012 by John Wiley & Sons, Inc.

and applications. "LVC" refers to the combination of three types of distributed simulations and applications into a single distributed system:

live–real, physical assets, including soldiers, aircraft, tanks, ships, and weapon systems

virtual–simulators of physical assets that provide real-world operator interfaces and humans in the loop, such as aircraft simulators, tank simulators, etc.

constructive–pure simulations either controlled by human beings (called "semi-automated forces") or run entirely without human intervention (called "closed simulations")

The goal of combining these three types of distributed systems is to combine real, virtual, and constructive assets into one seamless and coherent environment operating in real time. These LVC systems present new challenges, ones not found in conventional distributed simulation systems. Software architectures originally designed to support integrating systems for distributed simulation are not necessarily well suited to support the live component of LVC systems. This is due to the fact that when real, live systems are mixed with virtual reality and/or constructive simulations, the demands of the live systems dominate the resulting LVC system. Support for the "live" component of LVC systems is TENA's reason for being.

TENA is sponsored by the DoD's Test Resource Management Center (TRMC) and Joint Forces Command's Joint National Training Capability (JNTC). All TENA specifications and software are owned by the US Government and are provided free of charge to any organization that wishes to take advantage of TENA.

In the following sections, we discuss the various aspects of TENA in more detail. We start by discussing the requirements for range systems in general and then distributed real-time LVC systems in particular. Then we discuss the core capability of TENA contained in the TENA Meta-Model, TENA object models, the TENA Middleware, and the TENA Repository. Finally, we discuss the tools and utilities necessary for the efficient creation of a TENA execution.

TENA'S DRIVING REQUIREMENTS

An architecture is a bridge from requirements to design, in which the most important, critical, or abstract requirements are used to determine a basic segmentation of the system. These "driving requirements" dictate the structure of TENA and its content. There are driving requirements in the technical realm (providing guidance on how TENA systems should be constructed) and in the operational realm (providing guidance on what features TENA logical ranges must have). TENA's driving requirements are derived from a detailed analysis of DoD supplied source documents, as well as feedback from the range community itself.

The most important technical driving requirements are interoperability, reuse, and composability. These three terms are all related, but each has a different emphasis. Interoperability is the characteristic of an independently developed software element that enables it to work together with other elements toward a common goal. There are many degrees of interoperability, varying on a continuum. For the purposes of the range community, the most important degree of interoperability is termed "semantic interoperability" (Tolk et al., 2008), interoperability built upon a foundation consisting of a common language and context for communication.

Reuse is the ability to use a software element in a context for which it was not originally intended, and so is focused on the multiple uses of a single element. Composability is the ability to rapidly assemble, initialize, test, and execute a system from members of a pool of reusable, interoperable elements. Composability can occur at any scale—reusable components can be combined to create an application, reusable applications can be combined to create a system, and reusable systems can be combined to create an enterprise.

To achieve interoperability, one must have a common architecture, an ability to meaningfully communicate (including a common language and a common communication mechanism), and a common understanding of the context (including the environment and time). In addition to these features, reuse requires well documented interfaces and composability requires a repository that can contain the composable elements as well as metadata about how they can be assembled.

TENA's operational driving requirements focus on supporting a distributed LVC exercise, called a "logical range," throughout the entire range event lifecycle. TENA must support testing and training in a network-centric warfare environment, rapid development of applications and logical ranges, integration with modeling and simulation (M&S) systems, integration with a wide variety of range systems, and the gradual deployment of TENA onto the ranges. The most important of TENA's operational requirements is to provide a solid technical foundation for integrating live systems into distributed LVC environments.

REQUIREMENTS OF DISTRIBUTED REAL-TIME LVC SYSTEMS

The US DoD ranges and laboratories use systems of sensors to take measurements for the purpose of testing and/or training. Many of these sensor systems are embedded systems. The testing and training events occur in the real world, meaning that real missiles are launched, real tanks are driven, real aircraft are flown, etc.; so real measurements must be taken in real time and with high performance (e.g. latencies measured in milliseconds). The sensor systems are themselves inherently distributed, typically over a large geographic area. The sensor systems can include half a dozen to several hundred individual component sensors. So, DoD ranges are large-scale, distributed, real-time, and embedded systems. For testing and training activities, these large-scale sensor systems using live data are augmented with simulator systems and simulations yielding the final LVC system. Such an LVC system can be used, for example, to enable a pilot in a real aircraft in Nevada to fly with a pilot in a simulator in Florida on a mission to engage synthetic tanks that are being simulated in Texas.

Interoperability Challenges

Individually, the parts of these large-scale LVC systems are quite complicated. When the individual parts are combined to form an LVC system as a whole, the result is a large enterprise made up of a variety of complex interrelated systems.

There are many aspects of such an enormous enterprise, any one of which, if overlooked, could impair or even cripple the functioning of the entire enterprise. One aspect of interoperability is providing common low level communication protocols, enabling communication among systems in the enterprise. A second aspect of interoperability concentrates on commonality, or data agreement, about the nature and form of data exchanged.

Over the years, industry and academia have produced a variety of protocols and middleware products addressing one or both of these first two aspects of interoperability, such as Distributed Interactive Simulation (DIS) (IEEE, 1995) and the High-Level Architecture Runtime Infrastructure (HLA RTI) (Kuhl et al., 1999). The US DoD testing and training ranges and laboratories have made use of both DIS and HLA RTI as part of their approach to creating LVC systems.

More recently, the test and training range community have embraced TENA (Powell et al., 2002; Noseworthy, 2005) to support their LVC activities. TENA has addressed the first aspect of interoperability by building upon the existing body of work on low level data communication protocols (Noseworthy, 1996).[1] TENA has focused on the second aspect of interoperability by founding the fundamental concepts of its architecture on formal computer enforced data format agreements rooted in the model-driven software principles of Model Driven Architecture (MDA) (MDA Group, 2003). Yet even satisfying these two aspects of interoperability (common protocols and data agreements) has proven to be insufficient to ensure interoperability throughout an enterprise as large and complex as an LVC system. Common protocols and data format agreements are only the first steps along the path to interoperability.

The Effect of "Live"

The nature of purely simulated environments differs from the nature of live environments in two critical ways. First, in a purely simulated environment the flow of time need not be constrained by reality. Second, in a purely simulated

---

[1]CORBA Specification, http://www.omg.org/cgi-bin/doc?formal/04-03-12, Object Management Group, March 2004, version 3.0.3; Event Service Specification, http://www.omg.org/docs/formal/01-03-01.pdf, Object Management Group, March 2001, version 1.1; Notification Service Specification, http://www.omg.org/docs/formal/02-08-04.pdf, Object Management Group, August 2002, version 1.0.1.

environment the cost of failure (e.g. due to a software defect) is generally much lower than the cost of failure in a live environment. When live systems are mixed with virtual reality and/or constructive simulations, the demands of the live systems dominate the resulting LVC system.

Time, in a purely simulated environment, can be slowed, accelerated, and paused; and the exercise can be restarted, rewound, and even replayed from the beginning. In an LVC system, the distributed applications supporting the live components do not have that same ability. Live components operate in real time. Thus, the distributed applications that support those live components must also operate in real time. In an LVC system, none of the time management or pause/restart/rewind features can be tolerated by the live components, and thus cannot be tolerated anywhere within the LVC system as a whole. The nature of an actual missile test simply does not allow for a virtual or constructive component in the LVC system to pause or "slow time" for any reason. The unforgiving nature of the real-time live components requires that the virtual and constructive components meet the demands of a real-time system.

Software defects can cause computer systems to fail or give erroneous results. The increased complexity of distributed computer systems tends to increase the likelihood that a software defect will cause at least some part of the system to malfunction. This principle holds for distributed simulation systems as well. In a purely simulated environment, such as those created with virtual reality and/or constructive simulations, the cost of failure may be little more than the inconvenience of restarting one or more of the simulations. Certainly, the time lost may be significant; however, pure simulations are inherently less expensive than a live event with real assets. Indeed, cost savings is one of the primary reasons to simulate real systems, instead of simply using the real systems themselves. While being forced to restart a missile interception simulation due to a software defect is undesirable, being forced to redo an actual missile interception test due to a software defect is unacceptable. This discrepancy in the cost of failure places demands on the software architectures used to support LVC simulations that are not placed on the software architectures used to support only virtual and/or constructive simulations. As a result, little or no attention to preventing software defects was paid during the design of those other architectures. In fact, merely detecting software defects in simulations using any of the other popular software architectures can be such a challenge that many defects are discovered very late. Worse still, some defects can go entirely unnoticed, producing erroneous simulations results.

Decentralized Development

By their very nature, LVC systems cannot be composed of a single application replicated throughout the entire LVC system. This fact is a result of the wide differences in the nature of the applications that support live systems, the applications that create virtual reality simulators, and the applications that create constructive simulations.

Multiple organizations, teams, and individuals contribute to various parts of the large-scale LVC system, often at widely different periods of time. The development of the software applications is performed in a distributed fashion—distributed both geographically and temporally. In addition, the development is nearly always decentralized, lacking a common authority to provide a uniform and consistent development process. Achieving interoperability between the myriad complex pieces of software used in LVC systems in the face of such a chaotic software development environment is extraordinarily difficult. Yet, the success or failure of the LVC system rests squarely with the ability of these software pieces to successfully interoperate.

Formal, computer enforced agreements describing the nature and form of data exchanged in a large-scale LVC system are necessary to provide a common understanding of how and what data are to be communicated, and furthermore to ensure that that understanding is then implemented in every application comprising the system.

Without the use of formal, computer enforced agreements detailing the nature and form of the data exchanged, it is extremely difficult to ensure that the applications in an LVC system are abiding by those agreements. TENA provides (and requires the use of) a mechanism to create these formal, computer enforced data exchange agreements. Other distributed simulation software architectures have no such mechanism.

TENA OVERVIEW

An overview of TENA is shown in Figure 20.1 (Powell et al., 2002, 2003). TENA recognizes five basic categories of software.

1. *TENA applications (Range Resource Applications and TENA Tools)*—Range Resource Applications are range instrumentation or processing systems (software applications) built to be compliant with TENA and are the heart of any TENA execution. TENA Tools are generally reusable TENA applications, stored in the repository and made available to the community, that help facilitate the management of a logical range through the entire range event lifecycle.

2. *Non-TENA applications*—range instrumentation/processing systems, systems under test, simulations, and C4ISR systems not built in accordance with TENA but needed for a test or training event.

3. *The TENA common infrastructure*—those software subsystems that provide the foundation for achieving TENA's goals and driving requirements. These include the TENA Repository, as a means for storing applications, object models, and other information between logical ranges; the TENA Middleware, for real-time information exchange; and the Event Data Management System, for storing scenario data, data collected during an event, and summary information. Currently, the TENA Middleware is the most

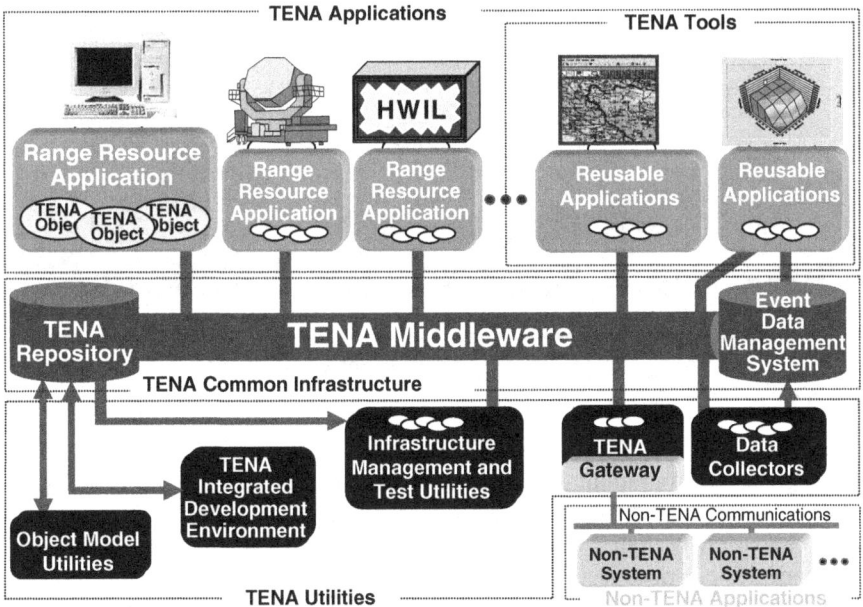

Figure 20.1 An overview of TENA.

mature of the three elements, being in its sixth released version. The TENA Repository has reached its initial operating capability but plans are in place for a substantial upgrade in the near future. The Event Data Management System has only been prototyped and has not yet been adopted formally as the TENA standard data management solution.

4. *The TENA object models* — the common language used for communication between all range resources and tools. The set of objects used in a logical range is called the "logical range object model (LROM)" and may contain TENA standard object definitions as well as non-standard (user-defined) object definitions.

5. *TENA utilities* — applications specifically designed to address issues related to usability or management of the TENA logical range.

The most important aspect of the TENA architecture, at the center of the diagram in Figure 20.1, is the TENA Middleware. The TENA Middleware uses Unified Modeling Language (UML) based model-driven automatic code generation to create a complex application "under the covers." This greatly reduces the amount of software that must be handwritten and tested. The TENA Middleware combines the programming abstractions of distributed shared memory, anonymous publish-subscribe, and model-driven distributed object-oriented programming into a single intuitive middleware system. Thus, the TENA Middleware weaves together a unique combination of model-driven, code generated software

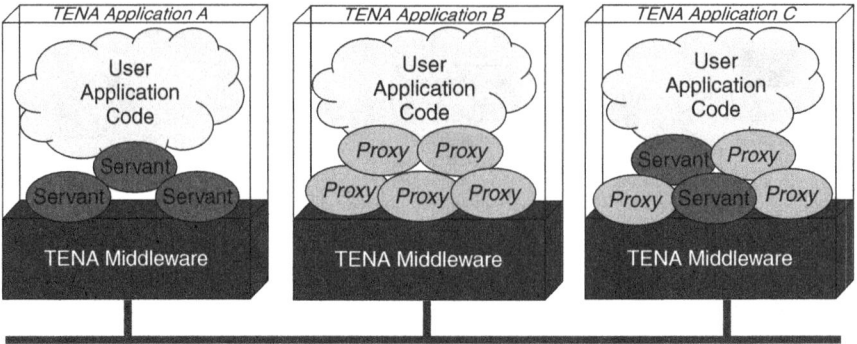

Figure 20.2 TENA applications, servants, and proxies, linked together.

with high level programming abstractions and an API designed to detect programming errors at compile time rather than runtime, where the cost of an error could be extremely high.

From the perspective of its users, TENA is a peer-to-peer architecture for logical ranges. Applications can be both clients (consumers of information) and servers (producers of information) simultaneously. In their role as servers, applications serve TENA objects called "servants." In their role as clients, applications obtain "proxies" as a result of subscription, each proxy representing a particular servant. Only servers can write to their servant objects' publication state. The TENA Middleware, the auto-code generated TENA object model code, and the user's application code are compiled and linked together (see Figure 20.2).

THE TENA META-MODEL

A meta-model is "a model that defines an abstract language for expressing other models".[2] All modern programming languages have a meta-model. The TENA Meta-Model defines several fundamental abstract concepts, constructs, and the relationships between them. The TENA Middleware supports these abstractions.

A TENA object model is a model constructed using specific TENA Meta-Model concepts and forming specific relationships between those concepts. A TENA object model can be represented in two ways. The first is using a UML class diagram. The second is using a text based representation called the TENA Definition Language (TDL).

Figure 20.3 shows a simple example of a TENA object model represented in both UML and TDL. TDL is a formal, readily understandable representation of an agreement describing the nature and form of data exchanged in a large-scale

---

[2]Common Warehouse Metamodel—Glossary, http://www.omg.org/docs/formal/01-10-25.pdf, Object Management Group, November 2001.

```
package Example
{
  enum Team { Team_Red, Team_Blue, Team_Green };

  local class Location
  {
    float64 xInMeters;
    float64 yInMeters;
  };

  class Vehicle
  {
    string name;
    Team team;
    Location location;
  };

  message Notification
  {
    string text;
  };
}; // package Example
```

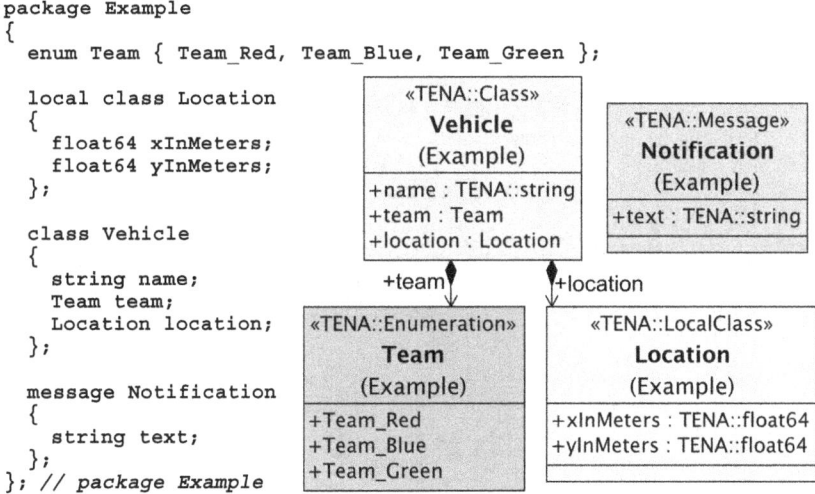

Figure 20.3   Example TENA object model shown in TDL and UML.

TENA LVC system. The TENA Object Model Compiler (OMC) compiles TDL, such as the TDL depicted in Figure 20.3, into C++, thereby automatically generating code that enforces the described agreements. All applications in the LVC logical range use this automatically generated code, even if the applications' development is distributed and decentralized.

It is important to keep in mind that a TENA object model is a collection of abstract ideas. Those abstract ideas have corresponding concrete C++ implementations in the TENA Middleware.

The TENA Meta-Model consists of a number of constructs, each of which can be used by other constructs. The entire TENA Meta-Model is illustrated in Figure 20.4. Each construct is identified as a UML class-like item, and each relationship is denoted on the diagram by a UML association line. For example, the line connecting the box labeled "Class" to the box labeled "Local Class" means that in TDL a "class" can contain a "local class." From this diagram one can understand every aspect of the TENA Meta-Model. Some important highlights are described in the following subsections.

Stateful Distributed Object (SDO)

A Stateful Distributed Object (SDO) (also called a TENA Class and denoted in Figure 20.4 as simply "Class") is an abstract concept formed by the combination of a Common Object Request Broker Architecture (CORBA) distributed object with data or state (Rumford et al., 2001; Noseworthy, 2002). The state consists of the data attributes of the SDO. The state is disseminated via anonymous publish-subscribe and cached locally at each subscriber.

CORBA has long provided the illusion that a distributed object's methods exist on a local object in the native programming language of the application.

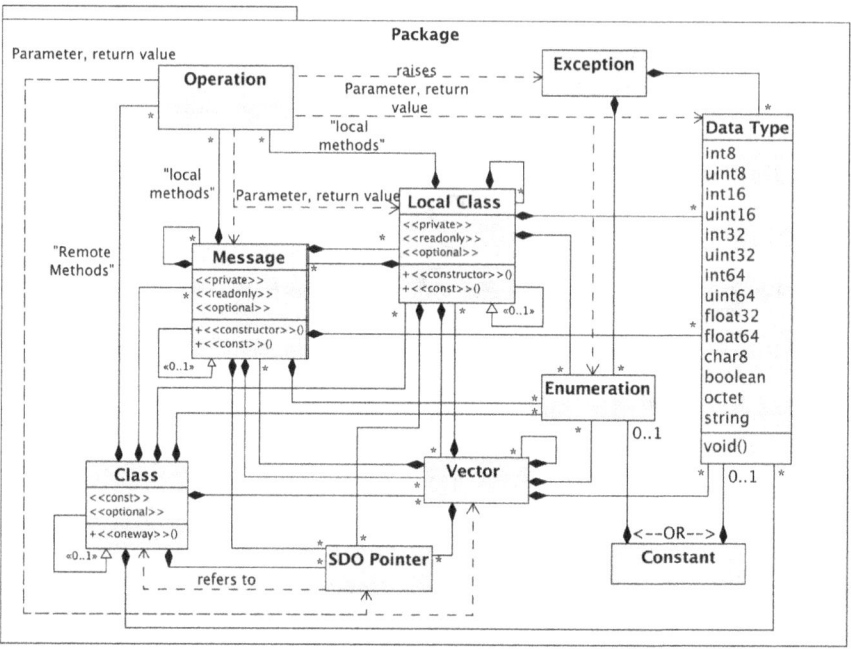

Figure 20.4   A UML representation of the TENA meta-model.

Unbeknownst to the application programmer, the distributed object's methods may in fact involve a remote method invocation to another application on the network.

The SDO extends this concept to include not only methods but also data attributes. The TENA Middleware provides the illusion that both an SDO's methods and state can be accessed locally, as if the SDO was a local object in the application.

An SDO's state is disseminated via anonymous publish-subscribe to applications based on subscription characteristics, such as the type of the SDO. Subscribers can read the state of an SDO as if it were a local object. Every time the remote application modifies an SDO's attribute, the SDO's state is disseminated to interested subscribing applications so that the local caches in the subscribing applications are updated.

The state making up an SDO can consist (see Figure 20.4) of fundamental data types, local classes (described below), enumerations, vectors, and SDO pointers; and these constructs can contain other constructs. In addition to state attributes, every SDO may have remote methods. As with attributes, any of the fundamental data types or complex constructs from the TENA Meta-Model may be used as method parameters or return types, with the notable exception of an SDO itself. This is because an SDO cannot be passed by value. SDOs can inherit (singly) from other SDOs. Inheritance in TDL, as in all OO languages, means inheriting both the methods and the attributes of the base class SDO. However, unlike in C++, inheritance in TENA is not necessary to implement polymorphism—different method implementations for SDOs can be used on an instance-by-instance basis throughout a logical range. The details of this feature are beyond the scope of this introduction.

Combining the concepts of a CORBA distributed object with the distributed shared memory and anonymous publish-subscribe programming paradigms to create an SDO results in a programming abstraction possessing the benefits of all these systems. An SDO

- supports the remote method invocation (RMI) concept that is very natural to distributed object-oriented system programmers
- provides direct support for disseminating data from its source to multiple destinations
- supports reads and writes of data as if they were any other local data—a concept familiar to virtually every modern programmer
- enables model-driven auto-code generation which eliminates the tedious and error-prone programming chores common to distributed programming
- has an Application Programmer's Interface (API) that is easy to understand.

SDO Pointer

In the TENA Meta-Model, a pointer to an SDO behaves in the same way as pointers to objects in the meta-models of other programming systems. An SDO pointer refers to or points to a particular SDO instance (including a "null" instance). Naturally, which particular SDO instance is being pointed to can be changed. An SDO pointer can be de-referenced to yield the SDO to which the pointer referred.

Local Class

A local class in the TENA Meta-Model is similar to an SDO in that it too is composed of both methods and attributes. However, the methods and attributes of a local class are always local with respect to the application holding an instance of the local class. Invoking a method on an SDO generally involves a remote method invocation to a remote machine where the method's computation is actually performed. Invoking a method on a local class, however, always results in the method's computation being performed locally in the invoking application. Similarly, the attributes of a local class are always local to the application holding the instance of the local class. This is in sharp contrast to the attributes of an SDO. Even though SDO attributes are accessed as if they were local to the application, in fact their values may have been set in a remote application and cached in the local application after having been disseminated. The parameters or return types of methods on a local class follow the same rules as those for an SDO—everything in the TENA Meta-Model is allowed except an SDO. Since local classes are always passed by value, they are legal parameters and return types. Local classes may also inherit (singly) from other local classes.

The local class concept greatly improves the power and expressiveness of the TENA Meta-Model. While the SDO concept is adept at modeling many real-world problems, often situations arise where it is impractical to use RMI or remotely modifiable attributes. The methods of local classes ("local methods") can be used to standardize both the interface to and the implementation of algorithms of importance to the event designers.

Message

A message in the TENA Meta-Model is identical to a local class with one important distinction. A local class can only be indirectly disseminated to a remote application via an RMI on an SDO or as an attribute of an SDO. A message can be directly disseminated to remote applications, i.e. it can be sent and received. In Figure 20.3, the "Notification" object is a message.

The TENA Meta-Model provides messages to allow application developers to model events that do not persist in time. An SDO models a concept in the distributed application that has a non-zero lifetime and whose state may be modified over that lifetime. A message models a one-time event that typically needs to be delivered to multiple applications.

Vector

Often, real-world systems can be modeled best using objects with a variable number of one type of attribute. For example, a real-world vehicle does not always have the same number of passengers. For these types of situation, the TENA Meta-Model offers a construct called "vector." Vectors are used to model the concept of varying cardinality of attributes. The TENA Meta-Model supports vectors of fundamental data types, local classes, messages, and SDO pointers.

Const Attribute

The TENA Meta-Model allows SDO attributes to be constant. This means that the corresponding attribute(s) can be initialized at the time the SDO is instantiated, but may not be modified subsequently. A Const attribute is an important modeling construct that has the additional benefit of providing for implementation optimizations—Const attributes need not be resent to subscribers when the other attributes of an SDO are updated.

Optional Attribute

Optional attributes provide direct support in TENA's high level data exchange agreement to model the concept of data values that may or may not exist at the time the SDO or message needs to be sent. Attributes that are not expressly optional are required. Optional attributes help prevent the common practice of programmers placing incorrect values (like "0") into attributes when they do not know the accurate value to use.

TENA OBJECT MODEL COMPILER (OMC)

As explained above, the concepts and constructs in the TENA Meta-Model are abstract. The example object model shown in Figure 20.3 is a depiction of abstract ideas. The TENA OMC generates concrete C++ constructs based on the particulars of a given object model. These C++ constructs are linked with the TENA Middleware.

Strong typing in the TENA Middleware prevents the accidental misuse of an object at compile time. This enforcement increases the likelihood that an application that compiles will run successfully. Careful crafting of the TENA Middleware API allows the C++ compiler to detect and prevent errors such as trying to use one object where another is needed. The TENA Middleware API provides clear, consistent, and unambiguous memory semantics. By using modern C++ programming practices, the TENA Middleware prevents at compile time the common software defects of failing to deallocate memory or corrupting the free store by deallocating the same block of memory twice. The TENA Middleware makes every effort to detect potential programming errors at compile time. When compile time detection is not possible, runtime detection is employed (e.g. detecting when an application fails to set a required attribute or attempts to read an unset optional attribute). The TENA Meta-Model is designed to make TENA easy to use. The TENA Middleware's API is designed to be *hard to use wrong*. The combination enables the rapid development of robust applications.

The TENA OMC bridges the gap between the abstract concepts described in TENA object models and the concrete C++ applications that use the TENA Middleware. The source code automatically generated by the TENA OMC from TDL is linked directly with the TENA Middleware, customizing it based on the particular object model(s) used. The TDL can be written by hand using a text editor or it can be generated from the MagicDraw UML tool[3] using the TENA UML profile and the TENA plug-in written especially for that tool.

By using model-driven automatic code generation, TENA saves software developers the considerable burden of creating the equivalent software by hand. In addition to saving the time and money required to produce the equivalent object model software by hand, generating the object model software with the TENA OMC eliminates untold numbers of software defects that might otherwise be introduced.

---

[3]For information on the MagicDraw UML tool, go to http://www.magicdraw.com/. The TDL MagicDraw Plugin can be found at: https://www.tena-sda.org/display/Tools/MagicDraw+Plugin+TDL+Generator/.

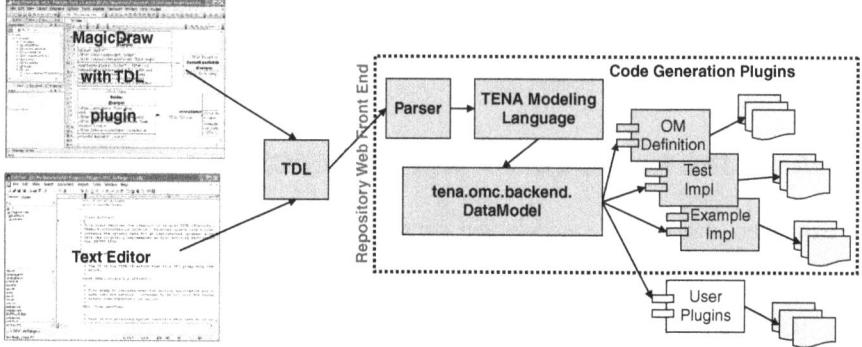

Figure 20.5   The extensible TENA object model compiler.

The TENA OMC provides a flexible and extensible framework that uses object models to generate C++ products that are verified to be correct. These products go beyond the autogeneration of object model code. The TENA OMC generates other products as well, such as example TENA applications, test programs, and documentation. Furthermore, by extending the OMC, members of the TENA community use these same object models to generate products tailored to their specific purposes, e.g. database loggers and automatically generated gateway applications that can easily bridge legacy architecture systems with more modern TENA-based systems. The overall design of the OMC is illustrated in Figure 20.5.

TENA STANDARD OBJECT MODELS

The set of TENA standard object models enable semantic interoperability among range resource applications by providing a standard "language" that all range resource applications use to communicate. The TENA standard object models can be thought of as a range-community-wide set of interface and protocol definitions encapsulated in an object-oriented design. Defining a complete TENA standard object model set will not be an easy or short-term process. It will require the work of many range engineers over many years to standardize the various types of information that are being used today. A "bottom-up" approach, based on the definition of small reusable "building block" objects that encode the most important elements of the range community information domain, is being used to manage the TENA object model development process. The bottom-up approach provides early standardization of many key building block objects and enables work to proceed incrementally towards range integration long before a comprehensive set of standard object models can be completed. Due deference to all of the work done on the DIS (IEEE, 1995) project to create a large body of information content relevant to the simulation world has been paid in the creation of the TENA standard object models.

The set of TENA standard object models is being developed using a prototype-based, iterative approach. Only object definitions that have been tested at multiple ranges during many logical range executions can be advanced for standardization. The object model used for a given logical range execution, called the Logical Range Object Model (LROM), can contain object definitions that are either custom designed for that particular logical range or are in the various stages of standardization. In this way, the range community is not forced to use object definitions that may be inappropriate for their logical range, but TENA still promotes the standardization that is necessary to create interoperability. After object definitions have been tested they can be put forward to the TENA Architecture Management Team (AMT) as candidates for formal standardization. The AMT is responsible for studying these definitions and de-conflicting them with any other object definitions that have already been put forward for standardization. When the AMT approves the object definitions, they are considered approved as TENA standards. When a substantial fraction of the range community information domain is standardized, interoperability between range resource applications will be greatly enhanced.

As of this writing in April 2011, the following TENA standard object models have been approved:

- TENA-Platform-v4—"Entity" and track information
- TENA-PlatformDetails-v4—other subsidiary entity information
- TENA-PlatformType-v2—encodes the type of an entity, including how that type is represented in the DIS Entity Type encoding scheme
- TENA-Embedded-v3—embedded weapons, sensors, and air transponder classes
- TENA-Munition-v3—describes a munition as a type of platform
- TENA-SyncController-v1—an object model for synchronizing sensors and targets
- TENA-UniqueID-v3—an encoding of the common site, application, and object ID scheme
- TENA-TSPI-v5—time–space position information, including position, velocity, acceleration, orientation, angular velocity, and angular acceleration. The implementation for these local classes includes the ability to convert from one coordinate system to another
- TENA-Time-v2—an object model for encoding various forms of time, and an implementation for converting between the different representations
- TENA-SRFserver-v2—an object model used to serve specific Spatial Reference Frames (SRFs) for a given execution
- TENA-AMO-v2—the Application Management Object (AMO) is used to enable remote control of a given application. It is used by exercise management tools, and considered a mandatory object for each TENA application to publish

- TENA-Engagement-v4—contains the Fire, Detonation, and Engagement-Results messages, used to manage combat interactions between platforms
- TENA-Exercise-v1—an object model to describe the entire exercise and its participants and organization
- TENA-GPS-v3—a model of a GPS sensor and the data it produces
- TENA-Radar-v3—an extensible model of a radar that allows users to model the radar at a number of different levels of detail

As an example, Figure 20.6 illustrates the most recent version of the TENA-Platform-v4 standard object model using UML.

Coordinate Conversions in TSPI

The invention of local classes was the direct result of the community's inability to come to a consensus on what coordinate system to use for the standard "Position" object. Some users wanted geocentric (as DIS prescribes). Some wanted geodetic (latitude and longitude). Many wanted some form of local tangent plane, as their ranges were all built on that coordinate system. Since there was no real possibility of an agreement on which coordinate system was universally applicable, the TENA project decided to allow users to use whatever coordinate system they wanted, with the conversions being done by the object model implementation itself. This decision required the ability to standardize both the interface to, and the implementation of, a set of software classes, and distribute these interfaces and implementations to all interested participants. The mechanism the project used to implement this requirement was the local class.

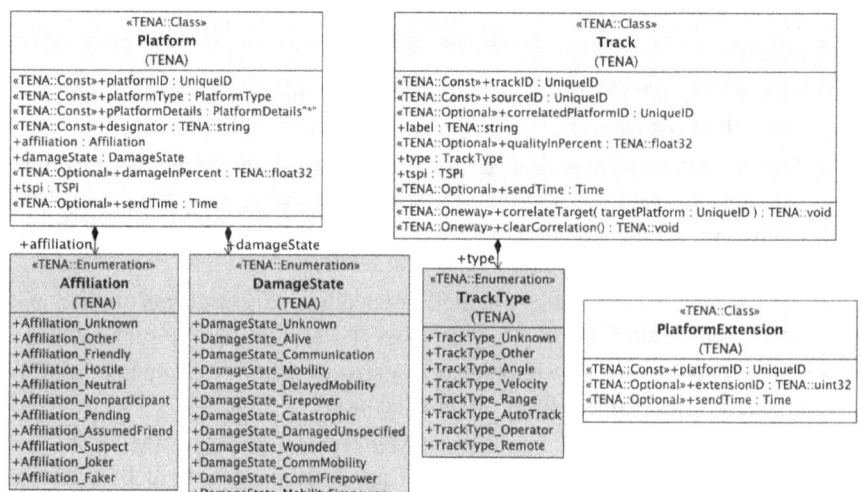

Figure 20.6   The TENA-Platform-v4 standard object model in UML.

In the TENA standard object models, a Platform SDO contains a TSPI local class which itself contains the Position local class. When the owner/server of that particular Platform receives updated position information from a sensor and wants to update the Platform object's position accordingly, it writes the position into the Platform's Position local class using a specified Spatial Reference Frame (SRF). The updated Platform SDO's state is then transmitted to all of its proxies spread throughout the exercise. When a client application wishes to read the Platform proxy's Position object, the client provides an SRF object corresponding to the coordinate system in which it wishes to receive the information. The appropriate local method then decides whether the reader wishes to read the position in the same SRF as the writer wrote it or in a different SRF. If the reader wants the position in the same SRF as was written by the writer, the Position object is simply handed to the reader without undergoing any translation. If the reader wants the object in some different SRF from what was written, the local method performs the translation automatically and provides an appropriate Position object in the SRF that the reader requested. In this scheme, neither the reader application nor the writer application are aware that they are participating in an exercise with other applications using different coordinate systems—everything is handled automatically by the encoded local methods. Translations are only done when necessary and only done by the reader application—this scheme is called "lazy translation."

The TENA Standard TSPI object model provides translation of coordinate systems in all of its constituent parts: time, position, velocity, acceleration, orientation, angular velocity, and angular acceleration. For the Position local class, the allowed SRFs are geocentric, geodetic, local tangent plane, and local spherical tangent plane. Other constituents of TSPI have other SRF possibilities. The coordinate conversion algorithms used are provided by the Synthetic Environment Data Representation and Interchange Specification (SEDRIS) project's Spatial Reference Model (SRM) implementation.[4]

TENA REPOSITORY

Fostering software reuse is a key goal of TENA. Towards that end, the TENA Repository and website services serve as a clearinghouse of information, reusable software components, and applications. More than 6200 registered users currently use the TENA Repository and website services.

The TENA Repository provides access to the various TENA products such as the TENA Middleware, the TENA OMC, TIDE, the TENA standard object

---

[4]For more information on SEDRIS go to http://www.sedris.org/. The SRM in particular is described at http://www.sedris.org/srm.htm. The SRM software is available at http://www.sedris.org/dwn4trpl. htm. Information on TENA's use of SEDRIS for coordinate conversions can be found at https://www. tena-sda.org/display/MW/TENA+TSPI+-v5+User+Guide.

models, and other information. In addition, the repository provides access to numerous software components developed by members of the TENA community.

The TENA website also includes wiki spaces, helpdesks, documentation, and email reflectors. These features are a single cohesive set of services designed to foster collaboration to mitigate the challenges of developing interoperable applications in a distributed and decentralized fashion. This repository and website services address an important aspect of the successful creation of large-scale LVC simulations. The TENA Repository and website services can be accessed at http://www.tena-sda.org/.

TENA MIDDLEWARE

At the core of TENA, the TENA Middleware is the high performance, real-time, low latency communication infrastructure used by range resource applications and tools during execution for all communication regarding objects in the LROM. The TENA Middleware is linked into every TENA application along with the LROM object definitions. The TENA Middleware is built to support the TENA Meta-Model and is thus the communication mechanism for all objects in the TENA object model. The purpose of the TENA Middleware is to provide range users with a unified API to support SDOs, messages, and other aspects of the TENA Meta-Model.

Figure 20.7 shows a high level schematic of the TENA Middleware. The Middleware is based on two popular open-source software packages: the Adaptive Communication Environment (ACE), which provides a platform-independent set of low level C++ classes and patterns to abstract away from the underlying operating system; and the ACE ORB (TAO), a full featured open-source implementation of the CORBA standard.[5] In the diagram, the box labeled "TENA objects," which are C++ objects generated from the provided TDL object model, inherit from an object framework defined in the middleware. Users interact with the middleware using a unified API, allowing them to access objects, publish, subscribe, and receive messages from the Middleware in a consistent fashion. As of this writing, the current version of the TENA Middleware is 6.0.2.

TENA TOOLS AND UTILITIES

Each interoperability architecture brings hurdles as well as advantages, and TENA is no exception. To minimize the hurdles, TENA provides a number of tools and utilities to make the job of creating and integrating TENA applications as easy as possible. Below, we describe five of these tools, which aid range developers in creating their software (TIDE), verifying that it works properly

---

[5]Information about ACE can be found at http://www.cs.wustl.edu/schmidt/ACE.html and information about TAO can be found at http://www.cs.wustl.edu/schmidt/TAO.html.

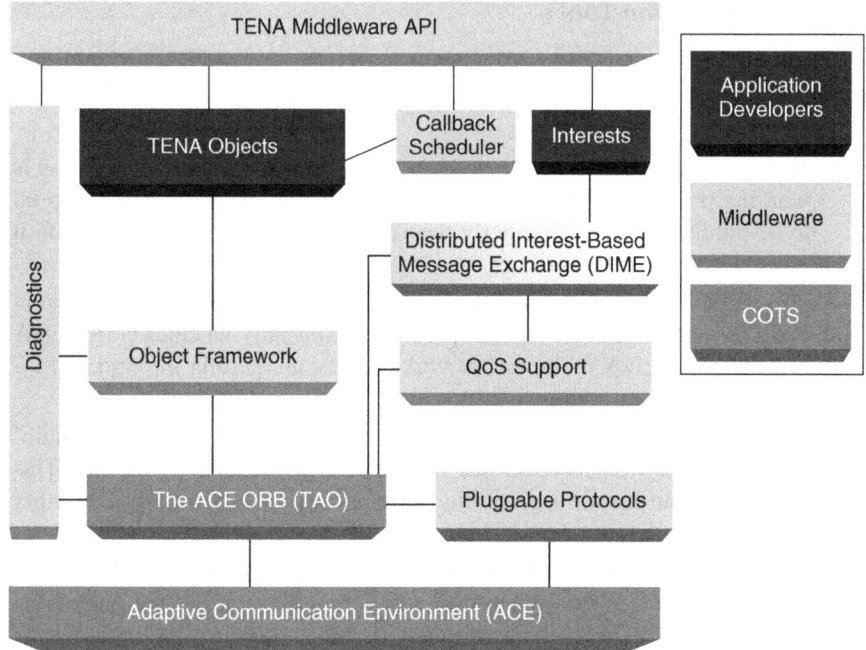

Figure 20.7   High level design of the TENA Middleware.

(IVT), monitoring its execution (TENA Console), bridging it to other architectures (Gateway Builder), and interacting with other applications in multiple security domains (SimShield). All but the last are freely available to any TENA user; SimShield, on the other hand, is a commercial product.

TIDE

The Eclipse-based TENA Integrated Development Environment (TIDE) facilitates the development and maintenance of TENA applications. Using TIDE, TENA developers can create object model components or search for existing object model components. TIDE can be used to generate example applications that use the object model(s) selected by the TENA developer. TIDE can also be used to semi-automate the upgrading of TENA applications to use newer versions of the TENA Middleware or the updating of TENA applications to take into account changes introduced by new revisions of object models. Thus TIDE can greatly reduce the time and effort required to create and maintain TENA applications. Finally, TIDE has the ability to expedite the creation of TENA applications by injecting helpful code snippets directly into TENA applications as they are being written.[6]

---

[6] See https://www.tena-sda.org/display/TIDE/Home.

Interface Verification Tool

The Interface Verification Tool (IVT) can be used as a surrogate TENA publisher or subscriber in testing TENA applications. The IVT can perform four basic functions:

- Verify TENA application interfaces—to see whether a given application is working properly and is properly publishing the information it is supposed to be publishing, and is properly receiving and handling the information it is supposed to be subscribing to.

- Test network functionality—to see whether the network is functioning properly with regard to TENA operations, especially whether both TENA reliable and TENA multicast network packets are getting through end-to-end.

- Generate scenario/platforms—in this use case the IVT generates a number of Platform objects and has them move in predefined patterns. This type of simple scenario generation capability is used to make sure other applications are seeing everything they are supposed to see.

- Provide real-time monitoring and analysis of events—in the role of subscriber, the IVT can show the user what other applications are publishing and can therefore alert the user to any possible problems with the event.

TENA Console

The TENA Console is a utility used to monitor the status of a TENA event. Diagnostics embedded in the Middleware, and thus in every TENA application, report statistics and alert messages directly to the TENA Console. Multiple instances of the Console can be run at any location during the event. The Console is a separate stand-alone Java-based application that is delivered along with the Middleware. Figure 20.8 shows the Console while running. Each tab shows various aspects of the execution, with the tab illustrating the connected applications being shown.

The capabilities of the TENA Console include status reporting, alert reporting, application and network health monitoring, and embedded diagnostics control. TENA Consoles can monitor application heartbeats to detect unresponsive applications and alert console operators. The TENA console operator can manually ping applications and execution managers to detect unresponsiveness. Network monitoring is accomplished with continuous inter-application pinging when necessary. The Console is capable of monitoring both Reliable (TCP) and Best Effort (UDP Multicast) communication between applications. Best Effort monitoring will only monitor active multicast addresses (i.e. will not cause unnecessary multicast group joins). The TENA Console has the ability to remotely terminate unresponsive applications. Applications can abnormally terminate or become disconnected from the execution, leaving behind "ghost" objects that need to be cleaned up. The TENA Console allows operators to be notified about (as well as proactively check for) unresponsive applications, and instruct all applications to terminate "ghost" objects.

Figure 20.8 TENA Console, set to the "applications" tab.

Gateway Builder

The Gateway Builder is an interoperability tool designed to significantly reduce the time, effort, and cost of integrating LVC applications that use different inter-operability architectures into distributed training exercises and test events. The Gateway Builder streamlines the integration process and reduces the time and effort of creating gateways. Unlike traditional gateways that are hard-coded for a specific use, Gateway Builder is a flexible, extensible, graphically driven tool that automatically generates gateways to bridge simulation and live protocols.

The Gateway Builder approach begins with the definition and import of various object model specification files. The Gateway Builder user selects two interoperability architectures, each with their own object model, to create a map defining the translation logic between the different architecture domains. Along with a variety of predefined mapping types, the Gateway Builder enables the user to import conversion functions and make use of user-defined configuration and look up files to support the mapping functionality. After the mapping process is complete, the Gateway Builder map is generated to produce custom gateway instances. The generated components include the scripts necessary for the installation, build, and configuration of the generated gateway application.

The key element in the technical approach to the Gateway Builder is its reliance on code generation techniques to reduce development effort. Other design considerations are reuse, modularity, flexibility, and independence of any particular object model. The Gateway Builder serves as a framework for managing and controlling the event integration process. Mappings, object models, and transformation functions from multiple exercises are stored, viewed, and leveraged.

The Gateway Builder is designed to integrate live and synthetic systems into training solutions, and supports a wide variety of protocols, including various versions of TENA, HLA, and DIS.[7]

SimShield

SimShield provides the capability to label, segregate, protect, and exchange data between TENA systems executing at different sensitivity or classification levels. SimShield is made up of two components: the Policy Editor and the Trusted Bridge. The Policy Editor is an easy-to-use graphical user interface that permits security and TENA domain experts to enter and review the reclassification rules that govern the intercommunication between single level logical ranges. Once approved, the administrator can install and implement the configured rule set on the Trusted Bridge in a matter of minutes, which then enforces the installed rule set. SimShield meets the data format and real-time performance requirements for distributed operations, exercises, and training.[8]

TECHNICAL DECISIONS AND THE ADVANTAGES OF THE TENA APPROACH TO INTEROPERABILITY

When an interoperability architecture is designed, a number of questions immediately present themselves. There is no generally applicable correct answer to these questions. Each must be answered based on the specific requirements of the domain in which the architecture is to be used. As we have said before, the TENA domain is one of live systems interacting with other live systems as well as virtual and constructive simulations. In this domain, the needs of the live systems dominate the architectural choices that need to be made.

Technical Decisions

The first decision that needs to be made is whether to specify an on-the-wire protocol or an API standard for a software infrastructure. Before TENA, the designers of DIS chose an on-the-wire standard and the designers of HLA chose an API standard. TENA also chose an API standard because an API standard allows for a common software infrastructure, that is tested and verified, to be distributed to the entire range community. Having a single software infrastructure with an API designed around high level programming constructs makes the creation of applications cheaper, easier, and less prone to error. An API standard also allows for future technological advancements in data transmission to be more seamlessly and cost-effectively incorporated into the infrastructure.

---

[7]See https://www.tena-sda.org/display/MSR/Home for more information about the Gateway Builder.
[8]See http://www.trustedcs.com/products/SimShield.html for more information on SimShield.

The second decision was whether to create a centralized client/server architecture or a peer-to-peer architecture. In a centralized client/server architecture, all data produced by applications are routed through the central server, which decides which applications it will forward the data to. A centralized client/server architecture has some advantages in that it allows both the centralization of data collection and the centralization of information routing. A peer-to-peer architecture, on the other hand, allows information to be both created and consumed by any individual application, with data being routed directly from the information source to any interested application. This direct routing of information makes a peer-to-peer architecture more efficient than a client/server architecture as it minimizes the data transmissions necessary to perform the required communication. The live systems on test and training ranges currently operate as a peer-to-peer set of systems, with each application sending information directly to the intended recipients. Since the ranges have grown up in an organic ad hoc fashion over the years, and each system sends and receives data based on how it was designed, a peer-to-peer architecture was the more obvious fit to this domain. Also, a peer-to-peer architecture is the more general approach. If a peer-to-peer architecture can be adequately created, a client/server architecture can be easily emulated. A client/server architecture, on the other hand, cannot adequately emulate a peer-to-peer architecture.

The third decision was whether to use a runtime interpreter to interpret the information being transmitted to each application or to compile the object definitions into the application code. TENA chose the latter approach so as to maximize the capability of catching software errors at compile time rather than at runtime. The decision to compile object models into the heart of the application led to the technical solution of auto-code generation. While technically difficult to master at first, once auto-code generation is used to create some aspect of a software application (like the object model definitions), it becomes easy to expand into other areas. This decision led inexorably to the auto-code generation of user starting-point code, test programs, example programs, and even the ability of users to auto-generate almost their entire software application.

The fourth decision was whether to design an object model that represents just a single spatial reference frame or multiple reference frames. DIS chose a single, fixed reference frame (geocentric). As discussed above, the designers of TENA chose to incorporate a technical solution to this conundrum and allow multiple reference frames to exist at the same time in the same exercise with automatic conversion between them. This decision allowed individual ranges to choose the coordinate system best suited to their test and training needs, while simultaneously creating the technical capability for TENA users to standardize other algorithm interfaces and implementations, not just those related to coordinate conversion.

The final architectural decision was to make the TENA Middleware hard to use wrong. In the domain of live systems, where a software error could cost millions of dollars or even be threatening to human life, the most overriding requirement is to make the software hard to use wrong—to minimize the chance of serious software defects. Every design decision that was made on the TENA Middleware API was made with this in mind. So far as could be accommodated under this constraint, the software was also made as easy to use as possible, presenting to the user high level object-oriented programming constructs. There are times when users might complain about some aspect of the TENA interface, because it makes them do work (like writing the C++ "Try... Catch" code, or providing valid initialization data) that other architectures do not make them do. However, the final product is always superior software, as reliability is enforced by the architecture through the Middleware API and the generated code. One must always remember that while it is possible to build software that is easy to use right, it is sometimes critical to build software that is hard to use wrong. All too often, software that is easy to write turns out to be easy to use wrong.

Advantages of TENA

TENA's technical approach brings a number of advantages to users who are looking for a solution to their interoperability problems. These advantages are primarily directed at the live range community, but could also be effective for other M&S activities. Nothing in its design excludes TENA from consideration as an interoperability solution for other M&S domains, even those that historically have had very little interaction with the live aspect of LVC integration.

The first advantage of TENA is that all TENA software and support are freely available to any user who wishes to take advantage of them. TENA is directly funded by the US Government to meet specific range community needs, but users are welcomed from other communities as well. TENA has users throughout the United States, Europe, Australia, and Asia, and is continually growing and improving as time passes.

The second advantage is that TENA's technical approach emphasizes reliability and its consequent cost-savings by building enforcement mechanisms for the data contracts into the API itself. Because of this design, many errors are caught at compile time rather than at runtime. TENA software is hard to use wrong and thus it is more difficult to create a TENA application that will fail during a test or training event.

The third advantage is that the auto-code generation capability makes creating a reliable TENA application as simple as possible. The TIDE tool manages the installation and configuration of the TENA Middleware and Object Model software, as well as assisting in upgrading and maintenance of the users' code. The code generator creates auto-generated starting points for the user—C++ files that have all the tedious aspects of programming already filled out for the user, with simple comments that the user should insert his code in specific places. The existence of the auto-generated starting point files means that a TENA user never starts with a blank page—a lot of work has already been done for him. This allows the user to rapidly develop real-time distributed LVC applications, all tied together by the TENA Middleware and the user's object models. Once his program is written, the auto-generated test programs allow the user to perform unit testing on his software application with almost no effort on his part to create a test harness. Since unit testing is easy, and most software errors are detected either by the compiler or the auto-generated test programs, integration testing is usually performed in a relatively short period of time using the IVT. Auto-generated gateways built by the Gateway Builder tool allow TENA users to quickly integrate their software with other architectures, such as DIS and HLA.

The fourth advantage is that the TENA Middleware software and all the auto-generated object model and test software have been thoroughly tested and verified in the TENA test lab. Thousands of individual unit tests and hundreds of system tests are performed on each version of the Middleware and Object Model Compiler. These tests ensure that, to the best of TENA's ability, all of the software defects have been removed from the TENA software before release. In the case where defects in the TENA software are found, an extensive helpdesk resource is available to users to ask and answer questions and get information about known problems. The TENA Repository also has extensive documentation, programmer's guides, release notes, and collaboration capabilities that give users a common documented view of the state of the TENA universe at any one time.

Each of the technical decisions and advantages above was made to respond to the specific needs of TENA's government customer. In meeting those needs, TENA has become the most capable and sophisticated interoperability solution in existence for the LVC community.

RELATED ACTIVITIES AND FUTURE DIRECTIONS

TENA is not a finished product. Currently, there are plans to expand TENA's applicability in the range community beyond the current context. The TENA in a Resource Constrained Environment (TRCE) program is extending TENA into the domain of embedded systems, unreliable networks, and smart phones. The Joint Mission Environment Test Capability (JMETC) program is using TENA to create large-scale persistent mission environments for the joint testing community.

TRCE

The mix of hardware, communications systems, and technologies in current test and training networks has led to considerable variability in the quality and speed of the network links and the types of computational platforms required to support test and training objectives. The TENA Middleware must be enhanced to reliably and seamlessly support these types of networks and computational environments to truly provide a common interoperability architecture. The TRCE project develops technologies for inclusion in the TENA Middleware to support this broad range of hardware and network link constrained environments.

Currently TRCE has focused on two of these objectives: improving TENA's support for variable quality and low data rate network links including wireless networks, and expanding TENA's support for handheld and embedded instrumentation computational platforms including smart phones and tablet devices.

A relay-node capability has been prototyped that allows TENA to transmit data through very constrained network links (down to 2400 baud) and through variable quality networks. TENA has also been ported and demonstrated to work on iPhones, iPads, Android phones, and Gumstix embedded processors.[9] These capabilities will be deployed into the mainstream TENA Middleware in a subsequent version for release to the entire TENA community.

JMETC

JMETC is a DoD corporate approach for linking distributed facilities for the creation of a joint mission environment for testing. JMETC enables customers (developers of new weapon systems) to efficiently evaluate their warfighting capabilities in a joint context. A subsidiary goal of JMETC is to provide compatibility between testing and training so that range resources built to perform testing can also be used in a training context and other range resources built for training purposes can be used in a test context.

The core of JMETC is a reusable and easily reconfigurable infrastructure, which consists of the following products.

- Persistent connectivity—a permanent network used and accredited for test capabilities. The JMETC network currently reaches over 60 test and training sites in the United States. While the technical challenges of creating a dedicated classified network are substantial, the ability to permanently accredit the JMETC network for classified operations is the one feature that saves the most time and money for JMETC's customers, as this process can be very long and expensive if done on an ad hoc basis.

- Middleware—JMETC uses the TENA Middleware for all of its information dissemination needs.

- Standard interface definitions and software algorithms—JMETC uses the TENA standard object model definitions and implementations, augmented by certain other object models designed specifically for the joint mission environment testing performed by JMETC users.

- Distributed test support tools—JMETC uses the TENA tools and utilities described above, but augments them with over 20 other tools that have been developed by the test community. All of these tools are provided free of charge to JMETC users.

- Data management solutions—JMETC deploys a prototype data management solution for both data initialization and runtime data collection. JMETC expects to use the TENA Standard Event Data Management System when that capability is deployed.

- Reuse repository—JMETC uses and enhances the TENA Repository and website services.

[9]See http://www.gumstix.com/.

JMETC provides a common joint testing process and a customer support team for users so as to maximize their productivity and make distributed testing more effective and more cost-efficient. More information on JMETC can be found at http://www.jmetc.org/.

FUTURE DIRECTIONS

LVC systems succeed or fail on how well their constituent parts can interoperate. Interoperability can only be assured when there are formal computerized contracts detailing the characteristics of the individual systems, the data exchanged among them, and the nature of the enterprise activity as a whole. These contracts must then be computer enforced to ensure that they are actually followed throughout the enterprise.

TENA has done considerable work on enforcement of data exchange contracts, but significant research into the means to accurately characterize individual systems as well as the nature of an entire LVC activity as a whole remains, along with research into how best to enforce the characterization contracts once they are created. The existence and enforcement of these contracts would, for the first time, make it possible to address the fundamental interoperability question that plagues those that need to perform LVC activities—"Will this software system interoperate within that particular activity?"—in a way that assures the correctness of the answer. Research into additional technologies will seek to provide capabilities to support the successful planning, execution, analysis, and history of LVC activities, such as test events and training exercises for the US DoD.

SUMMARY

The inherent nature of the software and systems used to carry out LVC test events and training exercises implies that such enterprises are composed of multiple heterogeneous systems, applications, and hardware platforms, geographically distributed at multiple sites. The large-scale distributed nature of these enterprises tends to further increase the number of distinct systems involved. In all known LVC events, multiple organizations contributed to various parts of the enterprise at different periods of time. This application development is nearly always decentralized, lacking a common authority providing consistent development processes and design decisions for each system.

When real, live systems are mixed with constructive and/or virtual reality simulations, the demands of the live systems dominate the resulting LVC system. Software architectures intended to support the creation of distributed simulations are not particularly well suited to meet the unforgiving demands of live distributed systems. TENA is the software architecture primarily used by the US DoD testing and training community. The needs of that community run the gamut of large-scale, real-time, distributed and embedded software systems. Oftentimes, these applications are combined to form large-scale LVC systems. The TENA

Middleware is designed to enable the rapid development of real-time distributed LVC systems. The TENA Middleware combines distributed shared memory, anonymous publish-subscribe, and model-driven distributed object-oriented programming paradigms into a single distributed middleware system. This unique combination provides a programming experience that is both simple and powerful for a broad variety of real-time distributed applications.

Future warfighting concepts, relying heavily on network-centric warfare capabilities, need a new technical foundation for both test and evaluation and warfighter training, based on interoperability and reuse. The nation cannot afford to rebuild the entire test and training range capability, so a strategy must be created to ensure that needed legacy capabilities are adapted, and future investments are designed, to be interoperable in support of these future warfighting scenarios. TENA embodies this strategy by prototyping, developing, and standardizing the software and interfaces needed to achieve a more profound interoperability throughout the range community. TENA is based on state-of-the-art research in advanced information technology and many lessons learned in large-scale distributed real-time systems development, and will provide the necessary foundation for developing sophisticated and powerful testing and training systems in the future.

ACKNOWLEDGEMENTS

The TENA architecture and software described above was developed by SAIC Inc. and other contractors under contract to the US Government (contracts 1435-04-01-CT-31085, DASG60-02-D-0006-40, and DASG60-02-D-0006-122). The authors would like to thank all the members of the TENA Team who have contributed many fine ideas and solutions over the years, but especially to the person with the vision to see TENA implemented and deployed across the world, the Director of the TENA Software Development Activity, George J. Rumford.

REFERENCES

IEEE (1995). *IEEE Standard for Distributed Interactive Simulation—Application Protocols (Set)*. IEEE, standard No. 1278.1 and 1278.1A.

Kuhl F, Weatherly R and Dahmann J (1999). *Creating Computer Simulation Systems: An Introduction to the High Level Architecture*. Prentice Hall, Englewood Cliffs, NJ.

MDA Group (2003). http://www.omg.org/docs/omg/03-06-01.pdf, version 1.0.1.

Noseworthy JR (2002). IKE 2—implementing the stateful distributed object paradigm. In: *Proceedings of the 5th IEEE International Symposium on Object-Oriented Real-Time Distributed Computing (ISORC 2002)*. IEEE, Atlantic City, NJ.

Noseworthy JR (2005). Developing distributed applications rapidly and reliably using the TENA middleware. In: *Military Communications Conference Proceedings 2005 (MILCOM 2005)*. IEEE, Atlantic City, NJ.

Noseworthy JR (1996). A novel and efficient distributed shared memory algorithm for strict causal consistency. ECSE Dept. Ph.D. thesis, Rensselaer Polytechnic Institute, Troy, NY.

Powell ET et al. (2002). The TENA Architecture Reference Document, http://www.tena-sda.org/documents/tena2002.pdf. The Foundation Initiative 2010 Program Office.

Powell ET, Lucas J, Lessmann K and Rumford GJ (2002). The test and training enabling architecture (TENA) 2002 overview. In: *Proceedings of the Fall 2002 Simulation Interoperability Workshop*. http://www.sisostds.org.

Powell ET, Lucas J, Lessmann K and Rumford GJ (2003). The test and training enabling architecture (TENA) 2002 overview and meta-model. In: *Proceedings of the Summer 2003 European Simulation Interoperability Workshop*. http://www.sisostds.org.

Rumford GJ, Vuong M, Bachinsky ST and Powell ET (2001). Foundation initiative 2010: the design of the second TENA middleware prototype. In: *Proceedings of the Fall 2001 Simulation Interoperability Workshop*. http://www.sisostds.org.

Tolk A, Turnitsa CD and Diallo SY (2008). Implied ontological representation within the levels of conceptual interoperability model. *Int J Intell Decision Tech, Special Issue* **2**, 3–19.

Chapter 21

Combat Modeling using the DEVS Formalism

Tag Gon Kim and Il-Chul Moon

INTRODUCTION

Modeling combats is an important, yet difficult task. While combat modeling is a critical contributor (Piplani et al., 1994; Department of Defense, 2006) to the key areas of military science, such as training (Zavarelli et al., 2006), analysis (Cramer et al., 2008), and acquisition (Manclark, 2009), combat modeling is difficult (1) because the modeling requires complex knowledge background, i.e. defense domain knowledge as well as the modeling and simulation knowledge (Sung et al., 2010) and (2) because the modeling result is often required to satisfy real-world settings in terms of applicability, realism, precision, and error resistance (Roza, 2004). Therefore, inherently, combat modeling in the real world often becomes a joint work between the subject matter experts for supplying the domain knowledge; the modeling and simulation experts for representing the domain knowledge as simulation models; and the software engineering experts for implementing the models. Because of this emphasis on cooperation, we need a modeling methodology that supports the joint work between the three types of experts (Kim et al., 2010). In detail, the support should include a clear and complete communication method between the three experts, a facilitated implementation from the modeling result, and a reusability of the developed models in other related scenarios. To briefly illustrate a methodology providing such supports in the modeling process, this chapter introduces the DEVS formalism (Kim and Zeigler, 1987; Zeigler et al., 2000) and its implementation framework from the combat modeling perspective (Mittal et al., 2007; ACIMS, 2007; Kim et al., 2010).

Engineering Principles of Combat Modeling and Distributed Simulation, First Edition.
Edited by Andreas Tolk.
© 2012 John Wiley & Sons, Inc. Published 2012 by John Wiley & Sons, Inc.

This chapter walks through a case study of modeling a naval air defense scenario with the DEVS formalism and an implementation framework to show the design and the implementation of a sample combat model. First, the walk-through shows how the DEVS formalism is utilized to formally and transparently represent a combat model. From the theoretic perspective, the DEVS formalism is essentially a formal description of a discrete event system, and the formalism is well applicable to describing weapon system operations as well as human command and control behavior. For example, the digital systems of weapon systems are easily describable by the DEVS formalism since the formalism is an extended version of finite state machines that is used to design digital systems. Additionally, human tactical behavior is adaptable to the discrete event system when we regard humans as agents accepting outside stimuli and generating responses. Second, the walk-through demonstrates how a DEVS implementation framework corresponds to the DEVS formalism and facilitates model implementation. With the theoretic foundation of the DEVS formalism, a DEVS implementation framework, i.e. DEVSim++ (Kim and Park, 1992; Kim et al., 2010), enables modelers to easily implement models as software artifacts. The framework provides templates that the models specified by the DEVS formalism naturally fit in. Third, the walk-through explains a simulation procedure using a reusable and general purpose simulation engine included in DEVSim++ and provides a sample battle experimentation using this DEVS based approach.

MODELING FRAMEWORK USING DEVS FORMALISM

Before starting the walk-through of a naval air defense scenario, this section describes the DEVS formalism and the DEVSim++ framework used for combat modeling. First, we review the combat modeling and simulation practice with the formalism support; then we introduce further details of the DEVS formalism and the DEVSim++ framework in the combat modeling context.

Combat Modeling and Simulation with Formalism

Modeling and simulation with the support of formalism enables a practice that is more transparent, faster, and more reusable. There are various types of formalism, i.e. the DEVS formalism, the Petri-Net, the System Dynamics, etc., and modelers choose a type of formalism that fits the modeling objective and the context. For instance, a combat situation can be viewed as a discrete event system exchanging signals and physical interactions between subsystems, which will lead to using the DEVS formalism.

Each type of formalism has its own representation basis, and the conceptual models of combats are representable by following the basis of the chosen formalism. The DEVS formalism is built upon set theory and the functional representation, and on the other hand the system dynamics is represented as a number of differential equations. This chapter provides the DEVS models specified by the textual notations used in set theory and the functional representation. Additionally, a DEVS model is representable as a DEVS graph diagram.

Implementing the represented models in the chosen formalism is easily achievable by utilizing an implementation framework supporting the formalism. For example, a conceptual model in the DEVS formalism is implementable by using DEVSim++ (Kim and Park, 1992; Kim et al., 2010), DEVSJava (ACMIS, 2007), DEVSML (Mittal et al., 2007), etc. to produce an implemented model that is executable by a generic simulation engine. At this stage, even though a model uses the same formalism, the implementation result can be different. For instance, DEVSim++ and DEVSJava implements a model by extending the C++ and the Java class templates, respectively, designed to meet the formalism specification. On the other hand, DEVSML utilizes a script format which looks similar to the notations of conceptual models.

Finally, simulating the implemented model depends on the choice of an implementation framework. One of the most important advantages of using formalism is the separation between the simulation engine and the model. Therefore, once a modeler finishes the implementation of models by using the formalism implementation framework, the modeler is also able to apply the generic simulation engine provided by the implementation framework. Figure 21.1 illustrates the formalism supports at the modeling, the implementation, and the simulation stages.

Combat Modeling by the DEVS Formalism

Introduced by Ziegler (1984), the DEVS formalism is a formal specification of a discrete event system. As illustrated in Figure 21.2, the discrete event system is a general case of the discrete time system which has a uniform time interval and the finite state machine that does not have internal state transition that happens when a time advancement of a state is expired. The importance of the discrete event system in combat modeling is the prevalence of modeling entities that have the characteristics of the discrete event system. For instance, most of the digital systems in the fire control systems and the communication control systems are finite state machines. Also, the periodic scan of the opposition forces by radar and sonar systems is a typical example of discrete time systems. Also, discrete event systems are easily adaptable to express human activities, i.e. the command and control process. Therefore, modeling a discrete event system is a key of combat modeling, and the DEVS formalism, which is specialized in expressing a discrete event system, is a key tool in combat modeling.

This formalism enables a large discrete event system to be specified by hierarchically decomposing the system into modules. After the decomposition, we have two types of specifications: the hierarchy structure of the system modules and the internal structure, or the state transition, of the system modules. These two specifications of the decomposed discrete event systems correspond to the coupled model and the atomic model respectively in the DEVS formalism.

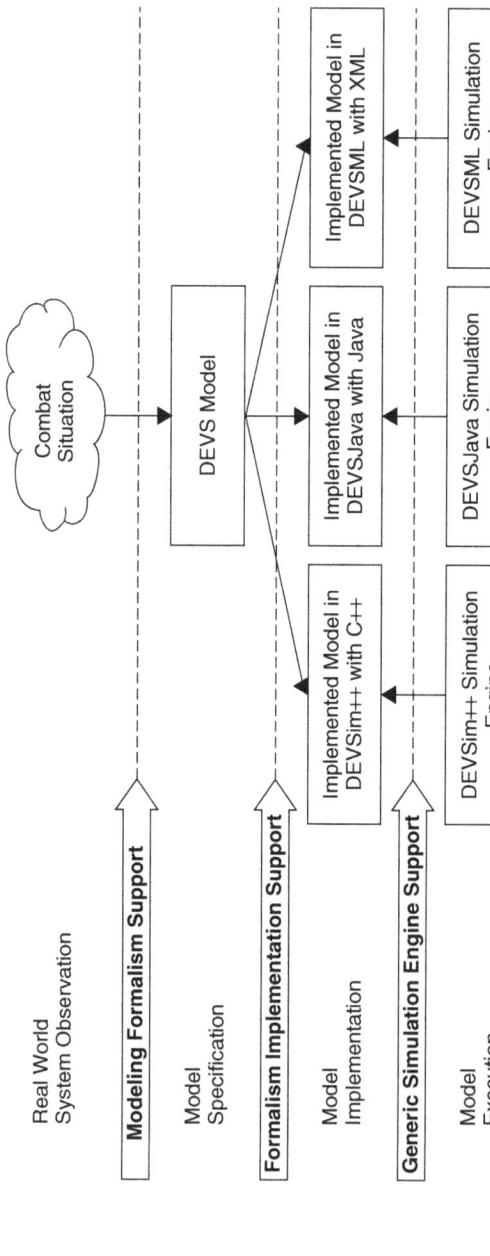

Figure 21.1 Modeling and simulation with the DEVS formalism.

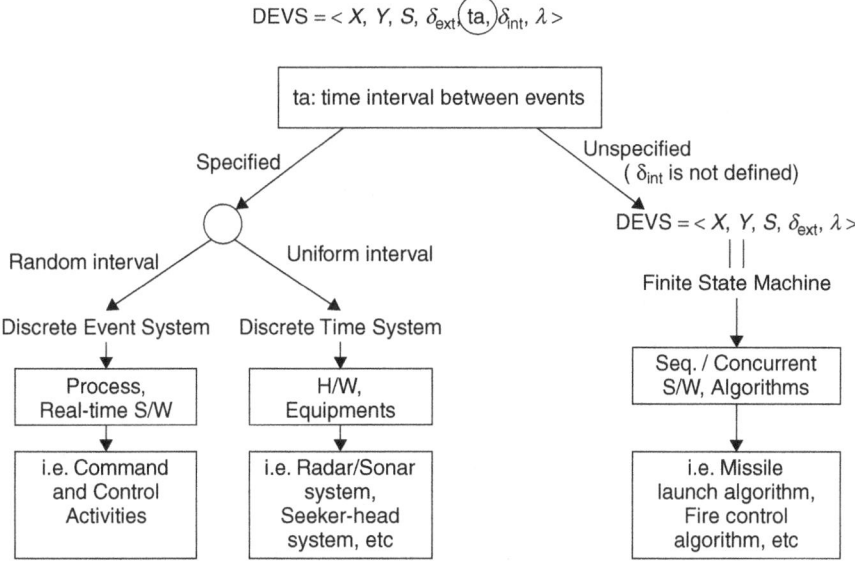

Figure 21.2   Modeling of combat systems in the DEVS formalism.

From the combat modeling perspective, this system decomposition and modular representation of the DEVS formalism is analogous to the engagements of opposing units and the unit constructions. For instance, a combat scenario of two opposing platoons can be viewed as a discrete event system whose sub-modules are two platoons interchanging engagement events. In the light of the DEVS formalism, the scenario is decomposable into two opposing platoon modules. Subsequently, each platoon module is also decomposable into multiple soldier modules. This decomposition of the scenario resembles the hierarchy structure of the system modules, or the coupled model of the DEVS formalism. Then, the soldier modules become the decomposed system modules, or the atomic model of the DEVS formalism. After this modeling concept of the coupled and the atomic models, we proceed to the detailed modeling of the coupled models of the unit structures and the atomic models of the individual soldiers. First, a coupled model is specified as the following 7-tuple in the set-theoretic context.

$$CM = \langle X, Y, M, \text{EIC}, \text{EOC}, \text{IC}, \text{SELECT} \rangle$$

where X is a set of input events, Y is a set of output events, M is a set of all component models,

$$\text{EIC} \subset \text{CM}.X \times \text{U}_i M.X_i \qquad \text{external input coupling}$$
$$\text{EOC} \subset \text{U}_i M.Y_i \times \text{CM}.Y \qquad \text{external output coupling}$$
$$\text{IC} \subset M.Y_i \times \text{U}_j M.X_j \qquad \text{internal coupling}$$
$$\text{SELECT} : 2^M - \emptyset \rightarrow M \qquad \text{tie-breaking function}$$

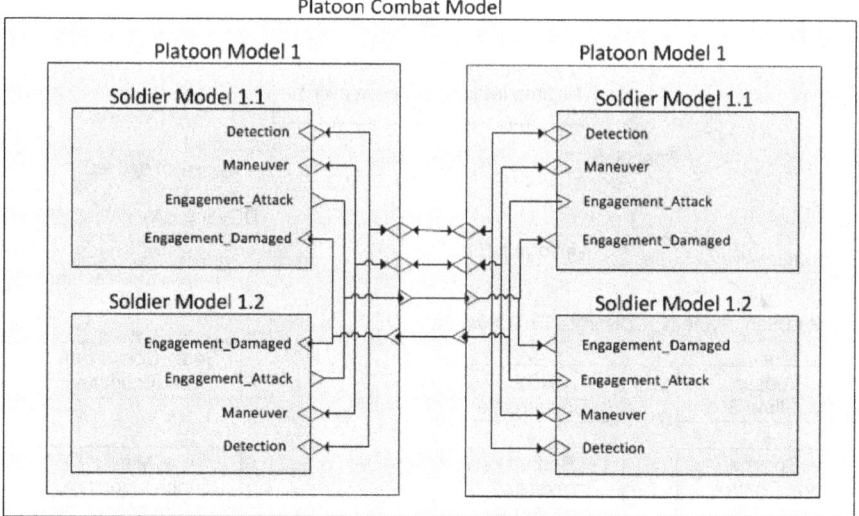

Figure 21.3   A DEVS diagram representation of a coupled model, or a platoon model.

Given this textual representation, we can illustrate a coupled model as the DEVS diagram in Figure 21.3. More detailed explanation about the DEVS diagrams are provided by Song and Kim (2010). Additionally, we specify the example platoon engagement scenario as the coupled model specifications textually in Figure 21.4.

Second, under the assumptions of the hierarchical structure of the coupled model, we identify an atomic model, or a soldier model, to specify. According to the DEVS formalism, an atomic model is specified as the following 7-tuple.

$$\text{AM} = \langle X, Y, S, \delta_{\text{ext}}, \delta_{\text{int}}, \lambda, \text{ta} \rangle$$

where X is a set of input events, Y is a set output events, S is a set of sequential states. δ_{ext} is $Q \times X \to S$, an external transition function, where $Q = \{(s, e) | s \in S \text{ and } 0 \le e \le \text{ta}(s)\}$, the total state set of M, δ_{int} is $S \to S$, an internal transition function, λ is $S \to Y$, an output function, and ta is $S \to R_{0,\infty}$, a time advance function.

Just like the coupled model, we can graphically represent this textual specification with an atomic level DEVS diagram visualizing the state transition of the atomic model. However, the DEVS diagram for an atomic model is not a simple component coupling diagram such as the diagram for a coupled model, so we provide a brief explanation. Figure 21.5 shows a simple state transition diagram that follows the DEVS atomic model diagram. First, the states of an atomic model, or S_0 and S_1 in S, are illustrated by circular nodes in the diagram. Additionally, the node also specifies the time advancement when the transition occurs, or $\text{ta}(S_0)$ and $\text{ta}(S_1)$. Then, the diagram has two types of directed edges to show two types of state transition events: the internal and the external events. The

```
PlatoonCombatModel=<X, Y, M, EIC, EOC, IC, SELECT>

  X = Ø, Y = Ø

  M={PlatoonModel1, PlatoonModel2}, EIC = Ø, EOC = Ø

  IC={(PlatoonModel1.Detection, PlatoonModel2.Detection),
     (PlatoonModel2.Detection, PlatoonModel1.Detection),
     (PlatoonModel1.Maneuver, PlatoonModel2.Maneuver),
     (PlatoonModel2.Maneuver, PlatoonModel1.Maneuver),
     (PlatoonModel1.Engagement_Attack, PlatoonModel2.Engagement_Damaged),
     (PlatoonModel2.Engagement_Attack, PlatoonModel1.Engagement_Damaged)}
PlatoonModel1=<X, Y, M, EIC, EOC, IC, SELECT>

  X={Detection, Maneuver, Engagement_Attack}

  Y={Detection, Maneuver, Engagement_Damaged}

  M={SoldierModel1.1, SoldierModel1.2}
  EIC={(this.Detection,SoldierModel1.1.Detection), (this.Detection,SoldierModel1.2.Detection),
     (this. Maneuver,SoldierModel1.1. Maneuver), (this. Maneuver,SoldierModel1.2. Maneuver),
     (this. Engagement_Attack,SoldierModel1.1. Engagement_Attack),
     (this. Engagement_Attack,SoldierModel1.2. Engagement_Attack)}

  EOC={(SoldierModel1.1.Detection, this.Detection), (SoldierModel1.2.Detection, this.Detection),
     (SoldierModel1.1.Maneuver, this.Maneuver), (SoldierModel1.2.Maneuver,this.Maneuver),
     (SoldierModel1.1.Engagement_Damaged, this.Engagement_Damaged),
     (SoldierModel1.2 Engagement_ Damaged, this.Engagement_Damaged)}

  IC={(SoldierModel1.1.Detection, SoldierModel1.2.Detection),
     (SoldierModel1.2.Detection, SoldierModel1.1.Detection),
     (SoldierModel1.1. Maneuver, SoldierModel1.2. Maneuver),
     (SoldierModel1.2. Maneuver, SoldierModel1.1. Maneuver) }
```

Figure 21.4   A textual DEVS specification of a coupled model, or a platoon model, in mathematical notation for the DEVS formalism.

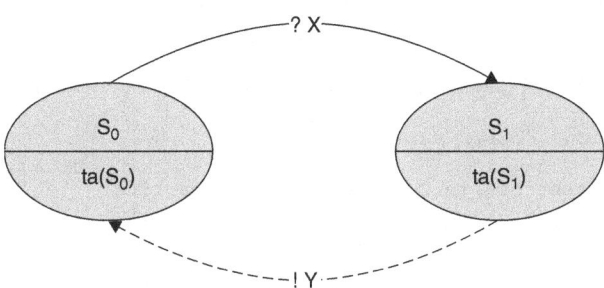

Figure 21.5   A graphical representation of a DEVS atomic model.

solid line represents an external event, which is specified by the question mark on the edge, for a state transition, or δ_{ext}, and the dotted line shows an internal event for a state transition, or δ_{int}. Finally, the ouput function of the transition is marked by the exclamation mark, or !, on the edge. Further information on the DEVS diagram can be found from Song and Kim (2010).

Once we apply this specification method to the soldier model, a modeler can design the soldier model and represent the model as Figure 21.6 graphically and as Figure 21.7 textually.

To put it simply, the DEVS coupled model in combat modeling is the combat scenario and the combat entity structure, and the DEVS atomic model in combat

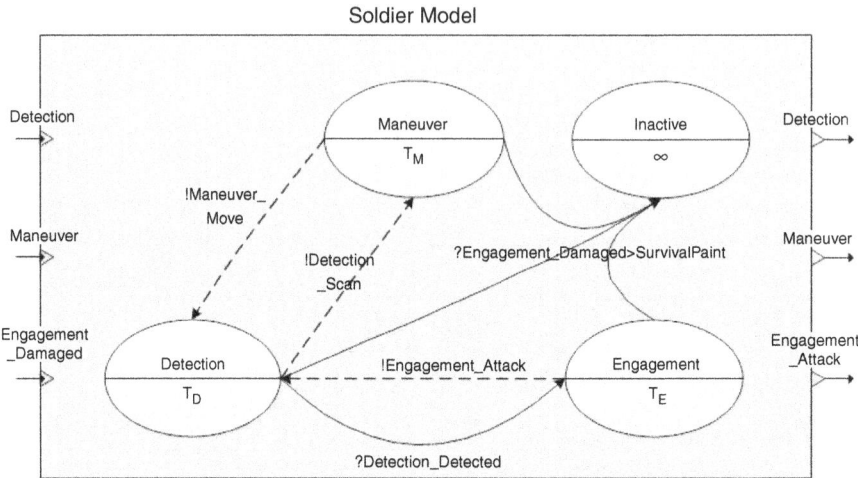

Figure 21.6 A DEVS diagram representation of a DEVS atomic model, or a soldier model.

```
SoldierModel=< X, Y, S, δₑₓₜ, δᵢₙₜ, λ, ta>

  X= {Detection, Maneuver, Engagement_Damaged}

  Y={Detection, Maneuver, Engagement_Attack}

  S={Maneuver, Detection, Engagement, Inactive}

  δₑₓₜ(Maneuver, Engagement_Damaged > SurvivalPoint)=Inactive
  δₑₓₜ(Detection, Engagement_Damaged > SurvivalPoint)=Inactive
  δₑₓₜ(Engagement, Engagement_Damaged > SurvivalPoint)=Inactive
  δₑₓₜ(Detection, Detection_Detected)= Engagement

  δᵢₙₜ(Maneuver)=Detection
  δᵢₙₜ(Detection)= Manuever
  δᵢₙₜ(Engagement)=Detection

  λ(Maneuver)= Move Message through Maneuver port
  λ(Detection)= Scan Message through Detection port
  λ(Engagement)= Attack Message through Engagement_Attack port

  ta(Maneuver)=Tₘ
  ta(Detection)=T_D
  ta(Engagement)=T_E
  ta(Inactive)=infinite
```

Figure 21.7 A textual DEVS specification of an atomic model, or a soldier model.

modeling is the combat entity that is not decomposable any further. However, there is still the question of determining whether an entity is a further decomposable or not. Fundamentally, this question falls into the modeling scope and the resolution, but also there is a useful guideline. Let us imagine that we model a soldier in the previous platoon combat example. Figure 21.8 shows two possible modeling results. We assume that modeled soldiers are capable of maneuver, detection, and engagement behaviors. From these behavior enumerations, the first model specifies that a soldier has three states: the maneuver, the detection, and the engagement states. Since a system has only one state at a certain time,

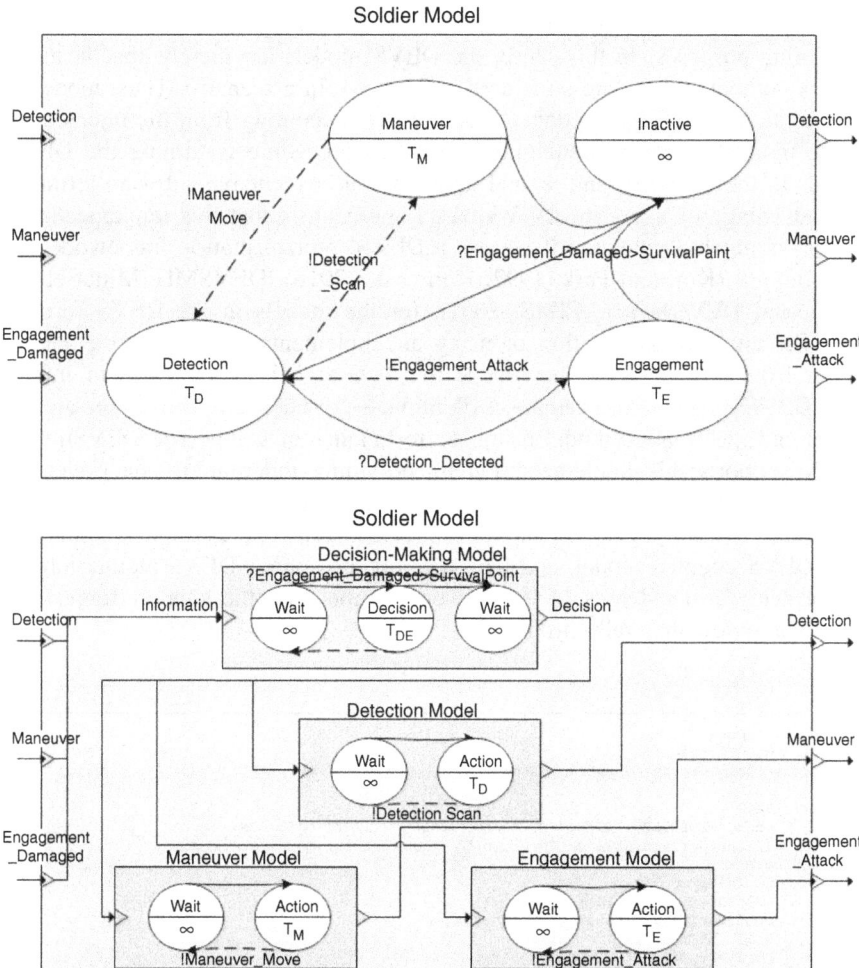

Figure 21.8 Two alternative models of the soldier in the platoon combat scenario: (a) a DEVS atomic model that cannot model behaviors simultaneously; (b) a DEVS coupled model that can model the simultaneous behaviors.

this soldier cannot perform multiple behaviors at the same time, which is not true if we model details of combat scenarios, i.e. a soldier running and shooting simultaneously. Hence, if the modeling details should cover such simultaneous behaviors, we need to specify the behavior as corresponding atomic models as well as the soldier as a coupled model—see the second model in Figure 21.8.

This platoon example shows how to apply the DEVS formalism to the military unit structure and the unit behavior. However, it should be noted that this is only one suggestion of the modeling detail that might vary by different modelers.

Implementing Combat Models by DEVSim++ Framework

The purpose of combat modeling is eventually simulating the models for analytic or training purposes. In this sense, the DEVS models are merely specifications of how combat entities are structured and behave in a scenario. Thus, modelers need to simulate DEVS models to see the results coming from the interactions of the models. For the simulation, a required procedure is turning the DEVS models in the diagrams and textual notations into executable software artifacts. One advantage of using the DEVS based approach is that this implementation process is much accelerated if we use a DEVS implementation framework, i.e. DEVSim++ (Kim and Park, 1992; Kim et al., 2010), DEVSML (Mittal et al., 2007), and DEVSJava (ACIMS, 2007), for the models in the DEVS formalism. The enabled acceleration of using an implementation framework mainly comes from two factors. The first is the well organized templates to implement DEVS models. For example, DEVSim++ provides templates of the atomic model and the coupled model using the techniques of C++. The DEVSim++ library supports this implementation by providing inheritable code skeletons. Figures 21.9–21.11, and 21.12 show two examples of the templates from the DEVSim++ library with the model implementation: one example for the platoon DEVS coupled model, and the other for the soldier DEVS atomic model. We present further details of the implementation with the support framework when we review the walk-through.

```
1    #pragma once
2    #include "atomic.h"
3    class CSoldierModel : public CAtomic
4    {
5    private:
6         enum {MANEUVER, DETECTION, ENGAGEMENT, INACTIVE} m_Status;
7         // further variable definitions to represent the model attribute
8    public:
9         CSoldierModel();
10        ~CSoldierModel(void);
11        bool ExtTransFn(const CDEVSimMessage&);
12        bool IntTransFn();
13        bool OutputFn(CDEVSimMessage&);
14        TimeType TimeAdvanceFn();
15   };
```

Figure 21.9   A part of the header source codes implementing the soldier atomic model in C++ using the DEVSim++ library.

```
1    #include "StdAfx.h"
2    #include "SoldierModel.h"
3    CSoldierModel::CSoldierModel()
4    {
5         AddInPorts(3, "Detection", "Maneuver", "Engagement_Damage");
6         AddOutPorts(3, "Detection", "Maneuver", "Engagement_Attack");
7         m_Status = MANEUVER;
8         // further model initialization steps
9    }
10
11   bool CSoldierModel::ExtTransFn(const CDEVSimMessage &message)
12   {
13        if ( m_Status == DETECTION )
14        {
15            if ( message.GetPort() == "Detection" )
16            {
17              m_Status = ENGAGEMENT;
18              return true;
19            }
20        }
21        // further routines handling external event transitions as specified in δext
22        return true;
23   }
24
25   bool CSoldierModel::IntTransFn()
26   {
27        if ( m_Status == MANEUVER )
28        {
29            m_Status = DETECTION;
30            return true;
31        }
32        // further routines handling internal event transitions as specified in δint
33        return true;
34   }
35
36   bool CSoldierModel::OutputFn(CDEVSimMessage &message)
37   {
38        if ( m_Status == ENGAGEMENT )
39        {
40            CEngagementInformation *message =
41              new CEngagementInformation(/* include variables of target model id and
42              the amount of damage */ );
43            message.SetPortValue("Engagement_Attack",message);
44        }
45        // further routines handling model outputs as specified in λ
46        return true;
47   }
48
49   TimeType CSoldierModel::TimeAdvanceFn()
50   {
51        if ( m_Status == MANEUVER )
52        {
53            return Tm;
54        }
55        // further routines handling model outputs as specified in ta
56        return Infinity;
57   }
58
59   CSoldierModel::~CSoldierModel(void)
60   {
61   }
```

Figure 21.10   A part of the source codes implementing the soldier atomic model in C++ using the DEVSim++ library.

The second factor is the increased reusability of engines, models, and algorithms (Sung et al., 2005; Bae and Kim, 2010). Since we implement models as the template provided by the selected implementation framework, the implemented models can be reused in other coupled models. For instance, the soldier model in the example is currently coupled in the platoon context,

```
1   #pragma once
2   #include "coupled.h"
3
4   class CPlatoonModel : public CCoupled
5   {
6   public:
7        CPlatoonModel(void);
8        ~CPlatoonModel(void);
9   };
```

Figure 21.11 A part of the header source codes implementing the platoon coupled model in C++ using the DEVSim++ library.

```
1   #include "StdAfx.h"
2   #include "PlatoonModel.h"
3
4   CPlatoonModel::CPlatoonModel(void)
5   {
6        CSoldierModel *soldier1 = new CSoldierModel();
7        CSoldierModel *soldier2 = new CSoldierModel();
8
9        // Adding models specified in M
10       AddComponent(2, soldier1, soldier2);
11
12       // Adding coupling relations specified in IC
13       AddCoupling(soldier1,"Detection",          soldier2, "Detection");
14       AddCoupling(soldier2,"Detection",          soldier1, "Detection");
15       AddCoupling(soldier1,"Maneuver",           soldier2, "Maneuver");
16       AddCoupling(soldier2,"Maneuver",           soldier1, "Maneuver");
17
18       // Adding coupling relations specified in EIC
19       AddCoupling(this,"Detection",              soldier1,"Detection");
20       AddCoupling(this,"Maneuver",               soldier1,"Maneuver");
21       AddCoupling(this,"Engagement_Attack",      soldier1,"Engagement_Attack");
22       AddCoupling(this,"Detection",              soldier2,"Detection");
23       AddCoupling(this,"Maneuver",               soldier2,"Maneuver");
24       AddCoupling(this,"Engagement_Attack",      soldier2,"Engagement_Attack");
25
26       // Adding coupling relations specified in EOC
27       AddCoupling(soldier1,"Detection",          this,"Detection");
28       AddCoupling(soldier1,"Maneuver",           this,"Maneuver");
29       AddCoupling(soldier1,"Engagement_Damaged", this,"Engagement_Damaged");
30       AddCoupling(soldier2,"Detection",          this,"Detection");
31       AddCoupling(soldier2,"Maneuver",           this,"Maneuver");
32       AddCoupling(soldier2,"Engagement_Damaged", this,"Engagement_Damaged");
33   }
34
35   CPlatoonModel::~CPlatoonModel(void)
36   {
37   }
```

Figure 21.12 A part of the source codes implementing the platoon coupled model in C++ using the DEVSim++ library.

but the soldier model itself is reusable when other coupled models require a similar soldier model. Additionally, the algorithms of the model behavior are reusable. For example, the soldier's detection behavior might be similar to the firefighter's detection behavior; then one behavior algorithm associated with a state of a model is reusable in another model's state. Finally, the simulation engine of a selected implementation framework is reusable. As far as a model is implemented by following the template of the DEVSim++ library, the model is executable by a general simulation engine provided by DEVSim++. This means that modelers are freed from implementing model execution parts repeatedly, which has to be done when modelers use not the implementation framework but generic programming languages without framework supports. Figure 21.13 depicts such reusability at the engine, the model, and the algorithm levels. We discuss this reusability issue further in the next subsection.

DISCUSSIONS ON MODELING PROCEDURE USING DEVS FORMALISM

This section discusses further on using the DEVS formalism and an implementation framework. The first discussion is about the reusability of using a formalism oriented approach rather than a direct implementation approach. The combat modeling field has been a practical area that emphasizes the real-world applications, so some modelers are keen to use direct implementation of scenarios with the object-oriented approach. We highlight the generational reuse technique from the object-oriented approach and the compositional reuse technique from the DEVS based approach.

The second discussion is how subject matter experts and modeling and simulation experts work together through an implementation framework. Nevertheless, combat modeling still requires extensive involvement of subject matter experts, and learning the formalism and an implementation framework would be a difficult task for some of the military experts. Therefore, we provide a co-modeling approach in which the modeling expert and the subject matter expert work together for combat modeling.

Having stated the objectives of this section, readers may skip it if they want to focus on actual combat modeling outputs and walk-throughs with less interest in the detailed characteristics of modeling procedures using the DEVS formalism.

Reusability from Object-Oriented and DEVS-Based Methodologies

Biggerstaff and Richter (1989) categorizes reusability techniques into two types: generational reusability and compositional reusability. The generational reusability is the patterns of generating software. One such generational reusability is the inheritance of the object-oriented approach (Stroustrup, 1988; Zeigler, 1990; Hill, 1996). A class is inheritable by another class, and the inherited class provides the patterns, such as the lists of methods and attributes in the inherited class, to the inheriting class. The compositional reusability is composing multiple entities to create a larger model. For instance, a DEVS coupled model is the result of compositions of DEVS models.

Since the DEVS based implementation often relies on the object-oriented approach, the DEVS implementation framework built upon the OO paradigm language, i.e. DEVSim++ and DEVSJava, utilizes both reuse techniques (Choi and Kim, 1997). As described in Figure 21.14, the combination of the DEVS based implementation and the object-oriented approach (Choi and Kim, 1997) enables

492

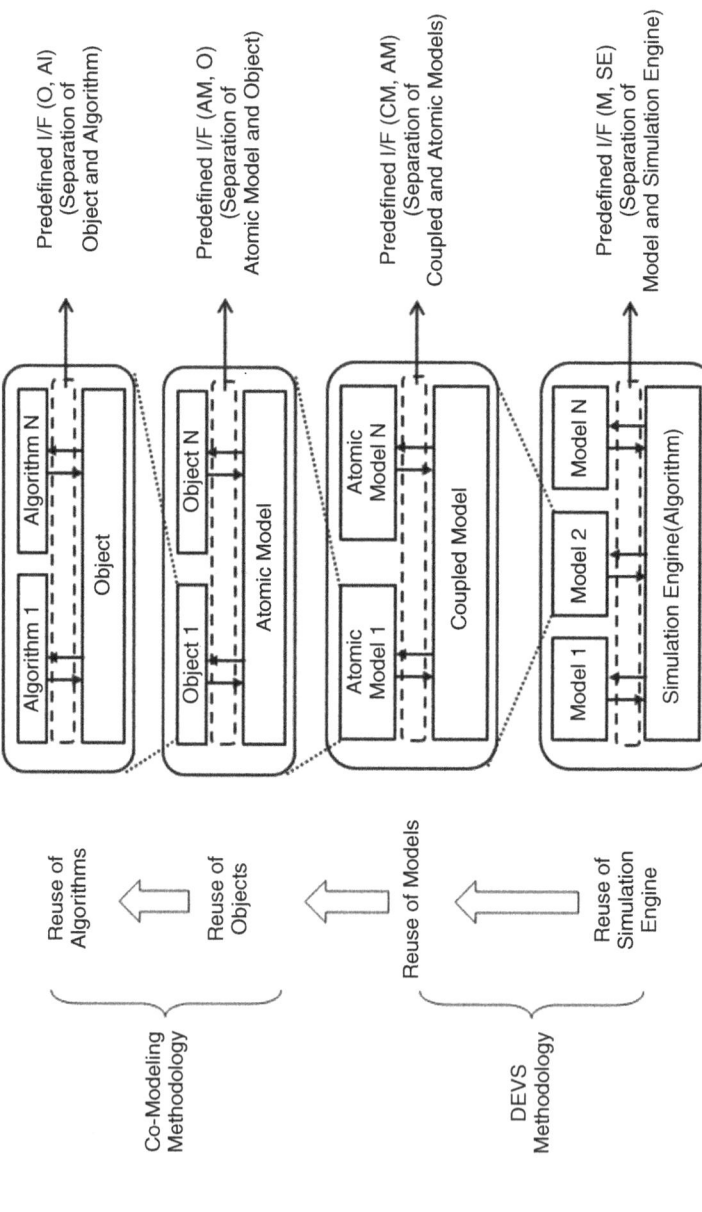

Figure 21.13 Illustration of the reusability at the simulation engine, model, object and behavior algorithm levels when using the DEVS based approach.

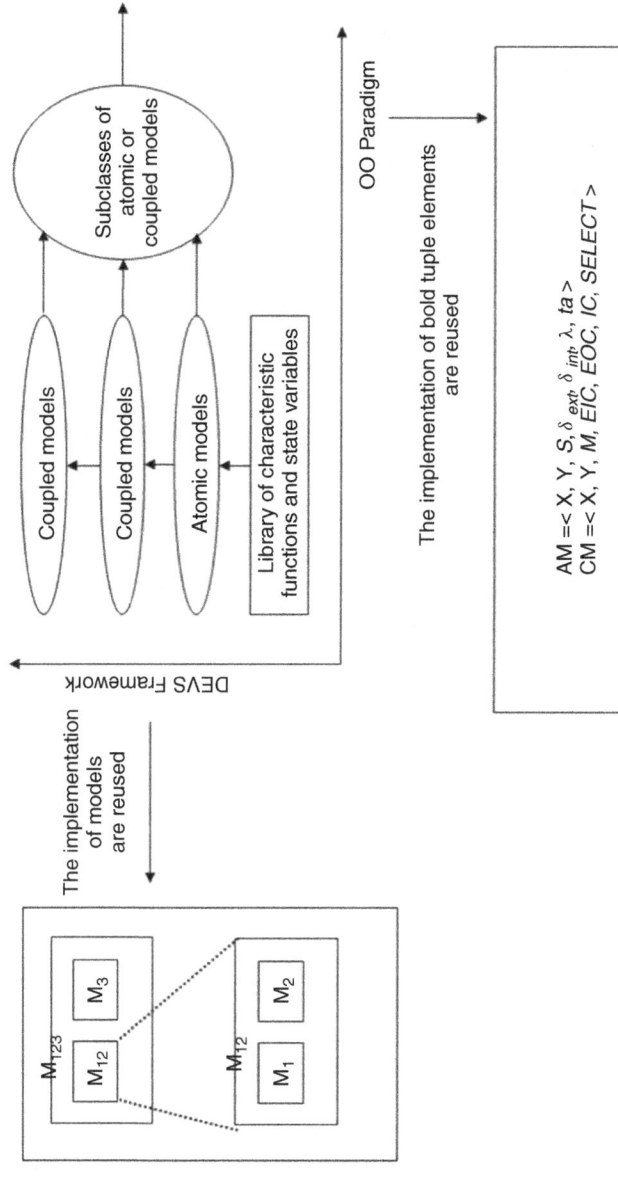

Figure 21.14 Illustration of two dimensional reusabilities: utilizing the composition-wise reusability of the DEVS framework and the inheritance-wise reusability of the object-oriented paradigm.

493

two different types of reusability. The first kind of reusability is the reusability of behavior and states, or the internals, of the models. This reusability is mainly achieved through utilizing the inheritance of the object-oriented approach. For instance, when we implement a commanding officer model by inheriting the soldier mode, we can reuse all the soldier model's behavior and states, and we can implement only the behaviors and states that are different from those of the solder models. The second kind of reusability is the reusability of implemented models in composing new models. This reusability is mainly achieved through the composition of the DEVS based implementation. For instance, when we compose a company model, the pre-implemented platoon model can be reused by coupling multiple platoon models into a single company model.

Co-modeling Methodology for Combat Modeling

Besides the reusability, the DEVS based implementation enables a better cooperative modeling approach, or co-modeling (Sung et al., 2005, 2010), between the subject matter experts and the modelers. Given a combat scenario, modeling and simulating this domain-specific scenario requires deep domain knowledge and modeling and simulation expertise. If we were to assume that a modeler has in-depth knowledge in software technologies, i.e. programming a simulation model, the modeler needs information about military science, i.e. extensive data about weapons, strategy, tactics, etc. The formal implementation of the DEVS based framework allows partitioning the overall model implementation into the implementation of the model structure and the implementation of the detailed model behavior. Therefore, modelers focus on the implementation of the discrete event system structure of a combat scenario. Concurrently, the subject matter experts concentrate on the implementation of the behavior of the modeled entities. For example, the modelers implement how the state of a soldier model transitions between the detection, the maneuver, and the engagements; and the subject matter experts implement the advance patterns when the soldier model is in the maneuver state. See Figure 21.15 for the illustration of the cooperation.

CASE STUDY: NAVAL AIR DEFENSE

This section provides the walk-through of the combat modeling by the DEVS based approach. First, we introduce an example combat scenario: a naval air defense scenario at the engagement level. Second, we illustrate how we model the scenario as a hierarchical discrete event system; and how we model the atomic models in the hierarchy. Finally, we show a sample battle experimentation design and its result.

Modeled Scenario: Naval Air Defense

We present a walk-through of combat modeling by introducing a case study of the naval air defense scenario. The naval air defense is a sequence of activities to make enemy warplanes and missiles ineffective or to reduce their threat level

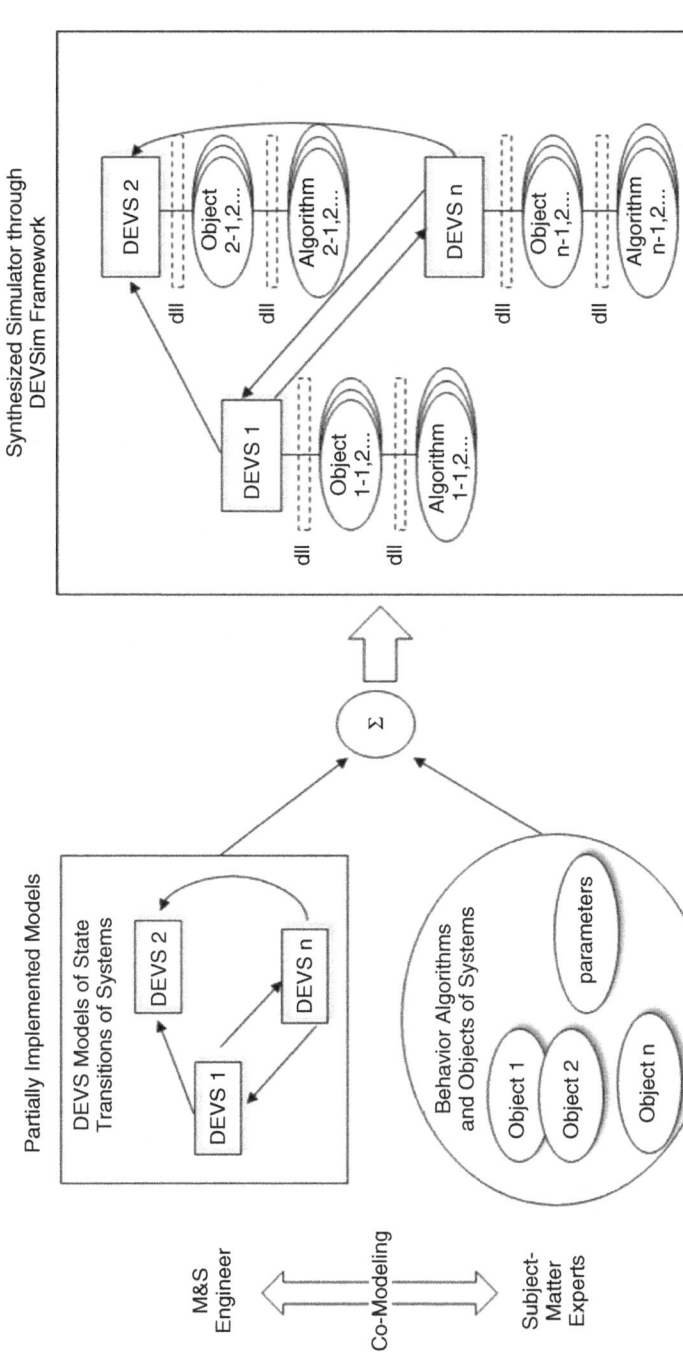

Figure 21.15 Illustration of the co-modeling approach that partitions the modeling by the subject matter experts and the modeling and simulation engineers.

against friendly warships, and such sequence of activities is specified as the doctrine of naval air defense (Neary, 2008).[1] The defense of the warship is difficult mainly due to its geographical characteristics, e.g. no concealment against threats. To compensate for this difficulty, modern navies have developed anti-air weapon systems, which resulted in the multilayer defense concept. This concept assumes that many warships with various detection abilities defend themselves with respect to the distance of the threats by stages. The warship tries to mitigate threats from long distances to short distances, according to the ranges of various weapon systems on board. At the same time, warships maintain a certain formation to optimize the opportunities to intercept the threats.

To simplify the walk-through example, we limit ourselves to model a few features in the naval air defense described in a doctrine of the Republic of Korea Navy. First, our simulation limits air threats to missiles, specifically. We assume that the missiles fly directly toward a warship. Then, we model on-board weapon systems with different capabilities in terms of detection and counter-measures. Figure 21.16 illustrates a snapshot of the naval air defense scenario. In the figure, multiple warships defend themselves with respect to the distance of the threats, or the red missiles, by stages. The warships try to mitigate the threats from long distances to short distances, according to the ranges of three weapon systems, i.e. a long range weapon, or surface-to-air missiles (SAMs); a medium range weapon, or rolling airframe missiles (RAMs); and a short range weapon, or closed-in

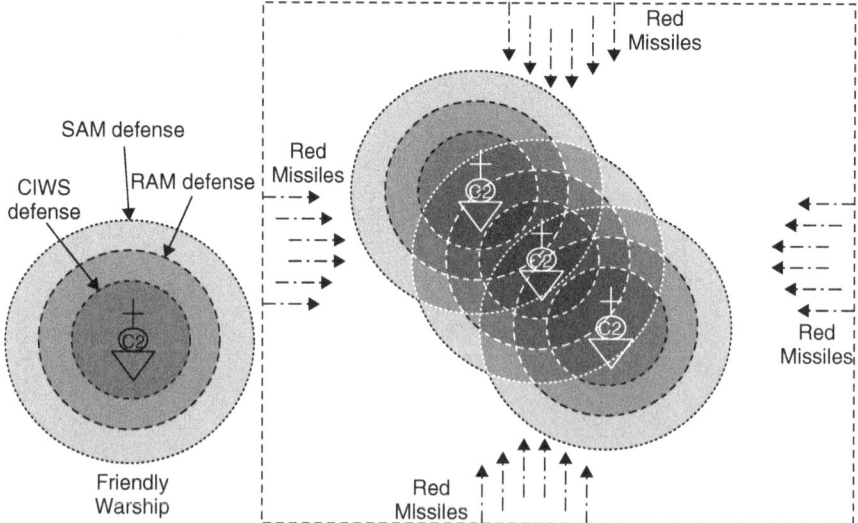

Figure 21.16   Snapshot of the naval air defense scenario.

[1] Interviews with Air Defense Experts at the Naval Education and Training Command of the Republic of Korea Navy, 2010.

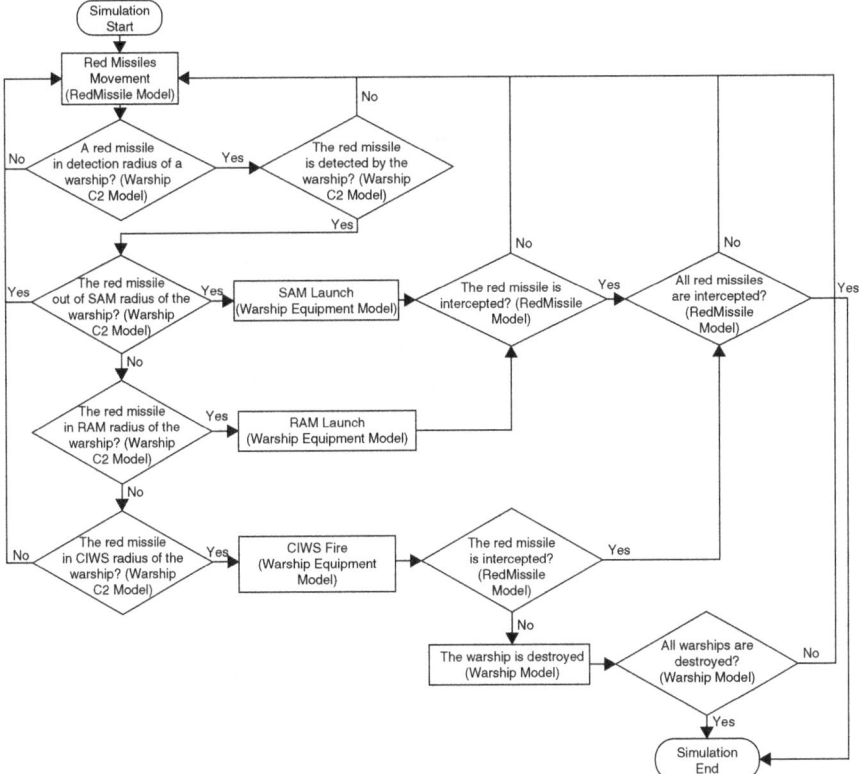

Figure 21.17 Simplified flowchart of the naval air defense scenario.

weapon systems (CIWS). The warship command and control (C2) personnel command and operate the warship weapon system.

Whereas Figure 21.16 is the snapshot illustration of the scenario, Figure 21.17 is the flow of the naval air defense procedure. This flow is also extracted from the navy doctrine, and the flow contains model behaviors as well as a simulation process. For example, the flow specifies that a warship will launch counter-measures which have the longer range first, i.e. SAMs, and the shorter range later, i.e. CIWS. This behavior needs to be formally specified within the model behavior. Moreover, the flow suggests that there should be interactions between the red missiles and the warship weapon systems to represent the possibility of interceptions. This means that the model needs to couple the weapon systems and the red missiles to model the interception, which should be specified as the model coupling in the DEVS formalism. Also, the flow says that a simulation will terminate when either warships or missiles are annihilated. This needs to be implemented within the simulation model, not as a modeled entity but as a control monitoring the simulation progress. Finally, we enumerate the modeled features of the target scenario in Table 21.1 (Kim and Sung, 2007).

Table 21.1   The Input Parameters and the Output Performance Indexes of the Naval Air Defense

Type	Name	Description
Input parameters	Number of warships	The number of warships composing fleet (default = 3)
	Number of red missiles	The number of red missiles in the threat forces (varied in the experimental design)
	Warship position	The X, Y, Z coordinates of warships composing fleet (warships' default coordinates ⟨150 km, 170 km⟩, ⟨160 km, 170 km⟩, ⟨170 km, 150 km⟩)
	Red missile position	The directions of red missiles in the threat forces. The red missiles are located in four directions relative to warships before engagement starts. For each direction, the number of red missiles is the same (four default directions: north, west, east, south)
	Red missile speed	The missile speed of red missiles in the threat forces. The speed of all missiles is the same (varied in the experimental design)
	RADAR radius	The maximum detection radius of RADAR on warship (default = 1000 km)
	CIWS radius	The maximum radius of CIWS warship (default = 3.2 km)
	RAM radius	The maximum radius of RAMs on warship (default = 10 km)
	SAM radius	The maximum radius of SAMs on warship (varied in the experimental design)
Performance index	Survival rate of warships	The ratio of the number of survival warships to the total number of warships when an interoperation simulation is terminated
	Intercept rate of the red missiles	The ratio of the number of intercept red missiles to the total number of red missiles when an interoperation simulation is terminated

Development of Combat Models

After the combat scenario is introduced and studied, we build a discrete event system model in the DEVS formalism. To build the system model, we identify potential models, interactions, and underlying parameters. To list a few, the items below are easily identified as potential models from the scenario:

- warship
- red missile
- weapon systems, i.e. SAMs, RAMs, and CIWS
- warship command and control (C2)

These identified items are the key elements of the combat scenario, and there are obvious coupling relations between them. For instance, a warship has weapon systems, so the warship model should contain the weapon system models. The warship C2 controls the weapon systems, so the warship C2 model is coupled with the weapon system models; and the warship C2 model is naturally contained by the warship model. The warships engage the red missiles, so the warship models and the red missiles should be coupled to pass the simulated engagement events. After recognizing these key models and their coupling relations, we hierarchically structure the models by the specification of the DEVS coupled models, and we present the DEVS diagram from this scenario in Figure 21.18.

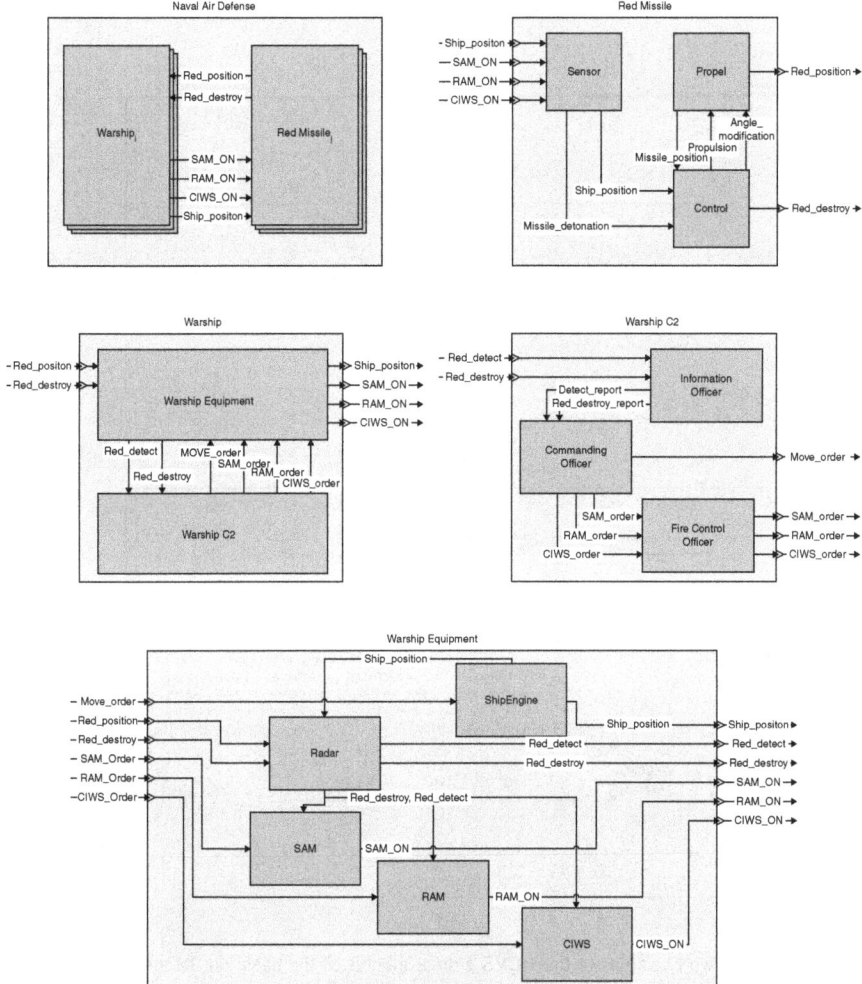

Figure 21.18 Diagrams of the DEVS coupled models of the naval air defense scenario.

With the DEVS coupling structure, we now identify what should be the inputs and the outputs of the designed models. Hence, we move toward designing the atomic model that is the model which is not decomposable any further at the targeted modeling abstraction. For instance, the propel model of the red missile model corresponds to the propulsion part of an anti-ship missile which is a sufficient detail in this naval air defense model. Therefore, we model this propulsion component as an atomic model and represent its behavior as the state transition diagram or the DEVS atomic model diagram—see Figure 21.19. The DEVS diagram specifies that the propel model enters the active state once the model receives the propulsion event, which represents the launch of the missile. Similarly, we also represent the behavior of the commanding officer model in Figure 21.19. The diagram of the commanding officer model specifies the events that start the decision making and the ordering cycle of the C2 activity. In the

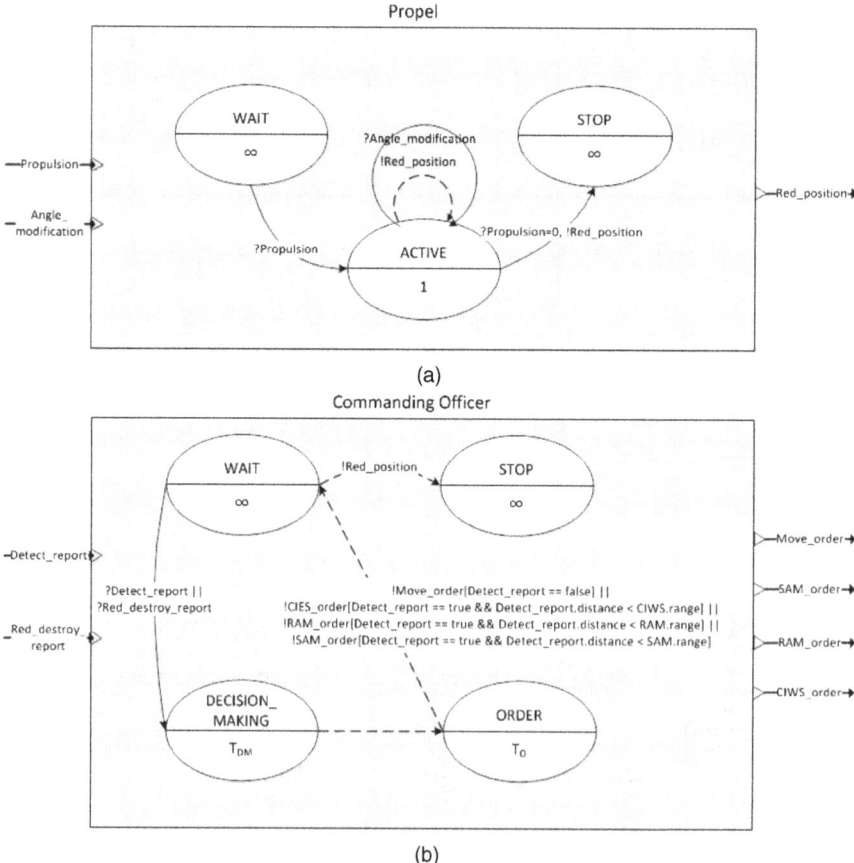

Figure 21.19 Two examples of the DEVS atomic models of the naval air defense scenario: the propel atomic model of the red missile coupled model (a) and the commanding officer atomic model of the warship C2 coupled model (b).

real world, the counter-measures are launched according to the distance of the approaching threat, so the output function of the ordering state has a detailed output condition that simulates the weapon selection flow in Figure 21.17. That being said, whether the decision-making and the ordering cycle is further decomposable or not is fundamentally dependent on the modeler's target abstraction level, and currently they are not modeled as a simultaneous behavior.

Now, we have the state transition level descriptions of the combat modeling, but we need further modeling of what should be the behavior of the system when it is in a certain state. For instance, we need to model the behavior of the propel model of the red missile when it enters the activation state. The behavior would specify the maneuver of the missile, or the approaching trajectory toward a target warship. Furthermore, the output function of the activations state requires the message creation of the approaching information for the activation of the external event function of other coupled models. This trajectory simulation is considered to be the low level detail of the hardware operation in this modeling level, for which we need further information from subject matter experts. In the DEVS modeling approach, this detailed system behavior needs to be coded as an algorithm that is invoked in either external event function, internal event function or output function. Therefore, this algorithm level modeling is delegated to the modeling work at the implementation stage, not at the model development stage.

Implementation of Combat Models

This section introduces the walk-through of the implementations corresponding to the DEVS coupled model, the DEVS atomic model, and the algorithms modeled in the previous section. We proceed with the implementation by utilizing the DEVSim++ implementation framework, and we present the code level implementation to show how the diagrams are realized as software artifacts.

First, we develop codes to represent the DEVS coupled model in the DEVSim++ framework. Figure 21.20 is the C++ code implementing the warship model using the DEVSim++ framework. Coding a coupled model consists of two parts. The first part adds the sub-modules that the coded coupled model contains in the hierarchy; refer to line 11 in Figure 21.20. The contained sub-modules are specified as M in the formalism of the DEVS coupled model. The second part specifies the coupling relations between a pair of a sub-module and another sub-module; or a sub-module and the coded coupled model that is represented *this* in the code; refer to lines 17–32 in Figure 21.20. The coupling relations are specified as EIC, EOC, and IC in the formalism of the DEVS coupled model.

Second, we implement the DEVS atomic model that contains the state transition design. Figure 21.21 is the DEVSim++ implementation of the propulsion atomic model of the red missile coupled model. We enumerate the possible discrete event states that are already specified as S in the atomic model design; see line 12 in Figure 21.21. After setting up the state variable, we implement the state transition functions that are invoked by either external or internal events,

```
1   #include "WarShip.h"
2   #include "WarShipC2.h"
3   #include "WarShipEquip.h"
4   ............
5
6   CWarShip::CWarShip(/*.... Required parameters */)
7   {
8        CModel* pWarShipC2 = new CWarShipC2(id, attackRange);
9        CModel* pWarShipEquip = new CWarShipEquip(id, detectRate);
10
11       AddComponent(2, pWarShipC2 ,pWarShipEquip);
12
13       AddInPorts(2, "red_position", "red_destroy");
14       AddOutPorts(4, "ship_position", "sam_on", "ram_on", "ciws_on");
15
16       //EIC
17       AddCoupling(this, "red_position", pWarShipEquip, "red_ position ");
18       AddCoupling(this, "red_destroy", pWarShipEquip, "red_destroy");
19
20       //EOC
21       AddCoupling(pWarShipEquip, "ship_position", this, "ship_ position ");
22       AddCoupling(pWarShipEquip, "sam_on", this, "sam_on");
23       AddCoupling(pWarShipEquip, "ram_on", this, "ram_on");
24       AddCoupling(pWarShipEquip, "ciws_on", this, "ciws_on");
25
26       //IC
27       AddCoupling(pWarShipC2, "move_order", pWarShipEquip, " move ");
28       AddCoupling(pWarShipC2, "sam_ order ", pWarShipEquip, "sam_ order ");
29       AddCoupling(pWarShipC2, "ram_ order ", pWarShipEquip, "ram_ order ");
30       AddCoupling(pWarShipC2, "ciws_ order ", pWarShipEquip, "ciws_ order ");
31       AddCoupling(pWarShipEquip, "red_destroy", pWarShipC2, "red_destroy");
32       AddCoupling(pWarShipEquip, "red_detect", pWarShipC2, "red_detect");
33   }
```

Figure 21.20   A part of the source codes implementing the warship coupled model in C++ using the DEVSim++ library.

and the functions are specified as δ_{ext} and δ_{int} in the atomic model formalism. Since these state transition functions need to be invoked by the DEVSim++ simulation engine, the functions should be coded as the publicly accessible method in the atomic model object; refer to the implementation from line 19 to 42 in Figure 21.21. Moreover, as the state changes, the model broadcasts its outputs through the defined output posts, or λ in the DEVS atomic model formalism, and this output operation is specified by the method implemented from line 44 to 52 in Figure 21.21. Finally, the discrete event system has its time advance function, or ta in the DEVS formalism, to specify the next internal event creation whose alternative meaning is the time-out of the current state, and we implement this time advance function as a method from line 54 to 63 in Figure 21.21.

Third, we implement the algorithm that changes the model information, such as the location of the red missile, in the active state of the propulsion atomic model. This algorithm describes the most detailed description of the system behavior, so this algorithm is better implemented with the subject matter expert's support. Therefore, we separate the implementation of the system behavior from the state transition implementation of the coded model. From line 31 in Figure 21.21, we implement a function call that utilizes a dynamically linked library implementing the maneuver algorithm. Figure 21.22 shows the implemented algorithm in a separate code library.

```
1    #include "Propel.h"
2    #include "libRedMissileObj.h"
3    ....
4
5    CPropel::CPropel(int id)
6    {
7        SetName("Propel");
8
9        AddInPorts (4, "propulsion", "angle_modification");
10       AddOutPorts (1, "red_position ");
11
12       m_Status = WAIT;
13
14       this->id = id;
15   }
16
17   CPropel::~CPropel(void){ }
18
19   bool CPropel::ExtTransFn(const CMessage &message)
20   {
21       if ( message.GetPort()  ==  "propulsion")
22       {
23           if ( message.getSpeed() == 0) m_Status==STOP;
24           else m_Status==ACTIVE;
25           // Perform supporting initializations
26       }
27
28       else if (message.GetPort() == "angle_modification")
29       {
30           //Invoke an maneuver function implementing a detailed movement algorithm
31           red_missile->RNextPosition(posX, posY, posZ, shipPosX, shipPosY, shipPosZ,
32           speed, 1);
33           ............
34       }
35       return true;
36   }
37
38   bool CPropel::IntTransFn()
39   {
40       if (m_Status == ACTIVE)
41           m_Status = ACTIVE;
42       return true;
43   }
44
45   bool CPropel::OutputFn(CMessage &message)
46   {
47       if (m_Status == ACTIVE)
48       {
49           CRedMissileObj* red_missile_pos = new CRedMissileObj(posX, posY, posZ, id);
50           message.SetPortValue("red_position ", red_missile_pos, true);
51       }
52       return true;
53   }
54
55   TimeType CPropel::TimeAdvanceFn()
56   {
57       if (m_Status == ACTIVE)
58           return 1;
59       else if (m_Status == WAIT)
60           return Infinity;
61       else if(m_Status == STOP)
62           return Infinity;
63       return Infinity;
64   }
```

Figure 21.21 A part of the source codes implementing the propel atomic model in C++ using the DEVSim++ library.

```
1   void CRedMissileObj::RNextPosition(double x, double y, double z, double ship_x,
2   double ship_y, double ship_z, double speed, double time)
3   {
4       distance = sqrt (((x - ship_x)*(x - ship_x)) + ((y - ship_y)*(y - ship_y)) +
5       ((z - ship_z)*(z - ship_z)));
6
7       x_distance = x - ship_x;
8       y_distance = y - ship_y;
9       hypo_distance = sqrt (((x - ship_x)*(x - ship_x)) + ((y - ship_y)*(y - ship_y)));
10
11      redZCosinetheta = (z / distance);
12
13      redSpeed = speed;
14      redXPos = x - (redSpeed*time*(x_distance/hypo_distance));
15      redYPos = y - (redSpeed*time*(y_distance/hypo_distance));
16      redZPos = z - (redSpeed*time*redZCosinetheta);
17  }
```

Figure 21.22 A part of the source codes implementing the movement trajectory algorithm of the propel atomic model in C++ using the DEVSim++ library.

The key of this implementation procedure is implementing the formally designed model, i.e. DEVS models, by an implementation framework supporting the formalism, i.e. DEVSim++. Compared with the object-oriented modeling and programming approach, this formal approach loses the flexibility of the general software implementation. On the other hand, the formalism supports better reusability and transparency in the model development. The models of the system structure, the system state transition, and the system behaviors are represented as the DEVS coupled model, the DEVS atomic model, and the algorithm for a specific state of an atomic model. Utilizing the DEVSim++ implementation framework supporting the DEVS formalism, these models are implemented with a clear separation that facilitates the reusability and the cooperation of modelers and subject matter experts.

Simulation of Combat Models

The general purpose simulation engine in the DEVSim++ executes the implemented models, so there is little demand for modelers to develop the codes related to the simulation execution. This utilization of the simulation engine is allowed by the formal implementation of the models. Under the assumption that the simulator is executed in a stand-alone machine, the simulation engine controls the flow of the simulation execution, which is fundamentally how to call functions in models and to propagate messages between models.

At the initialization stage, the simulation engine instantiates the implemented models. Then, the first task of the simulation engine is calling the time advancement functions of models and finding the minimum time advancement (see Figure 21.23). After finding the minimum, the engine assumes that the simulation time is increased to the minimum time advancement if the model with the minimum time advancement has no stacked external events. If the model has external events, the model will call the external transition function to change the states by the incoming event messages. Once the simulation time is increased, the

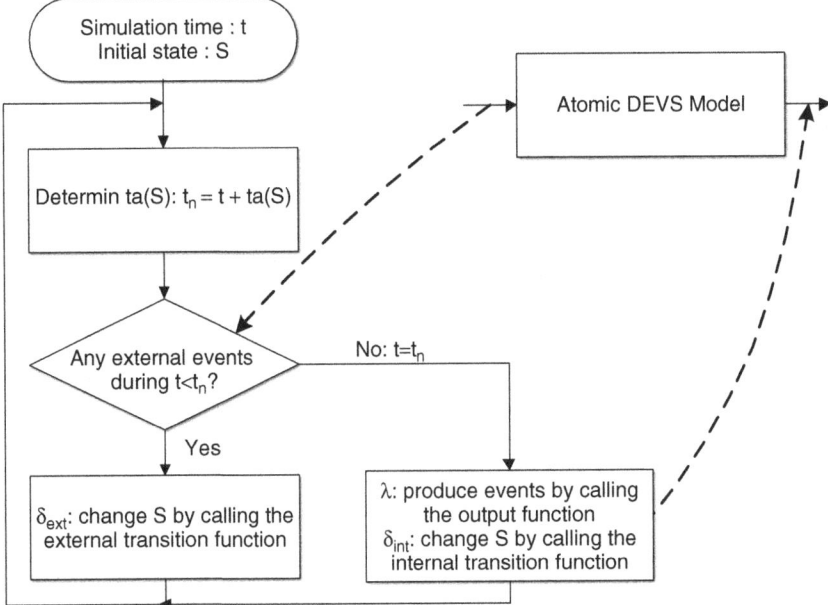

Figure 21.23 A simplified flowchart of simulation execution of a DEVS atomic model.

called model will produce events by running the output function after the internal state transition, and the engine calls the internal transition function. Finally, the produced events will be delivered to coupled models as described in Figure 21.24, and the simulation engine will request the next minimum time advancement.

Rather than the simulation execution being controlled by the simulation engine, the design, setup, and analysis of the battle experiment are the key responsibilities of the combat modelers when they aim to use the simulator for an analytic purpose. The battle experiment design is ultimately related to the question of what-ifs, e.g. whether a warship would better survive if the speed of the red missile is lower than we expected or not. For instance, Table 21.2 is an example of the battle experiment design. The design enumerates the parameters, the number of red missiles, the speed of the red missiles, and the intercept radius of a SAM; then for each parameter the design specifies the range of the variation.

Once the experiment is designed, modelers needs to implement the experimental setup depend on the battle experiment design specifying which parameters, i.e. the speed value used in the maneuver algorithm of the active state of the propel atomic model, vary with what range. The varied parameters are the member variables in the implemented model object or the static variables in the algorithm library. The parameter variation is organized by varying three different parameters at three different levels each, which makes total 27 experiment cases, see Table 21.2 for the details of the experimental design. Then, we repeat executing simulations

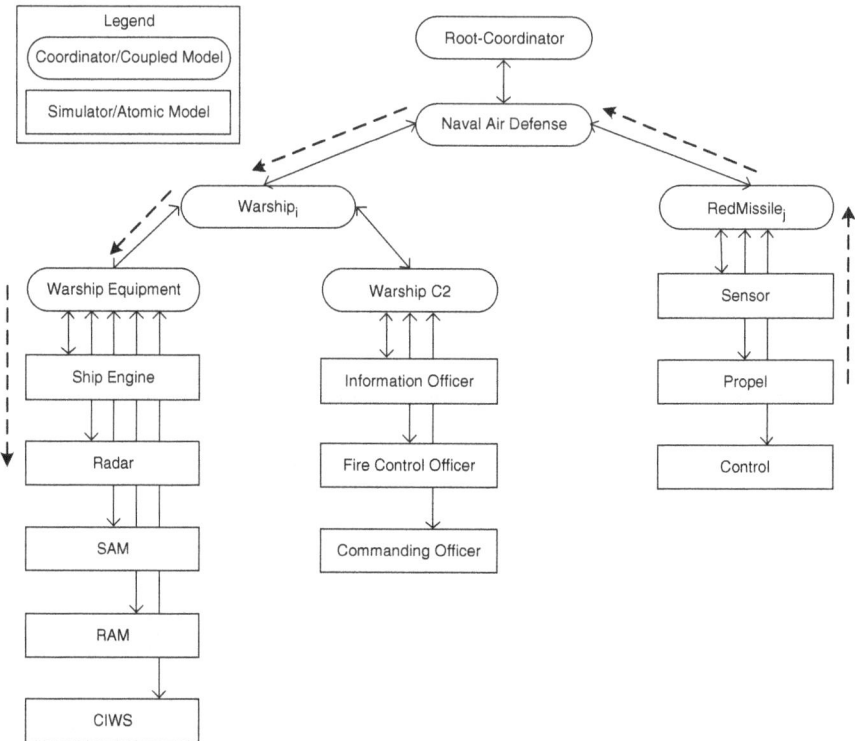

Figure 21.24 A snapshot of simulation execution; the dotted line shows a sample event message passing produced by the output function of a DEVS atomic model.

Table 21.2 Battle Experiment Design of the Naval Air Defense, and 10 Replications are Done (= 270 Executions)

Experiment variable name	Experiment design	Implications
Number of red missiles	8, 16, or 24	Low, medium, or high level of air threats
Red missile speed	300, 480, or 680	Slow, medium, or fast speed of red missile
Missile interception radius of a warship SAM	90000, 110000, or 130000	Short, medium, or long interception radius of warship
Total number of battle experiment cells	27 cells (= 3 × 3 × 3 cases)	Total number of battle experiment cells according to experiment variables

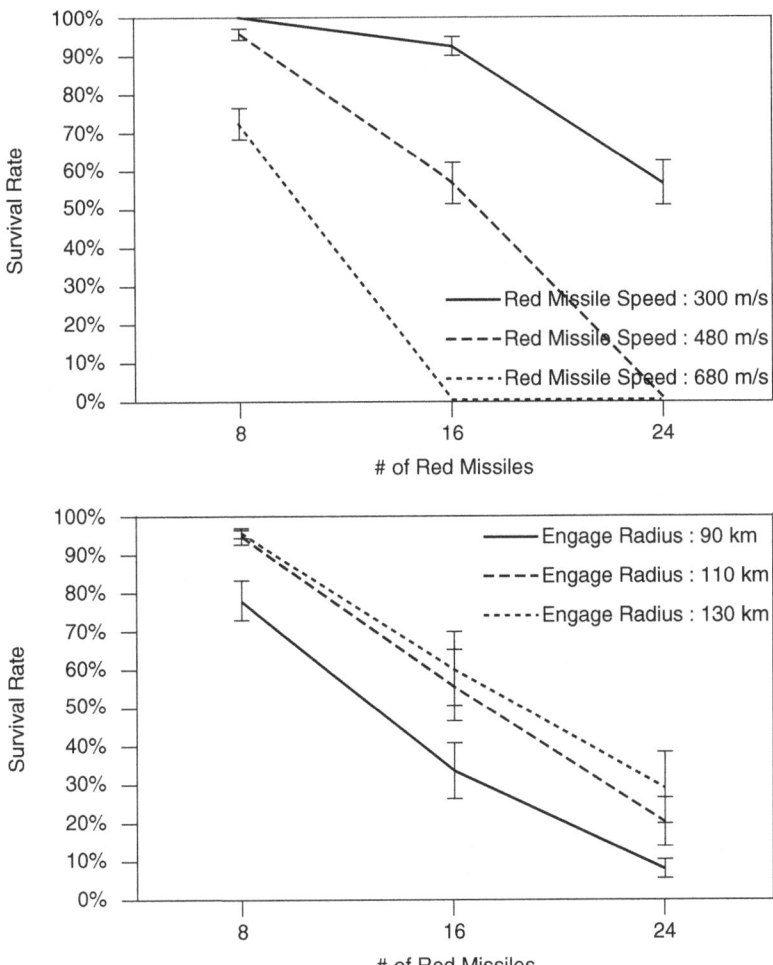

Figure 21.25 Performance index trend of various parameters of the naval air defense.

with each parameter variation case for ten times to reduce the variance of the stochastic simulation runs.

The analysis of the battle experiment starts by reviewing the trend of the simulation result, i.e. the survival rate of warships, in relation to the varied parameters. This analysis is performed by drawing a line chart or by the response surface analysis. Figure 21.25 is one example of the line charts visualizing the trend. As the number of parameter increases, it becomes difficult to visualize the statistical significance and the contribution of each varied parameter. Hence, we perform a meta-model analysis which is building an alternative model that explains the simulation model result approximately and is easily interpretable by

Table 21.3 Regression Analysis of the Sensitivity and the Robustness by Building a Meta-Model

	Standardized coefficient
# of reds	−0.679
Red speed	−0.567
Engage radius	0.211
Adjusted R^2	0.803

observing various statistical indicators, such as P values and standardized coefficients. Table 21.3 is an example of performing the meta-model analysis where we build a regression model. For further information on this battle experiment design, Carley (1999) provides details of the experiment procedure as well as the reasoning behind the statistical examinations.

The presented combat model is initially validated and verified by relying on the subject matter experts in the modeled domain. Yet, it is neither verified nor validated by comparing the simulation result to historic data because such historic data are extremely rare in the combat modeling domain. The subject matter experts examine the response surface and the statistical analysis results of the simulation outcomes to find discrepancies between the model results and their expertise.

CONCLUSION

This chapter briefly introduces the formalism, the framework, and the application of DEVS based combat modeling with a walk-through of producing a combat model of a naval air defense scenario. This DEVS based approach aims to achieve the theoretic grounding and the efficient implementation of combat models. Some may see this approach as a less efficient modeling methodology due to the inflexibility in its representation and implementation and its initial learning time. However, the DEVS formalism is a complete representation of a discrete event system, and most combat scenarios can be described as a discrete event system. Therefore, using the DEVS formalism will provide a complete and transparent representation of combat models, once modelers are experienced using the formalism. Furthermore, the models in the DEVS formalism are easily implemented by following the templates of a DEVS implementation framework, i.e. DEVSim++. Finally, the implemented models are executed by the simulation engine in the framework.

This chapter contains only a brief walk-through, so readers are encouraged to follow suggested reading materials for more practical and detailed information. Since this chapter deals with a comprehensive procedure of building a combat model in the DEVS formalism, details are spread across multiple lines of publication. First, the theoretic research on the DEVS formalism is well introduced in Zeigler (1984), Kim and Zeigler (1987), and Zeigler et al. (2000). Second, more information on developing and representing the DEVS based model is provided by Zeigler et al. (2000), Kim et al. 2010, Song and Kim (2010), and Sung et al. (2010). Third, readers interested in the implementation of DEVS based models are encouraged to follow ACIMS (2007), Mittal et al. (2007), and Kim et al. (2010). Fourth, the applications of combat models developed by the DEVS based approach are illustrated in Seo et al. (2008), Sung et al. (2009), Choi et al. (2010), Kim et al. (2010, 2011). Finally, the methodologies and examples of battle experiments are introduced in Carley (1999), Moon and Carley (2007), and Kim et al. (2011).

FURTHER READING

Piplani LK, Mercer JG and Roop RO (1994) *Systems Acquisition Manager's Guide for the use of Models and Simulations*, Report of the DSMC 1993–1994. Defense Systems Management College, Fort Belvoir, VA. Elsevier.

Zeigler BP and Hammonds PE (2007) *Model and Simulation-based Data Engineering*. Elsevier Science and Technology Books. Academic Press, New York.

REFERENCES

ACIMS (2007). http://www.acims.arizona.edu/SOFTWARE.

Bae JW and Kim TG (2010). DEVS based plug-in framework for interoperability of simulators. In: *Spring Simulation Multiconference 2010, Orlando, FL*. Proceedings of the 2010 Spring Simulation Multiconference (SpringSim'10). ACM, New York, NY, pp. 147–153.

Biggerstaff T and Richter C (1989). Reusability framework, assessment, and directions. In: *Software Reusability*, edited by T Biggerstaff and AJ Perlis. ACM Press, New York, NY, pp. 1–41.

Carley KM (1999). On generating hypotheses using computer simulations. *Syst Eng* **2**, 69–77.

Choi Y and Kim TG (1997). *Reusability Measure of DEVS Simulation Models in DEVSim++ Environment*. SPIE, Orlando, FL, pp. 244–255.

Choi CB, Bae JW and Kim TG (2010). Challenges of teaching modeling and simulation theory to the domain experts in a blended learning environment. In: *International Conference on Information Technology Baswed Higher Education and Training, Urgup, Turkey*, pp. 376–382.

Cramer MA, Beach JE, Mazzuchi TA and Sarkani S (2008). Understanding information uncertainty within the context of a net-centric data model: a mine warfare example. *The International C2 Journal* **3**, 1–55.

Department of Defense (2006). Office of the Under Secretary of Defense for Acquisition. Technology and Logistics. Acquisition Modeling and Simulation Master Plan. Systems Engineering Forum. Web Documents retrieved from http://www.acq.osd.mil/se/docs/AMSMP_041706_FINAL2.pdf. April 17, 2006.

Hill DRC (1996). *Object-Oriented Analysis and Simulation*. Addison-Wesley, Reading, MA.

Kim TG and Park SB (1992). The DEVS formalism: hierarchical modular systems specification in C++. In: *1992 European Simulation Multiconference, York, UK*, pp. 152–156. http://smslab.kaist.ac.kr/paper/CF/CF-18.pdf

Kim TG and Sung CH (2007). Objective-driven DEVS modeling using OPI matrix for performance evaluation of discrete event systems. In: *Proceedings of the 2007 Summer Computer Simulation Conference (SCSC). Society for Computer Simulation International, San Diego, CA, USA*, pp. 305–311.

Kim TG and Zeigler BP (1987). The DEVS formalism: hierarchical modular system specification in an object oriented framework. In: *Proceedings of the 1987 Winter Computer Simulation Conference*. IEEE Press, Piscataway, NJ, pp. 669–566.

Kim JH, Choi CB, Moon I and Kim TG (2010). DEVS-based doctrine validation of fleet anti-air defense. In: Proceedings of the 2010 Spring Simulation Multiconference (SpringSim'10). ACM, New York, NY.

Kim JH, Moon I and Kim TG (2011). *New Insight into Doctrine via Simulation Interoperation of Heterogeneous Levels of Models in Battle Experimentation. Simulation*. DOI: 10.1177/003754971141477.

Kim TG, Sung CH, Hong SY, Hong JH, Choi CB, Kim JH, Seo KM and Bae JW (2010). DEVSim++ toolset for defense modeling and simulation and interoperation. *J Def Model Simul* **8**, 129–142.

Manclark J (2009). Air Force Test and Evaluation Presentation. US Air Force T&E Days. http://www.aiaa.org/pdf/industry/presentations/manclark.pdf

Mittal S, Risco-Martin JL and Zeigler BP (2007). DEVSML: automating DEVS execution over SOA towards transparent simulators. *Proc SpringSim* **2**, 287–295.

Moon I and Carley KM (2007). Modeling and simulation of terrorist networks in social and geospatial dimensions. *IEEE Int Syst (Special Issue on Social Computing)* **22**, 40–49.

Neary CJ (2008). Navy Surface Tactical Missiles. AIAA Strategic and Tactical Missile Systems Conference. http://www.aiaa.org/pdf/industry/presentations/Neary.pdf

Piplani LK, Mercer JG and Roop RO (1994). *Systems Acquisition Manager's Guide for the use of Models and Simulations. Report of the DSMC 1993–1994*. Defense Systems Management College, Fort Belvoir, VA.

Roza ZC (2004). *Simulation Fidelity Theory and Practice*. Delft University Press, Delft, The Netherlands.

Seo KM, Sung CH and Kim TG (2008). Realization of the DEVS formalism in MATALB/Simulink. In: *Proceedings of Grand Challenges in Modeling and Simulation (GCMS-2008), Edinburgh, Scotland*, Jun 16–19, pp. 251–256.

Song HS and Kim TG (2010). DEVS diagram revised: a structured approach for DEVS modeling. In: *Proceedings of European Simulation and Modelling, Belgium*, Oct 25–27, pp. 94–101

Stroustrup B (1988). What is object-oriented programming? *IEEE Software* **5**, 10–20.

Sung CH, Hong SY and Kim TG (2005). Layered structure to development of OO war game models using DEVS framework. In: *Summer Computer Simulation Conference (SCSC '05), Philadelphia, PA*, pp. 65–70.

Sung CH, Hong JH and Kim TG (2009). Interoperation of DEVS models and differential equation models using HLA/RTI: hybrid simulation of engineering and engagement level models. In: *Proceedings of the 2009 Spring Simulation Multiconference (SpringSim '09)*. Society for Computer Simulation International, San Diego, CA. Article 150, 6 pages.

Sung C, Moon I and Kim TG (2010). Collaborative work in domain-specific discrete event simulation software development. In: *Proceedings of the 2010 19th IEEE International Workshops on Enabling Technologies: Infrastructures for Collaborative Enterprises (WETICE '10)*. IEEE Computer Society, Washington, DC, pp. 160–165.

Zavarelli J, DeChiaro SA, Fournier J, Schweickert DA and Zislin A (2006). Live virtual constructive experiments for C2 evaluation. In: *Proceeding of 11th International Command and Control Research Technology Symposium*, Paper I-090. http://www.dsci.com/document/Article_ICCRTS.pdf.

Zeigler BP (1984). *Multifaceted Modeling and Discrete Event Simulation*. Academic Press, London.

Zeigler BP (1990). *Object-Oriented Simulatzon with Hierarchacal Modular Models*. Academic Press, San Diego, CA.

Zeigler BP, Kim TG and Praehofer H (2000). *Theory of Modeling and Simulation*. Academic Press, Orlando, FL.

Chapter 22

GIS Data for Combat Modeling

David Lashlee, Joe Bricio, Robert Holcomb, and William T. Richards

INTRODUCTION

This chapter will provide a very basic understanding of how synthetic environments are typically generated for use in military constructive, virtual, Internet based, and embedded simulations. It briefly describes synthetic environments and some instruments used to gather and describe key data types, and provides some descriptions of key data structures. The understanding of source data for synthetic environment generation and implications for synthetic environment representation are also discussed. It is also important to have visibility on organizations that collect and process data since understanding what to expect from them determines whether or not a plan a for a dedicated data collection mission is necessary or not.

The intended use of synthetic environments for combat modeling is a key determinant in setting fidelity and resolution requirements and it is reflected in data collection, processing, and software runtime considerations. As an example, if a simulation is to be connected with other simulations and exchange entity/state data special considerations need to be taken to make sure that "ground truth" is maintained.

The views expressed in this chapter reflect the views of the authors alone, and do not necessarily reflect the views of any of their organizations. In particular they do not reflect the views of the US DoD Modeling and Simulation Coordination Office or the US Navy.

Engineering Principles of Combat Modeling and Distributed Simulation, First Edition.
Edited by Andreas Tolk.

Finally, this chapter will briefly describe types of runtime applications that use synthetic environments to include geographical information systems and describe some advanced processes for generation of synthetic environments.

Geospatial information systems (GIS) are used extensively by the military during real-world operations such as combat and in many operations other than war (OOTW). These systems, commonly referred to as command and control systems, have access to large databases of GIS data of many flavors and formats. Commanders can select the layers of data that they wish to look at to help them make decisions concerning their current military operations. However, when you move over to the world of combat modeling and simulation, the simulation may require different data, or the same data, but in a very different format. The reason for the difference is that the computer, which is doing some of the decision making for the simulation, defines "optimal" differently than a human will when it comes to perceiving information.

Generally this large amount of data has to be cut down and shaped into a format that represents the God's eye view or "ground truth" of the environment in the simulation. Unlike command and control systems that are often just presenting data to humans for analysis and do not have a real-time requirement, a simulation is managing many stationary and non-stationary objects, referred to as entities. Some of these entities are computer controlled and need to be able to perceive each other and the environment while making decisions that appear believable to the humans interacting with them. Often the simulation has the requirement to manage all the data and entities at a fast enough rate to be considered real time. Because of this increased load on the resources of the computer, optimizations in data formats are made for performance reasons. Otherwise, the computer that is running the simulation will be unable to keep pace with real-time speeds, due to the large amount of data it is keeping track of. This difference in data requirements is a distinction between the two types of systems and mimics the differences between the real-world command and control systems' data and the real world. This chapter will discuss what GIS data are, the general formats they come in, how GIS is used in a simulation, and how to create a terrain database for a simulation. It will also discuss how to mitigate or fix some of the problems with the data prior to and once imported into a simulation system and a few of the many issues that a terrain developer will encounter when building a terrain for a simulation exercise.

TYPES OF SYNTHETIC ENVIRONMENTS

The term "synthetic environment" refers to a representation of a dimensional space that may or may not represent an actual location within the world. Due to advances in computer graphics it is possible to create very visually convincing environments, where every element of that environment is synthetically generated; however, the environment has no correlation with any place that exists in the world. This type of fictional environment is known as an "unprecedented synthetic environment." It is also possible to generate visually convincing environments based on real-world data gathered from various types of sensors. This type of synthetic environment is referred to as "precedented." Regardless of whether generating a fictitious location using artists' tools or by using empirical methods from sensor data collection, the 2D or 3D representation of an environment is considered synthetic.

Unprecedented Synthetic Environments

Unprecedented synthetic environments are generated without using empirical data gathered from a geographic location. It is important to note that an artist recreating a specific geographic location from memory is a form of collecting empirical data and the synthetic representation does not fall within an unprecedented synthetic environment. In combat modeling, a 2D unprecedented synthetic environment may be created to support concept exploration where it has been determined that, during design of an experiment, a fictitious environment would both suffice and present opportunities for attenuating environmental characteristics. A 3D unprecedented synthetic environment may be used to stress specific cognitive attributes of a virtual simulation assisted training system.

Unprecedented synthetic environments are typically generated using software tools that have emphasis on artistic principles and procedural generation, where main software concentration is visual appeal (and they use sophisticated physics based algorithms to achieve perceptually convincing environments). There is no unifying standard that allows for combination of unprecedented synthetic environment representation and spatial metadata attributes to support concept experimentation from a design of experiment perspective. This raises issues especially when applying metadata to an unprecedented model to support design of experiment goals and objectives in a standardized way.

Precedented Synthetic Environments

A *precedented synthetic environment* includes some or all environmental data elements that were empirically gathered from a specific geographic location. The types of data gathered from measurement instruments depend on the nature and objectives of analysis, experimentation, training, or test and evaluation that combat model enables. It is important that such measurements and descriptions are aligned with combat model runtime needs in order to meet design of experiment objectives.

It is important to note that although a precedented synthetic environment might contain data collected from sensors, any results obtained by combat model interaction with a synthetic environment representation are not empirical results since a synthetic environment is an abstracted representation of the geographic area and uses assumptions and systemic representations within a specific domain for its representation (Table 22.1).

Table 22.1   Synthetic Environment Comparisons

Issue	Unprecedented synthetic environments	Precedented synthetic environments
Data sources	Artists' tools and procedural algorithms for data generation runtime application	Geological surveys, satellite sensors, aerial sensors, LIDAR
Level of fidelity	Generally very high in selected areas due to the terrain artists' ability to add detail where needed. Procedural algorithms aid in adding complexity	Can be low to very high depending on the data collection method. Satellite imagery may produce low resolution images and LIDAR may produce very detailed models of the environment
Development time requirements	Generally very high due to the need to have an artist produce much of the content manually. Procedural algorithms help reduce the time required to create high quality environments	Generally lower than unprecedented synthetic environments since sensors collect most of the data and these may be used directly by simulations. Some sensor data will need post processing to be usable, which can increase the development time requirement

GEOSPATIAL DATA COLLECTION

This section provides an overview of the data collection mechanisms used to collect environmental data. The output of the environmental data collection becomes source data for both 2Dand 3D GIS applications that will ultimately constitute the synthetic environment representation. Before an overview and discussion of GIS it is important to have an understanding of how environmental data are generated and the types of sensors that generate such data.

According to Elanchi and Van Kyl (2006) remote sensing is defined as the acquisition of information about an object without being in physical contact with it. Data and information are acquired by electromagnetic, acoustic or potential. Remote sensing is commonly associated with electromagnetic techniques such as low frequency radio waves through microwave, submillimeter, far infrared, near infrared, visible, ultraviolet, x-ray, and gamma-ray regions of the spectrum. Figure 22.1 provides a graphical depiction of the electromagnetic spectrum.

The type of remote sensing data acquired depends on the type of information being sought. With advances in technology, the trend is to have lower cost and more capable sensors, where the output of data resolution is increased (higher frequencies and laser based sensors), sensors themselves become smaller, and they consume less power (Table 22.2).

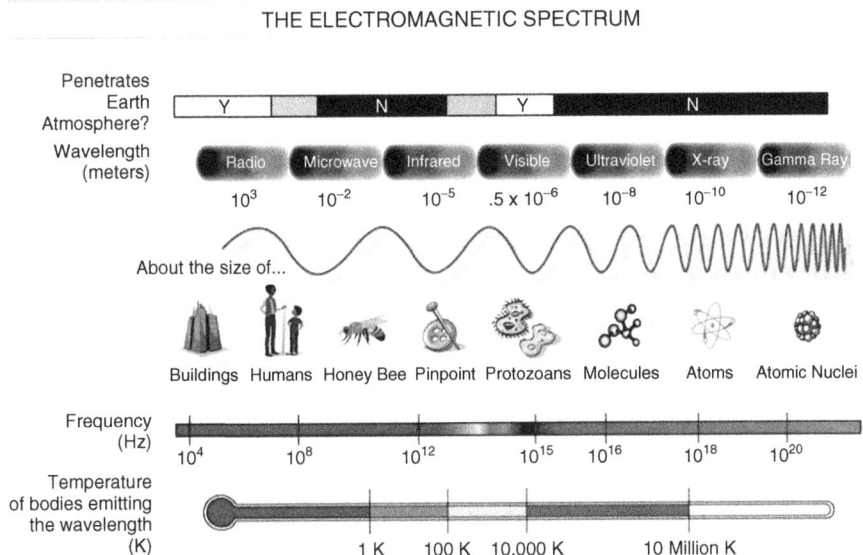

Figure 22.1 Electromagnetic spectrum. *Source: NASA,* 2011.

In Figure 22.2 note the difference in resolution representing the same geographical region using different wavelength spectra. The higher the frequency, the more power the sensor requires and the image resolution becomes higher.

The Basic Types of Data

GIS data are the primary starting point for building the virtual environment in many simulations used for training. In order to make the training useful it is necessary to make it as realistic as possible. To do this, real-world data are collected from GIS systems and then formatted for use by the simulation systems used for training. Generally, simulations will utilize GIS data regarding elevation, soil types, bodies of water, road networks, buildings, and anything that is fixed in place and will not move during a simulation exercise. Sometimes a simulation will also utilize environmental GIS data such as humidity, wind speed and direction, illumination, salinity, turbidity, and even solar radiation.

All of the different GIS data need to be combined into a usable format that the simulation can understand during a simulation exercise. In order to do this, you have to understand the basic types of GIS data that simulations normally deal with. The most common data types are raster data, triangular information networks (TINs), and vector data. Each has its unique uses and abilities as well as challenges.

Table 22.2 Types of Remote Sensing Date

Type of data needed	Type of sensor used	Examples of sensors
High spatial resolution with wide coverage	Imaging sensors, cameras	Large format cameras (50+megapixels), Seasat imaging radar, Buckeye System
High spectral resolution over limited areas of along track lines	Spectometers, spectroradiometers	Multispectral imaging radiometers Lansat multispectral mapper and thematic mapper (1972–1999), SPOT (1986–2002), Galileo NIMS (1989)
Limited spectral resolution with high spatial resolution	Multispectral mappers	
High spectral and spatial resolution	Imaging spectometers	Spacebone imaging spectrometer (1991), Hyperion (2000)
High accuracy intensity measurement along line tracks or wide swath	Radiometers, scatterometers	SeaWinds (2002) scatterometers
High accuracy intensity measurement with moderate imaging resolution and wide coverage	Imaging radiometers	SMOS (2007)
High accuracy measurement of location and profile	Altimeters, sounders	TOPEX/Poseidon(1992)
3D topographic mapping	Scanning altimeters and inferometers	Shuttle Radar Topography Mission (2000)
3D feature generation	Active sensors LIDAR/LADAR, imaging sensors and cameras	Leica Geosystems, TOPCON, Bell Aerospace

Modified from Elanchi and Van Kyl (2006).

Raster Data

Raster data, or 2D gridded data, are data that are broken up into uniform cells with each cell forming a box; each cell contains a small amount of data that describes something about the area the cell overlaps in the real world. For example, an image of the ground surface, such as in Figure 22.3, is a raster data set and each pixel in the image is a cell in the raster representing the color of the terrain at that particular point. However, raster data sets could alternatively indicate the relative elevation of each pixel within the "picture"; these data can then be

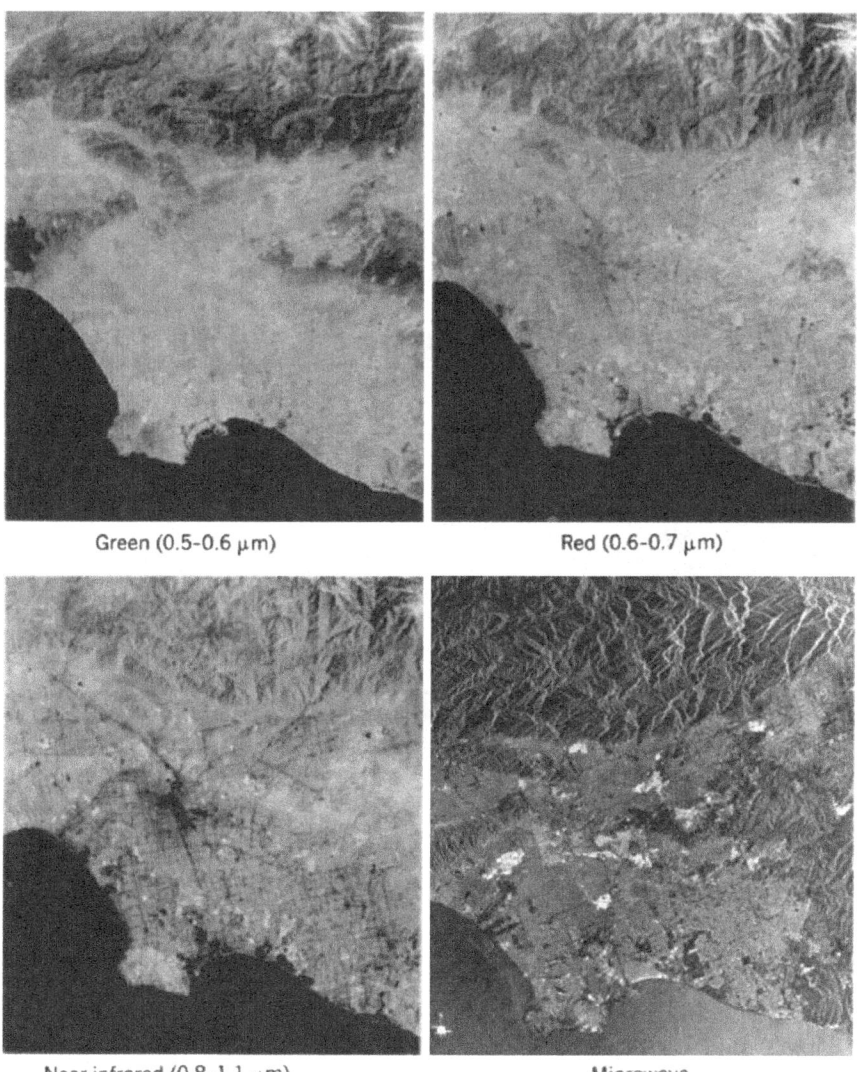

Green (0.5-0.6 µm)   Red (0.6-0.7 µm)

Near infrared (0.8-1.1 µm)   Microwave

Figure 22.2   Multispectral data collected from sensors. *Source*: Elanchi and Van Kyl (2006).

displayed as a bitmap allowing the viewer or the simulation to easily see the changes in elevation. For example, as in Figure 22.4, a single cell could contain information about the average elevation of the part of the terrain that the cell overlaps. It could also indicate the average number of trees in an area or the type of soil at a particular location.

Raster data are very versatile, but do have some limitations. One of the biggest drawbacks to raster data is that it is difficult to capture subterranean

Figure 22.3 VBMP 2002 image raster, 1 meter resolution data.

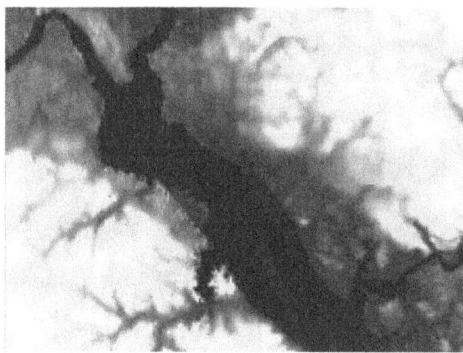

Figure 22.4 DTED level 1 data, approximately 90 meter resolution; colors represent elevation height.

Figure 22.5 Vector data 1:250,000 scale from a VPF data set.

features such as caves or vertical features such as cliffs because representing an overhang is not possible with a 2D image. However, this can be mitigated by using another type of data such as polygon data or voxel data (Figure 22.5). Another limitation of raster data is that it can be very memory intensive. Even if the raster image compresses very well for storage on disk, it will still require to be expanded in computer memory in order to be used. This can be mitigated by some type of paging scheme that only loads the parts of the raster data that are required and unloads the parts that are no longer needed.

Examples of raster data sets are

- Digital Terrain Elevation Data (DTED)
- Digital Elevation Model (DEM)
- Shuttle Radar Topography Mission (SRTM)
- Joint Photographic Experts Group (JPEG)
- Bitmap (bmp)
- American Standard Code for Information Interchange Grid (ASCII Grid)

Voxels

Voxels are a form of raster data that forms a 3D cube as seen in Figure 22.6. Unlike traditional raster data that can only show the topmost surface of an object, a voxel can be used to record subsurface features, the underside on overhangs, and even the material composition of the ground well below the surface. One example of this is a simulation called Ocean, Atmosphere, and Space Environmental Services (OASES) (O'Connor et al., 2001) which not only uses voxels to model the atmosphere but also the oceans and outer space (Richards, 2008). This ability to show more information about an object in great detail also creates a higher cost in storage and access times. Due to this increased level of complexity, voxels have historically not been used in large-scale simulations. However, with advances in current computer hardware, using voxels is becoming more practical.

Triangulated Irregular Network (TIN)

Elevation data can also be found in a triangulated irregular network or TIN format. A TIN is simply a series of triangles that forms a minimal set of polygons

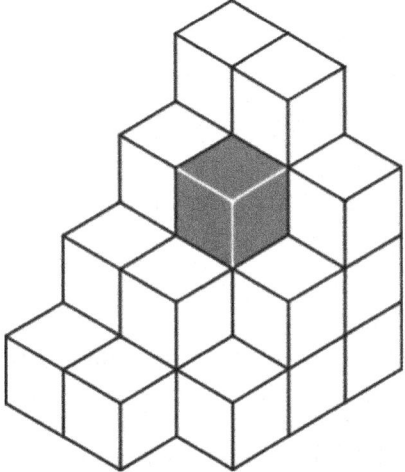

Figure 22.6   Voxel data.

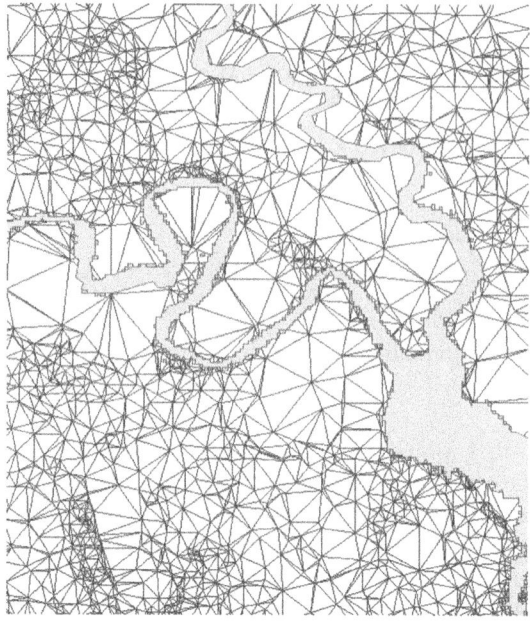

Figure 22.7 TIN elevation data with a water polygon superimposed.

that accurately portrays the shape of the terrain, as shown in Figure 22.7. The variable nature of a TIN allows a more accurate depiction of a surface than with a grid, but is not always suitable for all GIS applications. Each triangle has three vertices that store the X, Y, and Z positions of each corner. Similar to a raster image, each triangle can be drawn to produce a 3D landscape. However, unlike raster data, TINs can also be used to show subterranean surfaces and overhangs. Another difference is that each vertex can be shared between the triangles that touch it, so it is possible to get a reduction in the amount of data that is required to store the same information as a raster image in some cases. If the terrain is relatively flat, such as in a desert, it could be represented with fewer triangles resulting in a greatly reduced data size for the same terrain compared with raster data. If the terrain is very irregular, more triangles will be required, which may result in larger data sets than the equivalent raster data. TIN data generally have lower fidelity than raster data, but provide benefits in processing speed due to the design of CPUs and GPUs in computers today. It is not uncommon to combine TIN elevation data with raster imagery to create realistic terrain that can be processed in real time, as shown in Figure 22.8.

Vector Data

Vector data are the digital representation of geographical information. These data are formatted so that a computer can retrieve the information from a database and scale and redraw it accurately every time it is accessed. The geographical features this data type contains fall into one of three general classes: points,

Figure 22.8   TIN elevation data with raster imagery draped on top.

polygons, or lines. A point feature is simply a positional location on the map. A line or linear feature will have a start and an end point and may have a series of points in between these two that define the bends in the line. A polygon feature will have a center point followed by a series of points that define the perimeter of the polygon with the last point on the polygon overlapping the first point after the center point to close the polygon (Clarke, 2001, p. 20).

Additionally, each of these feature classes will usually have metadata associated with it that define a unique set of attributes for that feature. For example, a road may have the width, number of lanes, and type of construction (asphalt, hard packed dirt, gravel, concrete, etc.) associated with the linear feature. Most vector data also use a set of predefined codes to aid in feature identification as well as standardization between systems. Two well known standards are Feature and Attribute Coding Catalogue (FACC) (National Geo-spatial Agency, 2007) and the Environmental Data Coding Standard (EDCS). For example, a street or road vector datum would have an FACC code of "AD030," uniquely identifying it as a paved road (US National Imagery and Mapping Agency, 1998). However, this simple code is not the only attribute that is needed to fully describe the object. Fields such as height, width, elevation, name, material composition, color orientation, density, and the like help to further identify the object and indicate how the object will interact with the non-static objects (entities) within the simulation. Car entities, for example, would use the attributes of a road to determine how fast they can drive and which lane to drive in and produce believable behaviors for the entity.

The process of creating vector data called vectorization is a time consuming process involving significant human interaction. Basically, a person will take a paper map or digital image and scan it in either by hand or using an automated scanner. Once scanned, the object is given an ID number and a list of attributes appropriate to the object. The attributes are normally saved in one table and the points are stored in another. This process also introduces a fair amount of error into the data which can be caused by the resolution of the data, errors or poor quality in the original data, or even the steadiness of the human's hand while setting or editing the points or attributes for the object (Clarke, 2001, p. 122).

For example Figure 22.9 shows a digitized image being displayed in a GIS tool in which the user is defining an outline of a small lake. The accuracy of the outline is determined by the resolution of the image data, which for this example is 1 ft.

So at best each point along the outline is accurate to within 1 ft of the actual point in the real world. Additionally, as in Figure 22.10 where the centerline of the road is drawn in, the accuracy of the line is affected by the intervening terrain data, in this case the tree canopy.

So for this data set, under the canopy the error of the data is probably greater than 1 ft (Richards and Hibler, 2005). Finally, point features are then added to the data set for each of the houses in the image. Again the intervening terrain affects these data and it can be difficult to determine the exact center of the houses. Also with objects that have a height to them it can be difficult to accurately indicate the actual ground position due to the angle of the image and the time of day or season. The taller the object the greater the effect this has on its true

Figure 22.9 Screen shot of a pond being vectorized. Aerial Imagery © 2002 Commonwealth of Virginia.

Figure 22.10   Screen shot of features being vectorized. Aerial Imagery © 2002 Commonwealth of Virginia.

Table 22.3   Examples of Vector Data

Point features	Linear features	Polygonal
Trees, telephone poles, route labels, place names, points of interest, spot elevations, depth soundings, road signs, cell phone towers, schools, police stations, hospitals, churches	Road networks, fences, borders, streams, tunnels, bridges, dams, contour lines, power lines, sewer lines	Lakes, rivers, forests, soil types, populated areas, building footprints,geo-political areas

ground position unless one corner where the object touches the ground is visible on the image. The final result of this process is a set of vector data with a line, a polygon, and several points in it that can then be used to draw a simple street map or be used in a simulation.

3D Models of Objects

Another data format for simulations is the 3D object model such as a building or individual tree as shown in Figure 22.11. Most 3D models consist of a collection of 3D vertices (X, Y, and Z coordinates), information about how they are connected, and the polygon surface information such as which image to apply to the polygon. Generally most terrain databases will only deal with fixed objects but the idea behind all the 3D models is basically the same whether they are

Figure 22.11   Screen shot from Google SketchUp showing a house, a tree, and an office building.

fixed or mobile. The model is usually contained in a set of dedicated data files that contain the mathematical data and images of the surfaces of the object being described. Usually within a GIS database there is a vector file containing points that indicate the location of a 3D model, its orientation within the database, and the name and location of the file used to visualize the object. A few examples of the many 3D model file formats are Open Flight, Goggle Sketchup, and 3DS from Autodesk. Open Flight seems to be emerging as an impromptu industry standard for 3D terrain and model data within the simulation community.

Layers, Tiles, and Coordinate Systems

Before we start to discuss how to put all the data together, it is best to have a short discussion of layers, tiles, and projections. Layers are simple layers of data that are laid down one on top of another in a specific order to build the content of the terrain database. Tiles are arbitrary sections of the data that are grouped together allowing faster access to the data sets that are currently being viewed or worked with. Finally, coordinate systems are the means of locating places and objects in space and describing the relationships between them.

Simply put, a layer is a set of data that is grouped together for convenience and can be used as an overlay on other data sets. For example, a set of photographs mosaiced together can be considered an imagery layer (Growe and Toenjes, 1997). Additionally, a set of elevation rasters generally makes up the elevation layer. The image layer is often laid on top of the elevation layer so as to give an accurate depiction of the lie of the land. This can then be used to visualize the simulation as was shown in Figure 22.8. Once this is done another layer such

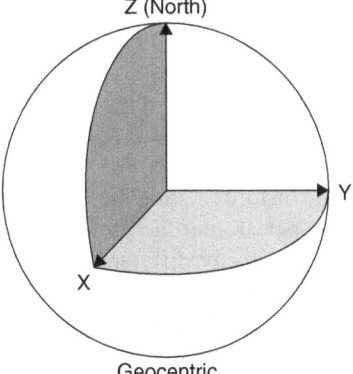

Figure 22.12 Geocentric coordinate system.

as the transportation network layer can be laid down on top of these two layers allowing the simulation to have a knowledge representation of how vehicles should move within the simulation.

Tiles on the other hand are essentially blocks of data that are cut into convenient sections to reduce the size of the data sets. This allows them to be managed, edited, and used by the simulation more efficiently. Tiling also allows for faster load times, and permits loading only the data that are required by the current view in the GIS tool or simulation software. When the terrain is very large, tiling may be the only means of using the terrain with simulation software. Since the terrain is not likely to be needed all at once, only the parts that are needed are loaded into the computer's memory and any tiles that are no longer needed are released. This allows the computer to most effectively use the limited memory in the system (Figure 22.12).

Finally, coordinate systems are the means by which we can relate different places in the world to one another. Most people are familiar with the Cartesian coordinate system from school. It has an X, Y, and Z axis and a point in space is represented as the distance from the origin along each of these three axes. The geocentric coordinate system is commonly used in simulation. It is a Cartesian coordinate system with the origin located at the center of the earth, with the Z axis coming out of the North Pole, the X axis along the Prime Meridian, and the Y axis at $90°$ east (Yang et al., 2000). The most commonly used survey defining the center and surface of the earth is the World Geodetic System taken in 1984 (WGS84) which is the same one used by GPS satellites today. The unit of measure in a geocentric coordinate system is the meter. This puts locations on the surface of the earth quite a distance away from the center. Using a geocentric coordinate, we can locate home plate of Fenway Park in Boston at $\{1529450.1792, -4466591.1310, 4274117.7724\}$.

There are several downsides to using a geocentric coordinate system. The first is that the representation is difficult for a person to understand by just looking at the X, Y, and Z coordinates. The second is that the direction of "up" varies depending on where the person is in relation to the origin. Also, there are some problems understanding where the center of the earth is exactly. This plays into consideration when converting from geocentric to another system that may be a projection.

Projections are a mapping of the 3D surface of a sphere to a 2D representation such as a map or computer screen. The best way to see this is to take a flat map of the world and attempt to wrap it around a sphere without allowing it to wrinkle; obviously this cannot be done without introducing distortion into the images. Just as there are many different ways to wrap the paper around the globe, there are many different projections. Since there are about as many different ways to project GIS data as there are individual states and countries around the world, most GIS tools automatically detect the native projection of the data, as it is read in, and will automatically convert it to a new projection of the user's choice.

Another very common coordinate system that people are familiar with is the geodetic coordinate system. The geodetic coordinate system specifies a latitude, longitude, and altitude for all objects. This is a polar coordinate system that represents a position as angles from the equator and Prime Meridian, and an altitude.

Universal Transverse Mercator is a grid based coordinate system. Imagine that the earth is a giant disco ball with each mirror being an individual tile. Each tile is a projection of the earth to a 2D map. This particular coordinate system exhibits more projection error near the edges of each of the tiles than in the center of the tile due to projection warping that occurs on the boundaries (US National Imagery and Mapping Agency, 1989).

When converting between coordinate systems, errors can be introduced in several areas. First there is the difference between the surveys, or data, that were taken to determine the center and surface of the earth. Keep in mind that the earth is not a perfect sphere but an oblate sphere that bulges in the middle. If one coordinate is in reference to one datum, then comparing it to another coordinate that references another datum, a serious error can be introduced since the origin and assumptions about the bulge of the earth will be different. Another area that can introduce error is that the floating point math that is used in a computer is not precise. Some numbers such as 0.1 cannot be perfectly represented using floating point numbers and have a very small amount of error. This error can accumulate throughout the mathematical operations that are performed to do the conversion, resulting in imperfect results.

Stitching the Data Together

While most command and control systems may use multiple layers of data at various levels of detail or resolution to allow a user to view the world at varying levels, a simulation can only use one set of data. The best way to think of this is that the simulation is simulating the real world and the real world is running at a scale of 1:1 with only one version of the Earth in its databank. However, the data sets that are available for simulation use will frequently be of multiple resolutions, variable quality, and usually not completely cover the area of interest. Sometimes this is intentional. An aviation simulation may not care about the location of every tree and only need the elevation data at enough of a resolution to represent mountain locations between airports, and then need very high resolution data at the airport. Therefore, it is up to the terrain developer to select the best quality data sets to completely cover the area of interest. Once all these data are collected and corrected they must be stitched together to give the simulation the best data set that it can handle and which meets the needs of the training exercise. It is always a tradeoff between performance and quality. Achieving both at the same time is difficult. There are several tools available to help with this process such as ESRI's suit of GIS tools, Global Mapper's Global Mapper tool, or TerraSim's Terra Tools just to name a few. While these tools will do most of the work for the developer, it is important to know the basics of this process and the decisions that will probably be encountered that will affect the resulting terrain database.

Basic steps to creating a terrain database for simulation training exercise

1. Identify the area that the simulation needs to cover
2. Select a data layer to work with. Elevation is usually the first one.
3. Determine the highest resolution that is required for the simulation.
4. Determine the resolution of the available data sets.

    (a) Compress or decompress the data as needed for raster data sets.

5. Re-project the data to match the projection required by the simulation.
6. Look for any holes in the data and find or create data to fill in the holes.
7. Inspect the data for errors and fix as needed.
8. Convert the data to match the simulation's resolution.
9. Start with the least desirable data set, and working up to the most desired data set

    (a) Convert the data to the simulation's format.
    (b) Determine the extent or footprint of the new data and use this to cut a hole in the older data.
    (c) Merge the old data with the new data.
    (d) Repeat this process until all data sets are pulled in.
    (e) Inspect the results to ensure the quality.

10. Have the simulation use the resulting data set and check for problems.
11. Have the customer inspect the data and check to see if it meets with their approval.

Putting It Together

To build a terrain database start with the elevation layer and work your way up to succeeding layers as needed and as required by the simulation that will receive the database. Normally the database will start with the worst resolution data set and work its way up to the best data set. The intermediate data sets will fill in the holes where the best data set does not cover the area of interest.

For example you are developing a terrain for a simulation that operates best at a 10 meter elevation scale. The area of coverage is a 50 mile square centered on the Hampton Roads area in southeastern Virginia. The available elevation data sets are DTED1 with 90 meter grid cells and DTED2 with 30 meter grid cells (US National Imagery and Mapping Agency, 1996). The process for combining these two data sets is fairly straightforward.

First define a polygon that matches the boundaries of the playbox. Use this polygon to cut the DTED1 data so that only the data needed for the playbox are pulled in and worked with from this point on. Now take the DTED2 data for the same area and cut it with the playbox polygon as well. Then take the resulting footprint of the remaining DTED2 data and use that to cut the DTED1 data, similar to using a cookie cutter and the DTED1 data are the cookie dough. Now once the DTED1 data are cut to the shape of the playbox and the DTED2 data combine the two sets. This should result in a new data set. Now that the data for the current layer is all combined into one data set, check it for errors and correct them as needed. This may require manual correction of the data or the use of an automated tool to fix the gaps in the data set. Repeat this process for each layer of data.

What Errors are Possible?

Now depending on the data that are being worked with there are a number of possible errors to look out for. Some are obvious and some can be rather subtle. One that crops up fairly often is the location of bodies of water such as the example in Figure 22.13. If you are using data sets from different resolutions of data you may end up with roads, buildings, and entire counties underwater. Or you may end up with a dry river or bay. Another problem you may have deals with the same body of water appearing twice in the created database as it was represented in two different data sets at two different resolutions. This can also happen with road networks (as in Figure 22.18 later) where it may be possible to have the same road or intersection appear multiple times in the database.

Water Flowing Uphill

There are several problems with using terrain data from different sources. One such example occurs during the process of stitching together elevation data and bodies of water. Due to the potential differences in the two sets of data, it is possible to get rivers that flow uphill, as in Figure 22.14. This problem not

Figure 22.13 Example of two data sets with mismatched water and land areas over the Yorktown, VA. The shaded area indicates land from VMAP0 data. Note that the Yorktown battlefield is under water according to VMAP0 data, Aerial Imagery © 2002 Commonwealth of Virginia.

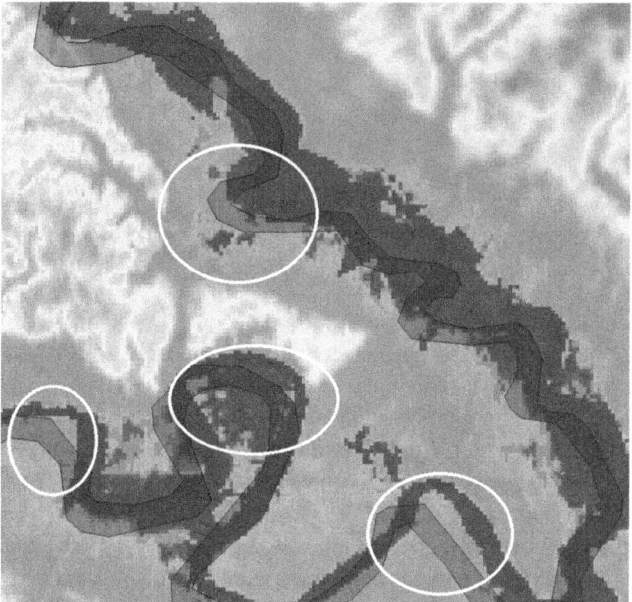

Figure 22.14 Examples of water flowing uphill.

only affects the simulation's fidelity but it also affects the look of the terrain when viewed by the user. Obviously water does not flow uphill under normal circumstances, which means that when these errors occur in the database they must be detected and fixed. There are several methods possible for dealing with this common situation. The first is to manually edit the data. For small databases or databases with few bodies of water this is feasible and requires no more than a terrain editing program to adjust the location of the bodies of water and their shorelines. Another simple method is to fill the digital elevation model (DEM) with water up to a given elevation. This method will produce a matching water

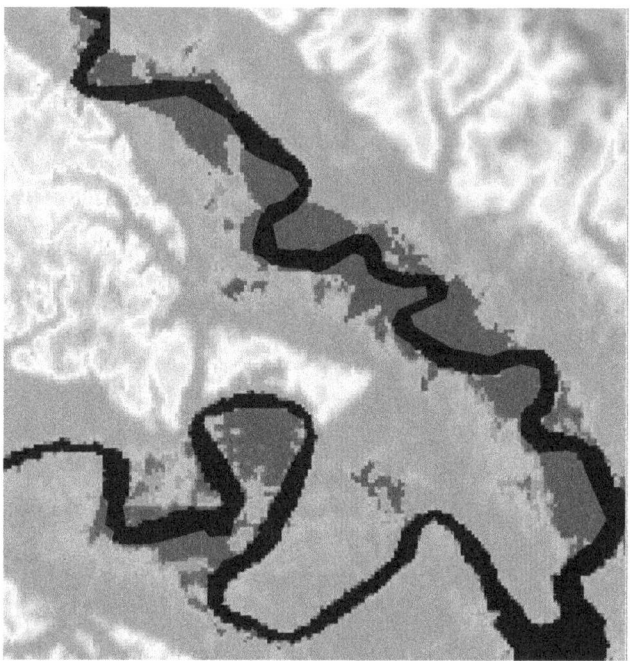

Figure 22.15   Using elevation data to show the main water body. The new water course is colored black.

body for the elevations, but it may also be unacceptable to the user as it could appear a bit granulated if the DEM is using a large cell size as is shown in Figure 22.15. Along these lines, another method allows for generating a water body that outlines the areas of no data for the DEM as many DEMs may not have elevation data for below the water surface. This method allows for the quickest processing time near coastal and tidal areas of the world.

However, for databases with many bodies of water or with very intricate shorelines and complicated elevation data, a more automated approach is needed. One such method is the D8 algorithm. Basically, the D8 algorithm follows the flow of water downhill from each grid cell and its surrounding eight neighbors. When the algorithm gets to a cell that has no higher neighbors, it is declared a pit (lake) and processing stops at each pit. Because of this, it is necessary to remove all the spurious flat areas and depressions from the data prior to processing (Garbrecht and Martz, 1999).

To alleviate the problem of pre-processing the data by hand to find the new water course, several other algorithms have been implemented. Of these, the Priority First Search (PFS) algorithm (Jones, 2002) shows promise. This method is a standard search algorithm used in computer science to search for the shortest path from a start location to a destination, in this example any pixel on the map to the lowest elevation pixel on the map as in Figure 22.16.

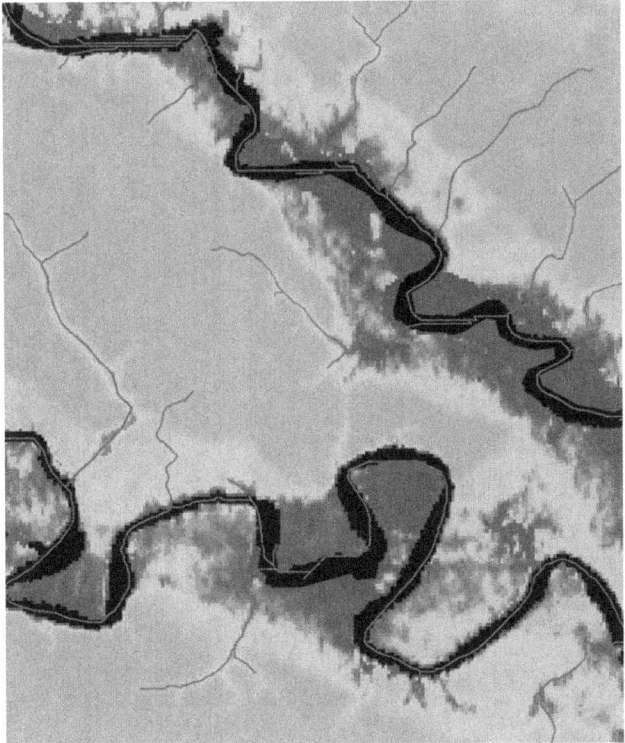

Figure 22.16   Results of a D8 algorithm to place streams and riverbeds in elevation data.

Fortunately, most GIS applications have these algorithms built in and there is no longer a requirement to develop your own code to produce the desired effect. However, whether you have developed your own algorithm or are using a commercial product, the end results will still have errors in them. For example the water course produced to generate Figure 22.16 matches up with the elevation data that it was generated from but, as can be seen in Figure 22.17, because of the difference in when the elevation data were taken and when the imagery was taken, there are still discrepancies between the data sets. This situation leaves the terrain developers in a situation where they will have to get the customer to determine which errors they are willing to accept prior to the start of an exercise. But they will also have to explain why the problem exists and the pros and cons of choosing one method over the other for fixing it and how long it will take, so that the customer can make an informed choice.

Disparity between Distributed Simulations

With simulations and simulators running at various different scales and with different coordinate systems, it is very likely that problems will crop up between

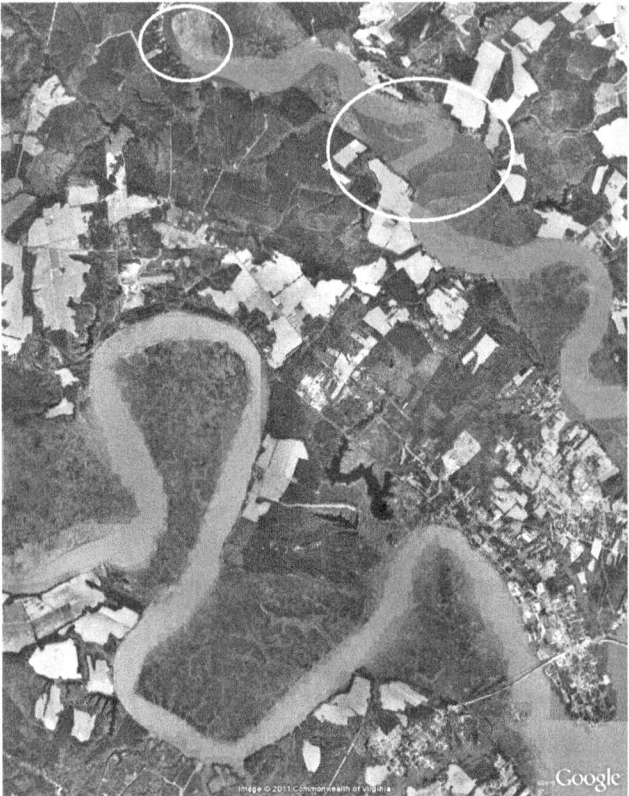

Figure 22.17 New water course overlaid on imagery data Image taken from Google Earth © 2010 Google.

them with respect to where units are located, how they move, and how they perceive entities from the other network simulations (IEEE, 1996). As an example of this take a look at a theoretical simulation running a theater level constructive simulation (TLCS), and a second theoretical simulation running a company level constructive simulation (CLCS). For the following example let us assume that the best available data for the TLCS are VMAP0 data, and the best available data set for the CLCS is available to it from VBMP data. If a TLCS player orders a tank platoon to move southwest via Mercury Boulevard in Figure 22.18, and a CLSC player orders a tank platoon to move northeast along Mercury Boulevard, they may not be able to see each other because they are using different sources for their data, which have Mercury Boulevard in different locations.

How do you fix these problems? The obvious solution is to pull the sets of data into the same GIS tool and make them physically match each other as much as possible and then export a single unified terrain data set. In this situation deleting the VMAP0 road network and replacing it with another data

Figure 22.18 Example of mismatching terrain data sets causing two units to not see each other. CLCS unit to the left and TLCS unit to the right. Aerial Imagery © 2002 Commonwealth of Virginia.

set's version of the road network is one possibility. Unfortunately, this may not entirely eliminate the problems due to different allowable input formats that may be required by the simulations. In those situations, it is best to work with terrain tools to achieve a correlated terrain between simulations. If it is not possible to create a correlated terrain, it may be necessary to have some human intervention to adjust the scenario or the behavior of the entities to account for differences of the terrain. Unfortunately, the question remains: which simulation has the correct terrain? Is it the CLSC with a 6 inch resolution or the TLCS with the 1:1,000,000 scale road networks? Ultimately this will only be answered by what data set best fits the training objective or the customer's requirements.

SUMMARY

This chapter introduced concepts needed to understand terrain data collection, data processing, terrain database generation, and some issues involved with terrain data and military simulations. Specifically, we introduce the concepts of raster data, vector data, TINs and 3D models, and how they work and complement each other. We also discussed some of the details of data layers, tiles, and coordinate systems. We briefly described the major fidelity issues with terrain data where it is possible to have water flowing uphill and a couple of methods for remedying that situation. Finally, we discussed in short one of the major problems with multi-resolution simulations and how multiple representations of the terrain can affect the outcomes of simulation exercises.

The following GIS tools were used to build and capture the images for this chapter:

- ArcMap: http://www.arcgis.com/home/
- Global Mapper: http://www.globalmapper.com

- Google Earth: http://www.google.com/earth/index.html
- Google Sketchup: http://sketchup.google.com
- VR-Vantage: http://www.mak.com

Other GIS tools that are also available and may be of use:

- Terrasim: http://www.terrasim.com/products/
- GRASS: http://grass.osgeo.org/

FURTHER READING

This chapter provides only a very high level overview on how terrain databases for combat simulations are produced. A production level understanding of this topic requires a good understanding in remote sensing, GIS, data models, 3D modeling. This is a domain that is in constant technological advancement that is reaching a high level of automation. Key challenges remain in the area of underlying data models and associated metadata related to terrain generation. The following is a short list of additional reading material:

- "What Every Computer Scientist Should Know About Floating Point Arithmetic," David Goldberg, March 1991, Computing Surveys, Association for Computing Machinery, reprinted at http://download.oracle.com/docs/cd/E19957-01/806-3568/ncg_goldberg.html
- Nielsen, Morten. "Spatial references, Coordinate systems, Projections, Datums, Elipsoids", May 5, 2007 http://www.sharpgis.net/post/2007/05/Spatial-references2c-coordinate-systems2c-projections2c-datums2c-ellipsoids-e28093-confusing.aspx
- Chunquan Cheng; Kazhong Deng; Jixian Zhang; Li Zhang;, "Block Adjustment of Airborne Imagery in Geocentric Orthogonal Coordinate System," *Information Engineering and Computer Science, 2009. ICIECS 2009. International Conference*, pp.1–4, 19–20 Dec .2009.

REFERENCES

Clarke KC (2001). *Getting Started with Geographic Information Systems*. Prentice Hall.

Elanchi C and Van Kyl J (2006). *Introduction to the Physics and Techniques of Remote Sensing* (2nd edition). John Wiley & Sons, New York, NY.

Garbrecht J and Martz LW (1999). Digital elevation model issues in water resources modeling. In: *Proceedings ESRI User Conference*, Paper 866.

Growe S and Toenjes R (1997). A knowledge based approach to automatic image registration. In: *International Conference on Image Processing (ICIP'97)*, Volume 3, p. 228.

IEEE (1996). *IEEE Standard for Distributed Interactive Simulation Application Protocols, IEEE Standard 1278-1.1995*. IEEE, Piscataway, NJ.

Jones R (2002). Algorithms for using a DEM for mapping catchment areas of stream sediment sample. *Comp Geosciences* **28**, 1051–1060.

NASA (2011). *My NASA Data+*. http://mynasadata.larc.nasa.gov/ElectroMag.html (last accessed 4 June 2011).

National Geo-spatial Agency (2007). *GEOINT Standards in the DoD IT Standards Registry*, version 07-3.0, November 6, p.14.

O'Connor MJ *et al*. (2001). Building on the realtime platform reference and envirofed FOMs. 01S-SIW-041 March.

Richards WT (2008). A study of environmental data usage in military simulations. In: *Proceedings of the ODU/VMASC Capstone Conference*.

Richards WT and Hibler D (2005). Automatically converting paper map images for use as notional terrain. In: *Proceedings: Inter-service/Industry Training, Simulation and Education Conference, Orlando, FL*, pp. 1360–1368.

US National Imagery and Mapping Agency (1989). *The Universal Grids: Universal Transverse Mercator (UTM) and Universal Polar Stereographic (UPS)*. Edition 1, TM 8358.2. NIMA, Washington, DC.

US National Imagery and Mapping Agency (1996). *Performance Specification – Digital Terrain Elevation Data (DTED)*. US Department of Defense document #MIL-PRF-89020B.

US National Imagery and Mapping Agency (1998). *Vector Smart Map Level 1 (VMAP1)*. US Department of Defense document # MIL-PRF-89033.

Yang Q, Snyder JP and Tobler WR (2000). *Map Projections Transformation Principles and Applications*. Taylor and Francis.

Chapter 23

Modeling Tactical Data Links

Joe Sorroche

INTRODUCTION

Tactical Data Links (TDLs), otherwise referred to Tactical Digital Information Link (TADIL) is a communication, navigation, and identification system that supports tactical data exchange between military systems. Tactical data are battlefield information about the location and type of battlefield resources. The Link 16 is an example of how this information is exchanged between battlefield resources. TDL is characterized by many standardized message formats and transmission characteristics such as position, type of aircraft, heading, altitude, and fuel and weapon status. The primary purpose of TDLs is to achieve compatibility and interoperability between command, control, and communications (C3) systems of the United States and other military forces. There are three primary functions of TDLs: air control, air defense, and surveillance.

There are two parts to TDLs: the messages and the networks. Some messages contain information about the transmitting platform. Other messages contain surveillance information. There are several types of TDLs that have specific message formats, and a unique network that carries these messages between users. These will be described later in this chapter.

HISTORY

Before World War II, the only way that TDL was exchanged between battlefield units was through radio systems. Information was exchanged only if there was an enemy aircraft, and if so the altitude and distance from a specified reference point were transmitted. This was an effort to eliminate casualties from friendly fire.

Engineering Principles of Combat Modeling and Distributed Simulation, First Edition.
Edited by Andreas Tolk.
© 2012 John Wiley & Sons, Inc. Published 2012 by John Wiley & Sons, Inc.

At the beginning of World War II in the Pacific, there was a high rate of aircraft loss due to friendly fire. Engineers developed a way to determine if the aircraft was friend or foe, which was known as Identification Friend or Foe (IFF). Each aircraft had an IFF radio transponder which transmitted a message that identified it as a friendly aircraft to others. However, there was still an unacceptable loss due to friendly fire because pilots forgot to turn on their IFF transmitter. One commander of a flight wing had a creative solution. The commander ordered signs be placed at the end of the runway that stated: Turn on your IFF or you will die! The rate of aircraft losses to friendly fire in that squadron was reduced to zero. When other commanders also displayed these signs, there was a significant drop in losses due to friendly fire.

The 1950s expanded on the concept of IFF and ways to add information in the IFF transponder message. The information would not only identify friend or enemy, but included location, direction, speed, altitude, and other relevant information required for attack or counter attack. It was decided to separate the information into two groups. One would identify friendly and enemy forces, and the other would determine the location information of each aircraft to ensure our aircraft had a tactical advantage in combat. And thus TDLs were born.

The 1960s saw the introduction of TADIL C/Link 14. It used very high frequency (VHF) to transmit information at a data rate of approximately 300 bits per second (BPS). However, the TADILs currently utilized for transfer of information were becoming insufficient due to technological advances. With the increased capacities of computers and increased capabilities of targets, the requirement for a faster, more robust system was needed. The 1970s saw the introduction of TADIL A, TADIL B, TADIL C, and the Army Tactical Data Link (ATDL). In the 1980s, TADIL J, otherwise known as Link 16, was introduced. In the decades that followed, other TADILs that were developed included Variable Message Set (VMF), Enhanced Position Location Radio Set (EPLRS), Situational Awareness Data Link (SADL), Link 22, Blue Force Tracker (BFT), and Integrated Broadcast Service (IBS), to name a few. Some of these are described in detail in this chapter.

Tactical Data Links Use

The use of TDLs among military forces is paramount in the reduction of friendly fire as well as in increasing the efficiency and fire power of military forces.

TACTICAL DATA LINK TYPES

There are several TDLs that have been developed since the late 1960s. These used the high frequency (HF) spectrum. As additional TDLs were developed, the VHF and ultra high frequency (UHF) spectrums were assigned, as shown in Figure 23.1.

As TDLs were developed, they were assigned to specific spectrums. Some of the most popular and widely used TDLs are described in the following sections.

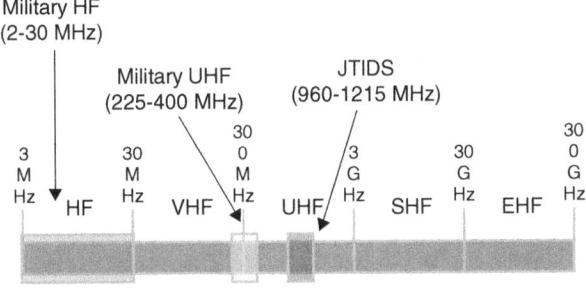

Figure 23.1   TADIL frequencies.

TADIL A/Link 11

TADIL A, or Link 11, was developed to exchange more information at a faster rate. Link 11 is a secure half-duplex TADIL radio link. Half duplex means that the radios can either transmit or receive, but not both at the same time. Link 11 is used by both NATO and the United States. Link 11 network participants exchange digital information among airborne, land-based, and shipboard tactical data systems. It is the primary means used to exchange data, which includes its own position, the position of other land or airborne units, both friend and foe, and the current status of its own unit, such as fuel, heading, altitude, and current mission. Link 11 can be used on either HF or UHF. However, the US Army uses only HF. Link 11 employs netted communication techniques and standard message formats for the exchange of digital information among airborne, land-based, submarine, and shipboard tactical data systems. It provides for the mutual exchange of information among net participants via HF or UHF radio. The M series messages used for Link 11 are shown in Table 23.1. Detailed descriptions of these messages can be found in Mil-Std-6011C.[1]

Each Link 11 message consists of two 30-bit frames. Each frame contains 24 bits of data along with 6 Hamming error detection and correction (EDAC) bits. Hamming error detection can detect two simultaneous bit errors at once. Hamming correction is an error correcting alogrithm that can correct single bit errors. The first 4 bits of the Link 11 tactical message contain the message label. The rest of the bits are defined by size and description fields specific to the message, for a total of 48 bits of tactical information.

Link 11 operates in a roll call mode among all participating units or pickets under the control of a net control station (NCS). In roll call, pickets are polled sequentially with messages, including roll call, NCS interrogation, picket reply, and NCS report. These messages can otherwise be called short broadcast messages. Each picket in the Link 11 network has a response time limit of 45 frames of 13.3333 ms duration. Pickets then transmit data upon receipt of a roll

---

[1]Mil-Std-6011C, Tactical Data Link 11/11B Message Standard, 5 March 2002.

Table 23.1   Link 11 Message Set

Message number	Message function
M.0	Test
M.1	Data reference position
M.81	Amplifying M.1
M.2	Air track position
M.82	Amplifying M.2
M.3	Surface track position
M.83	Amplifying M.3
M.4.A	ASW primary
M.84.A	Amplifying M4.A
M.4.B	ASW secondary
M.4.C	ASW primary acoustic
M.84.C	Amplifying M4.C
M.4.D	ASW bearing
M.84.D	Amplifying M4.D
M.5	Special points position
M.85	Amplifying M.5
M.6A	ECM intercept data
M.6B	ESM primary
M.86.B	Amplifying M.6B
M.6C	ESM parametric
M.86C	Amplifying M.6C
M.6D	EW coordination and control
M.86D	Amplifying M.6D
M.9A	Management information
M.9B	Pairing/association/correlation
M.9C	Pointer
M.9D	Link monitor
M.9E	Management
M.9F	Area of probability
M.9G	Data link reference point
M.10A	Aircraft control
M.11B	Aircraft mission status
M.11C	ASW aircraft status
M.11D	IFF/SIF
M.11M	EW/intelligence
M.811M	Amplifying M.11M
M.12	National/text/timing
M.812	Amplifying M.12
M.13	Worldwide national
M.14	Weapon/engagement status
M.15	Command

call message from the NCS. During the remainder of the time, each picket can receive reports from other members.

When a picket receives a roll call message from the NCS, the first message sent contains the message subtype set to "start code." The last message sent contains the message type "stop code." Messages sent between pickets have the message subtype set to "intermediate." The minimum response to a roll call message is a start code message and a stop code message in the signal message. If a picket does not answer its call within the roll call timeout period, it will be assumed that the picket did not receive the poll, and the NCS will poll the picket a second time. If the picket answered the first poll, the second poll will be ignored. If the picket still does not respond, the NCS will poll the next picket in the sequence.

Once all pickets have responded to the roll call message, the net cycle is considered complete. The NCS will then keep record of the net cycle time and fill in the net cycle time parameter. The following formula is used to compute the net cycle time:

$$\text{Net Cycle\_time} = [\text{current NCS transmission time} \\ - \text{previous NCS transmission time}]$$

in units of frames. If the NCS requires the time to wait before polling the next participating unit (PU), the following formula is used:

$$\text{Tx\_time} = [\text{callup} + 5 \text{ frames of wait time} + \text{PU reply}]/\text{net speed}$$

where callup = 8 frames. For fast mode, callup = 106.6664 ms, and for slow mode, callup = 176 ms. For fast mode, 5 frames wait time = 66.6665 ms. For slow mode, 5 frames wait time = 110 ms.

$$\text{picket reply} = (10 \text{ frames} + \text{data frames}) \\ \text{net speed} = 75 \text{ frames/s for fast speed and} \\ 45.4545454 \text{ frames/s for slow speed}$$

Therefore, for picket reply and all other messages,

$$\text{tx\_time} = [23 \text{ frames} + 2^*\text{nb\_msg}]/\text{net speed}$$

where nb_msg = number of messages of PU reply, or number of data messages.

Then, the NCS reports its own information. Once the NCS has broadcast its own information, each picket broadcasts link data in accordance with MIL-STD-6011C.[1]

TADIL B/Link 11B

TADIL B, or Link 11B, is an improved version of TADIL A/Link 11. Link 11B is a full-duplex, two-way, point-to-point link that provides for the serial transfer of data between units of the US Army, US Air Force (USAF), US Marine Corps (USMC), and the National Security Agency. Its participants are referred to as reporting units (RUs). Because it is point-to-point, each pair of RUs operates on a separate Link 11B channel, often referred to as a "B-Link." Data are forwarded among the RU pairs by forwarding RUs (FRUs). Link 11B employs the same message standard as Link 11. However, the equipment, some message protocols, and the data rate are different from those of Link 11, thus requiring special forwarding units to interface with Link 11 and Link 11B. TADIL B is not employed directly by Naval or Allied units.

Because Link 11 and Link 11B employ the same message standard and have virtually identical operational capabilities, they are often referred to together as Link 11/11B. An interface between them is provided by a forwarding participating unit (FPU). From an operational standpoint, they may be viewed as a single link.

The Link 11B message consists of 72 bits containing eight groups of 9 bits each. The start group uses 9 bits, but is not included in real-world Link 11B messages. The check group uses 9 bits while each of the six data groups uses 1 bit each to precede the 8 information bits.

Operations involving only Link 11 and Link 11B are referred to as Link 11/11B operations. These are fully described in MIL- STD-6011C.[1]

Link 11/11B Users

Link 11/11B is used by the US Army, USAF, USMC, and the National Security Agency using Control and Reporting Centre (CRC) SAM interfaces (CSI). Within the United States and some other NATO nations (e.g. France), Link 11B is used as the primary data link for ground based TACS (e.g. USAF MCE and USMC TAOC). The UK Royal Navy, Royal Marines and Royal Air Force in its ships, ship shore ship buffers (SSSBs), E-3D AEW, Nimrod MPA, and Tactical Air Control Centre use Link 11/11B. Within NATO, Link 11/11B is primarily used as a maritime data link. However, Link 11/11B is being adapted for theatre missile defense information exchange, so ground based (surface-to-air missile) SAM systems are or will be equipped with Link 11/11B. Within NATO and beyond, Link 11/11B is fitted in ships, airborne surveillance and with the Icelandic Air Defense System.

Link 11/11B is scheduled to be phased out of US assets by 2015. Because of this, the Link 11/11B equipment is being sold to new NATO members because they do not have sufficient funds to acquire Link 16 or Link 22 equipment.

TADIL J/Link 16

Joint Tactical Information Distribution System (JTIDS) was the program that developed Link 16. JTIDS is still used to identify the waveform and architecture of Link 16 and to differentiate between types of Link 16 terminals. TADIL J identifies the message set used in Link 16.

Link 16 is a communications, navigation and identification (CNI) system intended to exchange surveillance and command and control (C2) information among various C2 platforms and weapons platforms to enhance varied missions of each of the services. It provides multiple access, high capacity, jam resistant, digital data and secure voice CNI information to a variety of platforms. Link 16 is the primary NATO standard for TDL. NATO Standard Agreement (STANAG) 5516[2] and MIL-STD-6016D[3] describe the Link 16 message formats (Link 16 messages are also known as TADIL J messages) and Link 16 network instructions. Link 16 uses the JTIDS which is the communications component of Link 16. The terms Link 16 and JTIDS are frequently used interchangeably. The Multi-Function Information Distribution System (MIDS) is the NATO equivalent term for JTIDS.

Link 16 Message Set

Link 16 messages consist of network control messages, location of each RU, the location of other units, both friendly and hostile, C2 messages, and free text messages. Each message set has a specific functionality for the link. Table 23.2

Table 23.2 Link 16 Message Set

Message number	Message function
J0.X	Network control
J1.X	Network connectivity
J2.X	Own position location
J3.X	Track position location
J5.4	Acoustic bearing/range
J6.0	Intelligence information
J7.X	Track management
J8.X	Unit designator
J9.X, 10.X	Command/engagement status
J11.X	Target location/assignment
J12.X	Command/assignment
J13.X	Platform status
J14.X	EW parameters
J16.X	Image
J17.X	Weather over target
J28.X	Free text message
J31.7	No statement

[2]NATO Standard Agreement (STANAG) 5516.
[3]Mil-Std-6016D, Tactical Data Link 16 Message Standard, 12 December 2006.

contains the Link 16 message set and the description of each series. Additional message details can be found in MIL-STD-6016D.[3]

JTIDS messages are exchanged between units by transmitting them at the proper time. Each is assigned a time division multiple access (TDMA) time slot for transmission in the Link 16 network.

TDMA

TDMA is the method by which the transmission capacity available to the entire network is distributed among its members. A cyclical period of time is divided up into time slots which can be of different durations. Most time slots are allocated to specific units in the network. A unit transmits during its own time slot. All other units listen during this period, and they may or may not receive the transmission. Priority injection time slots may be available, which can reduce the length of time a unit has to wait before it is able to transmit high priority messages. If multiple units transmit in a priority injection time slot at the same time, the transmission might not be received. Because of this, the transmission is also repeated in the unit's own time slot.

JTIDS uses the principle of TDMA to divide network time and capacity into divisions called time slots. Each time slot is 7.8125 ms long with 128 time slots per second. Timeslots are organized into three interleaved sets (A, B, and C). An epoch is 12.8 min long, and is composed of 98,304 time slots. There are 112.5 epochs in a 24-hour day. Therefore, the current epoch, set and time slot number can be calculated from the current time. Operationally, groups of time slots are assigned to a common function known as a network participation group (NPG). Time slot assignments are published in a network data load (by a central net design agency), with participation groups identified by the time slot set, the offset of the time slot, and the time slot recurrence rate. The recurrence rate is expressed as an exponential power of 2 representing how often the time slot assigned to the NPG occurs within the set. TDMA architecture requires that each JTIDS participant, known as a JTIDS unit (JU), must know when its transmit time slots occur. JUs must be synchronized with a common network time to receive and transmit on the network. In JTIDS, one JU in a network is designated as the network time reference (NTR).

Link 16 Clock Synchronization

Clock synchronization can be accomplished using active synchronization, where the Link 16 terminal transmits round trip timing (RTT) messages to determine clock variance and when the epoch starts. Another way is passive synchronization, where the Link 16 terminal observes position and status messages from other terminals. Synchronization of the Link 16 radios has three phases: net entry, coarse synchronization, and fine synchronization. Before the net entry phase can begin, an NTR has to be designated. Only one radio in a network can be the

NTR. Net entry is the phase where the radio is attempting to listen for net entry messages. The NTR generates a J0.0 message every 12 s. Each radio attempting to enter the network will listen for a J0.0 message. Once the radio has received two consecutive J0.0 messages, it will move to the next phase. The next phase is coarse synchronization. This occurs after the completion of the net entry phase and allows the radio to transmit and receive data messages. The final phase is fine synchronization. Fine synchronization also allows the radio to transmit net entry messages to bring in other units attempting synchronization.

Frequency Hopping

Frequency hopping is a technique used by data links to provide jamming resistance. Jamming is the deliberate transmission of radio signals that disrupt communications by decreasing the signal to noise ratio of receivers. This is different from interference, where other radio frequencies happen to match the receiver frequency and interfere with reception. The jammer cannot follow the random frequency hops, so it is forced to spread its energy over the entire frequency range. Link 16 hops among 51 separate frequencies. Each hop is 13 μs for 76,923 hops per second. The hopping patterns are random and controlled by a crypto variable. The Link 16 frequency hopping concept is shown in Figure 23.2.

Network Participation Groups

Superimposed on the Link 16 network is a functional structure that allows the network to be adapted to specific C2 operational environments or needs. This adaptation is accomplished by allocating the network capacity among several virtual networks whose message transmissions are assigned to a single Link 16 function. JTIDS participants are then assigned to these networks, or functional

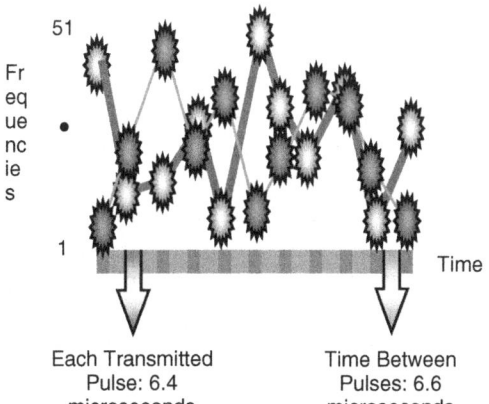

Each Transmitted Pulse: 6.4 microseconds

Time Between Pulses: 6.6 microseconds

Figure 23.2 Link 16 frequency hopping.

groups, as required by each of their missions and their capabilities. In addition to identification of friendly forces and position reporting, Link 16 functions include

- surveillance
- fighter-to-fighter
- air control
- imagery
- network enabled weapons
- ballistic missile defense
- electronic warfare
- mission management and weapons coordination
- two secure voice channels

There are also functional groups that support the operation of the network, including initial entry and RTT. All of these functional groups are known as network participation groups, or NPGs. The transmissions on each NPG consist of messages that support its particular function.

This structuring allows JUs to participate on only the NPGs necessary for the functions they actually perform. It is likely that all C2 platforms will operate on all applicable NPGs at all times. A maximum of 512 participation groups are possible. Of these, 30 have been allocated for subject oriented functions, and 22 of these are currently defined. The Air Force employs between 11 and 15 NPGs. NPGs 30 and 31 are specifically dedicated to Interim JTIDS Message Set messages, and are only used in NATO networks.

Each J series message is mapped to a specific NPG. The JTIDS Class 2 terminal allows the assignment of 64 time slot blocks and up to 32 participation groups from 512 possible selections. The NPGs have an external number, which is a number between 1 and 511, assigned by the network load, and an internal number, from 1 to 32. Each NPG is associated with a particular set of Link 16 messages. For example, the J2.x set of messages, which are precise position location indicators (PPLI), is transmitted on NPG 5 and 6.

Fighter platforms overcome the 12 s update rate for PPLI by transmitting J2.2 messages every 6 s and on NPG 5 and 6. Their time slots are staggered, so using both NPGs results in J22 messages transmitted every 2 s. Other slower air platforms can transmit their J2.2 messages every 6 or 12 s and still provide accurate location information. For J3.X series messages, their time slots can also be staggered such that J3.2 messages can be transmitted as often as required for accurate position information. Table 23.3 summarizes the NPGs and functionality.

NPGs support JU operational communications needs. They allow the network designer to separate the functions implemented in the J series messages. Network capacity is first allocated to NPGs, and then to the users that participate in that NPG.

Table 23.3   Link 16 NPGs and Functions

NPG	Function
2	RTT-A
3	RTT-B
4	Network management
5	PPLI, status A
6	PPLI, status B
7	Surveillance
8	Mission and weapons management
9	Air control
10	Electronic warfare
11	Imagery
12	Voice group A
13	Voice group B
14	Indirect PPLI
18	Network enabled weapons
19	Fighter to fighter net 1
20	Fighter to fighter net 2
21	BMD operations
29	Residual message
30	IJMS P messages
31	IJMS T messages

Communications Security

There are two types of communications security used in Link 16: Message Security Encryption Code (MSEC) and Transmission Security Encryption Code (TSEC). MSEC is used to encrypt the message data prior to the Reed–Solomon encoding, interleaving, and pulse generation. Reed–Solomon encoding is the scheme employed for encoding Link 16 message data and consists of the addition of error detection and correction. TSEC determines the amount of jitter, or a deviation in the time slot and the 32-chip pseudo-random noise variable. The TSEC crypto variable (CV), together with the net number and time slot number, also determines the frequency hopping pattern of the carrier.

Nets

The JTIDS waveform allows for 127 different nets, numbered 0 to 126. The net number, along with the TSEC CV and time slot number, determines the carrier frequency hopping pattern. These different hopping patterns keep the nets separate and distinct and allow multiple nets to operate at the same time. Multiple nets can be established by specifying different net numbers for a particular NPG, without changing the TSEC and MSEC CVs. If the MSEC and TSEC CVs are changed, then it is possible to have several variations of multi-netting, including blind relays. Multi-netting is defined as transmitting messages in the same time slots but on different NPGs and different net numbers from the main (default) net.

Stacked nets are also possible when creating a Link 16 network. A stacked net can be created by assigning the same group of time slots to the same NPG and with the same TSEC parameter but with different net numbers. The time slots must have the same set, initial slot number, and recurrence rate. They can have different net numbers and the same crypto keys, or different net numbers and different crypto keys, whichever is required for the network. The Link 16 voice nets and air control nets are examples of stacked nets.

Link 16 Use

Table 23.4 lists the Link 16 real-world platforms.

Table 23.4   Link 16 Platforms

US aircraft	Euro/other aircraft
F/A-18 Hornet	Saab Gripen
F-22 (receive only)	Dassult Mirage 2000-5F
F-15 Eagle	Dassult Mirage 2000D
F-16 Fighting Falcon	Eurofighter Typhoon
E-2C Hawkeye	Panavia Tornado
E-3 AWACS	Dassult Rafale
MH-60S/R NavalHawk family helicopters	Greek Embraer R-99A Airborne Early Warning and Control aircraft
E-8 Joint STARS	
EA-6B Prowler	
EA-18G Growler	
P-3C Orion	
EP-3C Aries	
Boeing RC-135 Rivet Joint	
UH 60 Black Hawk (planned)	
AH-64 Apache (planned)	
US ships	
US Carrier Battle Groups	NATO Frigates
Missile defense	
Arrow	
Patriot Missile	
Theater High Altitude Air Defense (THAAD)	
Short Range Air Defense (SHORAD)	
Joint Tactical Ground Station (JTAGS)	
Joint Land Attack/Cruise Missile Defense	

EPLRS/SADL

Situational Awareness Data Link (SADL) integrates US Air Force close air support aircraft with the digitized battlefield via the Army's Enhanced Position Location Radio Set (EPLRS). SADL is the integration of the modified EPLRS radio with an MIL-STD-1553B host interface and an aircraft avionics over the aircraft MIL-STD-1553B multiplex bus. It allows data from other SADL equipped fighters and ground EPLRS locations to be seen in cockpit displays.

SADL is a wireless radio network that operates in the UHF band, 420–450 MHz. There are eight center frequencies that SADL can operate on. Each center frequency is separated by 3 MHz. The operator can limit which center frequencies are used through a frequency map selection.

SADL radios frequency hop over eight center frequencies, providing jam resistance. The sequence is pseudo-randomly selected, and is controlled by the traffic encryption key (TEK). Frequency hopping occurs on a time slot by time slot basis. Therefore when an SADL radio uses a 2 ms time slot length, there are 512 hops per second. When the radio uses a 4 ms time slot length, there are 256 hops per second. Compared with Link 16 hopping rates, SADL is much slower. Link 16 hops at each pulse, so approximately 77,000 hops per second. Frequency hopping provides low probability interference (LPI) and low probability detection (LPD) which jammers cannot follow, and requires them to spread energy over the entire frequency range.

EPLRS also uses the principle of frequency hopping TDMA to divide network time and capacity into divisions called time slots. Each time slot is 1.953125 ms long with 512 time slots per second. An epoch is 64 s long, and is composed of 32,768 time slots. There are 44,236,800 epochs in a 24-hour day. Therefore, the current epoch, set, and time slot number can be calculated from the current time.

As mentioned in the section on Link 16 Message Set, the TDMA architecture requires that each EPLRS unit (EU) must know when its transmit time slots occur. EUs must be synchronized with a common network time to receive and transmit on the network. In an EPLRS network, one EU in a network is designated as the timing master initiator (TMI). The TMI transmits the timing pulse for all EUs in the SADL network. The TMI starts the network, but network timing is maintained by all network participants. Any radio can be designated the TMI, but only one radio in the network can be the TMI. For SADL aircraft, the flight lead (net shape 11) in the air–air mode is usually the TMI. For EPLRS ground networks, the network manager (ENM) in the air–ground mode and the gateway in the gateway mode are the TMIs. The TMI radio assumes its clock is correct just like the NTR in the Link 16 network. Radios that enter the net will attempt to match the clock drift of the TMI unit. Clock drift is the difference in time between the TMI and other SADL radios.

EPLRS/SADL provides Communications Security (COMSEC) and Transmission Security (TRANSEC) with a single CV—the traffic encryption key (TEK). It has an embedded secure data unit (SDU): KGV-13A programmable

COMSEC module. The TEK is used to modify various characteristics of EPLRS/SADL message processing and radio frequency transmission: baseband data encryption, data interleaving, frequency hopping, time slot scramble, and time slot jitter. The SDU must be loaded with at least one TEK in order to transmit or receive. There are three CVs associated with the radio: current traffic key, next traffic key, and initial key encryption key. The first two CVs are for the current crypto period and next crypto period respectively. The last CV is unique to every radio in the network and allows over the air re-keying. The TEK is required for net entry and data transmission.

Synchronization of the SADL radio has three phases: net entry, time unstable, and time stable. Before the net entry phase can begin, a network TMI has to be designated. Only one radio in a network can be the timing master. Net entry is the phase where the radio is attempting to listen for net entry messages. The next phase is time unstable. This occurs after the completion of the net entry phase, and allows the radio to transmit and receive data messages. The final phase is time stable. Time stable also allows the radio to transmit net entry messages to bring in other units attempting synchronization. Passive synchronization can also occur. This happens when the operator selects radio silent. Radio time stability is achieved after radio makes only minor adjustments to clock.

SADL addresses line-of-sight (LOS) issues by using the automatic relay functionality. This functionality is not operator selectable but is embedded in the programming of the radio. Therefore, SADL does not plan for specific platforms to "act" as dedicated relay assets like in most Link 16 networks. Relaying is supported on all communications services as an automatic function of the radio. Relaying allows for point-to-point, contention access multicast, and dedicated access multicast—up to six hops using relay chains that are self-forming and self-healing. Further, relaying provides automatic adaptation to dynamic radio frequency environments using multicast relaying without self-jamming (mutual interference). Two types of relays are available: dedicated and flood. A dedicated relay uses a separate dedicated time slot for each source time slot. For a 2 ms time slot, the relay occurs eight time slots after or 16 ms later. For a 4 ms time slot, the relay occurs eight time slots after or 32 ms later. Enabling flood relay allows all units who receive the message to relay it.

SADL Message Set

SADL uses the same J series messages as Link 16. However, SADL also uses two VMF messages for communication with Army ground units. The SADL message set is shown in Table 23.5. Real-world platforms that use SADL are the F-16 Fighting Falcon, Block 30, and the A-10C.

SADL Use

SADL is used by the United States only. The US air platforms that use SADL are the A-10C and F-16 Block 30 fighters.

Table 23.5   SADL Message Set

Link 16 to SADL	
J2.X	Own position location
J3.X	Track position location
J7.X	Track management
J12.X	Command/assignment
J28.2	Free text message
SADL to Link 16	
J2.2, J2.6	Air/land own position
J12.0	Pilot acknowledgements
J12.4	Pilot acknowledgements
SADL Air to Ground K Series	
K05.01	Position report summary
K05.19	Entity data summary

Link 22

Link 22, also known as NATO Improved Link Eleven (NILE), was developed as a replacement for Link 11. Link 22 was also designed to complement and interoperate with Link 16. Other features include an automated and simple management, which is less complicated than Link 11 and Link 16.

Link 22 is a (NATO) secure radio system that provides beyond line-of-sight (BLOS) communications. It interconnects air, surface, subsurface, and ground-based tactical data systems, and it is used for the exchange of tactical data among the military units of the participating nations. Link 22 can support NATO and Allied warfare tasking.

Link 22 employs a communications security (COMSEC) system, which is provided by the inclusion of an integral encryption/decryption device inside the Link 22 system. This cryptographic device (crypto) at the data link level is called the Link Level COMSEC (LLC). Link 22 transmission security is also available by the optional use of frequency hopping radios.

Tactical data are transmitted on Link 22 using J series messages. It uses the same field definitions as Link 16 to provide standardization between the two TDLs. Because J series messages are used, many Link 16 tactical messages can be transmitted without modification within the Link 22 network.

An operational Link 22 system is called a Link 22 super network. A Link 22 super network consists of only two units communicating with each other in a single NILE network. The most complex Link 22 super network would consist of 125 units, the maximum number with eight NILE networks. A unit participating within the Link 22 super network can be a member of up to four of the NILE networks.

Coverage beyond the Link 22 network is provided by the automatic relay of messages and the ability to adapt to changes automatically, without operator intervention, much like the EPLRS/SADL network. A unit will automatically retransmit a received message when necessary to ensure that the message is received by its addressees. The System Network Controller (SNC) calculates whether the relay is necessary based on its knowledge of the connectivity among units. The ability of a unit to relay can be affected by its relay setting. This setting's default is automatic relay, but the unit can be disabled from performing relay or designated as a preferred relayer. Relay is performed on a per message basis. Because messages are retransmitted only when necessary, this reduces the use of bandwidth.

Each NILE network can use either HF or UHF communications. HF communications are in the 2–30 MHz band, which provides BLOS communication, otherwise known as over the horizon. HF also provides direct LOS communications. UHF communications are in the 225–400 MHz band, which provides LOS communication only. Within each frequency band, either fixed frequency or frequency hopping radios can be used.

Link 22 has automated network management functions that require a minimal operator interaction. These functions are controlled by the transmission of network management messages. Each unit can define whether or not to automatically respond to or not to perform the network management functions.

At the tactical level, when a unit is congested, it can reduce the local traffic that it generates based on the provided congestion information. In addition, Link 22 has an automatic network message congestion management, and can do this in a number of different ways. The routing of messages takes congestion into account and will route messages using alternative paths to reduce message congestion. Link 22 has a dynamic TDMA (DTDMA) protocol. When DTDMA is enabled on a NILE network, it allows congested units to automatically request and receive additional capacity on a permanent or temporary basis. If DTDMA does not achieve the desired result, the unit managing a NILE network can change the configuration of the network to redistribute the available capacity or change the parameters of the network's media in use to increase the network's capacity. If these methods do not work, then a unit can contact the NILE network operator to decide which, if any, tactical messages received and queued for relay may be deleted.

Link 22 Users

The Link 22 program was initially conducted collaboratively by seven nations under the aegis of a Memorandum of Understanding (MOU). The original seven nations were Canada, France, Germany, Italy, the Netherlands, the UK, and the United States, with the United States acting as the host nation. Spain has replaced the Netherlands as a NILE nation. Finland has recently entered the NILE nations group. Other NATO nations have chosen either Link 16 or Link 11 as their preferred TDL.

Variable Message Format

Variable message format (VMF) was originally developed under Link 16. How-ever, during the development cycle, it became apparent that the information to be exchanged and the potential users of this concept were much larger than originally envisioned (Army only). Therefore, VMF was removed as a subset of TADIL J and developed as a separate standard, Mil-Std-6017. However, this standard contains the message set only; Mil-Std-2045-47001C defines the net-work and how these messages are exchanged. VMF messages, referred to as K series, contain much of the same information as is found in Link 11 and Link 16. The K series messages are too numerous to list here, but can be found in Mil-Std-6017.[4]

VMF is the message set currently used by the US Army. There are plans to expand this message set to NATO ground units.

Blue Force Tracker

Blue Force Tracker (BFT) was developed by the US Department of Defense. BFT provides military commanders and military forces with the location of friendly troops. In the military, blue denotes a friendly force and red denotes the enemy. BFT is based on the Force XXI Battle Command Brigade and Below (FBCB2) soldier and vehicle level workstations. However, BFT goes beyond the limitations of FBCB2. It also complements networks used for combat service support func-tions such as supply and maintenance. The two systems are most significantly limited by the range of two networks. The first is the UHF EPLRS network, which can transfer low speed data and global navigation satellite system (GPS) position and time information. The second is the VHF/FM SINCGARS radio, which is used for voice and low speed data within battalions and brigades. While it is jam-resistant, it is not encrypted and not authorized for classified information.

One limitation of BFT in urban areas is that GPS will not always penetrate buildings. However, the soldier wireless local area network of the intra-squad radio can provide connectivity. Each intra-squad radio that can receive the GPS information can propagate it to the rest of the network. It will be no more accurate than the position of the person receiving, but that is much better than no position. BFT is used by US air and ground forces as well as NATO coalition forces.

Integrated Broadcast Service

Integrated broadcast service (IBS) was built to replace the following legacy UHF tactical data links:

[4]Mil-Std-6017, Variable Message Format, 19 July 2006.

- TIBS—Tactical Information Broadcast Service
- TDDS—Tactical Related Applications (TRAP) Data Dissemination System
- TRIXS—Tactical Reconnaissance Intelligence eXchange System
- TADIXS-B—Tactical Data Information Exchange System—Broadcast

When TDDS, TIBS, and TRIXS UHF broadcasts were created, they supported near-real-time (NRT) needs of global, theater, and local users. However, interoperability and a standard message set did not exist between these systems. Therefore, in 1996, integration of these TDLs was directed by the US Government.

IBS integrates multiple intelligence broadcasts into a system of systems, and migrates tactical receive terminals into a single, related joint tactical terminal (JTT) family. IBS provides time critical situational awareness information to all IBS terminals and users on the network. IBS nodes can transmit information without requiring central node access or approval. The standard message set includes Link 16 and the Global Command and Control System set of messages.

IBS is used by US forces only.

TACTICAL DATA LINK TRAINING

As TDLs were developed, they became more complex, both the network and the information that was exchanged by participants in the network. It became necessary to train warfighters on how to develop TDL networks, as well as the methodology of what types of messages to exchange and when. The only available means of training at the time was during live fly events, or by simulating TDL networks. These simulations were not realistic because they did not simulate the network and, as such, provided limited training.

However, during the 1990s, simulators were being developed with higher fidelity training capability, and could be connected to one another for distributed training rather than train in "stand-alone." As simulations were connected to one another, protocols for distributed simulation were developed. The first distributed simulation protocol was the Aggregate Level Simulation Protocol (ALSP). This protocol was developed such that data packets exchanged on either a local area network (LAN) or a wide area network (WAN) contained the information necessary for simulations to participate in a training event. The second protocol developed was the Distributed Interactive Simulation (DIS), and the third is called the High Level Architecture (HLA).

Distributed Mission Training

As distributed simulation protocols were being developed, another effort was undertaken by the USAF: Distributed Mission Training (DMT). Distributed training provides a capability where multiple warfighters located at the same or different simulation facilities can participate in one training scenario, ranging from individual and team participation up to full theater level battles. The concept allows warfighters to train using almost any type of networkable training device including fighter, command and control, and cargo assets. Additionally DMT provides the ability to enhance training scenarios by creating additional assets not connected to the DMT network such as computer generated forces. These computer generated forces can be used to complete a training scenario if simulators are not available.

Later, adding live assets into the distributed training network was included in DMT. This combination of live, virtual, and constructive environments adds considerable training opportunities for joint and combined forces from their own location or a deployed training site.

Once these training capabilities were used on a regular basis, the next logical step was to create a TDL training capability. This training capability would require creating TDL models that would adequately provide command and control training.

Modeling Tactical Data Links

Modeling TDLs began with the Simulation Interoperability Standards Organization (SISO) Link 16 simulation standard effort. Link 16 was modeled first because there were immediate and operational requirements to model this TDL for Air Force C2 training. This effort consisted of using existing DIS models and incorporating them into one DIS and HLA model to enhance interoperability between DIS and HLA users and the US DoD services.

SISO-STD-002

One of the first widely accepted models for Link 16 is SISO-STD-002, the standard for Link 16 distributed simulation. SISO-STD-002 was developed and tested in the early 2000s, and incorporated five different DIS models. The standard was unique because it modeled both message exchange and the Link 16 network. Several US DoD standards organizations have adopted SISO-STD-002 as their model for Link 16. SISO-STD-002 has models for both DIS and HLA. This standard is also used as a template to model other TDLs. From this standard also came a way to quickly model TDLs for DIS and HLA.

Tactical Data Link Models for Distributed Simulation

TDL models for distributed simulation have been used for years to enhance C2 training. SISO-STD-002 provided the first widely used model for distributed training. From this model, models for other TDLs will be used in the proposed standard for modeling Link 11/11B and SADL, and is planned for use for IBS-I and IBS-S. These models will enable other simulation facilities and organizations to either model TDLs or quickly set up TDL message exchange while allowing for future growth without extensive changes.

The TDL models apply to protocol data unit (PDU) version 6 and earlier. Currently, IEEE 1278.1a is being updated and includes some changes to the Transmitter PDU that will impact the SISO-STD-002 and this proposal. Once IEEE 1278.1a is formally approved, SISO-STD-002 and TADIL TALES templates will be updated as well.

The TDL distributed models define a specific DIS signal and transmitter PDU structure, and HLA Base Object Models (BOMs) for data link modeling. These HLA BOMs define no new objects but a new object class that corresponds to the DIS transmitter PDU. The HLA BOM also defines a family of interactions that support all tactical data implementation that corresponds to the DIS signal PDU.

DIS Transmitter and Signal PDUs

The DIS distributed model shown at the end of this chapter shows how the DIS signal and transmitter PDUs are modeled for TDL modeling and message transmission. The TDLs will be distinguishable by TDL type. Additional enumerations can be added to the DIS Enumerated Bit Values for DIS document.

IEEE 1278.1a-1995 transmission rules for radios will apply. A transmitter/signal PDU pair or transmitter/signal/transmitter PDU wrap will be implemented as required.

Transmitter PDU Model

For the transmitter PDU, the static fields will be populated as defined in IEEE 1278.1a-1995. 64 bits of modulation parameter data are reserved for specific TDL modeling; therefore, the Length of Modulation Parameters field is set to 8 for 8 octets. If no TDL modulation parameters are defined, these will be defined as padding. If modulation parameters are added later, no changes are required unless a modeling fidelity increase is necessary. Another option is not to include the 64 bits of padding and set the modulation parameters field to zero. This would indicate no TDL modulation parameters are present, and would indicate TDL message exchange only. The transmitter PDU model is shown in Table 23.6. The yellow blocks indicate the additional TDL modulation parameters.

Signal PDU Model

For the signal PDU, the static fields will be populated as defined in IEEE 1278.1a-1995. There will be an additional 160 bits to define the specific link characteristics. If no specific link characteristics are defined, then the 160 bits will be padding for message transmission only. If modeling data are added later, no changes are required unless a modeling fidelity increase is necessary. The signal PDU model is shown in Table 23.7. The yellow blocks indicate the additional TDL modeling parameters.

Table 23.6 TDL DIS Transmitter PDU

Field size (bits)	Transmitter PDU fields			Description
96	PDU header	Protocol version	8 bit enumeration	
		Exercise ID	8 bit unsigned integer	
		PDU type	8 bit enumeration	
		Protocol family	8 bit enumeration	
		Time stamp	32 bit unsigned integer	
		Length	16 bit unsigned integer	
		Padding	16 bits unused	
48	Entity ID	Site	16 bit unsigned integer	
		Application	16 bit unsigned integer	
		Entity	16 bit unsigned integer	
16	Radio ID		16 bit unsigned integer	TDL Radio ID transmitting the signal
64	Radio entity type	Entity kind	8 bit enumeration	
		Domain	8 bit enumeration	
		Country	16 bit enumeration	
		Category	8 bit enumeration	TDL Enumeration IAW SISO DIS EBV Document
		Nomenclature version	8 bit enumeration	
		Nomenclature	16 bit enumeration	
8	Transmit state		8 bit enumeration	
8	Input source		8 bit enumeration	8—Digital data device
16	Padding		16 bits unused	
192	Antenna location	X component	64 bit floating point	
		Y component	64 bit floating point	
		Z component	64 bit floating point	

(*continued*)

Table 23.6 (*Continued*)

Field size (bits)	Transmitter PDU fields			Description
96	Relative antenna location	X component	32 bit floating point	
		Y component	32 bit floating point	
		Z component	32 bit floating point	
16	Antenna pattern type		16 bit enumeration	
16	Antenna pattern length		16 bit unsigned integer	
64	Frequency		64 bit unsigned integer	Set to left frequency
32	Transmit frequency bandwidth		32 bit floating point	Set to transmission frequency bandwidth, 3 dB down
32	Power		32 bit floating point	Power in dBm
64	Modulation type	Spread spectrum	16 bit Boolean array	Bit 1 set to 1 if frequency hopping used All bits set to 0 if not used
		Major	16 bit enumeration	TDL Enumeration IAW SISO DIS EBV Document
		Detail	16 bit enumeration	TDL Enumeration IAW SISO DIS EBV Document
		System	16 bit enumeration	TDL Enumeration IAW SISO DIS EBV Document
16	Crypto system		16 bit enumeration	TDL Enumeration IAW SISO DIS EBV Document
16	Crypto key ID		16 bit unsigned integer	
8	Length of modulation parameters		8 bit unsigned integer	Set as required
24	Padding		24 bits unused	
	Modulation parameters	TDL characteristic	Set as required for TDL modeling. For TDL message exchange only, these fields can be ignored	Integer, floating point, or enumeration

Table 23.7 TDL Signal PDU

Field size (bits)	Signal PDU fields		Description
96	PDU header	Protocol version	8 bit enumeration
		Exercise ID	8 bit unsigned integer
		PDU type	8 bit enumeration
		Protocol family	8 bit enumeration
		Time stamp	32 bit unsigned integer
		Length	16 bit unsigned integer
		Padding	16 bits unused
48	Entity ID	Site	16 bit unsigned integer
		Application	16 bit unsigned integer
		Entity	16 bit unsigned integer
16	Radio ID		16 bit unsigned integer	**Shall** contain the ID of the TDL radio or terminal transmitting the signal
16	Encoding scheme		16 bit enumeration	Bits 0–13 **shall** contain the number of TDL words not including the TDL header. Bits 14–15 **shall** contain the value 1 to indicate an encoding class raw binary data

(continued)

559

Table 23.7 (*Continued*)

Field size (bits)		Signal PDU fields	Description
16	TDL type	16 bit enumeration	TDL Enumeration IAW SISO DIS EBV Document
32	Sample rate	32 bit integer	This field **shall** be set to 0
16	Data length	16 bit integer	This field **shall** contain the length of tactical data in bits beginning after the samples field
16	Samples	16 bit integer	This field **shall** be set to 0
160	TDL signal PDU fields		Use 160 bits to describe TDL characteristics. Each field can be divided into 8, 16, or 32 bits of data, but must meet IEEE 1278.1a-1995 requirements for byte alignment. For message transfer only, the 160 bits can be defined as padding
	Message data	Array of octets	
	Padding	Signal PDU C2 padding to double-word boundary IAW IEEE 1278.1a	Padding (if needed) to increase total PDU size to a multiple of 32 bits

HLA Base Object Model

The Link 16 BOM added a family of interactions that support future TDL implementation of other data links. For this chapter, the acronym (LINK) will be used as a place holder for any TDL. For example, additional TDL interactions can be added to the (LINK)RadioSignal interaction, but can be designated as Link11RadioSignal or SADLRadioSignal. Complex data types and enumerations can also be added using the same method as for Link 16.

The TDL HLA BOM assumes that

1. the parent FOM contains all current DIS transmitter PDU records (not those associated with the PDU header) in accordance with IEEE 1278.1a as part of its object class hierarchy;

2. the parent FOM contains all current DIS signal PDU PDU records (not those associated with the PDU header) in accordance with IEEE 1278.1a as part of its interaction class hierarchy.

Conventions within the TDL distributed model BOM in OMT 1.3 format will follow those adopted by the Real time Platform Level Reference (RPR) FOM version 1.0 and 2.0. These conventions are as follows.

1. All names have the initial letter of each word capitalized.

2. All enumeration names end in the text "Enum" followed by a number. The number indicates the number of bits in the enumerated value.

3. All complex data type names end in the text "Struct."

Object Class Data

The TDL HLA BOM defines no new object classes. Instead the BOM will define a single complex data type (LINK)TransmitterStruct that corresponds to the TDL modulation parameters in the DIS transmitter PDU shown in Table 23.6. An attribute of this complex data type should be added to the object class in the parent FOM corresponding to the DIS transmitter PDU. Modulation parameters of the transmitter PDU shown in Annex A should map to the fields of the (LINK)TransmitterStruct complex data type attribute.

Parent object class fields are also modified such that they refer to the corresponding transmitter PDU fields (see TDL HLA BOM Assumption 1 above).

Note that for an RPR FOM implementation, an attribute of the (LINK)TransmitterStruct complex data type should be added to the RadioTransmitter object class.

Interaction Class Data

The family of interactions is a hierarchy in which the BOM's base class for this interaction is a generic class—the TDLBinaryRadioSignal interaction. This class is an empty class, contains no parameters, and is neither publishable nor subscribable. The specific parameters are properties of the various subclasses of this generic base class, and it is these subclasses that are published and subscribed to.

A TDL interaction will be added as a subclass of the TDLBinaryRadioSignal interaction, and contain the TDL network header parameters. The (LINK)MessageRadioSignal interaction contains the TDL message data. Additional interactions shown in Table 23.12 define the other types of Link 16 messages.

The Link 16 BOM design is such that the TDLBinaryRadioSignal interaction becomes a subclass of the parent FOM's equivalent of the DIS signal PDU.

Adding Tactical Data Link BOMs to the RPR FOM

Adding a TDL BOM to the RPR FOM consists of three steps: adding the TDL Link Radio Signal interaction, adding the (LINK)TransmitterData structure, and adding the parameters necessary including the associated enumerated and complex data types. Adding the interaction is the same for RPR FOM versions 1.0 and 2.0. The manner of adding the (LINK)TransmitterData structure differs between the two RPR FOM versions as shown in Tables 23.8–23.11.

The TDLBinaryRadioSignal class will be added as a subclass of the RawBinaryRadioSignal interaction class. This is done in order to allow access to the HostRadioIndex parameter in the RawBinaryRadioSignal interaction class. The HostRadioIndex parameter will tie the TDL message to a specific host and radio transmitter.

Table 23.8   TDL Complex Datatypes in RPR FOM 1.0

Complex datatype	Field name	Datatype	Cardinality
ModulationStruct	SINCGARModulation[52]	SINCGARSModulationStruct	0–1
	LINKTransmitterData[56]	LINKTransmitterStruct	0–1

Table 23.9   TDL Enumerated Values in RPR FOM 1.0

Identifier	Enumerator	Representation
RFModulationSystemTypeEnum16	Other	0
	Generic	1
	HQ	2
	HQII	3
	HQIIA	4
	SINCGARS	5
	CCTT_SINCGARS	6
	LINK LINK	8.XX

Table 23.10   TDL Complex Datatypes in RPR FOM 2.0

Complex datatype	Field name	Datatype	Cardinality
SpreadSpectrumStruct	SpreadSpectrum-Type	SpreadSpectrum-Enum16	1
	Padding	Octet	2
	SINCGAR-Modulation[52]	SINCGARS-ModulationStruct	0–1 (Spread-SpectrumType = SINCGARS-FrequencyHop)
	LINKTransmitter-Data[56]	LINKTransmitter-Struct	0–1 (Spread-SpectrumType = LINK_Spectrum-Type)

Table 23.11   TDL Enumerated Values in RPR FOM 2.0

Identifier	Enumerator	Representation
RFModulationSystemTypeEnum16	Other	0
	Generic	1
	HQ	2
	HQII	3
	HQIIA	4
	SINCGARS	5
	CCTT_SINCGARS	6
	JTIDS_MIDS	8
SpreadSpectrumEnum16	None	0
	SINCGARSFrequencyHop	1
	LINK_SpectrumType	2

The (LINK)RadioSignal interaction class is added as a subclass of a new interaction class, the TDLBinaryRadioSignal interaction, which itself is a subclass of the RPR FOM's RawBinaryRadioSignal interaction class, as shown in Table 23.12.

DIS to HLA Translations

For DIS to HLA translations, the SISO-STD-002 Appendix C provides guidance for DIS to HLA gateways for the Link 16 distributed model. There is also a guide in SISO-STD-005, the Link 11/11B standard for distributed simulation. As other TDLs are developed and standardized, these will provide translation guidance.

Table 23.12 TDL HLA Interaction Table

Interaction1	Interaction2	Interaction3	Interaction4	Interaction5
RadioSignal	Application SpecifcRadio-Signal DatabaseIndexRadioSignal EncodedAudioRadioSignal RawBinaryRadioSignal	TDLBinary-RadioSignal	LINKRadio-Signal	LINK_Message

Link 16 Network Model

As mentioned before, SISO-STD-002 models the Link 16 network as well as the message exchange. Because of this, SISO-STD-002 has different levels of fidelity. The standard allows for interoperability between fidelity levels, such that a level 3 fidelity participant can interoperate with a level 0 fidelity participant. The standard also allows for network latencies, and has methods to overcome simulation network limitations.

Transmitter PDU Model

The Link 16 transmitter PDU model is the same for all fidelity levels. All transmitter PDU modulation fields are required for all fidelity levels. The model is shown in Table 23.13.

The Link 16 Signal PDU models have different fidelity levels. One can choose a fidelity level that models message exchange only to time slots, encryption, and net entry and exit.

Level 0

Level 0 is the lowest level of fidelity. No Link 16 network characteristics are modeled in this fidelity level. The NPG and net fields are modeled in the signal message, but all other data in the JTIDS transmission header are not modeled, and are set to 255. Multiple messages are permitted in a single signal message. All messages within the signal message are assumed to be for the same NPG and net number with the same assumed packing. There is no message metering, and the maximum number of messages specified in the DIS standard is packed into the data area of a single signal message.

The Link 16 fidelity level 0 signal PDU model is shown in Table 23.14.

Link 16 time synchronization model for fidelity levels 0, 1, 2, and 3 does not model network entry and coarse and fine synchronization. Therefore, all units in the simulated network enter the network and automatically assume fine synchronization. For fidelity levels 0–2, the network synchronization ID is set to zero.

Table 23.13 Link 16 Transmitter PDU Model; All Fidelity Levels

Field size (bits)		Transmitter PDU fields		Description
96	PDU header	Protocol version	8 bit enumeration	
		Exercise ID	8 bit unsigned integer	
		PDU type	8 bit enumeration	
		Protocol family	8 bit enumeration	
		Time stamp	32 bit unsigned integer	
		Length	16 bit unsigned integer	
		Padding	16 bits unused	
48	Entity ID	Site	16 bit unsigned integer	
		Application	16 bit unsigned integer	
		Entity	16 bit unsigned integer	
16	Radio ID		16 bit unsigned integer	**Shall** contain the ID of the radio transmitting the signal
64	Radio entity type	Entity kind	8 bit enumeration	
		Domain	8 bit enumeration	
		Country	16 bit enumeration	
		Category	8 bit enumeration	21—Link 16 terminal IAW SISO DIS EBV Document
		Nomenclature version	8 bit enumeration	
		Nomenclature	16 bit enumeration	
8	Transmit state		8 bit enumeration	
8	Input source		8 bit enumeration	8—Digital data device
16	Padding		16 bits unused	
192	Antenna location	X component	64 bit floating point	
		Y component	64 bit floating point	
		Z component	64 bit floating point	

(*continued*)

Table 23.13   (*Continued*)

Field size (bits)	Transmitter PDU fields			Description
96	Relative antenna location	X component	32 bit floating point	
		Y component	32 bit floating point	
		Z component	32 bit floating point	
16	Antenna pattern type		16 bit enumeration	
16	Antenna pattern length		16 bit unsigned integer	
64	Frequency		64 bit unsigned integer	Mode 1 = 1131000000 (center frequency). For Mode 2 or 4, set to 969000000
32	Transmit frequency bandwidth		32 bit floating point	240000000 unless in communications mode 2 or 4, then 3000000
32	Power		32 bit floating point	Power in dBm
64	Modulation type	Spread spectrum	16 bit Boolean array	Bit 1 set to 1: frequency hopping for JTIDS communications mode 1. All bits set to 0: for JTIDS communications modes 2 or 4
		Major	16 bit enumeration	7—Carrier phase shift modulation
		Detail	16 bit enumeration	0—Other
		System	16 bit enumeration	8—JTIDS/MIDS
16	Crypto system		16 bit enumeration	0—Other
16	Crypto key ID		16 bit unsigned integer	0—Other
8	Length of modulation parameters		8 bit unsigned integer	8 = 8 octets
24	Padding		24 bits unused	
8	Modulation parameter #1	Time slot allocation mode	8 bit enumeration	Integer enumeration 0–4

Table 23.13   (*Continued*)

Field size (bits)	Transmitter PDU fields			Description
8	Modulation parameter #2	Transmitting terminal primary mode	8 bit enumeration	Integer enumeration 1—NTR 2—JTIDS Unit Participant
8	Modulation parameter #3	Transmitting terminal secondary mode	8 bit enumeration	Integer enumeration 0—None 1—Net position reference 2—Primary navigation controller 3—Secondary navigation controller
8	Modulation parameter #4	Synchro-nization state	8 bit enumeration	Integer enumeration 2—Coarse synchronization 3—Fine synchronization
32	Modulation parameter #5	Network synchronization ID	32 bit unsigned integer	TSA level 0–2, set to 0 TSA level 3, 4, set to 32 bit random number

Fidelity Level 1

Level 1 is similar to fidelity level 0 except that there is metered message data in the signal message. When the fidelity level 1 modeling is implemented, one time slot's worth of information is in one signal message. The NPG and net number fields are modeled, and the rest of the network modeling information is set to 255. Net entry is a low fidelity model. An NTR is assigned to the simulated network. The NTR issues a J0.0 message, typically in time slot A-0-6, at a rate of every 12 s. Modulation parameter 4 (synchronization state) in the transmission message is set to fine synchronization after reception of the J0.0 initial net entry message from the NTR. The first data message sent by a JU entering the network is the JU's PPLI.

The Link 16 fidelity level 1 signal PDU model is shown in Table 23.15.

Fidelity Level 2

Level 2 models the Link 16 time slots and also models metered data with no encryption. Link 16 messages are assigned to individual time slots and transmitted according to the Link 16 network load, which assigns transmission time slots for all JUs in the network. The NPG, net number, and time slot identification fields are filled in. However, network encryption is not modeled and the TSEC and

Table 23.14 Link 16 Signal PDU Model Fidelity Level 0.

Field size (bits)	Signal PDU fields			Valid range	Description
160	Link 16 signal PDU fields				
	NPG number		16 bit unsigned integer	0–511	NPG number. Used to segregate information within a JTIDS/MIDS network. Creates virtual networks of participants
	Net number		8 bit unsigned integer	0–127	Network number. Used to create virtual sub-circuits within NPG for stacked nets or between NPGs for multi-net
	TSEC CVLL		8 bit unsigned integer	255	Not modeled, set to 0xFF
	MSEC CVLL		8 bit unsigned integer	255	Not modeled, set to 0xFF
	Message type identifier		8 bit enumeration		Determines whether normal JTIDS header and message, RTT A/B, RTT reply, JTIDS voice, LET JTIDS, or VMF follows. See Table 5.2.2 for message type identifier enumerations
	Padding		16 bits	0	Set to 0
	Time slot ID	Time slot number	Bits 0–16	98303	Not modeled, set to 0x17FFF
		Padding	Bits 17–23	0	Set to 0
		Epoch number	Bits 24–31	0–112	Not modeled, set to 0x17FFF
	Perceived transmit time	Integer part	32 bit unsigned integer	4294967295	Not modeled, set to 0xFFFFFFFF
		Fraction part	32 bit unsigned integer	4294967295	
	Message data		Array of 32 bit unsigned integers		Link 16 messages

Table 23.15 Link 16 Signal PDU Model, Fidelity Level 1

Field size (bits)		Signal PDU fields		Valid range	Description
160	Link 16 signal PDU fields	NPG number	16 bit unsigned integer	0–511	NPG number. Used to segregate information within a JTIDS/MIDS network. Creates virtual networks of participants
		Net number	8 bit unsigned integer	0–127	Network number. Used to create virtual sub-circuits within NPG for stacked nets or between NPGs for multi-net
		TSEC CVLL	8 bit unsigned integer	255	Not modeled, set to 0xFF
		MSEC CVLL	8 bit unsigned integer	255	Not modeled, set to 0xFF
		Message type identifier	8 bit enumeration		Determines whether normal JTIDS header and message, RTT A/B, RTT reply, JTIDS voice, LET JTIDS, or VMF follows. See Table 5.2.2. for message type identifier enumerations
		Padding	16 bits	0	Set to 0
	Time slot ID	Time slot number	Bits 0–16	98303	Not modeled, set to 0x17FFF
		Padding	Bits 17–23	0	Set to 0
		Epoch number	Bits 24–31	0–112	Not modeled, set to 0xFF
	Perceived transmit time	Integer part	32 bit unsigned integer	4294967295	Not modeled, set to 0xFFFFFFFF
		Fraction part	32 bit unsigned integer	4294967295	
	Message data		Array of 32 bit unsigned integers		One time slot's amount of TADIL J words: 3, 6, or 12 words

569

MSEC fields are set to 255. The Link 16 fidelity level 2 signal PDU model is shown in Table 23.16.

Network latencies present a problem for Link 16 network modeling at fidelity level 2 and higher. Latencies in a Link 16 network are on the order of microseconds; however, latencies in a distributed simulation network are on the order of milliseconds. The section entitled Link 16 Model for Network Latency (later) presents models that address this issue.

Fidelity Level 3

Level 3 models the Link 16 encryption on transmission and message reception In addition, stacked nets, multi-nets, and crypto-nets can be modeled. All transmission information fields are filled in with meaningful data. The Link 16 fidelity level 3 signal PDU model is shown in Table 23.17.

Fidelity Level 4

Fidelity level 4 models everything from fidelity level 3 with the addition of the medium fidelity net entry and time synchronization procedures. All JTIDS header fields in the signal message are modeled with meaningful values. Medium fidelity synchronization procedures include net entry using RTT messages, time synchronization, and net exit.

Time Synchronization Modeling. Time synchronization modeling corresponds only to fidelity level 4. It is applicable to those systems where simulation of the fine synchronization methodology is paramount, potentially for high fidelity training, network testing, and network experimentation. Because the latency of WANs (latencies up to hundreds of milliseconds) is orders of magnitude higher than in a real Link 16 network (latencies up to 3 ms), this methodology will not meet the needs of sub-millisecond accuracy. Communities with the need for sub-millisecond accuracy will need to use a centralized server with a real-time operating system to simulate the microsecond intricacies of the JTIDS network. The term "high fidelity synchronization" is reserved for synchronization mechanisms that are able to model the sub-millisecond accuracy of the Link 16 network.

The accuracy of the synchronization mechanism should have an error rate less than the simulated time of propagation. The accuracy of the synchronization mechanism should be taken into account when modeling fine synchronization. The medium fidelity synchronization procedure is as follows.

1. The NTR begins by issuing net entry message pairs at a rate in accordance with the JTIDS terminal specification (typically in time slot A-0-6 at a rate of every 12 s).

2. A unique randomly generated key is filled into the network synchronization ID field. The primary JTIDS duty field should contain an NTR enumeration. At this point, the JU is considered to have achieved coarse synchronization.

Table 23.16 Link 16 Signal PDU Model, Fidelity Level 2

Field size (bits)		Signal PDU fields		Valid range	Description
160	Link 16 signal PDU fields	NPG number	16 bit unsigned integer	0–511	NPG number. Used to segregate information within a JTIDS/MIDS network. Creates virtual networks of participants
		Net number	8 bit unsigned integer	0–127	Network number. Used to create virtual sub-circuits within NPG for stacked nets or between NPGs for multi-net
		TSEC CVLL	8 bit unsigned integer	255	Not modeled, set to 0xFF
		MSEC CVLL	8 bit unsigned integer	255	Not modeled, set to 0xFF
		Message type identifier	8 bit enumeration		Determines whether normal JTIDS header and message, RTT A/B, RTT reply, JTIDS voice, LET JTIDS, or VMF follows. See Table 5.2.2 for message type identifier enumerations
		Padding	16 bits	0	Set to 0
	Time slot ID	Time slot number	Bits 0–16	98303	Set to network data load time slots
		Padding	Bits 17–23	0	Set to 0
		Epoch number	Bits 24–31	0–112	Set to network data load time slots
	Perceived transmit time	Integer part	32 bit unsigned integer	4294967295	Not modeled, set to 0xFFFFFFFF
		Fraction part	32 bit unsigned integer	4294967295	
	Message data		Array of 32 bit unsigned integers		One time slot's amount of TADIL J words: 3, 6, or 12 words

Table 23.17 Link 16 Signal PDU Model, Fidelity Level 3.

Field size (bits)		Signal PDU fields		Valid range	Description
160	Link 16 signal PDU fields	NPG number	16 bit unsigned integer	0–511	NPG number. Used to segregate information within a JTIDS/MIDS network. Creates virtual networks of participants
		Net number	8 bit unsigned integer	0–127	Network number. Used to create virtual sub-circuits within NPG for stacked nets or between NPGs for multi-net
		TSEC CVLL	8 bit unsigned integer	255	Set to TSEC CVLL provided by NDL
		MSEC CVLL	8 bit unsigned integer	255	Set to TSEC CVLL provided by NDL
		Message type identifier	8 bit enumeration		Determines whether normal JTIDS header and message, RTT A/B, RTT reply, JTIDS voice, LET JTIDS, or VMF follows. See Table 5.2.2 for message type identifier enumerations
		Padding	16 bits	0	Set to 0
	Time slot ID	Time slot number	Bits 0–16	98303	Set to network data load time slots
		Padding	Bits 17–23	0	Set to 0
		Epoch number	Bits 24–31	0–112	Set to network data load time slots
	Perceived transmit time	Integer part	32 bit unsigned integer	4294967295	Not modeled, set to 0xFFFFFFFF
		Fraction part	32 bit unsigned integer	4294967295	
	Message data		Array of 32 bit unsigned integers		One time slot's amount of TADIL J words: 3, 6, or 12 words

Table 23.18 Link 16 Signal PDU Model, Fidelity Level 4.

Field size (bits)	Signal PDU fields		Valid range	Description
160	Link 16 signal PDU fields	NPG number	0–511	NPG number. Used to segregate information within a JTIDS/MIDS network. Creates virtual networks of participants
		Net number	0–127	Network number. Used to create virtual sub-circuits within NPG for stacked nets or between NPGs for multi-net
		TSEC CVLL	255	Set to TSEC CVLL provided by NDL
		MSEC CVLL	255	Set to TSEC CVLL provided by NDL
		Message type identifier		Determines whether normal JTIDS header and message, RTT A/B, RTT reply, JTIDS voice, LET JTIDS, or VMF follows. See Table 5.2.2 for message type identifier enumerations
		Padding	0	Set to 0
		Time slot ID — Time slot number (Bits 0–16)	98303	Set to network data load time slots
		Padding (Bits 17–23)	0	Set to 0
		Epoch number (Bits 24–31)	0–112	Set to network data load time slots

Signal PDU fields sizes: NPG number — 16 bit unsigned integer; Net number — 8 bit unsigned integer; TSEC CVLL — 8 bit unsigned integer; MSEC CVLL — 8 bit unsigned integer; Message type identifier — 8 bit enumeration; Padding — 16 bits.

(continued)

Table 23.18 (*Continued*)

Field size (bits)	Signal PDU fields		Valid range	Description
	Perceived transmit time	Integer part	32 bit unsigned integer	0–4294967295	NTP time stamp format—NTP time stamps are represented as a 64-bit unsigned fixed-point number, in seconds relative to 0 h on 1 January 1900 Universal Time, Coordinated (UTC). The integer part is in the first 32 bits and the fraction part in the last 32 bits. The precision of this representation is about 200 picoseconds, which should be adequate for even the most exotic requirements
		Fraction part	32 bit unsigned integer	0–4294967295	
Message data			Array of 32 bit unsigned integers		One time slot's amount of TADIL J words: 3, 6, or 12 words

3. The JU will then transmit the appropriate RTT message (A or B). The synchronization state is set to coarse synchronization. The JU uses its own terminal perceived time in the perceived transmit time field.

4. The appropriate NTR/JU answers (in accordance with the JTIDS terminal specification), using the JU perceived time and the entity distance to calculate the perceived receive time. The RTT is then transmitted.

5. The transmitting JU fills its own terminal perceived time with the received transmit time field. The formula for filling in the receive time in the RTT reply is

$$RTT_{reply} = RT_{terminal} - t_{delay} + t_{propagate}$$

The t_{delay} is computed by

$$t_{delay} = RT_{time} - TT_{time}$$

where RT_{time} is the actual time held by the receiving/replying participant (derived from NTP, GPS, etc); $RT_{terminal}$ is the value of the simulated Link 16 terminal clock at the receiving/replying participant; TT_{time} is the actual time held by the transmitting participant (derived from NTP, GPS, etc); t_{delay} is the difference between the receiver's real-time clock at the time of receipt and the sender's real-time clock at the time of transmission (i.e. it approximates the emulation network latency), and $t_{propagate}$ is the propagation time of the radio frequency message in the simulated environment. This formula computes the perceived time of receipt by the receiving simulator with respect to the simulated terminal clock of the sender.

6. The originating JU then updates its own terminal time in accordance with the simulator model and the Link 16 fine synchronization procedures.

After the appropriate number of RTT exchanges have occurred (depending on whether the RTT A or RTT B method of synchronization was used and the internal terminal simulation model), the JU then will consider itself to be in fine synchronization, but continually issues RTT message pairs to maintain synchronization at rates specified within the JTIDS terminal specification. Once the terminal emulator model has met the requirements for fine synchronization, normal message transmissions commence in accordance with the Link 16 standard.

Communication Model for Simulations at Different Fidelity Levels

In the event that participants in a simulated Link 16 network set their respective simulations to operate at a different level of fidelity, the following model is available.

1. If a low fidelity network participant is in a simulated JTIDS network with a higher fidelity NTR, the network participant follows the low fidelity synchronization model. It skips the RTT synchronization process, changing directly to the fine synchronization state once it receives a J0.0 initial net entry message from the NTR or any Initial Entry JTIDS (IEJU).

2. If a higher fidelity network participant is in a simulated JTIDS network with a lower fidelity NTR, the higher fidelity participant either follows the low fidelity synchronization model or achieves fine sync with other high fidelity simulators. This can be accomplished by exchanging RTT-Bs or by passively synchronizing with other available high fidelity simulators. If no other high fidelity simulators are available to synchronize with, the high fidelity participant skips the RTT exchange and directly enters fine synchronization once the J0.0 is received.

3. If the NTR is a lower fidelity simulation and is unable to simulate full NTR duties, the NTR still has the ability to transmit net entry messages. The signal message is filled in with a J0.0 with zeroed time slot information. RTT emulation is not required of low fidelity NTRs.

4. A lower fidelity JU entering the net uses the network synchronization ID received from the NTR/IEJU in its transmission information PDUs. It then issues PPLIs at the assigned rate. This is nominally once every 12 s (equivalent to the A-0-6 time slot block), although this can occur at times of up to 24 s. All simulators regardless of fidelity accept another terminal's statement of synchronization capability if the network synchronization ID matches its own network synchronization ID.

In a lower fidelity synchronization simulation, non-reception of a PPLI message pair for 60 s indicates that the unit has fallen out of the simulated Link 16 network. Synchronization procedures are re-accomplished (i.e. reception of a PPLI message stating fine synchronization must occur before data from the JU will be accepted).

Link 16 Model for Network Latency

Each simulator that uses the SISO standard model for Link 16 simulations is assumed to be time synchronized to within ε milliseconds of a common reference time. The difference in the perceived time of any two emulators is in the interval $[-\varepsilon, \varepsilon]$. This is accomplished using the network time protocol, GPS clocks or any other suitable technology.

Each terminal maintains a local offset that is used to determine the time slot start and buffer retirement times. The locally perceived time is $t_p = t_c + t_l$ where t_c denotes the common reference time and t_l the local offset. A new time slot begins every 7.8125 ms measured from the initial perceived time. The kth slot begins at time $t_0 + 7.8125k$. The time slot occurring at time t_p is

$$(t_p - t_0)/7.8125$$

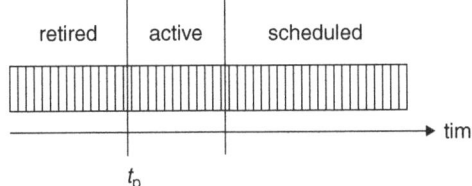

Figure 23.3 Retired, active, and scheduled time slots.

where t_0 denotes the initial perceived time. All Link 16 terminals also have a time slot expiration time denoted by t_B. The kth time slot expires at time $t_0 + 7.8125k + t_B$.

A time slot can be scheduled, active, or retired. An active time slot is one whose start time has passed (i.e. $t_p \geq t_0 + 7.8125k$), but which has not been retired (i.e. $t_p \leq t_0 + 7.8125k + t_B$). A retired time slot is a time slot that has been retired by the JU (i.e. $t_p > t_0 + 7.8125k + t_B$). A scheduled time slot is one that is neither active nor retired. Figure 23.3 shows the sets of retired, active, and scheduled time slots at a particular instance in time.

To provide an adequate model for latency, the Link 16 simulator should maintain an input buffer, an output buffer, a set of active time slots, and a set of retired time slots. When a time slot becomes active, the Link 16 simulator scans its output buffer for transmissions whose assigned time slot is equal to the active time slot. If more than one such transmission is found, the Link 16 simulator must select exactly one to be sent in the active time slot. The transmission is then removed from the buffer and sent as an emulation network message.

When a time slot is retired, the Link 16 simulator scans its input buffer for messages whose assigned time slot is equal to the retired time slot. The message whose transmitter location is closest to the receiver in simulation space is selected and delivered to the tactical data system. All other messages assigned to the retired time slot are discarded.

When the Link 16 simulator receives a transmission from the tactical data system whose assigned time slot is scheduled, it places that message into the output buffer. Otherwise, the message may be discarded or rescheduled for another available time slot. The actual behavior in this case depends on the terminal type being emulated.

When the Link 16 simulator receives a message from the emulation network and the time slot assigned to the message is not retired, that message is placed into the input buffer. Otherwise, the message is discarded.

This model uses a late time slot retirement scheme, employing an upper bound L on the network latency. When the sender has a transmission to send in the time slot starting at time T, the sender transmits a corresponding emulation network message when the time slot begins. The receiver delays the time slot retirement until time $T + B$, where $B > L > 7.8125$ ms. If a message is received prior to time $T + B$ and that message is intended for delivery in the time slot starting at T, then the receiver buffers the message. Once the time slot is retired, the messages received with that time slot number are sorted by transmission

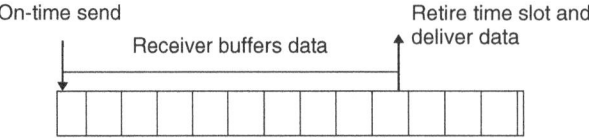

Figure 23.4 Late time slot retirement model.

distance in simulation space and the closest is taken as the message to have successfully arrived. Figure 23.4 depicts the beginning of a time slot and the corresponding time slot retirement by the receiver.

This model preserves the message order and contention properties of the Link 16 network when B greater than L. The end-to-end latency seen by end systems is increased by B. When B is smaller than L, the contention properties will not, in general, be preserved.

These models can be used when modeling Link 16 time slots, message metering, and net entry using RTT messages. For distributed simulations, it is usually not feasible to emulate higher fidelity Link 16 models, due to the high cost, unpredictable network latencies, changing network latencies, and difficulty in achieving time synchronization. Most USAF distributed simulation training organizations use fidelity level 1 model for Link 16 training, and that level fulfills their Link 16 training requirements. Higher levels should be implemented on a non-distributed basis, where Link 16 simulations are located in the same facility.

CONCLUSIONS

The TDL models developed for distributed simulation have made it possible to provide much needed training for US and NATO warfighters. These developments have provided a low cost method for TDL training in a distributed environment. SISO-STD-002[5] makes available a standard model for current and future TDLs. These models can be implemented quickly, and give the option, not requirement, for simulations to change when higher fidelity models are defined. The higher fidelity standards will have fidelity/interoperability between simulations, much like the SISO-STD-002[6] standard for Link 16[7].

[5]SISO-STD-002, Standard for Link 16 Simulations, 10 July 2006.

[6]SISO-STD-005, Standard for Link 11/11B Simulations.

[7]IEEE Std 1278.1-1995; IEEE Standard for Distributed Interactive Simulation—Application Protocols, 21 September 1995.

Chapter 24

Standards-Based Combat Simulation Initialization using the Military Scenario Definition Language (MSDL)

Robert L. Wittman Jr

INTRODUCTION

This chapter introduces the basics of combat simulation initialization and starts by identifying the logical data sets commonly associated with initialization. After this we move to a discussion of the challenges of initializing single systems and consistent initialization among federations of simulations and simulations and mission command devices. Next the Simulation Interoperability Standards Organization's (SISO) MSDL standard is introduced as a mechanism to describe and exchange military scenario initialization data in a common verifiable format. We conclude with a discussion on key research and development areas of interest to MSDL technologies.

For the purposes of this chapter we shall discuss initialization at three levels: model based initialization; military scenario based initialization; and federation based initialization. The next sections provide characterizations of the data sets providing the initial conditions at each level. The characterizations are intended

Engineering Principles of Combat Modeling and Distributed Simulation, First Edition.
Edited by Andreas Tolk.
© 2012 John Wiley & Sons, Inc. Published 2012 by John Wiley & Sons, Inc.

to provide the basic general nature of the data categories and are not exhaustive lists of either detailed model or scenario based initial conditions.

Model Based Initialization Level

The model based initialization level focuses on the types of data necessary to drive specific models of the environment, land and air platforms, and individual combatants and other life-forms, and the organizations to which the individuals belong. These models may vary in their level of detail in terms of both the variables that need to be populated necessary to bring them to life within the simulation and the parameters they can provide for interactions with other simulation based actors, data collection systems, and other federated simulations. The initialization data relevant to the physical dynamics of the systems are typically called model performance and characteristic data or model parametric data and include platforms (both equipment and human), data driving algorithms providing visual and infrared sensing, weapon delivery (hit or miss), mobility (how fast can I go on given terrain), vulnerability (what kind of damage do I take when hit), etc. This level also includes the basic runtime format data necessary for the environment to provide height above terrain requests, environmental feature requests, obstacle queries and other environment related outputs. For the purposes of this discussion the assets (number and type of sensors, weapons, communications assets, etc.) associated with a single actor (equipment, human or other life-forms, and units) are also considered part of the model based initialization data set.

Behavioral models and data sets are less standardized and include categories such as tactical movements such as bounding over-watch, movement formations, and more complex behaviors associated with assaults and room clearing activities.

Scenario Based Initialization

Scenario based initial conditions focus on the following:

- the organizations and actors that will be played within the simulation;
- the relationships between the organizations and actors;
- the initial status of the actors including their locations, health status, supply and munitions load for each actor (platform level for entity based simulations and unit level for aggregate simulations—those simulations that represent generalized strength, mobility, engagement, engineering, and other dynamics for the unit while only keeping a list of individual unit contributors for accounting purposes and updating the generalized strength, mobility, etc.);
- the planned missions and/or tasks for any of the associated actors;
- the geographic area where the simulation will take place; and
- the starting time and duration of the simulation.

The scenario level of initialization has been the focus of recent standards related efforts within the SISO resulting in the MSDL and as such provides the focal point for much of the remainder of this chapter.

Federation Based Initialization

Federation based initialization is the third data initialization category to be discussed. This category considers consistent initial conditions across all of the data types mentioned in the previous sections for simulations that will be joined and run as a federation of systems. Typically each system/federate within the federation manages the initialization categories identified above and may have unique data addressing data formats and content (more, less, or different) for the simulated capabilities they are providing to the federation.

Although beyond the scope of this section, this difference in data sets can impact the ability of the federation to provide a fair fight environment to actor interactions among two or more federates. A classic example of the fair fight issue highlights the impact of independent and different terrain model detail and techniques. In this example two entities are owned by two separate federates that are shooting at each other. One federate has a more detailed terrain representation with a hill separating the entities while the other federate does not represent the hill and can see and thus fire at the other entities giving them an unfair advantage. Although very simplistic, this issue with inconsistent modeling must be accounted for across all federate interactions to insure the federation meets the needs of the user.

As mentioned additional effort will be necessary to manipulate and provide an appropriate level of consistency when exchanging initialization or runtime data between simulation federates. This usually involves providing data mappings as part of the initialization data sets for unit and entity types: munitions, supplies, environmental, and other data (Lacy and Tuttle, 1998; Prochnow et al., 2005; Thakkar and Lindy, 2005; Blais et al., 2009; Sprinkle et al., 2011).

For the purposes of this discussion there is also a special case of federation based initialization that occurs in many simulation assisted exercises that accommodates exchange of information between the simulations and real-world mission command systems. Historically, the interface between the two systems is handled via translations between simulation formats and real-world message formats such Joint Variable Message Format (JVMF), United States Message Text Format (USMTF), and more recently Publish and Subscribe Service (PASS) and Data Distribution Service (DDS) topics. In general there is a much richer set of data sent from the simulation to the mission command devices providing message based and graphical situational awareness than orders generated by the commanders and their staff on mission command systems to the simulations. In some limited cases, such as for indirect fire control, there is an interface allowing commands to be input from the real-world devices. By and large in this setting the simulations and mission command devices are initialized independently with somewhat consistent organizational and environmental data sets.

Initialization Processes

A final short paragraph on initialization processes is necessary to round-out and wrap up this introductory section. The formality and coordination of the initialization process depends on the size and complexity of the exercise. For a simple single simulation exercise without external connectivity to other simulations or mission command devices the process can and should be defined within the simulation's documentation. For larger more complex simulation based events involving single or multiple real-world mission command devices as well as multiple simulation federates, a well-defined rigorous system of systems initialization process is necessary. This section does not intend to provide a one size fits all initialization process, but instead identifies the high risk areas that need to be addressed with the initialization process to support reduced costs and enhanced federation stability such as the following:

- agreement on environmental representation consistency and correlation boundaries;
- agreement on unit, platform, and life-form enumeration definitions and the enumeration mapping and tracking process;
- agreement on pre-runtime scenario change management and control. Scenario here includes order of battle, positions, health status, graphics overlays, environmental representation, actor ownership, mission command device interaction, etc;
- steps and magnitudes of change (e.g. is it allowable for one federate to change the order of battle and then reflect this change during simulation runtime or does this change need to be disseminated to all or some subset of the federates prior to runtime) for orderly pre-runtime changes within the scenario.

The focus here is on the orderly and consistent progression from pre-runtime to runtime across the federation to include simulation, mission command devices, and other federates.

BACKGROUND

This section begins to focus on the Military Scenario Definition Language (MSDL). It provides a short history of MSDL followed by an explanation of the primary design considerations for MSDL and how it evolved into an SISO standard. Finally a description of how MSDL is related to other standards will be provided.

The Origins of MSDL

The Close Combat Tactical Training (CCTT) program provided the first scenario related definition construct to what is now the MSDL language. CCTT is a vehicle platform virtual training environment that provides tools to support pre-exercise scenario development to simulation execution management and control to after action review. As part of this toolset the Commanders Exercise Initialization Toolkit (CEIT) provided an ability to specify the units, entities, and tactical graphics to be included within a CCTT training exercise. The tool leveraged Commercial-Off-The-Shelf (COTS) technology in the form of Microsoft Power-Point to construct the scenario and then save it in a format specific to CCTT for ingestion during its initialization process. These CEIT scenario files could then be saved and/or modified for later use or linked for a before and after discussion as part of the CCTT after action review (Wittman and Abbott, 2006).

From 1997 to 2001, as the concepts for the Army's next generation entity level simulation OneSAF were initiated and matured, the idea of reusing scenarios within and between simulations was discussed as having the potential to provide huge cost and time savings for a variety of reasons. These savings could be found when running a scenario multiple times using simulations of different fidelities for analytical purposes; for sharing the same scenario in a consistent manner across a number of simulations in a federation; as well as providing a way to leverage different scenario development tools and viewers to develop and manipulate the scenarios. During this period there was also a special interest in developing Army and DoD-wide modeling and simulation standards to promote cost savings and reuse from top levels of Army DoD leadership (Lacy and Tuttle, 1998).

This culminated in specific guidance provided by Walt Hollis, Deputy Under-Secretary Army for Operational Analysis (DUSA-OR), to the OneSAF Program Management to prioritize development of specifications, formats, process, and tools that could be matured into industry-wide standards. As a result of the OneSAF activities and the Army leadership's guidance the MSDL concept was born and matured into an international standard within the OneSAF program under the programmatic and technical leadership of OneSAF Program Managers: LTC Thomas Coffman (May 2000 to June 2003); LTC John "Buck" Surdu (June 2003 to July 2006) (MSDL Study Group Chair (2005–2006) and MSDL Product Development Group Co-Chair (2006–2007)); LTC Robert Rasch (July 2006 to December 2009) (MSDL Product Development Group Co-Chair (2007–current); the continuing and consistent guidance of OneSAF DPM John Logsdon (2000–current); and the technical leadership of the OneSAF Government's Chief Architect Cindy Harrison (1999–2003) and MITRE Chief Architect Robert Wittman (2000–current).

MSDL Design Characteristics

Designing and implementing reusable software components has been assessed as costing much more time and resources than development for single-use implementations. Part of this cost is due to accounting for non-project related use, extension, and understanding of the product. For sharing scenarios this meant identifying and assessing not only how to develop appropriate scenario development tools but how to specify an interface to the scenario that would allow maximum reuse across the modeling and simulation community with low introductory costs. During the design process four design characteristics were considered fundamental to the eventual success in transitioning the OneSAF developed MSDL into an open international standard. The four design principles are listed below: application independence; separation of data from code; separation of concerns; and leveraging commercial and industry standards are presented in turn in the following subsections.

Application Independence

This concept establishes the driving need for allowing different applications from different vendors to leverage and use the interface and/or data specification at a reasonably low introductory cost. The applications may include existing COTS (i.e. Microsoft PowerPoint, Excel, etc.) or Government-Off-the-Shelf (GOTS) applications (i.e. existing mission command systems such as the Army Battle Command Systems (ABCS), Command Post of the Future (CPOF), etc.), or new systems yet to be developed. These systems with relatively little modification (this of course is dependent on the system being modified) should be able to use the interface specification and import and/or export data in accordance with the specification.

Separation of Data from Code

Directly supporting the first concept is the design principle to separate software code from data. In addition the data must be specified such that they can stand alone without the need for libraries of code with embedded business rules to understand, leverage, or use the data produced or consumed in compliance with the specification. Of course helper utilities and libraries are always useful, but the main tenet is they should not be required for producing or using the specified data. What this means is that the data should be defined so that they can stand alone, can be validated prior to use, and are understandable and accessible to those that want to develop compliant data producing or consuming applications.

Separation of Concerns

MSDL's exclusive focus is restricted to information relevant to a military scenario and does not include application-specific, training-specific, exercise-control-specific, or other types of simulation initialization type information. This type of information might include the speed of simulation execution, or specifying which simulated entities will be represented in different simulations when the scenario is being used to drive a federation. It is expected that relationships between an MSDL document and initialization documents supporting other concepts such as model, simulation, and federate allocation can be included within existing MSDL elements or extended into an instance document using unique namespace entries to identify implementation-specific extensions to the standard.

Leverage Industry and Military Standards

From its beginnings the OneSAF program, under the guidance of OneSAF Chief Architect Cindy Harrison, prioritized the identification and reuse of existing model algorithms, data specifications, and software development standards from the commercial, modeling and simulation industry, and military standard domains. Three standards that have special significance to MSDL include the eXtensible Markup Language (XML), the US DoD Interface Standard Common Warfighting Symbology—2525B, and the Joint Consultation, Command and Control Information Exchange Data Model (JC3IEDM). A short description of the impact of each is provided in the following paragraphs.

Early on in the development cycle XML was identified as having a number of features supporting the desired design characteristics and as such it was selected to define the initial MSDL specification (Lacy and Tuttle, 1998; Abbott et al., 2007). First and foremost, XML had a large and growing user-base that was already familiar with its use. Second, it allowed for explicit definition of the data elements within MSDL such that it was easy to explicitly specify and separate code from data. Third, there was an available and growing set of commercially available schema definition, editing, and validation tools available (Lacy and Tuttle, 1998). Finally, a number of integrated parsing tools were beginning to become available for integrated use within a number of programming and scripting languages such as Java, C++, Visual Basic, and Perl.

Likewise early in the scenario definition development process MIL-STD 2525B was identified as supporting the necessary design considerations to specify a code independent definition of a military scenario. The MSDL conceptual data model and resulting XML based data model relied heavily on 2525B's primary focus of a military scenario. It provided the starting point for code independent specification of the units, equipment, installations, tactical graphics, and Military Operations Other Than War (MOOTW). Many of the data elements within MSDL came directly from the 2525B standard.

Finally, the JC3IEDM is an internationally defined and managed ontology-oriented set of products defining the range of information to be interchanged between coalition command and control (C2) systems (Wittman and Lopez-Cuoto, 2007). The JC3IEDM is owned and managed by Multination Interoperability Program (MIP) and the North Atlantic Treaty Organization (NATO) Management Board. A short history of the MIP gives some context to the importance of leveraging the JC3IEDM. The MIP was established in 1998 with representation from Canada, France, Germany, Italy, the United Kingdom and the United States with the very specific objective of enabling information exchange between cooperating but independent national C2 and mission command systems. Today there are 24 nations actively participating in the MIP program (Abbott et al., 2007; Wittman and Lopez-Cuoto, 2007). Because the JC3IEDM incorporates a host of international standards and provides a wide range of commonly used enumerations across the weather, military organizations, and other military domains the MSDL schema uses the same enumerations and in some cases the same JC3IEDM XML representation for its scenario concepts. To ensure identification and appropriate maintenance of the data elements leveraged from the JC3IEDM these data elements are called out using a JC3IEDM namespace within the MSDL schema. For the complete set of the JC3IEDM documentation see http://www.mip-site.org.

THE PRINCIPAL MSDL ELEMENTS

This section defines the nine principal data elements within the MSDL standard data model. These elements describe a military scenario according to the MSDL Product Development Group's (PDG's) military scenario definition: "A specific description of the situation and course of action at a moment in time for each element in the scenario. The description of the scenario conveys reality (what is true about the situation, such as the forces identified as participants in the situation) and perceived reality (what is considered to be true based on intelligence information)" (SISO, 2008, p. 13). The contents of the specification are expected to grow and extend over time to support other types of information such as equipment loading information, unit orders and individual soldier tasking, and individual platform readiness or damage state. The evolving Coalition-Battle Management Language (C-BML) now in trial use under the C-BML PDG will be used to supply order and task information. This section is not intended to replicate the MSDL specification, but to provide enough detail of the core data elements to give the reader insight into their definition and use. It should be noted the MSDL specification SISO-STD-007-2008 includes a comprehensive set of Altova XML-Spy generated graphics detailing the hierarchal nature of the MSDL schema. A limited sample set of these graphics are provided in this section to give insight into the comprehensive specification.

As mentioned above, and shown in Figure 24.1, there are nine foundational elements to the MSDL specification. They range in the data they hold from scenario meta-data to specific order of battle and tactical graphic representations to be interpreted and used during the scenario's execution. During the course of its development as a standard the MSDL schema underwent a number of substantial changes from the original schema offered by the OneSAF program. These changes included the addition of weather conditions as defined within the JC3IEDM and a reduction in the number of primary elements within the schema. These elements were removed from standards consideration either because they were too complex and contentious (and hence could be dealt with or more appropriately dealt with in later MSDL releases) or were being developed by another separate SISO PDG (the C-BML PDG provides a good example of this type of coordinated activity.)

The remainder of this section introduces and describes each of the nine elements.

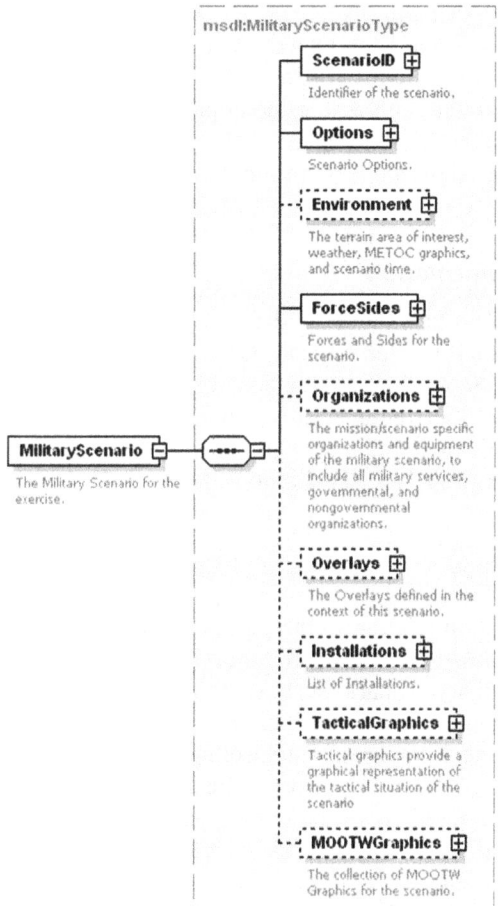

Figure 24.1   MSDL elements (SISO, 2008, p. 20).

ScenarioID Element

The first element within the MSDL specification provides formatted fields to hold information or meta-data about the MSDL instance document. The data model and resulting schema were reused from the Base Object Model SISO modelID specification. It contains four generalized types of summary information: (1) basic name and type data; (2) classification and restriction information; (3) purpose and domain use information; and finally (4) general reference information (SISO, 2008, pp. 20–26).

Name and Type Information

These data elements hold the basic identifying elements of the scenario including its name, the type of military scenario, the version of the military scenario, and the last modification date to the scenario document. The type element may contain an enumeration of "FOM" for Federation Object Model (FOM), "SOM" for Simulation Object Model (SOM), or "BOM" for Base Object Model (BOM) value, but can also include a string value with a short description of the MSDL type such as "Improvised Explosion Device (IED) related," "US Only Maneuver," "Multi-National Logistics," or other descriptive terms identifying the type of military scenario (SISO, 2008, p. 21).

Classification and Restriction Information

This set of data holds the security classification of the data within the document. It can hold the common security marking terms such as "Unclassified," "For Official Use Only," "Secret," "Top Secret," or designations unique to the users of the military scenario. Within this set there is also a field to hold release restrictions regarding specific individuals or organizations, such as "Restricted to NATO Nations," "Restricted to US Only," or "Distribution Unlimited" (SISO, 2008, p. 23).

Purpose and Use Information

This category of data holds information relating to the purpose and domain use of the military scenario. The purpose field allows for the entry of a general statement of the purpose for the military scenario such as "training scenario for Company A." The application domain element identifies one of the application domains, Analysis, Training, Test and Evaluation, Engineering, or Acquisition, to which the military scenario applies. There are also two fields supporting use descriptions. The first identifies use limitations for describing where the scenario is not appropriate and there is a final use history field on where the scenario has been employed (SISO, 2008, pp. 23–24).

Reference Information

Reference information identifies the general nature of the military scenario such as "Military Warfare," "Civil Affairs," or "NATO Operations" to provide context for a set of user defined keywords; example keywords include "entity level," "aggregate level," "engagement," etc. Point of contact information for the scenario is also included in this area and may include key personnel responsible for the creation or maintenance of the scenario. A final subsection allows for a list of references for the development of the scenario. References may include specific operations orders, road to war information, or other military relevant document references (SISO, 2008, pp. 24–26).

Options

The options element establishes the global parameters used within the military scenario. The global options include (1) the MSDL version description; (2) the organization detail contained within the scenario; and (3) the coordinate representation, and naming convention/standard used throughout the scenario. It should be noted that basic schema validation provided by World Wide Web Consortium (WC3) compliant tools will not automatically verify global selections throughout the instance document. Therefore if data validation to this level is necessary specific customized data checking scripts or applications should be created using schematron or XML related validation technology (SISO, 2008, pp. 26–29).

MSDL Version Description

Although this field sounds duplicative to the version field within the ScenarioID element it is quite different. This field allows MSDL document validation against a specific version of the MSDL schema and as such holds the MSDL schema version for which the instance document should be validated (SISO, 2008, p. 27).

Organization Detail

The organization detail field has two parts. The first part characterizes the aggregation level of the scenario as aggregate or entity. If the aggregation field is set to "true" the scenario is aggregate based and will only include unit or organizational level information and will not identify specific equipment or life-form components making up the organizations. If aggregate based the second, echelon, field should hold the echelon to which the aggregation is made. Standard entries for the echelon include "squad," "platoon," "company," "battalion," "regiment," etc. If the military scenario specifies all or some of the units down to the platoon level the echelon field would be platoon. Scenarios that include equipment level detail set the aggregation field to "false" (SISO, 2008, pp. 27–28).

Scenario Data Standards

The scenario data standards element includes a symbology standard structure and a coordinate data standard structure. The symbology standard structure allows for the standard name to be used such as "2525B" along with major and minor version identifiers. The final structure describes the coordinate system type to be used within the military scenario document. One of the following four types can be selected (SISO, 2008, pp. 28–29):

- Military Grid Reference System (MGRS) supports a MGRS grid zone consisting of a MGRS grid square, precision, easting, northing, and elevation fields
- Universal Transverse Mercator (UTM) system supports a UTM coordinate consisting of a UTM grid zone, easting, northing, and elevation
- Geodetic Coordinate (GDC) system provides fields for a latitude, a longitude and elevation
- Geocentric Coordinate (GCC) system provides fields for an x, y, and elevation.

Environment Element

The environment element describes the scenario time, the extents of the contiguous geographic area of interest within the military scenario, the weather across the geographic area, and the meteorological and oceanographic conditions across the environment (SISO, 2008, pp. 29–51).

Scenario Time

The scenario time provides the start time date and time of the scenario in yyyy-mm-ddThh.mm.ss Zulu format. This element provides a reference point for all relative time used later in the military scenario (SISO, 2008, p. 30).

Geographic Extents

The geographical area of interest is provided through an upper right and lower left coordinate along with a name for the area. Both the upper right and lower left structure allow an independent coordinate type selection. Again if it is necessary for an application to validate consistency with the coordinate type selected in the options structure a specialized validation application must be developed and applied (SISO, 2008, pp. 31–36).

Weather

The weather is provided across the extent of the area of interest and is specified for the start time of the scenario. If the weather is dynamic across the simulation execution this must be done post military scenario initialization via runtime updates or with the addition of a dynamic weather data set. The weather is a complex structure leveraging the constructs from the JC3IEDM weather data model and defines the following information types (SISO, 2008, pp. 36–46).

- *Atmosphere* — This structure holds the humidity, the height of an inversion layer to provide mixing of released material in the environment, pressure, and temperature variables.
- *Cloud cover* — This defines the cloud cover conditions across the area of interest. It provides the type of clouds, dimensions, coverage, and light refraction for use within the importing simulations. Examples of cloud types include "clouds," "radioactive cloud," "smoke," etc.
- *Icing* — Icing is described in terms of an icing enumeration type and a specific severity code. Examples include "clear icing," "mixed icing," etc.
- *Lighting* — The lighting structure provides a category representing the class of light, when the light type begins and ends for the light type, and a special moon phase code for the moon class of lighting. Light type examples include "daylight," "moonlight," "nautical twilight," etc.

- *Precipitation* —Precipitation is defined by an enumeration type of precipitation and the rate at which the precipitation is falling. Examples of precipitation types include "hail," "rain," "sleet," "snow," etc.
- *Visibility* —There can be many visibility items within the area of interest that are defined by a category code and a range. Example values of visibility types include "blowing snow," "fog/mist," "sandstorm," "smoke," etc.
- *Wind* —There can also be multiple wind instances within the scenario. They are also defined by a category code, altitude, direction, speed, and a special nuclear yield qualifier code. Some examples of wind types defined within the JC3IEDM wind types are as follows: "constant," "squalls," "gusting," "variable," "not known," etc.

Meteorological and Oceanographic (METOC) Factors

In addition to the JC3IEDM weather data structures a meteorological and oceanographic (METOC) structure reflecting the 2525B standard is also used within MSDL. As there is redundancy between these two weather structures it is expected that subsequent releases will resolve to a single weather structure either through combination of both or elimination of one of the structures. The METOC structure within MSDL supports the following entries (SISO, 2008, pp. 46–51).

- *Symbol identifier* —A symbol identifier code defines the type of METOC structure. The symbol identifier can be used to identify a wide range of meteorological structure including cloud coverage, precipitation, icebergs, sheet ice, shore lines, visibility sandstorms, etc. (Department of Defense, 2005, pp. C1–C2).
- *Unique designation* —used to uniquely identify a METOC structure.
- *Date time group* —provides a time to end the existence of the METOC structure as related to the start time of the scenario.
- *Quantity* —the number of METOC structures identified by the symbol identifier within the scenario.
- *Location and movement* —the location of the METOC structure as well as the direction and speed of the structure.

ForceSides Element

The ForceSides element defines the relationship between sides that are of interest within a scenario. Sides are at the highest level within the forces and sides hierarchy. Forces belong to sides and are expected within the importing simulation to inherit relationships between the sides. The side relationships can be asymmetric with side A defined to have a friendly relationship to side B while side B can be assigned a hostile relationship to side A at scenario start. It is up to the individual simulation to decide on how to initialize and process this information based on their support for multi-sided asymmetric relationships between units. Obviously once side A is attacked or fired on by side B one would expect side A to respond appropriately at the entity and/or aggregate level (SISO, 2008, pp. 51–53).

Once units and equipment are defined the units and equipment will have a relationship to a superior unit with the topmost belonging to a side or force. All forces should belong to a single side. For example "Coalition Forces" could be identified as a side. The side could contain forces for the United States, Great Britain, France, and Germany. Additional forces for each individual country's service branches could be created with allegiance to each of the forces representing each country.

ForceSides Name and Type Information

Each force and side is assigned a unique identification code and name within the ForceSides structure. An allegiance reference is provided for force designations that have allegiances to other forces or sides. In the example above all of the individual countries would have allegiance to the "Coalition Forces" side. The allegiance reference holds the unique identification code of the force or side with which they show allegiance. The ForceSides structure also provides an optional field to hold the military service of the force or side. A standards based country code enumeration can also be provided for each country that is identified. Again it can be omitted if not needed to adequately define the force or side (SISO, 2008, p. 51).

Association Information

The association structure allows this ForceSide structure to hold its relationship with all other ForceSide structures in the scenario. For every other ForceSide structure a relationship field is established to hold the relationship "friendly," "neutral," "hostile," "suspect," or "unknown." These relationships are intended to define the relationship at simulation start time and it is up to the importing simulation to evolve the relationships as appropriate during scenario execution (SISO, 2008, p. 52).

Organizations Element

The organizations element houses the unit and equipment of interest within the military scenario for all sides and forces. This is often times referred to as the task organization or as the order of battle for a scenario. Within MSDL both of the subordinate structures have their origins in the 2525B standard and as such both have the symbol identifier as the key element to identify unit and entity types. All units and entities within the MSDL file should be linked hierarchically through an appropriate organization based chain of command to the ForceSides structure (SISO, 2008, pp. 53–75).

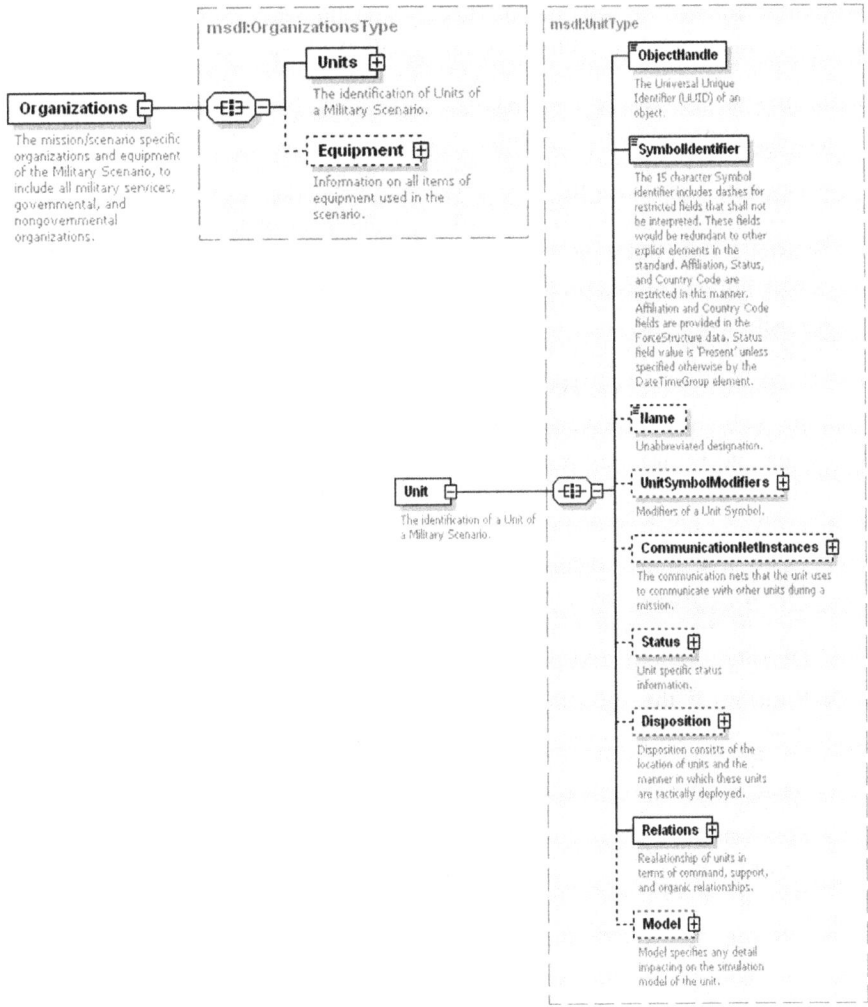

Figure 24.2 Organization units structure (SISO, 2008, pp. 53, 55).

Units

The unit structure is enclosed as a substructure within the organizations and units elements as shown in Figure 24.2 and can be used to define both military and non-military organizations that have a hierarchical relationship and are of interest within the scenario. If desired individual equipment, soldiers, and other life-forms can also be represented within the MSDL equipment structure showing the unique platform level assets of the organization (SISO, 2008, pp. 54–68).

Unit Identification. Unit identification is provided at the most basic level with a unique identifier, used as a reference for other subordinate units

594 Chapter 24 Standards-Based Combat Simulation Initialization

and equipment and as a name for display and human identification purposes. A symbol identifier holds unit type information as defined within the 2525B Appendix A—C2 Symbology: Units, Equipment, and Installations. As defined within 2525B the symbol identification code holds the following (Department of Defense, 2005, pp. A1–A2; SISO, 2008, pp. 55–58].

- Position 1 provides the coding scheme that indicates which overall symbology set to use for the item (unit, equipment including life-forms, and installations) of interest. For units, equipment, and installations an "S" for warfighting code is used (Department of Defense, 2005, p. A1).

- Position 2 holds the unit's affiliation. For the case within an MSDL file this field is restricted to a "–"; as an affiliation it is explicitly defined within a stand-alone element (Department of Defense, 2005, p. A1).

- Position 3 provides the battle dimension, indicates a symbol's battle dimension and is currently restricted to ground "G" within the MSDL standard (Department of Defense, 2005, p. A1). It is expected that MSDL will evolve to other domains over time.

- Position 4, status, indicates a symbol's planned or present status. Within MSDL, status is presumed to be "Present" until the standard evolves to support more comprehensive truth versus perceived values (Department of Defense, 2005, p. A2; SISO, 2008, p. 108).

- Positions 5 through 10 hold the function ID that identifies a symbol's function. Each position indicates an increasing level of detail and specialization and provides generalized types for unit, equipment, and installations (Department of Defense, 2005, p. A2).

- Positions 11 and 12, symbol modifier indicator, identify indicators present on the symbol such as echelon, feint/dummy, installation, task force, etc. (Department of Defense, 2005, p. A2).

- Positions 13 and 14, country code, identify the country with which a symbol is associated (Department of Defense, 2005, p. A1). This value is not used as it is provided within the ForceSide structure of MSDL.

- Position 15, order of battle, provides additional information about the role of a symbol in the battle space. The codes include A, air order of battle; E, electronic order of battle; C, civilian order of battle; G, ground order of battle; N, maritime order of battle; and S, strategic force related (Department of Defense, 2005, p. A2).

Communications. At the unit level communications nets are specified and provide the type of communication network owned by the unit, an identifier for that network, and the types of communication services provided by the network. The types of networks specified can be "COMMAND_NET," "OPERATIONS_INTELLEGENCE_NET," "ADMIN_LOGISTICS_NET," and "FIRE_SUPPORT_NET." The types of communication services provided can be defined as enumerations for data transfer (DATTRF), facsimile (FAX), identify friend or foe (IFF), image (IMAGE), Multilateral Interoperability Programme (MIP) Common Interface Service (MCI), message handling service (MHS), tactical data link (TDL), video service (VIDSVC), voice service (VOCSVC), and not otherwise specified (NOS) (SISO, 2008, pp. 58–59).

Status. The status field identifies the Mission Oriented Protective Posture (MOPP) level of the unit (Level 0, none; Level 1, overgarments and helmet; Level 2, to be determined; Level 3, chemical protective mask and hood are added; Level 4, butyl rubber gloves are added) and the weapons control status as an enumeration of free, tight, or hold (SISO, 2008, pp. 59–60).

Disposition. The disposition field provides the location of the unit either by the center of mass or lead position within the unit. This is specified as part of the location structure. The direction of movement, speed, formation, and formation position with a larger unit structure are also fields within this structure (SISO, 2008, pp. 60–63).

Relations. The relations structure provides the relationship between this unit and its task-organized superior. The structure holds a reference to a superior unit or to ForceSide structure. In addition to a reference to the superior unit, a coded enumeration (unit; ForceSide; or not specified) also provides the type of structure holding the superior. If the superior is a unit the unique identifier of the unit is housed in this structure and a coded command relationship type is also provided in terms including ORGANIC; ATTACHED; OPCON; TACON; ADCON; and NONE. An additional structure holds information showing the units that are supported by this unit. The structure provides the type of support from the enumeration set (GS, General Support; DS, Direct Support; R, Reinforcing; GSR, General Support Reinforcing, and NOT_SPECIFIED), a reference to the supported unit, and the role of support in terms of Priority of Effort Chemical, Engineer, Fires, and Intelligence (SISO, 2008, pp. 64–67).

Model. The model field identifies the expected resolution of the unit being represented within the simulation and is generically coded to "high," "medium," and "low." It is up to the importing simulation to map to and create an appropriate unit for instantiation within the simulation (SISO, 2008, p. 68).

Equipment

The equipment structure can be used to define both military and non-military equipment, personnel, and other life-forms that may be of interest within the military scenario. These equipment and life-form items represent the assets and are linked to the organization constructs by including a reference to the unit to which they belong. Currently equipment items can belong to only one organization (SISO, 2008, pp. 68–75).

Equipment Identification. Like units and other elements within MSDL equipment identification is provided with a unique identifier and a name for display and human identification purposes. A symbol identifier holds equipment type information as defined within the 2525B Appendix A—C2 Symbology: Units, Equipment, and Installations. As defined within 2525B, the symbol identification code holds the following (Department of Defense, 2005, pp. A1–A2; SISO, 2008, pp. 68–71).

- Position 1 provides the coding scheme that indicates which overall symbology set to use for the item (unit, equipment including life-forms, and installations) of interest. For units, equipment, and installations an "S" for warfighting code is used (Department of Defense, 2005, p. A1).

- Position 2 holds the unit's affiliation for the case within an MSDL file. This field is restricted to a "–" as its affiliation is explicitly defined within a stand-alone element (Department of Defense, 2005, p. A1).

- Position 3 indicates a symbol's battle dimension and is currently restricted to ground "G" within the MSDL standard. It is expected that MSDL will evolve to other domains over time (Department of Defense, 2005, p. A1).

- Position 4, status, indicates a symbol's planned or present status. MSDL status is presumed to be "Present" until the standard evolves to support more comprehensive truth versus perceived values (Department of Defense, 2005, p. A2; SISO, 2008, p. 108).

- Positions 5 through 10 hold the function ID that identifies a symbol's function. Each position indicates an increasing level of detail and specialization and provides generalized types for unit, equipment, and installations (Department of Defense, 2005, p. A2).

- Positions 11 and 12, symbol modifier indicator, identify indicators present on the symbol such as echelon, feint/dummy, installation, task force, etc. (Department of Defense, 2005, p. A2).

- Positions 13 and 14, country code, identify the country with which a symbol is associated. This value is not used as it is provided within the ForceSide structure of MSDL (Department of Defense, 2005, p. A2).

- Position 15, order of battle, provides additional information about the role of a symbol in the battle space. The codes include A, air order of battle; E, electronic order of battle; C, civilian order of battle; G, ground order of battle; N, maritime order of battle; and S, strategic force related (Department of Defense, 2005, p. A2).

Communications. An equipment item that has a communications device will reference the network the communications device is on and the owning unit of that communications network. This provides the importing simulation with information to establish that a platform (equipment item) has the ability to communicate on a network and that network is owned by a unit (SISO, 2008, pp. 71–72).

Disposition.   The disposition structure provides the initial location, direction of movement, speed and formation position of a particular equipment item (SISO, 2008, pp. 72–73).

Relations.   The relations structure provides the relationship of an equipment item to its organic superior and the unit or force/side that owns the equipment item (SISO, 2008, pp. 73–74).

Model.   The model field identifies the expected resolution of the unit being represented within the simulation and is generically coded to "high," "medium," and "low." It is up to the importing simulation to map to and create an appropriate equipment item for instantiation within the simulation (SISO, 2008, p. 75).

Overlays Element

The overlays element is used to group the intelligence elements or tactical graphics that are included within the military scenario. This element provides fields for an overlay name and unique reference. It also allows identification of the type of overlay that has been created and can be one of the following: "Operations," "Fire_Support," "Modified_Combined_Obstacle," "Intel," "Recon_Surveillance," "Obstacle," "Air_Defense," "Logistics," "A2C2," and "User_Defined." (SISO, 2008, pp. 75–76).

Installations Element

The installations element describes the detected installations as they stand at scenario start time as predetermined by the intelligence gathering process for force, side, or unit individually. The description of any corresponding actual instances, the reality aspect, is unspecified in this version of MSDL.

Installation Identification

Like the unit and equipment element identification, installation identification is provided with a unique identifier and a name for display and human understanding. A symbol identifier provides for the installation type as defined within the 2525B Appendix A—C2 Symbology: Units, Equipment and Installations. Installations can range from a technological research facility to a public water service facility. The symbol identifier provides the following information for the installation (Department of Defense, 2005, pp. A1–A2; SISO, 2008, pp. 76–81).

- Position 1 provides the coding scheme that indicates which overall symbology set to use for the item (unit, equipment including life-forms, and installations) of interest. For units, equipment, and installations an "S" for warfighting code is used (Department of Defense, 2005, p. A1).

- Position 2 holds the unit's affiliation for the case. Within an MSDL file this field is restricted to a "–" as its affiliation is explicitly defined within a stand-alone element (Department of Defense, 2005, p. A1).

- Position 3 indicates a symbol's battle dimension and is currently restricted to ground "G" within the MSDL standard. It is expected that MSDL will evolve to other domains over time (Department of Defense, 2005, p. A1).

- Position 4, status, indicates a symbol's planned or present status [A2]. Within MSDL status is presumed to be "Present" until the standard evolves to support more comprehensive truth versus perceived values (Department of Defense, 2005, p. A2; SISO, 2008, p. 108).

- Positions 5 through 10 hold the function ID that identifies a symbol's function. Each position indicates an increasing level of detail and specialization and provides generalized types for unit, equipment, and installations (Department of Defense, 2005, p. A2).

- Positions 11 and 12, symbol modifier indicator, identify indicators present on the symbol such as echelon, feint/dummy, installation, task force, etc. (Department of Defense, 2005, p. A2).

- Positions 13 and 14, country code, identify the country with which a symbol is associated. This value is not used as it is provided within the ForceSide structure of MSDL (Department of Defense, 2005, p. A2).

- Position 15, order of battle, provides additional information about the role of a symbol in the battle space. The codes include A, air order of battle; E, electronic order of battle; C, civilian order of battle; G, ground order of battle; N, maritime order of battle; and S, strategic force related (Department of Defense, 2005, p. A2).

Owner Location Orientation and Name Data

Several structures provide for the name of the installation, the installation location, and its orientation, and a reference identifier to the owner of the graphic (SISO, 2008, pp. 77–78).

Associated Overlays

The associated overlays field provides a link from the installation element to the overlays concerned with the recorded installation (SISO, 2008, pp. 80–81).

TacticalGraphics Element

The TacticalGraphics element describes the tactical action based information as known by a particular force, side or unit individually (SISO, 2008, pp. 91–92).

Tactical Graphics Identification

Like other MSDL elements, tactical graphic identification is provided with a unique identifier and a name for display and human understanding. A symbol identifier provides for the installation type as defined within the 2525B Appendix B—C2 Symbology: Military Operations. The symbol identifier provides the following information for the installation (Department of Defense, 2005, p. B1; SISO, 2008, p. 83–76).

- Position 1 provides the coding scheme that indicates which overall symbology set to use for the item (unit, equipment including life-forms, and installations) of interest. For units, equipment, and installations a "G" for the tactical graphics code is used (Department of Defense, 2005, p. B2).

- Position 2 holds the unit's affiliation for the case within an MSDL file. This field is restricted to a "–" as its affiliation is explicitly defined within a stand-alone element (Department of Defense, 2005, p. B1).

- Position 3 provides the operation group the tactical graphic belongs to and can be one of the following: T, Tasks; G, C2 & General Maneuver; M, Mobility/Survivability; F, Fire Support; S, Combat Service Support; and O, Other (Department of Defense, 2005, pp. B1–B2).

- Position 4, status, indicates a symbol's planned or present status. Within MSDL, status is presumed to be "Present" until the standard evolves to support more comprehensive truth versus perceived values (Department of Defense, 2005, p. B1; SISO, 2008, p. 108).

- Positions 5 through 10 hold the function ID that identifies a symbol's function. Each position indicates an increasing level of detail and specialization and provides generalized types for unit, equipment, and installations (Department of Defense, 2005, p. B1).

- Positions 11 and 12, symbol modifier indicator, identify indicators present on the symbol such as echelon, feint/dummy, installation, task force, etc. (Department of Defense, 2005, p. B1).

- Positions 13 and 14, country code, identify the country with which a symbol is associated. This value is not used as it is provided within the ForceSide structure of MSDL (Department of Defense, 2005, p. B1).

- Position 15, order of battle, provides additional information about the role of a symbol in the battle space and is defaulted to "X" for tactical graphics (Department of Defense, 2005, p. B1).

Owner and Affiliation Data

Two structures provide for the owner of the tactical graphic through a reference to the owner's unique identifier. The affiliation structure holds the threat HOSTILE, FRIEND, NEUTRAL, UNKNOWN perceived by the owner of the graphic associated with the intent of the graphic (SISO, 2008, p. 83).

AnchorPoints

The AnchorPoints structure holds the data points to connect the line segments associated with a tactical graphic. The start and end points of the line segments are placed in sequential order in the data structure (SISO, 2008, p. 83).

Associated Overlays

The associated overlays field provides a link from the installation element to the overlays concerned with the recorded installation (SISO, 2008, p. 83).

Symbol Class Information

Symbol class information holds structures to define the type and modifiers to a type of tactical graphic. Tactical graphic types include points; lines; areas; boundaries; nuclear, biological, and chemical events; and tasks (SISO, 2008, pp. 83–87).

MOOTW Graphics

The MOOTWGraphics element, shown in Figure 24.3, describes the detected MOOTW concepts as determined by the intelligence gathering process by each force, side, or unit individually. The description of any corresponding actual instances, the reality aspect, is unspecified in this version of MSDL (SISO, 2008, pp. 92–97).

MOOTW Identification

Like other MSDL elements, MOOTW graphic identification is provided with a unique identifier and a name for display and human understanding. A symbol identifier provides for the MOOTW graphic type as defined within the 2525B Appendix E—C2 Symbology: Military Operations Other Than War (MOOTW) Symbology. The symbol identifier provides the following information for the MOOTW graphic (Department of Defense, 2005, pp. E1–E2; SISO, 2008, pp. 93–94).

- Position 1 provides the coding scheme that indicates which overall symbology set to use for the item (unit, equipment including life-forms, and installations) of interest. For MOOTW concepts an "O" is used for the code (Department of Defense, 2005, pp. E1–E2).
- Position 2 holds the unit's affiliation for the case within an MSDL file. This field is restricted to a "−" as its affiliation is explicitly defined within a stand-alone element (Department of Defense, 2005, p. E1).
- Position 3 provides the symbol's primary category including V, VIOLENT ACTIVITIES; L, LOCATIONS; O, OPERATIONS; or I, ITEMS (Department of Defense, 2005, pp. E1–E2).

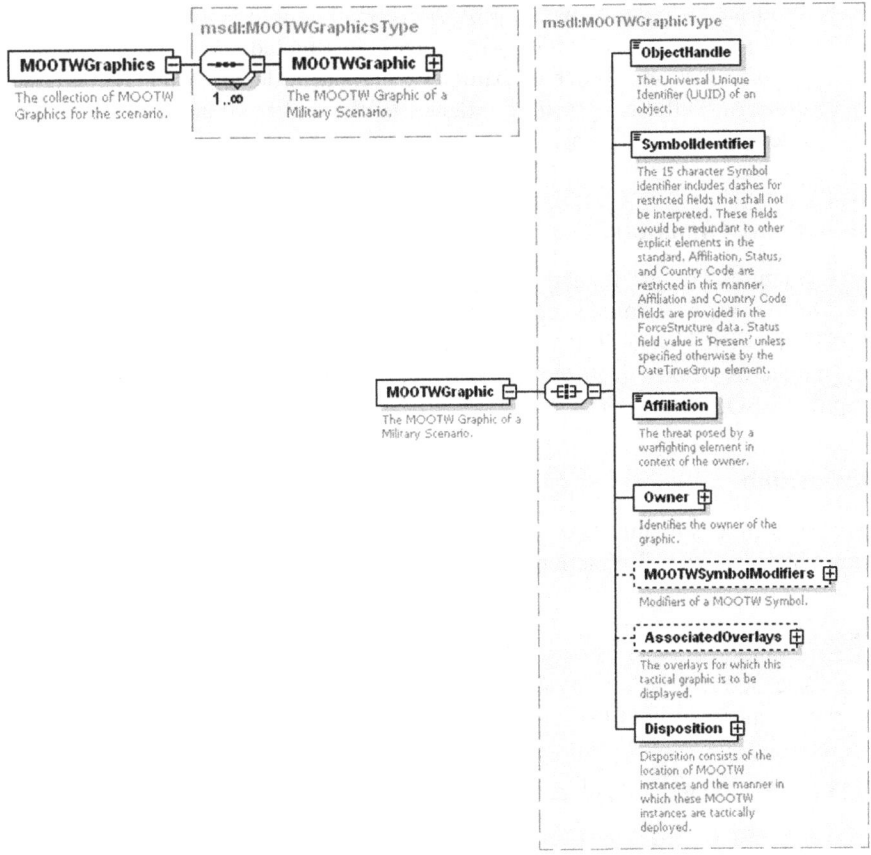

Figure 24.3 MOOTW graphics structure (SISO, 2008, p. 92, 93).

- Position 4, status, indicates a symbol's planned or present status. Within MSDL, status is presumed to be "Present" until the standard evolves to support more comprehensive truth versus perceived values (Department of Defense, 2005, p. E2; SISO, 2008, p. 108).
- Positions 5 through 10 hold the function ID that identifies a symbol's function. Each position indicates an increasing level of detail and specialization and provides generalized types for unit, equipment, and installations (Department of Defense, 2005, p. E2).
- Positions 11 and 12, symbol modifier indicator, identify indicators present on the symbol such as echelon, feint/dummy, installation, task force, etc. (Department of Defense, 2005, p. E2).
- Positions 13 and 14, country code, identify the country with which a symbol is associated. This value is not used as it is provided within the ForceSide structure of MSDL (Department of Defense, 2005, p. E2).

- Position 15, order of battle, provides additional information about the role of a symbol in the battle space. The codes include A, air order of battle; E, electronic order of battle; C, civilian order of battle; G, ground order of battle; N, maritime order of battle; and S, strategic force related (Department of Defense, 2005, p. E2).

Owner and Affiliation Data

Two structures provide for the owner of the MOOTW graphic through a reference to the owner's unique identifier. The affiliation structure holds the threat HOSTILE, FRIEND, NEUTRAL, and UNKNOWN perceived by the owner of the graphic associated with the intent of the MOOTW concept depicted by the graphic (SISO, 2008, p. 94).

Disposition

The disposition structure provides location, direction of movement, and speed with which the MOOTW action is occurring (SISO, 2008, p. 97).

Associated Overlays

The associated overlays field provides a link from the installation element to the overlays concerned with the recorded installation (SISO, 2008, p. 97).

FUTURE DIRECTION AND RESEARCH

There are a number of avenues for future growth, research, and sharing of the MSDL standard. This final section provides some insight into each of these areas by starting with maturation and evolution of the standard as managed by the SISO MSDL PDG. It then looks at an evolving mechanism for authoritative mission command order of battle data input from the Global Force Management Data Initiative (GFM DI) into the MSDL format for efficient sharing and initialization of mission command systems.

The MSDL PDG continues to pursue an iterative and incremental development strategy for MSDL with the group currently working on MSDL Version 2.0 with the following prioritized list (Wittman, 2008).

(a) Create a common simulation standard language framework that will support the harmonization and synchronization of the C-BML and MSDL. A current prototype activity moving forward in the NATO MSG-085 activity is to reference a C-BML instance document within an MSDL scenario file. The C-BML document provides the planned execution for the task organizations with the MSDL document. The C-BML document in its order set will reference the appropriate MSDL units and equipment items and tactical graphics as part of a comprehensive initialization package.

(b) Devise common unit and equipment enumerations that provide for greater interoperability and semantic understanding than simple text based names for units and equipment. The Distributed Interactive Simulation (DIS) standard and 2525B standards are both under consideration as providing a common, accepted way to uniquely identify unit and entity types.

(c) Logistics and supply attribution: create a data model that allows specification of logistic and supply data in support of the identified units and equipment.

(d) Develop air, space, intelligence, and naval domain extensions with specific unit, equipment, and tactical graphics support.

(e) Model perception support for side and force intelligence data entries accounting for perceived information within the scenario.

(f) Develop attack guidance matrices support that incorporates a data model that allows identification of targets and supports equipment to target pairings.

(g) Finally, the existing MSDL data model and its support for legacy, current, and future force military scenario definition will also be open for review and enhancement as necessary.

The Global Force Management Data Initiative (GFM DI) defines a comprehensive solution to automate and provide efficient access to the DoD force structure. Through a wide set of authoritative organization servers and a defined data model to access the information it allows the modeling and simulation community an awesome capability to access authoritative order of battle data. It is currently populated with authorized unit structures but is expected to support on-hand data in the foreseeable future with support for personnel, logistics, mission command, and readiness data [Sprinkle et al. 2011]. There is a current prototype activity supporting simulation initialization using the GFM DI through the Joint Forces Command's Joint Training Data System (JTDS) Order of Battle Service (OBS). It has been proposed that a direct service from the GFM DI Authoritative Services into MSDL would enhance a standards based capability and make the authoritative data available to the range of simulations and supporting military scenario tools that leverage the MSDL standard. Because both MSDL and the Global Force Management Data Information Exchange Data Model (GFMIEDM) have XML schema (XSD) representations the transformation from one to the other should be straightforward using industry standard (WC3) data transformation technologies (Sprinkle et al., 2011).

Finally, as more MSDL documents are created the ability to easily find and share the scenario information will greatly enhance the return on investment in stepping up to MSDL technologies. It is envisioned that Service Oriented Architecture (SOA) and web based (Google-like) search engines will be leveraged to identify and retrieve the military scenario focused data sets.

THE EXPANDING MSDL USER BASE

The number of MSDL users continues to grow both in the United States and internationally. Within the United States, OneSAF continues to leverage and develop both import and export mechanisms for OneSAF so that its scenarios can be shared among MSDL compliant simulations across many different projects. Federation based architectures and toolsets such as the Modeling Architecture for Training Research and Experimentation (MATREX) federation uses MSDL within their scenario development and initialization process to ensure consistent initialization across individual simulation federates. The NATO Modeling and Simulation Group MSG-085 focusing on simulation and C2 interoperability is leveraging MSDL in concert with the evolving Coalition-Battle Management Language (C-BML) standard to resolve a number issues with federation initialization encountered in an earlier C-BML focused effort. Most recently at an International Training and Education Conference (ITEC 2011) held in Cologne, Germany, a number of corporations: CAE, IABG, SAAB, and MITRE—a United States Federally Funded Research and Development Center (FFRDC) described efforts to mature MSDL production, editing, and import capabilities. MSDL is also a recommended standard within the NATO Modeling and Simulation Standards Profile.

CONCLUSION

This chapter introduced the basics of combat simulation initialization and the challenges associated with single and federation based simulation initialization. The MSDL standard was described and related to other military and commercial standards. The chapter concluded with a discussion of future work and research on the MSDL standard as well as a brief summary of the expanding user base.

REFERENCES

Abbott J, Blais C, Chase T, Covelli J, Fraka M, Gagnon F, Gupton K, Gustavsson P, Peplow K, Prochnow D and Wittman R (2007). MSDL the road to balloting. In: *Proceedings of the Fall 2007 Simulation Interoperability Workshop*. http://www.sisostds.org.

Blais C, Pearman J, Dodds R and Baez F (2009). Rapid scenario generation for multiple simulation: an application of the military scenario definition language. In: *Proceedings of the Spring 2009 Simulation Interoperability Workshop*. http://www.sisostds.org.

Department of Defense (2005). *Interface Standard Common Warfighting Symbology*. MIL-STD-2525B w/Change 1, 1 July.

Lacy LW and Tuttle C (1998). Interchanging simulation data using XML. In: *Proceedings of the Fall 1998 Simulation Interoperability Workshop, Orlando, FL*. http://www.sisostds.org, pp. 1110–1119.

Prochnow D, Fogus M, Vintilescu J and Borum B (2005). Initialization of distributed simulations: a better way? In: *Proceedings of the Fall 2005 Simulation Interoperability Workshop*. http://www.sisostds.org.

SISO (2008). *Standard for: Military Scenario Definition Language SISO-STD-007-2008, 14 October*. Military Scenario Definition Language Product Development Group. 3100 Technology Parkway, Orlando, FL 32826, http://www.sisostds.org/ProductsPublications/Standards/SISOStandards.aspx.

Sprinkle R, Nash D, Hasan O and Wittman R (2011). Standards-Based Simulation Initialization, Interoperability & Representation with GFM DI. In: Proceedings of the Fall Simulation Interoperability Workshop. http://www.sisostds.org.

Thakkar N and Lindy E (2005). Initialization of the MATREX environment using MSDL. In: *Proceedings of the Fall 2005 Simulation Interoperability Workshop*. http://www.sisostds.org.

Wittman R (2008). *Euro SIW 2008 MSDL PDG Minutes*. In: Digital Library for MSDL PDG, Meeting Records for PDG held June 17 Euro SIW at http://www.sisostds.org/DigitalLibrary. aspx?EntryId=29668.

Wittman R and Abbott J (2006). Keeping up with the military scenario definition language (MSDL). In: *Proceedings of the Spring 2006 Simulation Interoperability Workshop*. http://www.sisostds.org.

Wittman R and Lopez-Cuoto S (2007). Using OneSAF to explore simulation and JC3IEDM alignment. In: Proceedings of the 2007 Simulation Technology Conference (SimTecT 2007), Brisbane, Australia 4–7 June 2007, *SimTecT 2007*, Paper 55.

Chapter 25

Multi-Resolution Combat Modeling

Mikel D. Petty, Robert W. Franceschini, and James Panagos

INTRODUCTION

A multi-resolution combat model, as defined here, is a composite model that links a unit level combat model, i.e. a model that primarily represents combat in terms of military units such as battalions and brigades, with an entity level combat model, i.e. a model that represents individual battlefield objects or platforms such as tanks and helicopters.[1,2] In such a model portions of a single executing battlefield scenario may be simulated in each of the models and simulated events occurring in one of the models' simulations may have an effect on events in

---

[1] This definition implicitly delimits the scope of this chapter in several ways. First, multi-resolution models can be and have been developed for non-combat applications (Odum and Odum, 2000); however, the scope of this chapter is intentionally limited to combat models. Second, it implies that multi-resolution combat models operate only at two levels of resolution, unit and entity. There is no reason why this must be so, and more than two levels of resolution are possible (Franceschini, 1999). However, many multi-resolution combat models do operate at those two levels and hereinafter that structure will be assumed. Finally, it defines multi-resolution combat models as combining separate unit level and entity level models. Of course, those levels of resolution could both be present in a single model (Davis, 2005). However, a number of multi-resolution combat models have been implemented as defined by linking separate models, and hereinafter that architecture will be assumed.

[2] Instead of *multi-resolution*, some sources use the terms *cross-resolution* (Davis, 1993; Reynolds et al., 1997), *variable-resolution* (Davis, 1995b), *selectable-resolution* (Davis and Bigelow, 1998), or *multiresolution* (Bigelow and Davis, 2003).

Engineering Principles of Combat Modeling and Distributed Simulation, First Edition.
Edited by Andreas Tolk.
© 2012 John Wiley & Sons, Inc. Published 2012 by John Wiley & Sons, Inc.

the other. Both of the preceding statements are true of distributed simulations in general (see Petty and Gustavson, 2012); characteristics specific to multi-resolution combat models are that the linked models have representations at distinctly different levels of resolution (unit and entity) and the representations of the simulated battlefield objects (units and entities) may change resolution at runtime from entity level to unit level, and vice versa.

This chapter is a tutorial on the basic concepts and operations of multi-resolution combat modeling and a review of existing multi-resolution combat models as they have been developed, implemented, and used. Reasons for developing such systems are given, key terms and concepts of multi-resolution combat models are defined, typical software architectures for implementing them are described, the history and literature of multi-resolution combat model implementations are surveyed, and challenges and open research issues in their implementation and use are identified.

This introductory section begins with brief introductions to entity level and unit level combat models as preparatory background. It then motivates the topic of multi-resolution combat models by providing justifications for their development and concludes with a clarification of the disparate and overlapping terminology used to describe such systems.

Background: Entity Level Combat Modeling

Entity level combat models[3] represent individual combat vehicles (e.g. tanks or helicopters), platforms (e.g. radar sets), and personnel (e.g. infantry squads or soldiers) as distinct simulation entities. Necessary state information, such as location, velocities, and ammunition load, is maintained for that entity separately and each entity is capable of independent action. Entity level models model the physical phenomenology of interest (e.g. moving, sensing, and shooting) at the level of individual entities. The entity level models consider the performance characteristics of the specific entity, the effects of terrain (and, less often, other environmental factors) on its actions, and the entity's location with respect to other entities. Entity level models are often based on performance data for the specific entity type; those data may be gathered from testing, operational use, or design specifications.

Entity level movement models may be table- or parameter-driven, where a function of entity type and terrain surface type (for ground vehicles) determine maximum speed. They may also be based on more general physics equations describing the entities' movement capabilities. Entity performance parameters such as turn radius and acceleration may be considered.

---

[3]In this chapter, the noun "model" will usually mean a computer program instantiating a mathematical or logical representation of combat or of a combat system. In contrast, the noun "simulation" is the execution of a model over time. The unfortunately common use of "simulation" (noun) as a synonym for "model" (noun) is intentionally avoided.

Entity direct fire combat models typically use conditional probability tables that encode the probability of a hit (P_h) given a shot and the probability of a kill given a hit (P_k). A simple P_h table might have two dimensions, weapon system and range; more sophisticated models will include other dimensions, e.g. target velocity or target aspect. The table entries are probabilities; e.g. a hit is scored if a random number is less than the appropriate P_h table entry.[4]

Entity sensing at the entity level often revolves around the determination of intervisibility or "line of sight" (LOS); here the question is whether the LOS between two entities is blocked by intervening terrain. Specific algorithms for determining intervisibility vary widely.[5] A simple LOS algorithm may extend a single ray measured from a single sensor point to a single target point, consider only terrain obstructions, and return only "blocked" or "unblocked" results (Youngren, 1994). More complex LOS algorithms extend multiple rays from the sensor point to multiple points on the target, consider not only the terrain but intervening physical obstructions such as vehicles, buildings, or trees having various opacities, and return a numerical value indicating what fraction of the target is potentially visible (Longtin, 1994). The specifics of most LOS algorithms are highly dependent on the representation format of the terrain database.

An important class of entity level combat models is semi-automated forces (SAF) systems.[6] In multi-resolution combat models the entity level model is usually a SAF system. SAF systems generate and control multiple battlefield entities during a scenario execution. The runtime behavior of the SAF entities is largely controlled by the SAF system's software. The algorithms that generate the behavior are designed to react to the simulation situation and perform realistic and doctrinally correct actions (Smith and Petty, 1992; Petty, 2009). SAF systems can be used to generate both opposing and friendly entities in battlefield training scenarios. The intent is that trainees derive some positive training benefit by "fighting" against the SAF opponents and "cooperating" with the SAF-friendly forces. SAF entities are also used in analysis and experimentation applications, where the SAF entities populate the battlefield as needed for the trial or experiment. SAF systems are normally able to interoperate with other models using standard interoperability protocols (Petty, 1995).

Background: Unit Level Combat Modeling

Unit level combat models represent military units (e.g. a mechanized infantry company) as an aggregate without separately representing each individual entity

---

[4]The P_h and P_k tables can be combined into a single table giving the probability of a kill given a shot, i.e. P_{ks}.

[5]Intervisibility determinations are typically computationally expensive and can consume a significant fraction of the computational power of the entity level models' host computer. For that reason, research to reduce the cost, both algorithmically and heuristically, has been ongoing; see Petty et al. (1992), Rajput et al. (1995), Petty (1997) and Petty and Mukherjee (1997).

[6]SAF systems are also widely known as computer generated forces (CGF) systems.

(e.g. infantry fighting vehicle) within the unit. The position, movement speed and direction, status, and composition of a unit are maintained for the unit as a whole, and are often computed as the result of statistical analysis of the unit's actions. Unit movement in such models may be controlled by an operator, scripts, commands involving conventional map control measures such as routes and phase lines, or some combination of any of these. Movement speed is affected by a unit's movement capabilities and the effects of terrain.

The results of engagements between units in unit level models are produced using aggregate attrition models. Distinct activities such as target acquisition and lethality assessment are combined into the attrition calculations. Individual entities are not represented in these units, so the details of entity–entity engagements are not modeled. The contributions of the individual entities to the combat's outcome are accumulated over the entire unit (for homogeneous models) or over weapon system classes within the unit (for heterogeneous models). The attrition calculations produce results that affect the overall units.

A common and well known type of aggregate attrition models are Lanchester equations (Lanchester, 1956), which exist in various forms (Taylor, 1980a,b,1981; Fowler, 1996a).[7] Lanchester equations are differential equations describing the rate of change of Blue and Red force strengths X and Y as a function of time, with the function depending only on X and Y. One partly generalized version of the Lanchester equations has the following form (Davis, 1995a):

$$\frac{dX}{dt} = -K_y X^r Y^s \text{ and } \frac{dY}{dt} = -K_x Y^t X^u$$

where K_x and K_y are the attrition rate coefficients of the Blue and Red force, respectively; and r, s, t, and u are free, time-independent parameters that can be used to "tune" or customize the equations for particular situations (also see Dare and James, 1971). The equations may be extended in various ways, e.g. to include constant reinforcement-rate terms, as well as other effects (Helmbold, 1965). There are two special cases of the generalized form of the Lanchester equations: the "square law" corresponds to $s = u = 1$ and $r = t = 0$; the "linear law" corresponds to $r = s = t = u = 1$.

$$dX/dt = -K_y Y \text{ and } dY/dt = -K_x X \text{ (square law)}$$
$$dX/dt = -K_y XY \text{ and } dY/dt = -K_x XY \text{ (linear law)}$$

The square law is usually taken to apply to "aimed fire" (e.g. tank versus tank) and the linear law to "unaimed fire" (e.g. artillery barraging an area without precise knowledge of target locations).

The attrition rate coefficients K_x and K_y depend on factors such as the time to acquire a target, the time of flight of the projectile, the single-shot probability of

---

[7]In addition to Lanchester equations, other models used for aggregated unit level attrition calculation include the Quantified Judgment Model (Dupuy, 1985, 1998), Fire Power Scores and Force Ratios (Anderson, 1974), and the ATLAS ground attrition model (Kerlin and Cole, 1969).

a kill, terrain, weather, and others. Methods for determining attrition rate coefficients include derivation form historical battle data (Peterson, 1953; Dupuy, 1995), Bonder–Ferrell theory (Bonder, 1967; Bonder and Farrell, 1970), and Markov dependent fire models. Target acquisition requires algorithms for search, screening, and detection (Fowler, 1996b; Koopman, 1999; Washburn 2002). Factors include probability of detection, LOS (or probability of LOS), and area of search. In unit level models, these factors are usually averaged across weapons systems or combatants. Other factors that influence engagement outcomes may be more difficult to represent in the equations; these include defensive variables such as armor and anti-weapons systems, and proactive behavior such as maneuver and use of terrain cover.

Motivation for Multi-Resolution Combat Models

Unit level combat models have certain desirable characteristics. Their aggregate representations provide computational efficiency, with the result that they can simulate large scenarios in terms of geographic scope and size of military forces involved and are often able to execute much faster than real time. However, existing unit level combat models based on mathematical model abstractions such as Lanchester equations do not exploit the detailed performance data and higher resolution models that are available at the entity level.[8]

Entity level combat models have a different set of advantages. The level of resolution of their models of combat phenomenology, such as moving, sensing, and shooting, is at the level of individual entities, which is both more intuitively accessible to model users and more directly supportable by available test and operational data on entity performance than the relatively abstract equations of a unit level model. However, current pure entity level combat models can be computationally expensive, thereby limiting the number of entities that can be concurrently simulated, and some entity level models tend to produce attrition levels higher than those seen historically (Petty and Panagos, 2008).[9]

There has been a long-standing desire (at least since 1992) to combine or link examples of these two classes of model so as to realize the best features of both in a single system. At least 15 distinct combinations of unit level and entity level models in multi-resolution combat models have been implemented since 1992 (there were already several in 1995 (Stober et al., 1995)). Despite

---

[8]There is no claim in this chapter, explicit or implicit, that higher resolution models are necessarily higher fidelity. The two characteristics are orthogonal and may exist in any combination in a model (Davis, 2005; Banks, 2010).

[9]Possible phenomena present in actual combat and accounted for in the parameters of unit level attrition models such as Lanchester equations but not in entity level models that could explain this include target duplication, shooter non-participation, suppression effects, self-preservation, and suboptimal use of weapons and targeting systems.

their variety, some common architectural and modeling elements have emerged in these systems; these include the combination of a unit level model with an entity level model, the development of some type of interface module to mediate the operations and interactions between the two models, the use of a distributed simulation interoperability protocol to connect the two models and the interface module, and resolution change operations such as disaggregation, wherein a unit represented in and controlled by the unit level model is instantiated as a set of entities represented in and controlled by the entity level model, and aggregation, where the reverse occurs.

Research, development, and implementation of multi-resolution combat models have been justified on the basis of several factors.

1. *Modeling flexibility*. The ability to change the representational level and possibly the host model of simulation objects at runtime provides the simulation user with beneficial modeling flexibility (Davis, 1993; Powell, 1997; Davis and Bigelow, 1998) and explanatory power (Franceschini et al., 1999).

2. *Scenario scalability*. The ability to simulate less important portions of a scenario at a lesser level of resolution (and computational cost), while simulating more important portions at a greater level of resolution, allows larger scenarios to be simulated for a given amount of computational power (Cox et al., 1995; Franceschini and Petty, 1995b; Castro et al., 2002).

3. *Results fidelity*. Detailed entity level performance information has the potential to provide high fidelity combat results for unit level models (Franceschini and Karr, 1994; Davis and Bigelow, 1998).

4. *Contextual realism*. Small unit training exercises can be conducted in an entity level virtual battle that is set within the context of a larger unit level constructive battle, adding realism and motivation to the training (Franceschini and Karr, 1994). This idea generalizes from training to the broader notion that an ongoing scenario in a unit level simulation can provide context and direction for the behavior of entities in an entity level simulation (Davis and Bigelow, 1998).

5. *Training enhancement*. Training higher level commanders and their staffs based on unit level models can be enriched by supplementing unit level models' typically aggregate statistical interactions with entity level models' detailed entity-specific interactions (Downes-Martin, 1991; Franceschini and Karr, 1994).

6. *Sensor stimulus*. Multi-resolution combat models can provide stimulus, in the form of very large numbers of geographically distributed entities corresponding to an operationally realistic scenario, to entities that are moving fast and/or possess long-ranged sensor systems (Calder and Evans, 1994; Root and Karr, 1994; Schricker et al., 1998b).

Terminological Clarification

The terminology used in the research literature to describe the components of multi-resolution combat models has varied over time. Some sources describe them as linking unit level and entity level models,[10] others as linking aggregate and disaggregate models, and still others as linking constructive and virtual models. The first and second pairs of terms are simply close synonyms (units are aggregated entities, entities are disaggregated units). The third pair are not, however, simply alternative terms for the same concepts. Instead "constructive" and "virtual" refer to categories within the "live, virtual, and constructive" typology often used to partition models in defense related applications (Banks, 2010). In a constructive combat model, both the battlefield objects, e.g. which may be units or entities depending on the model's level of resolution, and the people making up or operating those battlefield objects, i.e. their personnel or crews, are simulated.[11] In contrast, in a virtual combat model, while the battlefield objects are simulated, the people operating them are by definition real; e.g. a real human pilot operating a helicopter simulator. While unit level combat models are almost always constructive, the terms are not synonymous, and unit level is more appropriate in reference to multi-resolution combat modeling because it refers to a resolution level rather than a category of model.

In the case of entity level and virtual, the latter term is even less appropriate in this context, as many entity level combat models are in fact constructive, not virtual (e.g. OneSAF). Nevertheless, the term "virtual" does relate to one of the motivations for multi-resolution combat modeling. Even in a multi-resolution combat model linking unit level and entity level models that are both constructive, the overall distributed simulation system can also include virtual simulators with human participants (e.g. crew members in a tank simulator) who benefit from the system's multi-resolution capabilities. For example, the participants may interact with entities under the control of the entity level model that have been disaggregated from a unit of the unit level model.

CONCEPTS AND OPERATIONS
OF A MULTI-RESOLUTION COMBAT MODEL

This section provides the conceptual framework of multi-resolution combat modeling, within which specific details and capabilities can be placed in context. A notional example of a multi-resolution combat model illustrates the essential structure and operation of such a system, and key terms are defined in relation to that example. A generic implementation architecture is described,

---

[10]These terms are used in this chapter because they seem most descriptive of the levels of resolution involved.

[11]While a constructive model may have a human operator, he/she is typically not representing or acting as the crew of a specific entity.

identifying the primary components of such a system. A set of operations that change the representational resolution of simulation objects at runtime are cataloged.

Key Terms and an Example

Units are military organizational groupings, such as battalions and brigades; they are normally composed of multiple subordinate units, e.g. a battalion may comprise three companies. *Entities* are individual military objects, such as tanks and helicopters; while entities can be decomposed into parts or assemblies, such decomposition is not normally pertinent in this context. *Object* is a generic term to mean either a unit or an entity.

Resolution is the amount of detail represented in a model. Typically an entity level model is considered *high resolution*, representing individual entities and their interactions in some detail. In contrast, most unit level models are relatively *low resolution*, representing units and their interactions more abstractly. Model resolution may be thought of as a scale from high to low, with a *level* of resolution designating a particular value on that scale.[12]

An *interaction* is an attempt by one object to modify the state of another, for example direct fire, logistics resupply, and communications are all interactions. An *intra-level interaction* is an interaction that takes places between objects at a single level of representation (entity level or unit level) during a simulation at that level. On the other hand, an *inter-level interaction* is an interaction between objects at different levels of representation.[13]

The terms *aggregate* and *disaggregate* can be nouns, verbs, or adjectives. As adjectives, an aggregate model represents objects that are composed of other objects (Rumbaugh, 1991); here that means a unit level model. On the other hand, a disaggregate model represents objects that are not (normally) further decomposable, here referring to an entity level model. The verbs *aggregating* and *disaggregating* and the nouns *aggregation* and *disaggregation* will refer to the processes of changing representational resolution of a set of objects from entities to a unit, or vice versa, respectively. *Aggregated* and *disaggregated* will be used as past-tense verbs, to indicate the completed processes of aggregation or disaggregation.

A multi-resolution combat model is a composite model that links a unit level combat model with an entity level combat model. Such a model can represent a unit in aggregate form in the unit level model and the entities that

---

[12]Note the inverse relationship between model resolution and the military organizational hierarchy when using the terms *high* and *low*. Usually, higher resolution models represent objects lower in the military hierarchy (entities), and lower resolution models represent objects higher in the military hierarchy (units).

[13]Instead of *inter-level interaction*, some sources use the term *cross-level interaction* (Powell, 1997; Reynolds et al., 1997).

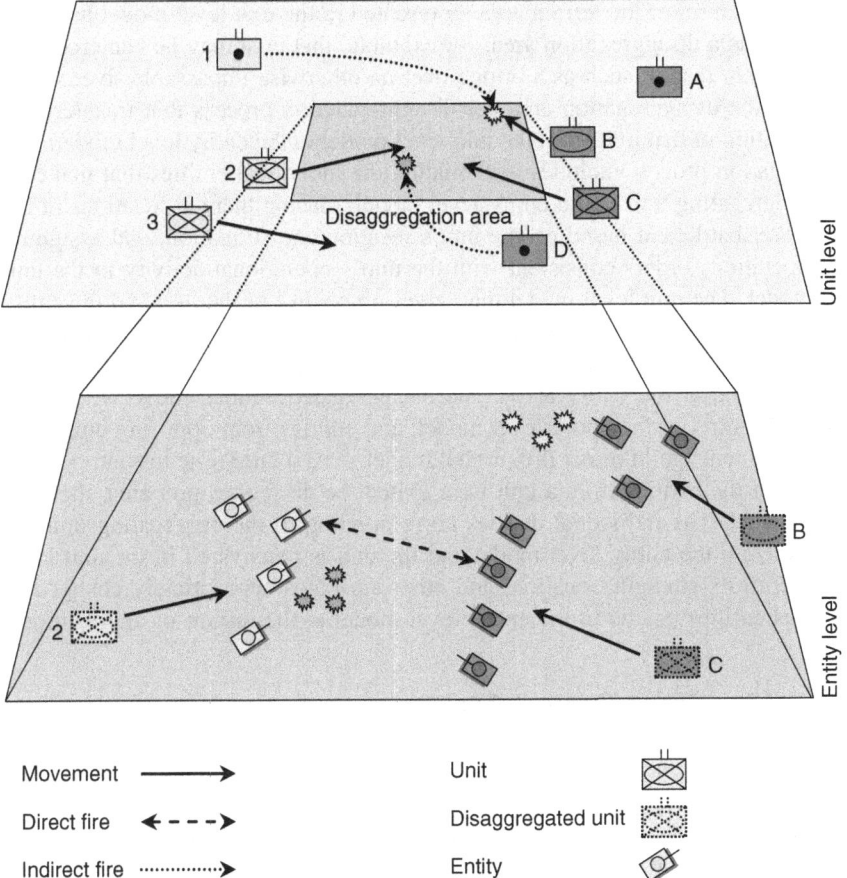

Figure 25.1   Conceptual structure of a multi-resolution combat model.

make up that unit in disaggregate form in the entity level model. Figure 25.1 illustrates the conceptual structure of such a system; it shows a schematic example of an instant in time during the execution of a notional scenario. The upper half of the figure shows an executing unit level model. In that model, the objects being simulated are units, which move and engage in combat using unit level models. The lower half of the figure shows an executing entity level model. In that model, the objects simulated are entities, which move and engage in combat using entity level models.

In the figure, the unit level model is executing a simulation involving two brigade sized forces composed of company and battalion maneuver units. Initially, none of those units, i.e. no entities belonging to those units, exist in the entity level model. The unit level scenario occurs in a large terrain area, a portion of which is also represented by the terrain database of the entity level model.[14] A subset of the terrain area represented in the unit level model has been designated as a disaggregation area; for example, that area may be centered on a crucial terrain feature such as a bridge over an otherwise impassable river. Units that enter the disaggregation area are disaggregated, a process that transfers the representation of that unit from the unit level model to the entity level model. The disaggregation process includes instantiating the individual entities that make up the disaggregating unit in the entity level model, placing them at locations in the entity level battlefield based on the unit's location and formation, and assigning them operations orders consistent with the unit's operational activity in the unit level model. The unit level model then gives up control of the unit and the entity level model takes over control of that unit's entities. After the disaggregation operation, the simulation execution resumes, with a portion of the scenario now being modeled at the entity level. The disaggregated entities move within and under the control of the entity level model, and entities from opposing units may interact, e.g. engage in direct fire, in that model.[15] At a later time instant, perhaps when all of the entities from a unit have exited the disaggregation area, that unit is aggregated. The individual entities corresponding to the aggregating unit are removed from the entity level model and the unit is reactivated in the unit level model, with its strength, location, and other attributes appropriately changed to reflect its entities' status in the entity level model at the instant of aggregation.

Typical Implementation Architecture

The typical implementation architecture of multi-resolution combat models of the type described here has three primary components: a unit level model, an entity level model, and an interface module that logically connects the two models. The interface module performs resolution translation and object state maintenance functions needed to support the resolution change operations and inter-level interactions to be discussed later.[16] Figure 25.2 illustrates the implementation architecture from two perspectives, logical and physical. Figure 25.2(a) shows the logical architecture; the unit level and entity level models each communicate with the interface module, which mediates between them. Unit level information describing the state and activities of the units it is simulating pass

---

[14]The terrain area of the entity level model may in fact coincide with the terrain area of the unit level model, though it is usually a subset.

[15]Not all of the entities being simulated in the entity level model must necessarily be the result of disaggregating a unit from the unit level model. Most multi-resolution combat models support the inclusion of entities that are not the result of disaggregation and are not eligible for aggregation. Crewed simulators may generate such entities.

[16]The interface module has different names depending on the specific implementation, including Simulation Interface Unit (Karr et al., 1993; Peacock et al., 1996), Virtual Simulation Interface Unit (Schricker et al., 1996), Agg/Disagg Unit (Petty and Franceschini, 1998a), and Aggregation/Disaggregation Unit (Williams et al., 1998).

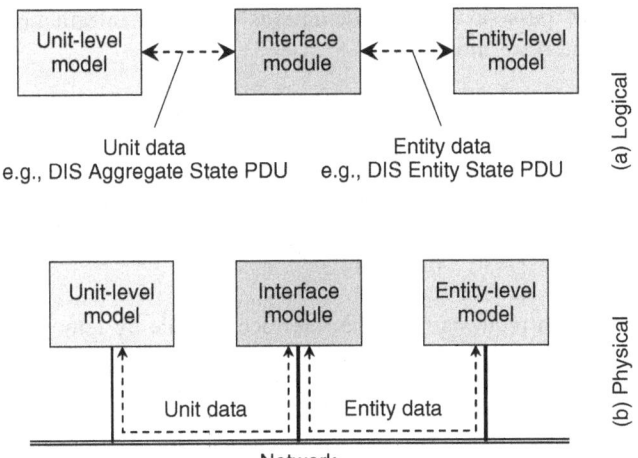

Figure 25.2 Logical and physical architectures of a multi-resolution combat model.

from the unit level model to the interface module. Similarly, entity level information describing the state and activities of the entities it is simulating pass from the entity level model to the interface module. Figure 25.2(b) shows the physical architecture. The two models are not actually connected through the interface module; instead, all three primary components are connected via a network. They send and receive information about units and entities, as well as messages relating to resolution change operations, using a distributed simulation interoperability protocol (SIMNET, Distributed Interactive Simulation (DIS), or High Level Architecture (HLA), as discussed earlier).

In multi-resolution combat model implementations, certain functions are typically assigned to specific components in the system.[17] The unit level model performs these functions:

1. Model units and simulate their actions over time.

2. Send periodic unit status updates to the interface module.

3. Respond to disaggregation requests from the interface module by replacing its normal representation of the disaggregating unit with a passive representation and relinquishing control of the unit.

4. Respond to aggregation requests from the interface module by replacing its passive representation of the aggregating unit with a normal unit and resuming control of the unit.

---

[17]Other assignments of these functions to the system components are possible and have been used. In particular, the monitoring of disaggregation and aggregation triggering functions by the interface module is not universal (Franceschini and Petty, 1995b). However, for simplicity, hereinafter it will be assumed that the functions are assigned as listed. In addition, these lists are not complete; additional functions will be discussed later.

5. Update the status of passive unit representations based on information received from the interface module.

The entity level model performs these functions:

1. Model entities and simulate their actions over time.

2. Send periodic entity status updates to the interface module.

3. Respond to disaggregation requests from the interface module by instantiating the entities making up the disaggregating unit in the entity level model.

4. Respond to aggregation requests from the interface module by removing the entities making up the aggregating unit from the entity level model.

The interface module performs these functions:

1. Translate unit status updates into entity status updates, and vice versa, as needed.

2. Monitor the trigger conditions for disaggregation, and, when met, issue disaggregation requests to the unit level and entity level models.

3. Monitor the trigger conditions for aggregation, and, when met, issue aggregation requests to the unit level and entity level models.

In most multi-resolution combat models, the unit level and entity level models are connected via a standard distributed simulation interoperability protocol, such as SIMNET, DIS, or HLA.[18],[19] DIS contains predefined standard messages within the protocol that may support the execution of a multi-resolution combat model (Calder and Evans, 1994; Franceschini and Petty, 1995a; Foss and Franceschini, 1996; Stober et al., 1996; Schricker et al., 1998a). By comparison, HLA does not include predefined messages (in HLA terms, objects or interactions) for multi-resolution combat modeling (or any other type of model, for that matter; this is consistent with its design intent); developers must define and include the necessary messages in the Simulation Object Models and Federation Object Models of the system.

Resolution Change Operations

A defining characteristic of multi-resolution combat models is the ability to change the representation level of simulated objects at runtime. A set of several resolution change operations have been defined, implemented, and found to

---

[18]The SIMNET, DIS, and HLA interoperability protocols have been described elsewhere in this volume (Petty and Gustavson, 2012); those details are not repeated here.

[19]Note that multi-resolution simulation and distributed simulation are not synonyms. A *distributed simulation* is one where the simulation computation is distributed across multiple computers and hosts connected by some communications infrastructure. Multi-resolution simulations may be distributed or not, and distributed simulations may be multi-resolution or not.

be useful. Two of these operations, aggregation and disaggregation, have already been mentioned. Because objects at different levels of resolution are normally related by composition (e.g. a unit is composed of multiple entities), resolution change operations typically involve decomposing, in some form, one lower resolution object into multiple higher resolution objects (disaggregation and its variants), or composing many higher resolution objects into one lower resolution object (aggregation and its variants). These operations are described here: disaggregation, aggregation, pseudo-disaggregation, pseudo-aggregation, partial disaggregation, and subset disaggregation.

Disaggregation[20]

Disaggregation is the operation by which representation of a unit is transferred from the unit level model to the entity level model. In the disaggregation process a unit, which was being simulated by the unit level model, is converted into the entities that compose it, which are then simulated by the entity level model. The disaggregation process has these steps.

1. The interface module detects that a disaggregation trigger condition has been met for some unit in the unit level model. It sends a disaggregation request to the unit level model for that unit.

2. The unit level model sends information about the disaggregating unit to the interface module; that information may include the unit's location, the types and numbers of entities in the unit, and the unit's current operational activity.

3. The interface module translates the unit level information into entity level information for each of the entities that make up the disaggregating unit. It sends an instantiation request to the entity level model for each entity.

4. The entity level model instantiates the disaggregating units' entities per the instantiation requests and commences actively simulating those entities.

5. The unit level model ceases actively simulating the unit that was disaggregated.

In step 3, specific entity level attributes for each entity must be inferred or interpolated for each entity in the disaggregating unit. This process may be straightforward or problematic, depending on the attribute; for example, the location of the disaggregating unit and its formation (which may be explicitly specified by the unit level model or inferred from the unit's current operational activity), along with doctrinal formation templates that specify for a given formation where a unit's entities will be relative to each other and the unit's center of mass, are used to determine the locations at which to instantiate the unit's

---

[20]Instead of *disaggregation*, some sources use the term *deaggregation*.

entities in the entity level model. The placement process may be affected by local terrain features, e.g. ground vehicles should not be placed in lakes or rivers (Franceschini, 1992; Clark and Brewer, 1994).

Often a passive representation of a simulation object at one level is maintained while it is actively simulated at the other level, a technique referred to as *ghosting*. For example, a unit that has been disaggregated and is thus being simulated as entities by the entity level model may still be represented as a unit in the unit level model. A ghost unit holds a place in the unit level model for the disaggregated unit should it eventually be aggregated, and to support any interactions that occur only between units at the unit level.[21] Usually the attributes of the passive, or *ghost*, representation are derived from those of the active representation, and are updated periodically as the simulation proceeds. For example, the locations of the entities of a disaggregated unit may be collected from the entity level network messages by the interface module and combined to produce a location for the unit. The aggregated unit information is periodically passed by the interface to the unit level model, where it is used to update the status of the ghost unit.

Aggregation

Aggregation is the reverse of disaggregation. It is the operation by which representation of a unit is transferred from the entity level model to the unit level model. In the aggregation process a unit, which was being simulated as a set of entities by the entity level model, is converted into a unit, which is then simulated by the unit level model. The aggregation process has these steps.

1. The interface module detects that an aggregation trigger condition has been met for some unit in the entity level model.[22] It sends an aggregation request to the entity level model for each entity of the unit.

2. The entity level model removes the aggregating unit's entities from the entity level model. It sends final attribute values for each entity in the aggregating unit to the interface module.

3. The interface module aggregates and translates the entity level information into unit level information for the aggregating unit. It sends an instantiation request to the unit level model for the aggregating unit.[23]

4. The unit level model instantiates the aggregating unit per the instantiation request and commences actively simulating the unit.

5. The entity level model ceases actively simulating the entities of the aggregating unit.

---

[21]Instead of *ghost*, some sources use the term *surrogate* (Powell, 1997).

[22]Of course, there are no units in the entity level simulation. What this means is that the entities that compose a unit have collectively met an aggregation trigger condition, e.g. their average location is outside the designated disaggregation terrain area.

[23]If the unit was not removed from the unit level model but instead maintained as a ghost unit, then an activation request, rather than an instantiation request, would be sent.

Pseudo-Disaggregation

In the pseudo-disaggregation operation, active representation of a unit remains in the unit level model, but a process similar to disaggregation is used to generate over time entity level information for the entities that compose that unit (Root and Karr, 1994). The unit level model continues to actively simulate the unit and it is involved in all actions and interactions in the unit level model as usual. The center of mass, movement speed, movement direction, and formation for a pseudo-disaggregated unit are passed from the unit level model to the interface module. In a process similar or identical to that of determining locations for entities during disaggregation, the interface module uses the information about the pseudo-disaggregated unit to produce locations, movement speed, and movement directions for the entities that compose the pseudo-disaggregated unit. Then either the interface module or the entity level model sends network messages (such as DIS Entity State PDUs) that provide the appearance that the individual entities of the pseudo-disaggregated unit are present in the entity level model. In fact they are not; they are not controlled by the entity level model and do not participate in the interactions of that model. Pseudo-disaggregation is useful to stimulate models of sensor or surveillance platforms without incurring the computational cost of disaggregating the unit and simulating the entities individually (Schricker et al., 1998b).[24,25]

Pseudo-Aggregation

In the pseudo-aggregation operation, active representation of the entities that compose a unit remains in the entity level model, but a process similar to aggregation is used to generate over time unit level information for the unit composed by the entities. The entity level model continues to actively simulate the entities and they are involved in all actions and interactions in the entity level model as usual. The locations, movement speeds, and movement directions for the entities of the pseudo-aggregated unit are passed from the entity level model to the interface module. The interface module aggregates and translates the entity level information into unit level information for the pseudo-aggregated unit. It sends this information to the unit level model, where it is used to update that model's representation of the pseudo-aggregated unit. In most respects, this operation is quite similar to the unit ghosting process.

---

[24] A process similar to pseudo-disaggregation is described in Powell (1997). It differs in that the entity level data generated for the individual entities of the pseudo-disaggregated unit are not actually output to the simulation network, with the result that those entities do not appear to exist in the entity level simulation. Instead, the generated entities' information is retained in a data structure internal to the interface module, where it can be consulted to resolve certain inter-level interactions. This process is not named in Powell (1997); it may reasonably be termed *internal pseudo-disaggregation*.

[25] Some entity level models operate entirely in a mode quite similar to pseudo-disaggregation, e.g. TACSIM (Smith, 1997).

Partial Disaggregation

Partial disaggregation allows the representation of a unit in a useful way at the entity level without completely disaggregating it.[26] In partial disaggregation the representation of the unit is not completely transferred to the entity level; consequently, the entities that are instantiated in a partial disaggregation are not fully functional at the entity level. Pseudo-disaggregation, described earlier, is a special case of partial disaggregation, where almost none of the representation of the unit has been transitioned to the disaggregate level. To generalize from pseudo-disaggregation to partial disaggregation, assume that the entity level model supports the instantiation of entities with a subset of the entity level model's capabilities, rather than the "all" of disaggregation or the "nothing" of pseudo-disaggregation. For example, an entity could be instantiated with inter-visibility and direct fire damage resolution capabilities at the entity level but not autonomous behavior and movement. Those capabilities not present in the partially disaggregated entities would be provided by the unit level model. Partial disaggregation has been implemented using ModSAF as the entity level model (Schricker et al., 1998b).

Subset Disaggregation

In subset disaggregation, the disaggregation operation is applied to a subset of the entities that compose a unit, i.e. not all of a unit's entities are instantiated in the entity level model, and the unit level model retains some representation of the unit for the entities not disaggregated.[27] To compare subset disaggregation to partial disaggregation, in partial disaggregation some of the capabilities for an entity are disaggregated, whereas in subset disaggregation some of the entities for a unit are disaggregated.

Implementation Issues and Challenges

Certain issues and challenges consistently arise in the implementation of multi-resolution combat models. Their resolution often varies across implementations. These issues are described here: triggering resolution changes, disaggregation overload and spreading disaggregation, reconciling simulated time, terrain correlation, results correlation, and inter-level interactions.

Triggering Resolution Changes

There are a number of possible criteria for triggering resolution change operations, especially disaggregation and aggregation.

---

[26]Some sources use the term *partial disaggregation* not as defined here but instead to mean what will herein be called *subset disaggregation* (Reynolds et al., 1997).

[27]Instead of *subset disaggregation*, some sources use the term *partial unit disaggregation* or *partial disaggregation* (Reynolds et al., 1997).

1. *Designated area*. Disaggregation occurs when a unit is located within a portion of the terrain area designated in advance as a disaggregation area. There may be multiple designated disaggregation areas. Aggregation occurs when the center of mass of the disaggregated unit, as determined from the entities in the unit, moves out of the disaggregation area. In order to reduce the likelihood of a rapid sequence of resolution change operations for a unit moving along the perimeter of a disaggregation area, the boundary of the disaggregation area can be defined to be slightly larger for triggering aggregation than for disaggregation.

2. *Proximity*. Disaggregation occurs when a unit is within a critical range of one or more hostile units or entities. The range may be the maximum detection range of any of the sensor systems in the hostile unit or entities. This criterion can lead to spreading disaggregation, which is described later.

3. *Intent to interact*. The aggregate unit intends to interact (e.g. employ sensors or conduct direct fire) with some other disaggregated unit (virtual vehicles).

4. *Operator initiative*. Disaggregation or aggregation occur when triggered by a human operator, who may do so at will. This criterion allows human intelligence to decide when a resolution change is appropriate, so anomalous behavior should not result, and is useful during testing and early development of a new multi-resolution combat model. However, reliance on human operators constrains the range of applications for the system.

5. *Commander's focus of interest*. A commander may wish to view different parts of the battlefield at different levels of granularity. As the commander's focus of interest shifts around the battlefield, units may be disaggregated or aggregated in response (Downes-Martin, 1991). The precise mechanism may be manual or automatic, depending on whether the commander's focus of interest is available to the system.

Disaggregation Overload and Spreading Disaggregation

In a multi-resolution combat model the number of entities instantiated as a result of disaggregation operations can become too large for some part of the system, perhaps the computational or storage capacity of the entity level model's host computer or the bandwidth of the network connecting the unit level and entity level models' host computers (Williams et al., 1998); the condition has been referred to as *disaggregation overload* (Trinker, 1994). Disaggregation overload may be the result of *spreading disaggregation*, which can occur when disaggregation of units is triggered by geographic proximity to disaggregate entities, one of the disaggregation triggers listed earlier (Petty and Franceschini, 1995). Spreading disaggregation, which is a chain reaction of disaggregations, may occur when each disaggregation operation creates new entities close enough to additional units to trigger their disaggregation as well.[28]

Reconciling Simulated Time

Unit level and entity level models may differ in their treatment of time. Entity level models may execute in real time, with the passage of time within the entity level matching that of the real world, so as to enable interoperation with crewed simulators or other human participants.[29] Unit level models are often time-stepped, with the simulation time advancing a fixed amount of time for each computational cycle of the model. The size of the simulated time step ordinarily has nothing to do with the time required to compute the events of that time step, and the models generally execute as fast as possible. In a multi-resolution combat model combining real-time and non-real-time execution, or even when neither model is real time but one executes more quickly with respect to simulated time than the other, the passage of time in the two models must be synchronized. If it was not, events in one model would proceed at a different rate than those in the other, potentially producing temporal and causality anomalies and serious realism breakdowns. The interface module can be used to reconcile time between the two models. The faster simulation can be paused periodically so that its simulated time does not advance past the slower simulation's time. During the pauses, the interface module collects information from the slower simulation (e.g. position and status of entities and any interactions from entities to units); when the slower simulation resumes execution, the interface module forwards the collected information to the faster model.

Terrain Correlation

Entity level models often represent the terrain of the entity level battlefield in great detail, with individual roads, buildings, trees, and bushes represented, and the possible locations an entity may occupy are essentially continuous over the terrain; such detail is appropriate for entity level modeling. In contrast, unit level models generally use terrain that has been partitioned in a regular grid of squares or hexagons, with the terrain of each grid element abstracted into one or more terrain attributes that apply to the entire element (e.g. hexagons in the Joint Theater Level Simulation (Kang and Roland, 1998)). This difference affects entity locations during disaggregation, as mentioned earlier. The difference in the representations can lead to terrain correlation issues. Terrain correlation is the general question of whether two (or more) different representations of the same terrain

---

[28]Instead of *spreading disaggregation*, some sources use the term *chain disaggregation* (Reynolds et al., 1997).

[29]Not all entity level models execute in real time, and of those that do, they may have the ability to execute as fast as possible in simulations without human participants. Nevertheless, reconciling real-time and non-real-time execution is an issue in some multi-resolution combat models.

have differences that cause discrepancies in the results of a simulation that uses both of them. Terrain correlation is an issue in distributed simulation in general and has been studied in that context (Schiavone et al., 1995; Petty et al., 1996). It has special considerations in multi-resolution combat modeling because, as noted, the unit level and entity level models will typically have terrain representations at different levels of resolution and different representations (e.g. polygons at the entity level and hexagons at the unit level), increasing the opportunities for significant discrepancies (Petty and Franceschini, 1998b).

For example, serious discrepancies could arise if a bridge spans an otherwise impassable river in the unit level terrain representation but does not exist in the entity level terrain representation. That situation would allow units to cross the river in the unit level model but prevent entities from crossing the river in the entity level model, with the consequence that the outcome of the scenario could be substantially and anomalously affected by whether units seeking to cross the river were disaggregated or not. Moreover, if the terrain can be changed in tactically significant ways during the course of the simulation, e.g. engineers demolishing a bridge, then those terrain changes must be propagated from the model in which they occurred to the other; this function is usually assigned to the interface module.

Results Correlation

It is generally considered desirable that in a multi-resolution combat model the outcome or results of a scenario be independent of the resolution level at which that scenario is simulated. In other words, the results of executing a scenario should be the same, or similar within a desired tolerance, regardless of which units were disaggregated and aggregated over the course of the simulation, and when those operations occurred; here this similarity is termed *results correlation*.[30] If the results do not correlate in this way, then differences between the unit level model's and the entity level model's representations of the same events are introducing discrepancies.

For example, suppose that in a multi-resolution combat model a tank company is ordered to perform a road march from its current location to a destination location. That movement may be simulated as a unit moving in the unit level model, or as the individual entities of the company moving in the entity level model. In each case the terrain database in the model at the level the movement is simulated will be accessed to get information needed to simulate the movement. It is reasonable to expect that the amount of simulated time required for the unit to complete the move in the unit level model will be consistent with the time required for the corresponding formation of entities to complete the move in the entity level model, but it may not be. Differences in terrain representations or

---

[30]Instead of results correlation, some sources use the terms *consistency of prediction*, or simply *consistency*, e.g. Davis (1995b), Powell (1997), Reynolds et al. (1997) and Franceschini et al. (2000).

movement models may introduce discrepancies, and those discrepancies could have significant effects on the scenario outcome.

Results correlation can result from several sources in the models' representations, some of which have already been mentioned: time, terrain, combat, sensing and detection, logistics, and others. It is an important topic in multi-resolution combat modeling research (Davis, 1995b; Reynolds et al., 1997; Franceschini, 1999).

Inter-Level Interactions

As previously stated, one design intent of a multi-resolution combat model is that simulated events occurring in one of the models' simulations may have an effect on events in the other. The usual means by which this may occur are the resolution change operations, e.g. a unit is disaggregated into entities, which then participate in the entity level simulation. However, an alternative to the resolution change operations exists. Inter-level interactions are interactions (as previously defined) between a unit and an entity directly across the resolution boundary without first disaggregating the unit into entities or aggregating the entities into a unit. Inter-level interactions can be used to avoid spreading disaggregation and the resulting disaggregation overload, mentioned earlier, and allow the system to include unit level or entity level models that cannot support resolution change operations.

Inter-level interactions were introduced in Franceschini and Petty (1995b) and Powell (1997), who explained the need for them and gave example scenarios to show how they might be used. Some types of possible inter-level interactions and brief summaries of how they might be implemented are listed; for extended discussion of all of these types of inter-level interactions as well as others, see Petty and Franceschini (1998a).

1. *Indirect fire*. Indirect fire is generally aimed at a geographical area where hostile entities are expected to be. Artillery volleys at the unit level can be automatically translated into individual artillery detonations at the entity level and applied against the entities there. Similarly, but with more difficulty, individual artillery detonations at the entity level can be aggregated into volleys and resolved at the unit level. Implementations in both directions exist (Karr et al., 1993; Karr and Root, 1994a).

2. *Direct fire*. Direct fire, aimed fire at a specific entity, is more problematic as an inter-level interaction than indirect fire, largely because of the effect of intervisibility on direct fire and the differences between intervisibility handling at the unit and entity levels. Inter-level direct fire can be converted to intra-level direct fire, either at the entity level by using pseudo-disaggregation or partial disaggregation of the unit to produce entities, or at the unit level by first using pseudo-aggregation or partial aggregation to produce a unit and then resolving the direct fire at that level.

3. *Operations orders*. Military forces are typically operating under a set of orders that define the mission and provide direction for the actions that

they take. If the entity level model has automated behaviors available to control the simulated entities, as is typically the case with SAF systems, the operations orders controlling the actions of a unit in the unit level model can be translated automatically or by the operator of the SAF system into the entity level behaviors when the unit is disaggregated.

4. *Communications*. Communications in reality are always from an entity to an entity, i.e. there is no "unit" in any physical sense. To resolve inter-level point-to-point communications in a multi-resolution combat model, determine which entity in the communicating unit would be communicating (e.g. the command vehicle) and a location for that communicating entity (by pseudo disaggregation, partial disaggregation, subset disaggregation, or by default, e.g. the center of mass of the unit) and resolve the communications in the entity level model using that location. For inter-level broadcast communications, the process is more complicated, as locations for all possible recipients of the communications may be needed. Deliberate jamming of communications and detection of non-communication emissions are similarly complicated in the inter-level context.

5. *Terrain updates*. During a simulation, units or entities may modify the terrain at their level in ways that are tactically significant at the other level, e.g. an engineering unit demolishes a bridge which should be impassable to entities for the remainder of the execution. Such modifications are best first resolved at the level they are initiated, i.e. as an intra-level interaction, and then translated into terrain updates suitable for the terrain representation at the other level, conveyed to that level, and applied there (Petty and Franceschini, 1998b).

IMPLEMENTATIONS AND APPLICATIONS

This section turns from general discussions of the architecture, operations, and issues in multi-resolution combat modeling to specific instances and uses of such models. A brief survey of the actual implementations of multi-resolution combat models and of the associated literature is provided, and some examples of practical applications of the models are described.

A surprisingly large number of multi-resolution combat models with the basic architecture described here, existing unit level and entity level models linked via a networked simulation interoperability protocol and implementing aggregation and disaggregation operations, have been implemented. Table 25.1 is a partial list of implemented multi-resolution combat models. Brief descriptions of the multi-resolution combat models in Table 25.1 follow.

Integrated Eagle/BDS-D

Arguably the first operating multi-resolution combat model of the type described in this chapter to be implemented, the Integrated Eagle/BDS-D system linked the

Table 25.1   Implemented Multi-Resolution Simulation Systems

System	Year	Unit level model(s)	Entity level model(s)	Interoperability protocol(s)	Reference(s)
Integrated Eagle/ BDS-D	1992	Eagle	CGF Testbed	SIMNET DIS	Franceschini, 1992; Karr et al., 1992, 1993; Franceschini and Karr, 1994; Karr and Root, 1994a,b; Karr and Franceschini, 1994; Franceschini and Petty, 1995b; Franceschini, 1995; Schricker et al., 1996
Eagle II	1992	Eagle	SIMNET SAF	SIMNET	Powell, 1993
Corps Level Computer Generated Forces	1994	Eagle	ModSAF	DIS	Calder and Evans, 1994; Calder et al., 1995a,b; Peacock et al., 1996
BBS/SIMNET	1994	BBS	SIMNET SAF	SIMNET DIS	Hardy and Healy, 1994
SOFNET-JCM	1994	JCM	SOFNET	DIS	Babcock et al., 1994
ABACUS/ ModSAF	1995	ABACUS	ModSAF	DIS	Cox et al., 1995; Weeden and Smith, 1996; Weeden et al., 1996
Eagle/ITEMS	1996	Eagle	ModSAF ITEMS	DIS	Schricker et al., 1996
Integrated Eagle/ ModSAF	1998	Eagle	ModSAF	DIS	Schricker et al., 1998b; Franceschini et al., 1999
Swedish Air Defense	1998	TYR FBSIM	ARTEVA	HLA	Sköld et al., 1998; Seiger et al., 2000
Battlespace Federation	2000	ARES	ModSAF	HLA	Beeker, 2000
JTLS-JCATS	2002	JTLS	JCATS	HLA	Bowers et al., 2002; Bowers and Prochnow, 2003a
ACTF-MRM	2004	CBS	JCATS	DIS HLA	Zabek et al., 2004
JWARS-JSAF	2005	JWARS	JSAF	HLA	Macannuco et al., 2005
Joint Multi-Resolution Model	2007	JTLS	JCATS	HLA	Cayirci, 2009
NATO Training Federation	2008	JTLS	JCATS	HLA	Cayirci, 2009

Eagle unit level model (developed by the US Army Training and Doctrine Command Analysis Center) with the Institute for Simulation and Training's Computer Generated Forces Testbed (IST CGF) using the SIMNET, and later DIS, protocol. The project's goal was the integration of the unit level and entity level models as a demonstration and proof of principle. Many of the basic architectural ideas (e.g. the interface module and its functions) and resolution change operations (e.g. aggregation and disaggregation) were first developed for this system (Franceschini, 1992, 1995; Karr et al., 1992, 1993; Franceschini and Karr, 1994; Karr and Franceschini, 1994; Karr and Root, 1994b; Franceschini and Petty, 1995b; Schricker et al., 1996). Eagle units were disaggregated into their component IST CGF entities. While the units were disaggregated, Eagle's simulation shifted to real-time execution to receive and incorporate the entity information being generated in real time. Combat occurred in the entity level model between individual entities. The first inter-level interactions were also developed; operations orders from the unit level model were sent to the entity level model's operator for execution and non-disaggregated units in Eagle could attack disaggregated entities in the IST CGF with indirect fire (Karr and Root, 1994a).

Eagle II

Developed independently around the same time as Integrated Eagle/BDS-D as a prototype multi-resolution system, Eagle II was similarly innovative. The essential ideas of linking a unit level model to an entity level model, providing disaggregation and aggregation operations, and implementing the multi-resolution operations in an interface module are all present in Eagle II (Powell, 1993). The Eagle II system included, in addition to the unit level and entity level models, a three-dimensional visualization of the battlefield. A proof of concept demonstration of the system took place in 1993 (Powell, 1993).

Corps Level Computer Generated Forces

The Corps Level Computer Generated Forces system linked the Eagle model with ModSAF, which was the most important SAF system at that time (Calder and Evans, 1994). Developed to support the Joint Precision Strike Demonstration the Corps Level Computer Generated Forces model generated a combined arms force of maneuver and artillery units at the corps level in order to study the effects of deep strikes on logistics and command, control, communications, and intelligence (Calder et al., 1995a,b). The model stimulated and interacted with tactical systems and their operators and was developed to include a number of specialized capabilities and components (Peacock et al., 1996). It was the first multi-resolution combat model to be used for practical applications.

BBS/SIMNET

The BBS/SIMNET system links the Brigade/Battalion Battle Simulation with SIMNET using the SIMNET Semi-Automated Forces (SAF) as the Computer

Generated Forces system (Hardy and Healy, 1994). This system was developed to learn about unit level and entity level interoperation and means to combine unit level battle staff training (BBS) with entity level team tactical training (SIMNET).

ABACUS/ModSAF

Developed in the UK, ABACUS/ModSAF linked ABACUS, a unit level constructive model used by the UK Army for command and staff training that normally represented units of company size, with the ModSAF entity level model (Cox et al., 1995). The stated goal of the system was to study methods, benefits, and challenges for interoperability between unit level and entity level models (Weeden and Smith, 1996). A 1996 demonstration of the system included (via DIS) two additional entity level models that were not involved in the aggregation and disaggregation operations (Weeden et al., 1996). As with many of the systems on this list, this system's architecture included an interface model that mediated the resolution change operations.

Integrated Eagle/ModSAF

Integrated Eagle/ModSAF was a further development of the Corps Level Computer Generated Forces system. Pseudo-disaggregation capabilities were added to the system in order to generate large numbers of realistically behaving entities (Schricker et al., 1998b). Integrated Eagle/ModSAF was used in the Dynamic Multi-user Information Fusion (DMIF) project in 1998 (Franceschini et al., 1999). DMIF investigated automated tools to process the massive amounts of information generated by sensing instruments in a theater level scenario and compared various implementations of those automated tools (called data fusion engines) to assess their effectiveness. Integrated Eagle/ModSAF's pseudo-disaggregation capability was used to generate sufficient numbers of entities to represent a theater level scenario, more than 10,000 individual battlefield entities. To accurately assess the fusion engines, the generated entities had to conduct themselves in a realistic manner. Integrated Eagle/ModSAF was also used in Army Experiment 6 (AE6) in 1999 (Franceschini et al., 1999). While the DMIF used the model in an analysis setting, AE6 served as a training exercise for command staff personnel. As with DMIF, pseudo-disaggregation was used to generate a large number of individual entities to create the context of a large scale scenario needed to train command staff personnel in a realistic scenario.

Swedish Air Defense

The Swedish Air Defense model was developed to test aggregation and disaggregation strategies (Sköld et al., 1998).[31] It combines three, rather than the more

---

[31]The name "Swedish Air Defense" is the authors' invention; the sources do not name the model.

typical two, models and levels of resolution (Seiger et al., 2000). TYR is a unit level model of air defense operations; it represents air attack and air defense units. FBSIM models individual aircraft and air defense entities. ARTEVA is a fine-grained model of bomb and missile impacts on single entities. In the system disaggregations occur from TYR to FBSIM, and if missiles are fired or bombs are released, a disaggregation from FBSIM to ARTEVA will occur to resolve the combat.

Battlespace Federation

The Battlespace Federation linked a unit level campaign model with the ModSAF entity level model using HLA. Developed as a testbed for simulation concepts, Battlespace Federation also included a virtual command post, a national strategic intelligence model, and a stealth viewer (Beeker, 2000).

JTLS-JCATS

The JTLS-JCATS model linked Joint Theater Level Simulation (JTLS), a unit level model widely used as a driver for command post training exercises (Kang and Roland, 1998), with Joint Conflict and Tactical Simulation (JCATS), an entity level model used for both training and analysis (Bowers et al., 2002; Bowers and Prochnow, 2003a). JCATS has internal multi-resolution capabilities, and so disaggregation is done with the model rather than in a separate interface module. Special features of HLA including object management, ownership management, and time management are used to support the system. Among other applications, JTLS-JCATS has been used for training emergency response to officials in an emergency response center (Bowers and Prochnow, 2003b).

ACTF-MRM

The Army Constructive Training Federation—Multi-Resolution Modeling system tem linked the unit level Corps Battle Simulation with the entity level Joint Conflict and Tactical Simulation. ACTF–MRM is designed to drive command post training exercises for commanders and their staffs (Zabek et al., 2004). It also includes the Combat Service Support Training Simulation Systems, a detailed logistics model, and other components.

JWARS-JSAF

The JWARS-JSAF system links the Joint Warfare System, a unit level model of joint theater warfare, with Joint Semi-Automated Forces, an entity level model developed from ModSAF that represents air, land, and sea entities (Macannuco et al., 2005). Intended primarily for analysis applications, JWARS-JSAF also supports human participation in a scenario using real-time execution.

Joint Multi-Resolution Model and NATO Training Federation

Like JTLS-JCATS, the Joint Multi-Resolution Model links the JTLS unit level model and the JCATS entity level model. The NATO Training Federation, which was developed from the Joint Multi-Resolution Model, also links those models; it supported a major NATO training exercise in 2008 (Cayirci, 2009).

CHALLENGES

Both theoretical and practical challenges exist with respect to the development and use of multi-resolution combat modeling (National Research Council, 1997). Research addressing these challenges has a long history, and the research results have informed the practicel implementations listed earlier. Specific challenges in multi-resolution combat modeling, which implicitly define research directions, include results correlation, resolution flexibility, inter-level interactions, resolution change triggers, and software complexity (Franceschini et al., 1999).

Results Correlation

As mentioned earlier, the aggregation and disaggregation operations introduce fundamental validity questions regarding correlation between levels of resolution (Davis, 1995b; Reynolds et al., 1997; Franceschini, 1999; Franceschini and Mukherjee, 1999). Research has been ongoing in this area; a variety of approaches have been investigated, including disaggregation methods designed to avoid introducing correlation issues (Franceschini et al., 2000) and predictive aggregation/disaggregation algorithms (Chua and Low, 2009). A systematic comparison of the results produced by multiple models at different levels of resolution on a set of identical combat scenarios revealed that differences in spatial representation, force aggregation, and time step can lead to different outcomes; this suggests potential sources of results correlation error in multi-resolution models (Hillestad et al., 1995). Detailed investigation into the fundamental issues associated with maintaining consistency between multiple levels of resolution led to several important observations about multi-resolution modeling, including the need to handle concurrent interactions at multiple levels of resolution and the potential for time step differences between levels to introduce inconsistencies (Reynolds et al., 1997; Natrajan, 2000). Multi-resolution entities, which can exist and interact at multiple levels of resolution concurrently and maintain an internal state that is consistent across those levels, were proposed as a solution. The useful notion of mapping functions, which map object attributes from one level of resolution to another, was introduced in the context of this research.

Resolution Flexibility

Most existing multi-resolution combat models exploit the military echelon hierarchy when defining resolution levels that pervade the simulation (i.e. every object, including the synthetic natural environment, is represented at the "company level" or the "entity level"). Increasing resolution flexibility aims at creating multi-resolution combat models that can be built and controlled on a per-data-item basis. For example, the resolution of the synthetic natural environment does not need to be directly tied to the resolution of military units maneuvering in it—it may be useful for a variety of environment resolutions to be available to a single military unit model. This is obviously a more complex problem than exploiting the military echelon hierarchy for resolution levels, but there would be a big payoff in terms of flexibility if such systems could be built.

Inter-Level Interactions

As mentioned earlier, existing multi-resolution combat models have some limited inter-level interactions between entities. The most common example is indirect fire from an aggregate artillery battery against a disaggregated company. However, the question remains whether other more complex inter-level interactions could be implemented in a useful way. For example, would it make sense to allow a tank (at the disaggregated entity level) to engage in direct fire combat with an aggregate company? However, neither implementing inter-level interactions nor understanding their impact on results correlation is simple (Reynolds et al., 1997). Unlike the familiar and well understood mathematical models of combat at the unit level and the natural resolution and data supported models of combat at the entity level, there is no theoretical or experiential basis for direct inter-level interactions. The most focused study of inter-level interactions to date found that the best available mechanism for implementing such interactions was generally some form of pseudo-disaggregation (Petty and Franceschini, 1998a).

Resolution Change Triggers

Finally resolution change triggers need to be explored more fully. Because a huge number of data items can be represented at different detail levels in a multi-resolution combat model, the user must be given sophisticated tools to be able to select a resolution level for individual data items efficiently and effectively. Many automated triggers suffer from the spreading disaggregation problem (also known as chain disaggregation), in which the disaggregation of one entity causes the disaggregation of other entities, and spreading across the simulation. In the worst case, the advantages of multi-resolution combat modeling are defeated because all of the entities will end up being represented at the disaggregate level. Research conducted in this area has developed and compared different disaggregation triggers, including operator selection, fixed geographic areas, and proximity thresholds between units (Williams et al., 1998). More work is needed on devising semi-automated or automated approaches that provide the user with useful ways of managing resolution changes without introducing effects like spreading disaggregation.

Software Complexity

Because they are implemented using distributed simulation interoperability protocols, such as DIS or HLA, to link the unit level and entity level models an extra layer of software complexity is introduced (Franceschini and Petty, 1995b). Because they are linkages of two different models that may be maintained by separate organizations, configuration management effort is potentially doubled.

RELATED RESEARCH

Related research into broader multi-resolution modeling has gone on as well.[32] A study based on the development of a series of increasingly abstract mathematical models for a specific military scenario (precision weapons used to defeat an attacking ground force) produced a number of findings relevant to multi-resolution modeling in general, including the power of characterizing the models with formal mathematics, the utility of distinguishing among phases of military operations when developing aggregate models, and the notion of hierarchical trees of variables where child variables in the tree are more detailed aspects of the information aggregated in their parent variable (Davis and Bigelow, 1998). Elements of the latter idea have been developed further into the concept of "integrated hierarchical variable resolution," and the simplifying assumption that all object representations within a single model are at the same level of resolution has been set aside to outline a more general multi-resolution structure (Davis, 2005). Starting from the analogy of using motion capture from a person to drive the movements of an animated person, space–time constraints were studied as a mechanism to use the behaviors of an object at one level of resolution, e.g. an entity, to drive the behavior of an object at another level of resolution, e.g. a unit (Reynolds, 2002). Multi-resolution modeling was found to provide "crucial enabling capabilities" for model validation when the latter is understood not as a demonstration of model correctness (or incorrectness), but rather as a categorization of the range of applications for which a model has been found to be useful (Bigelow and Davis, 2003). A proposed common framework for specifying components to better support interoperability was studied using a multi-resolution model as a test case (Muguira and Tolk, 2006). An examination of the relationships between composability and multi-resolution modeling concluded that ontological techniques to document the scope, representation, and structure of models may improve both (Davis and Tolk, 2007).

CONCLUSION

Multi-resolution combat modeling is an approach and an architecture for modeling combat scenarios when both large scenario size and high resolution entity

---

[32]Space limits in this chapter preclude any attempt at a comprehensive literature survey of multi-resolution modeling. What follows is at best an arbitrary sampling of a few of the important results. The scope of multi-resolution research in general goes well beyond the "unit level model linked to entity level model" system that was the focus of much of this chapter.

level modeling are needed. The value of the approach is evident from the number of distinct implementations of multi-resolution combat models and their continuing development and use. Research challenges remain which, when solved, will further enhance the flexibility and accuracy of such models.

ACKNOWLEDGMENTS

Early research and development into multi-resolution combat modeling was sponsored by the US Army Simulation, Training, and Instrumentation Command (now Program Executive Office for Simulation, Training, and Instrumentation) and the US Army Training and Doctrine Command Analysis Center. Preparation of this chapter was supported by the University of Alabama in Huntsville's Center for Modeling, Simulation, and Analysis. An anonymous reviewer of an early draft of the chapter provided valuable comments that substantially improved its content and coverage.

REFERENCES

Anderson LB, Bracken J, Healy JG, Hutzler MJ and Kerlin EP. *IDA Ground–Air Model I (IDAGAM I)*, Report R-199, Institute for Defense Analysis, Alexandria VA, 1974.

Babcock DB, Molnar JM, Selix GS, Conrad R, Castle M, Dunbar J, Gendreau S, Irvin T, Uzelac M and Matone J (1994). Constructive to virtual simulation interconnection for the Sofnet-JCM interface project. In: *Proceedings of the 16th Interservice/Industry Training Systems and Education Conference, Orlando FL, November 28–December 1*. National Training and Simulation Association.

Banks CM (2010). Introduction to modeling and simulation. In: *Modeling and Simulation Fundamentals: Theoretical Underpinnings and Practical Domains*, edited by JA Sokolowski and CM Banks. John Wiley & Sons, Hoboken, NJ, pp. 1–24.

Beeker E (2000). An HLA experiment in disaggregation. In: *Proceedings of the Spring 2000 Simulation Interoperability Workshop, Orlando FL, March 21–31*. http://www.sisostds.org.

Bigelow JH and Davis PK (2003). *Implications for Model Validation of Multiresolution, Multiperspective Modeling (MRMPM) and Exploratory Analysis*. RAND, Santa Monica, CA.

Bonder S (1967). The Lanchester attrition-rate coefficient. *Oper Res* **15**, 221–232.

Bonder S and Farrell RL (eds) (1970). *Development of Models for Defense Planning*. Report SRL 2147, TR-70-2, AD 714 667. Systems Research Laboratory, University of Michigan, Ann Arbor, MI.

Bowers FA and Prochnow DL (2003a). Multi-resolution modeling in the JTLS-JCATS federation. In: *Proceedings of the Fall 2003 Simulation Interoperability Workshop, Orlando FL, September 14–19*. http://www.sisostds.org.

Bowers FA and Prochnow DL (2003b). JTLS-JCATS federation support of emergency response training. In: *Proceedings of the 2003 Winter Simulation Conference, New Orleans LA, December 7–10*. IEEE, Piscataway, NJ.

Bowers FA, Prochnow DL and Roberts J (2002). JTLS-JCATS: design of a multi-resolution federation for multi-level training. In: *Proceedings of the Fall 2002 Simulation Interoperability Workshop, Orlando FL, September 8–13*. http://www.sisostds.org.

Calder RB and Evans AB (1994). Construction of a corps level CGF. In: *Proceedings of the Fourth Conference on Computer Generated Forces and Behavioral Representation, Orlando FL, May 4–6*, Institute for Simulation and Training, pp. 487–496.

Calder RB, Peacock JC, Panagos J and Johnson TE (1995a). Integration of constructive, virtual, live, and engineering simulations in the JPSD CLCGF. In: *Proceedings of the Fifth Conference on*

Computer Generated Forces and Behavioral Representation, Orlando FL, May 9–11, Institute for Simulation and Training, pp. 71–82.

Calder RB, Peacock JC, Wise BP, Stanzione T, Chamberlain F and Panagos J (1995b). Implementation of a dynamic aggregation/deaggregation process in the JPSD CLCGF. In: *Proceedings of the Fifth Conference on Computer Generated Forces and Behavioral Representation, Orlando FL, May 9–11*, Institute for Simulation and Training, pp. 83–91.

Castro PE, Antonsson E, Clements DT, Coolahan JE, Ho Y-C, Horter MA, Khosla PK, Lee J, Mitchiner JL, Petty MD, Starr S, Wu CL and Zeigler BP (2002). Modeling and simulation research and development topics. In: *Modeling and Simulation in Manufacturing and Defense Acquisition: Pathways to Success*, edited by PE Castro et al. National Research Council, National Academy Press, Washington, DC, pp. 77–102.

Cayirci E (2009). Multi-resolution federations in support of operational and higher level combined/joint computer assisted exercises. In: *Proceedings of the 2009 Winter Simulation Conference, Austin TX, December 13–16*. IEEE, Piscataway, NJ, pp. 1787–1797.

Chua BYW and Low MYH (2009). Predictive algorithms for aggregation and disaggregation in mixed mode simulation. In: *Proceedings of the 2009 Winter Simulation Conference, Austin TX, December 13–16*. IEEE, Piscataway, NJ, pp. 1356–1365.

Clark KJ and Brewer D (1994). Bridging the gap between aggregate level and object level exercises. In: *Proceedings of the Fourth Conference on Computer Generated Forces and Behavioral Representation, Orlando FL, May 4–6*, Institute for Simulation and Training, pp. 437–442.

Cox A, Maybury J and Weeden N (1995). Aggregation disaggregation research–a UK approach. In: *Proceedings of the 13th Workshop on Standards for the Interoperability of Defense Simulations, Orlando FL, September 18–22*, Institute for Simulation and Training, pp. 449–464.

Dare DP and James BAP (1971). *The Derivation of Some Parameters for a Corps/Division Model From a Battle Group Model*. Defense Operational Analysis Establishment, West Byfleet, UK, M7120.

Davis PK (1993). *An Introduction to Variable-Resolution Modeling and Cross-Resolution Model Connection*. RAND R-4252-DARPA.

Davis PK (1995a). *Aggregation, Disaggregation, and the 3:1 Rule in Ground Combat*. RAND MR-638-AF/A/OSD.

Davis PK (1995b). An introduction to variable-resolution modeling. In: *Warfare Modeling*, edited by J Bracken, M Kress and RE Rosenthal. Military Operations Research Society, Alexandria, VA, pp. 5–35.

Davis PK (2005). *Introduction to Multiresolution, Multiperspective Modeling (MRMPM) and Exploratory Analysis*. Working Paper WR-224, RAND National Defense Research Institute, Santa Monica, CA.

Davis PK and Bigelow JH (1998). *Experiments in Multiresolution Modeling*. RAND National Defense Research Institute, Santa Monica, CA.

Davis PK and Hillestad RJ (1993). Families of models that cross multiple levels of resolution: issues for design, calibration and management. In: *Proceedings of the 1993 Winter Simulation Conference, Los Angeles CA, December 12–15*. IEEE, Piscataway, NJ, pp. 1003–1012.

Davis PK and Tolk A (2007). Observations on new developments in composability and multi-resolution modeling. In: *Proceedings of the 2007 Winter Simulation Conference, Washington DC, December 9–12*. IEEE, Piscataway, NJ, pp. 859–870.

Downes-Martin S (1991). The combinatorics of vehicle level wargaming for senior commanders. In: *Proceedings of the Second Behavioral Representation and Computer Generated Forces Symposium, Orlando FL, May 6–7*, Institute for Simulation and Training.

Dupuy TN (1985). *Numbers, Prediction, and War: Using History to Evaluate Combat Factors and Predict the Outcome of Battles*. Dupuy Institute, Annandale, VA.

Dupuy TN (1995). *Attrition: Forecasting Battle Casualties and Equipment Losses in Modern War*. Nova Publications, Falls Church, VA.

Dupuy TN (1998). *Understanding War: History and Theory of Combat*. Nova Publications, Falls Church, VA.

Foss WR and Franceschini RW (1996). A further revision of the aggregate protocol. In: *Proceedings of the 14th Workshop on Standards for the Interoperability of Defense Simulations, Orlando FL, March 11–15*, pp. 727–738. http://www.sisostds.org.

Fowler BW (1996a). *De Physica Beli: An Introduction to Lanchestrial Attrition Mechanics*, Part I. IIT Research Institute/DMSTTIAC, Report SOAR 96–03.

Fowler BW (1996b). *De Physica Beli: An Introduction to Lanchestrial Attrition Mechanics*, Part II. IIT Research Institute/DMSTTIAC, Report SOAR 96–03.

Fowler BW (1996c). *De Physica Beli: An Introduction to Lanchestrial Attrition Mechanics*, Part III. IIT Research Institute/DMSTTIAC, Report SOAR 96–03.

Franceschini RW (1992). Intelligent placement of disaggregated entities. In: *Proceedings of the 1992 Southeastern Simulation Conference, Pensacola FL, October 22–23*, Institute for Simulation and Training, pp. 20–27.

Franceschini RW (1995). Integrated Eagle/BDS-D: a status report. In: *Proceedings of the Fifth Conference on Computer Generated Forces and Behavioral Representation, Orlando FL, May 9–11*, Institute for Simulation and Training, pp. 21–25.

Franceschini RW (1999). Correlation error in multiple resolution entity simulation, PhD Dissertation, University of Central Florida.

Franceschini RW and Karr CR (1994). Integrated Eagle/BDS-D: results and current work. In: *Proceedings of the 11th Workshop on the Standards for the Interoperability of Defense Simulations, September 26–30, Orlando FL*, Institute for Simulation and Training, pp. 419–423.

Franceschini RW and Mukherjee A (1999). A simple multiple resolution entity simulation. In: *Proceedings of the Eighth Conference on Computer Generated Forces and Behavioral Representation, Orlando FL, May 11–13*, Institute for Simulation and Training, pp. 597–608.

Franceschini RW and Petty MD (1995a) Status report on the development of PDUs to support constructive+virtual linkages. In: *Proceedings of the 12th DIS Workshop on Standards for the Interoperability of Defense Simulations, Orlando FL, March 13–17*, Institute for Simulation and Training, pp. 385–388.

Franceschini RW and Petty MD (1995b). Linking constructive and virtual simulation in DIS. In: *Distributed Interactive Simulation Systems for Simulation and Training in the Aerospace Environment*, SPIE Critical Review 58, Orlando, FL, pp. 281–298.

Franceschini RW, Schricker and Petty MD (1999). New dimensions in simulation. *Mil Train Tech* **4**, 46–50.

Franceschini RW, Wu AS and Mukherjee A (2000). Computational strategies for disaggregation. In: *Proceedings of the Ninth Conference on Computer Generated Forces and Behavioral Representation, Orlando FL, May 16–18*, Institute for Simulation and Training, pp. 543–554.

Hardy D and Healy M (1994). Constructive and virtual interoperation: a technical challenge. In: *Proceedings of the Fourth Conference on Computer Generated Forces and Behavioral Representation, Orlando FL, May 4–6*, Institute for Simulation and Training, pp. 503–507.

Helmbold RL (1965). A modification of Lanchester's equations. *Oper Res* **13**, 857–859.

Hillestad R, Owen J and Blumenthal D (1995). Experiments in variable resolution combat modeling. In: *Warfare Modeling*, edited by J Bracken, M Kress and RE Rosenthal. Military Operations Research Society, Alexandria, VA, pp. 63–86.

Kang K and Roland RJ (1998). Military simulation. In: *Handbook of Simulation: Principles, Methodology, Advances, Applications, and Practice*, edited by J Banks. John Wiley & Sons, New York, NY, pp. 645–658.

Karr CR and Franceschini RW (1994). Status report on the integrated Eagle/BDS-D project. In: *Proceedings of the 1994 Winter Simulation Conference, Orlando FL, December 11–14*. IEEE, Piscataway, NJ, pp. 762–769.

Karr CR and Root ED (1994a). Integrating aggregate and vehicle level simulations. In: *Proceedings of the Fourth Conference on Computer Generated Forces and Behavioral Representation, Orlando FL, May 4–6*, Institute for Simulation and Training, pp. 425–435.

Karr CR and Root ED (1994b). Integrating constructive and virtual simulations. In: *Proceedings of the 16th Interservice/Industry Training Systems and Education Conference, Orlando FL, November 28–December 1*, section 4-4. National Training and Simulation Association.

Karr CR, Franceschini RW, Perumalla KRS and Petty MD (1992). Integrating battlefield simulations of different granularity. In: *Proceedings of the Southeastern Simulation Conference 1992, Pensacola FL, October 22–23*. IEEE, Piscataway, NJ, Institute for Simulation and Training, pp. 48–55.

Karr CR, Franceschini RW, Perumalla KRS and Petty MD (1993). Integrating aggregate and vehicle level simulations. In: *Proceedings of the Third Conference on Computer Generated Forces and Behavioral Representation, Orlando FL, March 17–19*, Institute for Simulation and Training, pp. 231–240.

Kerlin EP and Cole RH (1969). *ATLAS: A Tactical, Logistical, and Air Simulation*. Technical Paper RAC-TP-338, Research Analysis Corporation.

Koopman BO (1999). *Search and Screening*. Military Operations Research Society, Alexandria, VA.

Lanchester FW (1956). Mathematics in warfare. In: *The World of Mathematics*, edited by JR Newman. Simon & Schuster, New York, NY, Vol. 4, pp. 2138–2157.

Longtin MJ (1994). Cover and concealment in ModSAF. In: *Proceedings of the Fourth Conference on Computer Generated Forces and Behavioral Representation, Orlando FL, May 4–6*, Institute for Simulation and Training, pp. 239–247.

Macannuco D, Snow C, Painter RD and Jones JW (2005). Multi-level resolution engagement modeling through a JWARS-JSAF HLA federation. In: *Proceedings of the Spring 2005 Simulation Interoperability Workshop, San Diego CA, April 3–8*. http://www.sisostds.org.

Muguira JA and Tolk A (2006). Applying a methodology to identify structural variances in interoperations. *J Def Model Simulat* **3**, 77–93.

National Research Council (1997). *Technology for the United States Navy and Marine Corps, 2000–2035: Becoming a 21st-Century Force, Volume 9 Modeling and Simulation*. National Academy Press, Washington, DC.

Natrajan A (2000). Consistency maintenance in concurrent representations, PhD Dissertation, University of Virginia.

Odum HT and Odum EC (2000). *Modeling For All Scales: An Introduction to System Simulation*. Academic Press, San Diego, CA.

Peacock JC, Bombardier KC, Panagos J and Johnson TE (1996). The JPSD Corps Level Computer Generated Forces (CLCGF) system project update 1996. In: *Proceedings of the Sixth Conference on Computer Generated Forces and Behavioral Representation, Orlando FL, July 23–25*, Institute for Simulation and Training, pp. 291–301.

Peterson R (1953). Methods of tank combat analysis. In: *Report of the Fifth Tank Conference*, edited by H Goldman and G Zeller. Report 918, AD 46000, Ballistic Research Laboratory, Aberdeen Proving Grounds, MD.

Petty MD (1995). Computer generated forces in distributed interactive simulation. In: *Distributed Interactive Simulation Systems for Simulation and Training in the Aerospace Environment*, edited by TL Clarke. SPIE Optical Engineering Press, Bellingham, WA, Vol. CR58, pp. 251–280.

Petty MD (1997). Computational geometry techniques for terrain reasoning and data distribution problems in distributed battlefield simulation. PhD Dissertation, University of Central Florida.

Petty MD (2009). Behavior generation in semi-automated forces. In: *The PSI Handbook of Virtual Environment Training and Education: Developments for the Military and Beyond; Volume 2: VE Components and Training Technologies*, edited by D Nicholson, D Schmorrow and J Cohn. Praeger Security International, Westport, CT, pp. 189–204.

Petty MD and Franceschini RW (1995). Disaggregation overload and spreading disaggregation in constructive+virtual linkages. In: *Proceedings of the Fifth Conference on Computer Generated Forces and Behavioral Representation, Orlando FL, May 9–11*, Institute for Simulation and Training, pp. 103–111.

Petty MD and Franceschini RW (1998a). *Interactions Across the Aggregate/Disaggregate Simulation Boundary in Multi-Resolution Simulation*. Technical Report, Institute for Simulation and Training, December 20.

Petty MD and Franceschini RW (1998b) *Special Topics in Multi-Resolution Simulation*. Technical Report, Institute for Simulation and Training, December 20.

Petty MD and Gustavson P (2012). Combat modeling using the high level architecture. In: *Engineering Principles of Combat Modeling and Distributed Simulation*, edited by A Tolk. John Wiley & Sons, Hoboken, NJ, 2012.

Petty MD and Mukherjee A (1997). The sieve overlap algorithm for intervisibility determination. In: *Proceedings of the 1997 Spring Simulation Interoperability Workshop, Orlando FL, March 3–7*, pp. 245–255. http://www.sisostds.org.

Petty MD and Panagos J (2008). A unit-level combat resolution algorithm based on entity-level data. In: *Proceedings of the 2008 Interservice/Industry Training, Simulation and Education Conference, Orlando FL, December 1–4*, National Training and Simulation Association, pp. 267–277.

Petty MD, Campbell CE, Franceschini RW and Provost MH (1992). Efficient line of sight determination in polygonal terrain. In: *Proceedings of the 1992 IMAGE VI Conference*, Phoenix AZ, July 14–17, Image Society, pp. 239–253.

Petty MD, Hunt MA and Hardis KC (1996). Terrain correlation measurement for the HLA platform proto-federation. In: *Proceedings of the 15th DIS Workshop on Standards for the Interoperability of Defense Simulations, Orlando FL, September 16–20*, Institute for Simulation and Training, pp. 691–702.

Powell DR (1993). Eagle II: a prototype for multi-resolution combat modeling. In: *Proceedings of the Third Conference on Computer Generated Forces and Behavioral Representation, Orlando FL, May 17–19*, pp. 221–230.

Powell DR (1997). Control of entity interactions in a hierarchically variable resolution simulation. In: *Proceedings of the Fall 1997 Simulation Interoperability Workshop, Orlando FL, September 8–12*, pp. 21–31. http://www.sisostds.org.

Rajput S, Karr CR, Petty MD and Craft MA (1995). Intervisibility heuristics for computer generated forces. In: *Proceedings of the Fifth Conference on Computer Generated Forces and Behavioral Representation, Orlando FL, May 9–11*, pp. 451–464.

Reynolds PF (2002). Using space-time constraints to guide model interoperability. In: *Proceedings of the Spring 2002 Simulation Interoperability Workshop, Orlando FL, March 10–15*, Institute for Simulation and Training. http://www.sisostds.org.

Reynolds PF, Natrajan A and Srinivasan S (1997). Consistency maintenance in multiresolution simulations. *ACM Trans Model Comp Simul* **7**, 386–392.

Root ED and Karr CR (1994). Displaying aggregate units in a virtual environment. In: *Proceedings of the Fourth Conference on Computer Generated Forces and Behavioral Representation, Orlando FL, May 4–6*, Institute for Simulation and Training, pp. 497–502.

Rumbaugh J (1991). *Object-Oriented Modeling and Design*. Prentice-Hall, Englewood Cliffs, NJ.

Schiavone GA, Nelson RS and Hardis KC (1995). Interoperability issues for terrain databases in distributed interactive simulation. In: *Distributed Interactive Simulation Systems for Simulation and Training in the Aerospace Environment*, edited by TL Clarke. SPIE Critical Reviews of Optical Science and Technology, Vol. CR58, SPIE Press, Bellingham, WA, pp. 281–298.

Schricker SA, Franceschini RW, Stober DR and Nida JD (1996). An architecture for linking aggregate and virtual simulations. In: *Proceedings of the Sixth Conference on Computer Generated Forces and Behavioral Representation, Orlando FL, July 23–25*, Institute for Simulation and Training, pp. 427–434.

Schricker SA, Franceschini RW and Petty MD (1998a). Implementation experiences with the DIS aggregate protocol. In: *Proceedings of the Spring 1998 Simulation Interoperability Workshop, Orlando FL, March 9–13*, pp. 768–776. http://www.sisostds.org.

Schricker SA, Franceschini RW and Adkins MK (1998b). Using pseudo-disaggregation to populate a large battlefield for sensor systems. In: *Proceedings of the Seventh Conference on Computer Generated Forces and Behavioral Representation, Orlando FL, May 12–14*, Institute for Simulation and Training, pp. 475–483.

Seiger T, Holm G and Bergston U (2000). Aggregation/disaggregation modeling in HLA-based multi resolution simulations. In: *Proceedings of the Fall 2000 Simulation Interoperability Workshop, Orlando FL, September 17–22*. http://www.sisostds.org.

Sköld S, Holm G and Mojtahed V (1998). Building an aggregation/disaggregation federation using legacy models. In: *Proceedings of the Fall 1998 Simulation Interoperability Workshp, Orlando FL, September 14–18*. http://www.sisostds.org.

Smith RD (1997). Disaggregation in support of intelligence training. In: *Proceedings of the 19th Interservice/Industry Training, Simulation, and Education Conference, Orlando FL, December 1–4*, National Training and Simulation Association, pp. 712–718.

Smith SH and Petty MD (1992). Controlling autonomous behavior in real-time simulation. In: *Proceedings of the Southeastern Simulation Conference 1992*, Pensacola FL, October 22–23, Society for Computer Simulation, pp. 27–40.

Stober DR, Kraus MK, Foss WF, Franceschini RW and Petty MD (1995). Survey of constructive+virtual models. In: *Proceedings of the Fifth Conference on Computer Generated Forces and Behavioral Representation, Orlando FL, May 9–11,* Institute for Simulation and Training, pp. 93–102.

Stober DR, Schricker SA, Tolley TR and Franceschini RW (1996). Lessons learned on incorporating aggregate simulations into DIS exercises. In: *Proceedings of the 14th Workshop on Standards for the Interoperability of Defense Simulations, Orlando FL, March 11–15*, Institute for Simulation and Training, pp. 391–396.

Taylor JT (1980a). *Lanchester Models of Warfare*, Volume I. Defense Technological Information Center (DTIC), ADA090843, Naval Post Graduate School, Monterey, CA.

Taylor JT (1980b). *Lanchester Models of Warfare*, Volume II. Defense Technological Information Center (DTIC), ADA090843, Naval Post Graduate School, Monterey, CA.

Taylor JT (1981). *Force-on-Force Attrition Modeling*. Military Operations Research Society, Alexandria, VA.

Trinker A (1994). *General Architecture for Interfacing Virtual and Constructive Simulations in DIS Environment*. Technical Report IST-TR-94-28, Institute for Simulation and Training, September 14.

Washburn AR (2002). *Search and Detection* (4th edition). Operations Research Section, INFORMS, Baltimore, MD.

Weeden N and Smith M (1996). Aggregation disaggregation research—a UK approach, Update. In: *Proceedings of the 15th Workshop on the Interoperability of Distributed Interactive Simulation, Orlando FL*, Institute for Simulation and Training, September 16–20.

Weeden N, Usher T and Healy M (1996). Demonstration of vertical and horizontal simulation interoperability between ABACUS, ModSAF, NavySF, and JANUS. In: *Proceedings of the 15th Workshop on the Interoperability of Distributed Interactive Simulation, Orlando FL*, Institute for Simulation and Training, September 16–20.

Williams TM, Sale N, Smith M and Shakoor AS (1998). Automated variable representation. In: *Proceedings of the Fall 1998 Simulation Interoperability Workshop, Orlando FL, September 14–18*. http://www.sisostds.org.

Youngren MA (1994). *Military Operations Research Analyst's Handbook, Volume I: Terrain, Unit Movement, and Environment*. Military Operations Research Society, Alexandria, VA.

Zabek A, Henry H, Prochnow D and Wright M (2004). The army constructive training federation multi-resolution modeling: the next generation of land component commander training at the unit of employment echelon of command. In: *Proceedings of the Fall 2004 Simulation Interoperability Workshop, Orlando FL, September 19–24*. http://www.sisostds.org.

Chapter 26

New Challenges: Human, Social, Cultural, and Behavioral Modeling

S. K. Numrich and P. M. Picucci

BACKGROUND AND MOTIVATION

The fall of the Berlin Wall in 1989 ushered in a new era in global security with the United States located at center stage reading from a radically revised script. No single nation could step up to take the place long held by the Soviet Union. No currently standing military could expect to defeat the United States in traditional armed combat. The rules had changed. Today's security landscape is dominated, not by a few large nations capable of fielding large military forces, but by a multiplicity of small nations struggling to build stable environments in the presence of multi-faceted threats. As internal conditions deteriorate as a result of social, economic, and environmental pressures, these nations become breeding grounds for insurgents and transnational groups whose tactics include terrorism. At the same time the distribution of power, political, economic and military, is becoming more diffuse. The strength and influence of the United States as a global power is inextricably linked to the system of alliances and partnerships built and maintained over the past decades in all areas of the world. As a result, the US military must be prepared to be a partner with other US institutions in supporting national security through promoting stability in key regions and providing assistance to nations in need. As a result of all of these trends, the US military has acquired a far more expansive mission set (US Department

Engineering Principles of Combat Modeling and Distributed Simulation, First Edition.
Edited by Andreas Tolk.
© 2012 John Wiley & Sons, Inc. Published 2012 by John Wiley & Sons, Inc.

of Defense, 2010a). In addition to maintaining the capacity to engage in major military conflict, the domain of most combat models, the US military must also be prepared to undertake counter-insurgency operations like that in Afghanistan against the Taliban (US Department of Defense, 2010b). US armed forces are also first responders in humanitarian assistance and disaster relief (HA/DR) missions from Indonesia and Japan in the Pacific to nearby neighbor, Haiti. To assist governments under stress, the US military has determined the capabilities it requires to provide support to stabilization, security, transition to a more stable national situation under the control of local government, and aid in essential reconstruction within the country. These have been formalized in a Joint Operating Concept (JOC) on the Military Support To Stabilization, Security, Transition and Reconstruction (STTR) Operations (US Department of Defense, 2006). The problem of adapting the military system to manage tasks not initially part of the military mission set was posed as the challenge question as part of defining the updated JOC of Logistics in 2010 (US Department of Defense, 2010c).

> *How can Joint Force Commanders and DOD integrate or synchronize and optimize joint, interagency, multinational, nongovernmental, and contracted logistics to simultaneously establish and maintain multiple Joint Force Commanders' operational adaptability and freedom of action in the design, execution and assessment of concurrent combat, security, engagement, and relief and reconstruction missions in an environment characterized by increasing complexity, uncertainty, rapid change, and persistent conflict?*

The JOC's problem statement contains most of the characteristics of modern military activity. It is joint, interagency, and multinational involving nongovernmental and contracted personnel in an environment in which the military is required to provide security in an environment of increasing complexity, uncertainty, rapid change and potentially persistent conflict. Under such circumstances, the notion of combined arms, once understood to be the effective use of air, land, and sea assets, is now augmented to include not just the unpiloted vehicles, but non-kinetic actions such as infrastructure development, economic relief, and nation-building.

Warfare is a tool for achieving an effect, as are all military operations. The desired effect of military domination is usually to gain political or material control of a region, its population and resources. For all its weaponry, military is human at its core; however, in today's expansion of military operations, that human core is becoming an increasingly dominant factor. The adaptive adversaries of irregular warfare gain their advantage by their ability to control a local population, to be enmeshed within and unidentifiable from it, and to compete with the local government for legitimacy in the eyes of the population. Irregular warfare, counter-insurgency, counterterrorism, and stability operations all focus on the population and the tools used to work within that population demand as complete an understanding of the human terrain as maneuver warfare does of the geophysical terrain. Therefore, combat models for today and the future if they are to be relevant to the full range of military operations must include socio-cultural and behavioral modeling to the extent that the human terrain impacts the choice of weapons and the success of the military operation. JOCs include the military capabilities needed for the military operation referenced and thus provide an initial guide to the types of models that might be required in extended combat models. For additional reading about the impact of human dynamics on the military environment, the reader is referred to the text *Operational Culture for the Warfighter* written for use at the Marine Corps University (Salmoni and Holmes-Eber, 2008). The text examines five cultural dimensions, five dimensions of the battlespace that can be examined and included into military planning and execution. It also points to the other social models that are needed for understanding the human dynamics of the battlespace. The dimensions include the physical environment and how a culture views such resources; the economic dimension and how a culture deals with the acquisition and distribution of physical and symbolic goods; the social structure and how a culture organizes its societal relationships; the political structure describing the way in which a culture approaches governance; and the dimension of beliefs and symbols, those factors that frame the local belief system and often have a profound effect on behavior. The missions, the operational environment, and the human environment are the foundational pieces on which the socio-cultural and behavior components of combat models should be based.

MISSION, AREA, LEVEL, AND OPERATOR (MALO) CONCERNS

Conceptual models, such as that provided by "Operational Culture," can provide direction to the warfighter but they do little to provide guidance on how to operationalize their components into existing combat models. Said operationalization requires the marrying of a variety of expertise across military operators, the modeling community, and social scientists. In an effort to facilitate this communication we have suggested, in previous work (Picucci and Numrich, 2010), that modeling and simulation efforts incorporating socio-cultural attributes (and as a consequence combat models that incorporate socio-cultural models) need to focus upon four characteristics: mission, area of responsibility, level of command, and operator expertise. The following discussion defines these characteristics and notes similarities and differences in their application to existing combat models.

Mission

The first characteristic, mission, goes beyond a simple definition of the intent to which the model is being used. At its most simple this characteristic expresses the fact that the socio-cultural concerns to the military will vary depending not only on the broad categorization of mission (HA/DR missions require different modeling and simulation capabilities than do counter-insurgency missions) but also on the characteristics and goals of the specific mission (not all HA/DR missions will emphasize the same concerns). Social sciences often employ fundamentally different modeling approaches to their fields of study depending on the perceived purpose of the work. Models that are geared toward prediction are not the same as those that are targeted at description, explanation, or evaluation.

Area (of Responsibility)

The second characteristic emphasizes that while much of the social sciences are concerned with the discovery of covering laws, it is generally acknowledged that said laws find unique expressions within different communities and populations. The discovery of a specific variable relationship in one state or community does not guarantee that the same relationship holds constant in another state or community with the natural implication that a model that works well as an evaluative tool for an HA/DR mission in Haiti will not necessarily be applicable to a similar mission carried out in Japan.

Level (of Command)

The third characteristic that must be incorporated into successful socio-cultural model development is an understanding of whether the model is being constructed for relevance at the tactical, operational, or strategic levels. This is made more difficult by the understanding that socio-cultural factors at both the micro and macro levels can impact decision making processes, perceptions, and beliefs. Turn-around time requirements, data availability, reach-back and subject matter expert support can vary widely from the strategic to tactical levels, thus impacting the type of modeling effort that should be pursued. The same model that explains macro level economic conditions sufficiently for strategic concerns may be entirely insufficient (or even misleading) if used to support a tactical model that relies on individual perceptions of those conditions.

Operator (Expertise)

The last characteristic reflects a need to understand who will be employing the tool: what their level of socio-cultural expertise and comfort is. The true challenge in this regard is the ability to develop tools that are useful to a wide range of expertise levels. A particular difficulty in this regard is the inclusion of socio-cultural modeling elements that are both understandable and transparent in their effects to those with limited socio-cultural experience as well as sophisticated enough in their integration such that highly proficient individuals will not dismiss the contributions as superficial or artificial.

Of course, these four components are interrelated. For instance, ongoing conflicts in Iraq and Afghanistan have impacted the availability of Foreign Area Officers (FAOs) in other theaters. As a consequence, the area of responsibility and operator expertise factors are interrelated to a high degree. There is also the potential for high correlation between the level of command and operator expertise as the individuals tasked with strategic issues have often had greater access to professional military education opportunities in this area than have individuals tasked with tactical problem sets.

Implications

While the four MALO components represent crucial nodes upon which any given model design team must reach a shared understanding, they also have more broadly based implications. One implication of understanding MALO's impact on the broader range of socio-cultural model inclusion is that models that incorporate socio-cultural elements will be less generalizable. Existing combat models can, and do, utilize non-socio-cultural elements across differing mission sets and areas of responsibility. For example the use of a logistics model to determine personnel and material availability operates in the same manner regardless of mission supported and is altered by the area of responsibility only by boundary conditions imposed by the theater (physical environment, security conditions, etc.). How the model calculates transit times and load levels is unaffected by these factors. The same cannot be said for most socio-cultural models. A task as ubiquitous as identification of local power brokers can require isolation of fundamentally different factors depending on whether one is seeking to identify individuals of influence relevant to an HA/DR mission or those relevant to a counter-insurgency effort. Likewise the direct and indirect power relationships within a society are often highly specific. Building a model generalizable across these domains introduces clear risks. At best, the result will be too abstract for operational employment and at worst the results from such a model could prove misleading. Therefore determining the bounds of this generalizability and verifying appropriateness of a particular instantiation are critical issues to the inclusion of socio-cultural elements within existing combat models.

Likewise, although the problem of scalability across levels of command is not unique to socio-cultural models, there are aspects of socio-cultural modeling that exacerbate the issue. This is most clearly indicated in the phenomena in which macro-structures exhibit behavior very different from the aggregation of their constituent micro-components (Urry, 2005). Not only does the behavior of individuals alter when in groups but the characteristics of that behavior differ depending on the identity of the group but also on what other individuals are in the group at a particular point in time.

A significant issue linked to varying degrees of operator expertise is the risk associated with skepticism toward this form of modeling. The skepticism with which most operators approach socio-cultural models raises the stakes for any implementation. This skepticism can take a number of forms. For those outside of their comfort/expertise area the issue is one of overcoming the bias that socio-cultural elements are not necessary to carrying out military tasks. A more sophisticated but no less dangerous skepticism is that advanced by those who already believe in the utility of socio-cultural understanding. Their skepticism stems from the belief that no model or simulation effort can adequately capture the nuances of culture in a manner that will make the model a useful tool. This puts a substantial burden on the model builders: not only must they incorporate enough transparency to demonstrate utility to the former group, but that transparency cannot come at the price of model complexity that will be necessary to placate the latter group. Furthermore, the pervasiveness of these beliefs means that any misstep along any of a number of points of failure can have far ranging repercussions, not just for the specific model but across the range of socio-cultural model integration efforts.

Non-MALO Concerns

Although the MALO construct serves to illustrate several of the issues inherent in the inclusion of socio-cultural modeling into combat models, there are other factors that further complicate the issue. Combat models are inherently behavioral; alterations in the battlespace are measured in changes in the performance of units within that space; and effects manifest as visible impacts on the behavior of units. By contrast much of socio-cultural modeling needs to focus on incremental changes not in behavior but in opinion, belief, and decision making calculus. While models can depict these kinds of shifts, doing so is rarely reflective of the real-world information operators can be presumed to have. Further, although behavioral changes may result from shifts in these kinds of variables they often do so only once certain "tipping points" have been reached (Gladwell, 2000). These points probably shift in response to perceptions of macro level indicators. It is important to note that perception is a key element (Granberg, 1993) such that even were a combat model to incorporate a number of "objective" socio-cultural indicators those indicators would have to be "interpreted" by the entities within the simulation before the probabilities of behavioral shifts were calculated.

Further difficulties arise from the fundamentally dynamic nature of the entities that would have to be instantiated within existing combat models. Tank, aircraft, or unit behaviors although constrained by external factors (weather, terrain, damage, etc.) change in the degree of capability they have, but their fundamental behavior set does not, generally, alter over the course of a model run. Within the socio-cultural realm this same stricture does not apply. Additionally socio-cultural phenomena are often characterized by the interplay of micro elements on macro conditions and the interpretation of macro conditions on micro behaviors (Mayntz, 2004). These issues largely mandate the inclusion of cognitively complex agents. Unfortunately the design of such agents is a difficult, time consuming task usually resulting in agents that are only appropriate to a specific mission, area, level, and operator expertise. This also impacts existing entities within combat models. If they are not modified such that their behaviors shift in response to changes at the macro and micro levels then the inclusion of socio-cultural modeling to provide those changes cannot alter modeled behaviors.

Compounding these concerns are issues regarding uncertainty both in data measurement and relationship. Given the complex interplay between micro and macro forces, even small degrees of uncertainty can have compounding effects on model behaviors. Finding ways to adequately display uncertainty and properly characterize model results under those conditions present real challenges to this effort. Most serious is the issue how to acculturate users, used to models with relatively low degrees of uncertainty and straightforward cause and effect mechanisms, to the value of appropriate use of socio-cultural informed models.

While the above discussion describes a daunting task, it is worthwhile to note that social science modeling does not have to be a complete real-world simulation. Holding it to that standard is a bridge too far as no modeling effort ever achieves direct one-to-one correspondence with reality. Scoping of the problem through detailed understanding of the MALO elements can bring modelers to a better understanding of what factors will need to be incorporated, which agents will require what levels of complexity, and allow the modelers to meet operator expectations.

UNDERSTANDING SOCIO-CULTURAL DATA

Data required for modeling socio-cultural factors is both similar to and yet different from data routinely used in combat modeling. Like data for combat modeling, socio-cultural data will come in either structured or unstructured formats. Structured data is information that has been reduced to numerical values found in tables or data sheets of various sorts. In combat modeling unstructured data are generally needed to describe how humans function in the combat environment and are most often found in manuals that describe military doctrine or the tactics, techniques, and procedures (TTPs) used in specific combat environments. Performance data used to specify the way entities function in kinetic simulations is most often formatted—lists of numerical specifications for such factors as speed, turn radius, a weapon's range and kill ratio. In both cases, the provenance of the data is clear and performance data are often certified by appropriate military authorities. The data that govern the performance are ascribed to a specific entity, a particular fixed wing aircraft, or a ground-launched weapon, and the way it is employed in the battle conforms to the TTPs for employment under the particular context. Under normal circumstances both performance and employment data are available and are sufficient for modeling the engagements specified in the scenarios.

Socio-cultural data come in both structured and unstructured forms, but there are no certifying authorities for their correctness and there are likely to be notable gaps in their availability. Engagements in traditional combat modeling involve situations in which one or more weapons of known capability are directed at one or more targets of known vulnerability. Statistical means are used to represent the uncertainties. More importantly, the engagements are military against military. Including a more comprehensive view of the human dynamics changes the problem from one-on-one to many-on-many, where the sides interacting include the local government, internal military forces, external military forces, leaders of internal factions, influences from external connections, and the local population. Each faction involved has long-term motivations, short-term goals, and perceptions of what a desired end-state might be—driven by ethnic, religious, or cultural perspectives. This translates into a situation in which the socio-cultural behaviors are not only entity specific, but highly context-driven in situations where behaviors are impacted by a multiplicity of factors that influence human perception and reaction. For centuries, the military has understood the necessity of understanding the mind and perceptions of the enemy. What is new is the need to understand the perceptions of the various factions of the population.

Thus socio-cultural modeling involves both the situation and the way the different factions in the population perceive both the situation and the evolution of the situation in time from their perspective. The situation might be specified by information found in standard statistical tables describing the economy, the availability of education or medical care, the availability of food, water or housing, even birth and death rates. In some cases, models might be used to provide dynamics not present in statistical tables. In nearly all cases of interest where there is a population at risk, essential data at the local level are apt to be outdated (last tabulated 10 or more years ago) or missing altogether.

The situation as measured by all means and techniques and quantified by nationally and internationally accepted standards is only part of the story. The way the different factions of the population perceive and react to the situation is at the core of working effectively in the human environment. A simple example can provide a graphic, immediately accessible picture of the power of perception.

Consider the following situation. The national government promises to provide a guaranteed four-hour period each day at a given hour of reliable electric power to every home in the nation. For many developing nations around the world, four hours of reliable electric power would be a boon; however, in major industrial nations having only four hours of reliable electric power a day would be a disaster. Further, let us consider that the promise is made with a guarantee that it would persist for the next four months. Again, for populations not accustomed to reliable electric power, the promise of having daily power for the next four months would be a major improvement in their lives and they would consider the government successful at least at the level of providing energy infrastructure. Back in the industrial nation, the failure to remedy the situation of only four hours of reliable power for a four-month period would be totally unacceptable and worthy of a no-confidence vote against the government.

The situation is identical, but the perception and consequent reaction are based upon expectations conditioned by experience and societal norms. When taking into account the human dimension, it is not sufficient to know the situational data; it is also critical to understand how the different groups in the area perceive that situation.

Perceptual Data

Irregular warfare is defined by Joint Publication 1–02 as "a violent struggle among state and non-state actors for legitimacy and influence over the relevant population(s)" (US Department of Defense, 2001). Acquiring data on the current perceptions of various relevant factions in a population is the first part of the problem, but having a functional understanding of those perceptions requires going a step beyond. Knowing perceptions is perquisite to being able to influence them to achieve an end, and the ability to influence perceptions necessitates an understanding of those long-term factors that condition the formation of perceptions. As a result, understanding perceptions has a historical and cultural context that may not be apparent in the data themselves.

Historical information provides a window on both the formation of perceptions and the behaviors driven by them. There is also a dynamic associated with perceptions. Conditions change, attitudes change, external influences change—and all of them drive perceptions and actions. Knowing how a faction reacted to similar circumstances in the past may not determine current actions, but as good Bayesians we believe that the past provides indicators. Cultural, societal, and religious norms are strong drivers of behavior and they tend to be more persistent. Understanding what a cultural, ethnic, or religious group values and how it deals with property and leadership, with crime and redress, provides a necessary foundation for working effectively with the population and their leadership.

With context firmly in hand, there are a number of means for probing the perceptions of the population. In a literate society, the media is a valuable resource for understanding the perceptions of the population, particularly the vernacular media. Translation brings with it several sources of bias. The very fact that something is being translated implies that it is intended for consumption by a foreign audience and selection criteria include both what the media moguls feel would interest the foreign audience and what they wish the foreign audience to understand from a particular story. This is particularly true when there is considerable government control of the media and the government has a purpose in controlling its output. Translations themselves induce a bias through inexpert ability to capture a sentiment in a foreign language. Thus the vernacular is the most important source for acquiring perceptual data.

Interviews also provide insight into the attitudes and perceptions of local populations; however, information acquired from these sources must also be carefully interpreted. Subject matter experts, expatriates and focus groups often have their own agendas and biases. Unless they are currently living in the society in question, their assessments are second-hand: valuable, but not necessarily correct in all aspects.

Surveys and polls are extremely valuable, but also subject to error. Populations at stress are often less interested in giving answers to pollsters than they are in just surviving. In repressive environments, people are not likely to be forthcoming about their own reactions and will at times give answers that they feel the pollsters are seeking. There are a number of reliable, professional polling groups who understand how to build questions to elicit the right information from a local population, questions presented by members of their local culture and phrased in ways that make sense in that culture.

Perceptions are not static and are influenced by whether the individual feels that his or her situation is improving or degenerating. This implies that both the situational data and perceptual data must be collected at regular intervals in crisis situations. Crisis situations normally exist in environments where the ability to collect data of any type is seriously hampered by a lack of security, by the state of the infrastructure, and by the ability to access representative parts of the population in spite of difficult conditions. Thus the need for continuous monitoring is greatest under circumstances least amenable to data collection of any sort.

Data Considerations Summarized

Including the human dimension in military tools and models increases demands already placed on data. In socio-cultural assessments, far more of the data must be extracted from unstructured text from diverse sources. Even the situational data are likely to have inconsistencies, lack current information, and be totally non-existent in areas of most critical concern. The human dimension involves human decision makers who function on the basis of perceived reality; therefore, perceptual data take on an importance not found in traditional force-on-force modeling.

In the end, all actions, kinetic and non-kinetic, change perceptions and influence a population. Similarly, what is said and not said at critical times can change perceptions. The ability to work effectively with factions in a population depends upon understanding the context in which their perceptions are formed and being able to work within that context to change perceptions that drive behavior and thus change the operational environment to accomplish the military mission, whatever its nature.

IN SEARCH OF A COMMON VOCABULARY

The ability to collect, store and use data effectively is dependent upon some type of framework, a common vocabulary or taxonomy, that permits all participants regardless of the individual backgrounds to understand and correctly interpret the information that is being exchanged among humans first and then among processors in a computational environment. There is no such common vocabulary among the various social sciences and the situation only becomes more complicated when the military are added to the conversation. The Defense Science Board summarizes the problem in the following way:

> *Technologies to support an understanding of human dynamics lie at the intersection of a broad set of disciplines: the social sciences (anthropology, sociology, political science, history, and economics), the biological sciences (neurobiology), and the*

mathematical sciences (computer science, graph theory, statistics, and mathematics). These typically independent disciplines have distinct histories, terminologies, methodologies (observational versus experimental) and evaluation approaches (quantitative versus qualitative), which sometimes lead to inconsistent practices, outcomes, and/or recommendations. (US Department of Defense, 2009)

A vocabulary used by one community will be imbued with the perspectives and methods of that community. This will be true in the vocabularies explored in the rest of this section. The first vocabulary discussed is drawn from a computational architecture and the second finds its origin in the military mission space, but both include both situational and perceptual data.

A Vocabulary to Support Analysis

Conflict Modeling, Planning and Outcomes Exploration (COMPOEX) was developed by the Defense Advanced Research Projects Agency (DARPA) in response to the need to plan complex military operations in environments that must be represented by large-scale systems of systems that include non-physical systems (e.g. political and military systems). All of these systems are considered to be adaptive and to exhibit emergent or unexpected behavior. The COMPOEX framework was designed to provide an adaptive and robust approach to planning based on a comprehensive study of both the structure and dynamics of the systems in the environment (Waltz, 2008).

The vocabulary built for COMPOEX begins with the military's deconstruction of influences in terms of political, military, economic, social, information, and infrastructure (PMESII). These factors are characterized as either attributes in the virtual or simulated world, or as political or social actors. Table 26.1 shows those factors that are actor-oriented and include perceptual data. In Table 26.2, the factors describe the operational environment and tend to be purely situational.

An Operational Perspective

The data categorizations from COMPOEX were created based on a military construct PMESII, but were developed by analysts who design capabilities that

Table 26.1 Actor-Oriented Data Sets—Partially Perceptual

Political	Regional influences
	National government
	Government institutions
	Local government
	Military organization
	Criminal network
	Non-governmental organizations (NGOs)
Social	Population segment attitude

Table 26.2   Factors Describing the Virtual World—Situational

Economic	National macro-economy
	Meso-economy
Infrastructure	Electrical power
	Telecommunications
	Water service
	Sanitation service
	Healthcare services
	Education services
	Manufacturing
	Agriculture
	Construction
	Food production and distribution
	Transportation networks
Information	Media sources
	Media channels
Military	Security by rule of law
	Military deployment
	Military engagement
	Insurgent targeting

support decision makers. The framework we are about to examine emerged from a concerted effort to understand the needs of Marines whose missions required them to work in foreign environments and to address those needs from the perspective of both the military operator and social science theory.

Operational Culture for the Warfighter (Salmoni and Holmes-Eber, 2008) is written from the perspective of anthropologists working with Marines who bring to the table use cases from around the world that span a broad spectrum of operations, including combat operations, counter-insurgency, stability and reconstruction, humanitarian assistance and disaster relief, and training and operating with foreign forces. The authors use three different descriptive models from anthropology in combination to create a framework designed to help the Marine understand, work with, and potentially influence peoples from different cultures. The three models are ecological, social structure, and symbolic, and are defined by the authors in the following way:

- *Ecological model*: focuses on the relationship between cultures and the physical environment
- *Social structure model*: examines the way the social structure of a group affects the roles, status, and power of various members
- *Symbolic model*: studies the beliefs, symbols, and rituals of a cultural group (Salmoni and Holmes-Eber, 2008)

Each of the three models looks at a socio-cultural group from a different perspective. The ecological model is strongly economic, dealing with how the group works with elements of the environment and the ownership and production of commodities of value, even hi-tech commodities. It looks at conflict as a logical outcome of competition for scarce resources. The manner in which a culture deals with commodities of value has a direct impact on how the people will respond to reconstruction, development, and relief operations handled by the military.

The social structure model's perspective is one of power and influence—how a society organizes its structures and relationships. From the perspective of this model, war and violence are a result of power struggles in which losing groups continually challenge the existing system to acquire better access to goods or to shift their positions in the power structure. The social structure model provides valuable insight into insurgencies and civil strife.

The developers of the symbolic model are oriented more toward ideology and take the position that culture is the product of thought and the construction of belief systems and values. Physical environments and social structures do not illuminate how a society functions; rather, these relationships emerge from the ideals and beliefs that guide the choices of the individual and group. The symbolic model helps explain the psycho-emotional reasons behind actions and conflicts in the society. Therefore, the symbolic model is particularly important in understanding conflicts involving ideological and religious motivations.

The models are important for providing a robust scientific foundation for understanding violence, its causes and potential means for resolution; however, they are too abstract to apply directly to the development of military plans and actions. The authors take these three models and expand them into five specific cultural dimensions that can be observed in areas where the Marines operate and incorporated into assessments of the local situation for purposes of planning. Table 26.3 lays out the expansion of the models into dimensions as expressed in Salmoni and Holmes-Eber's text but groups together several tables and texts.

In Table 26.3 the symbolic model points directly to perceptual information; however, in each of the other dimensions, the way the structures are developed forms and is formed by the world view of the people. Therefore, in working all of these dimensions, there are situations to be observed and reactions to the situations framed by the world view of the population. Corruption and nepotism can be observed and measured (Transparency International produces indices on corruption annually (Transparency International, 2010)), but working effectively with a population requires knowledge of how that society views what we call corruption. Appendix B in *Operational Culture for the Warfighter* is uniquely valuable in that it presents the types of information one can observe in specific tactical situations and relates them to the elements of the framework. For the modeler, this information is akin to having a textbook on sensors and their operational importance, only in this case the sensor is the human. The framework exposed by Salmoni and Holmes-Eber is expressly directed toward understanding the culture, important for the Marines who have to navigate the unknown and often dangerous waters of peace-making and peace-keeping in a foreign culture,

Table 26.3   Framework for *Operational Culture for the Warfighter* (Salmoni and Holmes-Eber, 2008)

Dimension	Explanation	Specific Components
Ecological model		
1: The physical environment	The way a cultural group determines the use of physical resources. Who has access, and how the culture views these resources in terms of access and ownership	Water, land, food, materials for shelter, climate and seasons (the way they stress and frame the use of the environment), fuel and power
2: The economic	The way people in a culture obtain, produce, and distribute physical and symbolic goods (food, clothing, cars, or cowrie shells)	Formal and informal systems Systems of exchange networks Relationship systems
Social structure model		
3: The social structure	How people organize their political, economic, and social relationships, and the way this organization influences the distribution of positions, roles, status, and power within cultural groups	How the culture handles: age, gender, class, kinship, ethnicity, religious membership
4: The political structure	The political structure of a cultural group and the unique forms of leadership within such structures (bands, councils, hereditary chiefdoms, electoral systems, etc.). The distinction between formal, ideal political structures and actual power structures	Political organization, cultural forms of leadership, challenges to political structures
Symbolic model		
5: Beliefs and symbols	The cultural beliefs that influence a person's world view; and the rituals, symbols, and practices associated with a particular belief system. These include also the role of local belief systems and religions in controlling and affecting behavior	History, memory, folklore Icons, rituals Symbols and communication Norms, mores, taboosReligious beliefs

but equally important for the modeler who needs to imbue his actors with realistic behaviors that are situationally, operationally, and culturally appropriate.

The Value of Vocabularies and Frameworks for Understanding

Why spend so much effort looking at different vocabularies and frameworks? Vocabularies are essential for framing problems and for developing the data structures for architectures. The fact that there are currently many distinct vocabularies that do not readily map into one another is an indication of the immaturity of the field of computational social science. The problem stated by the Defense Science Board (US Department of Defense, 2009) is real. We have cultural divisions across the disciplines that encompass the perspectives from which we study human dynamics, and those divisions must be bridged before we can create a holistic approach, a single vocabulary or framework for incorporating social sciences into military processes. Each vocabulary will be formed from the perspective of the developer, but regardless of the source, no vocabulary, framework or taxonomy for socio-cultural behavior can be functional without provision for both situational and perceptual data.

ROLE OF ARCHITECTURES

What is architecture? According to the US Department of Defense, architecture is the structure of components in a program or system, their interrelationships, and the principles and guidelines governing their design and evolution over time (DoD 5000.59-P, reference (g)) (US Department of Defense, 1998). Architecture enables us to break down a complex problem into its simpler component parts and then recombine those parts to reformulate the complete, end-to-end system of systems. The individual components play their roles alongside and often interacting with the other pieces. However, architectures depend upon data structures and metadata rooted in the vocabularies used to deconstruct the problems. Even in the presence of a commonly accepted vocabulary, establishing semantic understanding among the reassembled pieces is a significant problem in socio-cultural modeling as well as in traditional combat modeling.

Fundamental to any architecture is enabling the interaction of individual components or models through exchange of data—common meaning and format –otherwise termed lexical and syntactic interoperability. This is a minimal standard and does not guarantee a meaningful, integrated output. One of the earlier, widely used architectures for distributed simulations, the Distributed Interactive Simulation (DIS) protocol, IEEE 1278, did not specifically include any technical means for establishing consistent meaning; however, those using this protocol to create simulations engaged in a careful review of "enumerations," the specific list of variables to be transferred using the DIS protocol data unit

(PDU). Similarly, when using the HLA, the meaningful exchange of data is established using the Federation Object Model (FOM) which, like the enumerations, is created in the process of standing up a particular simulation event. In both cases, the architecture specifications enable lexical and syntactic consistency, but human processes are needed to create meaningful interoperability or semantic consistency.

The reason for bringing meaning and semantic consistency to the fore in this discussion is the need for common vocabularies to create semantic understanding across the disparate disciplines involved in socio-cultural modeling and simulation. If model builders from different disciplines talk past each other when examining the problem and use different methodologies to create their models, there is a significant possibility that combining the models through the exchange of formatted data will not create a meaningful system or synthetic world. Semantic consistency implies that across the entire group of models being brought together for the simulation there are no conflicting assumptions buried in the models. The process of ferreting out inconsistencies in assumptions involves building a common vocabulary for discourse and, finally, for creating the data structures for the architectures.

Both architectures to be discussed provide for lexical and syntactic consistency, but rely on human processes to establish semantic consistency. COMPOEX is designed for execution of models that run much faster than real time and does not provide for distribution across multiple, separate computers. The HLA is designed to enable the individual models to run on independent platforms, allowing latency to be the limiting factor in geographic distribution. Human, social, cultural, and behavioral (HSCB) data and models using and responsive to HSCB data can be integrated into both architectures. These are not the only architectures being used for military simulation; however, they have been used more widely than some other systems and they serve distinct communities of interest.

Architectures have a very specific role in addressing problems. An architecture is not a solution; it is an enabler to a solution process. In his discussion on the challenges that HSCB issues pose for modeling and simulation, Tolk (2009) calls attention to NATO's Code of Best Practices for C2 Assessment and its focus on the process of problem solving. The first and most critical step is the formulation of the problem and the accompanying strategy for its solution. Simulations involving diverse models are developed because the solution to a problem demands that the problem space be parsed into discrete pieces each of which can be solved individually and all of which are essential to the complete understanding of the problem. The architecture is a means of bringing together those specific disparate pieces that are germane to the solution of the problem at hand. Architectures are enablers and they are effective to the extent that they create an environment in which the appropriate modules can communicate and interact. Consideration of which architecture is to be used is secondary and follows upon determining what method and modules are appropriate to the solution of a particular problem.

COMPOEX: An Analytic Perspective

COMPOEX was developed to permit the analyst to explore many potential futures and to gain an understanding of how different circumstances impact the outcomes of specific actions. The use of the architecture by DARPA was to enable rapid course of action analyses for military planning. As indicated earlier, the vocabulary that provides the framework for COMPOEX is based on the PMESII concept of the environment, a choice that reflects the orientation of the military analyst. Tables 26.1 and 26.2 show how the human actors and virtual world are laid out in terms of the PMESII actions.

COMPOEX was also designed to enable multiple levels of resolution. In irregular warfare, the goals are set at the strategic level, but events at the local level impact national-level thinking. Any nation is a system of systems where local activities combine in some way to produce the national environment—politically, economically, socially and militarily. COMPOEX acknowledges this by enabling analyses at three levels: local, district or provincial, and national. Where there are models that produce local information that can be aggregated to produce results at the provincial or national level, the local models 'report to' the aggregation algorithm and the results of the aggregation are passed up to the higher level. Interaction among the models is enabled at the data level where models report their results and stand ready to execute when data used for input changes. COMPOEX synchronizes at distinct time steps, at which point the models report and seek new data. Models also have time synchronization that is specific to their algorithms. If, for example, an economic model is designed to use data accrued over a three-month period, it waits until that time is marked by the COMPOEX time synchronization to seek data and execute. If the time step for COMPOEX execution is one week, then a model that executes every three months would wait on the order of 12 time steps before execution and would report back either in the next cycle or at a time delay representative of the real-world process.

Figure 26.1 provides an architectural view of the information presented in Tables 26.1 and 26.2. It also adds the concept of multi-resolution execution. The figure is adapted from one shown by Hartley (2010) in his validation analysis of COMPOEX. The colors in the background separate different PMESII domains; however, the illustration groups the Political, Social and Economic domains together at the left, leaving the remaining three separate and in order as Infrastructure, Information and Military. Notice particularly the interaction of the three local economic models as they produce results that go both to the interaction plane as well as to the aggregation algorithm. The aggregation algorithm reports through the interaction plane up to the provincial area power struggles. When aggregate models use local data, they can access them through the interaction plane as do PS1 and E1 in the illustration. In Figure 26.1, the political and social actions are on the left and the virtual world computations begin with economics and continue to the right.

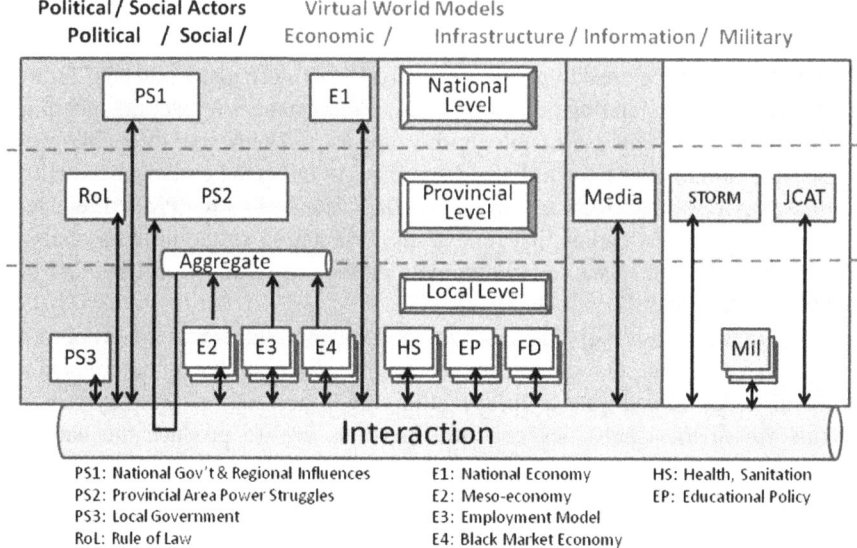

Figure 26.1   COMPOEX model framework.

In Figure 26.2, the concept of the COMPOEX architecture is turned on its side as if Figure 26.1 were rotated 90° clockwise, thereby placing the factors that compose the virtual world on the bottom and the interaction of the socio-political world on the top. Figure 26.2 emphasizes the different types of models that were used in the DARPA program and is adapted from Figure 2–6 in the paper by Waltz (2008). The network of nodes and lines in the upper portion of the figure represents interaction among the social and political actors engaged in a struggle for power and influence. The computational framework for these interactions is an agent based model developed by Taylor et al. (2006). The models that provide the virtual world computations come from various disciplines and methodologies, and include systems dynamics, Bayesian nets, Petri nets, Markov models and discrete time models. The agents interacting at the socio-political level react to the situational factors in the virtual world as they are presented through the conversion layer into the type of data that can be assimilated by the decision structure programmed into the agents.

The terminology of stocks, sources, and sinks is reflective of the systems dynamics approach used in bringing together diverse elements that comprise an interconnected system. The entire framework is grounded upon the theory of political power struggle. This theory is one of the three different anthropological models, the social structure model, used in developing the vocabulary for operational culture. Agent based models implement decision structures as part of the artificial intelligence that allows the agents to interact meaningfully with each other and with the environment. Implementing one model to motivate an agent's decisions is difficult, but has been done (Taylor et al., 2006). Using several models simultaneously would require a very complicated decision structure,

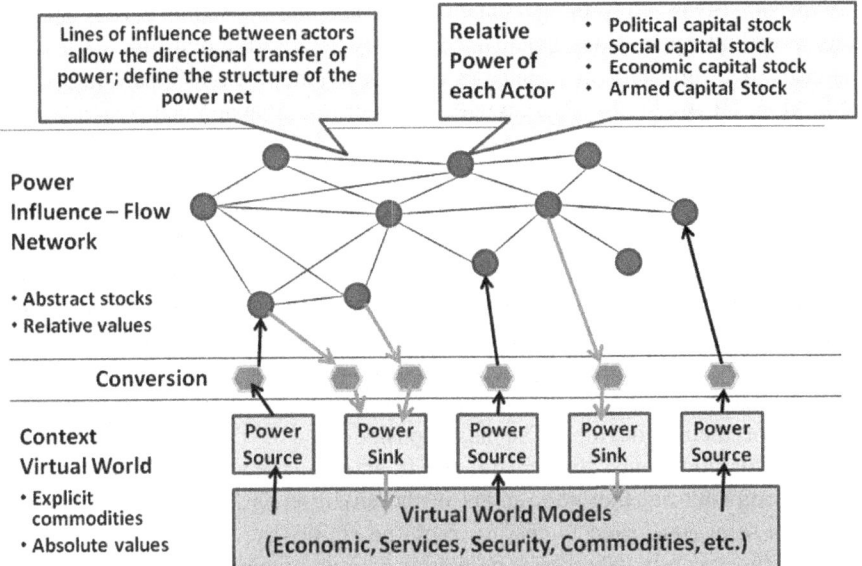

Figure 26.2 COMPOEX multi-resolution execution environment.

one in which an agent could at one point be making decisions as a political entity seeking power, and at another point the same agent's decision process could be dominated by a cultural or ideological tenet. In addition, the cultural orientation of specific populations would require that the rules in the decision structure be adapted to each new culture present in the environment. While this use of the COMPOEX architecture relies on the social structure model, the architecture itself is not limited by that model and could implement a different theoretical model by employing different heuristics governing the decision making in the agent based modes and by using an appropriate set of models in the associated virtual world.

COMPOEX was designed to create an environment in which models could be assembled to address a particular problem. Ideally, the architecture would be populated by a large number of models of different sorts and the analyst could compose his simulation based on the models that best fit his current application. While this, of course, is possible, it is not quite as simple as it sounds, because every new combination of models creates a new system that has to be tested for validity and sensitivity to the particular variables under scrutiny. It should be noted that this is true of all simulation environments, not just COMPOEX.

Adding HSCB Capability to an HLA Federation

In HLA-enabled simulations, there is a common synthetic world or battlespace in which entities of different types interact with one another and with the elements of the synthetic battlespace. The complications introduced by adding the human dynamics involves determining what entities to add and at what levels the interactions should take place. While the HLA is most often used in real-time simulation exercises to permit humans to interact with the synthetic environment in a natural fashion, HLA can also operate with different time management schemas that would allow faster than real-time execution. For this examination of the HLA, real-time execution will be assumed.

Traditionally, the battlespace is populated with Blue and Red forces—friendly and adversary entities. Irregular warfare requires that there be a focus on all the groups functioning in an operational area; thus, the first data that would come into play would be demographics. For each population group included, a representative number of individuals would have to be modeled together with their culturally determined decision making. This statement assumes that individuals and the groups to which they belong are not identical in terms of their behaviors. The next consideration is the decision structure, the artificial intelligence that creates the interaction between the human actor, his environment, and the other actors in the environment. HLA has often been used with semi-automated forces in which the entities can work without human intervention or can be taken over by a human as an operator directly controlling elements of the entity's behavior. Initial decisions in building HSCB federations would have to address how much intelligence would be built into which entities, and whether to permit human operators to take over certain entities.

The agents would have to be sensitive to the socio-cultural environment. Again, this is not exactly new for semi-automated forces. Vehicles are built to read terrain data and many respond to environmental data of various sorts, including dynamic changes in those data. The data exchange mechanisms of publish and subscribe allow this type of dynamism. The idea of separating the problem into environmental data and intelligent agents that react to those data is reminiscent of work done nearly a decade ago to develop environmental servers (Clark et al., 2002). The models that produced the situational data in the weather environment were hidden behind a server. Weather data could be supplied either by compute-intensive weather models running on massive computers elsewhere or live feeds from operational systems, providing data streams either prior to or during execution. The server would simply make the data from any source available to the simulated entities, which would then consume the data and behave accordingly. While the decision structure for human entities or agents executing in the environment would necessarily be more complex than that of a ship, tank, or aircraft, the concept of the server would remain the same, but in the case of simulations representing socio-cultural processes the environmental server might be used for economic or political environmental data.

The advantage of the server is that all entities get the same information, but the models providing that information could be replaced as better models are found. The server allows the environment to be dynamic without changing the way in which the entities access information. The application of servers to human behavior is not new and has been used by Silverman to allow human players in

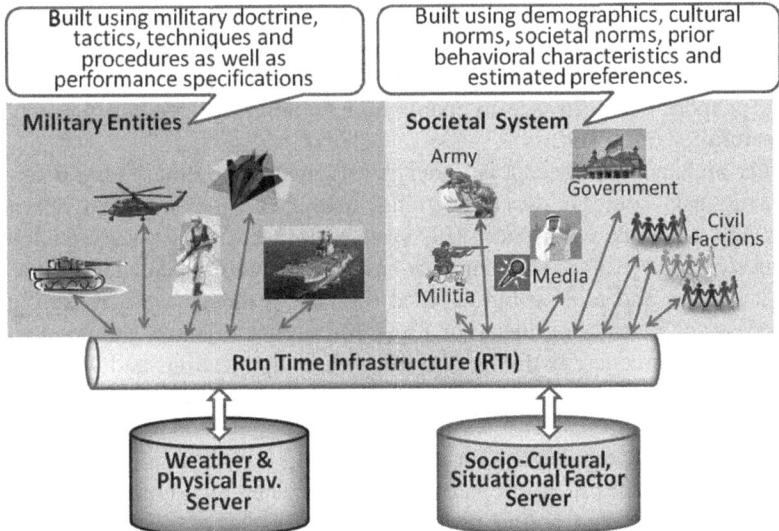

Figure 26.3   Socio-cultural server and societal agents in an HLA federation.

games to experience such factors as fatigue and stress, his behavior modification functions (Silverman et al., 2006). However, these behavior modifiers are distinctly different from situational factors in a socio-cultural environment. The notion of this type of application in an HLA federation is explored in greater detail by Numrich (2010). Figure 26.3 taken from Silverman et al. (2006) shows a notional concept of the HLA architecture separating the situational and perceptual data.

The HLA architecture should support the development of an agent–server system using the concept of situational and perceptual data. However, designers would have to exercise restraint in choosing what to simulate in this fashion, at what level of aggregation the simulation should be composed, and which theories of behavior should animate the agents. To create a federation that exhibits the consistency essential for validity, these considerations need to be addressed and documented at the outset.

Deciding About Architectures

Both architectures discussed above are capable of handling HSCB data and applications. They differ in their approaches because they come at the problem from different perspectives. These are not the only architectures currently being used to handle socio-cultural information. At least two other architectures are in use and under continued development, specifically to include HSCB data and models.

The Dynamic Information Architecture System (DIAS), originally designed to represent geospatially oriented natural systems, was linked with another object-based framework, Framework for Addressing Cooperative Extended Transactions

(FACET), to implement societal behaviors of individuals and groups interacting within the complex natural environment provided by DIAS (Christiansen, 2000). Both of these models were developed at Argonne National Laboratory and have been used for more than a decade to enable human entities to interact in a rich, dynamic natural environment.

A newer architectural concept is under development using the Web 2.0 services to wrap data sources, data processing, analytical modeling, and report generation into a flexible workspace. This system, SORASCS (Service Oriented Architecture for Socio-Cultural Systems), is based on the notion that a large number of data sources and tools could be used together if there were a relatively simple process for assembling them for use. This service-oriented architectural approach will accommodate both thin and thick client applications and will provide some common services and tools that can be added easily and seamlessly (Garlan et al., 2009).

There are multiple architectures for many reasons, but at least in part because they serve different types of users and applications. They all support syntactic and lexical consistency, but before there can be a semantic consistency within an architecture there has to be some level of agreement on vocabularies, the meaning of the terms used to describe the models and data that are integrated into the architectures. In an assembly of social scientists, there would be few if any who were willing to admit that the social science disciplines are ready to agree on a single vocabulary. Given this lack of agreement at the level of vocabulary, why then the press for an architecture?

The issue has been pushed by members of the military who are facing complex situations on a daily basis and who need tools to help them interpret the observable events in their environments and turn that understanding into effective action. The Air Force Office of Scientific Research and the Human Effectiveness Laboratory at the Air Force Research Laboratory chartered a study by the US National Academy of Sciences on "Behavioral Modeling and Simulation: From Individuals to Societies" published in 2008 (Zacharias et al., 2008). After an extensive review of the state of the art, the study board recommended an integrated, cross-disciplinary research program to include focused study on how to federate models and modeling components across levels of aggregation and detail. The report highlighted the need for semantic interoperability and the need for architectural testbeds for experimenting with different concepts of federation. Tolk (2009) urged the development of a community of interest to approach the issue of common vocabulary as well as an understanding of the nature and variety of models available to provide understanding of the societal factors involved in modern warfare.

While there are several architectures in which the human dynamics of combat modeling can be represented, none of them provides semantic consistency and all of them require considerable skill and knowledge on the part of the user and/or group building the federated system. Recalling the MALO principles, this fact would indicate that there are few systems that would be ready for routine military use. It may be that the best use of architectures today is to serve as a sandbox for experimentation—a means for working with and learning from the models we currently have and for developing design criteria for the next generation of models and architectures. We lack laboratories in which we can try experiments and test models and theories when we begin combining cross-disciplinary capabilities. Architectures can provide those synthetic laboratory environments for learning.

One of the terms that has emerged as the military simulation community shifts focus to examination of societal modeling is *hybrid modeling and simulation*. This term is properly associated with the composition of a simulation from different types of models and it should not be confused with the concept of hybrid warfare—a mixture of kinetic and non-kinetic means to achieve military objectives. The term *hybrid modeling* is not uniquely defined. A collection of models at different levels of aggregation but focused on a single topic or discipline, e.g. microeconomics models feeding a macroeconomic model, would be an example of hybrid modeling. Similarly, economic models and political models feeding the same environment and possibly sharing some of the same data would be a different sort of hybrid model in which data and models of different domains are linked to create a more holistic environment. An agent based model that has agents that represent individuals as well as societal structures like governmental departments or political parties would be an example of a hybrid model that uses the same computational framework but very different heuristics to represent specific parts of the society being examined. Just as there is no single approach to defining the socio-cultural environment, there is no generalized approach to creating a valid hybrid modeling environment. There is no formal or even consistent practice for determining which models to assemble, even if there were a preferred architecture for handling the assembled models. The manner in which socio-cultural factors are introduced into combat simulations, whether as heuristics behind the decision making of simulated leaders or commanders or population factions or as a means of establishing a dynamic situational context that interacts with and can be impacted by the actors in the simulated world, is not defined. The foundational challenge is to understand what can and cannot be modeled and whether introducing a partial solution helps solve or obfuscates the problem under exploration. Hybrid modeling that includes socio-cultural models as well as the tried and tested combat models is and will remain a challenge area for simulation for some years in the future.

CONCLUSIONS AND REMAINING CHALLENGES

Socio-cultural modeling and simulation requires that we bring together capabilities from a broad range of scientific disciplines, but we lack the connective tissue to make them work seamlessly together. The disciplines from which the capabilities arise have different vocabularies; they interpret events differently; their concepts of measurement, data, and models differ; and all of these differ from what the military user expresses as his needs. To frame the problems well, we must begin to engage in the type of cross-talk from which a common vocabulary arises. Salmoni and Holmes-Eber (2008) engaged in such a dialog in the development of operational culture. Language is the first problem and it has a direct impact on the data structures for architectures.

The National Academy study also indicated that there is a problem of realistic expectations on both the social science and military side of the discussions. The military, long accustomed to kinetic models that result from deterministic equations, expect generalizable, predictive models, but are confronted by social scientists who present descriptive models and explanations. Social sciences contribute valuable understanding to the context of irregular warfare, but it is neither predictive nor generalizable. Applicability is limited by

- the nature of the mission (M)
- the area of operation (geographic regions and their societal components) (A)
- the levels of command (strategic, operational, and tactical) (L)
- and the knowledge and skill level of the end user or operator (O)

These very real considerations, MALO, not only limit generalizability of models and simulation, but provide what may also be a useful framework for deconstructing problems. Picucci and Numrich (2010) discuss these concepts at length. Socio-cultural information is needed for many military missions and the mission can narrow the scope of the vocabulary, data, and models required to help the military user gain essential insight whether at home training for deployment on a particular mission or in the midst of a mission and planning for the next set of actions. A counter-insurgency mission does not require the same information and understanding as does a disaster relief mission. Responding to a tidal wave in Indonesia, a hurricane in Haiti or an infrastructure failure in a developed nation all require different approaches based upon the geography, the demographics, the motivations and capabilities of the population, and a determination of current needs in each case. The approach that worked yesterday in Iraq may not work today in Afghanistan. In fact, the actions that produced beneficial results last week in Kandahar may not work there this week because conditions have changed and the reactions of the population and the insurgency toward NATO forces have changed. The area of operation makes a difference, geographically and culturally, and the environment is dynamic, adaptive, and reactive to a multiplicity of factors exogenous to military activities.

The National Academy study and the use of multi-resolution modeling as a driving concept behind COMPOEX represent the recognition on the part of the scientific community that resolution and granularity are essential issues for data and for models. The technical community looking at human behavior looks at individual, group, and society. The military divides the problem along strategic, operational, and tactical levels of operation. Both concepts, while distinctly different, point to the need for multi-resolution modeling. Research has not as yet yielded a general method to aggregate from individual to group, because an individual does not always exhibit the same behaviors when in the presence of others. The extent to which local information can and should influence strategic and operational decisions is a hotly debated issue in many military circles under the guise of distributed versus central command and control. Both communities, military and research, understand the importance of aggregation, and both the scientific and military levels of aggregation will have to be represented in models that include socio-cultural factors.

The research communities in their rush to get capability to the military have not fully understood the limitations of resources available at different types of military installations. Research grade tools that require continual contact with the developer are not practical for deployed forces at any level. The military will always experience limitations in the availability of communication channels (readily accessible bandwidth) and qualified human resources. Architectures and the capabilities they bring together have to be designed with the end user in mind, and may never be deployable beyond a command headquarters where there is a skill base to use the application. When data acquisition is a problem, that problem cannot be handed over with tools to the operational user who has neither time nor resources to begin a quest for data. The end user in the field, on the other hand, has valuable data, but no simple means of entering them into a system where he or other users can then retrieve them. The community is still working on a methodology to better cope with these issues (Tolk et al., 2010).

Taking these MALO limitations into account can help both the social scientist and the military user to communicate more effectively about the nature of the problem and the potential for providing suitable tools designed to improve understanding about the socio-cultural environment of concern to the military.

Architectures running faster than real-time applications to enable analytic study and real-time architectures that allow the user to interact with the synthetic environment exist and can be used with HSCB data and models; however, the architecture that provides a sandbox for studying socio-cultural modeling as it applies to military needs is likely to be the most valuable architecture for the next decade as the research and military communities gain expertise in handling problems with socio-cultural implications.

The desire to bring socio-cultural information into simulations of the operational environment for the military brings new urgency to creating a robust understanding of hybrid modeling and its limitations. Whether the objective is to engage in multi-resolution modeling within the same technical domain (e.g. economics or human decision making) or to combine models from different technical domains to create a more holistic human environment, there are no generalizable methodologies for integrating such models or for assessing the adequacy of the resultant federation. Micro-economic models do not aggregate into a macro-economic model. Different methodologies grounded in different theoretical approaches are used in micro- and macro-economics. Similarly the behavior of the individual does not simply aggregate to form a group. Individuals not only behave differently when in a group, the same individual is apt to adjust his response in different groups based on his role in the group and relationships to other group members. There are numerous challenges facing the modeling community, but there, as the demand for socio-cultural modeling increases, research can bring new understanding, novel methods, and potential solutions. Success will require that military users, computational modelers, and domain experts in the social sciences engage in dialog to first specify their problems well based on military mission requirements and the demands of divers operational environments. Novel computational methods and engineering solutions alone cannot conquer this problem space which is inherently interdisciplinary and multi-faceted. Time spent in interdisciplinary and multidisciplinary contexts to clarify the problems of socio-cultural modeling will produce dividends in time and resources saved in the processes of meeting the many challenges facing researchers, modelers, and military in this critical domain (Numrich and Tolk, 2010).

REFERENCES

Christiansen JH (2000). A flexible object-based software framework for modeling complex systems with interacting natural and societal processes. In: *Proceedings of the 4th International Conference on Integrating GIS and Environmental Modeling (GIS/EM4): Problems, Prospects and Research Needs, Banff, Canada, September 2000*.

Clark DL, Numrich SK, Howard R and Purser G (2002). Meaningful interoperability and the synthetic natural environment. In: *Proceedings of the 2002 European Simulation Interoperability Workshop*. http://www.sisostds.org.

Garlan D, Carley K, Schmerl B, Bigrigg M and Celiku O (2009). Using service-oriented architecture for socio-cultural analysis. In: *Proceedings of the 21st International Conference on Software Engineering and Knowledge Engineering (SEKE2009), Boston, July*.

Gladwell M (2000). *The Tipping Point: How Little Things Can Make a Big Difference*. Little, Brown and Company.

Granberg D (1993). Political perception: explorations in political psychology. In: *Explorations in Political Psychology*, edited by S Iyengar and WJ McGuire. Duke University Press.

Hartley DS (2010). *DARPA COMPOEX VV&A*. http://home.comcast.net/~dshartley3/COMPOEX/compoex.htm (last accessed 9 August 2010).

Mayntz R (2004). Mechanisms in the analysis of social macro-phenomena. *Phil Soc Sci* **34**.

Numrich SK (2010). How might socio-cultural modeling fit into distributed simulations. In: *Proceedings of the Fall Simulation Interoperability Workshop*. http://www.sisostds.org.

Numrich SK and Tolk A (2010). Challenges for human, social, cultural and behavioral modeling. *SCS M&S Magazine*, Quarterly Magazine of the Society for Modeling and Simulation International, January.

Picucci PM and Numrich SK (2010). Mission-driven needs: understanding the military relevance of socio-cultural capabilities. In: *Proceedings of the 2010 Winter Simulation Conference, December*. IEEE, Piscataway, NJ.

Salmoni BA and Holmes-Eber P (2008). *Operational Culture for the Warfighter: Principles and Applications*. Marine Corps University Press, Quantico, VA.

Silverman BG, Gharathy G, O'Brien K and Cornwell J (2006). Human behavior models for agents in simulators and games: part ii—gamebot engineering with PMFserv. *Presence-Teleop Virt* **15**.

Taylor G, Bechtel R, Morgan G and Waltz E (2006). A framework for modeling social power structures. In: *Proceedings of Conference of the North American Association for Computational Social and Organizational Sciences, June 22–23*.

Tolk A (2009). Emerging M&S challenges for human, social, cultural and behavioral modeling. In: *Proceedings of the Summer Computer Simulation Conference*. IEEE, Piscataway, NJ.

Tolk A, Davis PK, Huiskamp W, Schaub H, Klein GL and Wall JA (2010). Challenges of human, social, cultural, and behavioral modeling (HSCB): how to approach them systematically? In: *Proceedings of the Winter Simulation Conference, Baltimore, MD*. IEEE, Piscataway, NJ, pp. 912–924.

Transparency International (2010). *Corruptions Perceptions Index (CPI)*, Berlin, GE. http://www.transparency.org (last accessed 6 August 2010).

Urry J (2005) The complexity turn. *Theor Cult Soc* **22**, 1.

US Department of Defense (1998). Under Secretary of Defense for Acquisition Technology, *DoD Modeling and Simulation (M&S) Glossary*. http://www.msco.mil/files/MSCO%20Online%20Library/DoD%205000.59-M%20-%20MS%20Glossary1%20-%2019980115.pdf (last accessed 6 August 2010).

US Department of Defense (2001). *Dictionary of Military and Associated Terms, Joint Publication (JP) 1-02, 12 April 2001* (as amended through 19 August 2009).

US Department of Defense (2006). *Military Support to Stabilization, Security, Transition and Reconstruction Operations Joint Operating Concept, Version 2.0*. http://www.dtic.mil/futurejointwarfare/concepts/sstro_joc_v20.doc (last accessed 13 May 2011).

US Department of Defense (2009). Defense Science Board, Office of the Under Secretary of Defense for Acquisition, Technology and Logistics, *Understanding Human Dynamics*. http://www.acq.osd.mil/dsb/reports/ADA495025.pdf (last accessed 20 May 2011).

US Department of Defense (2010a). *Quadrennial Defense Review Report (QDR)*. http://www.defense.gov/qdr/images/QDR_as_of_12Feb10_1000.pdf (last accessed 20 May 2011).

US Department of Defense (2010b). *Irregular Warfare: Countering Irregular Threats Joint Operating Concept, Version 2.0*. http://www.dtic.mil/futurejointwarfare/concepts/iw_joc2_0.pdf (last accessed 20 May 2011).

US Department of Defense (2010c). *Joint Concept for Logistics*. http://www.dtic.mil/futurejointwarfare/concepts/jcl.pdf (last accessed 13 May 2011).

Waltz E (2008). *Situation Analysis and Collaborative Planning for Complex Operations, 13th ICCRTS: C2 for Complex Endeavors*. Technical Paper 151, March 2008.

Zacharias GL, MacMillan J and Van Hemel SB (ed.) (2008). *Behavioral Modeling and Simulation: From Individuals to Societies*. Report of the National Research Council of the National Academies, National Academies Press, Washington, DC.

Chapter 27

Agent Directed Simulation for Combat Modeling and Distributed Simulation

Gnana K. Bharathy, Levent Yilmaz, and Andreas Tolk

This chapter covers three topics. First, multiple synergies between intelligent agent technologies and simulation are defined. Following the introduction of agents, their types, and agent architectures, we present a dichotomy on the use of agents for simulation and the use of simulation for agents. Second, the role and significance of the use of the agent paradigm in combat modeling and distributed simulation is delineated. Specifically, we examine how the agent metaphor can improve combat modeling, and in what domains agent technology can effectively enhance distributed simulation. Finally, we focus on examples of agent supported simulation and agent based modeling in the military domain for applications that support training and analysis for non-kinetic military and political applications. The cases described include a state level agent based model and a village level learning game. The framework on which such models are built is also introduced.

INTRODUCTION

The dynamic and distributed nature of combat simulation applications, the significance of exploratory analysis of complex phenomena, and the need for modeling the micro level interactions, collaboration, and cooperation among real-world entities is bringing a shift in the way systems are being conceptualized (Yilmaz and Ören, 2005). Using intelligent agents in simulation models is based on the idea that it is possible to represent the behavior of active entities in the world

Engineering Principles of Combat Modeling and Distributed Simulation, First Edition.
Edited by Andreas Tolk.
© 2012 John Wiley & Sons, Inc. Published 2012 by John Wiley & Sons, Inc.

in terms of the interactions of an assembly of agents with their own operational autonomy. The possibility to model complex situations whose overall structures emerge from interactions between individual entities and to cause structures on the macro level to emerge from the models at the micro level is making the agent paradigm a critical enabler in modeling and simulation of complex adaptive systems (Miller and Page, 2007).

Recent trends in technology as well as the use of simulation in exploring complex artificial and natural information processes (Denning, 2007; Luck et al., 2003) have made it clear that simulation model fidelity and complexity will continue to increase dramatically in the coming decades. The dynamic and distributed nature of combat simulation applications, the significance of exploratory analysis of complex phenomena (Miller and Page, 2007), and the need for modeling the micro level interactions, collaboration, and cooperation among real-world entities is bringing a shift in the way systems are being conceptualized. The emergent need to model complex situations whose overall structures emerge from interactions between individual entities and cause structures on the macro level to emerge from the models at the micro level is making the agent paradigm a critical enabler in modeling and simulation of complex systems (Epstein, 2006).

This chapter aims to provide a basic overview of potential uses of agents for simulation, as well as the use of simulation technologies to study simulation for agents. Agent systems are defined as systems that are composed of a collection of goal-directed and autonomous physical, human, and logical software agents situated in an organizational context to cooperate via flexible and adaptive interaction and cognitive mechanisms to achieve objectives that cannot be achieved by an individual agent. Agent Directed Simulation (ADS) is promoted as a unified and comprehensive framework that extends the narrow view of using agents simply as system or model specification metaphors (see Figure 27.1). Rather, it is posited that ADS is comprehensive in the integration of agent and simulation technology. ADS consists of three distinct, yet related areas that can be grouped under two categories as follows:

- *Simulation for Agents* involves the use of simulation modeling methodology and technologies to analyze, design, model, simulate, and test agent systems.
- *Agents for Simulation*: (1) agent supported simulation deals with the use of agents as a backend and/or front-end support facility to enable computer assistance in simulation based problem solving; (2) agent based simulation, on the other hand, focuses on the use of agents for the generation of model behavior in a simulation study.

BACKGROUND ON SOFTWARE AGENTS

Software agents are entities that (1) are capable of acting in purely software and/or mixed hardware/software environments; (2) can communicate directly with other

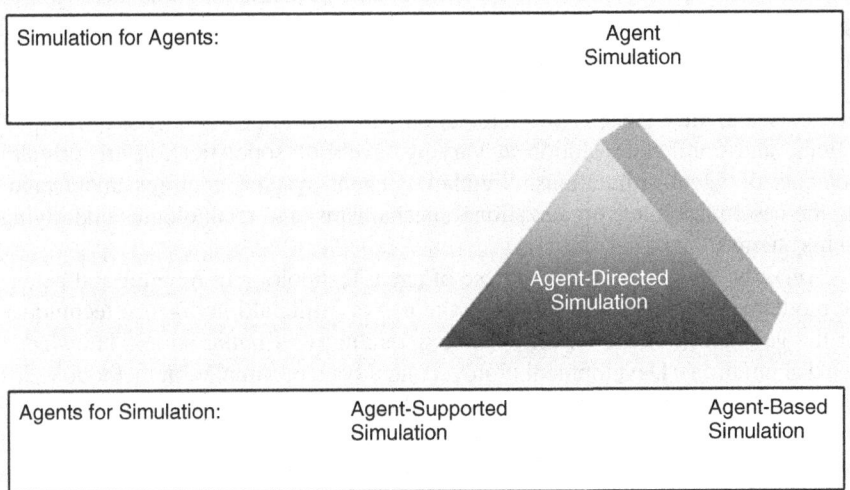

Simulation for Agents:	Agent Simulation	
	Agent-Directed Simulation	
Agents for Simulation:	Agent-Supported Simulation	Agent-Based Simulation

Figure 27.1   Dimensions of ADS.

agents; (3) are driven by a set of goals, objectives, and tendencies; (4) possess skills to offer services; (5) perceive their environment; and (6) can generate autonomous behavior that tends toward satisfying their objectives (Ferber, 1999; Weiss, 1999).

Software Agents

Agents are autonomous software modules with perception and social ability to perform goal-directed knowledge processing over time, on behalf of humans or other agents in software and physical environments. When agents operate in physical environments, they can be used in the implementation of intelligent machines and intelligent systems and can interact with their environment by sensors and effectors. The core knowledge processing abilities of agents include reasoning, motivation, planning, and decision making. The factors that may affect decision making of agents, such as personality, emotions, and cultural backgrounds, can also be embedded in agents. Additional abilities of agents are needed to increase their intelligence and trustworthiness. Abilities to make agents intelligent include anticipation (proactiveness), understanding, learning, and communication in natural and body language. Abilities to make agents trustworthy as well as assuring the sustainability of agent societies include being rational, responsible, and accountable. These lead to rationality, skillfulness, and morality (e.g. ethical agent, moral agent).

While many publications generally talk about agent based simulation as a collective term, it is useful to distinguish between three dimensions of ADS.

Agent simulation involves the use of simulation conceptual frameworks (e.g. discrete event, activity scanning) and technologies to simulate the behavioral dynamics of agent systems by specifying and implementing the behavior of autonomous agents that function in parallel to achieve objectives via goal-directed behavior. In agent based model specifications, agents possess high level inter-action mechanisms independent of the problem being solved. Communication protocols and mechanisms for interaction via task allocation, coordination of actions, and conflict resolution at varying levels of sophistication are primary elements of agent simulations. Simulating agent systems requires understand-ing the basic principles, organizational mechanisms, and technologies underlying such systems.

Agent based simulation is the use of agent technology to monitor and gener-ate model behavior. This is similar to the use of artificial intelligence techniques for the generation of model behavior (e.g. qualitative simulation and knowledge based simulation). Development of novel and advanced simulation methodologies such as multi-simulation (Yilmaz and Phillips, 2007) suggests the use of intelli-gent agents as simulator coordinators, where runtime decisions for model staging and updating take place to facilitate dynamic composability. The perception fea-ture of agents makes them pertinent for monitoring tasks. Also, agent based simulation is useful for having complex experiments and deliberative knowledge processing such as planning, deciding, and reasoning.

Agent supported simulation deals with the use of agents as a support facility to augment simulations and enable computer assistance by enhancing cogni-tive capabilities in problem specification and solving. Hence, agent supported simulation involves the use of intelligent agents to improve simulation and gam-ing infrastructures or environments. Agent supported simulation is used for the following purposes:

- to provide computer assistance for front-end and/or backend interface func-tions;
- to process elements of a simulation study symbolically (e.g. for consistency checks and built-in reliability); and
- to provide cognitive abilities to the elements of a simulation study, such as learning or understanding abilities.

For instance, in simulations with defense applications, agents are used as support facilities to

- fuse, integrate and deconflict the information presented by the decision maker;
- generate alarms based on the recognition of specific patterns;
- filter, sort, track, and prioritize the disseminated information; and
- generate contingency plans and courses of actions.

Complex System of Systems

The nature of the problems in combat systems engineering is increasingly becom-ing distributed. Supervision of telecommunication, supply chain, and logistics networks, as well as information management subsystems, is distributed across the nodes of the network. Hence, a large number of operators and programs need to work collectively in a cooperative manner to control such systems to achieve desired objectives. Individual operators and programs often have a partial view of the overall system, and their actions should be efficiently coordinated so that the overall system can react in a desired manner.

The following observations regarding the nature of modern and complex system of systems call for mechanisms to distribute activities and intelligence in terms of interaction between relatively independent and autonomous agents.

- *Problems are widely distributed*: Decisions and control actions are not only taken by system entities, but also individuals (actors) carrying out functions at various levels. The allocation of roles to decision makers and the propagation of performance criteria to influence local objectives are critical in specifying the behavior of the system.

- *Systems are becoming goal-directed and adaptive*: In well adapted systems goals and constraints are often implicit and embedded in the process. In a dynamic environment, an effective realization depends on self-organizing and adaptive mechanisms that are in place to change properties of the process to meet the current needs.

- *System processes evolve over time*: Processes are frequently modified to update the structure and mechanisms to keep some measure related to the relevant performance objective near an optimum. Control of adaptation, however, is distributed across all components and subsystems. A useful and credible model for the analysis of the process and prediction of responses to changes in the circumstances must reflect the mechanisms underlying the evolution of the process.

- *Socio-technical systems are human centered*: What people actually do, how they communicate and collaborate, how they solve problems, resolve conflicts, and learn behavior matters in the outcome of a process. Hence, representing activities requires modeling communication, collaboration, teamwork, conflict resolution, and tool and technology usage.

- *Software intensive systems engineering is moving toward designs using concepts of autonomous and interacting entities*: The history of software intensive combat systems development is moving toward the use of abstractions that suggests designs in terms of assemblies of entities, which are widely distributed and autonomous. Objects with autonomy and adaptiveness are presented as the basic elements of design in such systems analysis and design methodologies.

A GENERIC AGENT ARCHITECTURE FOR COMBAT SYSTEMS

Typically, agents supporting other applications may be categorized into four types, namely (1) information agents that are designed to gather and monitor

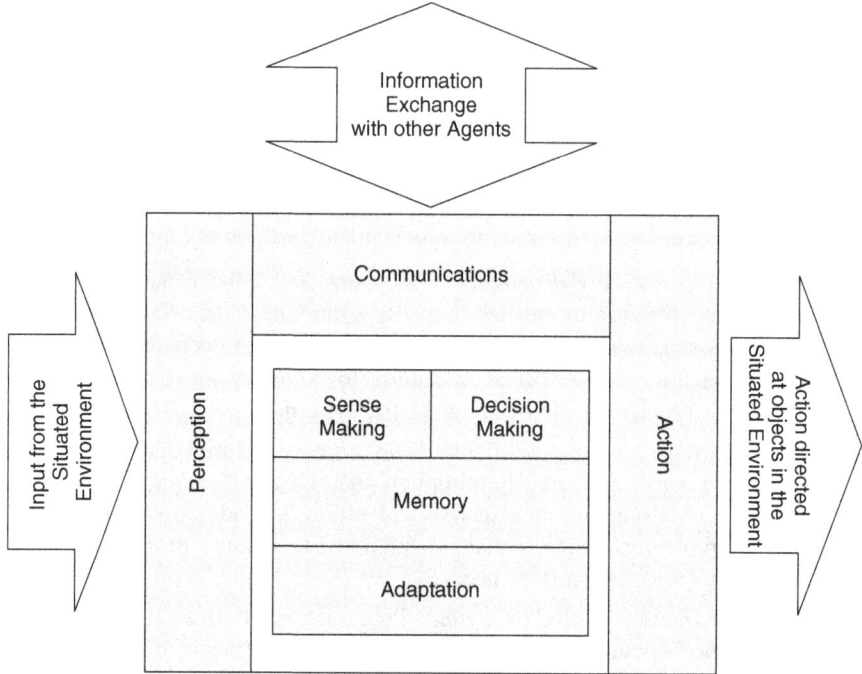

Figure 27.2   A generic agent architecture.

information from the real world; (2) modeling agents which specialize in esti-mating the real-world state; (3) planning agents which produce plans or courses of action from the current state of the world; and (4) transactional and brokering agents who ensure interoperability. Regardless of the type, there is a common structure to high level agent architecture.

The architectural frame for generic agents shown in Figure 27.2 was proposed by Moya and Tolk (2007) in support of discussing how the agent character-istics can be realized. It is kept simple on purpose, as we do not want to prescribe solutions, but just make developers aware of domains that need to be taken into account. If a domain is not covered in a solution, it should not be covered by purposeful design. As such, Figure 27.2 is a guideline for combat system engineers and simulation and agent developers. It is neither complete nor exclusive.

There are three external and four internal architectural domains identified in Moya and Tolk (2007). The external domains comprise those functions needed within an agent to interact with agent's environment:

- The *perception domain* observes the environment. Using its sensors, the agent receives signals from his environment and sends this information to the internal sense making domain.

- The *action domain* comprises the effectors. If the agent acts in his environment, the necessary functions are placed here. He receives the task to perform tasks from the internal decision making domain.

- The *communication domain* exchanges information with other agents or humans. If the agent receives information, it is sent to the internal sense making domain. He receives tasks to send information from the internal decision making domain.

The internal domains categorize the functions needed for the agent to act and adapt as an autonomous object. The four domains identified here are the following:

- The *sense making domain* receives input (sensors and communication) and maps this information to the internal representation that is part of the memory domain. This domain comprises potentially data correlation and data fusion methods, data mediation capabilities, methods to cope with uncertain, incomplete, and contradictory data, etc.

- The *decision making domain* supports reactive as well as deliberative methods, as they have been discussed in this chapter. It uses the information stored in the memory domain and triggers communications and actions.

- The *adaptation domain* may be connected with perception and action as well, but that is not a necessary requirement. The comprised function group updates the information in the memory domain to reflect current goals, tasks, and desires.

- The *memory domain* stores all information needed for the agent to perform his tasks. It is possible to distinguish between long-term and short-term memory, different methods to represent knowledge can be used alternatively or in hybrid modes, etc.

In accordance with the above structure and in addition to specific capabilities required for the given application, agent systems should also incorporate three fundamental capabilities, namely observation, state estimation, and plan and control. Autonomous agents need to observe the state of their world (including contexts) to operate. Often, the observations also include noise as well as signals. The agents also need to estimate the future state space by combining observations of the state of the world and historic information. Frequently, techniques such as Kalman filtering and maximum likelihood estimation solutions to probabilistic expressions of positions are employed. The agent execution typically modeled as combinatorial optimization problems, or in simpler or continuous stage spaces, through dynamic programming algorithms.

This architectural frame is not intended to replace more concrete architectures. It is a blueprint that needs to be adapted and made concrete by applications extending and enhancing this hub based on the requirements for combat systems engineering projects. For example, the critical components of planning, understanding, and anticipation enabling the agent to reason need to be detailed in the decision making component. Similarly, how the learning feature is realized needs to be captured when implementing a solution based on this blueprint.

AGENT SUPPORTED SIMULATION FOR COMBAT DECISION SUPPORT SYSTEMS

In this section, we introduce a case study that illustrates the use of agent technology as a support facility enabling symbiotic decision support. Decision science involves understanding cognitive decision processes, as well as methods and tools that assist decision making. Combat strategy problems are typically characterized by significant uncertainty. Proper simulation based decision support methodologies that are consistent with the way experts use their experience to make decisions in field settings could improve modeling courses of action (COAs), simulating them faster than real time, and then performing COA analysis. Real-time decision making requires models and tools that allow interaction with the system of interest. A symbiotic decision support system is defined as one that interacts with the physical system in a mutually beneficial way. It is highly adaptive and not only performs "what if" experiments to control the physical system but also accepts and responds to data from the physical system. Intelligent agents are used in this study to filter, perceive, and understand the state of the system to drive the simulation system to explore outcomes of potential COAs. Figure 27.3 depicts the architecture of an agent supported simulation based decision support system that illustrates the knowledge generation and machine learning components.

In this study, a multi-resolution coordinated mission for unmanned air vehicles (UAVs) is used to gain experience and identify the means for principled design of such systems. The Recognition Prime Decision Module (RPDM) and COA generator constitute the situation awareness component that involves perception, understanding, and anticipation functions. The COA simulator, which is based on multi-simulation, explores the outcomes of potential COAs recommended by the control agents. We use multi-UAV (MUAV) simulation (see Figure 27.4) as an experimental frame that drives the multi-resolution simulation framework that constitutes the tactical and high resolution simulations. Tactical federate is the component with which the decision maker interacts to (1) monitor situation models generated by the system and (2) use multi-simulation in decision making. The interaction between MUAV and tactical federate is handled by a separate component that clusters the targets to identify battalions and establish teams based on an agent based contract net protocol, which is a distributed task allocation algorithm. The tactical federate uses the information passed from the MUAV to instantiate and to configure a team with a strategy.

The strategy is identified by using a Bayesian network that constitutes the anticipation element of the decision making lifecycle supported by intelligent agents. The team, high resolution, and engagement federates along with MAK Stealth 3D Game Engine constitute the high resolution simulation. Situation

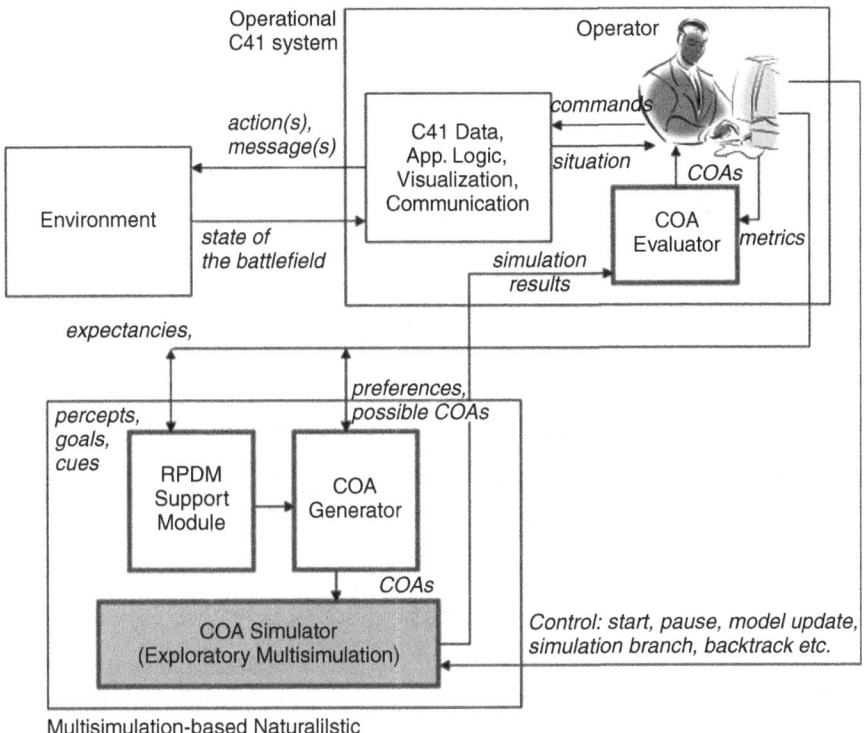

Figure 27.3   Agent supported decision support system.

awareness and controller agents make recommendations for updating the multi-simulation. Situation awareness is defined as the perception of elements in a particular environment within time and space, the comprehension of their meaning, and the projection of their status in the near future. Situation awareness, as depicted here, provides a set of mechanisms that enable attention to cues in the environment. The understanding, diagnosis, and COA generation processes in this study are based on the following phases coordinated by agents: (1) target clustering; (2) situation model (target graph) generation for each UAV; (3) consensus model generation for collective understanding of the situation; and (4) expectancy and COA generation.

AGENT BASED SIMULATION FOR NON-KINETIC DECISION SUPPORT SYSTEMS

Rationale for Non-Kinetic Modeling

Understanding the behavior of individuals and society is becoming increasingly important for the military as well as policy makers. The literature tells us that

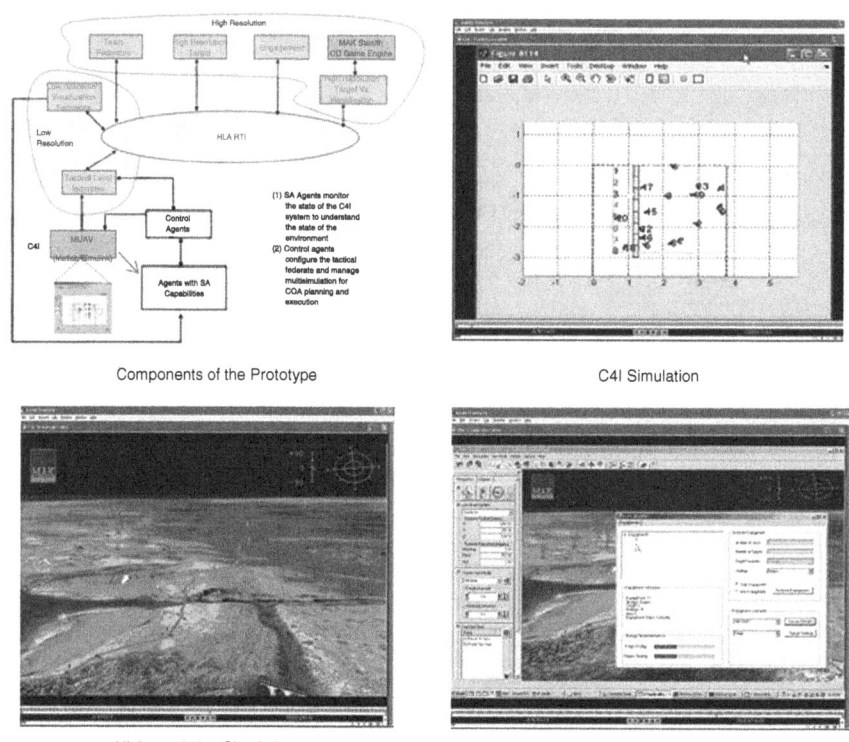

Components of the Prototype   C4I Simulation

High-resolution Simulation   Creating a New Engagement

Figure 27.4 A multi-resolution target engagement scenario.

"the aim of counter-insurgency (and community policing) is not solely to destroy groups at the enemy or criminal end of the spectrum, but also to progressively shift individuals and groups closer to the friendly end." Examples are given in Kilcullen (2006), Petraeus et al. (2006), Nagl (2002), and Chiarelli et al. (2005). Specifically, as a field example, General McChrystal, recent past Commander of the International Security Assistance Force (ISAF) for Afghanistan, routinely issued distinct orders and clear guidance on this subject when he stated, "The conflict will be won by persuading the population, not by destroying the enemy" (Fisher, 2009).

"Blue-on-Blue" or force-on-force warfare is rare and is relegated to the realm of the past, when combat implied fighting a known enemy with guns. Today, the US military is expected to be as much a transformative agent "winning the hearts and minds of people" as a fighting force. Even a US battlefield commander tasked with defeating insurgents is advised to follow doctrinal field manuals 3–24 (Counter-Insurgency or COIN) and 3–07 (Secure, Stabilize, Transition, and Reconstruction, SSTR). He or she must be aware of the hearts and minds of all factions—friendlies, neutrals, and oppositional—if he/she is to succeed in defeating forces against progress and bring about greater goods. The "adversary" includes not just insurgents, but also many non-kinetic elements as well; and ultimate victory is not in eliminating the "enemy" but in bringing the "adversary" to the negotiating table and ultimately transforming them into partners in progress.

Therefore, the forces must be trained to diagnose the source of grievance and appreciate the root causes of conflicts. The forces are often called on to contribute to nation building tasks, including building up institutions to provide public goods and services, the lack of which has been potentially also causing discontent. In policing terms, the counterpart to this is to engage and build up community groups like neighborhood watch, anti-drug programs, and so on since the police have too few resources alone to prevent and mitigate the many situations that might develop.

The adversary to instilling the rule of law and transitioning to host-nation reconstruction is potentially any corrupt or biased individuals or groups who perpetrate tribalism or anarchy in place of progress toward equitable sharing of jobs, services, and resources—any individuals or groups who propagate conflicts, and/or destabilizing patterns.

This means the battlefield commander must become an inquirer into the social design of the area of operation and its ethno-political factions. And this is the reason for modeling the stakeholders (factional groups, leaders, followers, institutions, etc.) and how they might be influenced as outlined.

What Kind of Agent Framework would Support Non-Kinetic Decision Making?

Wouldn't it be ideal if decision makers, strategists, and soldiers in the military get access to decision support and training systems that help explore (e.g. find out which strategy works best) and learn the social systems they are coming into contact with?

Currently, modeling of social systems can be delineated into a number of distinct schools (or research tracks) of modeling, depending on the tools and techniques employed. These include conceptual modeling, logic modeling, statistical modeling, social network analysis, agent based modeling (cellular automata), intelligent agent based modeling, systems dynamics, Bayesian modeling, and discrete event modeling. As can be noted, two of the techniques mentioned above are agent based, i.e. agent based modeling is not a monolithic technique.

The approach (or tools adopted) differs depending on the "granularity of the model and capability of the individual agents," ranging from "spatial models with simple rule based cellular automatons" (often referred to as thin, highly abstracted, dumb, light agents) to cognitively detailed, thick agents (Pahl-Wostl, 2002b). While our primary concern in this chapter is with cognitively detailed, intelligent agents, we will also briefly overview thin or abstracted agents.

Before we delve into suitable agent based modeling systems, let us look at the key dimensions of agent based modeling. (As we shall see in the case we present, more than one tool/technique is employed within a given model, but for the sake of dimensionalization and the ensuing discussion in this section, we shall presume single techniques are employed.)

Agent based models can be categorized according to their respective structural features. For example, the National Research Council's commissioned study on behavioral modeling and simulation recently pointed out five dimensions along which agent based models can be categorized. The five dimensions are (1) level of cognitive sophistication (agency); (2) level of social sophistication (social networks); (3) heterogeneity and number of agents; (4) the use of a grid (and we would add to that a sense of space and mobility); and (5) the means of agent representation (rules versus equations) (Zacharias et al., 2008; Pew and Mayor, 1998).

This is also a suitable place to introduce the abstracted, simpler agent based models and thick agent models. Abstracted models are introduced as a baseline at the light end of the agent cognitive complexity spectrum (details to come later the chapter).

In ensuing sections of this chapter, we briefly outline a number of models, focusing on one example model that we are most familiar with (framework developed by Silverman et al. (2007b) at the University of Pennsylvania for non-kinetic modeling). We will not engage in the relative merits of one framework versus another and the discussion should not be misconstrued as such, because (1) all the frameworks are evolving continuously; (2) our familiarity is skewed as we know some frameworks better than others; and (3) at least one of the authors is associated with the frameworks that we focus on, thereby posing a bias in our evaluation.

Abstracted (Simpler) Agent Frameworks: Cognitive and Social Sophistication

Agent based models built using abstracted, simple agents with a large number of relatively abstracted (simpler) agents tend to be low in cognitive sophistication, but (owing to their numbers) have an advantage in social sophistication in a single dimension over models made using only highly cognitively sophisticated agents. Relatively small numbers of parameters on a large number of agents provide the heterogeneity. Some of these tools tend to provide for rudimentary spatial intelligence.

There are many significant surveys that discuss abstracted (also relatively simpler) agent frameworks (Sekenko and Detlor, 2002; Tobias and Hoffman, 2004; Castle and Crooks, 2006; Railsback et al., 2006; Nikolai and Madey, 2009). Railsback et al. compare the capabilities of a small but key sample of NetLogo, Mason, Repast, and Swarm through a template model built at multiple levels. At each level, the authors add more capabilities and evaluate the toolkits in characteristics such as environmental issues, model structure, agent scheduling, file input and output, random number generation, and statistical capabilities. Castle and Crooks (2006) evaluate Swarm, Mason, Repast, Star-Logo, NetLogo, Obeus, AgentSheets, and AnyLogic with a focus on geospatial capabilities as well as capabilities related to ease of learning, customization and use, pedigree and support. Nikolai and Madey (2009) complement these by taking a broader view based on audit or inspection of available documentation (rather than hands-on experimentation and exploration). Nikolai and Madey consider "language required to program a model and to run a simulation, operating system required to run the toolkit, type of license that governs the toolkit, primary domain for which the toolkit is intended, and types of support available to the user." We refer the readers to the above surveys for additional details (note that there are also a number of other surveys of "simple" agent based model tools).

Some of these tools (e.g. Repast, MASON) also provide ability to customize the agent capabilities, which means, within limits, capabilities can be enhanced.

While these frameworks are relatively simple (and abstracted) in the cognitive dimension, they have some degree of social sophistication, albeit (often) in a single dimension of being connected to other agents through various neighborhood dynamics.

In addition to general purpose simulation toolkits, there have been specialized developments in the military domain. Of particular success in the military domain are ISAAC and Project Albert (and their derivatives).

ISAAC (Irreducible Semi-Autonomous Adaptive Combat), and its derivative EINSTein (Enhanced ISAAC Neural Simulation Toolkit), were developed as conceptual, toy models to explore combats as self-organized complex adaptive systems. Specifically, the tools help complement top-down Lanchesterian models by investigating, albeit at a very high level of abstraction, how "high level emergent behaviors" result from a host of "low level interaction rules" and what tradeoffs can be observed among notional variables. ISAAC was a proof of concept where agents are the basic combat units, equipped with doctrine, mission, situational awareness, and adaptability. The battlefield is represented by a two-dimensional grid. EINSTein is an enhancement with particular emphasis on usability, explorability, and adaptability, making EINSTein a conceptual laboratory for what-if explorations (especially in command and control topology and relevance of information) and, it is hoped, unearthing any holy grails of combat dynamics at a very high level.

The Project Albert was an international initiative sponsored by US military (US Marine Corps, TRADOC et al.) to explore the supportive role of simulation in military planning and decision making, especially in identifying critical human factors and shaping study questions, and it resulted in a proof of principle urban experiment simulated with two of the Project Albert Agent Based Modeling and Simulation (Map Aware Non-uniform Automata (MANA) (Lauren and Stephen 2002) and Pythagoras) and the legacy combat model JANUS.

MANA, initially inspired by ISAAC (Ilachinski, 1997, 2004), was built for New Zealand defense purposes but was quickly adapted elsewhere, including at Monterey's Naval Postgraduate School. Consisting of abstracted (also relatively simpler) agents, easy to use interface, ease of scenario building, fast execution, data farming capability to allow parameter space exploration, and ease of adaptability in Delphi language, MANA has come to synonymize a light agent platform for combat. It is also amenable to exploring network-centric warfare issues. The terrain modeling, for which it uses color-coded bitmaps, is editable during the simulation.

Unlike MANA, Pythagoras, another agent based model environment designed to support Project Albert, allows for relatively more sophisticated representation behaviors based on desires, motivators and detractors, soft rules to create individual decision differences, sense of affiliation to create groups (red, green, and blue) and relationships (friendly, neutrals, or enemies among them), concepts of hierarchy (leader, follower) and hierarchical influence on autonomy, and simulation events/triggers to change behavior sets. Still, as a simulation meant for a military context, it allows for weapons, sensors, communication devices, and terrain. Similarly, JANUS or JSAF are operated on a very large scale and take advantage of the human players to supplement the role-playing agents (i.e. person-in-the-loop).

Cognitive Sophistication and Agency

Of the dimensions of agent based modeling, the first three dimensions are the most salient and also concur with the three dimensions (agency, social networks, and functional heterogeneity) proposed by researchers (Pahl-Wostl, 2002b) to conceptually capture complexity in agents.

Need for Cognitive Sophistication

Let us look at the real world that one is trying to model as a system. There are many ways of classifying systems, and different classifications have different uses. According to Ackoff and Emery (1972), a system could be classified into four types depending on whether the essential parts of a system or the whole can display choice and therefore have purpose. An example of the classification is given in Table 27.1. Through persuasive argument, Ackoff shows that purposeful parts are one of the primary characteristics of both social and ecological systems.

Given that our study here relates to modeling social systems pertaining to the military, purposefulness (or lack of it) is an important consideration. In particular, the purposeful act of agent choice (and purposeful behavior which derives from it) is at the heart of a decision making and agent interaction, and therefore could shed light on how the effectiveness of policies, strategies, and tactics should be

Table 27.1 Classification of Systems According to Ackoff (1990)

	Parts	Whole	Example
Deterministic	No choice	No choice	Clock
Ecological	Choice	No choice	Nature, market
Animate	No choice	Choice	Person
Social	Choice	Choice	Corporation

measured. This characteristic of purposefulness in agents must be embodied in the model to actually be able to simulate the system accurately.

Thus, one key aspect or sub-dimension of cognitive sophistication in agency could be broadly summarized as purposefulness. Having purposeful agency is a qualitative change required in agents. Other aspects of cognition include memory, learning, reasoning, and so on. In these areas, continuous improvement is being made.

Sibbel and Urban (2004) attribute the cause of failure of several modeling endeavors in economic and organizational planning to ignoring human decision making and behavior, and point out the need for including these in modeling through agent technology.

In one case of nitrate pollution by farming agents, Reusser et al. (2004) found that choice of agent rationality outweighed any other types of uncertainty such as those deriving from parameter uncertainties or stochastic effects. Similarly, using abstracted (also relatively simpler) and cognitively detailed (intelligent) models of vampire bats, Paolucci et al. (2006) demonstrated the benefits of modeling agents as cognitively complex entities.

Examples of Intelligent Agent Frameworks with Cognitive Sophistication

ACT-R (Atomic Components of Thought or Adaptive Character of Thought), SOAR (State, Operator, and Results), and PMFserv are examples of agents having cognitively detailed agency.

ACT-R currently is one of the most comprehensive cognitive architectures that initially focused on the development of learning and memory and nowadays increasingly emphasizes the sensory and motor components (i.e. the front-end and backend of cognitive processing) (Anderson, 1983, 1990, 1993; Anderson et al., 2004). SOAR is another sophisticated cognitive architecture that can be used to build highly sophisticated cognitive agents. SOAR is a computational implementation of Newell's unified theory of cognition and focuses on solving problems (Newell, 1990). Agents using SOAR architecture are capable of reactive and deliberative reasoning and are capable of planning.

More recently, various efforts have been made to improve and complement these purely cognitive agents by including some aspects of the affective phenomena that are typically intertwined with human cognition. The effects of emotions on decision making and, more broadly, human behavior and the generation of emotion through cognitive appraisal are most frequently computationally implemented (Hudlicka, 2003).

Examples of cognitive agents with affective capabilities include Silverman et al.'s (2007a) PMFserv (precursor to StateSim and Non-Kin described in this chapter) and MAMID. Other frameworks are also working to improve their affective capabilities, but we shall not discuss them here.

MAMID (Methodology for Analysis and Modeling of Individual Differences) is an integrated symbolic cognitive-affective architecture that models high level decision making. MAMID implements a certain cognitive appraisal process to elicit emotions in response to external stimuli and evaluates the effects of these emotions on various stages of decision making (Hudlicka, 2003).

PMFserv and its derivatives are models of an agent's cognitive-affective state and reasoning abilities that is applied to profile the traits, cognitions, and reasoning of individual leaders, followers, and others. They also manage the markups of the objects, artifacts as well as other contextual settings they perceive and reason about. Built on top of PMPserv are social intelligence, institutional features that characterize the socio-political systems. One can use this model to experiment on and study Diplomatic, Informational, Military and Economic (DIME) (Paolucci et al., 2006) actions that might alter Political, Military, Economic, Social, Informational, Infrastructural (PMESII) effects, and to assess different courses of action on the stability of the groups and actors. StateSim was intended for design inquiry and course of action experimentation. Planners can enter their DIME courses of action in one of two ways—as a US or multinational PMFserv agent taking these actions according to a schedule or set of triggers, or via the Monte Carlo dashboard. Further details on PMFserv and its derivatives are forthcoming in the chapter.

In general, highly sophisticated cognitive agents are modeled based on a computational implementation of one or another overarching theory of human cognition. This approach requires modeling the entire sequence of information processing and decision making steps human beings take, from initial stimuli detection to responses via specific behavior.

Modeling a social system with decision making processes requires integration of several hitherto fragmented models and theories in social sciences. At Silverman's laboratory, such theories are operationalized into performance moderator functions (PMFs). "A performance moderator function (PMF) is a micro-model covering how human performance (e.g. perception, memory, or decision-making) might vary as a function of a single factor (e.g. sleep, temperature, boredom, grievance, and so on)" (Silverman et al., 2007a). PMFserv synthesizes several best available PMFs to characterize micro decision making in the agents. The macro behavior (be it within the individual, groups, or the landscape as a whole) emerges from these micro behaviors. Let us look at the dimensions that matter to agents.

Building Cognitively Detailed Agents and PMFserv Layer

In order to present an agent system, especially how the National Research Council's key dimensions (cognition, social sophistication, number of agents, etc.) of agent capabilities are actually incorporated in an agent system, in this section we introduce PMFserv, a COTS ("Commercial off the Shelf") human behavior emulator that drives agents in simulated game worlds and the "thinking, feeling" decision making, cognitively detailed agents that are central to this framework. This software was developed over the past 10 years at the University of Pennsylvania as architecture to synthesize many best-of-breed models and theories of human behavior in the context of social systems. The unscripted, autonomous agents inhabiting the social system then make decisions in context to drive events in the world.

The modeling framework is best described through a series of layers, built on top of each other. We will start with cognitive sophistication in PMFserv layer, followed by social intelligence in FactionSim, a super module built on top of PMFserv to serve as a generic simulator for social interactions, and finally the further layer that brings in institutional framework to assist in building virtual social systems. Elsewhere, Silverman et al. (2009a, 2010) have discussed how the unifying architecture and how different subsystems are connected. Below, we summarize the salient features of PMFserv.

- *Synthesizing theories:* PMFserv is a model of models that synthesizes a number of scientific theories and models from the literature. It is an environment that "facilitates the codification of alternative theories of factional interaction and the evaluation of policy alternatives" (Silverman et al., 2010). It includes a plug-in architecture that facilitates turning on and off different models and trying new ones. None of these PMFs is "home-grown"; instead they are culled from the literature of the behavioral sciences. "Dozens of best-of-breed PMFs within a unifying mind–body framework are integrated to develop a family of models where micro decisions lead to the emergence of macro behaviors within an individual. These PMFs are synthesized according to the interrelationships between the parts and with each subsystem treated as a system in itself." (Silverman et al., 2009a) An interesting set of psychological, anthropological, sociological, and economic theories have been employed in this framework.

- *Cognitive agents:* PMFserv has decision making agents with cognitive-affective state and reasoning abilities. This enables modelers to represent human behavior in models by profiling the traits, cognitions, and reasoning of individual leaders, followers, and others.

Cognitive agents could be described as the primary agents that occupy this world as cognitively very detailed, endowed with values including short-term goals, long-term preferences, standards of behavior, and personality.

The agent decisions are made by a cognitive appraisal process using multi-attribute utility theory combined with the OCC (Ortony, Clore, and Collins, 1988) or (Ortony et al., 1988) model of emotions to determine how a situation and potential actions might impact an agent's own physiology, personality, cultural values, and relationships. The environment provides contexts. According to Gibson (1979) ecological psychology, these contexts carry and make decisions available for consideration. These agents make decisions based on a minimum of two sets of factors:

- the system of values that an agent employs to evaluate the decision choices and
- the contexts that are associated with the choices.

$$\text{decision utility} = f(\text{values, contexts})$$

The values guide decision choices and, in our case, have been arranged hierarchically or as a network. The contexts sway the agent decisions by providing additional and context-specific utility to the decisions evaluated. The contexts are broken up into micro-contexts. Each micro-context just deals with one dimension of the context (e.g. the relationship between the perceiver and target and so on).

With a given set of values, an agent evaluates the perceived state of the world and the choices it offers under a number of micro-contexts, and appraises the extent to which its values are satisfied or violated. This evaluation finally sums up as utility for decisions. Thus, the decision utilities for all action choices available are calculated as a combination of internal values and external contexts. The decision utility may be sectioned into having different concern types, each of which might be correlated with different sets of emotional arousals.

Value Trees

The agent's aspirations (both short and long term), standards of behavior including cultural values and personality traits are represented through multi-attribute value structures called Goal (short-term concerns), Standards (codes of conduct or methods willing to undertake to attain goals) and Preference (long-term desires for the state of the world) (GSP) trees. Each node in the GSP tree is weighted with Bayesian probabilities or importance weights. The details of the tree can be found elsewhere (Silverman and Bharathy, 2005; Silverman et al., 2007a).

A Preference Tree is one's long-term desires for world situations and relations (e.g. no weapons of mass destruction, stop global warming, etc.) that may or may not be achieved in the scope of a scenario and time horizon, but are nonetheless important as guiding values. The Standards Tree defines the methods an agent is willing to take to attain his/her preferences, and what code others should live by as well. The Standards Tree nodes that Silverman et al. (2007a) use merge several best-of-breed personality and culture profiling instruments such as, among others, Hermann traits governing personal and cultural norms (Hermann, 2005), standards from the GLOBE study (House et al., 2004), top-level guidelines related to economic and military doctrine, and sensitivity to life (humanitarianism). "Personal, cultural, and social conventions render inappropriate the purely Machiavellian action choices ('One shouldn't destroy a weak ally simply because they are currently useless'). It is within these sets of guidelines that many of the pitfalls associated with shortsighted artificial intelligence (AI) can be sidestepped" (Silverman et al., 2007a). Standards (and preferences) allow for the expression of strategic mindsets. Finally, the Goal Tree holds short-term needs the agent seeks to satisfy each turn (e.g. vulnerability avoidance, power, rest, etc.), and hence drives progress toward preferences. At individual level, the Goal Tree also includes traits covering basic Maslovian type needs. At the leadership level, the Goal Tree expresses the attitude to risk through the duality of growing/developing versus protecting the resources in one's constituency.

Parameter sets and models enable individual differences to be reflected in the model.

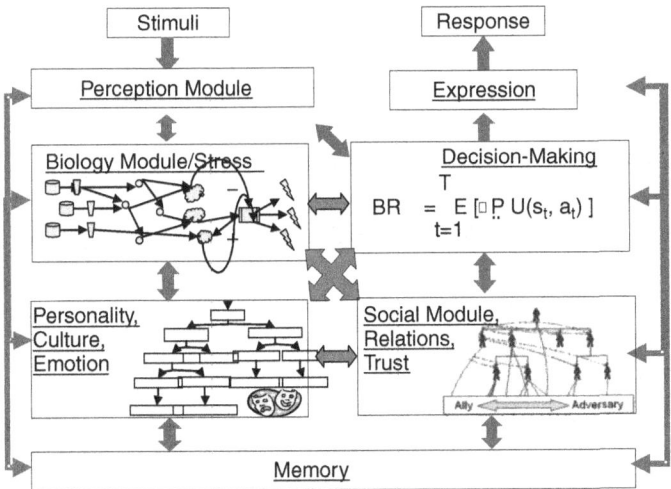

Figure 27.5 PMFserv agents: highlighting special features in decision making.

The agent architecture conforms to the generic architecture described in Figure 27.2. Special features of PMFserv agents are described in Figure 27.5, which expands on the module for decision making, including the influence of personality, culture and emotions, social interactions and physiology to account for the constraining role of stress in decision making. Under optimal stress, agents are vigilant and behave in a bounded rational fashion, but when stress levels (physical, emotion, or time stress for example) are too high or too low, the agents can be pushed into sub-optimal decision making patterns.

Decision Making Loop. The decision loop in the agents is best summarized by citing previous publications (Silverman et al., 2006, 2007a). For each agent, "PMFserv operates what is sometimes known as the Observe, Orient, Decide, and Act (OODA) loop. PMFserv runs the agent's perception (observe) and then orients the entire physiology and personality/value system to determine levels of fatigue and hunger, injuries and related stressors, grievances, tension buildup, impact of rumors and speech acts, emotions, and various mobilizations and social relationship changes since the last tick of the simulator clock. Once all these modules and their parameters are oriented to the current stimuli/inputs, the decision making/cognition module runs a best response algorithm to try to determine or decide what to do next. The summary is shown diagrammatically in Figures 27.5 and 27.6.

Interaction of, and Tradeoff among, Agent Dimensions in the Design

Should agents be abstracted (also relatively simpler) or should they be complex? Some interesting analytical results are indeed obtained when complexity at the

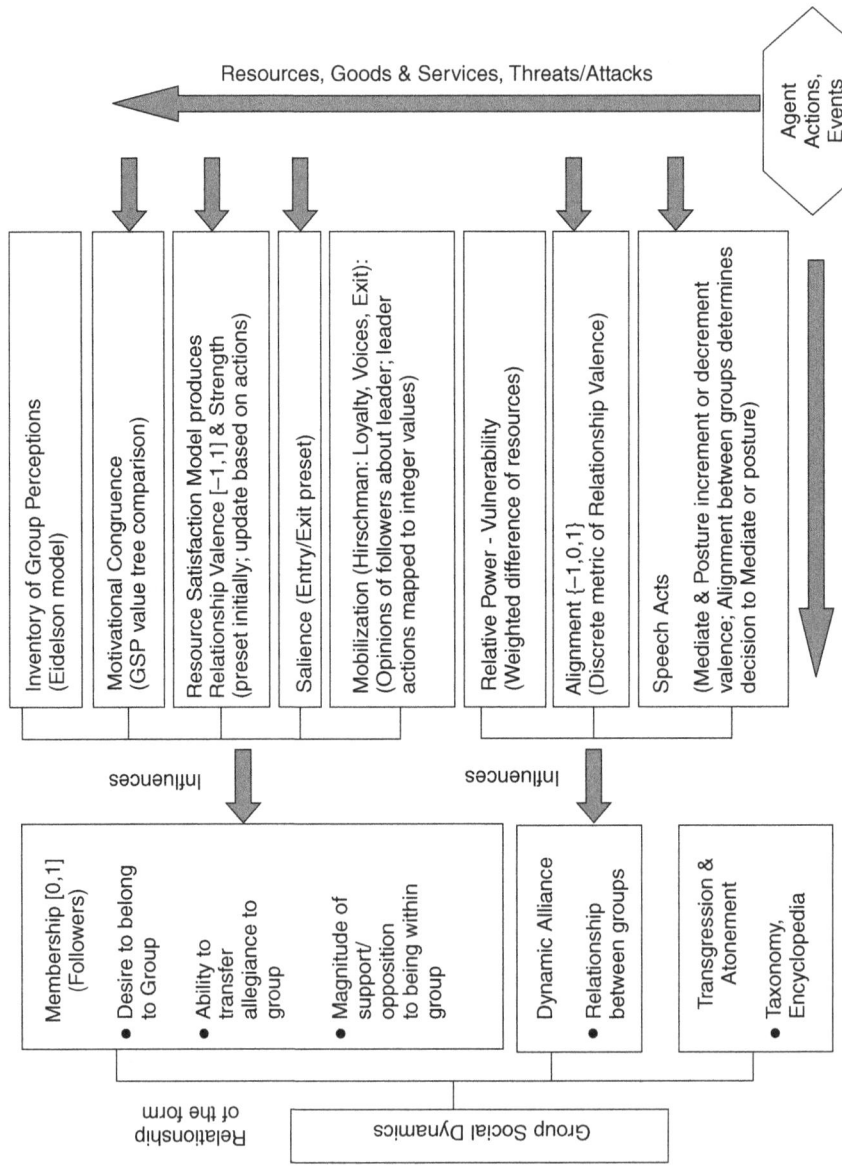

Figure 27.6 Social modules in FactionSim.

macro level is produced by simple micro level dynamics. We do not argue against the idea that agents should be kept as simple as appropriate or be based on the so-called keep-it-simple-and-stupid (KISS) principle. However, in modeling social systems, we argue that there is a place for complex agents.

Thus, there are agent technologies for representing behaviors in the military mission context through rule based engines. On the other hand, modeling non-kinetic contexts would benefit from a wider repertoire of human cognition and social sophistication. While high level abstract models may be sufficient for force-on-force operations, understanding the non-kinetic "humanscape" requires complex agents.

Artificial life is limited by its abstraction, and there is limited correspondence with the dynamics in a real society. These shallow agent models of artificial life can provide a very high level of understanding and insight into the decision process, particularly the effects of social interactions such as the neighborhood effect. These models have been exploited for illustrative and descriptive purposes, but there is a limitation to their application.

The models are largely single issue based and are primarily descriptive. While they are rich in connectivity along a single or limited issue dimension, cellular automata cannot account for details of cognitive and behavioral characteristics that are determinants of decision making in rich human subjects.

There is significant level of abstraction required to model any issue. This often means, in simple systems such as cellular automata, one can at best represent the higher level issues. If one were to trace back or drill down into the events that are occurring and provide explanations, it cannot be done below the level of abstraction.

There are of course more complex cellular automata models with options to set several variables/parameters (e.g. PSI—Lustick et al., 2004), thereby giving richer space. However, these enhanced models with increasing complexity are on their way to becoming more complex and richer models than cellular automata started out to be.

The assumption of rationality or predictability of the individual actors or agents is inherent in these models. In reality, the actors at best are fixed-rational, if not irrational. The models bring about complexity through the interaction of very simple parts. In reality, both parts as well as interactions bring about this complexity.

The system is likely to be sensitive to overly simplifying assumptions, abstractions, investigation of single issues at the twilight zone between chaos and order. Besides, traditional modeling approaches mainly stemming from technically oriented application domains more and more turn out to fail in supporting the social system models adequately.

Does this mean that one should build models with thick, cognitively detailed agents, rendering abstract agency obsolete? Owing to finite project resources available, in practice the dimensions are not independent, but are intricately linked to each other. In general, the more cognitively sophisticated the agents in an agent based model, the smaller the number of agents the model can accommodate, given the computational constraints of processing multiple cognitively sophisticated agents in a timely manner (Zacharias et al., 2008). This in turn affects the second dimension, the social sophistication. For example, the sophistication of the agents used in most cognitive agents means that these models are limited to no more than 10–20 agents.

The second dimension, the level of social sophistication, is also intricately linked to the third dimension, the number of agents, as well as to the first dimension, the level of cognitive sophistication. In general, the level of social sophistication is relatively low for small or large agent models while relatively high for mid-sized agent populations (Zacharias et al., 2008). This relationship is intuitive given that sophisticated social behavior requires some level of cognitive sophistication while cognitive sophistication beyond a certain level is limited by the aforementioned computational constraints.

An illustration of single-dimensional social sophistication with little cognitive sophistication is the residential segregation model devised by Nobel laureate Thomas Schelling. In Schelling's model, agents could have either black or white identities and their decision making was limited to a single decision concerning whether or not to stay in a particular neighborhood, based on a simple rule concerning the neighborhood's color composition (i.e. when the percentage of neighbors of the opposite color exceeds a certain threshold, move; otherwise, stay). Schelling's model can be easily implemented in agent based modeling toolkits such as Swarm, Repast, and NetLogo. The agents that are used in these modeling toolkits are usually low on the cognitive sophistication dimension and tend to follow simple rules and use some sort of cellular automata.

Intelligent Agent Frameworks: Cognitive and Social Sophistication

Unlike abstracted (simpler) agents who derive their social sophistication through a heterogeneous pool of a large number of agents giving a rich landscape where neighborhood effects interplay to bring about some emergence, cognitively sophisticated models will have an edge in social sophistication owing to their ability to deliberate in multiple dimensions that are of concern to social interactions. Thus, there is a tradeoff between the number of agents which help form larger networks, and cognitive sophistication which can bring multidimensional networks. The optima appear to be closer to either an intermediate-sized population model with moderately cognitively sophisticated agents or a hybrid model (using both cognitively sophisticated and abstracted (also relatively simpler) agents).

In Silverman (2010) cognitive-affective PMFserv models, a variety of techniques are employed to circumvent this issue and provide social intelligence. First (and overtly), a hierarchy of agent complexity is employed. The agents that need to deliberate decisions in details are cognitively very detailed and computationally expensive. Next come the agents who have been dumbed down slightly to make archetypal agents. These in turn are further expanded into numerous abstracted (also relatively simpler) agents in cellular automata to represent a physical and social landscape. The agents of these different complexities are dynamically interconnected so that leader agent decisions affect follower agents' actions, and vice versa. Unless the processing speed of computers increases dramatically over the next few years, this method seems to be one of the more reasonable ways to get around the problem of using cognitively sophisticated agents to build artificial social systems.

More importantly, social and contextual intelligence is built into agents by design. Let us see how this is done.

Adding Social and Contextual Intelligence

Groups/Factions and FactionSim Layer

The previous section overviewed the modules of a cognitive agent and some of the components that give it a social orientation. In this section we summarize additional modules that turn the cognitive agent into a socio-cognitive one.

Specifically, added on top of PMFserv is the FactionSim environment that captures a globally recurring socio-cultural "game," focusing upon inter-group competition for control of resources, further burdened with historic grievances and prejudices. This is game theory environment without the extreme parameter austerity and simplification, and also allows for PMESII campaign framework.

The environment is based on the premises that a vast majority of conflicts throughout history could largely be attributed to "greed" (i.e. the desire to acquire or control resources) and "grievance" (i.e. wanting to settle scores) among groups (Silverman et al., 2007b, 2009b, 2010) and groups, at the very least, consist of leaders and followers.

These games include not only resource factors (as in the case of game theory), but also socio-cultural and economic contextual factors such group affiliations, norms, sacred values, and interrelational practices, and individual value systems including personality, standards of behavior, goals and aspirations, etc.

FactionSim is a model of the social and organizational roles that exist in an area of operation (e.g. multi-state, state, sub-state, or region/village) and that may be played by the PMFserv agents. The agents belong to factions, which have resources, hierarchies of leadership, followers. The factions that agents belong to, as well as the agents themselves, maintain dynamic relationships with each other. The relationships evolve, or get modified, based on the events that unfold, blames that are attributed, etc.

Central to socio-cognition is faction or group. An agent could belong to one or more groups, the politically salient entities, headed by leader(s), and possessing resources and members. The groups can be arranged in a hierarchy with one or more sub-groups belonging to a super-group.

As explained in Silverman et al. (2008), the resources serve as indicators or proxies of (1) economic assets of the group in all forms including all factors of production and public goods that contribute to it; (2) level and type of security available to impose rule of law applied in the group, as well as will on other groups; and (3) political capital or popularity and support for the leadership as voted by its members.

This socio-political intelligence has been summarized in the above sub-model diagram (Figure 27.6), and handles governmental, institutional, and group management actions carried out by leaders as well as followers. In order to provide social dynamics, the agents have a membership model, a group-affinity model, group transfer dynamics, and sense-making about various political orientations and actions in historic contexts.

For example, the membership model describes how they also have a desire to belong to one or more factions with different levels of affinity, and "computes numerous averaged metrics using inputs from members of each group such as resource satisfaction levels, group collective beliefs (vulnerability, injustice, distrust), mobilization and votes, and motivational congruence (members versus leader)" (Silverman et al., 2007a). Group alignment and familiarity models compute numerous relationship related metrics such as alignment, strength of relationship, authority/legitimacy level, in-group influence, cross-group alignments and influence levels, and familiarity. A social transgression is an offense violating social norms or rules, and can include *faux pas*, taboos, materialistic (vandalism, fraud) or physical violence. Transgressions and atonement models affect how social and cultural norms might be violated, how that is accounted for, and manage how some of the transgressions could be atoned. This is a culture-specific module. In addition to internally held values, most societies also have social obligations or cultural norms that individuals are expected to conform to. These are well described in Knight et al. (2008). In addition, other modules contributing to social and group intelligence include the security model (skirmish model, similar to urban Lanchester), power–vulnerability computations (Johns, 2006), skirmish model (force size, training, etc.), economy model (Harrod–Domar model (Harrod, 1960)), black market, undeclared market (Lewis, 1954; Schneider and Dominik, 2000), formal capital economy, political model (loyalty, membership, mobilization, etc.) (Hirshman, 1970), institution sustainment dynamics, follower social network—cellular automata (Axelrod, 1998; Epstein, 2006; Lustick et al., 2004), small world theory/information propagation (Milgram, 1967), etc.

Adding Institutional Contexts: StateSim Layer

StateSim is a model of a state (or cross-state or sub-state) region and all the important political groups, their ethnic (and other) conflicts, economic and security conditions, political processes, domestic practices, and external influences. StateSim adds plug-ins and models atop FactionSim including a population model, economic services models, and the actual institutional agencies that allocate public goods and services (or not) to the factions in that region of interest. In doing so, StateSim is dependent on, and hence encompasses, PMFserv and FactionSim layers and provides the capability for modeling states, sub-states, and sets of states. StateSim seeks to recreate a given ethno-political situation and community in an artificial society framework and permits user/analysts to explore what issues are driving a given community and what might influence it.

As in the real world, institutions in the virtual world provide public goods and services, albeit imperfectly owing to being burdened with institutional corruption and discrimination. The economic system currently in FactionSim is a mixture of neoclassical and institutional political economy theories. The institutional models describe how different public goods are distributed (often imperfectly) among competing (or cooperative) factions. The public goods themselves are tied to the amount of resources for the institutional functions, including the level of inefficiency and corruption.

Institutions are used as a mediating force which controls the efficiency of certain services and can be influenced by groups within a given scenario to shift the equitableness of their service provisions. Political sway may be applied to alter the functioning of the institution, embedding it in a larger political-economy system inhabited by groups and their members. However, the followers of each group represent demographics on the order of millions of people. To handle the economic production of each smaller demographic, a stylized Solow growth model is employed: Solow (1956).

Each follower's exogenous Solow growth is embedded inside a political economy which endogenizes the Solow model parameters. Some parameters remain exogenous, such as savings rate, which is kept constant through time. As savings rates are modeled after the actual demographics in question and the time frame is usually only a few years, fixing the parameter seems reasonable.

Each follower demographic's production depends on their constituency size, capital, education, health, employment level, legal protections, access to basic resources (water, etc.), and level of government repression. These factors parameterize the Solow-type function, in combination with a factor representing technology and exogenous factors, to provide a specific follower's economic output. The economic output of followers is split into consumption, contribution, and savings. Consumption is lost, for the purposes of this model. Savings are applied to capital, to offset depreciation. Contribution represents taxation, tithing, volunteering, and other methods of contributing to group coffers. Both followers and groups have contributions, with groups contributing to any supergroups they belong to. Contributions, transfers, and spoils of other actions (e.g. attacks) are the primary source of growing groups' economy resources.

The unit of interaction is the institution as a whole, defined by the interactions between it and groups in the scenario. An institution's primary function is to convert funding into services for groups. Groups, in turn, provide service to members. In turn, each group has a level of influence over the institution, which it leverages to change the service distribution. Influence can be used to increase favoritism (for one's own group, for example) but it can also be used to attempt to promote fairness.

Groups, including the government, provide funding and infrastructure usage rights. Institutions are controlled by one or more dominant groups and there can be multiple competing institutions offering the same services. The distribution of services is represented as a preferred allotment (as a fraction of the total) towards each group. Institutions are also endowed with a certain level of efficiency, reflecting dollars lost in administration or misuse. The types of institutions currently modeled are public works, health, education, legal protections, and elections. For example, public works provide basic needs, such as water and sanitation; legal protections represent the law enforcement and courts that enforce laws; and the electoral institution establishes the process by which elections are performed, and handles vote counting and announcement of a winner.

Thus, StateSim adds plug-ins and models atop FactionSim including a population model, economic services models, and the actual institutional agencies that allocate public goods and services (or not) to the factions in that region of interest.

Metrics Designer and Calculator.   Finally, the description will not be complete without saying a few words about the metric system (Silverman et al. call it SAMA). In order to measure, calculate, and summarize the performance of the virtual world, there is a layer of stand-alone metric system on top of the database that tracks and stores output and can use lower level parameters to calculate higher level abstract parameters (e.g. Events of Interest or EOIs). The metric system is hierarchically organized, and by combining different lower level indicators and trends, higher level summary metrics of interest (e.g. key instability types such as rebellion, insurgency, and post-conflict metrics such as governance and development) may be computed.

This description of agents and their world is an oversimplification, but might serve the purpose of introducing the model. In-depth introduction to the framework, including the workings of the modules and the modeling methodology, have been published earlier (Silverman et al., 2005, 2006, 2007a, 2008, 2009b, c, d, 2010). Elsewhere in other publications (Silverman et al., 2005, 2009a, 2010), Silverman et al. have discussed how these different functions are synthesized, and reviewed how agents are profiled and how their reasoning works to make decisions etc. The models in PMFserv are social system models with particular emphasis on actor behavior. The models are built at different levels, one at a higher level of state or province or district and another at a lower level such as village. Both could be deemed to contribute to non-kinetic combat modeling.

Handling Heterogeneity (Parameter Space) and Number of Agents

Modeling Methodology and Handling Socio-Cognitive Model Load

In StateSim, heterogeneity and numbers of agents are furnished in two ways. StateSim typically profiles tens of significant ethno-political groups and a few dozen named leader agents, follower archetypes, and institutional actors. These cognitively detailed agents, factions, and institutions may be used alone or atop of another agent model that includes tens of thousands of lightly detailed agents in population automata.

Second, and more importantly, the individual differences among agents are not random. These models are knowledge based systems, and to a significant extent the modeling activity involves eliciting knowledge from subject matter experts as well as extracting knowledge from other sources such as databases and event data and consolidating the information to build a model of the social system.

In recent years, modeling methodologies have been developed that help to construct models, integrate heterogeneous models, elicit knowledge from diverse sources, and also test, verify, and validate models. In more complex agent models such as StateSim, the model building process (KE process) has been designed to satisfy the following functional requirements: (1) integrate individual social science models; (2) gather and systematically transform empirical evidence, tacit knowledge, and expert knowledge into data for modeling; (3) construct the model with the aim of reducing human errors and cognitive biases (e.g. confirmation bias) in constructing the model; (4) verify and validate the model as a whole; (5) run the model and explore the decision space; and (6) maintain the knowledge base over time.

An extensive web interview, which elicits subject matter expert inputs to construct the country models, has been designed. The subject matter expert input is triangulated against other sources of input such as databases like World Value Survey and events generated by automatic and manual scraping of webs and documents.

The details of the process are beyond the scope of this paper but can be found elsewhere (see, for example, Silverman and Bharathy, 2005; Silverman et al., 2009a, 2010; Bharathy and Silverman, 2011a, b).

Spatial Dimension and Mobility

While the first three dimensions identified by the NRC seem to provide the most germane distinctions, we shall also look at the fourth one, namely sense of space and mobility. We extend the fourth dimension, the use of a grid, to include sense of space and mobility. In combat models, this is an important feature. As with the social dimension, the spatial dimension also gets richer if there are large numbers of agents to represent and study. On the other hand, if the agent can reason well about space, that adds to its capabilities too.

Combat-specific frameworks such as MANA excel at spatial dimension and mobility; likewise, several other non-military models also provide this feature. For example, a large majority of abstracted (also relatively simpler) agents come equipped with grids and boids come with mobile agency. While space and mobility are important, any intelligence framework can be adapted to provide this. Owing to relative predominance and given that modelers no longer have to make an either-or choice regarding grids, this feature also seems to be of muted distinction.

For example, in StateSim, a compromise is found by using cognitively detailed agents, factions, and institutions atop of another agent model that includes tens of thousands of lightly detailed agents in population automata (Lustick et al., 2010). However, neither spatial considerations nor mobile agents are regular features in StateSim (mobile agents are now being developed for a current project). Where required, rudimentary mobility of agents and spatial sense for cognitively detailed agents are implemented through add-on models.

Knowledge Representation Paradigm

The distinction between the use of rules and the use of equations in agent representation seems to be increasingly blurred given the increasing proliferation of the combined use of rules and equations and given that equations can arguably be construed as a particular kind of rules. For example, StateSim uses both rules and equations for agent representation.

Integrated, Hybrid Modeling Paradigms with Intelligent Agents

Employing the right level of cognitive depth of agents is both a granularity issue and a qualitative issue (see Sibbel et al., 2004; Reusser et al. 2004 and Paolucci et al., 2006). Social systems are really characterized by both purposeful parts with cognition and personality (individual differences or heterogeneity), as well as complex relationships between them. Yet, intelligent agents are computationally expensive. There are practical difficulties in creating an entire, dense society using exclusively intelligent agents. Frequently, it is not necessary also. It would give more flexibility, if the decisions of the few archetypical intelligent agents were to be copied and multiplied by simple cellular automatons. It is our argument that different modeling paradigms can be, and have to be, integrated to best exploit their strengths and minimize their weaknesses.

In order to provide for the key social phenomena in non-kinetic military and socio-political systems relevant to non-kinetic military strategies, it is our thesis that the framework should include purposeful components such as agents with bounded rationality, rich cognitive processes, personality, value systems, heterogeneity among actors, social networks, as well as external institutional and economic frameworks which provide contexts for agents to operate. Models developed in such a framework can help provide insight into the behavior of social systems, particularly for conflict laden environments. By adding such models to the game space, realistic training games might be designed and exploratory, analytic, and decision support tools developed.

Let us see some examples of how intelligent agents are used (where necessary alongside simple agents) to model non-kinetic military and political systems.

General Overview of Hybrid Models

Cognitively detailed, intelligent agent models by nature are complex systems, and are often large collections of model of models. Such "model of models" collections are starting to show up in the HSCB field. They include COMPOEX (BAE Systems, 2009), StateSim (Silverman et al., 2010), NOEM (Salerno, 2008; Levis, 2009), SEAS (Chaturvedi and Mehta, 1999), CAESAR III model (Levis, 2009) to name a few. Some models have also been described in Zacharias et al. (2008).

While each of these models involves organization and country level simulation, each model has a markedly different focus. For example, COMPOEX and NOEM are projects that focus primarily on macro level issues, such as infrastructure and large-scale economics. The bulk of models in these collections are based on difference equations, such as resource flows. The COMPOEX suite addresses campaign planning issues through these tools, while the NOEM model concentrates on the national infrastructure. Both model collections have many qualitatively different models of individual infrastructure, economic, and institutional factors. On the other hand, the CAESAR III model focuses on micro level and meso level issue social systems. CAESAR III represents information processing and decision making of organizations as a function of the agents and social structure, as would be relevant to military C2 concerns. This model does not directly enforce macro-dynamics but instead relies upon models of agent cognition and information processing. The StateSim and SEAS models are a mixture of both micro level models and macro level models, having elements in common with both. StateSim and SEAS are agent based simulations that also explicitly synthesize models of the institutions, groups, and macro-economic factors. As a result, these have explicit models of macro-dynamics but also many cognitive sub-models within each agent.

Each model collection synthesizes many different interacting models, enabling emergent behavior. They also contain stochastic and path-dependent elements. Beyond this, they are all purposeful systems that attempt to complete goals within the system space. Some of them even have goals for many or all of the subcomponents (e.g. agents) of the system as well, leading to the kind of complexities seen in game theory and other fields. These approaches are necessary if one wants to capture enough of a social system's dynamics to make informed analysis and forecasting. While these models seem different, they are all complex systems.

Example Cases from Predominantly Cognitively Deep Agent Hybrid Models

Case A: Exploring Conflicts in a Political Unit (e.g. State/ District)

Using PMFserv agents, FactionSim, and StateSim institutional frameworks, Silverman et al. have developed a comprehensive, integrated, and validated agent based country modeling generator to monitor, assess, forecast, and most importantly explore national and sub-national crises to support decisions about allocating economic, political, and military resources to mitigate them: see Silverman et al. (2010).

As we have seen earlier, a StateSim model has several "moving parts." It helps the user/analysts to profile the stakeholders and their cognitions (PMFserv), the groups and networks the stakeholders move within (FactionSim), the formal and informal state institutions surrounding the stakeholder groups, and a suite of end-to-end lifecycle tools for managing the models and computational studies.

The resulting StateSim is a model of a state (or cross-state or sub-state) region and all the important political groups, their ethnic (and other) conflicts, economic and security conditions, political processes, domestic practices, and external influences. This model has a virtual recreation of the significant agents (leaders, followers, and agency ministers), factions, institutions, and resource constraints affecting a given country and its instabilities. These influential and important actors, who are required to deliberate over and make key decisions, are represented through cognitively detailed agents. The actors in this category include leaders at various levels and their archetypical followers. In order to save computational resources, Silverman et al. represent less salient actors (those who do not have to deliberate over key decisions) through a larger population model consisting of dumber agents making decisions on a simplistic scale of support versus opposition. For additional details, see Silverman et al. (2010).

The factions that agents belong to, as well as the agents themselves, maintain dynamic relationships with each other. The relationships evolve, or get modified, based on the events that unfold, blames that are attributed, etc. As in the real world, institutions in the virtual world provide public goods services, albeit imperfectly owing to being burdened with institutional corruption and discrimination. StateSim is an environment that captures a globally recurring socio-cultural "game" that focuses upon inter-group competition for control of resources (e.g. security/economic/political resource tanks).

StateSim seeks to recreate a given ethno-political situation and community in an artificial society framework. Doing so permits user/analysts to explore what issues are driving a given community and what might influence it. For instance, this includes going through the three phases of COIN doctrine (FM 3–24), the four phases of SSTR (FM 3–07), parts of the Civil, Military Operations (CMO) Training Manual, and the four steps of JIPOE.

As an example, one can readily envision an insurgency or rebellion existing in a world where there are inadequate resources to go around, inequality, discrimination, weak and corrupt institutions, and a number of factions or clans with a range of cultural and religious norms, aspirations, and grievances. It would be helpful to model and simulate such behaviors for the purposes of understanding the causes, influences, and their interactions. The intention of this modeling is not only to neutralize the enemies by force, but to understand the root cause behind the grievances in the larger population and progressively shift individuals and groups closer to the friendly end. To this end, a social system model with a set of simulated factions and insurgent agents operating in the context of inadequate, if not failed institutions is useful and appropriate. Through analyzing, modeling, and simulating such a social system, one comes to understand ways of co-opting and mobilizing the populace toward the rule of law and obviating and obstructing the need for insurgency. In fact, strict focus on militarily reacting to insurgent attacks can be counter-productive even for the military objective, as collateral damage to potentially woo-able factions would push these factions in favor of the insurgents.

Instead, a social system model such as StateSim can help with non-kinetic approaches to help diagnose the source of grievance, attempt to ameliorate the root causes, and foster the buildup of institutions, eventually strengthening the institutions for self-sustainment.

Validation

To date, StateSim has been applied in Iraq, Palestine, and four countries (Bangladesh, Sri Lanka, Thailand, and Vietnam) in the PACOM area of responsibility to model (forecast) emergence of state instabilities (insurgency, rebellion, domestic political violence, repression, etc.) which are macro-behaviors that emerge from the micro-decisions of all the participants, including the diverse leaders and followers. An example model has been schematically shown in Figure 27.7 and represents Bangladesh. The subsequent figure, Figure 27.8, shows how a higher level instabilities are computed from lower level decisions, events and state parameters.

StateSim has been subjected to a large correspondence test—320 specific forecasts were attempted in a 2008 DARPA challenge grant. Specifically, each of five metrics were forecast (rebellion, insurgency, domestic violence, political crisis, and repression) in each of four Asian countries separately for each quarter of 2004–2007. These $5 \times 4 \times 16$ quarterly model forecasts were compared with ground truth assembled by a third party and were found to have better than 80% accuracy on average. In a DARPA review, they discarded positive forecasts lower than 65% likely and negative forecasts greater than 35% likely. The remaining cases included only 59% of the total forecasts, but in that subset they were 94% accurate.

One can use this model to experiment on and study operations that might influence a region's instabilities and to assess different courses of action on the stakeholder groups and actors.

Experiment Support and End-to-End Data System

The StateSim library of models has passed significant tests of performance ($>80\%$ accuracy) and adaptability. Currently, the automation of modeling and analysis workflow is not complete but exists in multiple chunks of semi-automated pieces. The workflow starts with a web based instrument for eliciting all our parameter needs from cultural Intel SMEs: it includes data extraction for triangulation of

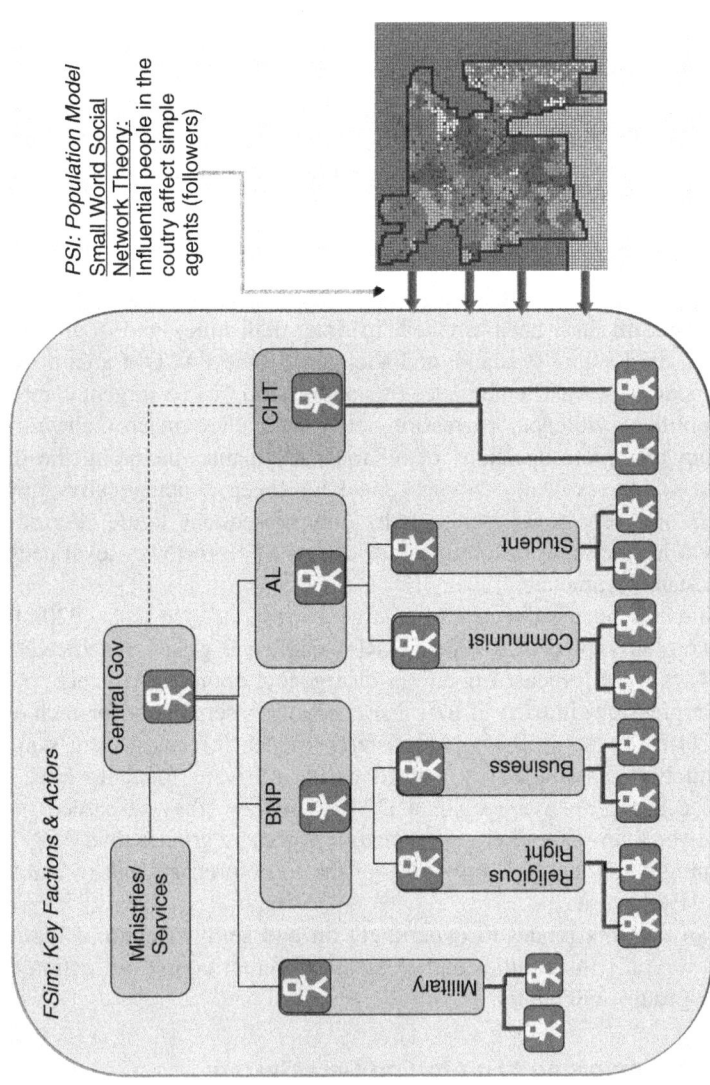

Figure 27.7 Overview of the components of StateSim.

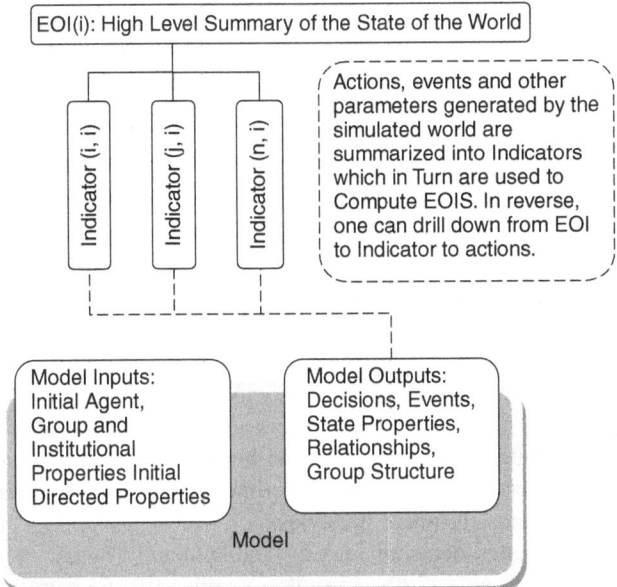

Figure 27.8   Metrics for the state of the word.

parameter estimates; that then populates and generates a StateSim model of a given region or community; that permits analysts to run it singly or under a Monte Carlo model controller for design of parameter experiments; that captures model outputs into an Oracle database; that includes viewers to aggregate low level actions into high level metrics; and that permits analysts to drill back down from high level metrics to individual causes including conditions and actors that took actions as well as a dialog engine that permits one to interview the actors about their rationale. For Monte Carlo runs and statistical analysis, one uses external applications (e.g. SAS, JMP, etc.) to assess the confidence intervals surrounding the forecasts. StateSim will not be exploited commensurate with its potential until the workflow automation is completed.

Case B: Non-Kinetic Military: NonKin Village: Cross-Cultural Trainer

Silverman et al. (2009c) created another model called NonKin at the village level using the above framework. NonKin (Non-Kinetic) Village is a small autonomous society that brings life to virtual factions and agents in an emergent sim-village. It is reconfigurable for a number of cross-cultural training goals. NonKin combines state of the art theories and models from the social sciences into a flexible (replaceable) "model of models" that realistically profiles a given culture and set of personalities (based on numerous validity studies in the Middle East, SE Asia, and Africa). Nothing is scripted, and it is all set up by tuning social science models to the society of interest.

Once set up, NonKin presents the user with many networks to unravel (kinship, ego/political, economic, insurgent, etc.) and with individual artificial intelligences playing roles in each network that cause them to carry out daily life and missions. The agents have personalities, follow social norms, are sensitive to cultural transgressions, live out economic lives (black market and formal market), and form grievances about different clans, factions, and groups. They also have an explanation engine so they can dialog about all of this with a player. As one advances the clock and comes back for repeated visits, more and more of this gets exposed. One needs to be patient to unravel it. Social system model based games such as mock villages can be employed to provide immersive experience in learning to be sensitive to stakeholder issues, foreign cultures, and norms. In particular, it puts trainees under choice situations as opposed to instruction based learning. The player(s) interact(s) with autonomous socio-cognitive agents and tries to find out who lives there; who are the clan leaders and followers; what are the norms, grievances, and dilemmas of this world; and how to convince wavering villagers, insurgent supporters, and members of "crime" families to convert to more peaceful, democratic, and legitimate lifestyles.

In games such as these, better decision making and interactivity are two key contributions of Silverman et al.'s (2009c) agents to games. If one were to look into the history of games, the earliest games were all scripted. This strong tendency to script the behavior has continued even with agent based technology. Even in a number of recent games, the agents were either human-played avatars or hand-scripted finite state machines of limited capability. In essence, in these games, humans must play (directly or indirectly through scripts) all roles and supply behaviors and utterances.

In recent years, designers have been enhancing a feature or two of the agents; the enhanced feature is typically planning capability, but in some cases cognition. However, the majority of these agents do not tend to have well rounded capability. We find procedural dialogs or manually scripted dialogs for entirely scripted scenarios are the norm. Even the agent systems with relatively advanced dialog features offer choices by scripting a very large repository of branches of dialog graphs with the assumption that such dialogs can be exhaustively specified *a priori* (e.g. using dialog graphs). However, ultimately, this approach ends up limiting the realism, story, and the possibilities (Silverman et al., 2009c; Bickmore and Cassell (2005). In rare cases, agents such as those in PMFserv tend to offer a relatively balanced set of capabilities including cognition, affect, and social sophistication, as well as a large number of agents. In NonKin models, declarative dialogs have been demonstrated in preliminary scale and form (Silverman et al., 2009c). In this experimental dialog generation, only high level agent parameters are employed to dynamically generate context-specific dialogs at runtimes. This could be implemented with or without natural language capabilities. In the approach, which does not depend on NLP capability, a library of dialog fragments is required. In addition, this library in NonKin includes interaction models, which are similar to a plan based view of dialog graphs or workflows for scenes. This is made possible as these agents are autonomous and use a rich base of models which understand context. One profiles the agents and organizations of a social system by populating various model parameters.

The interactive dialogs are context-triggered dialogs, often questions and answers. As the dynamics shift and the parameter values evolve, the agents use an explanation engine to automatically explain their goals, standards, preferences, relationships, grievances, and so on (Silverman et al., 2009c).

On the other hand, further enhancement of dialog capability will come when the context sensing autonomous agents are infused with natural language ability, but this is at least a step further away and requires sufficiently advanced natural language processing algorithms to be incorporated.

Pattern of Life.   Once the agents are profiled, they are able to take over life activities and decision making for the avatars they are plugged into. Each villager is different and continually runs its Observe, Orient, Decide and Act (OODA) loop to determine what daily life functions to perform in order to satisfy its Maslovian needs, e.g. safety, security, socializing and relationships, prayer, grievance atonement, etc. This permits the scenes to be unscripted. Agents are based on models with parameters that change as the situation evolves. Thus, agents are adaptive and react to what you do in the world.

The village (the Figures 27.9, 27.10, 27.11 give a snapshot of the village in game form) is marked up with various places of interest including that of habitation, employment, prayer, infrastructure, market, etc. Using affordance theory, these are marked up with the goods and services they offer and the functions or actions one can perform there (e.g. transactions, praying, socializing, eating, sleeping, etc.). This is both theoretically consistent with ecological psychology and efficient for software engineering and maintenance. The villagers scan these structures for what is possible and then compare that to their need for sustenance, belonging, employment, etc. They then make choices that satisfy their internal needs and values. The results are the emergent behavior and the pattern of daily life.

The economic and other resources and services of their respective groups are managed by the leader agents, who using public goods attempt to influence the villagers. Some leaders are interested in the greater good, some run the black

Daily Life Action	Component Steps	Affording Structures
Buy Food	Leave current location Go to location Buy 1 food unit	Economic structures with food inventory
Eat	Go home Eat 1 food unit	Food on hand or food at home
Go to Work/School	Leave current location Go to location Work/Attend class	Economic structure as place of employment, school house
Pray	Leave current location Go to location or home Pray	Residence or religious structure during pray time/day
Sleep	Go home Sleep	Residence from hours 2100-0600
Socialize	Find family/friends Socialize	Any structure with friends or family, passers-by on street
Attack	Choose target Acquire weapon Move to site Conduct attack	Any group's structures

Figure 27.9   A snapshot of village daily life.

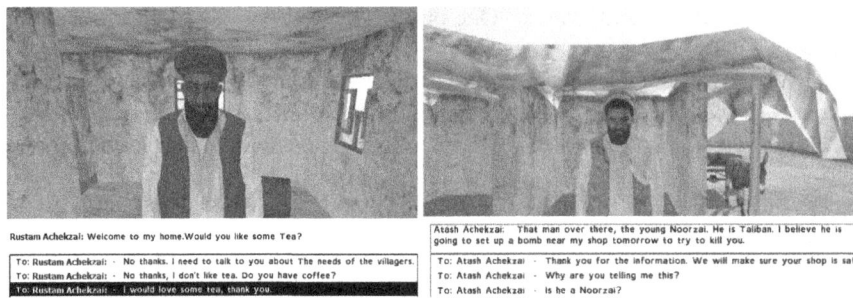

Figure 27.10 Trainee (soldier) playing the character exploring village culture and life (Courtesy: Silverman et al., 2009c).

(a)   (b)

(c)   (d)

Figure 27.11 (a) Commander giving you orders to go on patrol; (b) going on patrol and stopping at Malik's compound; (c) spotting a suspect with the help of an Afghan National Army member who is on your patrol; (d) dead young male after you shot him during your patrol.

market, and some are highly corrupt and self-serving. One faction, due to its religious values, might run insurgent and kinetic operations. A player can take actions and also influence the village.

Effects and Outcomes. All agents in the game can explain their internal socio-cognitive states, group grievances, relations/alignments, fears, and wants. By conducting patrols and "walking the beat," players become familiar to the villagers, improve relations, and learn about cultural sensitivities and how operations and courses of action can help and/or cause unintended and emergent effects at the tactical level. NonKin can be run *faster than real time* between patrols (e.g. 2 weeks later, 3 months later) to see how actions translate into effects.

Training Goals.   Training developers use a commander agent to issue daily orders to the player. The goal is to push the player through a training program such as, for example, the three stages of counter-insurgency methodology (survey the social landscape, make friends/co-opt the agenda, and foster self-sustaining institutions). Other training goals are to teach the player about cross-cultural awareness, conflict reduction, and inter-group cooperation. NonKin has a multi-echelon capability and can alternately support small unit training in immersive villages as well as staff and commander training at HQ.

Assessment Tools.   There are numerous assessment components and tools in NonKin that foster feedback, after action review, reflective learning, and so on. As explained in Silverman et al. (2009c), "the player learns to use the given resources judiciously and in a culturally sensitive way to achieve desired outcomes."

Server for 3D Worlds.   NonKin runs on a remote server and can be bridged to drive the agents in any given 3D world. In the past it has been used for VBS2, Unreal, RealWorld, JSAF, OneSAF, etc. One can also use NonKin in an unembodied mode to conduct Monte Carlo style experiments on courses of action and policies that might influence this world.

Case C: Integration of Village and District Level Simulations

One of the recent interesting developments has been integration of both village and district through a communication layer. While such integration with a cellular automata based population model is carried out routinely, this is more extensive as both the worlds are cognitively detailed and can be drilled down at multiple levels. When StateSim is present, a village is capable of leap-ahead simulation. In this mode, a real-time village is temporarily suspended while StateSim simulates over weeks or months. After a leap-ahead has occurred, the macro level simulation updates the village based upon the macro level changes. These effects show up in conversations with villagers after the sim-ahead.

Also, through StateSim, a higher level external actor (e.g. ISAF) can attempt to influence factional leaders (agents) and in turn villagers. The purpose of this exercise is to explore different possible uses of StateSim models and NonKin Village (as explained earlier) to explore a multi-resolution model for brigade/battalion staff training on the higher echelon end and immersive small unit training on the lower echelon end.

Thus local actions in the village might ripple up to alter the behavior of groups and this should be discoverable by the staff trainees. Likewise, courses of action suggested by the staff and commanders need to be carried out by the small unit trainees, but the human "climate" they encounter might be paved one way (or the other) by staff choices. This is an attempt to explore how multi-resolution modeling may help multi-echelon trainees to experience the coordination of tactical and operational DIME actions and PMESII effects. As we write this, preliminary tests have been undertaken successfully, but the project still is ongoing.

Software frameworks and the simulation environment described above allow for multiple dynamic simulation approaches to be integrated into a single model and models at different levels combined. By judiciously integrating models, and simultaneously engaging cognitively detailed agents, abstracted (also relatively simpler) agents, systems dynamics, discrete events, and continuous equations, *as needed*, hybrid models can be easily created to more accurately capture a system's structure and dynamics. This integration is in the right direction.

As a result, the judicial combination of models at different resolutions as well as within models different technological solutions such as large cellular automata and few archetypical intelligent, cognitively detailed agents can simulate social systems through a combination of socially and culturally sensitive decision making and subtle political dynamics. This way our socio-political system simulation does not sacrifice behavioral complexity at the level of the individual but is well able to model the complexity of the larger demographics and network effects themselves. This approach will result in a simulation that is much more realistic and responsive than more traditional models.

CONCLUSIONS

Agent systems, including both agent supported and agent based systems, have come a long way since their inception a decade or two ago, and especially have made significant advances in recent years. The future looks promising.

Sub-national conflicts between states and non-state actors are predicted to increase in frequency and scale in the near future. It is vital for the US military to model and simulate how to influence these conflicts since they have the potential to lead to potentially greater anti-American sentiment and threaten American interests and security. Fortunately, a number of interesting and relevant social science theories and models are now available.

In this chapter, we postulate and demonstrate with some case examples an integrated multi-resolution agent framework that allows users to easily model and evaluate regions of interest. Overall, this integration of cutting-edge social science analysis of all key social, political, economic, cultural, and environmental details with state-of-the-art technology will result in a powerful and easy-to-use platform that allows decision makers to reconfigure and operate the simulator for new scenarios and regions of the world. This integrated platform will also allow planners to test "what if" questions about the effects of actions and interventions.

Agent Supported Applications

We anticipate that the use of agents will increase as the technological context extends with emerging trends, and critical drivers such as augmented cognition, semantic web, web services and service oriented computing, grid computing, ambient intelligence, and autonomic computing become more pervasive.

Simulation based design of such systems will require seamless integration of agents and/or agent technologies. We expect that the use of simulation for the design of agents will become a new avenue for research. Analytic and heuristic methods in specifying the deliberative architecture of agents are common. Simulation can play a significant new role in the design of agents in such a way that software agents use simulation to make decisions before acting on the environment. Two different uses for agent based technology are presented. In the first case study, an agent supported decision-making system provided a framework for situation awareness, perception, and understanding capabilities. In the other, an agent framework was used to build social systems for non-kinetic applications.

In complex human-made systems, there is a demand for automated but intelligent systems that can make complex decisions utilizing uncertain, dynamic, distributed, and heterogeneous information from multiple sources. The intelligent agents are used by themselves or in a hybrid model in combination with other techniques such as expert systems, neural networks, fuzzy logic, and genetic algorithms. There is a growing demand for these systems in many areas and we have described a military application in agent supported modeling: a case of multi-UAV (MUAV) simulation has been described with a multi-resolution simulation framework with Recognition Prime Decision Module (RPDM) and COA generator with provisions for knowledge generation and machine learning. The intelligent agents help support strategy and a decision making lifecycle, develop understanding, carry out diagnosis, and generate COAs.

In military operations, as in other applications, agents that would support decision making are critical. For example, agents that collect, organize, and disseminate data, information, and knowledge to increase transactional capabilities, understanding, and wisdom are invaluable. The key advantage of agents lies in their ability for autonomous, complex, dynamic and flexible, high level interactions. We believe that the research described above will advance the case for using agent systems in critical operations.

Agent Based Modeling

In the agent based modeling application, models of sub-national conflicts and instabilities are depicted. The sub-national conflicts between states and non-state actors, or among non-state actors, are predicted to increase in frequency and scale. It is vital to model and simulate how to influence these conflicts since they may be directly or indirectly tied to our interests and security. The principal impact of Silverman et al.'s (2009c, 2010) research is an integrated multi-resolution agent framework that allows users to easily model and evaluate these theories. With such a framework, different theories at different resolutions (state, group, individual) can be tested and compared. It might prove useful in representing, evaluating, and training in the COIN (Counter-Insurgency)/IW (Irregular Warfare)/SSTR (Security, Stabilization, Transition, and Reconstruction) capabilities. Two cases were also briefly summarized, one (StateSim) at a strategic level and Non-Kin operating at a tactical, village level. The most salient aspects of this framework are (1) the synthesis of social science theories and selection of best-of-breed models for agent decision processes and interactions; (2) rigorous validation efforts carried out to date; and (3) its end-to-end transparency and drill-down capability from the front-end model elicitation (web interview, database scraping) to the backend metrics views and through indicators to events and even to the ability to query the agents involved in the events.

When populating these virtual worlds with agents and institutions, Silverman et al. (2009a) triangulated three sources: (1) existing country databases from the social science community and various government agencies and non-governmental organizations; (2) information collected via automated data extraction technology from the web and various newsfeeds; and (3) surveys of subject matter experts. This approach has yielded the best possible approximations for almost all parameter values of the models (see Bharathy and Silverman, 2011a, b; Silverman et al., 2010). The cognitive agent approach supported by adequate validation can offer nearly the same performance as statistical models; in addition, it brings to bear a greater transparency, explainability, and means to draw understanding of the underlying dynamics that are driving behaviors. The three cases introduced in this paper are (a) the state, province, or district level policy model of StateSim; (b) the village level NonKin game; and (c) integrated StateSim–NonKin game.

StateSim simulates political, economic, social, and related interactions at the regional level. Village leaders and followers also exist in this level. StateSim is a "network of networks" that realistically models larger regional dynamics such as regional faction leaders, macro-economic environments, and corruptible institutions that perpetuate economic and social inequities between groups. StateSim has shown over 80% accuracy in recreating real-world community decision making based on numerous DARPA-sponsored validity studies in the Middle East, SE Asia, and Africa.

In games such as NonKin Village, the artificial intelligence based, unscripted agents perceive contexts and respond guided by their value system. Unscripted, autonomous agency is a useful trait for games. Use of validated models (Bharathy and Silverman, 2010, 2011a; Silverman et al., 2009c) gives confidence that the immersive experience is close to reality (Pietrocola and Silverman, 2010). In the integrated StateSim–NonKin model, NonKin Village has been integrated with StateSim in an attempt at multi-level echelon integration.

Silverman et al. (2009a) have developed a generic game simulator which can be employed to relatively rapidly model social systems with social, economic, or political phenomena. Two of the applications of this model are creating role-playing games for immersive training or studying through computational experiments about the issues at stake.

In doing so, this research also attempts to create artificial intelligence models of human beings and, more specifically, leader and follower agents based on available first principles from relevant disciplines in both natural and social sciences (Silverman et al., 2007a). Realism in artificial intelligence agents is an important prerequisite for analytic models such as StateSim, and helps analysts explore the world.

Even for models behind immersive training (NonKin), realism enables one to see more clearly how to influence and elicit cooperation from various actors. A related benefit of having realistic agents based on evidence from video and multi-player online games is that, if the agents have sufficient realism, players and analysts will be motivated to remain engaged and immersed in role-playing games or online interactive scenarios (Silverman et al., 2007a).

While realistic leader and follower behaviors are valuable, the challenge is that social sciences that form the backbone of the models remain fragmented and narrow specialties, and few of their models have computational implementations (Silverman et al., 2007a). It is therefore pertinent that Silverman's laboratory has a thrust in synthesizing relevant first principles and best-of-breed social science models, exposing their limitations and showing how they may be improved.

The idea of using realistic agent based simulation to test competing theories in the social sciences is an approach complementary to other existing approaches including experimentation. Especially when the availability of data is limited or the quality of data is poor, or when experimentation using human subjects is either difficult or impossible, simulations may be the best choice. Simulators such as PMFserv can serve as virtual Petri dishes where almost unlimited varieties of computational experimentations are possible and where various theories can be implemented and tested to be computationally proved (i.e. to yield "generative" proofs). These are only some of the virtues and possibilities of having a versatile simulator like PMFserv.

As we examined earlier, agent-specific features such as cognitive, affective, social sophistication and those such as diversity and variety brought about by heterogeneity and/or number of agents compete with each other for computational and developmental resources. These features also enhance each other in the sense that the capacity of agents in one dimension enhances their ability to make use of other features as well. This positive and balancing relationship among features is important in the design of agency. The models described here feature such balances.

According to Rittel and Webber (1973), social problems ("wicked problem" is the term used to characterize these problems, but Ackoff calls them "messes") have no definitive formulation and are "never solved, only re-solved over and over," due to "incomplete, contradictory, and changing requirements." The first step to understanding these problems is examining multiple perspectives. The solution often lies in designs that attempt to satisfice, optimize multiple, individual agendas, but those that would transcend individual agendas and dissolve the problems (Ackoff, 1994).

We believe that that these wicked problems can best be understood through experiential learning. The effective testbed for wicked problems is a social system model—still better, a social system model embedded in a game. The models described in this chapter provide a step toward this goal.

As a final remark, the authors would like to point out that the intention of this chapter was not to give a comparative review between agent frameworks. The extensive use of one particular framework is rooted in the fact that the authors had full access to and knowledge of this framework to utilize it to give examples of how agents can be used in support of combat modeling challenges. For some additional information on how agents can be used in distributed simulation, the interested reader is referred to Tolk (2005).

ACKNOWLEDGEMENTS

Several past and present members of the Ackoff Collaboratory for Advancement of Systems Approach (ACASA) at the University of Pennsylvania contributed to development of the main software framework described in this paper. They are, in the chronological order of getting involved, Barry Silverman, Kevin O'Brien, Jason Cornwell, Michael Johnes, Ransom Weaver, Gnana Bharathy, Ben Nye, Evan Sandhaus, Mark Roddy, Kevin Knight, Aline Normoyle, Mjumbe Poe, Deepthi Chandasekeran, Nathan Weyer, Dave Pietricola, Ceyhun Eksin and Jeff Kim.

REFERENCES

Ackoff RL (1994). *The Democratic Corporation: A radical prescription for recreating corporate America and rediscovering success*. Oxford University Press, New York.

Ackoff RL and Emery FE (1972). *On Purposeful Systems*. Tavistock, London.

Anderson JR (1983). *The Architecture of Cognition*. Harvard University Press, Cambridge, MA.

Anderson JR (1990). *The Adaptive Character of Thought*. Lawrence Erlbaum Associates, Hillsdale, NJ.

Anderson JR (1993). *Rules of the Mind*. Lawrence Erlbaum Associates, Hillsdale, NJ.

Anderson JR, Bothell D, Byrne MD, Douglass S, Lebiere C and Qin Y (2004). An integrated theory of the mind. *Psychol Rev* **111**, 1036–1060.

Axelrod R (1998). *The Complexity of Cooperation: Agent-Based Models of Competition and Collaboration*. Princeton University Press, Princeton, NJ.

BAE Systems (2009). *COMPOEX (COflict Modeling, Planning, and Outcome Experimentation) Reference Manual*, 16 January. Washington, DC.

Bharathy G and Silverman B (2010). Validating agent based social systems models. In: *Proceedings of the Winter Simulation Conference*. IEEE, Piscataway, NJ.

Bharathy G and Silverman B (2011a). Holistically evaluating agent based social systems models. *Simulation: Transactions of the Society for Modeling and Simulation International*. Under Final Review.

Bharathy GK and Silverman BG (2011b). Applications of social systems modeling in political risk management. In: *Risk Management in Decision Making: Intelligent Methodologies and Applications*, edited by J Lu, L Jain and G Zhang. Springer, New York, NY.

Bickmore T and Cassell J (2005). Social dialogue with embodied conversational agents. In: *Advances in Natural Multimodal Dialogue Systems*, edited by J van Kuppevelt et al. Springer-Verlag, Berlin, pp. 23–54.

Castle C and Crooks A (2006). *Principles and Concepts of Agent-Based Modelling for Developing Geospatial Simulations*. Working Paper 110, University College London.

Chaturvedi AR and Mehta S (1999). Simulations in economics and management: using the SEAS simulation environment. *Commun ACM,* March 1999, pp. 60–61.

Chiarelli PW, Michaelis PR and Norman GA (2005) *Armor in Urban Terrain: The Critical Enabler,* March–April.

Denning PJ (2007). Computing is a natural science. *Commun ACM* **50**:7, 13–18.

Epstein JM (2006). *Generative Social Science: Studies in Agent-Based Computational Modeling*. Princeton University Press, Princeton, NJ.

Ferber J (1999). *Multi-Agent Systems: An Introduction to Distributed Artificial Intelligence*. Addison-Wesley: London.

Fisher I (2009). The airstrike: protecting the people or destroying the enemy ? *New York Times At War Blog*, 4 September 2009.

Gibson JJ (1979). The ecological approach to visual perception. Houghton Mifflin, Boston.

Harrod RF (1960). A second essay in dynamic theory. *Econ J* **70**, 277–293.

Hermann MG (2005). Who becomes a political leader? Leadership succession, generational change, and foreign policy. *Ann Meeting Internl Studies Assoc, Honolulu, HW.*

House RJ, Hanges PJ, Javidan M, et al. (2004). *Culture, Leadership, and Organizations: The GLOBE Study of 62 Societies*. Sage Publications, Thousand Oaks, CA.

Hirschman AO (1970). *Exit, Voice, and Loyalty*. Harvard University Press, Cambridge, MA.

Hudlicka E (2003). Modeling effects of behavior moderators on performance. In: *Proceedings of BRIMS-12, Phoenix, AZ*.

Ilachinski A (1997). *Irreducible Semi-Autonomous Adaptive Combat (ISAAC): An Artificial-Life Approach to Land Warfare*, Center for Naval Analyses Research Memorandum CRM 97-61.

Ilachinski A (2004). *Artificial War: Multiagent-Based Simulation of Combat*. World Scientific Publishing: Singapore (ISBN: 9789812388346).

Johns M (2006). Deception and trust in complex semi-competitive environments. Dissertation, University of Pennsylvania.

Kilcullen D (2006). Counter-insurgency redux. *Survival* **48**, 116.

Knight K, Chandrasekaran D, Normoyle A, Weaver R and Silverman B (2008). *Transgression and Atonement*. In: Dignum, V. (ed.) AAAI Workshop Proc., Chicago.

Lauren MK and Stephen RT (2002). Map-aware nonuniform automata (MANA) a New Zealand approach to scenario modelling. *J Battlefield Tech* **5**.

Levis AM (2009). Model Presentation at the AFOSR Annual Investigators Progress Review, January.

Lewis WA (1954). Economic Development with unlimited supplies of labour. *Manchester School* **28**(2), 139–191.

Luck M, McBurney P and Preist C (2003). *Agent Technology: Enabling Next Generation Computing A Roadmap for Agent Based Computing*. Agentlink.

Lustick IS, Miodownik D and Eidelson RJ (2004). Secessionism in multicultural states: does sharing power prevent or encourage it? *Am Polit Sci Rev* **98**, 209–229.

Lustick I, Alcorn B, Garces M and Ruvinsky A (2010). *From Theory to Simulation*. Annual Meeting of the American Political Science Association, Washington, DC.

Milgram, S. (1967). The small world problem. *Psychol Today*, May, 60–67.

Miller JH and Page SE (2007). *Complex Adaptive Systems: An Introduction to Computational Models of Social Life*. Princeton University Press, Princeton, NJ.

Moya LJ and Tolk A (2007). Towards a taxonomy of agents and multi-agent systems. In: *Proceedings of the 2007 Spring Simulation Multiconference* **1**, 11–18.

Nagl JA (2002). *Counterinsurgency Lessons from Malaya and Vietnam: Learning to Eat Soup with a Knife*. Praeger, Westport.

Newell A (1990). *Unified Theories of Cognition*. Harvard University Press, Cambridge, MA.

Nikolai C and Madey G (2009). Tools of the trade: a survey of various agent based modeling platforms. *J Artif Societies Soc Simul* **12**.

Ortony A, Clore GL and Collins A (1988). *The Cognitive Structure of Emotions*. Cambridge University Press, Cambridge.

Pahl-Wostl C (2002b). Agent based simulation in integrated assessment and resources management. In: *Integrated Assessment and Decision Support. Proceedings of the 1st Biennial Meeting of the International Environmental Modelling and Software Society*, edited by A Rizzoli and T Jakeman, Vol. 2, pp. 239-250. URL: www.agentlink.org/roadmap/al2/roadmap.pdf (Downloaded: April 2011)

Paolucci M, Conte R and Di Tosto G (2006). A Model of Social Organization and the Evolution of Food Sharing in Vampire Bats. *Adaptive Behavior* **14**, 223–238.

Petraeus H, Nagl J and Amos J (2006). *U.S. Army/Marine Corps Counterinsurgency Field Manual*. FM 3-24, MCWP 3-33.5. Washington DC.

Pew RW and Mavor AS (1998). *Modeling Human and Organizational Behavior: Application to Military Simulation*. National Academy Press, Washington, DC.

Pietrocola D and Silverman BG (2010). Taxonomy and method for handling large and diverse sets of interactive objects for immersive environments. In: *19th Annual Conference on Behavior Representation in Modeling Simulation, Charleston*.

Railsback SF, Lytinen SL and Jackson SK (2006). Agent-based simulation platforms: review and development recommendations. *Simulations* **82**, 609–623.

Reusser D, Matt H and Claudia P-W (2004). Relating choice of agent rationality to agent model uncertainty — an experimental study. IIEMS Conference paper 2004. http://www.iemss.org/iemss2004/.

Rittel HWJ and Webber MM (1973). Dilemmas in a general theory of planning. *Pol Sci* **4**, 155–169.

Salerno J (2008). Social modeling-it's about time. In: *Proceedings of the Second International Conference on Computational Cultural Dynamics*, edited by VS Subrahmanian and A. Kruglanski. AAAI Press, Menlo Park, CA.

Schneider F and Dominick E (2000). Shadow economies: size, causes, and consequences. *J Econ Lit* **38**, 77–114.

Serenko A and Detlor B (2002). *Agent Toolkits: A General Overview of the Market and an Assessment of Instructor Satisfaction with Utilizing Toolkits in the Classroom*. Working Paper # 455.

Silverman BG (2010a). Systems social science: a design inquiry approach for stabilization and reconstruction of social systems. *J Intell Decision Tech* **4**, 51–74.

Silverman BG and Bharathy GK (2005). Modeling the personality and cognition of leaders. In: *14th Conference on Behavioral Representations in Modeling and Simulation*. SISO.

Silverman BG, Bharathy GK and O'Brien K (2006). Human behavior models for agents in simulators and games: Part II —gamebot engineering with PMFserv. *Presence*, **15**, 2.

Silverman BG, Bharathy GK, Eidelson R and Nye B (2007a). Modeling factions for 'effects based operations': part i—leaders and followers. *J Comput Math Organ Theor* **13**, 379–406.

Silverman BG, Bharathy GK, Smith T, Eidelson R and Johns M (2007b). Socio-cultural games for training and analysis: the evolution of dangerous ideas. IEEE Systems, *Man Cybernetics* **37**, 1113–1130.

Silverman BG, Bharathy GK, Nye B and Smith T (2008a). Modeling factions for 'effects based operations': part ii—behavioral game theory. *J Comput Math Organ Theor* **14**, 120–155.

Silverman BG, Bharathy GK and Kim GJ (2009a). The new frontier of agent-based modeling and simulation of social systems with country databases, newsfeeds, and expert surveys. In: *Agents, Simulation and Applications*, edited by A Uhrmacher and D Weyns. Taylor and Francis.

Silverman BG, Bharathy G and Nye B (2009b). Gaming and simulating sub-national conflicts. In: *Computational Methods for Counter-Terrorism*, edited by S Argamon and H Newton. Springer-Verlag, Berlin.

Silverman BG, Weyer N, Pietrocola D, Might R and Weaver R (2009c). NonKin village: a training game for learning cultural terrain and sustainable counter-insurgent operations. In: *Agents in Games and Simulations Workshop, AAMAS Conference, Budapest, HU, May 11, 2009* (to appear in JAAMAS).

Silverman B, Pietrocola D, Weyer N, Weaver R, Esomar N, Might R and Chandrasekaran D (2009d). NonKin village: an embeddable training game generator for learning cultural terrain and sustainable counter-insurgent operations. In: *Agents for Games and Simulations: Lecture Notes in Computer Science*, Vol. 5920/2009, 135-154, DOI: 10.1007.

Silverman BG, Bharathy GK, Nye B, Kim GJ and Roddy PM (2010). Simulating state and sub-state actors with statesim: synthesizing theories across the social sciences. In: *Modeling and Simulation Fundamentals: Theoretical Underpinnings and Practical Domains*, edited by J Sokolowski and C Banks. Wiley, New York, NY.

Solow RM (1956). A contribution to the theory of economic growth. *Q J Econ* **70**, 65–94.

Tobias R and Hofmann C (2004). Evaluation of free Java-libraries for social-scientific agent based simulation. *J Artif Societ Soc Simulat* **7**.

Tolk A (2005) An agent-based decision support system architecture for the military domain. In: *Intelligent Decision Support Systems in Agent-Mediated Environments*, edited by GE Phillips-Wren and LC Lakhmi. Frontiers in Artificial Intelligence and Applications, IOS Press, Vol. 115, pp. 187–205.

Weiss G (1999). *Multiagent Systems: A Modern Approach to Distributed Artificial Intelligence*. MIT Press, Cambridge, MA.

Yilmaz L and Ören TI (2005). Agent-directed simulation. *Trans Soc Comput Simulat Int*, special issue, Vol. 81.

Yilmaz L and Phillips J (2007). The impact of turbulence on the effectiveness and efficiency of software development teams in small organizations. *Software Process: Improve Pract J* **12**, 247–265.

Zacharias GL, MacMilan J and Van Hemel SB (eds) (2008). *Behavioral Modeling and Simulation: From Individuals to Societies*. National Academy Press, Washington, DC.

Chapter 28

Uncertainty Representation and Reasoning for Combat Models

Paulo C. G. Costa, Heber Herencia-Zapana, and Kathryn Laskey

The commander is compelled during the whole campaign to reach decisions on the basis of situations that cannot be predicted. The problem is to grasp, in innumerable special cases, the actual situation which is covered by the mist of uncertainty, to appraise the facts correctly and to guess the unknown elements, to reach a decision quickly and then to carry it out forcefully and relentlessly. (Helmut von Moltke, 1800–1891)

CAPTURING THE FOG OF WAR

Military missions in a net-centric environment require distributed, autonomous agents to share information and cooperatively assess unpredictable situations to support real-time decisions in a rapidly changing environment. Combat models replicate this environment at various levels of detail and for different purposes. In some cases, such as in support of the commanders' decision making process, they must be capable of reasoning about complex situations involving entities of different types, related to each other in diverse ways, engaging in complex and coordinated behaviors.

As an example, combat models are increasingly applied to asymmetric warfare, in which elusive, secretive, and decentralized threats engage in loosely coordinated, difficult to detect behaviors. Capturing their behaviors through observations gathered by a diverse collection of sensors is a daunting task fraught

Engineering Principles of Combat Modeling and Distributed Simulation, First Edition.
Edited by Andreas Tolk.
© 2012 John Wiley & Sons, Inc. Published 2012 by John Wiley & Sons, Inc.

with uncertainty. In this scenario, each agent[1] accesses only a partial view of the global situation to perform its local task. However, not only must the commander have a more informed view, but other agents may also need access to non-local information that is important to their given tasks. In order to account for these aspects, a combat model should consider the command and control structure and keep track of what information each major player has access to in the simulation. More specifically, these models must have a means to account for uncertainty at different data fusion levels, such as the ones defined by the Joint Directors of Laboratories (JDL) (White, 1988; Steinberg and White, 1999), listed in Table 28.1. At the lower levels, probabilistic methods have become standard. The Kalman filter and its many variants have their mathematical basis in the theory of stochastic processes (Hall and Llinas, 1997; Hall and McMullen, 2004). Sensors for detecting various kinds of objects are typically characterized by receiver operating curves that allow false positive and false negative probabilities to be traded off as detection thresholds are adjusted. Classification systems are characterized by confusion matrices, which represent misclassification probabilities for various entity/event types. Incorporating these into a simulation is a complex but well understood task.

For JDL levels 2 and above the situation is a bit different, because issues such as the need for interoperability, representation of relationships among entities, and other more complex patterns of knowledge must be taken into account. This increased complexity requires greater modeling sophistication to produce the needed results. Thus, further data are required, more information sources from different types are involved, incomplete input becomes the rule, and uncertainty thrives in its various forms. In an environment where uncertainty is ubiquitous, the trend for more sophisticated combat models will naturally produce representational challenges that can only be addressed with formalisms that are expressive

Table 28.1   The JDL Data Fusion Levels (White, 1988)

Data fusion level	Association process	Estimation process	Product
L3 Impact assessment	Evaluation (situation to actor's goals)	Game-theoretic interaction	Estimated situation utility
L2 Situation assessment	Relationship (entity-to-entity)	Relation	Estimated situation state
L1 Object assessment	Assignment (observation-to-entity)	Attributive state	Estimated entity state
L0 Signal assessment	Assignment (observation-to-feature)	Detection	Estimated signal state

[1]The word "agent" here is used to indicate the lowest unit being simulated. Depending on the model resolution and purposes it can be a brigade, a battalion, or even an individual troop.

enough to capture essential domain regularities and consistently support plausible reasoning. Most knowledge representation formalisms either cannot properly capture domain knowledge with uncertainty at JDL levels 2 and above, or can do so only in controlled environments (limited scalability and expressiveness). More expressive, uncertainty capable formalisms are needed to meet the challenges faced by modern combat models.

Recent years have seen rapid advances in the expressive power of probabilistic languages (e.g. Jaeger, 1994; Koller and Pfeffer, 1997; Mahoney and Laskey, 1998; Pfeffer, 2000; Milch et al., 2005; Richardson and Domingos, 2006; Laskey, 2008). In this chapter we shall focus on the application of specific Bayesian methods to support information fusion in combat models with uncertainty. These methods are explained in the next section, where we present the concepts of Bayesian networks (BNs) and multi-entity Bayesian networks (MEBNs). To illustrate how these technologies can be applied to information fusion in combat models, we give an example of BN modeling with a very simple military vehicle recognition. This simple model is then extended to capture some characteristics common to models at JDL level 2 and above (i.e. relationships between entities). Finally, we focus on MEBNs, which address the representational limitations posed by BNs, and illustrate the details of the technique with an extended version of the use case model.

BAYESIAN REASONING FOR COMBAT MODELING

Probability theory provides a natural way to capture domain knowledge with its associated uncertainty. The value of probability theory is as much qualitative as it is quantitative. As Judea Pearl states: "The primary appeal of probability theory is its ability to express *qualitative* relationships among beliefs and to process these relationships in a way that yields intuitively plausible conclusions." Further, he argues, "The fortunate match between human intuition and the laws of proportions is not a coincidence. It came about because beliefs are not formed in a vacuum but rather as a distillation of sensory experiences. For reasons of storage economy and generality we forget the actual experiences and retain their mental impressions in the forms of averages, weights, or (more vividly) abstract qualitative relationships that help us determine future actions" (Pearl, 1988, p. 15, emphasis in original). In this passage, as in much of his work, Pearl is clearly discussing probabilistic representations of kinds of knowledge that would be classified at higher levels of the JDL hierarchy. With probabilistic representations, it is natural to construct models capable of inferential reasoning. Reasoning from evidence to hypothesis is accomplished by means of Bayes's rule and is often called Bayesian reasoning.

In this section, we briefly introduce the concept of graphical probability models, a powerful formalism for representing complex probability models involving many uncertain hypotheses. In such models, graphs are used to express qualitative dependence relationships, and local distributions involving only a few variables at a time provide quantitative information about the strength of the relationships.

We start with the most popular technique, BNs (Pearl, 1988), and then move to MEBNs (Laskey, 2008), which extend BNs to a greater level of expressive power.

Bayesian Networks

BNs provide parsimonious means to express a joint probability distribution over many interrelated hypotheses. A BN consists of a directed acyclic graph (DAG) and a set of local distributions. Each node in the graph represents a random variable. A random variable denotes an attribute, feature, or set of hypotheses about which we may be uncertain. Each random variable has a set of mutually exclusive and collectively exhaustive possible values. That is, exactly one of the possible values is or will be the actual value, and we are uncertain about which one it is. The graph represents direct qualitative dependence relationships; the local distributions represent quantitative information about the strength of those dependences. The graph and the local distributions together represent a joint probability distribution over the random variables denoted by the nodes of the graph (Costa et al., 2006).

Bayesian networks have been successfully applied to create probabilistic representations of uncertain knowledge in diverse fields. Heckerman et al. (1995) review an extensive set of applications of BNs. A more recent collection of application papers can be found in Pourret et al. (2008). The prospective reader will also find comprehensive coverage of BNs in a large and growing literature on this subject (e.g. Pearl, 1988, 2000; Neapolitan, 2003).

While standard BNs provide a powerful and versatile modeling tool, they are designed for a limited class of problems—those involving reasoning about a fixed, known set of uncertain hypotheses and evidence items. An example is the problem presented in our case study, which involves identifying the type of a single vehicle at a single time from a given fixed set of evidence items. However, real-world problems often involve a variable number of entities of different types interacting with each other in complex ways. As an example, a modern battalion-level combat model requires representing and reasoning about many entities (e.g. tanks, multi-purpose vehicles, platoons, etc.), each exhibiting type-specific structure and behavior (e.g. tanks have a characteristic allowable range of speeds for each kind of terrain), along with individual characteristics (e.g. a particular tank is proceeding on a given piece of terrain). A BN can be used to model the expected behavior of an individual tank in a given battlefield situation. In a simulation involving many tanks, one might make a copy of the BN for each tank in the simulation. However, this solution fails to capture interactions among the tanks, and thus will be unable to capture important battlefield dynamics such as tanks in a group executing a common plan. More generally, BNs have propositional expressive power, and are thus limited to problems such as the simple vehicle identification in the case study, where each vehicle is considered in isolation. Models intended to capture variations of this situation (e.g. observing multiple vehicles at the same time and inferring whether they form a group, or simulating the behavior of a group) require a more expressive formalism that can represent multiple entities of interest (e.g. many vehicles, many reports coming in at different times), how those entities are related (e.g. are those vehicles a group?), how situations evolve in time (e.g. is vehicle "bravo 4" accelerating?), uncertainty about entities (e.g. spurious reports), and other sophisticated knowledge patterns observable in the battlefield.

Multi-Entity Bayesian Networks

MEBNs extend the propositional expressiveness of BNs to achieve the representational power of first-order logic (Peirce, 1885; Frege, 1967). A MEBN represents the world as a collection of interrelated entities, their respective attributes, and relations among them. Knowledge about attributes of entities and their relationships is represented as a collection of repeatable patterns, known as *MEBN fragments* (MFrags). A set of well defined MFrags that collectively satisfies first-order logical constraints ensuring a unique joint probability distribution is a *MEBN theory* (MTheory) (Costa and Laskey, 2006). To be considered a complete, consistent component of a combat model, a group of MFrags should form an MTheory.

An MFrag represents uncertain knowledge about a collection of related *random variables* (RVs). RVs, also known as "nodes" of an MFrag, represent the attributes and properties of a set of entities. As in a BN, a directed graph represents dependences among the RVs. The RVs within an MFrag may contain arguments, called *ordinary variables*. These ordinary variables serve as placeholders for which domain entities can be substituted. For example, in Figure 28.1, **Speed** (obj, t) is a RV with two arguments, obj and t. These two arguments can be replaced by a specific object (e.g. Vehicle2) and a specific time (T1) to obtain **Speed** (Vehicle2, T1), the speed of Vehicle2 at time T1. Thus, an MFrag provides a template that can be instantiated as many times as needed for a given situation. The resulting BN fragments can be composed into a BN called a *situation-specific Bayesian network* (SSBN), constructed from the MTheory to reason about a specific situation. Each SSBN can contain blocks of repeated structure (instances of the same MFrag), with different numbers of blocks for different situations. This ability to represent variable sized models with repeated structure is a powerful extension to the expressive power of standard Bayesian networks. Because an SSBN is just a regular BN, traditional BN algorithms can be applied to it with no special adaptations.

MEBN categorizes RVs into three different types (see Figure 28.1 for a graphical representation):

- *Resident nodes* are the RVs whose distributions are defined in the MFrag. In a complete MTheory, each RV has its local probability distribution

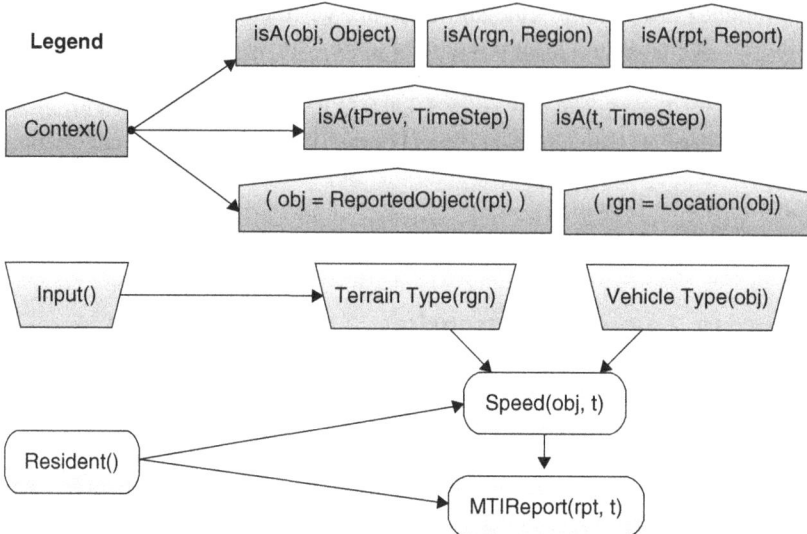

Figure 28.1 Structure of an MFrag. Directed arrows going from parent to child variables represent dependences between them. Nodes can contain a list of arguments in parenthesis, which are replaced by unique identifiers when the SSBN instantiation process is triggered.

uniquely and explicitly[2] defined in its home MFrag. The possible values of a resident node can be either instances of an existing entity or an ad hoc list of mutually exclusive and collectively exhaustive values.

- *Input nodes* are "pointers" referencing another MFrag's resident node. Input nodes provide a mechanism to allow re-use of RVs between MFrags.[3] Input nodes influence the probability distribution of the resident nodes that are their children in a given MFrag, but their own distributions are defined in the MFrags in which they are resident. In a complete MTheory, every input node must point to a resident node in some MFrag.

- *Context nodes* are logical (i.e. true/false) RVs representing conditions that must be satisfied for the distributions defined in the MFrag to be valid. Context nodes can represent several important types of knowledge patterns, and uncertainty about these patterns. For example, the context nodes in Figure 28.1 imply that the distribution of **MTI**(*rpt*, *t*) depends on the value of **Speed** (*obj*, *t*) when *obj* is the entity about which the report *rpt* is reporting. The problem of associating objects to reports is called *data association*. In cases where objects are close together, it may be uncertain

---

[2]If local probabilistic distribution is not explicit, then a default distribution should be assumed.

[3]It is possible to model "immediate" recursion by pointing input nodes as parents of a resident node within the same MFrag. Special care should be taken to avoid cycles in the resulting SSBN.

which object the report is referring to. This problem is known as *association uncertainty* or reference uncertainty. The RV **ReportedObject**(*rpt*) has as its possible values all the objects that could be the subject of the report *rpt*. We can represent association uncertainty by defining a probability distribution for **ReportedObject**(*rpt*) in its home MFrag. If we can infer from available knowledge that the value of a context node for an MFrag is satisfied, the probability distribution of the MFrag will be applied in the inference model. If there is no assignment of entities to variables in which the context nodes are satisfied, then the default distribution will be used instead. If it cannot be determined whether the context nodes are satisfied, the context nodes become virtually a parent of all the resident nodes in the MFrag, and the uncertainty is explicitly carried through the processing of the query.

A MEBN provides a compact way to represent repeated structure in a BN. An important advantage of MEBNs is that there is no fixed limit on the number of RV instances which can be dynamically instantiated as needed. MFrags can be viewed as representing modular "chunks of knowledge" about a given domain. Since an MTheory is a consistent composition of such "chunks," MEBN (as a formalism) is suitable for use cases addressing knowledge fusion (Laskey et al., 2007).

Case Study: the Vehicle Identification Bayesian Network

This section introduces BN modeling through a case study on fusing reports from several sources. The task is to receive reports from a moving target indicator (MTI) radar, an imaging sensor, a weather system, and a GIS database, and to apply this information to infer the type of an object hypothesized to be a military vehicle. Fusion is performed via Bayesian inference, using a model constructed from a combination of expert knowledge and data. The model is highly simplified, but is intended to be illustrative of a typical category of level 1 fusion problem, the identification and classification of individual entities. We also discuss briefly the use of the same BN model for other tasks such as prediction and simulation.

The next subsection describes how to formalize the fusion problem by specifying a joint probability distribution for hypothesis and evidence, and applying Bayes's rule to infer the probable vehicle type from the evidence. The next two subsections show how to compactly encode such a joint distribution as a BN, describing the BN structure and then describing the local probability distributions. Finally, we describe how to compile the BN using standard BN inference software, enter evidence, and obtain results from the model.

Joint Probability Distribution for Hypothesis and Evidence

We begin by defining RVs to represent the uncertain quantities of interest. Formally, an RV is a function mapping a *sample space*, representing the entire set of possibilities for the problem at hand, to an *outcome space*, representing the possible values of the uncertain quantity. For our problem, the sample space is the set Ω of all possible military situations. We define five RVs: V denotes the vehicle type, M denotes the MTI report; I denotes the imagery report; W denotes the weather report; and G denotes the report from the GIS database. Each of these RVs maps a situation $\omega \in \Omega$ to the value of the RV in that situation.

For example, suppose we are concerned with whether the object in question is a tracked vehicle, a wheeled vehicle, or a non-vehicle or false alarm. That is, the possible values of the RV V are v_T denoting a tracked vehicle, v_W denoting a wheeled vehicle, and v_N denoting a non-vehicle. In each situation, exactly one of these three possibilities will actually be the case. That is, if $\omega \in \Omega$ is the actual situation, we have $V(\omega) = v_T$ if the object is a tracked vehicle; $V(\omega) = v_W$ if the object is a wheeled vehicle ω; and $V(\omega) = v_N$ if the object is a false alarm. Formally, the RV V is a function

$$V : \Omega \to T_V = \{v_T, v_W, v_N\} \tag{1}$$

mapping a military situation ω to one of the elements of T_V.

A *probability distribution* for an RV assigns probabilities to subsets of the RV's possible values in a manner that satisfies the axioms of probability theory. Briefly and informally, the probability axioms state that (1) each probability is a number between 0 and 1; (2) the probability of the entire outcome set is 1; and (3) the probability of the union of two disjoint sets is the sum of their probabilities.[4] For example, suppose our probability distribution assigns a 50% probability to a wheeled vehicle and a 40% probability to a tracked vehicle, i.e. $\Pr[V = v_T] = 0.5$ and $\Pr[V = v_W] = 0.4$. From the axioms, we can deduce that there is a 90% chance that the object is a tracked or wheeled vehicle, i.e. $\Pr[V \in \{v_T; v_W\}] = 0.9$. We can also deduce that there is a 10% chance of a false alarm, i.e. $\Pr[V = v_N] = 0.1$.

The other four RVs, I, M, W, and G, are likewise defined as functions mapping Ω to their respective outcome sets. The output of an imaging sensor is an array of analog or digital pixels. For simplicity, we assume our model receives not the pixel array itself, but the output of an image processing system that classifies the image as a tracked vehicle, wheeled vehicle, or non-vehicle.[5] We denote these outputs as i_T, i_W, and i_N, respectively. An MTI sensor produces a sequence of returns, typically processed into tracks, where each track includes a trajectory of positions and velocities. For the purposes of our example, we make the extreme simplification that the MTI output has been processed into a report

---

[4]When an RV has infinitely many possible values, this axiom must be extended to include in finite sums. When an RV takes on values in a continuum, there is no way to consistently assign probabilities to all subsets of possible values; therefore, probabilities must be assigned to a collection of subsets, called σ-algebra, for which it is possible to consistently define probabilities.

[5]Probabilistic modeling of imagery at the pixel level is an active area of research, as is the use of BNs to integrate such models with data from other sources.

with one of four values: $M(\omega) = m_F$ reports a fast-moving object; $M(\omega) = m_M$ reports a a medium-velocity object; $M(\omega) = m_S$ reports a slow-moving object; and $M(\omega) = m_N$ indicates that no object has been detected at the location in question. We assume two possible values for the weather report: w_S is a report of sunny, clear weather, and w_O is a report of overcast, cloudy conditions. Finally, we assume the GIS report has four possible values: g_R for on-road; g_S for off-road smooth; g_H for off-road rough; and g_V for off-road very rough. Given these assumptions, the four evidence RVs are formally defined as functions:

$$
\begin{aligned}
I &: \Omega \to T_I = \{i_T, i_W, i_N\} \\
M &: \Omega \to T_M = \{m_F, m_M, m_S, m_N\} \\
W &: \Omega \to T_W = \{w_S, w_O\} \\
G &: \Omega \to T_G = \{g_R, g_S, g_H, g_V\}
\end{aligned} \tag{2}
$$

These five RVs can be combined into a multivariate RV, also called a random vector, that represents all the modeled features of the situation. The outcome space for this random vector is the Cartesian product $T_V \times T_I \times T_M \times T_W \times T_G$ of the five outcome sets. This random vector maps each situation $\omega \in \Omega$ to a vector $(v, i, m, w, g) \in T_V \times T_I \times T_M \times T_W \times T_G$. If we specify a probability distribution for this random vector, called a *joint* probability distribution, then we can recover probability distributions for lower-dimensional random vectors by an operation called *marginalization*. For example, the marginal distribution for (V, W) is obtained by marginalizing (summing) over the RVs I, M, and G as follows:

$$
\Pr[(V, W) = (v, w)] = \sum_{i \in T_I} \sum_{m \in T_M} \sum_{g \in T_G} \Pr[(V, I, M, W, G) = (v, i, m, w, g)] \tag{3}
$$

While it is often fairly straightforward to obtain marginal distributions for individual RVs from domain experts and/or data, direct specification of a high-dimensional joint distribution is typically prohibitively difficult. While marginal distributions for individual RVs are uniquely determined from the joint distribution, the reverse is not the case. There are typically many joint distributions consistent with a given set of single-variable marginal distributions. One might try to narrow the possibilities by specifying marginal distributions for small subsets of RVs, but this typically results in inconsistencies. In the next section a BN specifies a uniquely defined joint probability distribution from a set of local distributions, each involving only a few RVs at a time. This ease of specification is a powerful advantage of BNs.

Suppose we are given a joint distribution on a set of RVs, and we are interested in a hypothesis RV H. The distribution for H is obtained by marginalizing over the other RVs. The marginal distribution is denoted $\Pr[H]$, a notational shorthand for the set of marginal probabilities $\{\Pr[H = h] : h \in T_H\}$. Now, suppose we obtain evidence that the RV E has value e. We need to update our information about H to incorporate the new information about E. To do this, we use the *conditional* distribution of H given $E = e$, written $\Pr[H|E = e]$. This conditional distribution is given by

$$\Pr[H = h|E = e] = \frac{\Pr[H = h, E = e]}{\Pr[E = e]} \tag{4}$$

Equation (4) can be rewritten, using the identities of probability theory, as

$$\Pr[H = h|E = e] = \frac{\Pr[E = e|H = h]\Pr[H = h]}{\Pr[E = e]} \tag{5}$$

Equation (5) is called *Bayes's formula* or *Bayes's rule*. It is the fundamental equation of uncertainty dynamics, describing how uncertainty is updated as evidence is received. Bayes's rule is useful because a model is often specified in terms of a prior probability on the hypothesis variable and conditional distributions on evidence given the hypothesis variable. For such a model, equation (5) can be used to update the probability distribution for the hypothesis once evidence has been received.

In particular, suppose we have received reports that the imaging sensor reports a tracked vehicle; the weather is clear; the MTI reports medium speed; and the GIS reports that the object is on the road. Upon receipt of these reports, the probability that the object is a wheeled vehicle changes from $\Pr[V = v_W]$ to

$$\Pr[V = vw|I = i_T, M = m_M, W = w_S, G = g_R]$$
$$= \frac{\Pr[I = i_T, M = m_M, W = w_S, G = g_R|V = vW]\Pr[V = v_W]\Pr[V = v_W]}{\Pr[I = i_T, M = m_M, W = w_S, G = g_R]} \tag{6}$$

As discussed above, a BN uniquely and consistently specifies a joint distribution over a set of RVs by a graph representing a qualitative dependence structure, together with a set of local distributions quantifying the strength of the relationships. In the next two sections we show how to define a BN for the case study problem, and argue that the BN is a natural representation for the uncertainties in the problem.

Building the BN Structure

After defining the hypothesis and evidence RVs, the next step in building a model for our case study is to describe the qualitative relationships among the RVs. Typically, these qualitative relationships are obtained from domain experts, data, and general background knowledge such as the laws of physics and the axioms of geometry. We suppose the outcome of this process can be summarized as follws.

1. Tracked vehicles can travel at a maximum speed of 6 kph on very rough terrain where wheeled vehicles cannot travel.

2. Tracked vehicles are more likely to be off-road than wheeled vehicles. On rough off-road terrain, tracked vehicles usually travel at 8–12 kph, and wheeled vehicles usually travel at 5–9 kph. On smooth off-road terrain, wheeled vehicles usually travel at 10–20 kph and tracked vehicles usually travel at 5–15 kph. On roads, wheeled vehicles usually travel at 35–115 kph and tracked vehicles usually travel at 15–60 kph.

3. An MTI sensor provides approximate position and velocity for vehicles that are moving, but cannot see stationary objects.

4. An imaging sensor usually distinguishes vehicles from other objects, and usually reports correctly whether a vehicle is tracked or wheeled. Cloud cover can interfere with the ability of the imaging sensor to distinguish vehicles from other objects.

5. The GIS system usually reports accurately whether a location is on the road, off-road on smooth terrain, off-road on rough terrain, or off-road on very rough terrain. However, there are occasional errors in the terrain database.

This problem can be modeled using the BN depicted in Figure 28.2. Drawing such a diagram is usually the first thing one does when modeling with BNs, since representing domain relationships in a directed graph tends to be not only natural but also helpful to understand and organize the information. The resulting graph structure conveys the basic domain knowledge being captured by the expert. It can be verified by the expert and used to help others understand the relationships. That is, just by looking at the BN structure and node names one can realize the implications of the model.

To make the BN easier for humans to understand, it is customary to give meaningful names to the RVs and their possible values, in contrast to the mathematical notation used above. There, RVs are denoted by the first letters of the names shown in Figure 28.2. That is, V corresponds to *Vehicle type*, I corresponds to *Imaging report*, and so on. Note that the BN of Figure 28.2 contains two RVs not defined in the previous section, which considered only evidence and hypothesis RVs. These RVs, *Speed* and *Terrain*, denote features of the situation that are not directly observed but mediate the relationship between the hypothesis of interest and the evidence. Representing these RVs explicitly in the

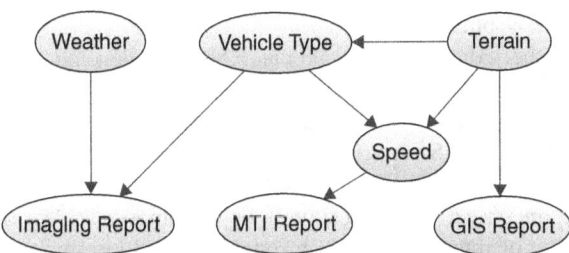

Figure 28.2 Vehicle identification BN.

BN has two advantages: it simplifies the structure of the BN, and makes specification of the probability distributions more natural. Following our established convention, when using mathematical notation these RVs are denoted as S and T, respectively.

The BN in Figure 28.2 includes three "report nodes" that are used to enter incoming evidence from the imagery, MTI, and GIS sensors. They are leaf nodes, meaning that they have no children and at least one parent node (i.e. a node that directly influences the distribution of its child nodes). Dependence, conditional dependence, and independence relationships among the nodes are all explicit in the structure. As an example, the imaging report is influenced by the weather conditions (e.g. cloudy weather might influence the sensor discrimination power) and vehicle type (e.g. a tracked vehicle is more likely to cause a tracked vehicle report than a wheeled vehicle), but it is independent of the terrain type.

There are many technical aspects of BN modeling, knowledge of which greatly improves one's ability to capture the nuances of the domain being modeled. Concepts such as d-separation (Pearl, 1988), intercausal dependence, independence of causal influence (Heckerman and Breese, 1996), context-specific independence (Boutilier et al., 1996), partitions, local decision trees, and others can be leveraged to produce rather sophisticated representations. Covering all these aspects is outside the scope of this chapter, but the interested reader will find a large body of research on the subject (e.g. Pearl, 1988; Neapolitan, 2003; Darwiche, 2009; Gamez et al., 2010; Jensen and Graven-Nielsen, 2010; Kjaerul and Madsen, 2010).

Defining the Joint Probability Distribution

For this simple example, we have made some assumptions regarding the background information to support the definition of the joint probability distribution of the model explained above, which would have been the result of data analysis in a real case. The local distributions for each of the RVs are described in turn below, along with the assumptions made to obtain the distributions.

Weather and Terrain

We first define local distributions for the weather and terrain RVs. These are root nodes in the BN, and so their distributions are specified directly, without conditioning on the values of any other RVs.

It is assumed that the weather conditions will be *clear* 80% of the time, and thus *cloudy* 20% of the time. That is, $\Pr[W = w_S] = 0.8$ and $\Pr[W = w_O] = 0.2$.

The *Terrain* node is necessary due to error in the GIS system. It represents the actual terrain type, of which the GIS report is noisy evidence. It is assumed that in the area of interest there is a 40% chance an object of interest will be on the road, a 30% chance an object of interest will be on smooth off-road terrain, a 20% chance an object of interest will be on rough off-road terrain, and a 10% chance an object of interest will be on very rough off-road terrain. That is, $\Pr[T = t_R] = 0.4$; $\Pr[T = t_S] = 0.3$; $\Pr[T = t_H] = 0.2$; and $\Pr[T = t_V] = 0.1$.

Vehicle Type

This RV represents the main hypothesis. Its prior distribution is conditioned on its parent, *Terrain type*. That is, the probability that a vehicle is of a given type depends on the terrain on which it is located.

1. *Road.* On the road, 60% of objects of interest are wheeled vehicles, 35% are tracked vehicles, and 5% are non-vehicles. That is, $\Pr[V = v_W|T = t_R] = 0.60$; $\Pr[V = v_T|T = t_R] = 0.35$; and $\Pr[V = v_N|T = t_R] = 0.05$.

2. *Off-Road Smooth.* On off-road smooth terrain, 50% of objects of interest are wheeled vehicles, 45% are tracked vehicles, and 5% are non-vehicles. That is, $\Pr[V = v_W|T = t_S] = 0.50$; $\Pr[V = v_T|T = t_S] = 0.45$; and $\Pr[V = v_N|T = t_S] = 0.05$.

3. *Off-Road Rough.* On off-road rough terrain, 35% of objects of interest are wheeled vehicles, 55% are tracked vehicles, and 10% are non-vehicles. That is, $\Pr[V = v_W|T = t_H] = 0.35$; $\Pr[V = v_T|T = t_H] = 0.55$; and $\Pr[V = v_N|T = t_H] = 0.10$.

4. *Off-Road Very Rough.* On off-road very rough terrain, 10% of objects of interest are wheeled vehicles, 70% are tracked vehicles, and 20% are non-vehicles. That is, $\Pr[V = v_W|T = t_V] = 0.10$; $\Pr[V = v_T|T = t_V] = 0.70$; and $\Pr[V = v_N|T = t_V] = 0.20$.

Note that according to this distribution, there is a positive probability that a wheeled vehicle is in off-road very rough terrain, whereas the domain expert indicated wheeled vehicles cannot travel in such terrain. We shall see later, in the description of the distribution for the *Speed* distribution, that wheeled vehicles in off-road very rough terrain are assumed to be stuck and therefore are not moving.

Imaging Report

The probability distribution for this RV is defined conditional on the values of its parents *Weather* and *Vehicle type*. Once again, we have made some assumptions to obtain the distribution.

1. *Clear* conditions provide a 96% chance of correct identification for any of the three vehicle types, and the mis-identification probabilities are split evenly between the two incorrect hypotheses. That is, $\Pr[I = i_x|W = w_S, V = v_x] = 0.96$ and $\Pr[I = i_x|W = w_S, V \neq v_x] = 0.02$ for $x \in \{W, T, N\}$.

2. *Cloudy* conditions decrease the likelihood of a correct imaging report to 40%. Again, mis-identification probabilities are split evenly between the two incorrect hypotheses. That is, $\Pr[I = i_x|W = w_O, V = v_x] = 0.40$ and $\Pr[I = i_x|W = w_O, V \neq v_x] = 0.30$ for $x \in \{W, T, N\}$.

Vehicle Type	Wheeled		Tracked		NonVehicle	
Weather	Clear	Cloudy	Clear	Cloudy	Clear	Cloudy
Tracked	0.02	0.3	0.96	0.4	0.02	0.3
Wheeled	0.96	0.4	0.02	0.3	0.02	0.3
NonVehicle	0.02	0.3	0.02	0.3	0.96	0.4

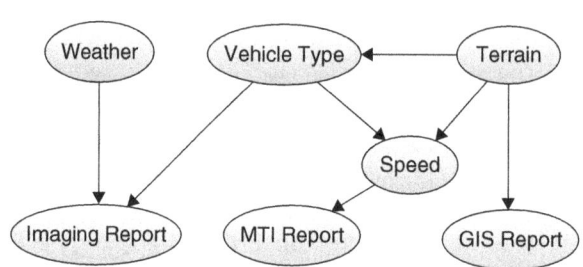

Figure 28.3   CPT of node *Imaging report.*

Figure 28.3 depicts the resulting *conditional probability table* (CPT) as modeled in the UnBBayes open-source probabilistic reasoning tool suite. In the figure, the combination of the states of the parents of a node (parent node titles in dark gray) appears at the top, while the node's states appear as rows in the table (node state titles in light gray). As a result, each column must add up to one.[6]

Speed and MTI Report

The *Speed* RV illustrates a common modeling technique used in BNs, which is to use a discrete RV to represent an inherently continuous measurement (i.e. the velocity of an object). Most BN packages implement inference algorithms requiring discrete-valued RVs, which explains why many resort to discretization. In this case, a few assumptions were made to simplify the *Speed* node's CPT while keeping the model fairly realistic. An overarching assumption about the speed data provided is that these are not precise, steady state speeds but ranges of those most commonly seen from historical patterns of operation. Speeds are broken into ranges that determine significant breaks in the data provided. Specifically, four states are given:

- *Fast*—this state, s_F in mathematical notation, corresponds to speeds of greater than 60 kph.
- *Medium*—this state, s_M in mathematical notation, corresponds to speeds in the range 16–59 kph.

---

[6]Other Bayesian software packages, such as Norsys Netica™, use a different convention with node states at the top, which results in probabilities within a given row adding up to 1.

| Vehicle Type | Wheeled | | | | Tracked | | | | NonVehicle | | | |
Terrain	Road	OffRoad Smooth	OffRoad Rough	OffRoad Very Rough	Road	OffRoad Smooth	OffRoad Rough	OffRoad Very Rough	Road	OffRoad Smooth	OffRoad Rough	OffRoad Very Rough
Stationary	0.1	0.1	0.1	1	0.1	0.1	0.1	0.1	0.7	0.7	0.7	0.7
Slow	0.1	0.3	0.7	0	0.3	0.4	0.5	0.75	0.15	0.2	0.25	0.3
Medium	0.3	0.5	0.1	0	0.5	0.4	0.3	0.1	0.1	0.05	0.05	0
Fast	0.5	0.1	0.1	0	0.1	0.1	0.1	0.05	0.05	0.05	0	0

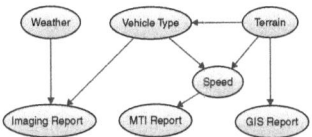

Figure 28.4   CPT of node *Speed*.

- *Slow* —this state, s_S in mathematical notation, corresponds to speeds in the range 1–15 kph.
- *Stationary* —this state, s_N in mathematical notation, corresponds to speeds of less than 1 kph, or a non-moving object.

Using these states, we define the RV

$$S : \Omega \rightarrow T_S = \{s_F, s_M, s_S, s_N\} \tag{7}$$

We further assume that each vehicle type is actually capable of speeds in ranges not identified in the description. Specifically,

- A *wheeled* vehicle in *OffroadVeryRough* terrain is stuck and therefore *stationary* 100% of the time.
- In other conditions, vehicles are assumed to be stationary 10% of the time.
- Vehicles spend time in speed bands below those described as they accelerate to speed, and vehicles may sprint at speeds above those described. Specific conditional probabilities are given in Figure 28.4.
- Non-vehicles are usually *stationary* with a 70% probability in all terrain conditions. Non-vehicles never have *fast* speed when off-road. Other probabilities for non-vehicles are given in Figure 28.4.

The CPT for node *Speed* is depicted in Figure 28.4.

Finally, the *MTI Report* RV has the same states as the *Speed* RV, and its CPT was designed to reproduce the accuracy of the sensor. As an example, when a vehicle's speed is *slow* then the MTI will return *stationary* 5% of the time, *slow* 80% of the time, *medium* 10% of the time, and *fast* 5% of the time.

GIS Report

If the GIS system were 100% accurate then this node would be redundant. We assume the GIS system is usually accurate. Specifically, we assume roads are recognized correctly 99% of the time, and the other three terrain types are recognized correctly 90% of the time. The remaining percentage is divided between adjacent terrain types. Specifically:

- $\Pr[G = g_R | T = t_R] = 0.99; \Pr[G = g_S | T = t_R] = 0.01; \Pr[G = g_H | T = t_R] = 0; \Pr[G = g_V | T = t_R] = 0;$
- $\Pr[G = g_R | T = t_S] = 0.05; \Pr[G = g_S | T = t_S] = 0.9; \Pr[G = g_H | T = t_S] = 0.05; \Pr[G = g_V | T = t_S] = 0;$
- $\Pr[G = g_R | T = t_H] = 0; \Pr[G = g_S | T = t_H] = 0.05; \Pr[G = g_H | T = t_H] = 0.9; \Pr[G = g_V | T = t_H] = 0.05;$
- $\Pr[G = g_R | T = t_V] = 0; \Pr[G = g_S | T = t_V] = 0; \Pr[G = g_H | T = t_V] = 0.1; \Pr[G = g_V | T = t_V] = 0.9.$

The Joint Distribution

The BN model described in this section defines a joint distribution over the seven RVs. This joint distribution is obtained as follows:

$$\Pr[(T, W, V, S, I, M, G)]$$
$$= \Pr[T]\Pr[W]\Pr[V|T]\Pr[S|V, T]\Pr[I|W, V]\Pr[M|S]\Pr[G|T] \qquad (8)$$

Compiling and Using the Model

After all CPTs have been defined and entered into the BN tool, then the graphical model can be compiled. Compiling a BN means applying a standard inference algorithm to compute marginal distributions for all the RVs in the BN. To see the impact of evidence, one can set a RV to a given value and apply the inference algorithm. The resulting beliefs reflect the application of Bayes's rule to the joint distribution, computing the conditional distribution of each node given the evidence nodes.

To illustrate how one could use the graphical representation to make inferences with the model, we present two distinct situations and show how the model responds. In the first case, let us suppose we must infer whether an object is a tracked vehicle, a wheeled vehicle, or a non-vehicle given the following reports:

1. The imaging sensor reports a tracked vehicle.
2. The weather is clear.
3. MTI reports medium speed.
4. GIS reports that the vehicle is on the road.

Figure 28.5 depicts the BN compiled in UnBBayes. Note that the evidence RVs are shown in gray, and have 100% probability of the observed values. In this example, the model infers a tracked vehicle relying most heavily on the imaging sensor. With the speed of 28 kph falling within the usual range for tracked vehicle travel on roads, there are no data provided to contradict the imaging sensor report.

In the second example, the information provided is not as consistent as the previous set.

1. The imaging sensor reports a wheeled vehicle.

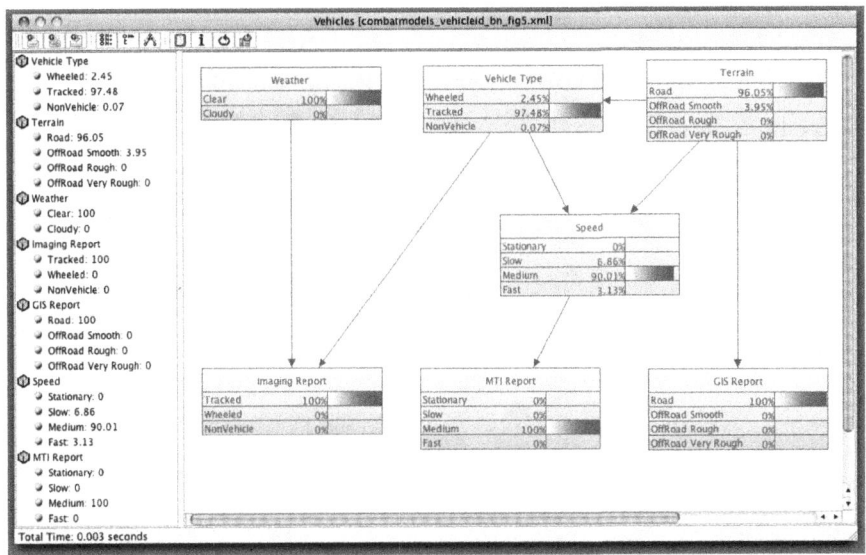

Figure 28.5   Compiled BN with evidence.

2. The weather is cloudy.

3. MTI reports slow speed.

4. GIS reports that it is on very rough terrain.

The compiled BN with the above evidence entered is depicted in Figure 28.6.

This example illustrates a disparity between sensors. Remembering that wheeled vehicles cannot move in *OffroadVeryRough* terrain and that the maximum speed for tracked vehicles in the same conditions is 6 kph, there is a strong inference that the vehicle is tracked. The imaging sensor contradicts this finding, but its effect is diminished due to the cloudy conditions. Therefore, the correct inference is a high probability for a tracked vehicle.

Even in its simplified BN version, the issues behind the vehicle identification problem can be easily generalized to a class of similar problems classified as JDL level 1. That is, data on specific parameters of a given object are used to infer the object properties, type, etc. BNs provide a flexible and straightforward means of capturing these parameters and performing principled inferential reasoning to increase knowledge about the associated properties.

This BN model could also be used to predict the speed of a vehicle of known type, by conditioning on the terrain and vehicle type and calculating the distribution for speed. The model could also be used to simulate behavior (i.e. speed in this case), and to simulate sensor reports. Although this model is highly simplified, and would have to be enhanced to be useful in real applications, it is clear that BN models can be used for a variety of purposes in combat modeling. We have focused on fusion, but BNs could also be used to represent the distributions used to simulate behavior of entities in combat simulations.

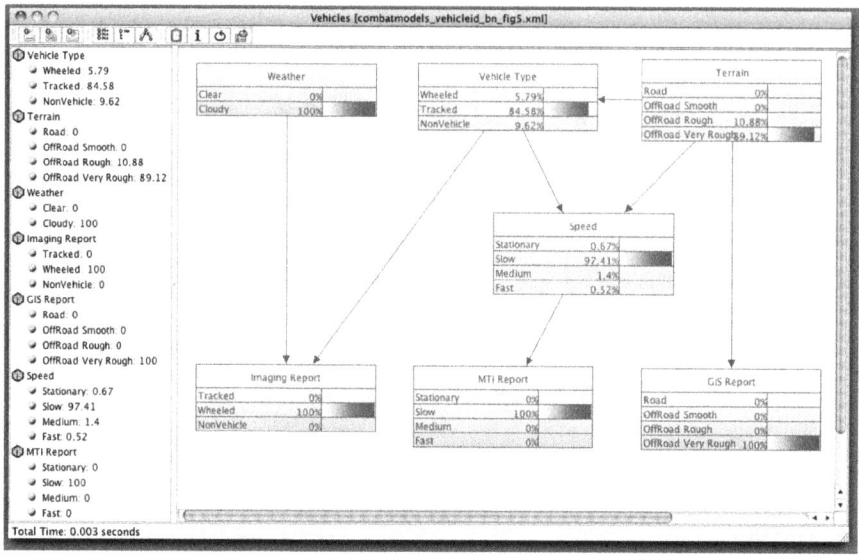

Figure 28.6   Compiled BN with conflicting evidence.

Case Study Extended: the MEBN version

Modeling with BNs is relatively straightforward. With some basic instruction in probability theory and BNs, many domain experts can learn to build models such as the one in Figure 28.2 by applying their domain knowledge directly to the situation described in the previous sections. Of course, the quality of the model will depend on the modeler's mastery of two aspects: knowledge about the domain and understanding of the technique. The first will directly impact whether the model captures the nuances of the situation with enough detail and fidelity, while the latter will help in producing the best model to answer the questions it is supposed to answer (i.e. knowing about advanced characteristics of BNs such as how to represent context-specific independence (Boutilier et al., 1996) will help to build better models). Still, less sophisticated models that are "good enough" for most purposes (e.g. as a tool to communicate details of the domain to non-experts) can be devised with relatively little difficulty.

While BNs are attractive as a natural and straightforward modeling tool, their limited expressive power makes them unsuitable for complex problems. The model in Figure 28.2 is good enough to assess the type of one vehicle given the three reports one gets about this vehicle and information on weather. For different situations and even slight modifications to the model's characteristics (e.g. more than three reports, reports from other types of sensor, different numbers of vehicles, etc.) one would have to build a new BN to capture the changes. In other words, the model shown in Figure 28.2 can be viewed as an instantiation of the modeler's knowledge about a specific situation. Any changes to this specific situation (other than changing the outcomes of the reports as for the two scenarios above) require building a new model.

MEBN modeling goes a level up in terms of abstraction. That is, MEBN can model variations in the situation such as changes in the numbers of entities and reports. Therefore, the modeler does not have to build an entirely new model to handle different numbers of vehicles or reports. In other words, upon receiving data from a specific situation (e.g. number of vehicles, reports, etc.) and a query set (with at least one element), a MEBN reasoner uses the information contained in the MTheory to instantiate the SSBN that best addresses the queries in the set. This is possible because of the greater expressiveness of MEBN. On the other hand, the leap in expressive power means that building an MTheory is a different type of effort from building a BN. Whereas building a BN involves applying domain knowledge directly to solve a specific problem, building an MTheory requires the modeler to think about repeatable patterns of domain knowledge that can be composed at runtime into a BN and then formalize these repeatable patterns as MFrags.

To cope with MEBN's greater expressiveness and higher demands on model design, there is the need for a formal methodology for designing MTheories. Figure 28.7 presents the Probabilistic Ontology Modeling Cycle (POMC), which is part of the Uncertainty Modeling Process for Semantic Technologies (UMP-ST) proposed in Carvalho et al. (2010) and then improved in Carvalho (2011). The POMC is intended to provide a methodology for building models in MEBN, and we shall use it to illustrate how the same example in the last section can be modeled as an MTheory.

As stated in Carvalho et al. (2010), the topmost circle (in blue) addresses the requirements for the model and involves defining the input that the model will be receiving. For simplicity, the only query to be addressed by the vehicle identification MEBN model of this section will be to identify the types of vehicles that are observed by the available sensors. Note that this assumption implies the possibility of various combinations of one or more reports coming from one or more sensors and related to one or more vehicles. This initial step also involves defining what evidence will be available to the model in order to reach its goals. In this example, the evidence list will be the same as used in the BN version (i.e. weather conditions and sensor input), but without the limitation on the number of sensors. In fact, the BN depicted in Figure 28.2 can be thought of as a special case of possible input combinations (i.e. three sensors, one report per sensor, all reports referring to the same vehicle).

The next three circles to the right (in green) refer to the analysis and design of the model. The first step in this phase (i.e. the first green circle) is to define what are the entities considered in the MTheory, including their attributes and relationships. In analyzing the vehicle identification BN, the obvious entities that

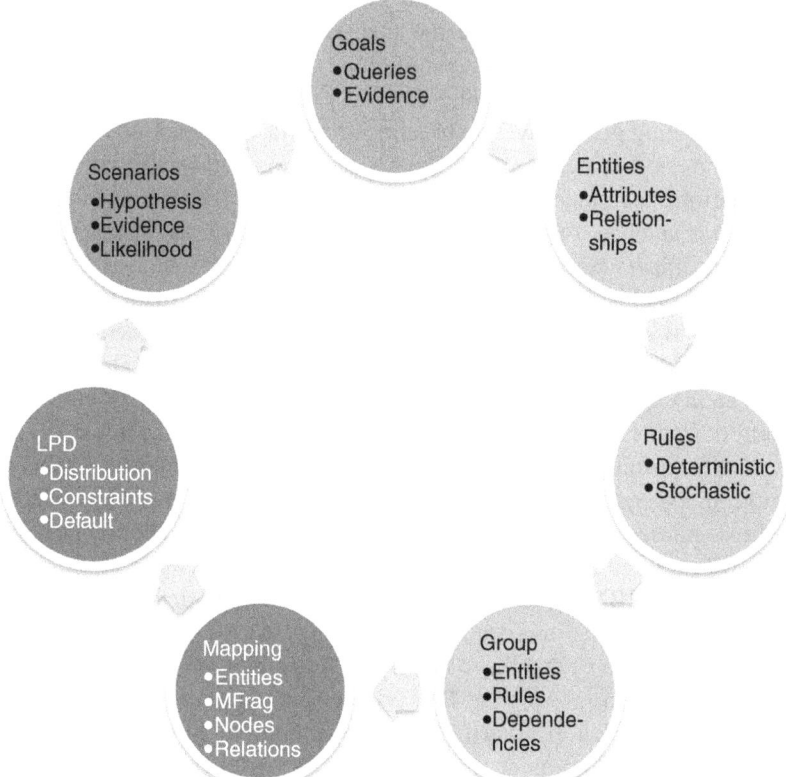

Figure 28.7 The Probabilistic Ontology Modeling Cycle (Carvalho, 2011).

come to mind are vehicles[7] and reports. That is, other types of entities can be listed as being part of the "vehicle identification" domain, but only those directly related to the purpose of the model should be listed. As an example, sensors play a role in this domain (e.g. generate reports), but for the purpose of assessing the type of a vehicle only the reports are needed, regardless of how they are generated. On the other hand, terrain type and weather conditions are attributes considered for inferential purposes so their associated entity/entities must be part of the model and included in our list. For our example, we will define an entity "region" with attributes of weather and terrain type. Finally, to account for the possibility of receiving multiple asynchronous reports we will also include the entity "time step" in our list.

---

[7]For semantic correctness, we used the term *object* so to avoid having a *vehicle* entity being classified as a *non-vehicle*.

The second step of the analysis and design phase is to assess the domain information that will be the basis for inference. That is, we need to define a set of rules reflecting the processes that happen in that domain. The last step of the analysis and design phase (i.e. the third green circle) involves an initial, notional structuring of the information collected in the previous two steps, assessing how the entities affect and are affected by the domain processes, what are the dependences observed in this interaction among entities and rules, and using those dependences to group related entities and rules accordingly.

The next phase is devoted to the model's implementation. There are two distinct steps to this phase, depicted as red circles in the POMC: the mapping and the local probability distributions (LPDs). Although the concept of building the actual model is tool agnostic, its realization is not. Thus, in this section we illustrate this phase by implementing the MEBN vehicle identification model using UnBBayes-MEBN.

The first implementation step (mapping) is tightly related to the result of the last step of the previous phase, since the grouping of the rules and entities according to the dependences observed among them will be the major factor driving the design of the model's structure. Once the entities are defined, we must describe their probabilistic properties and relationships through a collection of interdependent RVs (input nodes, resident nodes, and context nodes) and ordinary variables (for which individuals in the given situation will be substituted), all within the MFrag-based structure of a MEBN model. From a general point of view, the MFrag components in a probabilistic ontology can encode several ideas or patterns of a domain, as summarized in Table 28.2.

This table applies to MFrags encoding generic knowledge that applies to all individuals meeting their context constraints. Specific information about particular individuals is typically represented as findings, which are not described here.

Considering the mappings presented in Table 28.2, we can identify the following MFrags and variables in the vehicle identification model. See Figure 28.8 to visualize the MFrags, including dependences and contexts.

- *Reference_MFrag*: This MFrag gathers RVs representing references between different entities (e.g. the object reported on by a sensor, the region in which an object is located). The states of these resident nodes are instances of a type of entity (e.g. regions in the case of object locations). The current UnBBayes implementation defines the distribution for reference RVs as uniform on the entities involved in the situation. Future versions will have a richer representation. Because our case study does not involve reference uncertainty, the distribution for reference RVs is immaterial.
  - **Location**(*obj*): associates an object with the region in which it is located. States are instances of *Region*.
  - **ReportedRegion**(*rpt*): associates a report with the region on which it is reporting. States are instances of Region.

Table 28.2   Some knowledge patterns an MFrag can capture

Ideas/patterns	Description
Entity's properties	E.g. the type of an *Object* (tracked vehicle, wheeled vehicle, non-vehicle), weather of a *Region* (clear, cloudy)
Physical/abstract parts	If an entity is thought as structured, then we can represent its parts (e.g. GIS data, image data, and MTI data for *Report*)
References to entities	This is usually represented as an RV having an entity *A* as argument and entity *B* as its possible value (e.g. *ReportedRegion*(*rpt*), whose *rpt* is an ordinary variable having *Report* as its type, and *ReportedRegion*'s possible values are individuals of *Region*). N-ary references can be represented by using multiple arguments. This type of relationship can also be represented by a Boolean RV with two or more arguments, so that its value is true when the entities associated with its arguments are related. Thus, the possible values will be predefined, and the probability distribution can be defined by its LPD
Relationships	A non-functional relationship (e.g. an object is in the field of view of a sensor) can be represented by a Boolean RV having value True when the entities associated with its arguments are related, and False otherwise. The probabilities associated with those states are defined via LPDs
Dependencies	These are represented as arrows, pointing to resident nodes, and described by an LPD, which determines the probability of an RV given its dependences
Restrictions	Context nodes can represent restrictions in the domain

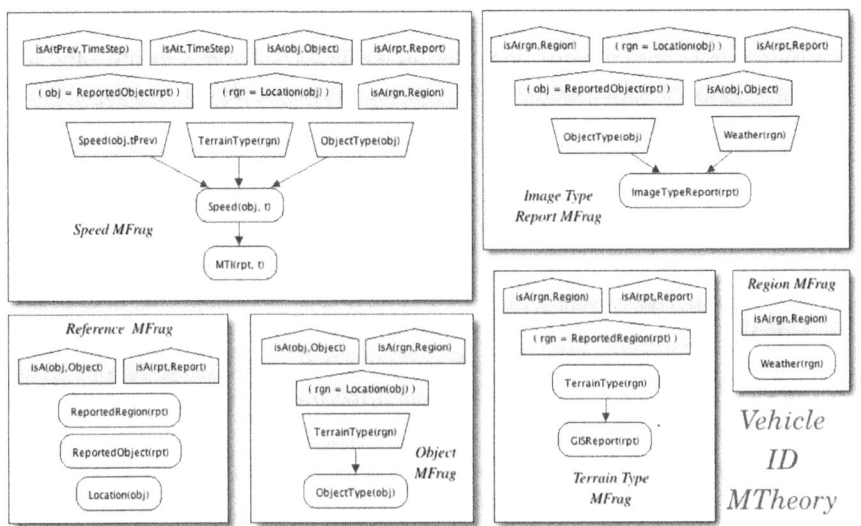

Figure 28.8   The vehicle identification MTheory.

- **ReportedObject**(*rpt*): associates a report with the object it is reporting on. States are instances of Object.
- **isA**(*obj, Object*): This is a context node indicating that the ordinary variable *obj* is of type *Object*. This context node indicates that instances of Object meeting the other context constraints can be substituted for the ordinary variable "obj" in any of the RVs of this MFrag. Because they represent the same concept, explanations of other "isA" nodes are omitted.

- *Speed_MFrag*: This MFrag represents speed of an object and its influence on the MTI report.

  - **Speed**(*obj, t*): represents the speed of an object at a given time step. States are elements of the set {*Stationary, Slow, Medium, Fast, Very-Fast*}.[8] Note that an additional speed category has been added to the ones presented above. This is in order to represent patterns of change in speed over time in a planned enhancement of the model, although it is not a vital aspect for this example.
  - **MTI**(*rpt, t*): represents the speed reported by the MTI at a given time step. Distribution is conditioned on the speed of the object generating the report. States are {*Stationary, Slow, Medium, Fast, VeryFast*}.
  - *rgn* = **Location**(*obj*): This is a context node stating that *rgn* is the location where *obj* is located. Context nodes will be explained only in the first MFrag in which they appear.
  - *obj* = **ReportedObject**(*rpt*): This is a context node stating that *obj* is the subject of report *rpt*.

- *ImageTypeReport_MFrag*: This MFrag represents reports from images (e.g. satellite images).

  - **ImageTypeReport**(*rpt*): represents the type of an object inferred from an image. States are elements of the set {*Tracked, Wheeled, NonVehicle*}.

- *Object_MFrag*: This MFrag represents properties of an object. In our simplified example, we represent only the type of the object. Other characteristics, such as shape, surface texture, or material composition, might be represented in a more sophisticated model.

  - **ObjectType**(*obj*): represents the type of the analyzed object. This is the query node (goal) for our example. States are elements of the set {*Tracked, Wheeled, NonVehicle*}.

- *TerrainType_MFrag*: This MFrag represents the terrain type of a region and the GIS report on terrain type. For our simple example, we represent the type of terrain and the GIS report about terrain type. Note that these

---

[8]We assume *Slow* is less than 15 kph, *Medium* is around 16–59 kph, and *Fast* is above 60 kph. We may consider "*VeryFast*" as fast as over 100 kph.

- ∘ **TerrainType**(*rgn*): represents the actual terrain type of a region. States are elements of the set {*Road, OffRoadSmooth, OffRoadRough, OffRoad-VeryRough*}.
- ∘ **GISReport**(*Report*): represents the terrain type obtained from a GIS report. States are elements of the set {*Road, OffRoadSmooth, OffRoad-Rough, OffRoadVeryRough*}.
- ∘ *rgn* = **ReportedRegion**(*rpt*): This is a context node stating that *rgn* is the region reported by the report *rpt*.

- **Weather_MFrag**: This MFrag represents the weather property of a region.

  - ∘ **Weather**(*rgn*): represents the weather condition (obtained from weather reports). States are elements of the set {*clear, cloudy*}.

Figure 28.9 shows the MTheory depicted in Figure 28.8 implemented in UnBBayes.

Note that **Speed_MFrag** contains a temporal recursion,[9] as the distribution of **Speed**(*obj*, *t*) depends on **Speed**(*obj*, *tPrev*). MEBN has enough expressiveness to represent temporal recursion. Inquisitive readers may ask how this recursion

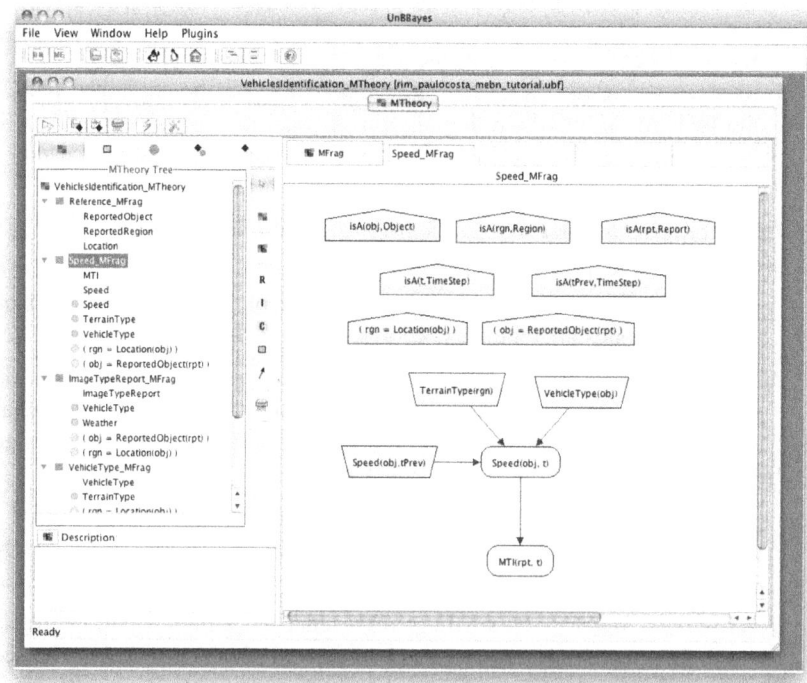

Figure 28.9 The vehicle identification MTheory in UnBBayes.

---

[9]Temporal recursion means a distribution for an RV with a temporal argument whose distribution depends on RVs at a previous time *t*.

is handled. In particular, there is no context node to enforce the constraint that *tPrev* precedes *t*, nor is an initial value shown to ground the constraint. A limited variety of linearly ordered recursion is handled automatically and implicitly by UnBBayes. The ordinary variables *t* and *tPrev* are of type *TimeStep*, which has been defined as a linearly ordered class. Thus, UnBBayes handles this recursion automatically.

After defining the structure of the model, the next step of the implementation phase is to define the LPDs for each MFrag. UnBBayes-MEBN provides a flexible way for declaring the LPD of resident nodes. This is done by using a special-purpose script, whose grammar is described in Figure 28.10. The *varsetname* is a dot-separated list of ordinary variables, and it refers to parent nodes containing all of those ordinary variables as arguments. *MIN, MAX* and *CARDINALITY* are respectively a minimal function, a maximum function, and a function to count the number of parent combinations, having *varsetname* as arguments and satisfying *b_expression*. If no script is provided, UnBBayes-MEBN assumes a uniform distribution (all values are equally probable). The current version of the LPD script does not include a way to define LPDs of reference nodes; thus, they use a uniform distribution.

The last phase of the POMC focuses on testing the model. Its only step (purple circle) is to build the scenarios that will be used to evaluate and validate the model. To do this, we specify the set of relevant entities in the situation. We assume a fixed, known set of entities in any given situation, although the number of relevant entities may vary from situation to situation.[10] If the situation includes the same number of entities of the same types as our example in the last

Figure 28.10 Dynamic LPD follows this grammar. The window at the top right corner shows how LPD script is edited in UnBBayes-MEBN.

---

[10]Although MEBN can represent *existence uncertainty*, in which previously unknown entities may be hypothesized, we do not consider existence uncertainty in our case study.

section (i.e. three reports, one object, one region, one time step), then a query on the type of object will trigger an SSBN construction process that will yield an SSBN that is identical to the BNs in Figures 28.5 and 28.6.[11] Otherwise, for any number of reports, objects, regions, etc., the SSBN construction process will result in the smallest SSBN needed to answer the query.

As an example, Figure 28.11 shows the SSBN that is obtained for a problem in which two objects are observed in a region reported by the GIS to be off-road very rough terrain; the weather is cloudy; the first object is reported by the imaging sensor as wheeled and by MTI as traveling slowly; and the second object is reported by the imaging sensor as tracked and by the MTI as traveling at medium speed. These reports taken collectively indicate a high likelihood of off-road very rough terrain, and consequently a high likelihood that both objects are tracked vehicles. Because of the cloudy conditions and the conflicting evidence, the model has concluded that the imaging sensor is likely to have reported erroneously.

A full tutorial on how to construct a MEBN model is outside the scope of this chapter, but the interested reader can refer to Matsumoto et al. (2011) or to the PR-OWL website at www.pr-owl.org.

As mentioned above, applying different data to the MTheory in Figure 28.9 would yield a different SSBN from that obtained in Figure 28.11.

Although we have shown that our MTheory can represent models including multiple vehicles, so far our example has remained at level 1 of the JDL hierarchy. The MTheory in Figure 28.9 allows for a flexible system that can handle a

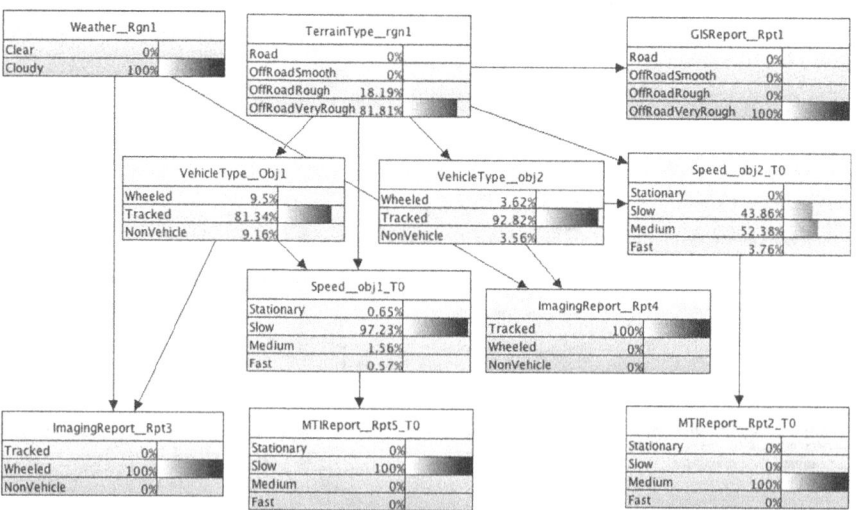

Figure 28.11 The Vehicle Identification SSBN.

---

[11]As noted above, there are minor differences in the local distributions of the BN model of the previous section and the MEBN model of this section, but the structure of the SSBNs will be the same for this case.

varying number of entities, but fails to address relationships between the entities and cannot leverage their interactions to make stronger inferences. Much of the MEBN representational power lies in its ability to address data fusion at JDL level 2 and above. We now extend our example to illustrate fusion at level 2. Military organizations are organized in units. Each unit consists of a group of personnel acting in various roles and using equipment characteristic of their unit type. To keep the model simple, we consider only a very basic kind of unit model. We introduce the concept of a *vehicle group* consisting of vehicles conducting an operation together. We assume two types of vehicle group: *TrackedGroup* and *WheeledGroup*, each of which contains mostly vehicles of its respective type, although occasionally a group may contain vehicles not of its type. We defi ne a **GroupType** RV to represent the type of a hypothesized group of vehicles. This RV has states in the set {*TrackedGroup, WheeledGroup, Not-AGroup*}, where the third state represents vehicles hypothesized to be a group which are in fact not a group. Figure 28.12 shows an MTheory that extends our example to include vehicle groups.

In the model, the LPD for the type of a group is conditioned on the terrain in which it was observed (e.g. a *WheeledGroup* is more likely to be observed on roads) and directly influences the vehicles observed (e.g. four tracked vehicles observed together are more likely to be part of a tracked group). This model illustrates just one simple example of the various types of relationships that can be captured by an advanced Bayesian representational framework such as MEBN, but it is sufficient to show that the logic provides a principled means to perform data fusion in combat models at JDL level 2 and above. Figure 28.13 shows the SSBN that results from a query about three different objects, two of them having the same data used previously in Figures 28.5, 28.6, and 28.11.

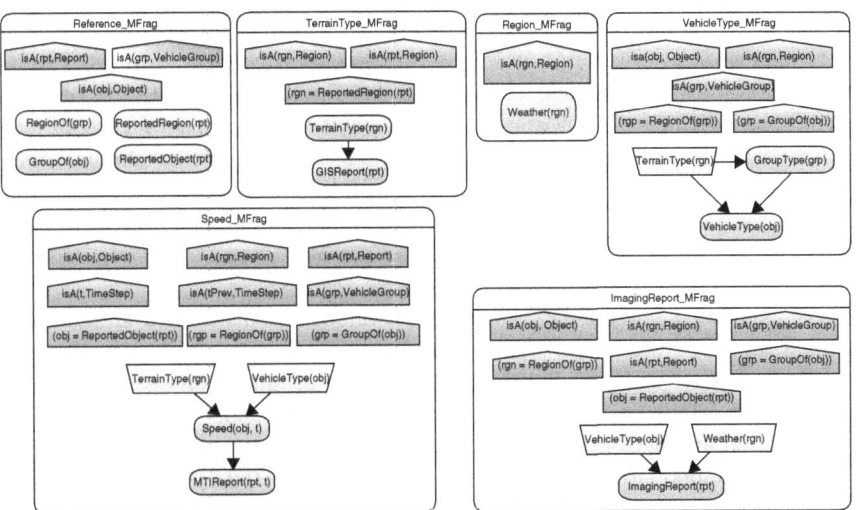

Figure 28.12 The vehicle identification MTheory—JDL2 version.

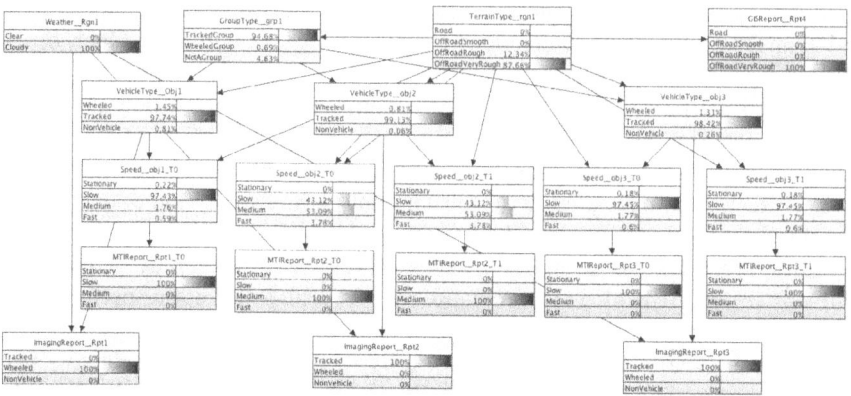

Figure 28.13 The vehicle identification SSBN—JDL2 version.

The SSBN in Figure 28.13 has strong evidence of a tracked group being observed, and because vehicles 1–3 are thought to be part of a tracked group their individual likelihoods of being tracked vehicles are also higher than they would be if the same data had been considered for the three vehicles separately. Wright et al. (2001) used a similar, although more sophisticated, model of vehicle groups to perform level 2 fusion. Their work demonstrated empirically the ability of level 2 models to improve the robustness of a fusion system to noisy information (e.g. dropped or crossed tracks; incorrect typing) about individual vehicles.

Capturing these kinds of relationships between entities and formalizing them using a logical framework is the purpose of ontologies, which represent knowledge about a domain in a structured and sharable way, ideally in a format that can be read and processed by a computer. When an expressive Bayesian logic framework is used to provide this formalization then the result is a probabilistic ontology (PO) (Costa, 2005; Costa and Laskey, 2006). POs expand the possibilities of standard ontologies by introducing the requirement of a proper representation of the statistical regularities and the uncertain evidence about entities in a domain of application. The probabilistic ontology language PR-OWL (Costa and Laskey, 2006; Carvalho, 2011) extends the ontology language OWL, providing the ability to represent uncertainty about the attributes of and relationships among entities represented in OWL ontologies.

As mentioned before, the increasing prevalence of distributed simulations, decision support systems, predictive analysis, and other capabilities for combat modeling points to a clear trend toward employing advanced probabilistic representations. Probabilistic ontologies can provide principled support for uncertainty representation and reasoning in distributed combat models. As illustrated above, this includes support for data fusion spanning multiple JDL levels. In addition, POs can represent the complex kinds of knowledge needed to support stochastic simulation of the behavior of entities in a combat simulation.

CONCLUSION AND FUTURE WORK

Future military missions in a net-centric environment will require distributed, autonomous agents to share information and cooperatively assess unpredictable situations to support real-time decisions in a rapidly changing environment. Situations change over time through dynamic evolution of the world, and also through actions taken by agents. Bayesian evidential reasoning is a theoretical model for individual-agent situation awareness that has achieved considerable practical success. A Bayesian agent expresses beliefs about the current situation as a probability distribution over situations. The agent also has a dynamic model of how the situation evolves, a model of the effects of actions on situation evolution, and a model of observations it can obtain. Using these models, the agent updates its beliefs about the situation as it takes actions and makes observations.

In order to cope with the increasing demands of simulation and decision support, combat models must be able to provide a principled methodology for dealing with the fog of war in a complex environment. A new generation of expressive probabilistic languages has recently emerged, allowing much richer representations of situations involving entities of different types, involved in a variety of relationships, and executing complex processes. In this chapter, we presented Bayesian techniques that are at the edge of the technology and have been applied in distinct military applications such as high level fusion and pattern recognition. Further, as a principled, mathematically coherent means to represent uncertainty about the attributes of, relationships among, and behavior of interacting entities of different types, probabilistic ontologies have the potential to support a powerful uncertainty representation and management capability for distributed stochastic combat simulations.

ACKNOWLEDGEMENTS

Paulo Costa and Kathryn Laskey would like to recognize the invaluable contribution from the many students of GMU's OR-719 course (Computational Models for Probabilistic Reasoning) who helped to improve and fine tune the vehicle identification models described in this chapter. Our special thanks go to Rommel Carvalho and Shou Matsumoto for their great work with the development of the MEBN version of the tutorial, and to Marcelo Ladeira and his UnBBayes team for their continuing support with the software.

Heber Herencia-Zapana was supported by the National Aeronautics and Space Administration under NASA Cooperative Agreement NCC-1-02043.

REFERENCES

Boutilier C, Friedman N, Goldszmidt M and Koller D (1996). *Context-Specific Independence in Bayesian Networks*, pp. 115–123.

Carvalho RN (2011). Probabilistic ontology: representation and modeling methodology. PhD Dissertation, George Mason University, Fairfax, VA.

Carvalho RN, Laskey KB, da Costa PCG, Ladeira M, Santos LL and Matsumoto S (2010). UnBBayes: modeling uncertainty for plausible reasoning in the semantic web. In: *Semantic Web*. InTech, Rijeka, Croatia, pp. 1–28.

Costa PCG (2005). Bayesian semantics for the semantic web. PhD dissertation, George Mason University, Fairfax, VA.

Costa PCG and Laskey KB (2006). PR-OWL: a framework for probabilistic ontologies. In: *Frontiers in Artificial Intelligence and Applications*. IOS Press, Baltimore, MD, Vol. 150, pp. 237–249.

Costa P, Laskey K and Laskey K (2006). Probabilistic ontologies for efficient resource sharing in semantic web services. In: *Proceedings of the Second Workshop on Uncertainty Reasoning for the Semantic Web, Athens, GA*.

Darwiche A (2009). *Modeling and Reasoning with Bayesian Networks* (1st edition). Cambridge University Press, Cambridge, MA.

Frege G (1967). Concept script, a formal language of pure thought modeled upon that of arithmetic. In: *From Frege to Godel: A Source Book in Mathematical Logic, 1879-1931*, edited by S Bauer-Mengelberg. Harvard University Press, Cambridge, MA.

Gamez JA, Moral S and Cerdan AS (2010). *Advances in Bayesian Networks* (1st edition). Springer-Verlag, Berlin.

Hall D and Llinas J (1997). An introduction to multisensor data fusion. *Proc IEEE* **85**, 6–23.

Hall DL and McMullen SAH (2004). *Mathematical Techniques in Multisensor Data Fusion*. Artech House, Boston, MA.

Heckerman D and Breese J (1996). Causal independence for probability assessment and inference using bayesian networks. *IEEE Transactions on Systems, Man and Cybernetics, Part A: Systems and Humans* **26**, 826–831.

Heckerman D, Mamdani A and Wellman MP (1995). Real-world applications of Bayesian networks. *Commun ACM* **38**, 24–26. Jaeger M (1994). Probabilistic reasoning in terminological logics. In: *Proceedings of the 4th International Conference on Principles of Knowledge Representation and Reasoning (KR'94)*. Morgan Kaufmann, Bonn, Germany.

Jensen FB and Graven-Nielsen T (2010). *Bayesian Networks and Decision Graphs* (2nd edition). Springer, New York, NY.

Kjaerulff UB and Madsen AL (2010). *Bayesian Networks and Influence Diagrams: A Guide to Construction and Analysis* (1st edition). Springer, New York, NY.

Koller D. and Pfeffer A (1997). Object-oriented Bayesian networks. In: *Thirteenth Conference on Uncertainty in Articial Intelligence (UAI-97)*, San Francisco, CA.

Laskey KB (2008). MEBN: a language for 1st-order bayesian knowledge bases. *Artif Intell* **172**, 140–178.

Laskey K, Costa P, Wright E and Laskey K (2007). Probabilistic ontology for net-centric fusion. In: *Proceedings of the Tenth International Conference on Information Fusion, Quebec, Canada*.

Mahoney SM and Laskey KB (1998). Constructing situation specific belief networks. In: *Proceedings of the Thirteenth Conference on Uncertainty in Artificial Intelligence* (UAI 97), Providence, Rhode Island, pp. 370–379.

Matsumoto S, Carvalho RN, Costa PCG, Laskey KB, Santos L and Ladeira M (2011). *Introduction to the Semantic Web: Concepts, Technologies and Applications*. iConcept Press, Annerley, Australia, p. 23.

Milch B, Marthi B, Russell S, Sontag D, Ong DL and Kolobov A (2005). Blog: Probabilistic models with unknown objects. In: *Proceedings of the 19th International Joint Conference on Artificial Intelligence*. Morgan Kaufmann, Bonn, Germany, pp. 1352–1359.

Neapolitan RE (2003). *Learning Bayesian Networks* (illustrated edition). Prentice Hall, Upper Saddle River, NJ.

Pearl J (1988). *Probabilistic Reasoning in Intelligent Systems: Networks of Plausible Inference* (1st edition). Morgan Kaufmann, Bonn, Germany.

Pearl J (2000). *Causality: Models, Reasoning, and Inference*. Cambridge University Press, Cambridge.

Peirce CS (1885). On the algebra of logic. *Am J Math* **7**, 180–202.

Pfeffer A (2000). Probabilistic reasoning for complex systems. PhD, Stanford University.

Pourret O, Naim P and Marcot B (2008). *Bayesian Networks: A Practical Guide to Applications* (1st edition). Wiley, New York, NY.

Richardson M and Domingos P (2006). Markov logic networks. *Machine Learning* **62**, 107–136.

Steinberg BC and White F (1999). Revisions to the JDL data fusion model. *Sensor Fusion: Arch Algorith Appl* **3719**, 430–441.

White FE (1988). A model for data fusion. In: *Proceedings of the First National Symposium on Sensor*, Chicago, IL.

Wright E, Mahoney SM and Laskey KB (2001). Use of domain knowledge models to recognize cooperative force activities. In: *Proceedings of the 2001 MSS National Symposium on Sensor and Data Fusion, San Diego*, CA.

Chapter 29

Model-Based Data Engineering for Distributed Simulations

Saikou Y. Diallo

INTRODUCTION

The goal of this chapter is to provide a framework for engineers and practitioners who are faced with the challenge of engineering suitable data to support two or more interoperating distributed simulations. Within this context, this section focuses on how, through model based data engineering, to ensure that the data to be exchanged exist, are semantically accessible and do not result in inconsistencies that might lead to false results. There are several applicable engineering approaches to data interoperability but we focus on model-based data engineering (MBDE) because of its military roots, its maturity, and its level of detail.

MBDE uses the same steps as data engineering, which was first introduced in the North Atlantic Treaty Organisation (2002). While in data engineering the focus is on preparing a system for integration within a federation, MDBE is centered on capturing the data requirements of the federation in the form of a common reference model (CRM). In order to do so, the notions of requirements, needs, and capabilities are introduced and the need to capture data dependencies is formulated. The steps of MBDE are expressed as a series of algorithms that together form a heuristic that approximate a CRM. In the next section, we frame the data interoperability problem as it relates to combat modeling.

Engineering Principles of Combat Modeling and Distributed Simulation, First Edition.
Edited by Andreas Tolk.

DATA INTEROPERABILITY CHALLENGES FOR COMBAT MODELING

In many instances, military users are presented with a vignette describing a scenario that needs to be trained, or a situation that has to be rehearsed. In many of those cases, there is a requirement to use existing combat models in a federated environment to foster reuse or simply to comply with budget restrictions. The main idea behind this requirement is that models can be composed and recomposed to represent multiple situations, thus reducing the need to create new models for new situations. Within the context of interoperating distributed simulations, there exist standards such as the Distributed Interactive Simulation (DIS) (IEEE, 2002) and the High Level Architecture (HLA) (IEEE, 2000) that are developed to connect simulations (communicate data and synchronize actions) in a distributed environment. In addition, there is a Federation Development and Execution Process (FEDEP) that governs how simulations can be prepared to be integrated in a federation (IEEE, 2003). This process has been generalized into the Distributed Simulation Engineering and Execution Process (DSEEP) (IEEE, 2003). These standards cover both the technical and the semantic aspects of capturing and exchanging data but an additional engineering process is required for the information exchange to occur without inconsistencies. An engineering method such as MBDE is necessary in use cases such as the following.

- *Simulation interoperability:* The combination of two or more simulation systems or combat models in a distributed environment might arise from the requirement to combine two or more simulation systems in order to create a system of systems. In this case, information must be exchanged from one simulation to another and MBDE is required to ensure that only the necessary and sufficient information is exchanged and that the federation behaves consistently as a whole. In addition, with current standards, it is often required to use a standardized or a *de facto* common data model (RPR-FOM, JLVC FOM, MATREX FOM), and MBDE can be used to map internal representations and supporting interfaces to the standardized data model.

- *Reuse of data:* In some cases, it might be necessary to reuse initialization data from another exercise with the necessary changes. In this case, MBDE is required to understand and align the data sets required to support the current scenario with the data that are being reused. The goal is to ensure consistency and avoid data duplication which can introduce ambiguities such as homonyms and synonyms. In addition, MBDE ensures a consistent interpretation of data.

- *Multiple data sources:* In the case of command and control systems or in terms of scenario development, it is often necessary to merge multiple authoritative data sources at multiple echelons in order to generate a full picture of the exercise. In that case MBDE ensures that the merging process is done consistently and that data are interpreted unambiguously by the participating systems.

In this chapter we use the simulation interoperability use case as the basis for discussion and illustration. Nonetheless, MBDE is applicable to the two other use cases mentioned above.

Before we move forward, there are three major assumptions the reader must be aware of and should avoid making when engaging in the process of MBDE.

- *Availability assumption:* data that need to be exchanged in a federation are available at least in one of the simulations, they can be communicated, and it is of secondary importance how they are generated internally within a simulation. This assumption also means that the functions that output data are universally aligned. This is most certainly not the case in general as for instance several simulations implement different line-of-sight algorithms. The reader is encouraged to read Davis and Anderson (2004) for a more substantial discussion on fair-fight issues.

- *Semantic alignment assumption:* The meaning of data is universally agreed upon within the federation. This is not necessarily true. For instance, when two data models refer to a geographical location, model one refers to location in terms of longitude–latitude while model two refers to location in terms of latitude, longitude and elevation.

- *Independence assumption:* Every datum can be produced independently by a simulation. This assumption ignores groupings of data that only have meaning when taken as a whole. As a simple example, the position of a tank can be expressed in one, two, or three dimensions which equates to one, two, or three groupings of data. The way one simulation models its position as opposed to another might result in some serious mismatches between a simulation's representation of the world and the federation's representation of the world.

Overall, it is essential to recognize that a federation of models is also a model. As such, it is designed to answer a modeling question. This question could be how to fulfill a training objective, how to identify the best system to acquire, or simply how to identify the best course of action. This observation is essential in the way one approaches the data interoperability problem. In general, there are two ways to approach a federation. One approach is to start from the top by developing a federation model and then identify systems that can fulfill the model. Another way is to first identify the systems that will participate in the federation and then generate a federation model based on the systems. Both approaches involve the following steps.

- *Identify the research question:* The initial action is to identify the research question as it guides what data to select and acts as a tie-breaker when two or more data are relevant. For instance, a tank model that is used in training and a tank model that is used in testing are both useful in a given scenario, but the decision on which to use is driven by the research question.

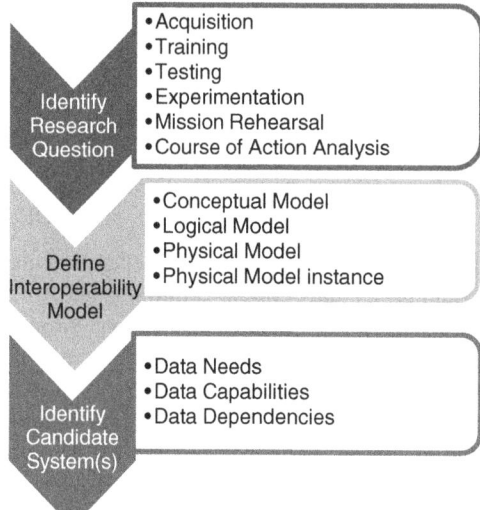

Figure 29.1 Top-down approach to data interoperability problem.

- *Define the interoperability model:* The interoperability model is the model that answers the research question. It is specified according to data modeling theory but usually takes the form of a relational model (Codd, 1970). It is first captured as a conceptual model which is used to generate a logical model. The logical model is normalized and turned into a physical model which is implemented as a physical model instance. It is important to note that in some cases existing models can be reused and some models are standards in given domains. For instance, the Joint Command Control and Communication Interexchange Data Model (JC3IEDM) is a standard data model for the North Atlantic Treaty Organization (NATO) Command and Control information exchange.

- *Identify candidate systems:* At this point, the data requirements for the federation are captured and it is time to select one or more candidate simulation systems that implement parts or the entire interoperability model. If there is a simulation that can do it, the work is done at the data level. If there is a need for more than one—possibly a combination of live, virtual, and constructive simulations—an engineering process is required to identify the data needs and capabilities of each simulation. In addition, special attention has to be paid to the data dependencies (relationships between data) within each simulation.

Figure 29.1 shows the suggested top-down approach and the questions or activities that need to be considered under each task.

MBDE works either way as it is generalized to support a federated approach. MBDE is an engineering approach that shows how to identify the needs and requirements and highlights the gaps between the simulation's capabilities and the requirements expressed by the interoperability model (Tolk and Diallo, 2010).

It also highlights differences in scope and resolution and shows where aggregation and disaggregation might be needed. In the next section, we introduce MBDE and its main steps.

MODEL-BASED DATA ENGINEERING

Before we introduce MDBE, it is essential to define some key terms from Diallo (2010).

- *Elements.* are real or imaginary things. An element in our case is the real military object such as a tank, a plane, or a bullet. An element can be modeled either as an entity or a property. For instance, one can model a tank as an object with several properties or simply model a tank as an entity capable of moving which reduces the tank to its ability to move.
- *Entities.* are abstractions of elements. An entity is a model of the real tank, plane, or bullet. Entities are found in a combat model as simplified versions of elements.
- *Properties.* are the characteristics of an element. In general, properties are modeled as attributes of an entity and those attributes have a counterpart in the real world.
- *Symbols.* are the representations of elements. Symbols can be numbers, strings, images, text, or a combination of symbols.
- *A domain.* is a collection of unique symbols. The domain is the set of elements that belong to a context. Every element is uniquely identifiable within a given domain. The domain can be either enumerated ($\{a, b, c, d\}$) or specified (positive integers).

Having defined these key terms, we now introduce MBDE. MBDE comprises four main steps as informally described in data engineering. We shall examine each in the next subsections.

Data Administration

The goal of data administration is to identify the format and physical location of data elements and their value domains when appropriate. For unstructured or semi-structured data a form of semantic enrichment is needed to generate structured data. Data administration results in an unambiguous definition of data elements, a classification of data in terms of entities, properties, and values or value domains, and documented data in the form of metadata. It is recommended that both data and metadata be captured in a machine language such as XML Schema (XSD) and recorded in a metadata registry administered following the ISO/IEC 1179 standard of metadata registry (ISO/IEC, 2003). The algorithm in Table 29.1 can be used for data administration.

At the end of this process every datum that is required to answer the modeling question is captured. It is important to note that entities, properties, and symbols

Table 29.1   Data Administration Algorithm for the Federation Model

```
1. Begin
      a. For every datum required to answer the modeling
         question
            i. Identify the format
           ii. If the datum is unstructured or semi-
               structured
          iii. Structure datum
           iv. Tag datum with Metadata and go to step vii
            v. If the datum is structured
           vi. Identify the location
          vii. Classify datum as entity, property or symbol
         viii. Identify the domain
2. End
```

Table 29.2   Data Administration for Candidate System

```
1. Begin
      a. For every candidate system
            i. Identify the datum that it can produce
           ii. If the datum is unstructured or semi-
               structured
          iii. Structure datum
           iv. Tag datum with Metadata and go to step vi
            v. If the datum is structured
           vi. Identify the location
          vii. Classify datum as entity, property or symbol
         viii. If it is a symbol identify its domain
           ix. Identify datum that is needed to produce it
            x. Classify as entity, property or symbol
           xi. Check if it is the last datum
          xii. If no go to Step i
         xiii. Else go to Step 2
2. End
```

are separated in three distinct categories. Consequently, every element is an entity, a property, or a symbol. The act of relating entities to properties and properties to symbols happens during the data management process. We can now apply the algorithm to each subsystem with additional steps as shown in Table 29.2.

At the end of this process every datum that the system can produce is captured along with its data requirements. As a simple example, if a tank with property {speed} is identified as a requirement in the federation and a tank simulation is identified as a candidate model, this algorithm requires that we identify what the tank simulator needs in order to calculate speed. This could be the properties

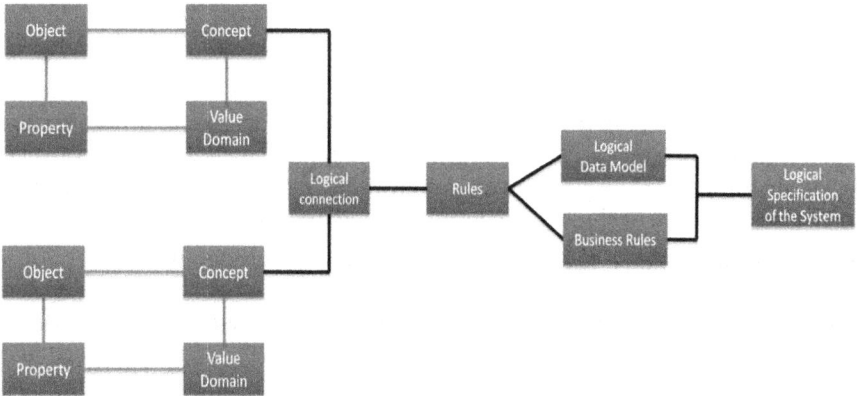

Figure 29.2   Data management process and artifacts.

{distance} and {time}. As a result we have the entity {tank} and the properties {speed} as requirement and {distance, time} as needs. In the next process, we will show how these data are further categorized in order to make the system participate in the federation.

Data Management

The goal of data management is to identify the logical relations between data elements and all relevant business rules including how to form meaningful grouping of elements otherwise known as composites (Tolk and Diallo, 2010). The application of data management results in the identification of all elements including their structure and context. Data management produces a logical data model that should also be captured in machine understandable language.

Figure 29.2 shows the data administration process that results in the specification of a logical data model. The data gathered in the data administration process must be linked together to form logical units of data that can be evaluated in context. We shall formally define what context and resolution is later in this chapter. For the moment, let us focus on generating logical units of data for the federation and the candidate systems. For this purpose, we define the following basic relationships:

- *hasProperty:* the grouping of entities with properties. This grouping is simply a relation that states "entity E has a property P";
- *hasSymbol:* the groupings of properties with symbols. This grouping captures the sentence "property P has value V."

In addition, it is necessary to capture the data dependences that are required to capture the interoperability model. In general, two elements are functionally dependent if a change in one implies a change in the other. For simplicity and to stay in the scope of combat modeling, there are two types of change that can occur.

1. *Existential dependence (E-dependent):* the creation or elimination of an element results in the creation or elimination of another. For instance the creation of a tank model results in the creation of a crew and, conversely, the destruction of an airplane implies the destruction of a crew. In this chapter we will limit E-dependences to entities.

2. *Transformational dependence (T-dependent):* a change in the value of a property implying the change in the value of another. To keep with our example, the change in the status value of a crew member (the pilot) from "alive" to "dead" implies changing the status of the plane, its speed and altitude. Again, in this chapter we will limit T-dependences to properties.

These dependences generate two additional groupings that we simply call *isE-dependent* and *isT-dependent*.

Having defined the groupings, we can now apply the algorithm in Table 29.3 to capture all of the groupings for the interoperability model.

It is important to note that the generation of the E-dependent and T-dependent groupings is entirely a function of the modeling question and therefore has to be done by inspection. The data administration process if done in a machine understandable way can serve as a core set of data that can be combined and recombined to generate a data model that is suitable for a given modeling question. As a simple example, let us assume that we want to create an interoperability model that involves a tank with properties {speed, size, type, position}. Following the algorithm, we decide that the creation of a tank results in the creation of four soldiers with a location property that is equal to that of the tank. This results in a {tank, soldier} E-dependent grouping and we use an enumeration of types as symbols for the type property and integers with domain miles per hour for the speed property. Further, we discount the idea of dismounted operations which results in the T-dependent grouping {position, location}. Let us now assume that

Table 29.3 Interoperability Model Generation Algorithm

```
1. Begin
     a. For every entity
          i. Generate E-Dependent grouping
               1. For every property
                    a. Generate T-Dependent grouping
                    b. if property applies to entity add
                       to has Property grouping
                         i. For every symbol
                              1. If symbol applies to
                                 property add to has
                                 Symbol grouping
2. End
```

we want to express that a soldier is co-located with the tank as in the "soldiers are mounted on the tank." This would result in a T-dependent grouping of {location, tank}. However, since we have restricted T-dependence to properties, we would have to make sure that the interoperability model includes a location property or an algorithm that allows for the derivation of a location in order to accommodate the notion of mounted or dismounted. The objective of dependences is to allow the modeler to express modeling decisions with the choice of groupings. This allows for modularity in the design of the model and transparency in interoperability. We repeat this process until all our entities, properties, and symbols are identified and grouped.

Having generated the interoperability model we apply another algorithm to the candidate systems and generate their respective models. In addition, for the candidate systems we must be aware of the data needs given the data requirements provided by the interoperability model. As a reminder, the objective is to combine the candidate systems such that the resulting combination produces a data model that is equivalent to the interoperability model. We assume that the candidate model can produce at least one entity in the interoperability model. Therefore step a in the following algorithm (Table 29.4) applies to those entities and not the entire set of entities in the candidate model.

This algorithm generates for each candidate model a set of sets that encompass not only the entities properties and symbols that are required to answer the modeling question but also all the entities, properties, and symbols that the candidate model needs to produce such an answer. Staying with our previous interoperability model of tank and soldier, we identify a candidate model that can talk about tanks. However, its E-dependent grouping is empty, meaning it is independent and therefore does not imply the creation of soldiers. This means that we now need a second candidate model that talks about soldier. In this model there is an E-dependent grouping of {soldier, weapon} meaning every soldier has a weapon. Further, soldiers have a property location but it is T-dependent on

Table 29.4 Candidate Model Generation Algorithm

```
1. Begin
      a. For every candidate entity
            i. Generate E-Dependent grouping
                  1. For every candidate property
                        a. Generate T-Dependent grouping
                        b. if property applies to entity add
                           T-dependent grouping to
                           has Property grouping
                              i. For every symbol
                                    1. If symbol applies to
                                       property add to
                                       has Symbol grouping
2. End
```

status meaning if the status is "mounted" it has the position of the vehicle it is mounted on; otherwise it has its own position. This poses a challenge that has to be dealt with during data alignment but the role of data management is to highlight these gaps such that the data alignment process can be conducted without additional data management tasks.

Regardless of the approach (top-down or bottom-up) these dependences always exist. From a top-down perspective, MBDE involves $n+1$ models (n being the number of models plus the candidate model) meaning that the dependences are first specified in the interoperability model, and then in each candidate model respectively. These dependences are then aligned to form a coherent whole. From the bottom-up perspective, the dependences are first defined within each candidate model and then MBDE is applied to create dependences between the models in order to generate the interoperability model.

The challenge is how to satisfy the need of the candidate system such that it answers the modeling question (or part of it) while remaining consistent. As a simple example, let us take a modeling question about tanks and a candidate system with an E-dependent grouping of tank and crew. Because of the dependency, a crew must be instantiated with every tank, meaning that all of the properties of a crew must be satisfied in order for the candidate model to talk about tanks. Consequently, some system must provide this information or it can be defaulted to some known value. In MBDE these gaps become directly apparent and have to be solved during data alignment.

Data Alignment

The goal of data alignment is to identify the resolution and scope of data and identify gaps and variances between representations. Data alignment results in a list of resolution and scope issues that have to be resolved within the federation. However, it is essential to first define what scope and resolution is using the terms that we have defined in this chapter.

Definition 1: *The scope of an element is the number of unique groupings in which it appears.*

Definition 2: *The resolution of an element is the number of elements within its scope.*

In this case, we can talk about the scope or resolution of an entity as well as the scope and resolution of a property. In the case of this chapter, we have limited the type of grouping and therefore we can rewrite the definitions as follows.

Definition 3: *The scope of an entity is the number of unique hasProperty and isE-dependent groupings in which it appears.*

Definition 4: *The resolution of an entity is the sum of the cardinality of the hasProperty set and the cardinality of the E-dependent set.*

Definition 5:   *The scope of a property is the number of unique hasSymbol and is T-dependent groupings in which it appears.*

Definition 6:   *The resolution of a property is the sum of the cardinality of the hasSymbol set and the cardinality of the E-dependent set.*

With these definitions, we can now provide a trivial data alignment algorithm that compares the resolution and scope of a candidate system's model with the interoperability model. This algorithm will simply step through all the groupings starting from the elements and ending at the symbols. Resolving the mismatches in resolution and scope depends largely on the goal of the federation and the resources available. There are several approaches but this discussion falls out of the scope of this chapter. Moving forward, we shall assume that resolution and scope issues have been resolved and we can proceed to the data transformation process. However, before we proceed there is an additional issue that must be considered: namely that of mandatory and optional elements. Mandatory elements are equivalent to the elements that are required to answer the modeling questions and optional elements are the members of the E-dependant groupings that are not required to answer the modeling question. As mentioned before, the interoperability model is the CRM. From a bottom-up point of view, the CRM is the model that represents the information exchange needs and capabilities of the systems given a set of requirements. Let us now formulate the mandatory and optional problem as follows.

- *Mandatory in the CRM and optional in systems:* In this case there is a potential for an information gap in the federation. The CRM rejects the E-dependent grouping from the system as incomplete and a decision has to be made to either add fictional data to complete the grouping or not to support that grouping. The latter decision can also have a cascading effect if the grouping is logically related to other groupings in the CRM, specifically if the business rules of the CRM require the existence of the grouping as a necessary precondition for other groupings.

- *Mandatory in systems and optional in the CRM:* In this case there is a potential for a gap in the system depending on the flow of information. The CRM can be extended to mandate the data element. However, the consistent application of such an algorithm results in the CRM mandating every element that is mandatory in every participating system which makes the CRM specific to the particular federation rather than the community of interest. The other approach would be for the system to make assumptions about incomplete data but that would result in a mismatch of information between the CRM and system and negate the value of using a CRM in the first place.

- *Mandatory in some systems and optional in others:* This alternative is the most complex, especially since the CRM is considered as a system. In this case, the federation members have to decide how to change their systems and the CRM to fit the information exchange needs of the federation. This approach is equivalent to that of manual semantic alignment which the CRM was built to avoid.

It is also important to distinguish between source and target or sending and receiving systems. Figure 29.3 shows that when the CRM is the target system, there is a conflict only when mandatory elements of the CRM intersect with optional elements of the sending system. In every other pairing, the information can be exchanged without additional conflict resolution.

When the CRM is the source system as shown in Figure 29.4 there is a conflict when the optional elements of the CRM intersect with the target system's mandatory elements. In every other case, there is no additional conflict resolution. In general, these two cases can be reduced to a "mandatory in target, optional in source" problem. However, the CRM is simultaneously source and target which adds an additional dimension to the problem. We shall address this issue in more detail in the section dealing with CRM construction.

Having discussed these issues, we move to the data transformation.

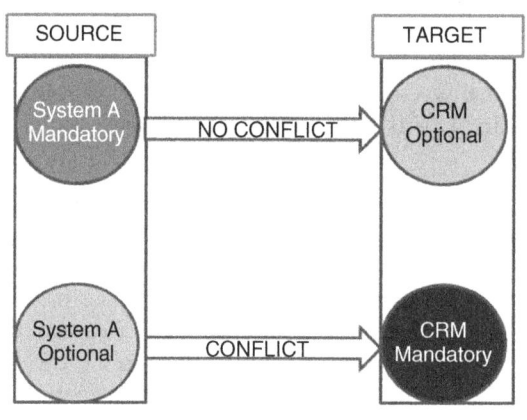

Figure 29.3 Intersection of mandatory and optional elements with the CRM as target.

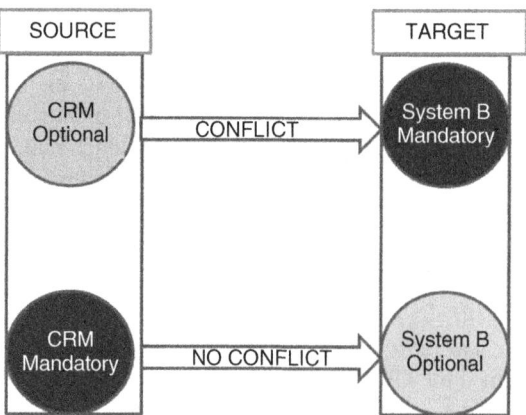

Figure 29.4 Intersection of mandatory and optional elements with the CRM as source.

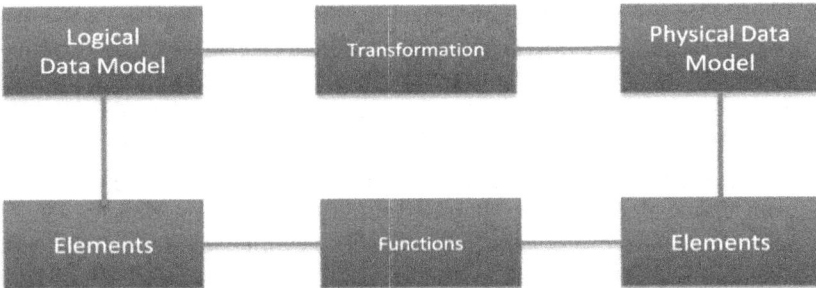

Figure 29.5 Data management process and artifacts.

Data Transformation

The goal of data transformation is to identify and implement mapping functions that generate valid sentences of the system as specified in data management. Data transformation generates a computable model of the logical data model which is called the physical data model. Data transformation ensures consistency between the logical and physical views of data and serves as the basis for an implementation or a physical data model instance.

It is important to note that data transformation should be structure preserving and should not result in any additional elements and dependences. In order to be more specific, additional rules are introduced.

Rule 1: *Every relation is maintained during data transformation.*

Rule 2: *An element preserves its set membership during data transformation.*

Rule 3: *Every dependency is maintained during data transformation.*

Rule 4: *Additional dependences shall not be introduced during data transformation.*

These rules serve as guidance in determining how to resolve conflicts when there is more than one candidate function. If more than one function fulfills the rules, they are considered equivalent for a given transformation and either can be chosen.

Data transformation in a perfect world is simply the generation of a physical data model but in reality we have to deal with legacy systems. Practically, this means that data transformation is often reduced to mapping properties and mostly values. A simple example would be to align units of measure, coordinate systems, or unit sizes. However, as MBDE has shown so far in data administration, management, and alignment, there are more issues to consider than just defining mapping functions for value sets. In fact there are two types of transformations.

- *Simple transformations:* Simple transformations exist when there is a direct correspondence (1:1) between elements. Tools can easily be used to discover and perform simple mapping. For complex heterogeneous models, however, simple transformations are the exception not the rule. In most cases we deal with complex transformations.

- *Complex transformations:* Complex transformation exists when one element in a source has more than one corresponding element in the target. This correspondence is generally referred to as *n:m* transformations with *n* and *m* both integers.

Both types of transformation exist at the element, property, and symbol level with the majority of transformations being of complex type. Let us consider each case separately.

- *Entity level:* Depending on the scope, resolution, and focus of the model, similar entities can be modeled in different manners. Therefore, an entity in the source model can be represented as an association of entities in the target model and vice versa.

- *Property level:* Identifying similarities between properties is complicated by the existence of synonyms and homonyms. Furthermore a property in one model can map to a composition of related properties in another.

- *Symbol level:* As a result of the complexity at the property level, the content of fields can also be expressed differently depending on the modeler's choices. These relationships have to be discovered and expressed explicitly during the mapping process.

In order to address this issue, we propose the application of the five steps depicted in Figure 29.6.

- *Designate source and target:* In the first step, all elements needed on the source and target sides must be identified. This includes elements and properties that need to be exchanged and therefore have to be mapped to each other. For each set of elements there must be a clearly defined source and target. As a result, the notion of source and target is not fixed. A source can become a target depending on the set being exchanged. The importance of this process was mentioned earlier.

- *Classify relationship:* In the next step, we identify the type of mapping. This mapping falls into one of two categories: 1:*n* mappings or *n:m* mappings (*n*:1 mappings and 1:1 mappings are trivial special cases). The three sources of complex mapping (elements, properties, and symbols) must be dealt with separately and thoroughly.

- *Identify as static or dynamic:* The definition of static and dynamic properties is necessary to support information exchange before runtime (static) and during runtime (dynamic). While static properties must only be aligned for the initialization phase, dynamic properties change their value during runtime and are the basis for information exchange during runtime. This

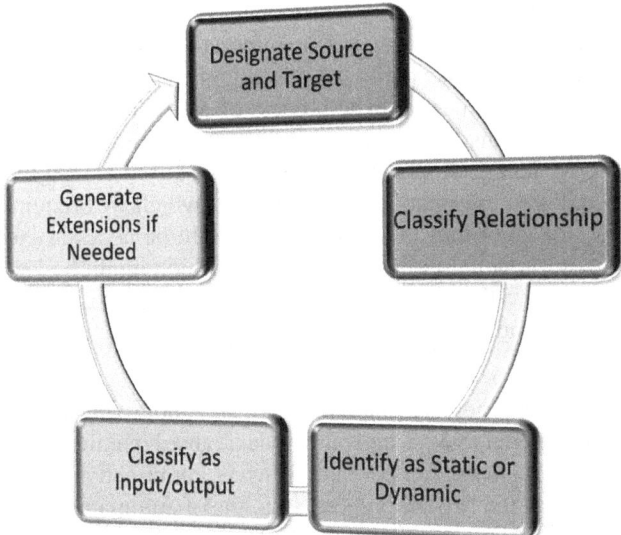

Figure 29.6   Data transformation mapping process.

step ensures that the right information exchange is consistent throughout the interoperability process. It is also important to note that the static elements correspond to the descriptive aspects of a model or its ontology while the dynamic elements represent the behavioral aspect of the model.

- *Classify as input or output:* A model can also be represented as a set of states which have input and produce output. For a given state, each model has a matching input/output pair that becomes a requirement for any other model that wants to interoperate with it while it is in that state. As a simple example consider a tank model that needs the location of a target and the line of sight as input in order to engage and another who needs to know the target type in addition. Both models can output a battle damage assessment if they are engaged and have three states (wait, detect, engage). If those two tanks are to fight in a battle through interoperability, it seems that the first model would have the advantage because its input set is a subset of that of the second model. However, there is no guarantee that it is the case just by looking at the data. The data questions such as needs, requirements, and capabilities as addressed by MBDE are necessary. The matching of states with respective input/output pairs is an additional step that helps the data modeler understand and predict the emerging interoperability space when the two models are coupled. While a full algorithm analysis is out of the scope of MBDE, the modeler must be prepared to address these issues before the data transformation process begins.

- *Generate extensions if needed:* As shown in the previous example, additional extensions to one or more models might be required to fulfill the state/input/output requirements. Parts of these extensions are formulated in the process of data alignment. However, it is important to keep in mind that data alignment is mostly concerned with the alignment of the models in terms of their syntax and semantics. The process of data transformation might require additional extensions to make the systems actually exchange data. Going back to our tank example, it might be necessary to extend the first tank model to add type information to its input set, which will require extensive changes to the algorithm. Alternatively, it might be necessary to default the tank type information in the second model which might be less resource intensive but more limiting in scope. The modeler has to make a decision based on the potential effect on the ability to answer the modeling question.

The issues identified in this section are not exhaustive and additional factors such as the pragmatics (how a model assigns meaning to data) and dynamics (how a model transitions from one state to another) of models while not directly related to data help guide the modeler in the MBDE process. In terms of automation data transformation is the process with the most potential in terms of data because it has well defined semantics at least for symbols.

SUMMARY

The context of interoperating distributed simulations presents multiple challenges at the data level. At the core of the problem is the ability to formulate a modeling question in the form of a data model that will serve as the blueprint for the federation. In this chapter, we propose MBDE as an engineering method to face these challenges in the form of high level algorithms that can serve as the basis for more specific algorithms based on the requirements of particular federations. While ideally MBDE is a top-down process, in practice it is an iterative process that has no predefined starting point but requires that each step be done at least once.

In data administration, we identify and specify the interoperability model as the representation of the modeling question we are trying to answer. If the approach is to build a federation, we use data administration to identify the capabilities and needs of each candidate system. In data management, we prepare the systems to be integrated such that the resulting information exchange is consistent across the federation. Consequently, we identify semantic groupings that function as atomic elements and specify how they relate to one another in the form of a logical data model. It is worth repeating that the resulting model might have more elements than the part of the interoperability model it is trying to fulfill and that in that sense it represents a particular system's representation of the interoperability model. In data alignment we identify disparities in scope and resolution between the data models and align them wherever possible. In data transformation we distinguish between simple transformations and complex transformations and show that the process is more than just the generation of a physical data model from a logical data model. In addition, we provide several rules, guidelines, and issues to consider when interoperating combat models.

MBDE as a whole is simply the creation of a CRM either explicitly as the interoperability model or implicitly by applying engineering principles to connect systems. In practice it means that for interoperability a process for MBDE always applies whether it is documented or not. It is worth noting that while data interoperability is the visible part of interoperability, alignment of pragmatics and dynamics are important in order to maintain a consistent behavior across a federation. The reader is encouraged to consider data modeling theory as an additional field that is very helpful for understanding some of the issues presented in this section. However, it is also important to remember that data modeling in modeling and simulation has its own requirements especially at the conceptual level. At the end of this process, the information exchange needs, capabilities, and requirements are clearly defined.

REFERENCES

Codd EF (1970). A relational model of data for large shared data banks. *Comm ACM* **13**, 377–387.

Davis P and Anderson RH (2004). *Improving the Composability of Department of Defense Model and Simulations*. RAND National Defense Research Institute, Santa Monica, CA.

Diallo S (2010). *Towards A Formal Theory of Interoperability*. Old Dominion University, Norfolk.

IEEE (2000). *High Level Architecture*. IEEE CS Press, Piscataway, NJ.

IEEE (2002). *Distributive Interactive Simulation*. IEEE Press, Piscataway, NJ.

IEEE (2003). *Federation Development and Execution Process*. IEEE CS Press, Piscataway, NJ.

ISO/IEC (2003). *ISO/IEC 11179*. ISO/IEC.

North Atlantic Treaty Organisation (2002). *NATO Code of Best Practice*.

Tolk A and Diallo SY (2010). Using a formal approach to simulation interoperability to specify languages for ambassador agents. In: *Proceedings of the Winter Simulation Conference*. IEEE, Piscataway, NJ.

Chapter 30

Federated Simulation for System of Systems Engineering

Robert H. Kewley and Marc Wood

INTRODUCTION

System of systems decisions are extraordinarily complex. It follows that combat modeling to support system of systems decisions is also complex. This chapter provides a framework to deal with that complexity. Simulations are typically built for engineering analysis of a system, not a system of systems. Therefore, simulation development in this environment involves architecting a federation of engineering models that represent the component systems. This federation is a system of systems in itself—all within an overarching system of systems engineering environment.

To some degree, simulation success will depend on the same factors that lead to success in more general system of systems engineering. The leadership team must make a very strong case for the value and importance of the federated capability. The operational experts must provide a clear description of the operational capability provided by the system of systems. Federation architects provide a simulation architecture that enables analysis of that system. Finally, simulation engineers must operate autonomously, and collaboratively, within the architectural framework and established standards to integrate their models. However, even in the most optimistic cases, system of system complexities will give rise to

Engineering Principles of Combat Modeling and Distributed Simulation, First Edition.
Edited by Andreas Tolk.
© 2012 John Wiley & Sons, Inc. Published 2012 by John Wiley & Sons, Inc.

unforeseen challenges. Flexibility, adaptability, frequent testing, and incremental development are the path to eventual success.

This chapter outlines an approach in three steps. The next section defines system of systems engineering and provides an overview of engineering and modeling and simulation approaches prescribed for that context. The following section highlights engineering principles, many of which are previously discussed in this book, that are particularly relevant in a system of systems context. Then a systems engineering methodology is prescribed for federation development. These sections are based in part on the experiences of the West Point Department of Systems Engineering in leading federation development for analysis of soldier systems. Many of the sample products were developed for that purpose.

SYSTEMS OF SYSTEMS ENGINEERING

System of systems engineering is an emerging practice in an increasingly connected world. The information revolution which has connected people, businesses, and societies has also enabled systems to become more connected—to accomplish things not possible with one system. According to the *Defense Acquisition Guidebook*:

> *A system of systems is a set or arrangement of systems that results when independent and useful systems are integrated into a larger system that delivers unique capabilities. (Department of Defense, 2010)*

They are further categorized by type—based upon the degree of centralized leadership and resourcing (Office of the Under Secretary of Defense, 2008). The methodologies in this chapter deal with *acknowledged* or *directed* systems of systems. Because the approach begins with capability and operational viewpoints for the system of systems architecture, the system of systems capability must be acknowledged or directed by a higher authority with at least some level of funding. *Virtual* or *collaborative* systems of systems emerge only by the voluntary consent of the participants. They can only be descriptively architected after the fact. When that occurs, they become *acknowledged*.

> *SoS engineering deals with planning, analyzing, organizing, and integrating the capabilities of a mix of existing and new systems into an SoS capability greater than the sum of the capabilities of the constituent parts. (Department of Defense, 2010)*

Because system of systems engineering is such a new field, there are three distinct lines of thought about the existence and relevance of system of systems engineering (Vilerdi et al., 2007). Some engineers recognize a fundamental difference between system of systems engineering and simple systems engineering. This school of thought calls for different management and engineering approaches to solve complex system of systems challenges. A second school of thought views system of systems engineering as simply another instance of systems engineering where the components are systems themselves. A third group is taking a wait and see approach as the discipline evolves.

This chapter falls in line with the *pro-difference* camp. System of systems engineering practices are driven by five distinguishing characteristics of a system of systems Sage and Cuppan (2001):

- Operational independence of the individual systems
- Managerial independence of the systems
- Geographic distribution
- Emergent behavior
- Evolutionary development

Additional differences include an evolving architecture, multiple groups of stakeholders with different perspectives, and distributed and unstable funding (Vilerdi et al., 2007).

In this environment, top-down prescriptive architecting is impossible. The component systems have their own architectures and, in many cases, little or no requirement or incentive to conform to the overarching system of systems architecture. Funding sources, test requirements, and delivery schedules vary across the component systems. Even the organizational culture and terms of reference vary across the component systems. Keeping in mind that the component systems have their own legitimate purposes and value, top-down approaches to system of systems engineering introduce costs and delays that eventually erode the value of the components. As system complexity grows, so do these costs and delays, eventually rendering the system of systems capability cost ineffective. The Future Combat Systems program was an effort by the US Army to design and develop a system of systems under the governance and oversight of a single program. The technical complexities of the program led to cost overruns, schedule delays, and the eventual cancellation of the program (Francis, 2009). In its wake, the Army has adopted a more traditional focus—giving more autonomy and priority to individual systems.

Given this daunting challenge, what hope is there for developing system of systems capabilities? While no accepted best practice for system of systems engineering has emerged, proposed solutions share some common traits. The Department of Defense issues guidance as to how the systems engineering process has different system of systems considerations (Office of the Under Secretary of Defense, 2008). This guidance outlines emerging principles that highlight the importance of organizational issues, shared technical management, open systems, and loose coupling. This approach is more formally described by management literature as federalism (Handy, 1992; Sage and Cuppan, 2001). It also draws from adaptation and evolution of complex systems (Ottino, 2004). Taken as a whole, these approaches are characterized as follows:

- a minimalist approach to centralized architecture and control;
- a shared understanding of the overarching system of systems functions;
- adherence to and enforcement of standards;

- trust and responsibility placed upon component system engineers—they are "dual citizens" committed to both component system performance and system of systems performance (Sage and Cuppan, 2001);
- incremental and iterative delivery of capabilities;
- robust testing and experimentation;
- design to enable future evolution.

Simulation practitioners have listed several roles for simulation support to system of systems. It is a valuable tool for prototyping adaptation and evaluating potential system of systems architectures (Sage and Cuppan, 2001). It has additional capabilities for understanding system emergent behavior and performing system of systems test and evaluation (Office of the Under Secretary of Defense, 2008). The National Defense Industrial Association identified four specific uses of simulations in support of system of systems engineering (Law, 2007):

- Provide dynamic interfaces and data to drive system testing
- Probe current and future capabilities, relationships, and architectures
- Support system performance evaluations
- Support operator in the loop training

They go on to point out that modeling and simulation has typically failed to live up to its potential in the system of systems domain. One of the key inhibitors is that models are built for a specific focus area. They cannot be federated easily, and the cost and difficulty of adaptation outweigh the potential benefits. Any methodology for simulation support to a system of systems must address this difficulty.

Descriptive approaches to modeling and simulation for a system of systems are beginning to emerge. The Missile Defense Agency has federated seven engineering models as the basis for their performance evaluation strategy. Consistent with system of systems engineering principles, they highlight collaboration as a key to modeling and simulation success (Repass, 2010). Another approach describes a model driven Discrete Event Systems Specification (DEVS) Unified Process for simulation based design, testing, and analysis of systems of systems (Mittal et al., 2008). This approach relies on the adoption of the DEVS formalism for specifying component models (Zeigler, 2003). The Army's Research Development and Engineering Command has developed a modeling and simulation unified architecture to support system of systems modeling across a variety of Army domains (Hurt et al., 2006). While these approaches are all relevant and successful in their domains, they can be aggregated into a more general approach.

The remainder of this chapter will elaborate on this approach. We will first discuss a series of engineering considerations for federated simulation. In light of these considerations, we then describe a systems engineering process for the development of federated simulations to support system of systems engineering. In a book on combat simulation, it is appropriate for this process to rely upon the architectural specification of the system of systems in the Department of

Defense Architectural Framework 2.0 (Department of Defense, 2009). The chapter concludes with a section on an appropriate technical management approach for success in federated simulation development.

This approach was formulated in the West Point Department of Systems Engineering, in collaboration with the Virginia Modeling, Simulation, and Analysis Center, over three years during which the department led the development of a federated simulation to support analysis of soldier systems for the Army's Program Executive Office—Soldier (Kewley et al., 2008a; Kewley and Tolk, 2009b). Descriptions and references to that project are included as examples. While our development team subscribed to a systems engineering approach from the outset, the details of that approach have evolved with experience. We will present the work using the latest instance of that process, recreating and updating original architectural views for consistency.

Case Study: Program Executive Office Soldier

The PEO Soldier Simulation Road Map is an effort by Program Executive Office (PEO) Soldier to develop within the Army a capability to model the effects of soldier equipment on unit-level effectiveness—focused at platoon and below. In November of 2003, Brigadier General James Moran, PEO Soldier, commissioned the West Point Department of Systems Engineering's Operations Research Center (ORCEN) to develop a model, or family of models, that would support PEO Soldier decision making with respect to soldier equipment. The ORCEN, working within the PEO, further defined the need as "PEO Soldier needs a simulation that allows the evaluation of platoon effectiveness based upon changes in Soldier tactical mission system (STMS) characteristics." Fulfillment of this need would bring the PEO in line with the Army's Simulation and Modeling for Acquisition, Requirements, and Training (SMART) program. The SMART program involves rapid prototyping using M&S [modeling and simulation] media to facilitate systems engineering so that materiel systems meet users' needs in an affordable and timely manner while minimizing risk (AMSO, 2002). Taking this need, the ORCEN evaluated a series of alternatives that ranged from creating a brand new simulation to adopting, in its entirety, an existing simulation. The team concluded that while developing a single model was cost and time prohibitive, no single existing model met the PEO's requirements. They recommended a federation of models including IWARS, OneSAF, and COMBAT^{XXI}. PEO Soldier accepted this recommendation and asked the ORCEN to lead the effort in building a team to develop this federation (Tollefson and Boylan, 2004).

While everyone understood the need for a federated modeling solution, the composition, type of integration, and level of detail for the federation were not so simple to agree upon. The ORCEN worked two parallel efforts from June 2004 until July 2005. First, they had to establish memoranda of agreement that would enable funding and collaboration within this project. This required significant negotiation between PEO Soldier, the Natick Soldier

Center (developer of IWARS), PEO Simulation Training and Instrumentation (PEO-STRI—developer of OneSAF), and Training and Doctrine Command Analysis Center—White Sands Missile Range (TRAC-WSMR—developer of COMBAT[XXI]). Second, they had to further refine the analysis requirements for the federation. In short, PEO Soldier did not have a list of analysis requirements; they had a list of equipment. The ORCEN worked with the PEO to categorize and streamline this list into a discrete set of modeling requirements that could be implemented by the members of the federation. Once these requirements were understood, it was easier for the modeling agencies to agree to develop these capabilities (Martin, 2005).

ENGINEERING REQUIREMENTS FOR FEDERATED SIMULATION

As described in the preceding section, cost and other constraints have thus far limited a top-down approach to developing a simulation capability that can be applied to a system of systems. However, a federated approach can provide the flexibility to maintain each system's integrity, while progressing toward the goal of effective system of systems simulation. Evolving system of systems capabilities demand a series of parallel training, acquisition, and fielding initiatives that must be fully integrated. Analysis of these interactions requires a robust and reconfigurable simulation paradigm (Kewley et al., 2008b). It must capture the engineering requirements and guide the conceptual development of an integrated, flexible system of systems simulation capability, which does not prohibit the continued development of each individual system within the federation.

The concept illustrated in Figure 30.1 is such a model. It provides a framework that will enable a team of engineers and analysts to rapidly conceptualize, develop, execute, and analyze data using federated simulation models. The outer ring shows a series of domains in which federated models have proven useful. The boxes just inside the outer ring represent current application areas from these domains. For each application, a different federation of models must be developed to support the question at hand. These federates must be held together, conceptually and technically, by the engineering capabilities, which form the core of the model in the center of Figure 30.1. These six engineering capabilities are the key research areas that will enable effective development of federated models:

- *Information Exchange System:* The capability to pass meaningful information between federates during the simulation run.
- *Environmental Representation:* The capability for federates to reference a shared and correlated environment in which entities interact.
- *Entity Representation:* The capability for federates to reference shared conceptually aligned information about entities in the simulation. Some of this representation is passed via the information exchange system.

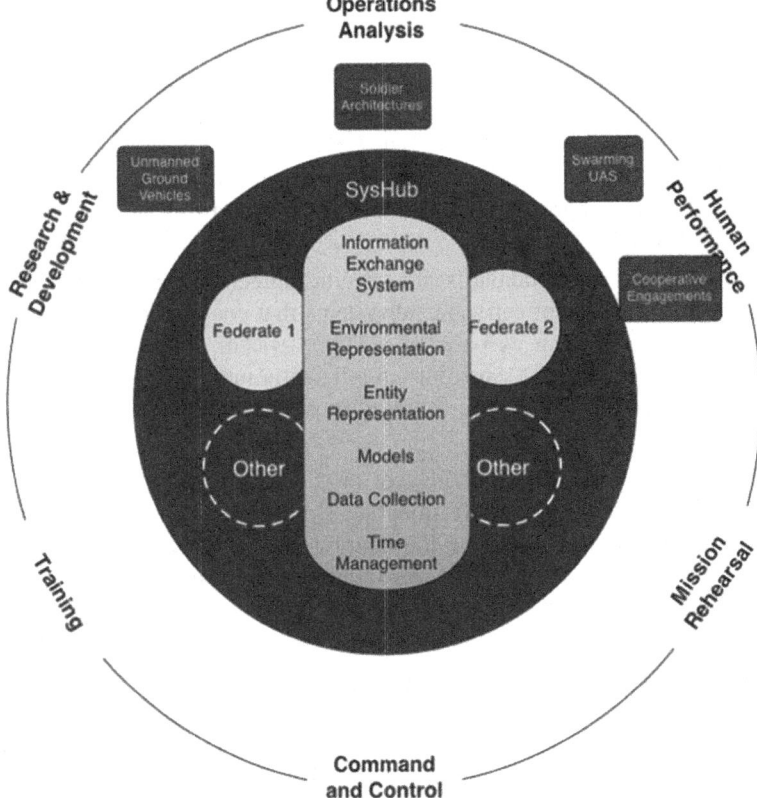

Figure 30.1 Engineering capabilities for federated simulations in support of systems of systems analysis.

- *Models:* Within the context of the analysis or training question, the internal models of each federate must be validated and coordinated across the federation.
- *Data Collection:* The capability to collect meaningful information from the simulation run in the context of the analysis question or training objective for which the federation was designed.
- *Time Management:* The capability for all federates to maintain a common time reference in order to synchronize simulated events.

Information Exchange System

An information exchange system must provide a basic level of interoperability between simulations. With respect to software systems, interoperability is the ability of two or more systems or components to exchange information and to use the information that has been exchanged (IEEE, 1990). It must have the following critical components:

- *Interoperability Model*. Two common models for providing interoperability are the Distributed Interactive Simulation (DIS) protocol (IEEE-SA Standards Board, 1998) and the High Level Architecture (HLA) (IEEE-SA Standards Board, 2000). These are established Institute of Electrical and Electronics Engineers (IEEE) standards for allowing simulations to share information at runtime.

- *Data Standard*. Data standards are specific agreements between agents responsible for different software subsystems that communicate with each other when functioning as parts of a larger system. The data model is crucial and must have the capability to efficiently integrate federates without creating additional trouble-shooting requirements to ensure they can communicate.

- *Middleware*. An effective middleware program will enable participating simulation developers to easily integrate simulations written in multiple programming languages without the requirement to encode and pass messages. The middleware should allow integration efforts to be focused on events dealing with the handling and interpretation of information exchange system messages internal to their model, reducing the time spent resolving inter-model issues.

- *Test Cases*. Test cases should be developed for each integration task. Developers can then compare the results with what they would expect from a stand-alone simulation. The test cases provide unambiguous messages and test capabilities to prevent costly misunderstandings and eliminate faulty assumptions.

- *Network Based State Data*. A common data picture is achieved when visibility of entity states exists for a federation of simulations. A network based data picture enables simulation-independent assessments.

- *Technical Integration Support*. A readily available and knowledgeable support team will reduce the time required to develop solutions for architecture and integration problems and provide valuable training to enable users to become more familiar with simulation tools.

Finally, expensive license fees and other charges can increase the overall cost as well as the complexity of management requirements. This must be an up-front consideration when selecting an information exchange system.

Case Study: Program Executive Office Soldier

The Army's experimentation program in support of transformation significantly challenged the existing federated simulation paradigm (Hurt et al., 2006). This program had a need for the aggregation of engineering level models in an environment that supported human-in-the-loop interaction and a robust command and control environment. The Army's Research, Development, and Engineering Command's (RDECOM) Modeling Architecture for Technology, Research, and Experimentation (MATREX) program took a systems engineering approach to attacking these challenges in a collective environment that spanned RDECOM in the context of system of systems analysis and integration. They developed not only an HLA based information exchange system, but also the tools and procedures required to support integration and build community acceptance (Gallant et al., 2009).

The Program Executive Office Soldier was able to utilize the MATREX information exchange system in its Simulation Road Map federated modeling solution. First, MATREX utilizes a proven data model, which two of PEO Soldier's federates, Infantry Warrior Simulation (IWARS) and One Semi-Automated Forces (OneSAF) had already completed development to support. It also includes an automated test case tool to distribute test cases, a robust middleware library which reduced debugging and encoding time, and a network-based battle command functionality (the MATREX Battle Command Management Services).

Finally, MATREX is a government-owned program, which allowed the project to avoid the additional cost of hiring a commercial simulation integration firm. Support from the MATREX program enabled the PEO Soldier project to move from an idea to a reality in a relatively short time. Without MATREX, the integration would have taken up to six additional months, and the solution would have been a fragile point solution based only on internal knowledge and the needs of the project. The MATREX tools proved to be a valuable information exchange system solution for the PEO Soldier simulation federation.

Environmental Representation

Environment encompasses natural and man-made physical elements of the battlespace occurring in the terrain, atmosphere, ocean, and space domains. In traditional modeling and simulation taxonomies, buildings and land based infrastructure fall under the domain of terrain. This typically includes features that are considered more persistent over time. In addition, those things such as weather and obscurants that change over much smaller time scales are also important elements that must be captured in the environmental representation.

In many simulation abstractions, entities interact with each other and with their environment. In federated simulations, it is not common for the models to share a single representation of the environment across the network. In most cases, each individual model has its own internal representation of that environment. If the results of the simulation are going to be valid and not obscured by differences in environmental representations, they must be sufficiently correlated for meaningful interchange and execution of a scenario.

Terrain databases created for different applications may or may not be created from the same source data. The source data may be of different resolutions and the techniques for sampling and integrating data may be different. Thus, it is not surprising that features such as roads, bridges, rivers, and forested areas on the map for different applications often do not nicely overlay each other. Additionally, if the surface relief is represented differently across applications, adjudication of line of sight and other interactions with the terrain become problematic. This is a particularly challenging area based on the fact that there often are different means of defining the relief (elevation over spatial extents of the area of interest) and extracting and representing features, which can cause major issues when federating simulations. It is essential for entities in the simulation to have the same notion of where environmental elements are located and to execute interactions using a correlated representation of that environment.

This can be even more difficult when terrain data models include features such as a road segment or surface element which can be categorized by type. Consider soil type, for example. Enumerations could be US Geological Survey (USGS) soil classifications or they might be more simply represented as sand, clay, mud, or gravel. If running a federated simulation scenario which considers ground vehicle mobility, it would be important to understand how the two different soil type enumerations might affect the underlying algorithms computing mobility as well as the capacity of the simulation to execute a fair fight.

The correlated generation of terrain information for simulation databases is vital to the success of federated simulation. There are three options for achieving correlated terrain representation. The first option is to have each model interact with terrain using copies of the same terrain data model and associated programming interface for terrain algorithms such as mobility factors or line of sight. A second option is to develop correlated data from the same geospatial data sources in a single tool. There are several commercial and government-owned tools that use a common set of raw geospatial feature data, elevation data, and imagery to generate simulation databases in multiple formats. The final option is to independently generate correlated databases using different tools. This option may introduce correlation errors due to differences between the tools. It is possible to manually correlate the resulting databases using common reference points, but this is a difficult and imprecise process. Regardless of the option used, it is important to validate the effects of terrain data on the scenario during the federation test phase. Checking the validity of observed movement rates, line-of-sight algorithms, and visual representations will give confidence in the accuracy and correlation of environments across the federated models.

Case Study: Program Executive Office Soldier

The Program Executive Office (PEO) Soldier Simulation Roadmap project utilized three federates, One Semi-Automated Forces (OneSAF), Infantry Warrior Simulation (IWARS), and the Combined-Arms Analysis Tool for the 21st Century (COMBAT^{XXI}) to achieve the desired simulation capability. In order to simplify terrain representation and minimize the chances for negative impacts on the terrain algorithms, the PEO Soldier project required each federate to use OneSAF's Environmental Runtime Component (ERC) as its underlying data model. All three federates natively referenced the ERC when accessing terrain information during simulation runs. The OneSAF team supported this strategy by providing C++ wrappers for the Java-based ERC. However, as new federates with different underlying terrain representations joined the federation, additional effort was required to build correlated representations using a terrain modeling tool.

Entity Representation

In a discrete event simulation, entities interact with each other and their environment so that the system changes over time. A simulation's state is defined as a collection of variables necessary to describe a system at a particular time, relative to the objectives of a study (Law, 2007). In combat modeling, the variables necessary to describe the system are the entities represented in the simulation and the state of an entity is the collective state of each of its attributes. Entities change their state during the course of the simulation through interaction with other entities or the environment. The challenge within the federation is timely and consistent tracking of each entity's state throughout.

In a federation, each supporting federate will probably use a different level of detail for the representation of each entity. Some models will require detailed entity representation with enough resolution to support calculations relevant to that entity's interaction with its environment and other entities. Other simulations may have a much more abstract representation with much less detail. In all cases, the resolution of the entity representation is dependent upon the question the simulation must answer, and the federation design must account for the differences.

An entity may also include a combination of static and dynamic attributes. The static attributes are set at simulation initialization and do not change over time. The type of weapon a particular soldier carries, for example, is often a static attribute. The dynamic attributes, on the other hand, do change. For example, a soldier's location and health status will change over the course of the fight, and these attributes will probably be very important to the entities represented in other federates. In many cases, these changes must be communicated to other federates via the information exchange system.

Three design steps are required for coordination among federates. Initially, developers (in coordination with operational experts) must identify a reference which describes the characteristics and capabilities of each entity that must be modeled. For example, if a simulation includes an M1A1 Abrams Main Battle Tank, there must be sufficient documentation of its movement speeds, ammunition and weapons characteristics, fuel capacity and other factors pertinent to accurately represent the entity in the simulation environment. The second step is for each federate to build a static and dynamic representation of the tank given the constraints of its internal data structures. Finally, for the dynamic attributes, each federate must define those attributes for which it requires notification when they are updated in another simulation. For example, one federate needing location information on an M1A1 Tank as it moves across the terrain may require the simulating federate to update its location on the terrain database. The federation designers must build a consolidated list of those attributes, and the information exchange system must have data structures for representing changes to those attributes as simulation time advances.

Case Study: Program Executive Office Soldier

Within the Program Executive Office (PEO) Soldier federation, the important entities to be represented included dismounted soldiers, combat vehicles, and unmanned aircraft. The Military Scenario Definition Language (MSDL) was used to define the units and entities that would participate in the battle. During scenario development, the Army's Training and Doctrine Command (TRADOC) sources defined the unit composition and Army Material Systems Analysis Activity (AMSAA) weapons characteristics and data identified entity capabilities. These sources provide a common shared reference for the supporting combat models to build a scenario. IWARS represented dismounted soldiers in great detail, with abstract representations of vehicles and aircraft. For OneSAF and COMBAT^{XXI}, the aircraft and vehicles were represented in detail while the dismounted soldiers were represented in less detail than for IWARS. For the dynamic attributes, the MATREX objects *IndividualCombatant, GroundPlatform*, and *AirPlatform* were used to represent these entities (MATREX, 2009). Each federate subscribed to these interactions and updated the internal states of these entities as required.

Models

Validating models in a federation is challenging and requires some particular considerations. A shallow application of interoperability might only seek to align models and federates by converting the simulation's data into the necessary input and output format specified by the federates. Although this can often be accomplished at the programmer level, this will achieve limited results. Support for interoperability protocols such as HLA and DIS often enable new federates to be plugged in with little or no programming or engineering, but the danger in this type of integration is that the federated model may not be designed to perform analysis in the new context. Just because a model has been validated in one context does not mean that it can automatically be used properly in a new context. Constructing valid federations requires additional work to ensure the model performs correctly in all contexts for which the simulation is designed to function.

The subject matter experts for the federated model must work with the simulation systems engineers to validate the model in its new context. Although we shall not address model validation in depth here, we would like to highlight a few guiding principles for validating federated models. First, it is assumed that the federated model has already been validated in some context. If so, the modeling assumptions from the original validation must then be checked within the new simulation context. It is imperative that this new context does not invalidate the original assumptions to a point where the model is not useful in the new context. Once this static validation is complete, further validation can take place during the federation test and evaluation phase. In this phase, data inputs and outputs from the model are checked against known quantities to ensure the model behaves properly. Finally, a subject matter expert in the domain to be modeled can visualize or check model runs to give face validity to the model in its new context.

Case Study: Program Executive Office Soldier

The Program Executive Office Soldier Simulation Roadmap federation's handling of IWARS soldiers mounted inside OneSAF vehicles highlights some important model validation issues. The implementation required IWARS to mark the soldiers as mounted in the IndividualCombatant information exchange and move the locations of the soldiers to match the location of the combat vehicle. In each federate, entities were not permitted to directly engage those entities that were marked as mounted and could only engage the vehicle. With respect to interoperability of the data structures, this approach worked perfectly. The entities moved around the battlefield and dismounted at the proper location. However, the new context violated the IWARS internal assumption that soldiers were always dismounted. The IWARS simulation, which owned the dismounts, had no internal model for assessing damage inside a vehicle. If a vehicle were hit, the federation would not be valid because it would not have been possible to assess damage against its occupants. Fortunately, OneSAF did have an internal model for damage to occupants of vehicles. When a vehicle was hit, OneSAF ran this damage model against all of its occupants, regardless of which simulation owned those occupants. If one of the IWARS soldiers was damaged, OneSAF sent a *DamageReport* interaction to IWARS so that IWARS could update the damage state of its mounted entities. Another validation problem was that these mounted soldiers were still able to acquire enemy targets and build situational awareness, even though they were completely enclosed in a vehicle. IWARS had to turn off the target acquisition algorithm for these mounted soldiers to replicate the fact that they could not see outside of the vehicle. This mount/dismount modeling challenge highlights the complexities that can arise when federating a simulation. In this case, federating IWARS and OneSAF introduced a new context that violated IWARS' assumption that soldiers would always be dismounted.

Data Collection

A federated simulation capability is constructed to answer a central question or series of questions. It does this by generating valid data that can then be analyzed in the context of the questions at hand. Successful analysis requires a clear definition of the questions to be answered and an understanding of the extent to which the questions can be accurately represented and answered by the tools available to the analyst.

There are two primary tasks involved in moving from questions to answers. First, the question must be translated to a form that can be represented in the simulation. To do this, the analyst must understand the modeling assumptions made by the designers of the simulation and phrase questions in terms of simulation inputs and outputs. The analyst must then establish criteria for translating the results into substantive answers to the original questions. Second, the simulation must be run and enough data collected to achieve the desired level of confidence. Once translated, the question establishes a requirement for data collection. If the data collection tools available to the analyst do not support the data requirements, tools must be adapted or developed to meet the need. If not feasible, the scope of the question must be reduced to fit the available data.

Current automated tools for data collection tend to fall into two broad categories. The first consists of stand-alone loggers which record raw data from DIS, HLA, or other data interchange systems. When processed data are required, it is often possible to construct small *ad hoc* tools to analyze and summarize the output of these lower level loggers. The second consists of integrated data collection and analysis systems that record information at a level of abstraction comparable with a simulation's internal object model. Integrated data collection systems are easier to use as long as they provide the data required.

In addition to automated tools, human observation is a valuable tool for data collection. In some cases, the most practical way of capturing data may be to have a human expert watch or participate in events as they unfold. If human psychological or physiological performance factors are an important part of the experiment, they must also be measured. Thus, visualization tools and immersive environments are also an important part of the analyst's data collection toolset.

Case Study: Program Executive Office Soldier

For the PEO Soldier project, analysts integrated three different data collection approaches. First, they leveraged the internal data collection systems for IWARS, One-SAF, and COMBAT[XXI] to collect data on events such as target acquisitions and weapons effects. A network based data collection tool collected HLA interactions and object updates in order to assess metrics such as soldier locations and unit attrition. Finally, developers created a custom data collection federate to monitor and record the situational awareness of soldiers on the battlefield. By this combination of tools, analysts were able to assess the necessary performance metrics and effectiveness metrics to answer critical questions.

Time Management

In order to execute successfully, time must be synchronized across federates so that the ordering of events and interactions mirrors the real system. Simulations are synchronous, so the federates must have a shared representation of time. There are a few different strategies to achieve this.

The first, and perhaps the simplest and most common, is to require federates to execute in real time. In this case, each federate keeps an internal clock synchronized with real time and sends messages and state updates at the time they occur. The federations' ordering of events is preserved with respect to real time. A modification of this approach is to have each federate run in a multiple of real time. This approach requires the least amount of modification to the participating federates; however, the repeatability of the simulation cannot be guaranteed. Additionally, there is no way to know if one federate speeds up or slows down with respect to real time, and there is a possibility for synchronization errors caused by a federate's ability to synchronize its internal clock with real time.

A second strategy is to implement some form of time management at the federation level. There are several approaches to this, each offering unique advantages and disadvantages (Carothers et al., 1997; Fujimoto, 1998; Taylor et al., 2003). These schemes all require tradeoffs with respect to performance, ease of implementation, and repeatability.

Case Study: Program Executive Office Soldier

Because the MATREX Protocore libraries had not yet implemented time management services at the federation level, the PEO Soldier project used real-time synchronization. Because of the complexity of IWARS, the performance penalty for real-time execution was fairly small. At the time, however, it was not possible for the current federation to guarantee repeatability in the time-ordering of its events.

SYSTEMS ENGINEERING PROCESS FOR THE DEVELOPMENT OF FEDERATED SIMULATIONS

The requirement for a systems engineering approach to federation development stems from the challenges of representing system of systems integration in a simulation environment. Typically, simulations exist that represent the operation of some of the subsystems. However, when these subsystems are integrated to form a new capability, a new simulation model is required to support analysis of system of systems effects. The federation developer must typically rely on a series of techniques from other disciplines and the expertise of supporting developers to design the federation. This section outlines a systems engineering approach that may be used to guide the process (Kewley and Tolk, 2009a).

The real challenge of federation development is to ensure that the final product is consistent with the purposes for which the simulation project was originally started. It should support analysis of different system of systems strategies, represented as simulation inputs. In so doing, there must be a mechanism for tying different simulation development activities to the requirements for certain system functions to be modeled and to certain outputs to be produced. A systems engineering approach to simulation development ensures that these ties to requirements are maintained throughout the process.

The process borrows from the model-driven architectures (MDA) approach to produce models of the simulation system on different levels—targeting different sets of stakeholders (Object Management Group, 2007). The capabilities and operational viewpoint activities produce an operational description of the system to be simulated. The operational products, analogous to the computation-independent model of MDA, are independent of the fact that a simulation model is being built. They consider the concerns of the operational stakeholders who must use or operate the system as it functions in the real world. The service viewpoint activities focus on the simulation architecture as a whole. These products, analogous to the platform-independent model of MDA, assign simulation functions and metrics to different simulation models. The primary stakeholders for system level activities are integration engineers and managers for each of the component models. The federation implementation activities focus on the detailed development of simulation components and the interfaces between them. These products, analogous to the platform-specific model of MDA, provide sufficient detail for software engineers to develop code that will interface with the overall simulation federation to provide the required results. In some cases, the software or test cases may be auto-generated from the technical specification.

While any system engineering architectural approach would work for simulation development, this book's focus on combat modeling makes the Department of Defense Architecture Framework (DODAF) a good choice for this example (Department of Defense, 2010). The steps of the architectural process and corresponding model views will be represented using the DODAF standard.

Capability Viewpoint Activities

Because a system of systems federation is such a difficult undertaking, the organization that proposes to attempt development should first clearly articulate the capabilities that this federation will deliver and how those capabilities will achieve organizational goals. DODAF's capability viewpoint is designed just for this purpose. Figure 30.2 shows the engineering activities necessary to identify federation capabilities.

- *Develop federation vision.* The senior leadership of the organization that owns, funds, or benefits from the system of systems capability should

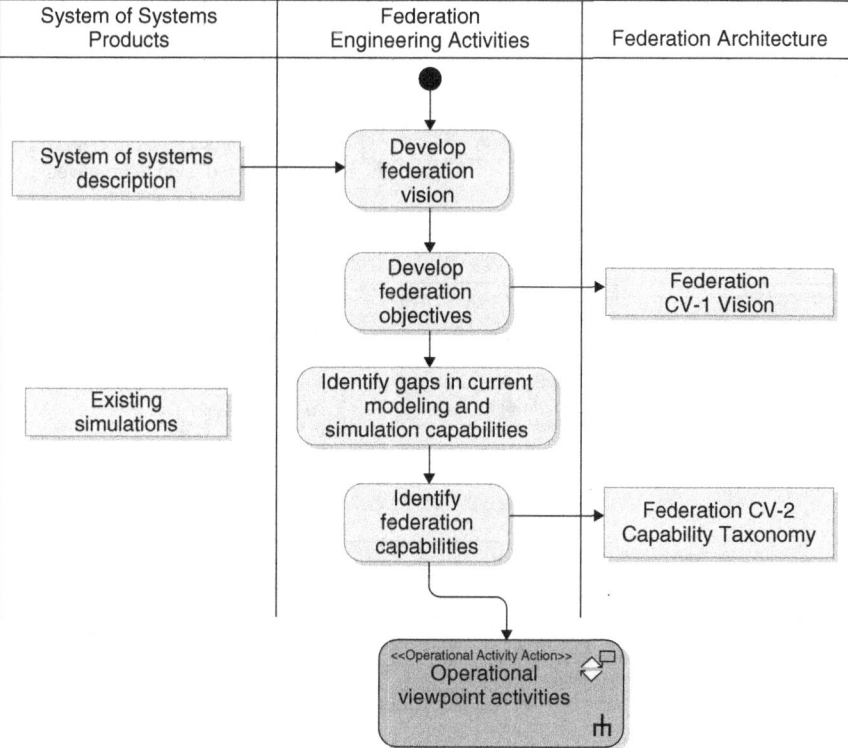

Figure 30.2 Capabilities viewpoint activities.

express how a federation of models will help them in the process to realize the desired system of systems capability. This vision will guide lower level engineers and leaders in further development of simulation objectives and capabilities.

- *Develop federation objectives.* Clearly state the objectives of the simulation development activity. Properly stated, these objectives will articulate not how good the simulation federation will be, but how that federation will support the decision processes, training processes, or implementation processes undertaken to realize the system of systems capability.

- *Identify gaps in current modeling and simulation capabilities.* Within DODAF, capabilities allow an organization to achieve their desired effects, or goals. Before defining capability requirements for a federation, the organization should thoroughly understand its current simulation capabilities and its ability to achieve federation objectives with those simulations. The differences between required capabilities and current capabilities are capability gaps.

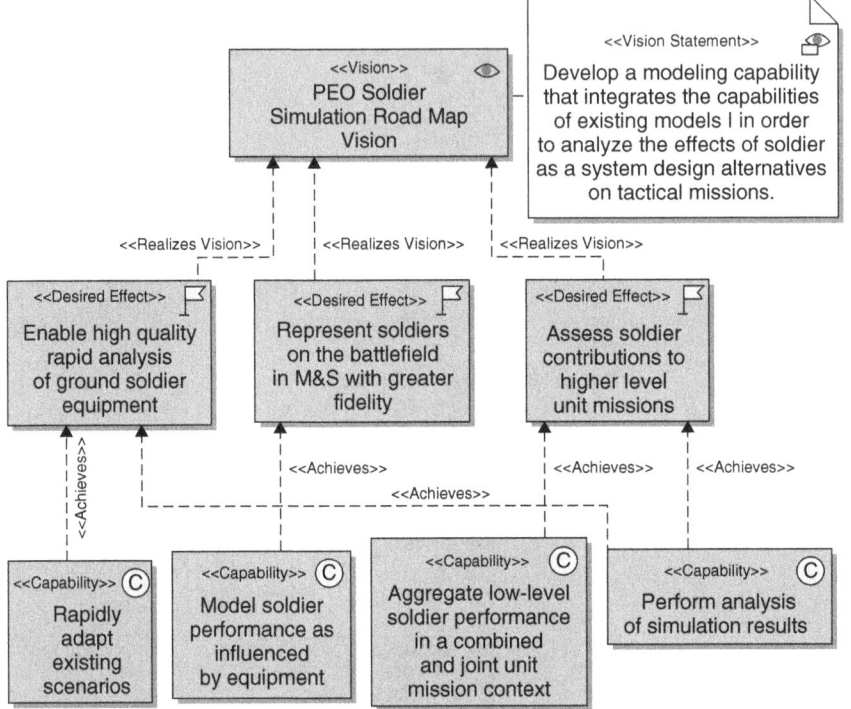

Figure 30.3 CV-1 vision for PEO Soldier federation.

- *Identify federation capabilities.* The capability gaps identified in the previous step naturally translate into the required capabilities for the federation that will be used to model the system of systems. These capabilities become the succinct high level design requirements for the federation.

The results of the capabilities viewpoint activities can be summarized in a Capabilities View-1 (CV-1) vision diagram. Figure 30.3 shows this view for the PEO Soldier federation. This is also an important decision point for the system of systems leadership team. Given the estimated timelines, costs, and risk involved with federation development, the leadership team must articulate to subordinate engineers whether the defined simulation capabilities, and their impact on organizational objectives, are worth the cost. If so, the leadership must resource simulation development with sufficient money, time, manpower, and risk mitigation measures to allow successful federation development.

Operational Viewpoint Activities

In performing the operational viewpoint activities, simulation engineers are focused on the problem definition phase of the systems engineering process

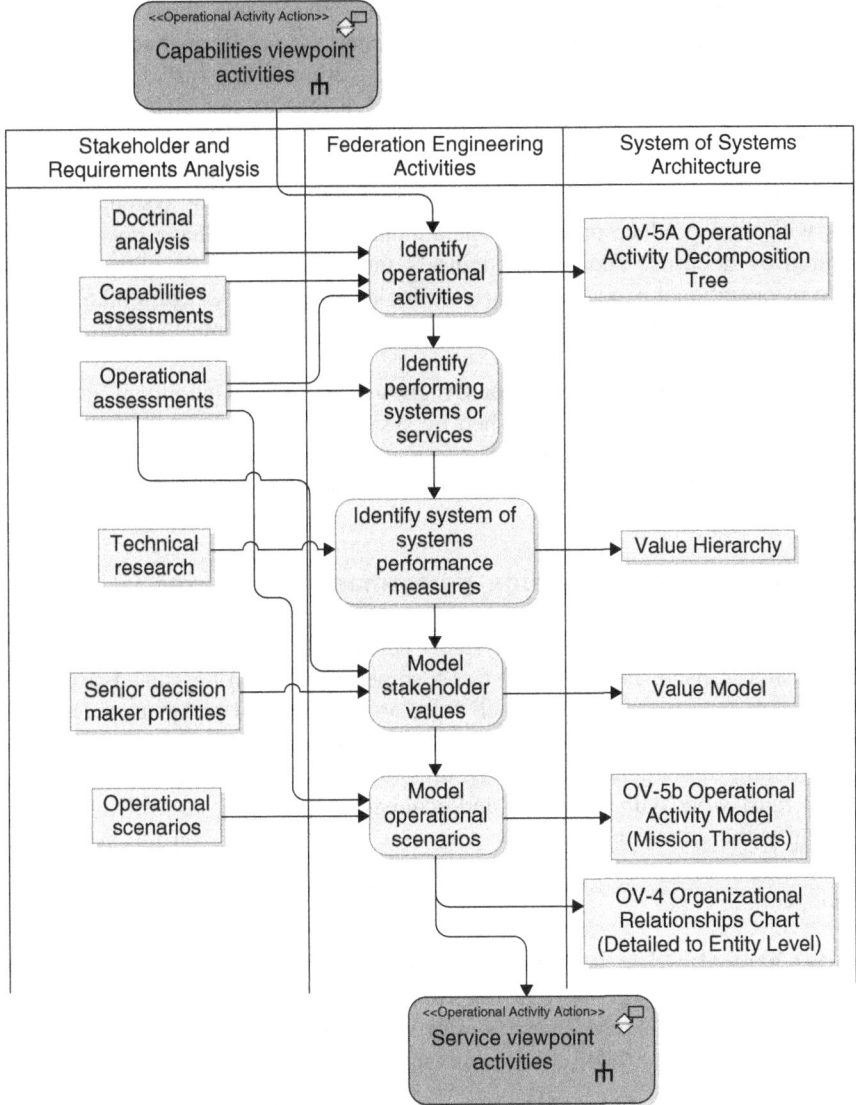

Figure 30.4 Operational viewpoint activities.

(Parnell et al., 2010). Their primary goal during this phase is to gain an understanding, documented as a set of DODAF views, of the system of systems to be modeled. Note that the views in this section do not define the simulation, but the system of systems to be modeled. Engineers should understand its users, its subsystems, and the value derived from its operation. The steps in Figure 30.4 represent the steps of this process.

Identify Operational Activities. In this phase, operational experts who employ the system of systems identify and define the operational activities performed by the system of systems. Often, the activities are further defined by operational doctrines. If this is the case, it is useful to consult doctrinal task lists such as the Universal Joint Task List or Army Universal Task List for commonly understood doctrinal activities (Headquarters Department of the Army, 2009; The Joint Staff, 2010). However, the emergent nature of systems of systems will often produce emerging operational activities that have not yet been captured by doctrine. Therefore, engineers must work carefully with the right group of operational experts to clearly define these activities. Figure 30.5 shows the operational activities for a dismounted command and control system.

Identify performing systems or services. The operational activities in the previous steps are accomplished by systems or services on the battlefield. Particularly with respect to system of systems activities, operational experts must define the performers that work together to accomplish the system of systems activity.

Identify system of systems performance measures. Given the identified operational activities and performers, operational experts must also help define the critical system of systems performance measures of interest for the decision at hand. For doctrinal tasks, some insights can be gained from the performance measures listed for each Joint or Army task (Headquarters Department of the Army, 2009; The Joint Staff, 2010). However, this exercise is not a simple listing of the performance measures from doctrinal manuals. Operational experts must work with engineers and acquisition authorities to drill into these to identify the critical performance measures that shed light on the system of systems engineering and acquisition decisions for which the federation is being developed.

Model stakeholder values. To further identify critical value measures, the engineers must work beyond doctrinal manuals to identify undocumented values. Values such as ease of use, simplicity, or the ability to transport or carry the system will reveal themselves in this analysis.

- *Identify stakeholders:* Identify stakeholders for the system of systems. These are not only users but also system owners, system developers, and the client for which the simulation study is being performed. Ensure the modeling detail captures enough information to answer the question at hand without incorporating so much detail that simulation development becomes overly difficult, expensive, and time consuming.
- *Elicit stakeholder concerns:* Using any combination of interviews, focus groups, and surveys, determine the fundamental concerns and values of the stakeholders identified in the previous steps.

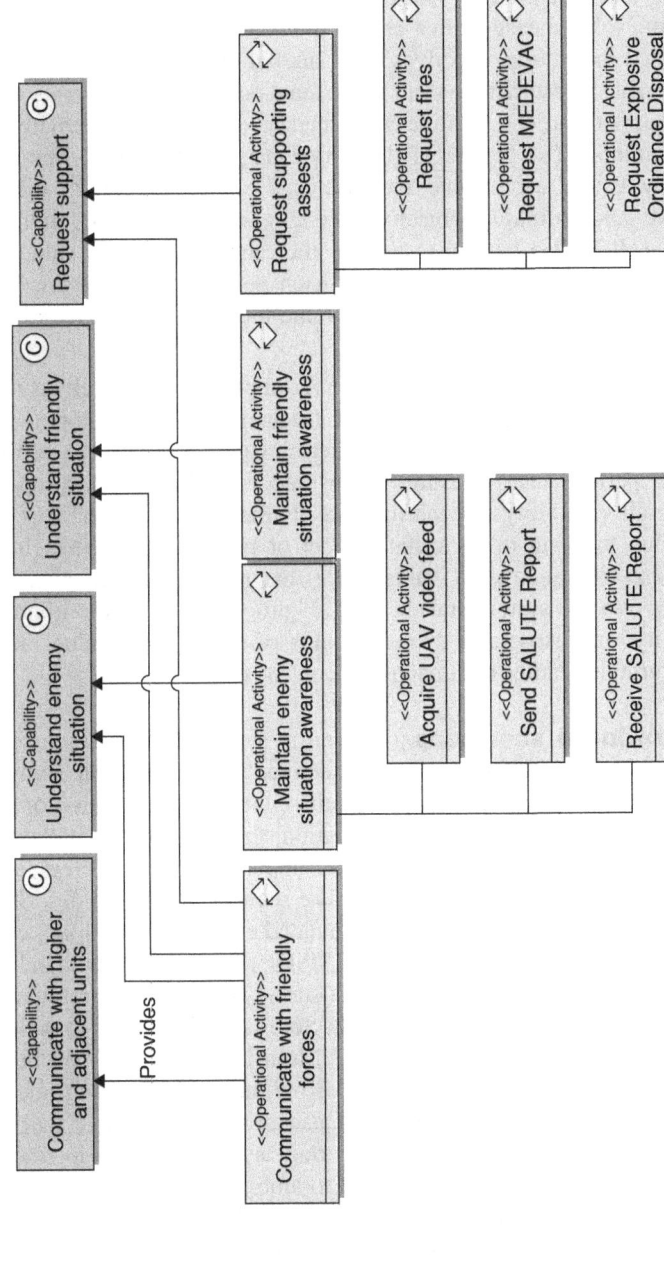

Figure 30.5 Operational activities for a dismounted command and control system.

- *Determine system of systems objectives:* Based on the operational activities and stakeholder concerns, determine the system of systems objectives that must be successfully met in order to deliver value back to the stakeholders.

- *Define value measures:* Once the objectives have been identified, determine value measures that can be used to see if the system of systems has indeed met those objectives. The simulation system, once developed, must be able to estimate system of systems performance in terms of these objectives so that the relative performance of different alternatives can be determined. The results of this phase may be represented as a value hierarchy where each value measure is a component of an individual objective, and individual objectives are components of the overall system objective. It is also helpful to specify the range of possible values, from least desirable to most desirable, for each performance measure (Parnell et al., 2010). Figure 30.6 shows the value hierarchy for a ground soldier command and control system.

- *Build system of systems value function:* A simple value hierarchy is not sufficient for direct comparison between alternatives. The development team must return to the stakeholders and determine the value curves and weights for each performance measure, taking into consideration the importance and overall variability of each measure (Parnell et al., 2010). This results in a value function that translates a set of performance scores for each alternative into an overall value score that represents the value of that alternative to the system stakeholders. Figure 30.7 shows the importance, variability, and associated swing weights of each of the value measures for the ground soldier command and control system.

Model operational scenarios.   Once the system of systems, its operational activities, and its values have been defined, the simulation study team must also define the scenario that represents the context for evaluation of system of systems performance. In a military simulation, this represents details such as terrain and weather, forces in the engagement, supporting friendly forces, and full definitions of the entities in the simulation. The scenario definition describes the mission, forces involved, and roles of the simulated entities. In a military context, the Military Scenario Definition Language (MSDL) is an excellent standard for this representation (SISO, 2008). Figure 30.8 shows the MSDL representation of the platoon level scenario used for the PEO Soldier project. This is a platoon-sized operation derived from the TRADOC Multi-Level Scenario 20 in which a Stryker platoon attacks to seize a series of buildings in a built-up area. The primary focus of the operation is the first squad of the platoon that will secure Objective A. Entity definitions are an important aspect of scenario definition. All too often, the names of entities can lead to ambiguous understandings of the actual capabilities represented. Entity definitions should be as specific as possible with references to authoritative sources that provide accurate data to simulation modelers who must represent these entities in their respective simulations. Finally,

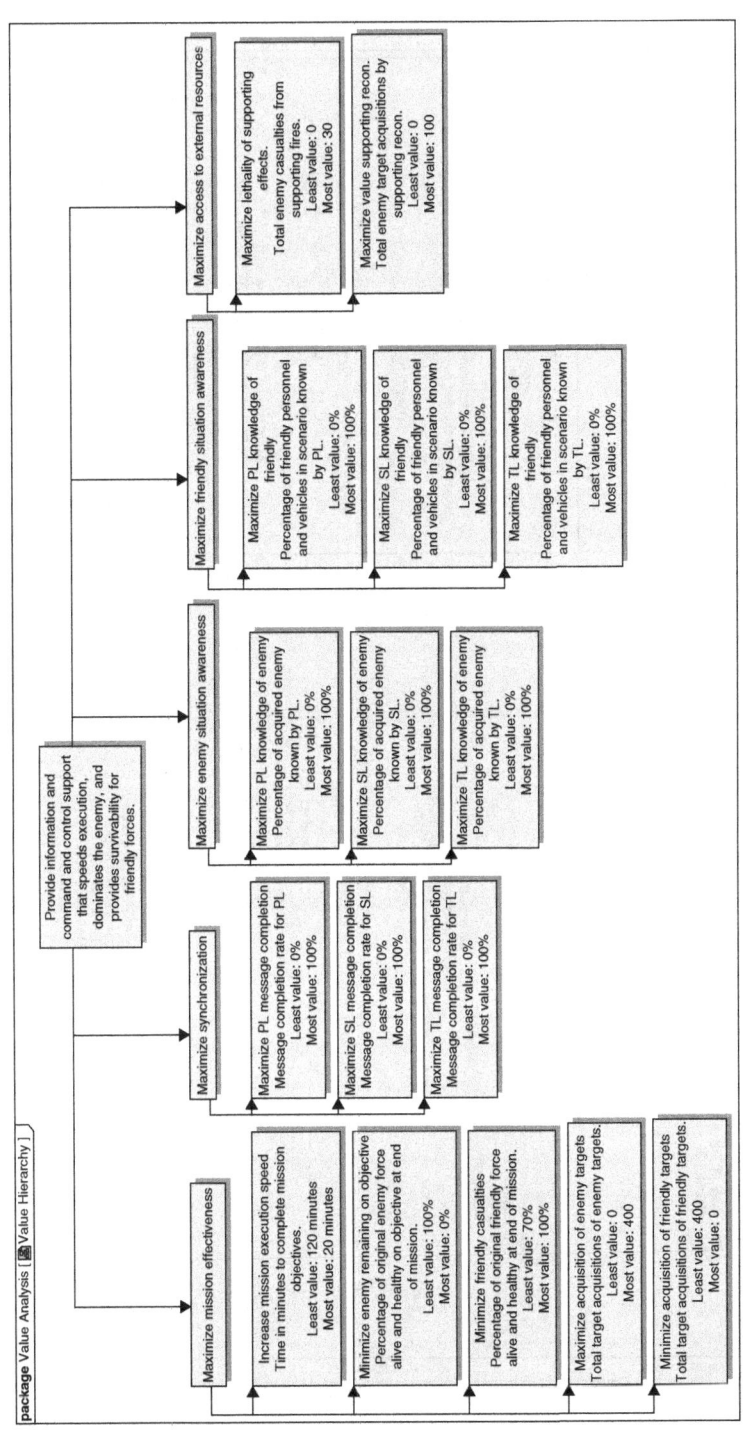

Figure 30.6 Ground soldier command and control system value hierarchy.

activity Swing weight matrix [Swing weight matrix]

	Most important	Very important	Important
High variability	: Increase mission execution speed Time in minutes to complete mission objectives. Least value: 120 minutes Most value: 20 minutes Weight: 100		: Maximize TL knowledge of enemy Percentage of acquired enemy known by TL. Least value: 0% Most value: 100 Weight: 30 : Maximize TL knowledge of friendly Percentage of friendly personnel and vehicles in scenario known by TL. Least value: 0% Most value: 100 Weight: 30
Medium variability	: Minimize friendly casualties Percentage of original friendly force alive and healthy at end of mission. Least value: 70% Most value: 100% Weight: 70	: Maximize acquisition of enemy targets Total target acquisitions of enemy targets. Least value: 0 Most value: 400 Weight: 40 : Maximize PL knowledge of enemy Percentage of acquired enemy known by PL. Least value: 0% Most value: 100% Weight: 40 : Maximize PL knowledge of friendly personnel and vehicles in scenario known by PL. Least value: 0% Most value: 100% Weight: 40 : Minimize acquisition of friendly targets Total target acquisitions of friendly targets. Least value: 0 Most value: 400 Weight: 40	: Maximize lethality of supporting effects. Total enemy casualties from supporting fires. Least value: 0 Most value: 30 Weight: 15 : Maximize SL knowledge of enemy Percentage of acquired enemy known by SL. Least value: 0% Most value: 100% Weight: 15 : Maximize SL knowledge of friendly Percentage of friendly personnel and vehicles in scenario known by SL. Least value: 0% Most value: 100% Weight: 15 : Maximize value of supporting recon. Total enemy target acquisitions by supporting recon. Least value: 0 Most value: 100 Weight: 15
Low variability	: Minimize enemy remaining on objective Percentage of original enemy force alive and healthy on objective at end of mission. Least value: 100% Most value: 0% Weight: 50	: Maximize PL message completion Message completion rate for PL Least value: 0% Most value: 100% Weight: 20	: Maximize SL message completion Message completion rate for SL Least value: 0% Most value: 100% Weight: 5 : Maximize TL message completion Message completion rate for TL Least value: 0% Most value: 100% Weight: 5

Figure 30.7 Swing weight matrix used to build system value function.

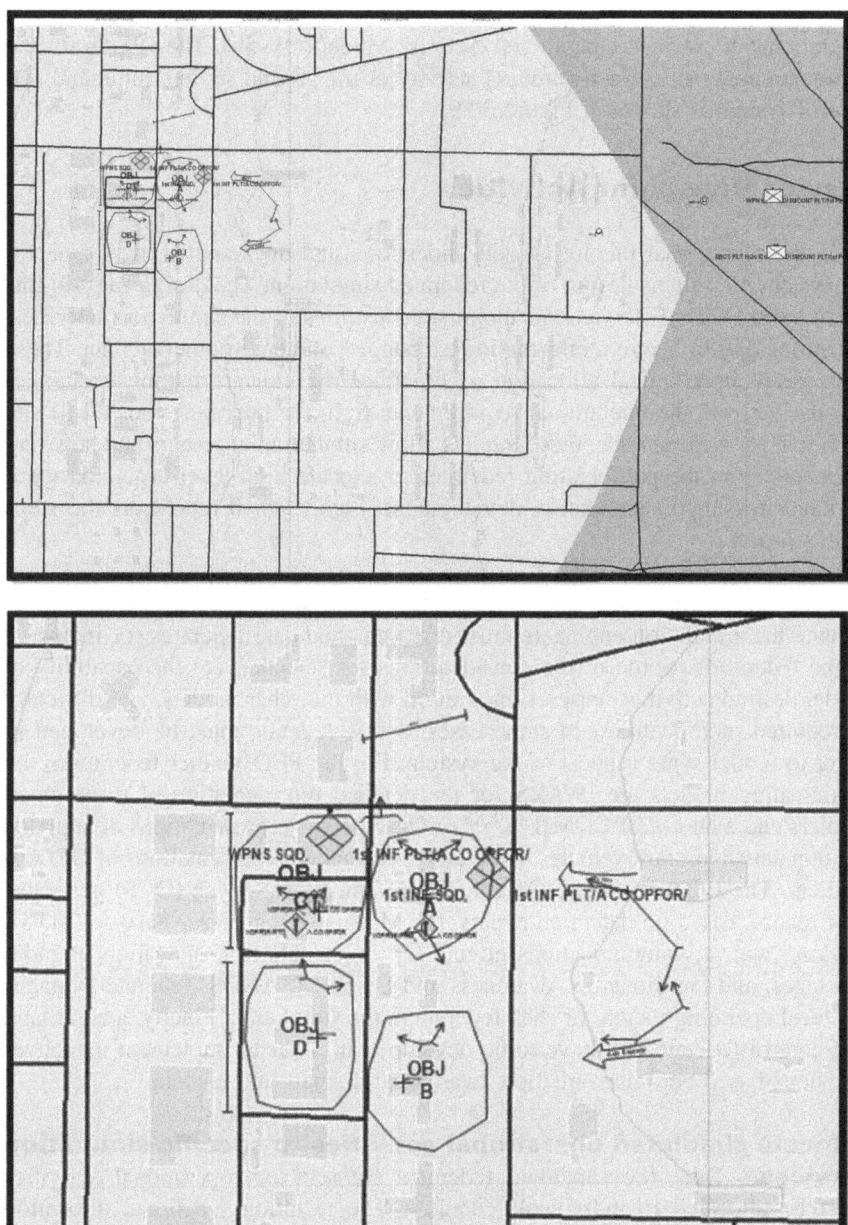

Figure 30.8 Military Scenario Definition Language representation of attack scenario.

within the scenario, the operational activities performed by the system of systems should be defined as a series of mission threads so that simulated events can be synchronized in the proper order. This model can be represented as a DODAF Operational View 5b—Operational Activity Model (OV-5b). The OV-5b for the maintain enemy situation awareness activity of the ground soldier command and control system is shown in Figure 30.9.

Service Viewpoint Activities

Once the operational picture is well understood and documented, it is time for the system of systems design of the federated simulation. The design steps in this phase build DODAF service viewpoint specifications of the simulation activities, federation data collection, information exchanges, and environmental data. These steps result in a logical allocation of functionality and information exchanges that derive from the operational requirements from the previous step. While not sufficient for writing code, these models allow simulation engineers and software engineers from the participating federates to allocate high level tasks and work packages to support simulation development. Figure 30.10 represents the steps of this process.

Select participating simulations.   Once the required functionality is known, the simulation engineers must research candidate federates for inclusion in the federation. Some of the considerations for selection are the capability to model desired activities, ease of integration with the other models, and difficulty of required modifications. In some cases, a new federate must be developed in order to model some aspects of the system. For the PEO Soldier federation, the participating models are IWARS for the detailed representation of dismounted soldiers and a choice of COMBAT[XXI] or OneSAF for representations of supporting unmanned aircraft, vehicles, fires, and less detailed representations of OPFOR soldiers. The BCMS federate Organic Communication Service (OCS) generates spot reports based on detection events, the Message Transceiver Service (MTS), working with a communications effects server, calculates propagation of radio messages, and the Situation Awareness and Display (SANDS) federate manages the local operating picture of each federate in the simulation. Finally, a command and control federate will have to be developed in order to implement maneuver decisions based upon the situation awareness of small unit leaders.

Allocate simulated operational activities to specific simulation services.   Once the candidate federates are selected, operational activities must be allocated to individual federates. The resulting functional allocation takes the OV-5b mission thread from the operational viewpoint and allocates specific activities to federates using swim lanes in the activity diagram. This allocation for the ground soldier command and control simulation is shown in Figure 30.11 as a DODAF Services Viewpoint-4 Services Functionality Flow

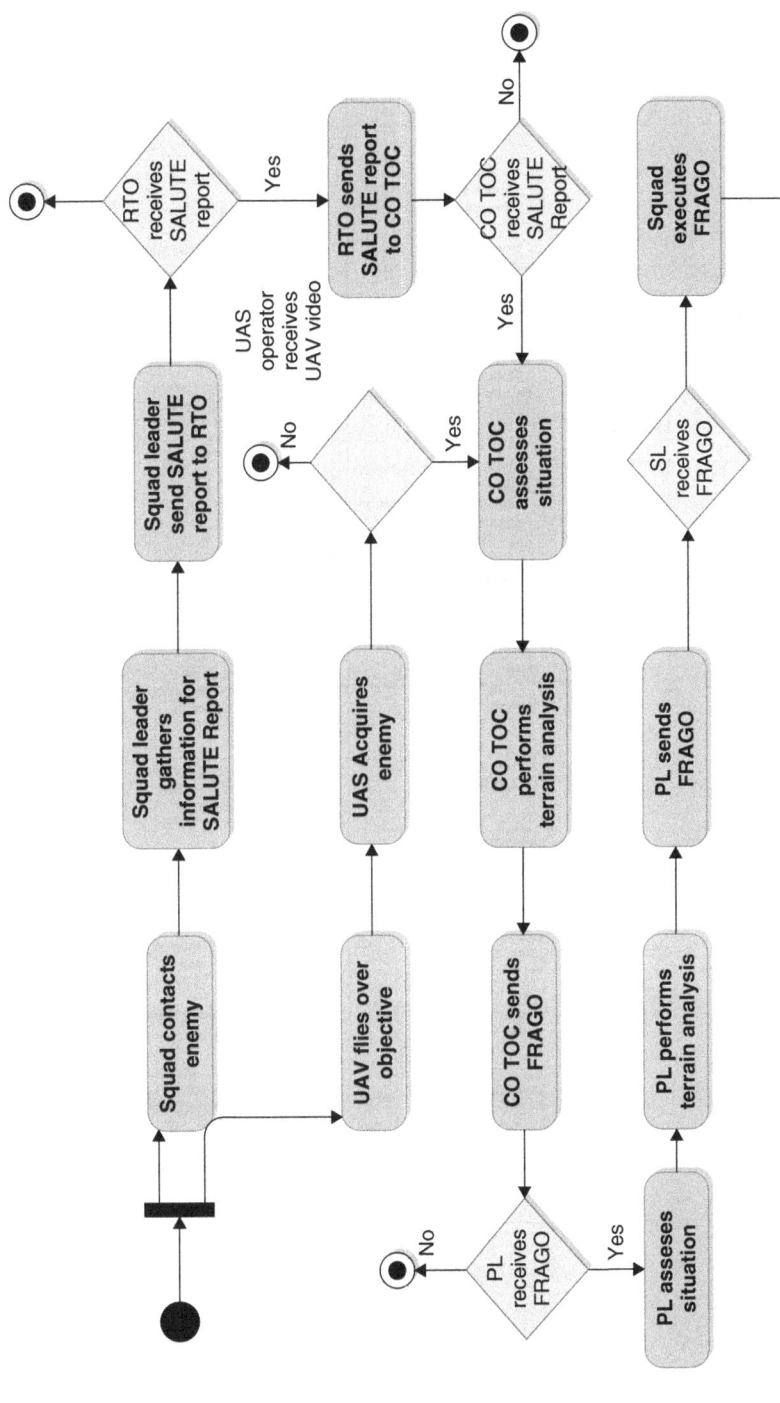

Figure 30.9 OV-5b Operational Activity Model for the maintain enemy situation awareness operational activity within a dismounted tactical mission.

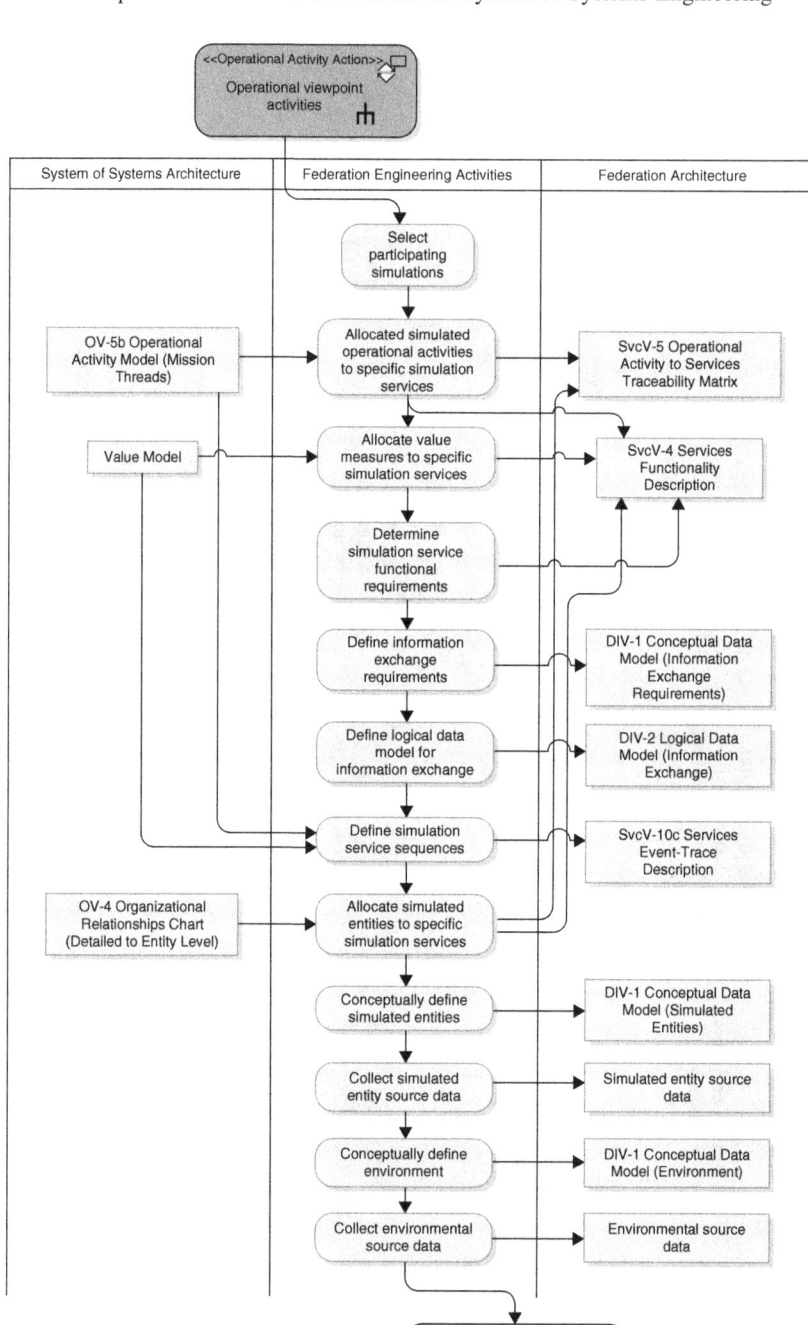

Figure 30.10   Service viewpoint activities.

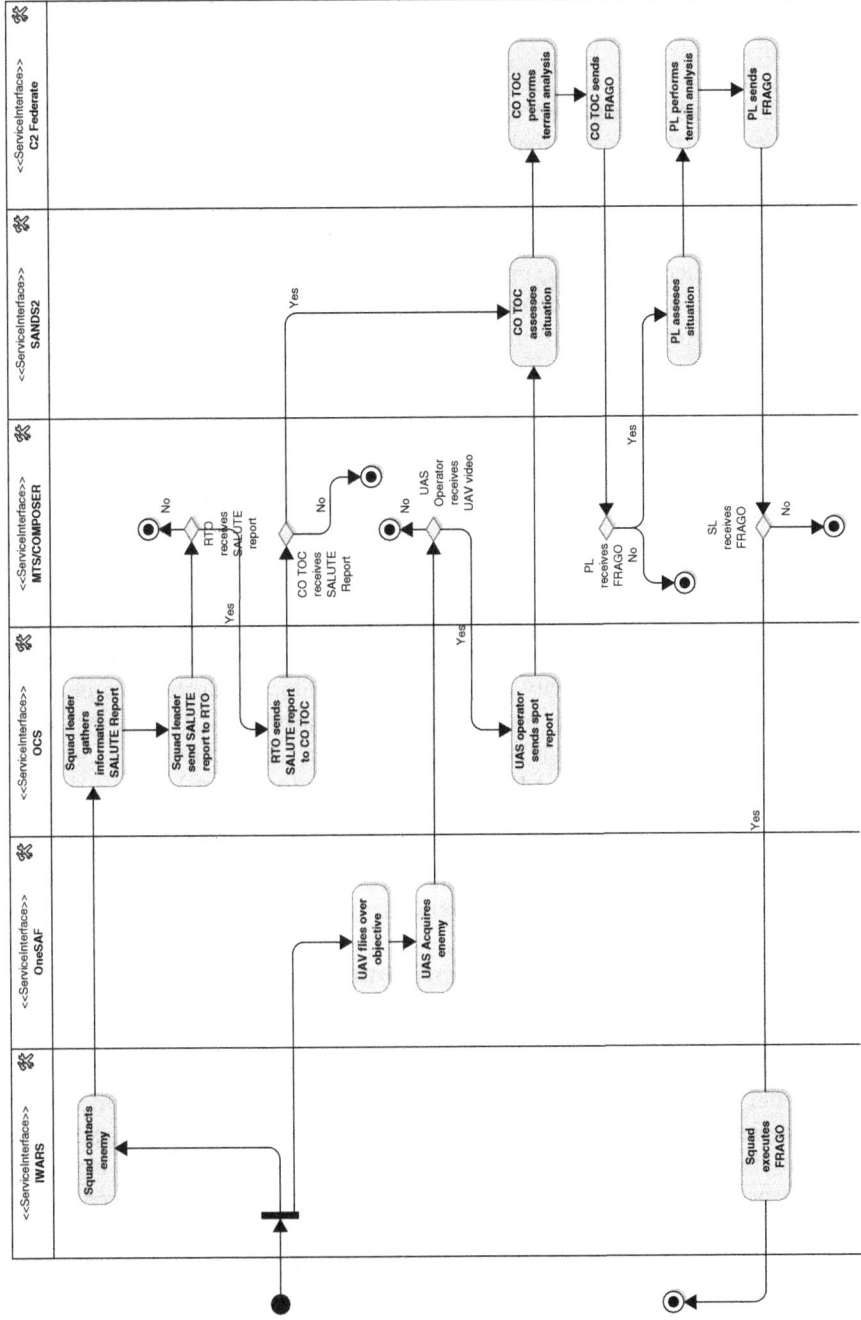

Figure 30.11 SvcV-4 showing allocation of situation awareness activities from the OV-5b to simulation services, represented as vertical swim lanes.

Description (SvcV-4). This view can also be used to generate a Services View-5 Operational Activity to Services Traceability Matrix (SvcV-5). For a federation, this allocation shows which simulation services model the actions within an operational activity.

Allocate value measures to specific simulation services. In a manner similar to the allocation of operational activities, the requirements to collect necessary performance data should be allocated to federates as well. In some cases, the required data may not exist in any one federate but will have to be collected from network interaction data collected by loggers. In the case of the PEO Soldier federation, most of the required data were available within the network data logger used to support the experiment. However, a custom data logger was required to capture the knowledge state of friendly forces over the course of the simulation.

Determine simulation service functional requirements. Once the modeling functions and data collection functions have been determined, the simulation functionality requirements may be formally specified for the models. These requirements documents may be used to support contracting with federate developers who must deliver models with the required functions.

Define information exchange requirements. In order for the federation to execute, data must be exchanged between the models. These requirements may be derived from the activity diagrams used to allocate functions to individual federates. Any time that a control line crosses a swim lane, there is typically a requirement for some amount of information to be passed in order to support that allocation. This results in a DODAF Data and Information View-1 Conceptual Data Model (DIV-1). Each entity represented in the DIV-1 is a conceptually aligned and meaningful information element that must be supported by the information exchange system. Figure 30.12 shows some of the data entities required for the simulation service shown in Figure 30.11.

Define logical data model for information exchange. As information exchange requirements are identified in the previous step, engineers must formally specify the data elements required to support that data exchange. These data requirements can be specified in a Data and Information View-2 Logical Data Model (DIV-2) as shown in Figure 30.13. This is a two-way process. It may be more efficient to delay this formal specification until the technical information exchange architecture is selected. In some cases, information elements from that architecture may be reverse engineered to provide the required information elements. In other cases, these elements must be defined from scratch before further specifying in the technical information exchange architecture. The DIV-2 adds detailed attributes and data types to the entities defined in the DIV-1. For the PEO Soldier federation, a hybrid approach was taken. Most of the entities in the DIV-1 Conceptual Data Model could be represented by using the

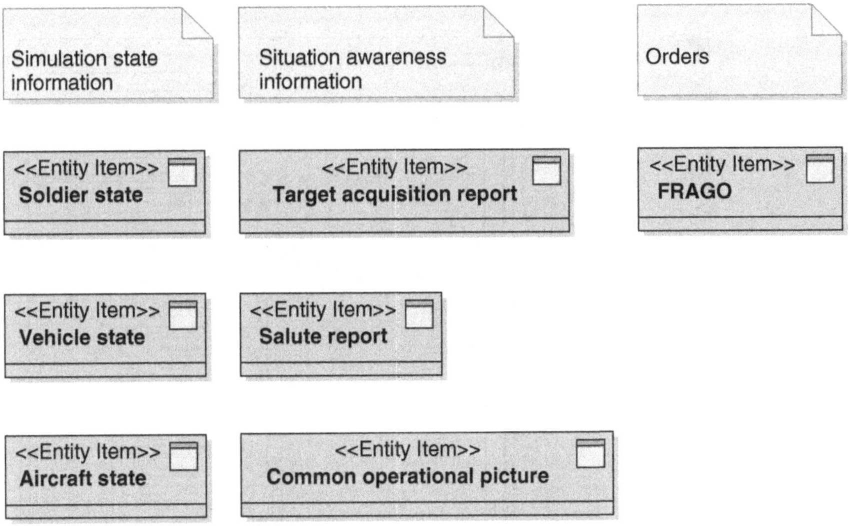

Simulation state information	Situation awareness information	Orders

<<Entity Item>> **Soldier state**	<<Entity Item>> **Target acquisition report**	<<Entity Item>> **FRAGO**

<<Entity Item>> **Vehicle state**	<<Entity Item>> **Salute report**

<<Entity Item>> **Aircraft state**	<<Entity Item>> **Common operational picture**

Figure 30.12   DIV-1 Conceptual Data Model showing the information exchange requirements for modeling dismounted situation awareness.

MATREX FOM (MATREX, 2009). However, additional work was done to allow the representation of a dismounted FRAGO (Kewley et al., 2010).

Define simulation service sequences.   Using the simulation services defined in the SvcV-4 and the data defined in the DIV-2, construct a DODAF Service View-10c Services Event-Trace Description (SvcV-10c) that shows the detailed sequence of simulation service events and information exchanges required to simulate a particular system of systems activity. This model is a common shared understanding between federation engineers and simulation developers about each service interface, the specific information passed in, the required functionality, and the specific information returned. This is all within the context of an already defined operational activity and supporting OV-5b. Figure 30.14 shows the SvcV-10c service functionality, interfaces, and information exchanges required to simulate the maintain enemy situation awareness operational activity defined in Figure 30.9.

Allocate simulated entities to specific simulation services.
Within the federation, the responsibility to maintain and publish entity state data must be allocated to specific federates. The necessary input information can be found in the scenario data defined in the operational viewpoint. This may be an MSDL file or a DODAF Operational View-4 Organizational Relationships Chart (OV-4). Each unit or entity defined in these models must be allocated to a specific federate that will maintain, update, and publish that entity's state data. Within the PEO Soldier federation, a specific XML format was defined that specified entity ownership from the MSDL data.

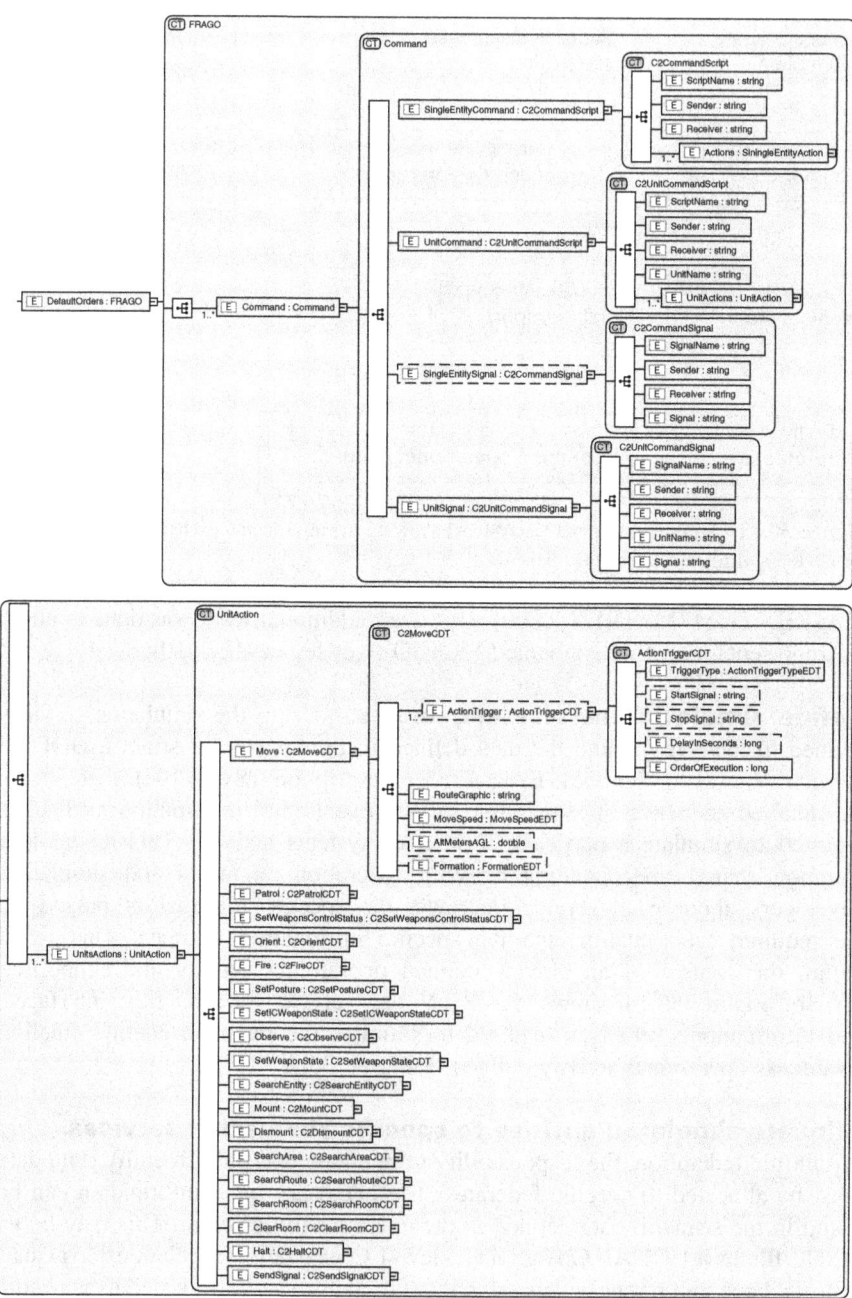

Figure 30.13 DIV-2 used to represent a dismounted FRAGO. This model was developed using an XML modeling tool.

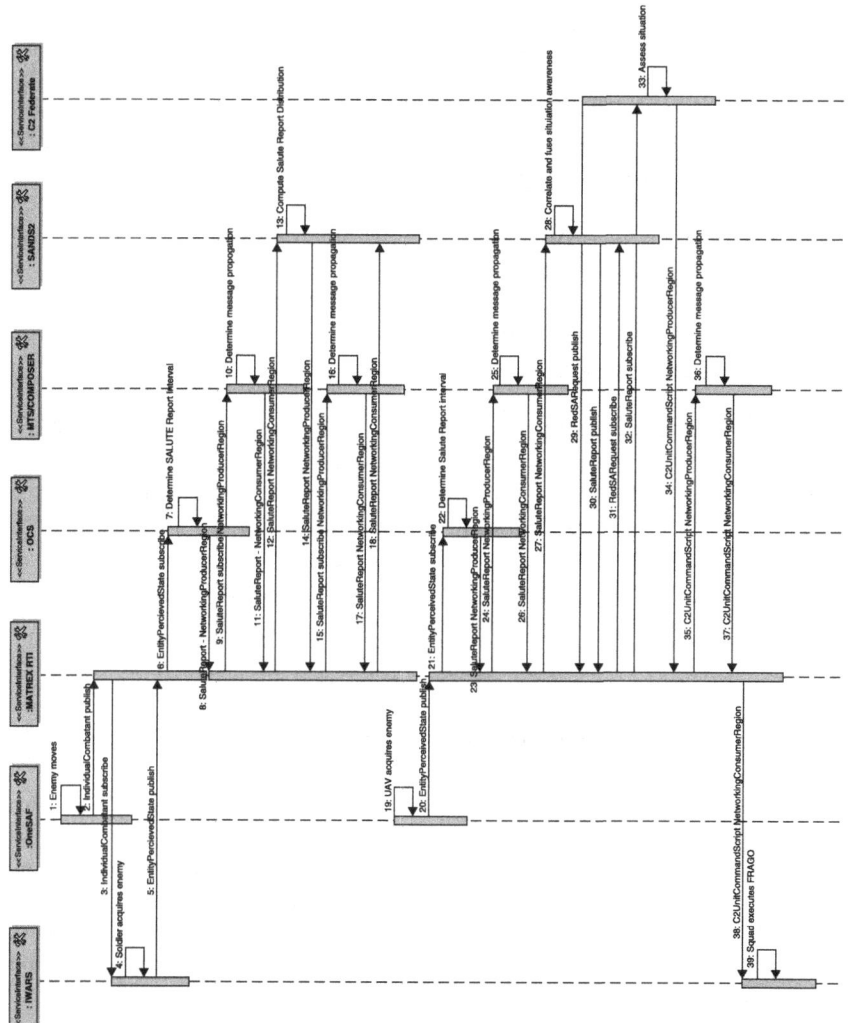

Figure 30.14 Svc-10c showing the specific sequence of simulation services and information exchanges required to simulate the maintain enemy situation awareness operational activity.

797

Conceptually define simulated entities. A meaningful federation requires a consistent and conceptually aligned representation of entities in different simulation models. For each entity in the scenario, the federation developers must provide a doctrinal or other reference that unambiguously defines the characteristics and capabilities of that entity. Participating federates will each use this common reference to build the necessary data to support their models.

Collect simulated entity source data. For each specific entity defined in the previous step, federation engineers must also go to authoritative sources to collect performance data for these entities. Typical data collected in this step include weapons lethality data, system survivability data, movement rates, sensor performance data, or communications capabilities.

Conceptually define environment. In addition to entities, the environment must be considered as well. Some of this information may be contained in the MSDL file generated for the scenario. Federation engineers must unambiguously define the terrain on which the simulation will take place, the weather, the time of day, and light data.

Collect environmental source data. This step represents the collection of source data necessary to appropriately represent the environment in the different federates. The environmental representation may not be the same for all federates. However, using the same source data will lead to correlated representations across the models. For the PEO Soldier federation, a terrain box at White Sands, New Mexico, was used. A collection of geospatial elevation data, imagery, and shape files was compiled to support terrain data generation using a commercial tool.

Federation Implementation Activities

Once the system has been designed and data have been collected, it is still necessary to do system development for all of the participating federates and for the overall federation integration architecture. These are all technical implementation activities that look to provide software engineers and programmers with sufficient information that will allow them to write code and deliver working federates within the overall specification. Figure 30.15 shows a diagram of the federation implementation activities and products.

Select information exchange technical standard. The simulation must exchange information across an architecture designed for this purpose. Simulation standards such as DIS or the HLA are possible choices. Another possibility is to use service oriented architectures, based on web standards, designed to support business or industrial systems. For the PEO Soldier federation, the MATREX High Level Architecture has been selected.

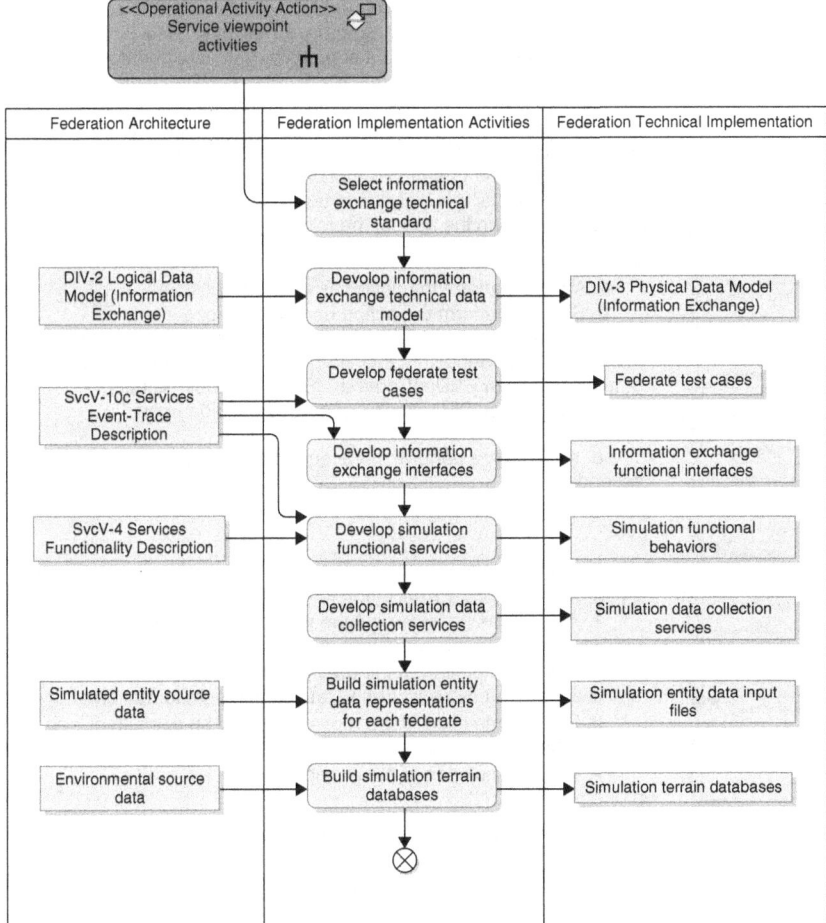

Figure 30.15   Federation implementation activities.

Develop information exchange technical data model.   The infor-mation exchange data model must be specified and represented in technical format selected in the previous step. In the case of HLA, this specification will be a federation object model (FOM). A web services architecture would require exten-sible markup language (XML) representations of the data. For the PEO Soldier federation, the MATREX FOM version 4.3 has been selected. This model con-tains most of the information elements needed to represent the entities of the DIV-1 in Figure 30.12. The FOM needed to be extended to add the FRAGO information exchange entity of the DIV-2 shown in Figure 30.13.

Develop federate test cases.   While simulation development is possible without test cases, they provide developers with a clear and understandable way to interpret and develop services in response to specified information exchanges.

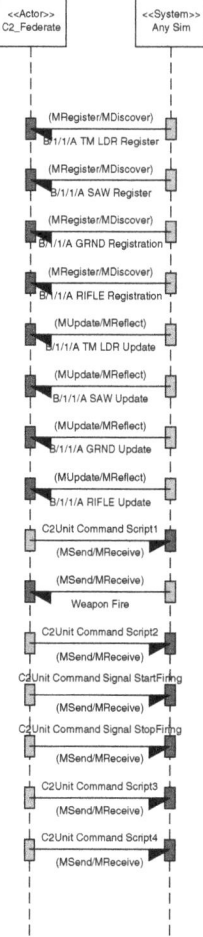

This is the most critical data structure for this test case. The required capability is to build a supporting data structure for individual tactical actions in this script. The receiving federate will have to scan through each of the 16 different task types in this data structure for possible behaviors to implement. Each behavior will include a data structure for that behavior and an ActionTrigger. The OrderOfExecution field in the ActionTrigger indicates the order in which these tasks must be performed. In terms of triggering the action, the following values are possible:

 a. ASAP - perform the action as soon as possible

 b. Signal - perform the action upon receipt of a signal string called StartSignal. This signal may come via two different interactions - C2SendHumanUnitSignal for human voice, visual, or tactile signls or C2UnitCommandSignal for radio signals.

 c. CompletionOfPrevious - perform the action upon completion of the previous action.

 d. AfterDelay - perform the action after DelayInSeconds seconds.

This interaction has six included tactical tasks for FireTeam-A/1st SQD INF/1ST PLT/A in the following order:

 1. As soon as possible, set the weapons control status to Hold

 2. Upon completion of the previous action, move along the specified route at a quick pace in an echelon right formation.

 3. Upon completion of the move, orient on an azimuth of 300 degrees in a line formation.

 4. Upon completion of the previous action, assume a prone position.

 5. Upon completion of the previous action, set the weapons control status to tight. This will allow engagement of nearby enemy forces.

Figure 30.16   Automated test case used for simulation developers to test their ability to receive a unit FRAGO via the information exchange system and implement appropriate operational activities.

Many development environments provide the capability to replicate services and information exchanges. For the PEO Soldier federation, the MATREX Automated Test Case tool allowed federation engineers to develop specific test cases for each federate that passed the required information and listened for appropriate responses. Figure 3.16 shows a test case given to developers for the PEO Soldier federation.

Develop information exchange interfaces.   Based on the information exchange requirements of the SvcV-10c, each simulation federate must develop the necessary interfaces to receive inputs and send outputs. For the PEO Soldier simulation, MATREX tools provided some assistance. Once the updated FOM was developed, the MATREX tools allowed the automatic code generation of Java and C++ stubs that would assist simulation programmers in building interfaces to the HLA RTI. This code generation saved these developers a great deal of time because they did not have to encode data before it was passed over the infrastructure. This task was handled automatically by the MATREX tools and generated code.

Develop simulation functional services.   In addition to receiving the necessary inputs and posting outputs, each federate needs to build the internal behaviors and representations necessary to accurately model the operational activities identified in the operational view. For example, in the PEO Soldier simulation, each federate needed to build the ability to execute unit or individual tactical tasks that were received via a FRAGO during the simulation run.

Build simulation data collection services.   In this step, the data collection requirements determined in the system level activity must be represented as output data from simulation federates or from data loggers tied to the federation. Depending upon the federate, these formats may be supported by standard database systems, or they may simply be text or log files that must be processed. The developers must build queries or tools that collect these raw data and summarize them as value measures from the value model built in the operational viewpoint. For the PEO Soldier federation, the most complicated data collection requirement was a capability to record the situation awareness state of each friendly entity for each minute of the battle. For this task, a custom MATREX federate was created to query situation awareness each minute on behalf of each friendly entity. It wrote a data file including the known friendly and enemy positions on the battlefield.

Build simulation entity data representations for each federate.   Using the entity allocation to simulation services defined in the services viewpoint, each federate must accurately represent its assigned entities in its own native input data formats. Using the conceptual definition of the entity and the collected source data, each simulation engineer must transform the source data into its own internal representation so that each simulated entity represents the conceptual definition of that entity as closely as possible. These internal representations may be databases, spreadsheets, XML files, or other file formats required by the participating simulations. In some cases, supporting tools for the simulation can ease this transition. In other cases, it is a laborious manual process using basic editors. For the PEO Soldier federation, most of the simulations could be directly read from the MSDL scenario file. An XML file specified the entities and types required by each simulation, and each model built the necessary data files required to simulate each entity in their respective models.

Build simulation terrain databases.   The source terrain data must also be converted into simulation-specific formats. In many cases commercial tools support this process for a variety of formats. In other cases, the terrain data may be read into the simulation models in existing geospatial format. The result of this step is correlated representation of the terrain across the federation. For the PEO Soldier federation, a commercial tool was used to generate correlated databases in the OneSAF ERC format and in an OpenFlight format for IWARS 3D viewer.

Dealing with Emergent Systems

In any system of systems, emergent capabilities will develop to potentially be integrated into the overall system. At first, these emergent capabilities will have their own stand-alone value—a value for which they were probably developed. However, as they enter into the system of systems environment, they will find ways to integrate within that architecture to deliver value in new ways that were not foreseen in either the design of the individual system or the overarching system of systems. With respect to a simulation environment, it may be advantageous to simulate these emergent properties and analyze their effects before committing significant resources to developing and maturing the system of systems capability.

For PEO Soldier, an emergent system is the Smartphone. A rapidly evolving commercial market introduced a small handheld radio that, given a supporting network, outperforms any radios developed specifically for dismounted soldiers. Ideally, a soldier could use the extensive collaboration, mapping, and messaging capabilities of the Smartphone for sharing of intelligence and situation awareness, for navigation, and for passing orders and requests. Unfortunately, the existing message sets and order formats used for dismounted command and control are well suited for voice communications and human understanding—not for automated reasoning and display of the information. In order for the Smartphone to deliver advanced command and control capabilities, some system of systems integration is needed to integrate this commercial technology with a manual orders and reporting system.

A potential solution is the Coalition Battle Management Language (C-BML) (SISO, 2010). Using this standard, a soldier could issue machine-readable reports, orders, and requests via a Smartphone interface. However, extensive testing of this interface is necessary in a number of environments to ensure feasibility and to improve the capability to a point where it could be fully developed and fielded. Some level of simulation based testing would be helpful. Therefore, Smartphones and C-BML need to be integrated into the simulation architecture.

A simulation federation deals well with emergent systems. The federation itself is a system of systems—with each federate its own system, designed to provide value in its own domain. However, by adopting federation rules and information exchange mechanisms, the federate can join the federation to provide system of systems insights above and beyond those provided by a single model. Figure 30.17 shows an adapted version of Figure 30.11 that integrates a Smartphone into the federation via C-BML services. Figure 30.18 shows the adaptation of Figure 30.14. These architectural changes drove system development and testing in order to produce a federation that allowed a Smartphone to

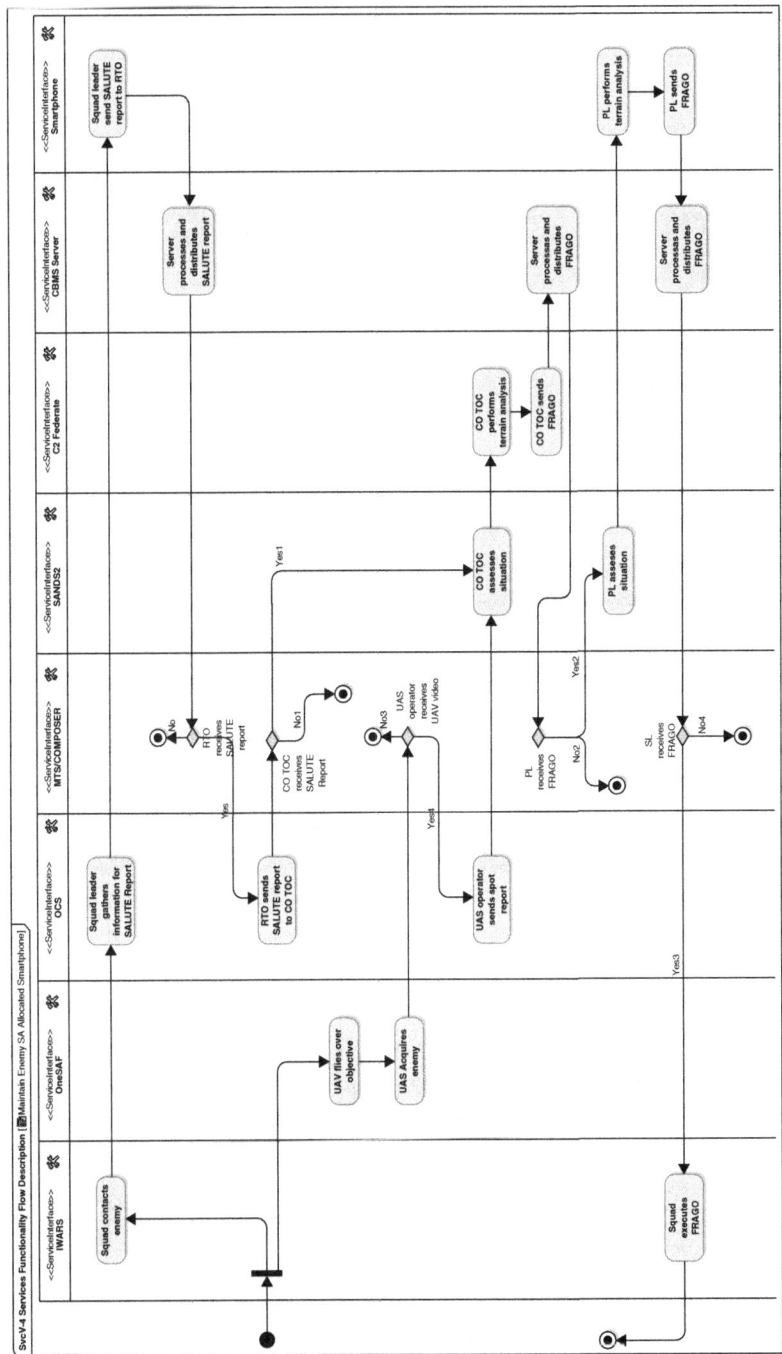

Figure 30.17 SvcV-4 showing the updated services functionality flow for integrating a Smartphone and C-BML services into the simulation architecture.

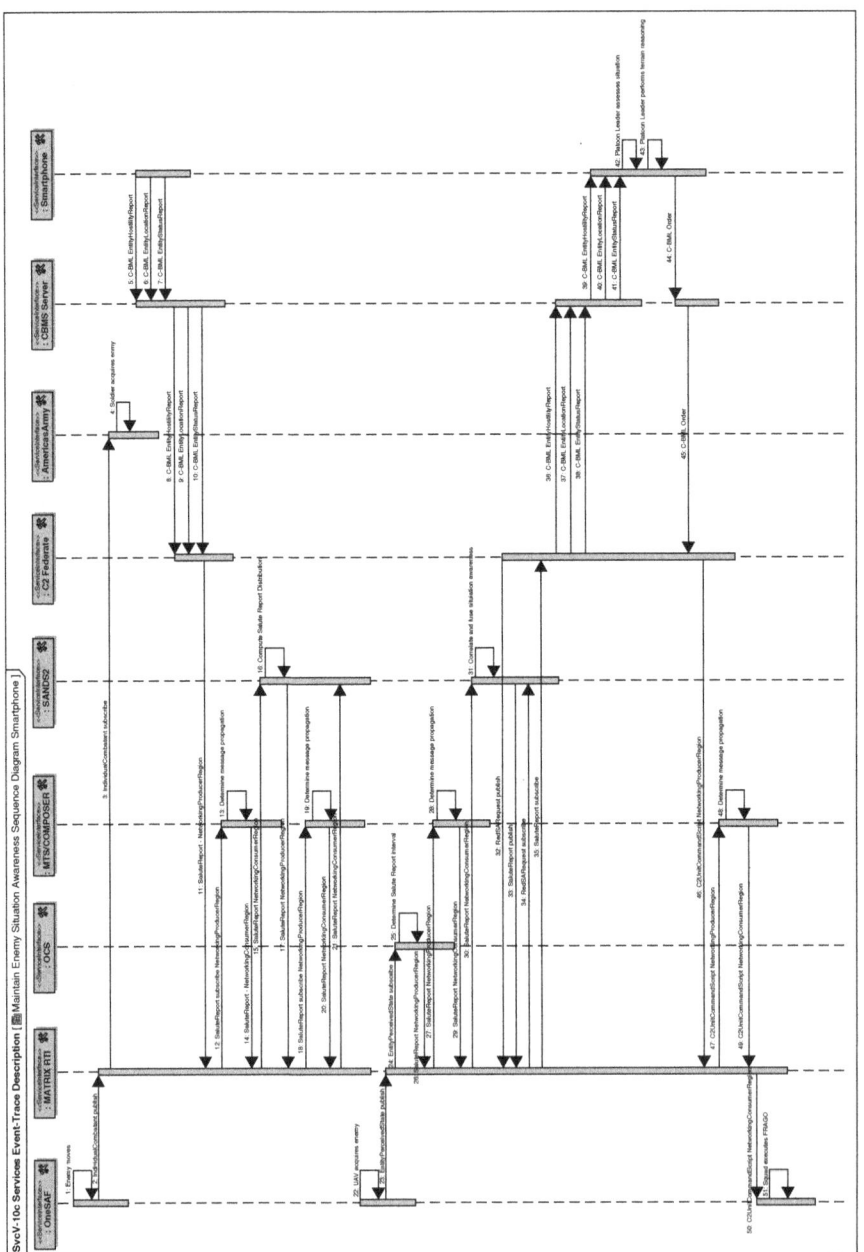

Figure 30.18 SvcV-10c showing the updated event trace description for integrating a Smartphone and C-BML services into the simulation architecture.

display current situation awareness based upon the reports issued during the simulated battle. The leader could assess the situation and issue an order via C-BML to be executed by the simulated entities.

This example shows how a systems engineering approach enables adaptation of the federation to deal with analysis of emergent systems. More complex adaptations would have to deal with changing the federation object model in order to accommodate more complex information exchanges that may be introduced by emergent systems. However, this same approach would serve federation engineers well when dealing with these changes.

Advantages of a Systems Engineering Approach

Delivery of the engineering products described at each step will give system developers all of the specifications they need to build software components that deliver the required functionality and interface with the federation architecture. The capabilities viewpoint will identify the purpose and required capabilities for the federation. The operational viewpoint will give them a conceptual context for integration. The services viewpoint will give them a semantic view of their components and an understanding of the overall simulation architecture. Finally the technical federation implementation products will specify their components in sufficient detail to allow them to interface with the selected technical infrastructure. A good systems engineering process requires stakeholders from all four of these steps to work together in a coordinated way.

There are three main advantages to using a systems engineering approach to federated simulation development. The first advantage is to ensure a clear line of logic from operational representations, to system level federation design, to coding and development. A second advantage is the separation of concerns permitted by modeling the system from different viewpoints. Operational experts can adjust the models in the operational viewpoint. System level experts can organize and specify the system using systems architecture tools, and code developers can work from technical specifications. The final advantage is that all of the systems engineering products support the engineering manager in implementation of the development and test plan. It breaks the complex federation into discrete pieces of functionality that can be developed, component tested, and integration tested in order to manage progress. This approach has helped during implementation of the PEO Soldier federation, saving a great deal of development time and effort that is typically spent rectifying poorly specified interfaces during integration tests.

TECHNICAL MANAGEMENT APPROACH FOR FEDERATED SIMULATION DEVELOPMENT

Given that a federated simulation system is a system of systems, a technical management approach appropriate for system of systems engineering is also appropriate for the development of a federation of simulations. Each simulation in the federation is built for its own purpose, and it delivers value for that purpose. Each simulation has its own independent management structure. Each simulation has its own funding, development, testing, and fielding cycle. Federation development must coexist with the development of each individual simulation, and it cannot obstruct the process by which individual simulations deliver value in their respective domains. It would be better if the federation activities, in addition to providing a system of systems analysis capability, also improved the value of the component simulations.

In this environment, the management approach will necessarily be collaborative. While some central management is necessary, the participants play a significant role, and they typically participate by consent, not by directive. The systems engineering process described in the previous section provides a series of technical steps to achieve federation success, but these steps are not sufficient. In addition to federating simulations, we are federating organizations. A few management principles need to be considered.

Aggressively Seek Shared Value

The lead engineer has the complex task of discovering and articulating shared value across a number of organizations. First, the overarching value of the federation and its support to a system of systems, which was identified in the capabilities view of the DODAF architecture (see the section on Capability Viewpoint Activities), must be clearly articulated. The federation must have a common higher purpose. It will be very difficult to achieve federation objectives if that system of systems objective is not resourced. If the federation lead engineer cannot get resources from the owner of the system of systems capabilities, the management team from the participating simulations will see that as an indicator of the relatively low importance of the system of systems and they will direct their efforts toward their individual funded objectives.

The lead engineer will also have to make some adjustments and compromises that show he or she clearly understands the independent value of the component simulations. Some of this can be done by providing resources to the developers of the component simulations that allow them to achieve system of systems objectives without using their own internal resources. The lead engineer will have to be sensitive to the critical scheduled events and releases for the component systems so that federation events do not conflict with key events for the participants. The more the lead engineer understands the objectives and priorities of the component simulations, the more he or she will be able to find and articulate capabilities that the federation will bring to these components that can be used in pursuit of their own objectives. In complex environments, systems stand to gain from some level of interdependence or synergy. In times of cost and schedule pressure, shared value is the glue that will hold the development team together.

Gain Senior Management Support

Once the lead engineer has a good understanding of the shared value proposition for both the federation and the component simulations, he or she must articulate this value to senior management in each organization to gain support. The system of systems capability and supporting simulation federation must have the support of a senior manager who can articulate its value to other senior managers and provide resources to achieve that capability. This senior manager at the system of systems level, along with the federation lead engineer, must then engage senior management for the participating simulations to agree on resources and gain support.

Gaining management support from participating simulations is not an easy task. These managers will be primarily concerned with their own challenges, resourcing, and schedules. They will answer first to the stakeholders who fund them for delivery of their own models. Federation development will be seen as additional work and resources that may conflict with their own objectives. This challenge will test the persuasiveness and charisma of the lead engineer. To be successful, the argument for support should include the following:

- clear articulation of the system of systems objectives and how federated simulation will support them;
- a persuasive argument, to include funding, from a senior manager who will benefit from the system of systems capability;
- an articulation of the value the participating simulations will obtain through participation in the federation;
- a minimal, open, standards based, and technically sound approach to federation architecture;
- a realistic initial project plan that demonstrates how federation development will not interfere with critical events for the participants.

Shared Technical Development and Management

Assuming shared value and senior management support is achieved, a team of engineers and developers will come together to realize a federation of simulations. This team is unlikely to work directly for the lead federation engineer. Additionally, since federation development is so challenging, they are likely to be some of the most capable engineers or developers in their own organizations. The lead engineer needs to understand his or her role in this setting. He or she is not capable of understanding all the aspects of federation as they apply to the component simulation, so deference to other members of the team must be given.

This approach is consistent with the ideas of subsidiarity and dual citizenship as addressed by Sage and Cuppan (2001). Subsidiarity places power at the lowest level of the engineering team. Once they understand the overarching objectives and general architectural approach, they are best equipped to make engineering decisions with respect to implementation, and their decisions are most likely to attain shared value, perhaps at some expense to federation value. This only works if each engineer is a "dual citizen" in two communities. They are responsible to their own organizations for maintaining the value of the component simulation, and they are responsible to the federation for doing their very best to achieve federation value within constraints.

While the lead engineer may have to resist over-engineering in order to allow the final implementation to evolve, he or she will have to actively intervene if it is clear that one of the participating engineers is not a dual citizen. The lead engineer can first appeal to the offender to refocus at some level on federation objectives. However, if this does not work, he or she must be willing to approach senior management for a different engineer on the team. In a federation environment of shared responsibility, nothing will impede progress more than conflicting agendas and a lack of respect for federation objectives.

CONCLUSION

Military forces around the world face the uncertainty of an evolving threat and more persistent warfare, coupled with the responsibility to anticipate the nature of conflict in the years ahead. They must continue to develop and integrate cutting edge technology in order to field, equip, and train forces to meet these challenges. While combat modeling and simulation has played an important role in the past, the field is on a steep growth trajectory as militaries around the world seek to develop and implement the tools of tomorrow.

Having already established a reputation for adding value, combat modelers must continue to drive advancements in the field. The stand-alone systems of yesterday are being replaced by integrated, federated, distributed systems. Whether utilized as an analytical tool in support of critical defense decisions or as a cost-saving, immersive training environment, simulation continues to be a relevant and tangible commodity.

The complexities of this world compel engineers to develop systems of systems to bring order to the defense domain, and it is imperative to utilize an integrated, systems approach in simulation. As federation offers the flexibility and cost-effectiveness to apply simulation to the system of systems, the chosen development methodology must be based upon sound engineering principles. This methodology must be executed via a systems engineering process, and managed in a way that shares value, gains senior leader support, and grows the federation without handicapping the federate. From this effort will emerge new simulation capabilities where the federation is greater than the sum of its individual federates. By leveraging the strengths of each system and eliminating the barriers to interoperability, the field of combat modeling and simulation as well as the whole of defense will benefit from this progress.

REFERENCES

AMSO (2002). *Planning Guidelines for Simulation and Modeling for Acquisition, Requirements, and Training*. Army Modeling and Simulation Office, Fort Belvoir, Virginia.

Carothers C, Fujimoto R, Weatherly R and Wilson A (1997). Design and implementation of HLA time management in RTI version f.0. In: *Proceedings of the 1997 Winter Simulation Conference*, edited by S Andradóttir, KJ Healy, DH Withers and BL Nelson. IEEE, Piscataway, NJ.

Department of Defense (2010). *Defense Acquisition Guidebook*. Pentagon, Washington, DC.

Department of Defense (2010). *Department of Defense Architecture Framework*, Version 2.02, Department of Defense, Washington, DC.

Francis PL (2009). *Defense Acquisitions—Issues to be Considered for Army's Modernization of Combat Systems, Testimony Before The Subcommittee On Airland, Committee on Armed Services, US Senate*. Technical Report, Government Accounting Office.

Fujimoto RM (1998). Time management in the high level architecture. *Simulation* **71**, 388–400.

Gallant S, Metevier C and Snively K (2009). Systems engineering for distributed simulation within MATREX. In: *Proceedings of the 2009 Spring Simulation Multiconference*. Society for Modeling and Simulation International, Vista, CA.

Handy C (1992). Balancing corporate power: a new federalist paper. *Harvard Bus Rev* **70**.

Headquarters Department of the Army (2009). *FM 7-15 The Army Universal Task List*. Department of the Army, Washington, DC.

Hurt T, McKelvey T and McDonnell J (2006). The modeling architecture for technology, research, and experimentation. In: *Proceedings of the 2006 Winter Simulation Conference*, edited by LF Perrone, FP Wieland, J Liu, BG Lawson, DM Nicol and RM Fujimoto, pp. 1261–1265. Winter Simulation Conference, Monterey, CA.

IEEE (1990). *IEEE Standard Computer Dictionary: A Compilation of IEEE Standard Computer Glossaries*. Technical Report, New York, NY.

IEEE-SA Standards Board (1998). *Standard for Distributed Interactive Simulation—Application Protocols*. Technical Report IEEE 1278.1A-1998.

IEEE-SA Standards Board (2000). *IEEE Standard for Modeling and Simulation (M&S) High Level Architecture (HLA)—Framework and Rules*. Technical Report IEEE 1516-2000.

Kewley R and Tolk A (2009a). A systems engineering process for development of federated simulations. In: *Proceedings of the 2009 Military Modeling and Simulation Conference*. Society for Modeling and Simulation International, San Diego, CA.

Kewley R and Tolk A (2009b). *PEO Soldier Simulation Road Map VI—Command and Control Modeling*. Technical Report DTIC: ADA509285. West Point Operations Research Center, West Point, NY.

Kewley R, Tolk A and Litwin T (2008a). *PEO Soldier Simulation Road Map V—The MATREX Federation*. Technical Report DTIC: ADA486791. West Point Operations Research Center, West Point, NY.

Kewley RH, Goerger N, Teague E, Henderson D and Cook J (2008b). Federated simulations for systems of systems integration. In: *Proceedings of the 2008 Winter Simulation Conference*. IEEE, Piscataway, NJ.

Kewley R, Turnitsa C and Tolk A (2010). Further exploration in primitives of meaning. In: *Proceedings of the 2010 Military Modeling and Simulation Conference*. Society for Modeling and Simulation International, Orlando, Florida.

Law AM (2007). *Simulation, Modeling, and Analysis* (4th edition). McGraw-Hill, New York, NY.

Martin G (2005). *PEO Soldier Simulation Roadmap: Initial Steps in Implementation*. Technical Report DTIC ADA435707, United States Military Academy Operations Research Center, West Point, NY.

MATREX (2009). *MATREX Federational Object Model*, Version 4.3. Modeling Architecture for Technology, Research, and Experimentation, Aberdeen Proving Ground, Maryland.

Mittal S, Zeigler B, Martin JR, Sahin F and Jamshidi M (2008). Modeling and simulation for systems of systems engineering. In: *System of Systems Engineering—Innovations for the 21st Century*. Wiley, New York, NY, Edited by Mo M. Jamshidi.

Object Management Group (2007). *OMG Model Driven Architecture*. http://www.omg.org/mda (last accessed 14 March 2008).

Office of the Under Secretary of Defense (2008). *Systems Engineering Guide for Systems of Systems* (1.0 edition). Pentagon, Washington, DC.

Ottino JM (2004). Engineering complex systems. *Nature* **427**, 399.

Parnell G, Driscoll P and Henderson D (eds) (2010). *Decision Making in Systems Engineering and Management* (2nd edition). Wiley, Hoboken, NY.

Repass T (2010). *Complexity of Simulating the Ballistic Missile Defense System*. Presented at the Modeling and Simulation Committee Meeting of the National Defense Industrial Association Systems Engineering Division. Presentation made by T. Repass (from the Modeling and Simulation Program, Directorate for Engineering, Missile Defense Agency) on August 17, 2010 in Rosslyn, VA.

Sage A and Cuppan C (2001). On the systems engineering and management of systems of systems and federations of systems. *Inform Knowledge Syst Manage* **2**, 325–345.

SISO (2008). *Military Scenario Definition Language*. Technical Report, Simulation Interoperability Standards Organization.

SISO (2010). *C-BML Guidelines Document*. Published by the Simulation Interoperability Standards Organization (SISO) located in Orlando, Florida.

Taylor S, Sharpe J and Ladbrook J (2003). Time management issues in cots distributed simulation: a case study. In: *Proceedings of the 2003 Winter Simulation Conference*, edited by S Chick, PJ Sanchez, D Ferrin and DJ Morris. IEEE, Piscataway, NJ.

The Joint Staff (2010). *Universal Joint Task List*. Authentication required. Joint Staff, Washington, DC.

Tollefson ES and Boylan GL (2004). *Simulation Roadmap for Program Simulation Roadmap for Program Executive Office (PEO) Soldier*. Technical Report DTIC ADA425648. United States Military Academy Operations Research Center, West Point, NY.

Vilerdi R, Ross A and Rhodes D (2007). A framework for evolving system of systems engineering. *J Def Software Eng*. **10**, 28–30.

Zeigler BP (2003). DEVS today: recent advances in discrete event-based information technology. In: *Proceedings of the 11th IEEE/ACM International Symposium on Modeling, Analysis, and Simulation of Computer and Telecommunications Systems, Orlando, FL*. Publication compiled for the MASCOTS Conference 2003. Page 148, Orlando, Florida, October 2003.

Chapter 31

The Role of Architecture Frameworks in Simulation Models: The Human View Approach

Holly A. H. Handley

Architecture frameworks are used by system engineers, acquisition professionals, and other stakeholders to communicate in a common language about complex systems (DoDAF, 2004). Frameworks collect and organize data for a system or system of systems through the use of a set of viewpoints and their corresponding products. This taxonomy can be used to populate simulation models created to analyze the behavior of the system under design, as well as to validate the performance of realized systems. The Human View was developed as an additional architecture viewpoint to specifically represent the role of humans in network enabled capabilities. It captures information about the human relationships in the system, including elements such as tasks, roles, constraints, social networks, training, and metrics. Simulations can extend the Human View to illustrate and capture the dynamic nature of human performance in the system environment; likewise executable models can include human based data to provide a more complete socio-technical simulation.

Engineering Principles of Combat Modeling and Distributed Simulation, First Edition.
Edited by Andreas Tolk.
© 2012 John Wiley & Sons, Inc. Published 2012 by John Wiley & Sons, Inc.

INTRODUCTION

"The Department of Defense Architecture Framework (DoDAF), Version 2.0, serves as the overarching, comprehensive framework enabling the development of architectures to facilitate the ability of Department of Defense (DoD) managers at all levels to make key decisions more effectively through organized information sharing across the Department, Joint Capability Areas (JCAs), Mission, Component, and Program boundaries" (DoDAF, 2010). An architecture framework defines a common approach for the development, presentation, and integration of data that compose the architecture description of a complex system. The use of a framework to capture and organize system data contributes to building more interoperable systems. Architecture frameworks also provide a mechanism for understanding and managing complexity, as system information is captured in a consistent taxonomy. The different viewpoints that compose the architecture framework can be used to capture multiple aspects of a complex system.

Network enabled capability (NEC) is a concept that "envisions the coherent integration of sensors, decision makers, effectors, and support capabilities to achieve a more flexible and responsive military" (NATO, 2010). NEC is a complex system of coalition forces operating in a network based environment. While networks can increase information sharing among coalition forces, the challenge of NEC is to ensure that the right information is being seen by the right people at the right time. The Human View was developed by a NATO panel as an additional architecture framework viewpoint to address problems defining the role of humans in networked systems. The goal of the Human View is to ensure that the human component is adequately considered in capability development and to provide a methodology for a rational integration of human and technology.

Since architecture frameworks provide system engineers a structured language to communicate about complex systems, the addition of the Human View provides a means to include human–system concepts in the dialogue. The Human View not only offers a method for representing human components of a complex socio-technical system for design purposes, but it also provides a taxonomic framework for data gathering and front-end analysis. "The selection of variables and metrics, the identification of relevant constraints, and the specification of performance shaping functions supports subsequent modeling and simulation" (NATO, 2010). Different levels of analysis and simulation can be performed with the Human View data, and when combined with data from the rest of the architecture framework (i.e. the operational and system viewpoints) a realistic evaluation of the human impact on the system performance can be made.

This chapter will first describe in more detail the concept of an architecture framework, including the addition of the Human View. It will then describe the relationship between the architectural data and simulation models, both for "as-is" and "to be" architectures. The use of the architecture information for three different levels of analyses will be explored, through the use of Human View examples. Finally, some general observations about the relationship between architecture frameworks and simulations will conclude the chapter.

ARCHITECTURE FRAMEWORKS

The DoDAF was created to ensure that architecture descriptions developed by various commands, services, and agencies were interrelatable between and among each organization's operational, systems, and technical architecture views (DoDAF, 2004). The framework provides a product focused method for standardizing architecture descriptions—the products are intended to represent consistent architectural information across the set of viewpoints describing an architecture. "One of the principal objectives is to present information in a way that is understandable to the many stakeholder communities involved in developing, delivering, and sustaining capabilities in support of a mission" (DoDAF, 2010). It does so by dividing the problem space into manageable pieces, first by different viewpoints, each with a particular purpose. The three major viewpoints include the operational view, which incorporates tasks and business processes, the system view, which describes the associated system platforms, functions, and characteristics, and the standards view, which conveys the set of rules that governs system implementation. The current version of DoDAF (version 2.0) also includes capability, data and information, project and services viewpoints (DoDAF, 2010). A description of these viewpoints is included in Table 31.1.

Each viewpoint contains a set of products that describe the information captured in that viewpoint. Products can be documents, spreadsheets, dashboards, or other graphical representations and serve as a template for organizing data in a readily understood format. The architecture products have become the focal point of architecture discussions and provide the means to integrate across architectures. When a system of systems is based on the same set of integrated products, the architectural elements can be aligned and referenced to each other. Most countries have adopted their own version of an architecture framework. The Ministry of Defence Architecture Framework (MoDAF) has been adapted by the UK Ministry of Defence (MoD) from the DoDAF. The original DoDAF views have been extended into seven MoDAF viewpoints. Along with the all, operational, systems services, and technical standards views, MoDAF adds the strategic view, which consists of views that articulate high level requirements for enterprise change over time, and the acquisition view, which consists of views that describe programmatic details to guide the acquisition and fielding processes. The Canadian Department of National Defense Architecture Framework (DNDAF) is also closely based on DoDAF. DNDAF provides a common view, operational view, system view, and technical view, all similar to the original four DoDAF views, but also includes strategic, capability, information, and security views. The overarching NATO Architecture Framework (NAF) has closely aligned itself with MODAF (NATO, 2007). While these frameworks are a major step in standardizing the information content of system architectures, the role of the human in the system has not been addressed in any of the national architecture frameworks.

Table 31.1   DoDAF Viewpoints and Descriptions (DoDAF, 2010)

DoDAF viewpoint	Description
All viewpoint	Describes the overarching aspects of architecture context that relate to all viewpoints
Capability viewpoint	Articulates the capability requirements, the delivery timing, and the deployed capability
Data and information viewpoint	Articulates the data relationships and alignment structures in the architecture content for the capability and operational requirements, system engineering processes, and systems and services
Operational viewpoint	Includes the operational scenarios, activities, and requirements that support capabilities
Project viewpoint	Describes the relationships between operational and capability requirements and the various projects being implemented
Services viewpoint	The design for solutions articulating the performers, activities, services, and their exchanges, providing for or supporting operational and capability functions
Standards viewpoint	Articulates the applicable operational, business, technical, and industry policies, standards, guidance, constraints, and forecasts that apply to capability and operational requirements, system engineering processes, and systems and services
Systems viewpoint	For legacy support, is the design for solutions articulating the systems, their composition, interconnectivity, and context providing for or supporting operational and capability functions

In 2002, NATO pursued a course of transformation termed Network-Enabled Capabilities (NEC)[1]. The NEC transformation focused on changes in three areas: people, processes, and technology. The challenge of NEC was "to achieve the proper mix of new human behaviors with organizational changes and innovative technologies" (Handley and Houston, 2009). In order to support this transformation, a NATO panel developed the NATO Human View. Its purpose was to augment existing architectural frameworks with additional information relevant to the human in the system The Human View contains seven static products that include different aspects of the human element, such as roles, tasks, constraints, training, and metrics. It also includes a human dynamics component to capture information pertinent to the behavior of the human system under design. Table 31.2 provides a description of the architecture products developed for the NATO Human View.

[1]http://www.nato.int/cps/en/SID-1F7151AF-2FE364A1/natolive/topics_54644.htm.

Table 31.2   NATO Human View products (Handley and Smillie, 2008)

Product	Name	Description
HV-A	Concept	A conceptual, high level representation of the human component of the enterprise architecture framework
HV-B	Constraints	Sets of characteristics that are used to adjust the expected roles and tasks based on the capabilities and limitations of the human in the system
HV-C	Tasks	Descriptions of the human-specific activities in the system
HV-D	Roles	Descriptions of the roles that have been defined for the humans interacting with the system
HV-E	Human network	The human to human communication patterns that occur as a result of *ad hoc* or deliberate team formation, especially teams distributed across space and time
HV-F	Training	A detailed accounting of how training requirements, strategy, and implementation will impact the human
HV-G	Metrics	A repository for human related values, priorities, and performance criteria, and maps human factor metrics to any other Human View elements
HV-H	Human dynamics	Dynamic aspects of human system components defined in other views

Simulation models can be used to realize the HV-H dynamics product; this allows the system engineer to analyze and evaluate alternative system configurations. "From this perspective, simulation forms part of the Human View by providing animation of state-transitions through which system components interact in pursuit of a mission" (NATO, 2010). Simulation can also be used to identify data discrepancies across views; enhanced simulations make it possible to cross-validate the data and examine their concordance and integrity within the Human View depiction.

THE ROLE OF ARCHITECTURE FRAMEWORKS IN SIMULATIONS

Framework products are graphical, textual, and tabular items that are developed for a given architecture description and describe characteristics of the system. The architecture framework was developed to provide a means to standardize the content of information collected about a system. Some attributes bridge two views and provide integrity, coherence, and consistency to architecture descriptions. However, the architecture framework in itself does not provide a means to analyze the content of the products or answer questions about the dynamic behavior of the architecture.

Many of the architecture products can be evaluated independently or in combination to find errors and predict some types of conflicts or behavioral problems for the system. A simulation, however, requires the information from one or more of the products to be entered into software so that the model can be "executed," that is, its performance can be evaluated over time. Dynamic models can be created for system designs where the behavior cannot be directly measured or observed, because the systems do not exist yet. The architecture products provide a useful and coherent framework within which to develop simulation models. Methodologies to convert the information from a set of architecture framework products into executable models has been developed and described in detail by Levis and Wagenhals (2000).

Architecture frameworks are often used to capture time-specific snapshots of system architectures. An "as-is" architecture representation captures information about a legacy system, while "to be" descriptions depict the development of new capabilities, or improvements to existing ones. Simulations may be desired to analyze existing systems for shortcomings and to then assess the impact of changing selected design variables; this is a simulation of an "as-is" architecture. Conversely, simulations may be required to predict system behavior of future systems yet to be designed and implemented; in this case the simulation model is based on the operational concept in order to evaluate design decisions for the "to be" architecture. " 'As-is' models can have an important function in informing high level 'to be' definitions. At times, it is impossible to generate ideas of new structures without first capturing what already exists" (Bruseberg, 2009). In both cases, a simulation model can be used to inform the system design and ensure that the capability that will be deployed meets the operational and system performance requirements.

Architecture frameworks have been explored as a basis for modeling and simulation in the military domain. The products provided by the architecture descriptions provide a systematic way to capture and analyze operational, system, and technical data; the framework provides a standardized and accepted representation of the military system. The interrelationships between the products provide a method to perform analysis on the system by tracing the flow of data between products and evaluating the time delay of functions and interactions; the results can then be compared with mission requirements.

The structured data also provide an input source for simulation models. DoDAF has been suggested as a way to capture Higher Level Architecture (HLA) relevant modeling and simulation information requirements (deSylva et al., 2004). HLA provides the ability for multiple computer simulations to exchange information; this federation of simulations is often necessary to model the complexity of military systems. By using DoDAF to structure the federation, the resulting models are at a level of detail consistent with the requirements of the program and are well suited to defining the federation objectives (deSylva, 2004). Similarly, architecture frameworks have been explored in conjunction with system of systems modeling. DoDAF can provide an overarching framework for combining multiple models to allow system of system analysis (Muguira and Tolk, 2006). Through its multiple viewpoints, DoDAF captures the multiple aspects required to represent the system of systems paradigm, as well as to address the interoperability concerns of utilizing multiple models. DoDAF can help ensure *composability*, or the alignment of the assumptions and constraints of the underlying models (Muguira and Tolk, 2006).

DoDAF has also been applied to support modeling at the capability level (Atkinson, 2004). A capability is defined as "the ability to achieve a desired effect under specified standards and conditions through combinations of ways and means (activities and resources) to perform a set of activities" (DoDAF, 2010). Architecture frameworks can provide an integrated model to capture, structure, and manipulate the large amount of data required to describe capabilities. The capabilities can be developed and evaluated within smaller functional domains, and then combined in a hierarchical manner to describe overall mission level capabilities (Atkinson, 2004). Systems are then mapped to the capabilities, and simulations over time can be used to evaluate performance with respect to metrics.

In all of these cases, the use of DoDAF makes "real-world" data available at a required level of detail for military modeling. It provides a standardized and accepted method to represent and structure the data for a system, system of systems, or capability level model. The set of DoDAF views and accompanying products provides the breadth of information required to populate a model, extract scenarios, identify constraints, and evaluate simulation results.

The Human View augments DoDAF by specifically collecting information about the parameters pertaining to the human in the system. It captures the human operator activities, tasks, communications, and collaborations required to accomplish mission operations and support operational requirements. Factors that affect, constrain, and characterize human behavior serve as the basis for performance prediction. Thus, simulations based on the Human View data support the added dimension of behaviors based upon the ways in which humans interact and impact the system performance.

TYPES OF ANALYSES

Three levels of analyses based on the Human View framework products can be explored (NATO, 2010). The first level is the analysis of a single product: the data contained in a distinct framework product are examined. The second level of analysis examines a combination of products: the data from one framework product are used to manipulate data in another framework product. This allows an independent variable defined by data in the first product to be varied and its effect measured on a dependent variable defined by data in the second product. The third level is an executable model of system performance. At this level, a simulation based on multiple products is executed over a varying domain. The first two levels of analyses can address issues about resources availability and conflict. The third level of analysis, the simulation, attempts to evaluate the system performance in response to varying demands and situational conditions (NATO, 2010).

Analysis of Single Products

A first order analysis can completed with a single product of an architectural description to provide a functional assessment of the data captured in that view. Individual views capture "snapshots" of different aspects of the system; in the case of the Human View, each product provides some perspective on the socio-technical attributes of the system. The different products are represented in different forms, i.e. figures, diagrams, or tables, which lend themselves to different methods of analysis. For example, the HV-C captures the human level activities of a system. These tasks can be described in terms of a sequence diagram, a temporal ordering of the tasks. This can give an indication of how a given sequence of tasks will perform, and the performance predictions for alternative sequences of tasks can be compared. A similar analysis can be used to examine the activities in the OV-5, Operational Activity Description. The activities can be evaluated under different time sequences of events which can be summed in order to calculate overall system performance. This ordered trace of operational activities is known as a "key thread" analysis.

Analyses with single products also provide many insights in comparing "as-is" and "to be" architectures. This method was used to evaluate changes in processes to improve military and civilian collaboration (Stevens and Heacox, 2008). During a cooperative exercise evaluating Defense Support to Civilian Authorities (DSCA) interaction processes, recommendations were made for change based on analysis of Human View framework products. The actual structures and interactions from the DSCA outcomes formed the basis of the "as-is" architecture, while the recommendations made after the exercise formed the basis of the "to be" architecture. By comparing the products, points of disparity between the two produced a list of suggested changes to be made for future DSCA operations. Four recommendations were made based on the comparison of the individual architecture products. The first recommended changes to tasking per organization based on an analysis of the HV-C (Tasks); this would help alleviate some of the burden placed on DSCA responders due to disparate standard operating procedures that led to task duplication and conflict. The second recommendation evaluated the HV-D (Roles); in this case the suggestion was to adhere to National Incident Management System (NIMS) naming conventions for quick identification of personnel in order to help establish communications. The third and fourth recommendations involved the coordination of distributed team members based on the analysis of the HV-E (Human Networks); these would define team requirements and interdependences for the responders, including associations between the field respondents and crisis command staff to share updates and information from activities. One of the major lessons from this exercise was that military and civilian entities operate quite differently. "However, these entities must support one another and engender compatible work processes in order to jointly respond to crisis situations in a successful manner" (Heacox and Stevens, 2008). Using the Human View framework products to capture the diverse information provided a way to view and compare joint military and civilian command and control operations and recommend improvements.

Analysis of Combinations of Products

Analyses performed with combinations of architecture products provide a second order evaluation regarding the interoperability of the data captured in each view. Since the performance of a system is often dependent on the relationship between the elements that compose the system, a key aspect of the architecture is the manner in which the information in one product interconnects with the data in other views. For example, the cross-product of HV-C Tasks with HV-D Roles demonstrates the impact the roles have on the task network, in that different roles bring different sets of knowledge, skills, and abilities (KSAs) to their assigned task. The correct match of role to task can improve the overall task process performance. Different task processes can be optimized by comparing and evaluating the underlying role attributes with regard to tradeoffs between task performance and operator workload. Alternatively, the analysis of the task network can form the basis to define the roles required for the system, i.e. defining competences for designated operators based on task requirements. A similar example is the allocation of operational activities described in the OV-5 to the system functions in the SV-4 (Systems Functionality Description). The system load balance can be evaluated and the need for reallocation of operational activities assessed.

Analyses with combinations of products can illustrate impacts to data in one product with respect to changes in other products. This methodology was used to evaluate proposed organizational changes to a Maritime Operations Center (MOC) in order to provide a more adaptive command and control structure (Handley et al., 2006). This project established a relationship between DoDAF views and personnel requirements as part of the effort to organize and standardize maritime operations. This also resulted in an initial architectural realization that was later developed into the Human View. The project analyzed 27 activities defined in the Decide Node Tree (OV-5) and 22 corresponding roles identified in the Organizational Relationships Chart (OV-4); three different versions of the OV-4 were developed based on the three different command structures that the organization was projected to assume. For each of the structures, the impact of the expansion and contraction of activities (functional responsibilities) and the corresponding workload was assessed. Additionally, the changes in command relationships of the candidate organizational structures changed reporting relationships for several of the roles. For example, as the organization "morphed" from one structure to the next, four of the roles changed cells, two roles became more specialized (assumed a narrower set of functional responsibilities), and two cells changed focus (assumed a different set of functional responsibilities). "Cells" are organizational units, similar to a "department"; the cell name reflects the functional responsibilities of the roles assigned to that cell. "Understanding the impact of different organizational structures on changes to role reporting relationships is important because maintaining relationships and patterns of interactions is necessary to fulfill responsibilities throughout the different organizational configurations" (Handley et al., 2006). The output of this analysis, comparing across two architectural framework products, provided three recommendations for the

DOTMLPF[2] Change Request (DCR) process regarding the impact of MOC organizational escalation and reduction on role responsibilities and relationships.

An Integrated Simulation Model

A comprehensive simulation model can be created by integrating information from multiple products of the architecture framework. Simulation models can evaluate the dynamic interoperability of the data and analyze the behavior of the system. For example, a simulation could include the tasks (HV-C) and their temporal relationships, roles (HV-D) with particular KSAs that are assigned to the task, and the expected performance criteria (HV-G) in order to address "what if" questions by focusing on the performance of the human process. This simulation can assist the systems engineer in analyzing and evaluating the human process dynamics under alternative system configurations.

Simulation models have previously been synthesized from architectural framework products and used to verify that the behavior of the architecture matches the desired system behavior (Wagenhals et al., 2000). Colored Petri nets were used to create discrete event system models of an architecture. The data collected in the framework products provided the necessary set of information to create an executable model that could reveal the logical, behavioral, and performance characteristics of the architecture. Typical models allow input parameters to be varied, constraints to be relaxed, and other variables to be explored to evaluate the effect on model outcomes. Wagenhals et al. (2000) developed an explicit methodology to generate discrete event models from architectural framework products to evaluate system behavior.

Using a similar methodology, a preliminary schema for the interaction of the Human View components was created, as shown in Figure 31.1. An event from the environment triggers a task (HV-C). The role (HV-D) responsible for the task begins processing it, coordinating with team members (HV-E) to exchange information during task processing. The way the task is processed may depend on characteristics of the actual person fulfilling the role (HV-B), including training completed (HV-F). Use of a system resource (HV-C), e.g. a computer interface, to complete the task is included in the model. Other constraints such as health hazards (HV-B) may moderate the performance of the task. Once the task is completed, metrics (HV-G) are used to evaluate performance. This modeling schema was implemented using the IMPRINT[3] simulation model; the model's input requirements can be mapped directly to the Human View data, as shown in Table 31.3.

Analyses performed with IMPRINT can provide the results required to evaluate the interaction of the data captured in the Human View products. Simulations

[2]Doctrine, Organization, Training, Materiel, Leadership and Education, Personnel and Facilities.
[3]Improved Performance Research and Integration Tool.

Table 31.3 Human View to IMPRINT Mapping (Handley and Smillie, 2010)

Product	Description	IMPRINT Data
HV-A Concept	A high level representation of the human component of the system	Hypothesis to be tested by the model
HV-B Personnel Constraints	Manpower Projections (HV-B1): Predicted manpower requirements for supporting present and future systems	*IMPRINT OUTPUT*: Number of desired MOS expected to be available per year
	Establishment Inventory (HV-B3): Current number of personnel by rank and job within each establishment	*IMPRINT OUTPUT*: Estimated number of soldiers needed
HV-B Human Factors Constraints	Health Hazards (HV-B5): Short- and long-term hazards to health that occur as a result of system operation, maintenance, and support	Stressors, such as heat, humidity, cold, wind, MOPP, and fatigue
	Human Characteristics (HV-B6): Operator capabilities and limitations with system operating requirements under various conditions	Personnel characteristics, such as ASVAB composite and cutoff
HV-C Tasks	Identify human level tasks	Functions/tasks decomposition
	Task-role assignment matrix	Task to operator assignment
	Tools required to accomplish a task	Tasks to system interfaces
	Information demands for specific tasks	Task demands (mental workload)
HV-D Roles	List of roles	Warfighters/operators
	Role responsibility	
	KSA competences	
HV-E Human Network	Role groupings or teams formed	Team functions
	Interaction types	Operator teams
	Team dependences	
HV-F Training	Training resources, availability, and suitability	Changing sustainment training frequency
	Training required to obtain necessary knowledge, skills, and ability	
HV-G Metrics	Human performance requirements	Mission level time and accuracy criterion
	Human task to metrics mapping	Task level time and accuracy standards
	Target values	*IMPRINT OUTPUT*: Crew performance, crew workload

MOS, Military Occupational Specialty; MOPP, Mission Oriented Protective Posture; ASVAB, Armed Services Vocational Aptitude Battery.

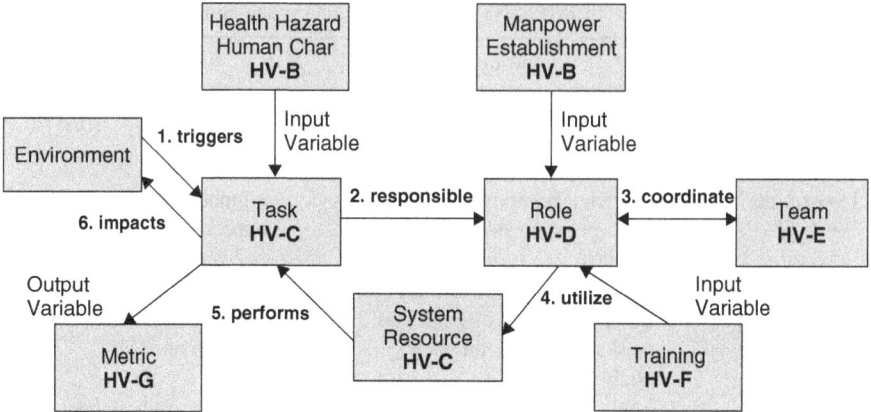

Figure 31.1   Human View dynamics schema (Handley and Smillie, 2010).

can be performed to provide expected levels of mission performance parameters. Task performance can be represented in terms of time to complete, percent steps correct, and task failure and its consequences, while role activity can be monitored through workload measures. After the baseline simulation of the system has been completed, additional simulations can be used to perform tradeoff analyses. For example, different role-to-task allocations will impact task performance and operator workload. Varying the assignment of system interfaces to tasks will affect channel conflicts (multiple tasks accessing the same resource) and thereby impact task performance and task failure. Creating a simulation model allows easy exploration of how changing parameters in one of the products impacts other aspects of the human data.

New design decisions can also be evaluated using the simulation model. When mapping new tasks onto existing roles, generally one of two methods is used. If the new tasks are similar to existing tasks, they are usually mapped to the role that already has the expertise (KSAs) on that type of task. In this case, the existing role now has more tasks, but requires less new training. The second method is to map new tasks to existing roles that are underutilized. In this case a better balance of workload among the roles is achieved; however, some roles may need additional training. Generally, the first method is used; however, an analysis performed using the simulation model can show the impact of the task mapping decisions. Changing skill levels can show how assigning a role with less experience would lead to a failure to complete key tasks, affecting system performance. Likewise, performance detriments due to overloaded roles can lead to delays and dropped tasks. Ultimately the goal of the simulation model is to show how changes to factors affecting the human elements of a system impact overall system performance (Handley and Smillie, 2008).

Simulations using information from other architectural viewpoints can also be completed. Prior to the development of the Human View, data from the operational view were used to configure a colored Petri net model in order to evaluate the impact of the insertion of a new technology into an established command and control process (Handley and Heacox, 2005). The simulation results provided a performance analysis of the improved timeliness of the Commander's Daily Update Briefing process using the Integrated Interactive Data Briefing Tool (IIDBT). Data were drawn from the following operational views: the Node Connectivity Diagram (OV-2), the Information Exchange Diagram (OV-3), the Organizational Relationships Chart (OV-4), the Activity Model (OV-5), and the Event-Trace Description (OV-6c). Additionally, system view architectural data were incorporated into the same model to evaluate the technical connectivity requirements to implement the system. This information was from the Systems Interface Description (SV-1) and Systems Communications Description (SV-2). The simulation results were used in the planning of the Trident Warrior 2005 exercise to select command and control processes for observation and evaluation.

CONCLUSION

Architecture frameworks can provide structured data to help define simulations in much the same way as they are used for the development of system architectures. They can provide a data source to populate executable models that can then be simulated to uncover the behavior and/or evaluate the performance of the system. Simulations can also provide a means of testing the relationships underlying the architecture framework; a simulation can confirm the integrity of the data captured in the framework products. The use of simulation models can assist in the exploration of the system design through both simple analyses and more complex simulations. Three different types of analyses were described: the analysis of the data of an individual product, analysis of a combination of products by using data in one product to manipulate the data in another product, and an executable model that incorporates data from multiple products in order to simulate the system behavior over time.

The Human View framework was developed to collect and organize human centered parameters in order to understand the way that humans interact and impact other elements of a system. Variables important to NEC such as technology, organization, and personal characteristics can be captured in the Human View products and traced through the data dependences to understand their impact on other system variables. In this way, the Human View can help facilitate the use of information technology to improve technology enabled collaboration. Simulation extends the Human View to capture the dynamic nature of human performance in a variable environment. Thus the Human View is not limited to an architectural description but becomes a basis for dynamic simulations.

REFERENCES

Atkinson K (2004). Applying design patterns to modeling and simulation. In: *Proceedings of the Fall Simulation Interoperability Workshop. Orlando, FL.* http://www.sisostds.org.

Bruseberg A (2009). *The Human View Handbook for MODAF*. Systems Engineering & Assessment. On behalf of the MoD HFI DTC, 2nd Issue, Bristol, UK.

DoDAF (2004). *Department of Defense Architecture Framework Working Group, DoD Architecture Framework Version 1.0 Deskbook*. Department of Defense, Washington, DC.

DoDAF (2010). Department of Defense Deputy Chief Information Officer, *DoD Architecture Framework Version 2.0*. http://cio-nii.defense.gov/sites/dodaf20/index.html.

Handley H and Heacox N (2005). *Commander's Daily Update Brief Process Baseline Version*. Technical Report, Pacific Science & Engineering Group, San Diego, CA.

Handley HAH and Houston NJ (2009). NATO human view architectures and human networks. In: *Proceedings of the 2009 MODSIM World Conference and Expo, Virginia Beach, VA*.

Handley HAH and Smillie RJ (2008). Architecture framework human view: the NATO approach. *Syst Eng* **11**, 156–164.

Handley HAH and Smillie RJ (2010). Human view dynamics—the NATO approach. *Syst Eng* **13**, 72–79.

Handley HAH, Sorber TJ and Dunaway JD (2006). *Maritime Headquarters with Maritime Operations Center, Concept Based Assessment Focus Area: Organization*. Human Systems Performance Assessment Capability Final Report, San Diego, CA.

Levis AH and Wagenhals LW (2000). C4ISR architectures. I: Developing a process for C4ISR architecture design. *Syst Eng* **3**, 225–247.

Muguira JA and Tolk A (2006). Applying a methodology to identify structural variances in interoperations. *J Def Model Simulat* **3**, 77–93.

NATO (2007). *The NATO Human View Handbook*, NATO RTO HFM-155 Human View Workshop Panel, Toronto, Canada.

NATO (2010). *Human Systems Integration for Network Centric Warfare*. RTO Technical Report TR-HFM-155, North Atlantic Treaty Organization, Research and Technology Organization.

Stevens C and Heacox N (2008). *Using NATO Human View Products to Improve Defense Support to Civil Authority (DSCA)*, 13th International Command and Control Research and Technology Symposium, Seattle, WA.

deSylva M J, Lutz RR and Osborne SR (2004). The application of the DoDAF within the HLA federation development process. In: *Proceedings of the Simulation Interoperability Workshop, Orlando, FL*. http://www.sisostds.org.

Wagenhals LW, Shin I, Kim D and Levis AH (2000). C4ISR architectures. II: structured analysis approach for architecture design. *Syst Eng* **3**, 248–287.

Chapter 32

Multinational Computer Assisted Exercises

Erdal Cayirci

INTRODUCTION

In computer assisted exercises (CAXes), computer support is provided for the following purposes:

- immersing the training audience (TA) in a realistic situation and environment;
- helping the exercise planning group and the exercise control staff (EXCON) to steer the exercise process (EP) towards exercise objectives.

Therefore, a CAX can be any type of exercise (i.e. live exercise, command post exercise), and CAX support tools are not limited only to military simulations. CAX support tools are involved in all stages of an EP to automate the processes, to reduce the duplication of work, to enhance the exercise environment, and to ensure that the EP flows towards the objectives. In this perspective, the CAX tools can be categorized into four classes:

- *Exercise planning and management tools:* These tools are used for the automation of processes, information management, and information exchange throughout an EP. They support the preparation and management of the scenario, as well as the main event and master incident lists (MEL/MIL).

Engineering Principles of Combat Modeling and Distributed Simulation, First Edition.
Edited by Andreas Tolk.
© 2012 John Wiley & Sons, Inc. Published 2012 by John Wiley & Sons, Inc.

- *Simulation systems and ancillary tools:* Simulation systems compute the possible results of the orders given by simulation operators. Ancillary software is also needed to prepare the simulation systems (e.g. database preparation tools).

- *Interfaces to command and control (C2) systems:* During a CAX, the TA uses operational C2 systems. Therefore, mediation-ware between the simulation software and C2 systems are needed.

- *Experimentation and analysis tools:* Tools are needed also for designing and managing experiments by using CAX data, for compiling and presenting the data collected by the simulation system, and for deriving information from these data.

The complexity of EP and CAX tools (i.e. CAX support systems) depends on the composition of the TA and exercise/training objectives (ETO). We shall explain TA and CAX support tools in more detail in the following sections. Although the complexity and length of EP varies from one exercise to another, an EP has typically four stages as depicted in Figure 32.1: specification, planning, execution, and analysis.

At the first stage, a command specifies the type, length, and objectives of the exercise. This command is typically one higher level to the TA and called officer specifying the exercise (OSE). The commander of the TA usually becomes officer conducting the exercise (OCE), and contributes to the specification phase especially when determining the ETO. Finally a third officer called officer directing the exercise (ODE) is also involved in the exercise specification process. ODE manages the exercise planning process during which the details like exercise scenario, exercise flow, communications and information systems, and real-life support issues are planned and prepared. In the third stage, the exercise is executed. Finally, EP is completed when the results from the exercise are analyzed and proper feedback is provided to the TA.

Overall the EP should be designed such that the ETO are achieved efficiently. We can define the efficiency of a typical EP with three parameters:

- full achievement of all ETO
- the length of the overall EP and especially of the execution stage
- the cost of the EP

Sometimes exercises may be conducted with different purposes, such as assurance and deterrence. In those cases, the definition of efficiency may be different. In this chapter, we focus on CAXes which mainly aim at training an

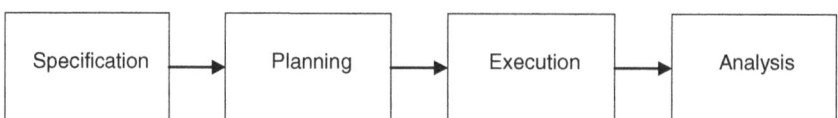

Figure 32.1   Typical exercise process.

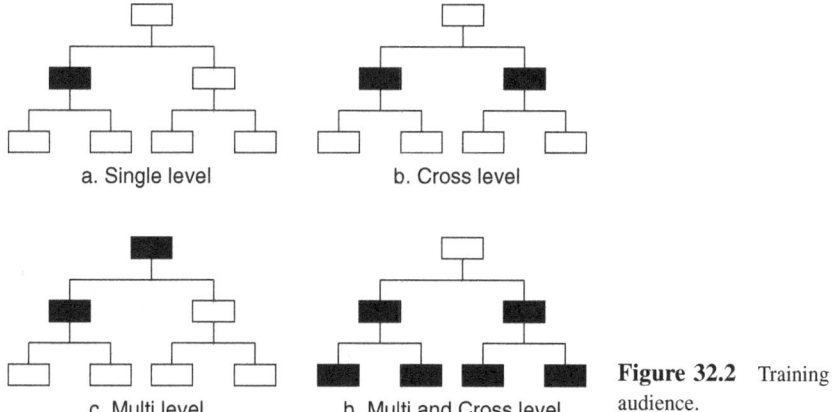

a. Single level

b. Cross level

c. Multi level

b. Multi and Cross level

Figure 32.2 Training audience.

audience and/or rehearsing for a mission. There is often a tradeoff between the length and the cost of an EP, and constraints may apply to these two parameters. For example a mission rehearsal may have a stringent length constraint, and this constraint may make it more expensive than a typical exercise with the same training objectives.

TRAINING AUDIENCE AND EXERCISE CONTROL STAFF

As stated before, the complexity of an EP and the exercise structure are based on TA and ETO (Cayirci 2007a, Cayirci and Marincic, 2009). TA is the focal point in an exercise structure. TA can be single level, multi-level, cross-level, and both cross- and multi-level as shown in Figure 32.2. A multi-level TA represents multiple levels of command trained at the same time in the context of a single scenario. Cross-level TA includes units or headquarters at the same level of command. When the units in a cross-level TA are from different services, the exercise becomes joint. A TA can have headquarters (HQ) and/or forces from different nations, which makes the exercise combined. In more and more exercises civilian national/international agencies and organizations like police, fire department, health agencies, and UN are involved. These civilian organizations usually become a part of EXCON and constitute the white or grey cell. They may also be a part of a TA. EXCON structure and the white/grey cell concept is explained later in this section.

TA can be co-located or various parts of the TA can be located in geographically remote sites (i.e. different cities, countries, or continents). The exercises that have TA components located at remote sites are called distributed exercises. Note that distributed simulation and distributed exercise are different things. A distributed exercise can be supported by a centralized simulation system or a centralized exercise can be supported by a distributed simulation. Locating client

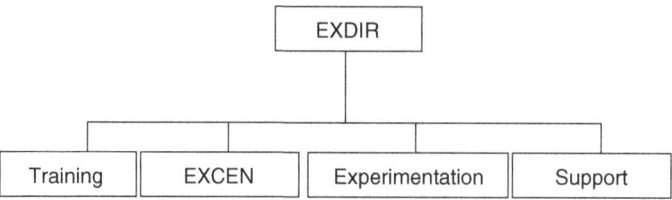

Figure 32.3 EXCON structure.

workstations of a simulation in remote sites does not make a simulation distributed. In distributed simulation computation for the simulations are carried out by multiple processes often on remote machines.

Another component in an exercise structure is the exercise control staff (EXCON). A typical EXCON model is shown in Figure 32.3. EXCON is guided by the exercise director (EXDIR). The training team (TT) consists of mentors, observer/trainers (O/T), subject matter experts and analysts. TT is deployed with TA, observes the TA, provides on site instructions and training, and collects inputs for after action review (AAR) and the evaluation of TA. Exercise center (EXCEN) is the organization responsible for the consistent and coherent flow of the exercise according to the ETO. EXCEN is explained in detail below. The experimentation team runs the experiments planned in conjunction with the exercise. Finally, the support team has elements like real-life support (RLS), visitor officer bureau (VOB), public information center (PIC), security office, and computers/communications support team.

The EXCEN functions shown in Figure 32.4 can be categorized into five broad classes as control centre (CONCEN), higher control (HICON), lower control (LOCON), and white/grey cell and situation forces (SITFOR). CONCEN monitors the current status of the exercise closely and steers it according to the ETO. HICON and LOCON represent the command levels/echelons that would normally be at the level above and below the TA respectively. LOCON and HICON consist of response cells (RCs). The number of RCs is dependent on the scenario and the TA. Each RC is made up of a number of planners, a number of simulation operators, and coordination staff. RCs are the main interface between simulation and exercise as explained later. White/grey cell is a response cell that is composed of subject matter experts or role players representing agencies,

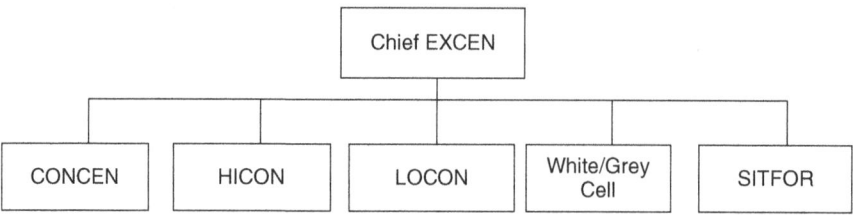

Figure 32.4 EXCEN model.

organizations, institutions, and individuals outside of the own or opposing force structure. SITFOR is the cell that manages the status of all the own and opposing forces in the scenario except for the ones represented by HICON and LOCON. When the opposing side is also played by a part of the TA, only the part of the opposing forces not controlled by the TA is managed by SITFOR.

CAX SUPPORT TOOLS

CAX support tools support EXCON and exercise planners to immerse the TA into a realistic situation and to steer the exercise towards its objectives. As we already explained in the Introduction, CAX support tools are categorized into four classes. The first of these categories is CAX planning and management tools.

Planning and Management Tools

CAX tools can support an EP starting at the specification stage. Software can provide a structured way to derive the ETO, missions, and operational tasks based on current documents such as training/exercise programs, training/exercise directives and guides, lessons learned databases, and the guidance by the OSE, OCE and TA Commander. Then this tool may compare the requirements with the capabilities of the available simulation systems, and can even figure out the required resources, e.g. person time, Communication and Information System (CIS) infrastructure, the cost of simulation, etc., according to some additional parameters such as the level of exercise and the size and the number of components in the TA. Since the number of available simulation systems is often small and it is easy to decide on the appropriate system, this tool can be more useful in determining the weaknesses of a simulation system and the approximate cost of using it for the exercise.

After the specification stage, the OCE should issue a planning guidance. The next key milestone in EP is the promulgation of an exercise plan (EXPLAN). In the preparation of EXPLAN many exercise support products including a scenario are prepared. The process for scenario development needs CAX support. In particular MEL/MIL development can be effectively supported by a tool. Moreover, the same tool can be used extensively for the management of an exercise because MEL/MIL constitutes the flow of an exercise and the workload on the TA.

Exercise Management Tools

In order to achieve the ETO, events and incidents are designed and injections are developed according to them before the exercise. An incident is a situation that stimulates actions by the TA. Related incidents are grouped into events. An incident is brought to the attention of the TA by using injections. Typically an event means several incidents and an incident is several injections. For example, embargo can be an event for an exercise. Then tracking and embarking

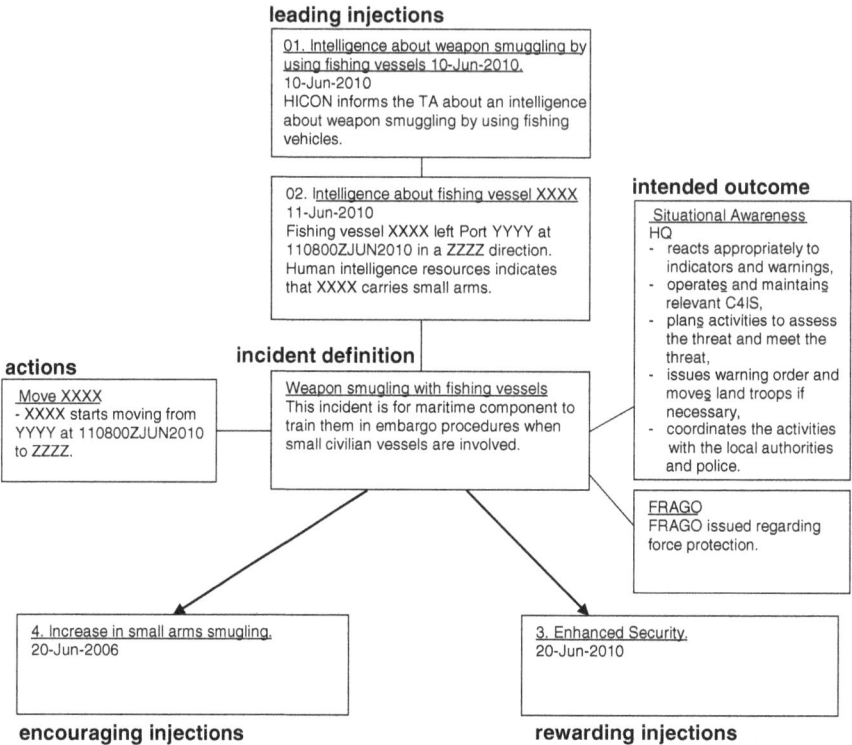

leading injections

01. Intelligence about weapon smuggling by using fishing vessels 10-Jun-2010.
10-Jun-2010
HICON informs the TA about an intelligence about weapon smuggling by using fishing vehicles.

02. Intelligence about fishing vessel XXXX
11-Jun-2010
Fishing vessel XXXX left Port YYYY at 110800ZJUN2010 in a ZZZZ direction. Human intelligence resources indicates that XXXX carries small arms.

intended outcome

Situational Awareness
HQ
- reacts appropriately to indicators and warnings,
- operates and maintains relevant C4IS,
- plans activities to assess the threat and meet the threat,
- issues warning order and moves land troops if necessary,
- coordinates the activities with the local authorities and police.

actions

Move XXXX
- XXXX starts moving from YYYY at 110800ZJUN2010 to ZZZZ.

incident definition

Weapon smugling with fishing vessels
This incident is for maritime component to train them in embargo procedures when small civilian vessels are involved.

FRAGO
FRAGO issued regarding force protection.

4. Increase in small arms smugling.
20-Jun-2006

3. Enhanced Security.
20-Jun-2010

encouraging injections   **rewarding injections**

Figure 32.5 An incident.

to a suspected ship can become an incident for the embargo event. A detection report about a suspected ship heading towards the restricted zone is an injection for the embarkment incident. Injections, incidents and events are listed in MEL/MIL, and are continuously modified during an exercise (i.e. dynamically scripted throughout the exercise) according to the performance of the TA.

An MEL/MIL management tool can provide a structured way to develop a well balanced MEL/MIL that covers all the ETO and TA fairly. First the ETO are entered into the system. Then every incident has to be linked to an ETO to ensure that they are designed according to the ETO. Moreover every incident is designed in a structured way as shown in Figure 32.5. There are leading injections and expected TA reactions (effects) for an incident. If the desired effects are reached for an incident, the TA is released with rewarding injections. Otherwise encouraging injections can be given. For example, a report about a violent crowd can be a leading injection for a crowd control incident. If the TA takes proper actions, then a report about the dispersion of the crowd can be given. Otherwise further violence can be reported as an encouraging injection.

MEL/MIL management is a key requirement that impacts not only the planning phase but also the execution phase of an EP. It is important also for the post exercise analysis and reporting phase. Therefore, a tool that can automate the

MEL/MIL scripting, provides the interfaces between the MEL/MIL scripts and the simulation, and collects the observations from the execution of the scripts is important for CAX support. This MEL/MIL tool may also have other supporting utilities, such as matrices that show the relation between the ETO and incidents, intensity created on the TA at a certain time by a certain incident, and geographical locations of injections on a map.

Scenario Management Tools

The definition and the content of a scenario change from one nation or an organization to another. A scenario is typically made up of modules and sub-modules, such as geo-military situations, country books, maps and geo data, order of battle for the forces involved in the exercise, roads to crises and other sub-modules that initiate the situation, planning guidance received from higher HQ, MEL/MIL, etc. Scenarios can be

- real, i.e. based on real geography, nations and situations,
- fictitious, often called generic,
- semi-fictitious, e.g. real geography but fictitious nations and events together with real nations and events.

In any case developing a scenario requires a long time (e.g. 12 months), the effort of a large team (e.g. five to ten people), and some additional resources such as geo data, maps and imagery. It also needs careful management and coordination to maintain the consistency of the scenario. A CAX support tool can shorten this process and ease many of the management and coordination challenges to keep the scenario consistent. For example, the ability to define a list of parameters may help in the process. This is a dynamic list, i.e. there is no limitation for the number of parameters, where scenario developers set parameters such as D day, the names of nations, etc. Then they use the names of parameters in the scenario modules. When they update the value for the parameter, it is automatically updated in all the modules where the related parameter is used. For example critical dates, locations, and names can be defined as parameters. Later when it is required to change them, changing the parameter value suffices. This saves time, and ensures that the change is done everywhere where it is required.

Such a tool also provides an effective way to generate scenario products from the fields in a setting. A developer can select the field from the available fields in the setting database, and ask the tool to generate a country book in any format (e.g. html, pdf, doc, etc.) from those fields. This tool can also import order of battle (ORBAT) data from the database of a simulation system. It is also possible to export ORBAT data edited by the tool to tools that support simulation systems, C2 systems, or directly support operations (OP). This helps the synchronization of many systems that need to work together consistently according to a common scenario.

The scenario tool should be in synchronization with the EP because many scenario modules are related to each other and are prepared during different stages of the EP by different exercise planning group members. Scenario tools must also support the management of multiple scenarios, querying scenario modules that fit a given set of specifications, merging scenario modules from different scenarios into a new scenario and scrutinizing the resulting scenario for a new setting. These capabilities can enable rapid scenario generation, which is a critical and emerging requirement for mission rehearsal trainings (MRT) and exercises (MRE). Rapid and expeditionary deployment of combined (multinational) forces requires MRE organized within days or weeks. The scenarios for such MREs must be prepared within several days. Note that a scenario is not only made of roads to crises and country books but also other modules like MEL/MIL.

CAX support tools may be useful not only for scenario management but also for content development for scenarios. For example, there are simulation systems that generate high resolution and extremely realistic 3D images. These simulation systems can be used to generate 3D imagery as if they are taken by spy planes or unmanned air vehicles.

Military Simulation Systems

Military simulations are categorized as live, virtual, and constructive as depicted in Table 32.1.

- **Live simulation.**   refers to a simulation that involves real people operating real systems. For example two pilots can be trained by using real aircraft in the air. In this case the aircraft and the pilots are real but the interactions between the aircraft are simulated and the simulation decides how effectively the pilots and the aircraft act against each other. Similarly, all the weapon systems can be equipped with emitters, and all the equipment and personnel can be equipped with sensors. If the weapons are aimed and fired correctly, the emission by the emitters can be sensed by the sensors, which indicate a hit and/or a kill based on some stochastic processes.

- **Virtual simulation.**   refers to a simulation that involves real people operating simulated systems. Examples for this are aircraft and tank simulators, where a simulator but not a real system is used to train a pilot or tank crew.

- **Constructive simulation.**   refers to a simulation that involves simulated people operating in simulated environments. Combat models that compute the possible outcomes of the decisions taken by headquarters fall in this

Table 32.1   Military simulations

Category	People	Systems
Live	Real	Real
Virtual	Real	Simulated
Constructive	Simulated	Simulated

category. In these simulations people and units are also simulated, as well as combat systems and the environment.

Among these classes military constructive simulations are often the only class of military simulation systems used in a CAX. As the technology matures live and virtual simulations federated with constructive simulations are more extensively used in CAXes. Before further elaborating on this topic, we first would like to clarify several issues related to this classification of live, virtual, and constructive.

First there has been a debate about the scope of live simulations. In live simulations the interactions between the real systems and people are simulated. A good example for this is fighter training by using real aircraft. Many experts believe that when a command and control centre joins an exercise by using the real C2 devices connected to a synthetic environment, it is also called a live simulation because C2 devices and the people using them are real. Although some experts do not agree with this, we believe that this is a correct interpretation. Therefore we can claim that the CAX community has been using live simulations for long time because many state of the art training centers are capable of feeding the C2 devices from the simulation systems.

Live simulations are also often mixed with live exercises. In essence a live exercise is a live simulation although it is not a computer simulation. However, the usages of live simulations are not limited to live exercises. In reality a live exercise can be supported also by a constructive simulation system, and a command post exercise can be supported by a live simulation. For example while one of the platoons is in the field running a live exercise, the other platoons of the company can be simulated in a constructive simulation system. Similarly live simulation systems like real command centers can be used in a CAX.

Similarly virtual simulations are often perceived as limited to things like aircraft, ship, and tank simulators. Virtual simulations can also be used to create realistic environments for command posts. Computer graphics, motion, fog and light generators can be used to create an environment that looks like a battlefield, and leaders can be asked to make their decisions in such an environment. Moreover, virtual simulations can be used to create products like video streams from unmanned aerial vehicles or 2D/3D imagery from satellites and/or aircraft. Especially when virtual simulations can interact with constructive simulations by using distributed simulation technologies, they become extremely useful for a CAX at any level.

We focus on constructive simulation systems in this chapter because usually a military constructive simulation system constitutes the core of CAX support. Constructive simulations are designed to find out the possible outcomes of the courses of actions taken by real people. They are constructed by many models, often based on stochastic processes, that calculate the results of interactions between the entities or units in a theatre. Constructive simulation systems can be classified into two categories according to their resolution, as summarized in Table 32.2.

Table 32.2   Military constructive simulations

Category	Level	Objects	Terrain
High resolution	Entity	Singular objects, e.g. a tank, a troop	High resolution, 200×200 km
Highly aggregated	Aggregate	Units, e.g. a battalion, a company	Low resolution, 4000×4000 km

- *High resolution simulations* are *entity level simulations* where singular military objects, e.g. soldiers, tanks, aircraft, are the primary objects represented. The resolution of terrain data is high, sometimes up to the plans of individual buildings. However, the simulated terrain is often limited to less than 200 km × 200 km. High resolution simulations are better suited for the tactical level. However, they are not only for tactical level simulation. They provide higher resolution that may be required also for operational and higher level purposes. Therefore, high resolution simulation systems should not be called tactical simulations.

- *Highly aggregated simulations* are *aggregate level simulations* where collections of military assets, i.e. units, are the primary objects represented. They use lower resolution terrain data but they can simulate in very large areas as large as continents. Similar to high resolution simulations, there is a tendency to call highly aggregated simulations operational level simulations, which is not correct. Aggregate level simulations may be very useful also for tactical purposes. Note that in civilian content tactical level is over operational level, which is opposite to the military hierarchy. In this chapter we use the military hierarchy in our definitions.

The constructive simulation systems can also be categorized based on their functionalities as follows.

- Service models are the simulation systems developed for the needs of a single service, i.e. army, navy, or air force.

- Joint models are either the simulation systems that fulfill the requirements of all services or federations made up of service models.

- Expert models are developed specifically to simulate certain functionalities such as logistics, intelligence, electronic warfare, homeland security, or space operations.

State of the art high resolution simulations can be used for simulating operations in regions as large as 2500 km × 2500 km. On the other hand, aggregate level simulations tend to be capable of simulating entities such as a single troop or a tank. Therefore, the gap between high resolution and highly aggregated systems is closing. Moreover, there are multi-resolution federations that integrate highly aggregated and high resolution simulations into a distributed simulation system typically by using HLA. Virtual and live simulation systems can also become federates in a multi-resolution federation.

There used to be a tendency to differentiate multi-resolution federations from live virtual constructive (LVC) federations (i.e. LVC federations and multi-resolution federations are often built separately). There are two main reasons for this.

- Virtual and live simulations have to run in real time. On the other hand multi-resolution federates typically need to be time managed. When a federation is time managed, anomalies in the movement of real-time systems (e.g. the movement of an aircraft may become not smooth) may be observed.
- Many virtual and live simulations are already federated by using DIS. There is a federation object model (FOM) called real-time platform reference (RPR) FOM based on DIS protocol data units (PDUs). Therefore many LVC federations use the RPR FOM. Since the RPR FOM does not support entity–aggregate interactions, it is not preferred for multi-resolution federations.

The modular FOM approach in HLA 1516–2010 provides new techniques to create a modular FOM that can include both the RPR FOM and other modules more appropriate for multi-resolution federations (DMSO 1998; IEEE 2000a,b,c). Reference federation architectures are already developed that can be used as both LVC and multi-resolution federations. However, most of the current multi-resolution federations are designed and implemented separately from LVC federations. Some examples for these implementations are joint multi-resolution federation (JMRF) and joint multi-resolution model (JMRM) in the United States, partnership for peace simulation network (P2SN) reference architecture by NATO partner nations, KORA and SIRA federation (KOSI) in Germany, and ALLIANCE in France. One recent endeavor is NATO Education and Training Network (NETN) reference federation architecture that integrates many highly aggregated/high resolution constructive and virtual systems. Bowers (2003) describes the foundations for the JMRM, as it was already perceived by Cayirci and Ersoy (2002). Reynolds and Srinivasan (1997) addressed the general problems of multi resolution modeling (MRM). In this book, more background information on such topics is published in chapter 25.

The NATO view on such MRM approaches is documented in Cayirci (2006, 2007b): NATO training federation (NTF) is a multi-resolution HLA federation and a part of NETN. Initial NTF has three combat models, namely joint theatre level simulation (JTLS), joint conflict and tactical simulation (JCATS), and virtual battlespace (VBS2). JTLS is a joint highly aggregated constructive simulation system. It best fits when the simulated units (simulation entities) are battalions, wings/air packages, i.e. multiple aircraft in an air mission, and ships (frigates, submarines, etc.). JCATS is a joint high resolution constructive simulation, where details like a single troop can be simulated by using high resolution terrain and environmental data. It is also possible to aggregate the simulated entities into

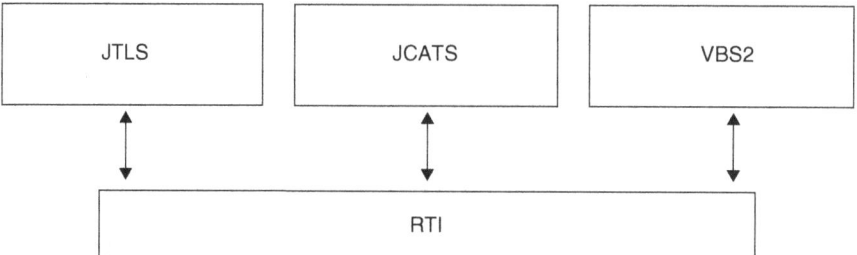

Figure 32.6   NATO training federation.

units and command them as aggregated units in JCATS. VBS2 is a very high resolution simulation system that can be used for very realistic visualization.

NTF is a very good example for reusability and interoperability of simulation tools. When the TA is lower than or equal to component command (corps) level, JCATS provides a better fidelity simulation. On the other hand, since it is high resolution, it needs many details for operating. Therefore, it is not viable to use purely JCATS in exercises where a high number of units and large areas are involved, e.g. corps and higher level exercises. JTLS is a very good match for those levels. However, many incidents of contemporary warfare and conflicts require high resolution planning also in high echelons, which is not available in JTLS. NTF connects JTLS and JCATS to close this gap. In NTF highly aggregated JTLS can be used as long as higher resolution simulation is not needed. When simulation resolution higher than the one that JTLS can provide is required, JCATS can be used. The outputs of these two simulations update the attributes of the units and entities in both simulations.

Mediation-Ware

CAX tools must replicate C4I environments during CAXs. In other words, simulation systems and all the other related software must be transparent to the TA. TA should carry out the exercise as if they are in an operation and commanding their subordinates by using C4I systems normally available to them. TA should receive orders and send their reports through C4I systems. This transparency can be achieved by the mediation tools between the simulation and C4I systems.

There are two ways for interaction between TA and simulation in this construct, and both of them are indirect. The first way is through the response cells who are acting as LOCON to the TA. TA gives an order and a response cell role plays as the headquarters or commanding officer of the subordinate unit that receives the order. Than the planners in the response cell coordinate and plan the execution of the order, and after applying some doctrinal C2 delays they pass the plans to the CAX operators who translate them to simulation orders. CAX operators receive both periodical and mission reports from the simulation system. They transfer these simulation reports to the planners in the response cell, who

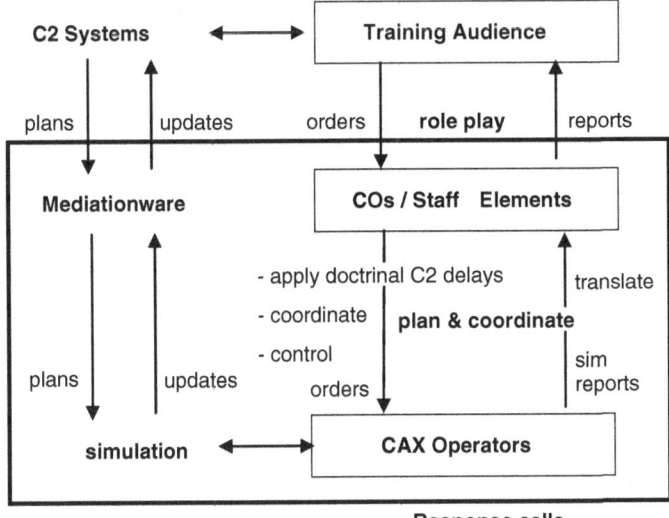

Figure 32.7 Simulation and the training audience.

translate them to reports for the TA. This interaction of TA with the simulation system is through *role play* by the response cell.

The second way is automatic interaction through C2 systems. Simulation systems can interact automatically with the operational C2 systems of the TA by using mediation-ware. For example, the air tasking orders developed in a C2 system can be translated into simulation orders and directly fed into the simulation. This interaction is from the C2 system to the simulation. The opposite direction is also possible. For example the simulated air missions can update the recognized air picture in a C2 device automatically and continuously.

Surdu and Poach (2000) envision even the seamless integration to allow the use of M&S services for the support of operation. Similar ideas and possible solutions are described in chapter 15.

MEL/MIL SYNCHRONIZATION

The bottom line is that the RCs are the part of this exercise construct that directly interact with simulations, but not the TA. In this approach simulations are used for the following reasons:

- to compute the possible outcomes, i.e. results, of the decisions made by TA.
- to simulate the entities and conditions not controlled by the TA or EXCON.
- to maintain a consistent white truth. Recognized operational pictures are derived from the white truth according to the intelligence capabilities and efforts of the sides and services.
- to stimulate C2 systems used by TA.

Although the simulation systems are designed mainly for the first two reasons, they can be used only for the last two in a completely scripted exercise, i.e. an exercise where the plans of TA are not simulated but the results of their decisions are prescribed based on forecasting and modified during the exercise according to the developing situation. Of course when TA is given injections based on the experience and intuition of EXCON, the risk that the injections are not coherent and realistic is higher. We call this risk a "negative training" risk because an HQ trained with unrealistic injections may plan considering that they can achieve the same results also in a real battlefield. Therefore, verified and validated simulation systems with validated databases are important and can reduce the "negative training" risk.

Contrary to the common belief, when there is an MEL/MIL, there is still need for simulation, and there should not be a tradeoff between an MEL/MIL and simulation. A good MEL/MIL is always needed to ensure that the exercise flows towards exercise/training objectives. However, the content of the MEL/MIL, the dynamic management of MEL/MIL, and the synchronization of MEL/MIL with the simulation are very important:

- MEL/MIL should not script the decisions of TA or hinder the decision and planning processes of TA.
- MEL/MIL should not script the results of the decisions taken by TA.
- MEL/MIL should not be fixed and dynamically maintained based on the performance of TA and ETO throughout the execution stage of an exercise.
- Situations can be created in simulations by controlling only situational forces. This suffices to synchronize simulation with MEL/MIL. However, the movement of situational forces should also be realistic.

CONCLUSION

Computer support is not limited to simulations in modern exercises. CAX tools include all kinds of systems that support EXCON and exercise planners to control the exercise for achieving ETO. A good MEL/MIL management tool, a set of simulations, and mediation-ware for C2 stimulation are needed for a typical CAX. When these tools are used by appropriate EXCON structure and procedures, the efficiency in achieving the ETO of an exercise can be maximized.

REFERENCES

Bowers A (2003). *Multi-Resolution Modeling in the JTLS-JCATS Federation*. Technical report, MITRE Cooperation.

Cayirci E (2006). *NATO Joint Warfare Center's Perspective on CAX Support Tools and Requirements*. ITEC'2006, May.

Cayirci E (2007a). *Exercise Structure for Distributed Multi-resolution NATO Computer Assisted Exercises*. ITEC'2007, May.

Cayirci E (2007b). *Distributed Multi-resolution Computer Assisted Exercises*. NATO Modelling and Simulation Conference 2007, NATO Report RTO-MP-MSG-056, Neuilly-sur-Seine, France.

Cayirci E and Ersoy C (2002). Simulation of tactical communications systems by inferring detailed data from the joint theater level computer aided exercises. *SCS Simul J* **78**, 475–484.

Cayirci E and Marincic D (2009). *Computer Assisted Exercises and Training: A Reference Guide*. Wiley & Sons, New York, NY.

DMSO (1998). *High Level Architecture Interface Specification*, Version 1.3NG. Washington, DC.

IEEE (2000a). *IEEE Standard for Modeling and Simulation (M&S) High Level Architecture (HLA)-Framework and Rules, Std 1516*. IEEE, Piscataway, NJ.

IEEE (2000b). *IEEE Standard for Modeling and Simulation (M&S) High Level Architecture (HLA) - Federate Interface Specification, Std 1516.1*. IEEE, Piscataway, NJ.

IEEE (2000c). *IEEE Standard for Modeling and Simulation (M&S) High Level Architecture (HLA)-Object Model Template (OMT), Std 1516.2*. IEEE, Piscataway, NJ.

Reynolds PF and Srinivasan S (1997). Consistency maintenance in multiresolution simulations. *ACM Trans Model Comput Simul* **7**, 368–392.

Surdu JR and Poach UW (2000). Simulations technologies in the mission operational environment. *SCS Simulat* **74**, 138–160.

Annex 1

M&S Organizations/ Associations

Salim Chemlal and Tuncer Ören

Within this book, many modeling and simulation organizations are mentioned as sponsors or proponents of activities, collaboration partners in development and standardization efforts, providers of repositories, and more. In support of his activities supporting collecting resources for the Body of Knowledge for Modeling and Simulation, Tuncer Ören compiles important information on publications, conferences, individuals, and organizations and publishes them on the website http://www.site.uottawa.ca/~oren/MSBOK/MSBOK-index.htm.

This annex is based on the list of modeling and simulation organizations as provided in June 2011, website: http://www.site.uottawa.ca/~oren/links-MS-AG.htm. We selected the most important entries of interest to a simulation engineer. The list supports combat modeling and distributed simulation and provides the reader with a short description and the link to more information.

This list is neither exclusive nor complete. In addition, new organizations may rise and others may be dismantled. Nonetheless, the following enumeration reflects the current state as used for writing this book.

We used government, industry, and academia as categories in which the organizations are enumerated in alphabetic order.

G: Government

 G1. AFAMS: Air Force Agency for Modeling and Simulation
 Nationality: United States
 URL: http://www.afams.af.mil/
 Description: AFAMS M&S capabilities enable the United States Air Force to organize, train, educate, equip, and employ current

Engineering Principles of Combat Modeling and Distributed Simulation, First Edition.
Edited by Andreas Tolk.
© 2012 John Wiley & Sons, Inc. Published 2012 by John Wiley & Sons, Inc.

and future air, space, and cyberspace forces for the full range of operations. Air Force M&S is a key enabler that accelerates and enhances existing air force warfighting capabilities.

G2. **AMSO:** US Army Model and Simulation Office
Nationality: United States
URL: http://www.ms.army.mil/
Description: AMSO aims to provide a vision, oversight, and management of M&S across the Army and to assist in developing a fully capable, coherent, and unified Army M&S strategy to organize and equip the Army with M&S capabilities in support of operating and generating force functions and institutional processes. The AMSO hosts the Army node of the DoD Model and Simulation Resource Repository (MSRR).

G3. **UK MOD AOF:** UK Ministry of Defence Acquisition Operating Framework
Nationality: United Kingdom
URL: http://www.aof.mod.uk/aofcontent/tactical/mands/
Description: MOD M&S is part of the Acquisition Operating Framework (AOF) and as such it is intended to improve the consistency of the MOD's application of policy and best practice and play an important role in delivering better solutions for defense in the future. The MOD simulation strategy applies to all UK Defence M&S employed in support of analysis, experimentation, capability acquisition and warfare development and, most importantly, front line operations and training.

G4. **MSCO:** Modeling and Simulation Coordination Office
Nationality: United States
URL: http://www.msco.mil/
Description: MSCO empowers the US Department of Defense with M&S capabilities that efficiently support the full spectrum of the Department's activities and operations. It performs those key corporate-level coordination functions necessary to encourage cooperation, synergism, and cost-effectiveness among the M&S activities of the DoD components. The MSCO is the Executive Secretariat for DoD M&S Management in fostering the interoperability, reuse, and affordability of cross-cutting M&S to provide improved capabilities for DoD operations.

G5. **DND/CF SECO:** National Defence and Canadian Forces Synthetic Environment Coordination Office
Nationality: Canada
URL: http://www.cfd-cdf.forces.gc.ca/sites/page-eng.asp?page=4238
Description: The DND/CF SECO is the focal point for M&S information, coordination, and guidance in the DND and CF. The

DND/CF SECO coordinates and sets the standards for operationally valid and verifiable models, synthetic environments, and distributed simulations that are used by multiple CF or DND branches.

G6. **DRDC SE:** Defence Research and Development Canada Ottawa's Synthetic Environments Facilities
Nationality: Canada
URL: http://www.ottawa.drdc-rddc.gc.ca/html/facilities_synthetic -eng.html
Description: The Synthetic Environments Battlelab, enabled by the Joint Simulation Network (JsimNet), is an experimental environment that can perform high-end computational processing and run unclassified simulations with advanced three-dimensional graphics for concept development and experimentation using exercise and visualization management tools. The Synthetic Environment Research Facility (SERF) aims to provide cost-effective simulations to study human interfaces, human–systems modeling and distributed mission training.

G7. **IDGS M&S COE:** Italian Defence General Staff Modelling and Simulation Centre of Excellence
Nationality: Italy
URL: https://transnet.act.nato.int/WISE/TNCC/CentresofE/MS/ Moreinform/File
Description: M&S COE supports Allied Command Transformation (ACT) efforts to transform NATO by providing subject matter expertise on all aspects of the M&S activities, improving training and education, amending capabilities and interoperability, assisting in doctrine standardization and supporting concepts development through experimentation.

G8. **KBSC:** Korean Battle Simulation Center
Nationality: Republic of Korea
URL: http://mss.cubic.com/MissionSupportServices/Operating Divisions/InformationOperationsDivisionIOD/TrainingExercise IOD/IODKoreanBattleSimulationCenterKBSC.aspx
Description: KBSC aims to conduct computer-driven service, joint and multinational training and battle simulation exercises for active US Army, US Air Force, US Navy, US Marine Corps, Republic of Korea and other allied military personnel. KBSC develops several large simulation exercises, such as Ulchi Freedom Guardian (UFG) exercise, a joint military exercise between South Korea and the United States. UFG is considered to be the world's largest and most complex distributed simulation-driven exercise.

G9. **MCMSMO:** US Marine Corps Modeling and Simulation Management Office
Nationality: United States
URL: https://www.mccdc.usmc.mil/mcmsmo/
Description: MCMSMO provides central coordination of Marine Corps Modeling and Simulation to support collaboration, interoperability, commonality and reuse of Marine Corps M&S tools, data and services. MCMSMO serves as the Marine Corps single point of contact on all Marine Corps M&S matters, and for coordination with the other Services, DoD, Joint Staff, and other agencies' M&S organizations.

G10. **MSIS:** US Modeling and Simulation Information System
Nationality: United States
URL: http://msis.dod-msiac.org/MSIS/Index.jsp
Description: The Department of Defense Modeling and Simulation Information System (MSIS) is operated by the Modeling and Simulation Information Analysis Center (MSIAC). MSIS is an electronic archive used to store metadata descriptions covering a variety of M&S tools, documents, databases, object models and contacts. A logon is needed to take full advantage of this repository.

G11. **NATO RTO:** NATO Research and Technology Organisation
Nationality: NATO
URL: http://www.rta.nato.int/
Description: The NATO RTO promotes and conducts cooperative scientific research and exchange of technical information amongst NATO nations and partners. NATO Modelling and Simulation Group (NMSG) is one of seven technical panels that were organized by the NATO RTO. NMSG supports cooperation among Alliance bodies, NATO Member and Partner Nations to maximize the effective utilization of M&S. Its primary mission areas include M&S standardization, education, and associated science and technology.

G12. **NMSO:** Navy Modeling and Simulation Office
Nationality: United States
URL: https://nmso.navy.mil/
Description: NMSO is chartered to promote the discovery and reuse of M&S resources, provide information on policy, data, and services and serve as the "action arm" of the Navy Modeling and Simulation Governance Board (GB). NMSO and MCMSMO serve as the single points of contact for M&S for the Navy and Marine Corps respectively.

G13. **PEO STRI:** US Army Program Executive Office for Simulation, Training, and Instrumentation

Nationality: United States

URL: http://www.peostri.army.mil/

Description: PEO STRI is the US Army's acquisition and contracting center of excellence for simulation, training and testing capabilities. It provides training and testing simulations, simulators, systems and instrumentation products and services to support the nation's security and other various military and civilian agencies.

G14. **TRAC:** US Army Training and Doctrine Command (TRADOC) Analysis Center

Nationality: United States

URL: http://www.trac.army.mil/

Description: TRAC conducts research on potential military operations worldwide to inform decisions about the most challenging issues facing the Army and the Department of Defense (DoD). TRAC develops and maintains a class of warfighting M&S referred to as force-on-force, ranging from individual objects to aggregated objects at corps level. TRAC M&S represent the Army's *de facto* standards for force-on-force M&S and are widely used by military, industry, and allies.

G15. **WEAO:** Western European Armaments Organization

Nationality: Various European Nations

URL: http://www.weao.weu.int/site/index.php

Description: WEAO is a subsidiary of the Western European Union (WEU) and has at the time this book was written 19 members: Austria, Belgium, Czech Republic, Denmark, Finland, France, Germany, Greece, Hungary, Italy, Luxembourg, Netherlands, Norway, Poland, Portugal, Sweden, Spain, Turkey, and the UK. The executing body is the Western European Armaments Group (WEAG). After 11 years serving the Defence Research and Technology (R&T) community in Europe, the Research Cell ceased its activities in August 2006. This website remains available as a source of information and documents, although without any further updates.

I: Industry

I1. **ETSA:** European Training and Simulation Association

Nationality: Various European Nations

URL: http://www.etsaweb.org/

Description: ETSA represents the European training and simulation community and provides an environment for users and suppliers to exchange opportunities, ideas, information, and strategies on training and simulation technology and methodology. It brings together all those who have a professional interest in improving the effectiveness of training and training related interoperability, standards, and codes of practice. It represents to governments and other users of training and simulation the non-partisan business interests of the industry.

I2. **ITSA:** International Training and Simulation Alliance
Nationality: United States
URL: http://itsalliance.org/
Description: ITSA is a worldwide network of training and simulation associations that share common goals; it promotes a better understanding around the globe of the importance of training and simulation technology for use in every profession and endeavor known to mankind. ITSA has associations in the USA (NTSA), Europe (ETSA), the Republic of Korea (KTSA), and Australia (SIAA).

I3. **MSIAC:** Modelling and Simulation Information Analysis Center
Nationality: United States
URL: http://www.dod-msiac.org/
Description: MSIAC is a Department of Defense Information Analysis Center; it assists DoD components/offices, other government agencies, academia, government contractors, and US industry in the analysis and dissemination of information within the DoD M&S areas. It is chartered to access, acquire, collect, analyze, synthesize, generate, and disseminate scientific, technical, and operational support information in the area.

I4. **NTSA:** National Training Systems Association
Nationality: United States
URL: http://www.trainingsystems.org/
Description: NTSA provides the training, simulation, mission planning, related support systems, and training services industries a recognized, focused, formal organization to represent and promote their business interests in the marketplace. The association provides industry forums to communicate the full capability and broad characteristics of all the elements of training systems and mission planning to include associated support services.

I5. **SAAB TCSC:** SAAB Tactical Combat Simulation Centre
Nationality: Sweden
URL: http://www.saabgroup.com/Air/Training_and_Simulation/ Complete_Training_Systems_Solutions/Tactical_Combat_Simula tion_Centre/
Description: SAAB TCSC is one of the leading providers of simulation and training solutions for military aviation where crews are trained in order to be experienced in advance and ahead of their operational mission to successfully perform in a complex joint operation. The TCSC is based around an HLA network allowing more pilot stations and C2 units to be added without any

re-configuration; other HLA compatible simulators can also be connected to the central systems.

I6. SIAA: Simulation Industry Association of Australia
Nationality: Australia
URL: http://www.siaa.asn.au/
Description: SIAA provides a focus forum for those involved with simulation technology in Australia to allow discussion and distribution of information and to further advance the research, development, and use of simulation technologies and practices in Australian society, industry, academia, and government.

I7. SISO: Simulation Interoperability Standards Organization
Nationality: International
URL: http://www.sisostds.org/
Description: SISO is an international organization dedicated to the promotion of M&S interoperability and reuse for the benefit of a broad range of M&S communities. SISO serves the global community of M&S professionals, providing an open forum for the collegial exchange of ideas, the examination and advancement of M&S related technologies and practices, and the development of standards and other products that enable greater M&S capability, interoperability, credibility, reuse, and cost-effectiveness.

A: Academia

A1. ACIMS: Arizona Center for Integrative Modeling and Simulation
Nationality: United States
URL: http://acims.eas.asu.edu/
Description: ACIMS is devoted to research and instruction that advance the use of M&S as means to integrate disparate partial solution elements into coherent global solutions to multidisciplinary problems with an emphasis on the need for sustained and continued development of the theoretical and conceptual foundations of M&S. ACIMS brings together a variety of researchers across various colleges of the State of Arizona in integrative M&S.

A2. AMSL: Auburn Modeling and Simulation Laboratory (Auburn University)
Nationality: United States
URL: http://www.eng.auburn.edu/csse/msl/
Description: AMSL promotes the study and development of new techniques using computer simulation; its areas of interest are simulation of computer networks, combat simulation, agent-based simulation, interoperability of networked systems, and software architectural design for interoperability.

A3. CMSA: Center for Modeling, Simulation and Analysis (University of Alabama in Huntsville)

Nationality: United States

URL: http://cmsa.uah.edu/

Description: CMSA is a research and development center at the University of Alabama in Huntsville. CMSA's work centers on M&S and systems engineering with special expertise on physics-based modeling, simulation interoperability and composability, discrete event simulation, spacecraft propulsion modeling, mathematical modeling and analysis, finite element modeling, and M&S education.

A4. IST: Institute for Simulation and Training (University of Central Florida)

Nationality: United States

URL: http://www.ist.ucf.edu/

Description: IST promotes the state of the art and science of M&S by performing basic and applied simulation research, supporting education in M&S and related fields, and serving public and private simulation communities. IST primarily focuses on applied research for military training in support of DoD and government subcontractors.

A5. LSC: Liophant Simulation Club

Nationality: Italy

URL: http://www.liophant.org/

Description: Liophant is an association located in Genoa University that brings simulation developers and users together. Liophant is devoted to promoting and diffusing the simulation techniques and methodologies, exchange of students, organization of international conferences, organization of courses and stages in companies to apply simulation to real problems.

A6. M&SNet: McLeod Modeling and Simulation Network

Nationality: International

URL: http://www.m-s-net.org/start/

Description: The M&SNet is a consortium of cooperating independent organizations active in professionalism, research, education, and knowledge dissemination in the M&S domain. It was established in 2003 by the Society for Modeling and Simulation International (SCS). The M&SNet aims to provide an organizational structure that will serve to integrate and enrich, within its organizations, M&S activities throughout the world. The M&SNet provides a framework within which organizations interested in M&S can interact, share expertise, and work on problems of common interest.

A7. MISS: McLeod Institute of Simulation Sciences

Nationality: International

URL: http://www.mcleodinstitute.org/

Description: The MISS is an initiative of the Society for Modeling and Simulation International (SCS). It consists of cooperating centers active in professionalism, research, education, and knowledge dissemination in the M&S domain; there are currently 26 MISS Centers around the world. The MISS aims to provide an organizational structure that will serve to integrate and enrich, within its Centers, the activities of M&S expertise throughout the world.

A8. MSREC: Modeling and Simulation Research and Education Center (Georgia Institute of Technology)
Nationality: United States
URL: http://www.msrec.gatech.edu/
Description: MSREC is an interdisciplinary research center at Georgia Institute of Technology that promotes cross-disciplinary research and development activities, including researchers in core M&S areas, supporting technologies, and innovative applications. The center offers education programs to train M&S practitioners, educators, and researchers.

A9. SCS: Society for Modeling and Simulation International
Nationality: International
URL: http://www.scs.org/
Description: SCS is the world's premier professional society devoted to M&S. SCS is chartered to advance the use of M&S to solve real-world problems, enhance simulation and allied computer arts in all fields, and facilitate communication among professionals in the field of simulation. SCS organizes meetings, sponsors and co-sponsors national and international conferences, and publishes the *SIMULATION: Transactions of the Society for Modeling and Simulation International* and the *Journal of Defense Modeling and Simulation* magazines.

A10. SMS Lab: Systems Modeling Simulation Laboratory at KAIST (Korea Advanced Institute of Science and Technology)
Nationality: Republic of Korea
URL: http://smslab.kaist.ac.kr/
Description: SMS Lab is devoted to research and development of M&S techniques and applications; it also supports an M&S education program at KAIST. The lab promotes several military projects with Korea Agency for Defence Development (ADD) and the development of simulation environments for war game training serving the Korean Navy, Marines, and Air Force.

A11. VMASC: Virginia Modeling and Simulation Center
Nationality: United States
URL: http://www.vmasc.odu.edu/

Description: VMASC is a multidisciplinary modeling, simulation and visualization collaborative research center of Old Dominion University. It furthers the development and applications of modeling, simulation, and visualization as enterprise decision making tools to promote economic, business, and academic development. It also supports the University's Modeling and Simulation degree programs, offering M&S Bachelors, Masters, and PhD degrees to students across the Colleges of Engineering and Technology, Sciences, Education, and Business.

This list is only a small subset of organizations. The web-based collection compiled in June 2011 for associations, organizations, or committees 105 entries, for centers or groups 34 entries, and for military organizations 27 entries. The interested reader is encouraged not only to visit the websites for updated information, but also to actively contribute to collect relevant information and updates and submit them to the website.

Annex 2

Military Simulation Systems

José J. Padilla

INTRODUCTION

Simulation systems known as CGF (computer generated forces) or SAF[1] (semi-automated forces) play a critical role in modern warfare by supporting activities like training, experimentation, acquisition, and analysis. Mainly, these activities can now be conducted at lower cost as they limit the unnecessary movement of troops and equipment and allow for the study of different scenarios efficiently. For instance, when training senior commanders in the execution of defense plans, simulation is used as an alternative to moving troops, planes, tanks, and ships. In addition, different scenarios that consider different troop levels and different types of planes reflecting different strategies can be studied in a short period of time. Simulation systems provide cost-effective, repeatable, and quantitatively analyzable means of "practicing" different scenarios. Scenarios range from joint/coalition strategic to tactical levels involving members from every branch and rank in the military hierarchy. In order to support missions, for instance, CGF simulations contain real-world terrain, human behavior (soldiers in a simulation get tired), weapons, and buildings among others. Although simulations systems can be standalone, they are used as a distributed system that runs different simulations simultaneously. For instance, one system simulates tanks, while another simulates planes, while yet another simulates soldiers. These three simulations

---

[1]SAF is differentiated from CGF in that in the former entities have some human input while in the latter they are autonomous. Behaviors of SAF and CGF rely on some level of artificial intelligence. Modern simulation systems support both.

Engineering Principles of Combat Modeling and Distributed Simulation, First Edition.
Edited by Andreas Tolk.
© 2012 John Wiley & Sons, Inc. Published 2012 by John Wiley & Sons, Inc.

are put together (in a federation) which would allow its users to train in a scenario that contains tanks, planes, and soldiers.

Simulation systems execute virtual or constructive simulations. In virtual simulations, real people interact with simulated systems allowing individuals to acquire a special skill such as flying an aircraft. In constructive simulations, simulated people interact with simulated systems. This allows a simulation to receive initial inputs without affecting outcomes. Virtual and constructive simulations can be used in tandem with live simulations (real people interacting with real people and/or systems) in what is called live virtual constructive (LVC) simulations.

Simulation systems, in their current technological form, can be traced back to the 1970s Lawrence Livermore Laboratory and its Janus program for conflict simulation using a graphical user interface. According to Shimamoto (2000), "the Livermore simulations have proved highly valuable to the military. They have been employed in Operation Just Cause in Panama and Operation Desert Storm in the Mideast, as well as for combat planning in Somalia, Bosnia, and other international trouble spots." The development of these simulations took off in the 1990s and they now exist in different forms to satisfy different needs. Currently, simulation systems are specialized by activities which usually translate into military branches. For instance, OneSAF is used mostly by the US Army for training, experimentation, and acquisition, while JSAF is used for joint training exercises.

In order to facilitate the study of simulation systems, this annex will provide a description of some of the most commonly used systems and a categorization according to their characteristics.

SIMULATION SYSTEMS

As previously mentioned, different simulation systems are used depending on particular needs. For instance, if the need is training for a mission where individuals from the Navy, Army, and Air Force are required to work together, there is a system for that. They seek to represent realistic scenarios where friendly and enemy forces are present for training or analysis purposes. They rely on physics-based models of missile trajectories and physiological factors in different environment and urban conditions. Visualization varies across simulation systems, from 2D considering high levels of granularity (brigade, for instance) to 3D considering low levels of granularity (soldier or squad level). These systems rely on frameworks or infrastructures to exchange data in a distributed environment. They are HLA[2] and DIS[3] compliant and are quickly relying on web services for interoperability. They use XML[4]-based languages to initialize scenarios (MSDL[5]) and

---

[2]High Level Architecture.
[3]Distributed Interactive Simulation.
[4]Extended Markup Language.
[5]Military Scenario Definition Language.

to exchange data between C2 and simulations and provide a common operational picture (BML[6] and C-BML[7]). In the following subsections, brief descriptions of the most commonly used simulation systems are presented. If the reader wants to know more, please refer to the simulation systems manuals.

Joint Semi-Automated Forces (JSAF)

JSAF was developed in the 1990s under the Synthetic Theater of War (STOW[8]) program. JSAF is a simulation system used to generate entity level units such as tanks, soldiers, munitions, ships, and aircraft in a synthetic environment. These entities can be controlled separately or organized into meaningful military units such as a platoon or a company. The entities seek to replicate their real-life counterparts, so they are affected by events such as getting tired (in the case of soldiers), obstacles due to terrain (trees), and limited line of sight (weather conditions). According to JSAF *User's Guide* (2003) JSAF entities exhibit realistic behaviors such as tanks driving along a winding road, ships following an intended course, units suffering combat damage, missiles exhibiting realistic trajectories, and ammo and fuel being depleted. Figure A2.1(a) shows a partial screen capture from JSAF as shown in the *User's Guide* (p. 78) and Figure A2.1(b) shows a screen capture of the main screen (JSAF Quick Guide, 2008, p. 2). Figure A2.1(a) shows the pull-down menu that allows users to select the desired level of granularity to be displayed. In addition, it shows the different tools available for creating and controlling artillery fire, checking gun ordnance inventories aboard units and selecting/firing guns, activating/deactivating radar, firing missiles, and controlling UAV (unmanned aerial vehicle) flight dynamics, among other things.

JSAF is an open environment where entities' attributes, tasks, and behaviors can be modified. This adds flexibility, but it may generate consistency issues with other simulations known to interoperate with it. For instance, we have two simulations, one for a tank and one for a plane. If within the tank simulation an attribute is added to the tank that it has a crew (before tank and crew were one entity), when the tank receives an impact from a missile launched by the plane, now the tank simulation may show a crew surviving the impact. This generates an inconsistency with the plane simulation as it expects the tank and crew to be killed.

JSAF is DIS and HLA compliant. DIS and HLA allow technical interoperability between simulations as highlighted in the previous example. DIS and HLA allow simulations to share data and synchronize actions in a federation on

---

[6]Battle Management Language.

[7]Coalition—BML.

[8]According to Lenoir and Lowood (2002), STOW was a large Department of Defense Program in M&S to "construct synthetic environments for numerous defense functions. Its primary objective is to integrate virtual simulation (troops in simulators fighting on a synthetic battlefield), constructive simulation (war games), and live maneuvers to provide a training environment for various levels of exercise."

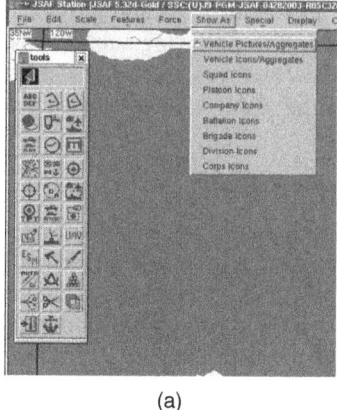

(a)

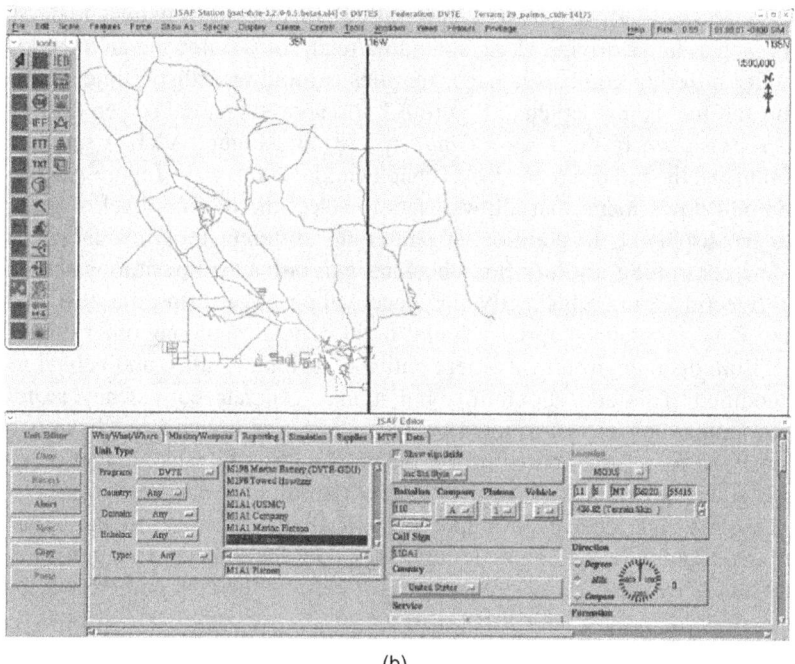

(b)

Figure A2.1 (a) JSAF partial screen capture (*User's Guide*); (b) JSAF main screen (*JSAF Quick Guide*).

a set of entities. In the previous example, the plane and tank are entities that are shared by the two simulations with probably their complete corresponding attributes, tasks, and behaviors.

JSAF is used by the Deputy Director J7 for Joint and Coalition Warfighting (DDJ&JCW), the Navy Warfare Development Command (NWDC), the Air Force Research Lab, and the US Marine Corps within the Deployable Virtual

Training Environment (DVTE) providing 2D strategic, operational, and tactical level support for joint navy and marines training in air, water, and land exercises. JSAF profile:

- Type of simulation: constructive
- Main use: joint
- Main purpose: training and experimentation
- Other users: Navy, Marines, and foreign governments such as Australia and the UK
- Level of support: strategic, operational, tactical
- Data model infrastructure support: HLA and DIS
- Display: 2D
- Military unit: up to brigade

One Semi-Automated Forces (OneSAF)

Born in the 1990s, OneSAF is a simulation system developed by US Army Simulation, Training, and Instrumentation Command to support training, experimentation, and acquisition activities ranging from the troop up to the command level. It is envisioned that OneSAF is the path to modernizing simulators across the US Army, especially for virtual trainers such as Aviation Combined Arms Tactical Trainer (AVCATT), Close Combat Tactical Trainer (CCTT), and the Common Gunnery Architecture (CGA). OneSAF has many JSAF features: replication of their real-life characteristics such as getting tired (in the case of soldiers), obstacles due to terrain (trees), and limited line of sight (weather conditions). One-SAF is HLA and DIS compliant in addition to supporting MSDL, JC3IEDM,[9] and BML. Unlike JSAF, OneSAF is Army centric (quickly being extended to support US Air Force, Joint, US Navy, and US Marine Corps simulation needs) and provides a wider range of support for land entities than for sea and air entities. This extended support is reflected in the higher detail fidelity of represented entities (buildings, land vehicles) and units (squad, brigade) and their range of behaviors. It can represent fully automated friendly and enemy forces, in addition to neutral.

OneSAF has had two incarnations: OneSAF Testbed (OTB) and OneSAF Objective System (OOS). OTB SAF is the successor of Modular SAF (ModSAF) and it provides an interactive, detailed, entity level simulation. According to the OTB SAF Overview (2000)

> OTB SAF entities can exhibit combat damage to their mobility and firepower according to the type of weapon used, the location and angle of incidence of the hit, and the range of the weapon. Similarly, an entity's weapons system exhibits realistic rates of fire and trajectories, and resource depletion is accurately simulated for both fuel and ammunition. Other simulated capabilities include intervisibility, target

---

[9]Joint Consultation Command and Control Information Exchange Data Model.

detection, target identification, target selection, fire planning, and collision avoidance and detection. These capabilities are based on, but not limited to, such appropriate realistic factors as range, motion, activity, visibility, arc of attention, direction, orders, and evaluation of threat.

Figure A2.2 shows a screen capture of OTB SAF (OTB SAF Overview, 2000). As can be seen, OTB SAF has a similar look to JSAF. The limitation with OTB SAF was that it required its user to be a good programmer and knowledgeable in military doctrine and tactics in order to modify entities and units' characteristics and behaviors.

OTB SAF was discontinued in 2006 and replaced by OOS SAF which had been in development since 2003. OOS SAF is a completely new system that allows composition of operations, systems, and control processes up to the brigade level. The current face of OneSAF is OOS SAF. In this incarnation, OneSAF is the next generation simulation system.

OneSAF is highly and easily scalable. Its premise of composition allows modelers the flexibility to represent different tasks and missions by modifying entities, units, and behaviors without the need to be programming or subject matter experts.

OneSAF profile:

- Type of simulation: constructive
- Main user: US Army

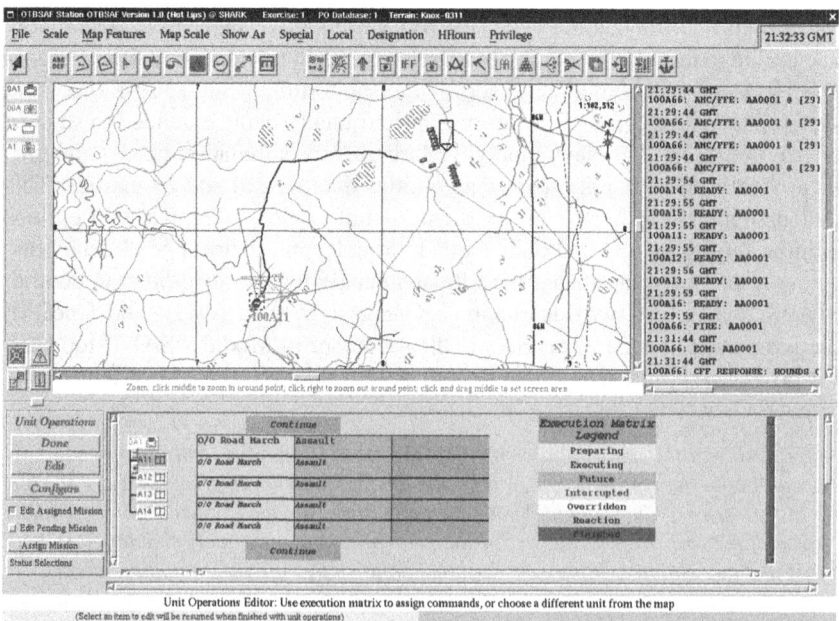

Figure A2.2   OTB SAF screen layout (OTB SAF Overview, 2000).

- Main purpose: training, experimentation, and acquisition
- Other users: Air Force, Joint, US Navy, US Marines, and foreign governments such as Canada, UK, and Australia
- Level of support: strategic, operational, tactical
- Data model infrastructure support: HLA, DIS, MSDL, BML
- Display: 2D and 3D
- Military unit: up to brigade

Joint Conflict and Tactical Simulation (JCATS)

Born in the 1990s, JCATS was developed by the Lawrence Livermore Laboratory and is currently sponsored by US JFCOM; it supports sea, land, and air exercises by the US Navy and US Marines. It is also used by the Department of Homeland Security (DHS) for training purposes. JCATS is similar to JSAF and OneSAF as it supports a large number of entities, generates realistic scenarios based on physics-based models, and is HLA and DIS compliant. Its advantage relies on its special support for training and rehearsing at the tactical level. Mission planning and rehearsal makes JCATS appealing to organizations such as DHS and the Department of Energy (DOE), among others. JCATS supports training and experimentation in a variety of environments and in special, urban environments where it provides a high level of detail of entities such as buildings (with doors, windows, and floor plans for instance), people (armed forces and civilians), and activities such as close air support and mount/dismount of vehicles. Figure A2.3 shows the JCATS workstation display (*JCATS: Simulation User's Guide*, 2003, p. 4-1).

US JFCOM uses JCATS in their LVC simulations in tandem with other systems such as VBS2 to provide different levels of detail in their training.

JCATS profile:

- Type of simulation: constructive
- Main user: joint
- Main purpose: training, analysis and experimentation
- Other users: US Navy, US Marines, DHS, DOE, NATO
- Main level of support: tactical
- Data model infrastructure support: HLA, DIS
- Display: 2D
- Military unit: up to battalion

VR-Forces

VR-Forces is a commercial of the shelf (COTS) simulation system developed by VT MÄK as an option to government solutions such as JSAF, OneSAF, and JCATS. It is used mostly for training at the tactical level providing a high level

Figure A2.3   JCATS workstation display (*JCATS: Simulation User's Guide*, 2003).

of detail of the battlefield. Like its government-sponsored counterparts, it is HLA and DIS compliant and, given that it has a programmable toolkit, modules can be built to make it complaint with XML-based languages. The toolkit can also be used to modify entities' characteristics and behaviors. It is used as a desktop solution by the US Air Force and the US Marines and by foreign governments such as Spain, Norway, and the Netherlands.

VR-Forces profile:

- Type of simulation: constructive
- Main user: all branches
- Main purpose: training and experimentation
- Other users: Spain, Norway, and the Netherlands
- Main level of support: tactical
- Data model infrastructure support: HLA, DIS
- Display: 2D and 3D
- Military unit: up to platoon

Virtual Battlespace 2 (VBS2)

VBS2 is a COTS simulation system developed by Bohemia Interactive Simulations to train and rehearse missions at the tactical level. Small combat teams

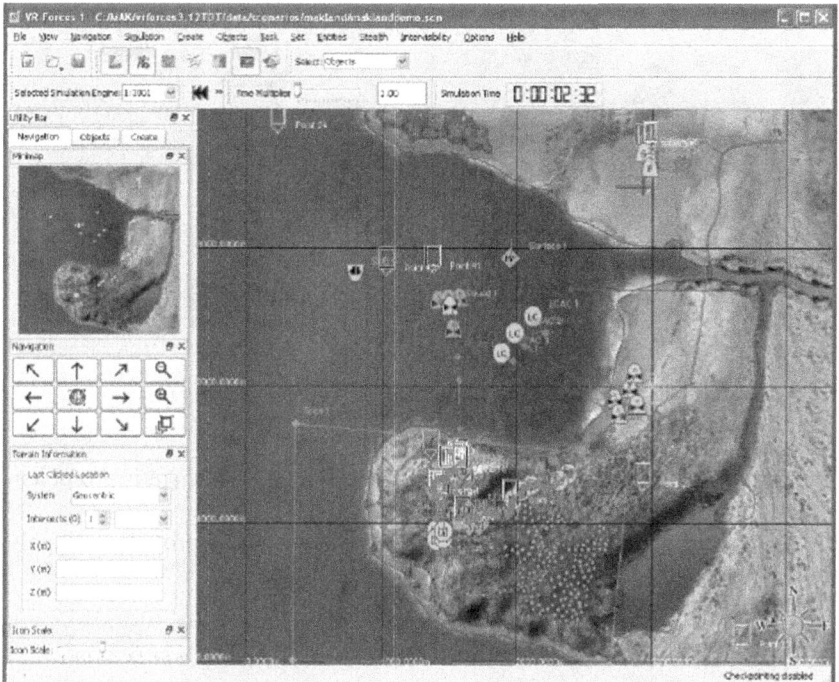

Figure A2.4   VR-Forces interface and demo scenario (*VR-Forces: Getting Started Guide*, 2009, p. 10).

rely on realistic environments for convoy training, fire support, visualization of weapons effect, and training in an urban environment. VBS2 has an extended database to simulate sea, air, and land vehicles and other miscellany. The database has objects (car batteries and water cooler, for instance), people (Afghan civilian, Iraqi police, or US civilian), vehicles (planes, ships, or armor vehicles from different nations), and weapons (a variety of handguns and rocket launchers). VBS2 is mainly used by the US Army. US JFCOM uses it in their LVC simulations. In an LVC simulation, a trainee uses a helmet with an eyepiece and a gun. The VBS2 simulation shows the trainee (as an avatar) within a squad. The trainee can shoot his/her gun and kill enemy forces within the simulation. USMC uses it as part of their DVTE program. Countries such as the UK and Canada use it as well as NATO.

Like previously mentioned simulation systems, VBS2 is HLA and DIS compliant allowing distributed simulation and connectivity with higher granularity simulation systems such as JSAF or JCATS. Figure A2.5 show the level of detail of some of the simulations using VBS2 (Bohemia Interactive Simulations, 2011). Figure A2.5(a), for instance, shows a screen shot of an interactive simulation to train a soldier how to greet and observe a person while looking for anything suspicious. In this case, the person has a large mid-section which may be indicative of explosives. Figure A2.5(b) shows the avatar of an Iraqi policeman

(a)   (b)

Figure A2.5   (a) Cultural awareness simulation; (b) water purification unit training.

communicating to the trainee's avatar (not in the screen) that there is a problem with the water purification unit.

As can be observed, unlike previous simulations VBS2 provides a highly detailed environment for individual or small team training.

VBS2 profile:

- Type of simulation: constructive, virtual
- Main user: US Army
- Main purpose: training and mission rehearsal
- Other users: US Secret Service, US Marine Corps, UK, Canada, and New Zealand among others
- Main level of support: tactical
- Data model infrastructure support: HLA, DIS
- Display: 2D and 3D
- Military unit: squad

Air Force Synthetic Environment for Reconnaissance and Surveillance Model (AFSERS)

The MUSE (Multiple Unified Simulation Environment)/AFSERS, developed by the Joint Technology Centre/System Integration Laboratory (JCT/SIL), is a simulation system mainly used for unmanned aerial vehicle (UAV) and intelligence, surveillance, and reconnaissance (ISR) training. Unlike previously mentioned simulation systems, the AFERS is focused on simulating systems that relate to air missions such as air vehicles (different UAVs, P-3, and U-2), landing systems, and radars. According to McClung and Jones (2002), the MUSE was developed as a Hunter UAV simulation and has evolved to include ISR platforms and sensor models (infrared), theatre and national capabilities, tactical communications and advanced mission planning. Like other simulation systems, AFSERS is HLA and

DIS compliant and it is used by the JCT/SIL in distributed simulation scenarios with other simulation systems such as AWSIM.

AFSERS profile:

- Type of simulation: constructive and virtual
- Main user: US Air Force
- Main purpose: training and mission rehearsal
- Other users: US Army, Joint, South Korea
- Main level of support: tactical
- Data model infrastructure support: HLA, DIS
- Display: 2D and 3D
- Military unit: battalion

Air Warfare Simulation (AWSIM)

AWSIM, developed in the 1980s by the Warrior Preparation Center in Germany, is the favored simulation system of the US and NATO Air Force for conducting simulation exercises in air warfare (air-to-air, air-to-surface, and surface-to-air combat) and space operations. Some of the simulations that AWSIM has are surface-to-air missiles (SAM), surface-to-surface missiles, (SSM), short range air defense systems (SHORAD), aerial refueling, and air-to-air engagement while considering variables such as speed, altitude and fuel consumption. According to Training Transformation Defined (2006) AWSIM was developed to

> train senior commanders and battlestaffs in the execution of wartime general defense plans that emphasize joint and conventional operations. . . . Today, it is the core model of the Air and Space Constructive Environment Suite used worldwide to train senior battle commanders and their staffs within the Air Force across the Department of Defense (DOD). It provides the opportunity to train for joint and combined prosecution of war using interactive computer simulations that replicate a realistic battlespace, incorporating various audiences through worldwide distribution.

AWSIM can be used with other simulation systems in order to conduct joint and coalition exercises. Perhaps the best known case in which AWSIM is used with other simulation systems is the Ulchi-Focus Lens (UFL), now known as the Ulchi-Freedom Guardian (UFG). The UFG is a large scale war fighting exercise and the world's largest command and control simulation driven exercise that takes place yearly between the South Korean and US Government military. The UFG's purpose is to train both military for the case of a North Korea attack. In the 2010 exercise, it was reported by the Xinhua News Agency (2010) that the exercise included about 55,000 South Korean soldiers and 30,000 US troops in South Korea and abroad. AWSIM is used in UFG exercises in conjunction with other systems such as the US Marine model Marine Tactical Warfare Simulation (MTWS) and US Army Corps Battle Simulation (CBS).

AWSIM profile :

- Type of simulation: constructive
- Main user: US Air Force
- Main purpose: training, mission rehearsal, experimentation
- Other users: NATO, NASA
- Main level of support: operational, tactical
- Data model infrastructure support: HLA, DIS, ALSP[10]
- Display: 2D
- Military unit: wing

Research, Evaluation, and System Analysis (RESA)

RESA, developed by SSC[11] San Diego, provides the US Navy with a simulation system for theater level naval warfare. According to Neyland (1997, p. 110), RESA was designed

> to support research and development and training of senior naval officers, focusing on command and control of battle group or force operations.... RESA has been used for requirement analysis, technology evaluation, concept formulation, system testing, system design, architecture assessment, operation plan evaluation, command and control training, joint operation interoperability, and distributed war gaming.

RESA is ALSP, DIS and HLA compliant which allow simulations to work in a distributed environment. RESA, being Navy focused, allows for the simulation of naval warfare such as submarines, sensors, warships, and C3[12] architectures as well as ground forces.

RESA profile:

- Type of simulation: constructive
- Main user: US Navy
- Main purpose: training, acquisition
- Other users: Joint, South Korea
- Main level of support: operational, tactical
- Data model infrastructure support: HLA, DIS, ALSP
- Display: 2D
- Military unit: –

---

[10] Aggregate Level Simulation Protocol.

[11] SPAWAR (Space and Naval Warfare Systems Command) Systems Center Pacific.

[12] Command, Control, and Communications.

Marine Air Ground Task Force (MAGTF) Tactical Warfare Simulation (MTWS)

Developed by SSC San Diego as well, MTWS's main purpose is to support tactical training exercises for the US Marine Corps. Despite the Marines using other simulation systems, MTWS provides the right mix of air, ground, maritime, and amphibious operations: MAGTF provides these combat operation elements in a wide range of tactical conditions while MARS (MTWS Analysis Review System) provides exercise review and analysis capabilities (Hardy et al., 2001).

MTWS profile:

- Type of simulation: constructive
- Main user: US Marine Corps
- Main purpose: training, analysis
- Other users: Joint, South Korea
- Main level of support: tactical
- Data model infrastructure support: HLA, DIS, ALSP
- Display: 2D
- Military unit: –

CATS TYR

CATS TYR was developed by C-ITS (back then named "Mandator") in collaboration with the Swedish Defence Research Agency (FOI) starting in 1996. The original purpose was to simulate and evaluate conventional wargaming with asymmetrical forces on a larger scale. Several core processes generated robustness and good scalability with multiple processors. Doctrines and models were supplied by FOI. Quite early support for civil units and PSO orders was introduced. Focus shifted towards operations other than war and disaster relief. Also the need for multi-resolution dictated that models would support smaller units as well.

When CATS TYR was put to use internationally additional features such as terrain, night vision capability, extended logistics, and decoys were added. In recent years optimization has enabled extended number of units as well as exceptional stability for the system.

TYR has served as the back-bone and core system in all VIKING CAX (Computer Assisted Exercises) events in an integrated environment containing other virtual and constructed simulation systems and C2 systems. VIKING events are international joint exercises, coordinated by the Swedish Armed Forces, that involve not only the military but also civilians and police forces. So far, there have been six VIKING events (99, 01, 03, 05, 08, and 11). VIKING 11 (conducted in April 2011) had the participation of about 2600 trainees representing Governments, the United Nations, Red Cross, and military forces among others. The event was distributed over nine remote sites, four in Sweden and five in other countries (Austria, Georgia, Germany, Ireland, and Ukraine).

CATS TYR is supported by a range of optional tools such as simplified scenario construction using MSDL and stand-alone AAR[13] tools with time sliders and "take home" packages, all integrated using HLA standards. Today, CATS TYR is developed into separate modules able to support other systems in a larger federation while supporting external models Fig. A2.6 shows a screenshot of CATS TYR (Karlström, 2011).

CATS TYR profile:

- Type of simulation: constructive
- Main user: Swedish Armed Forces in conjunction with police and civil organizations
- Main purpose: training and experimentation in PSO,[14] disaster relief, and conventional joint wargaming
- Other users: United Nations, other nations
- Level of support: strategic, operational, tactical
- Data model infrastructure support: HLA, MSDL, XML
- Display: 2D
- Military unit: –

GUPPIS/KORA OA (Korpsrahmen Simulationsmodell für die Offizierausbildung — Corps Frame Model for Officers Training)

KORA, developed by IAGB (Industrieanlagen-Betriebsgesellschaft mbH), is the software component of the GUPPIS[15] joint simulation system. It is used by the German Federal Armed Forces to train in air (surveillance, airspace control, air transport, among other capabilities), land (mounted and dismounted infantry among other capabilities), and naval (anti-surface and anti-air among other capabilities) warfare. In addition, KORA is also used for concept development and experimentation.

According to IABG (2010), the core of KORA is made up of a group of sub-models (Air Force, Navy, Logistics, Communications, Medical Services, and Reconnaissance, among others). Other components support the "generation of military structures and exercise scenarios, evaluation of exercises, editing of terrain data, processing of weapon system data and system administration" (p. 4). KORA is HLA and DIS compliant and is based on JC3IEDM standards.

---

[13] After Action Review.

[14] Peace Support Operations.

[15] Gefechtssimulationssystem zur Unterstützung von Plan/Stabsübungen und Planuntersuchungen in Stäbenvon Großverbänden und an Schulen/Akademien mit Heeresaufgaben (Combat simulation system for support of map and staff exercises within the HQs of major units or at schools and academies with army tasks).

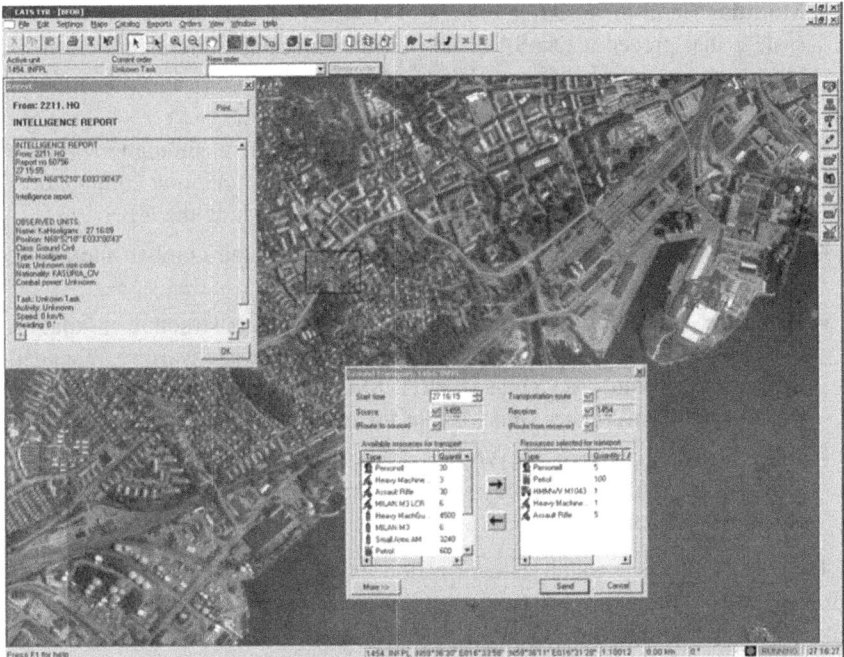

Figure A2.6 CATS TYR main screen.

KORA profile:

- Type of simulation: constructive
- Main user: German Federal Armed Forces
- Main purpose: training and experimentation
- Other users: Training Academies of the German Federal Armed Forces
- Level of support: strategic, operational, tactical
- Data model infrastructure support: HLA, DIS, JC3IEDM
- Display: 2D
- Military unit: –

Other Simulation Systems

- **Corps Battle Simulation—CBS:** a constructive simulation system that provides theater level training using ground battle scenarios
- **Joint Theater Level Simulation—JTLS:** a constructive, multi-sided simulation system used by joint and coalition forces with main focus the operational level of war

- **Modular Semi-automated Forces—ModSAF:** a constructive simulation system that precedes OneSAF and provides a scalable architecture allowing users to control different levels of military units in a wide range of scenarios
- **Warfighters Simulation—WARSIM:** a constructive simulation system that provides mission rehearsal capabilities for Army and joint commanders and their staff and supposed to replace systems such as the CBS
- **Janus Simulation System:** a 2D, multi-sided, ground combat interactive simulation system
- **Extended Air Defense Simulation—EADSIM:** a constructive simulation system that provides simulation of air, space, and missile warfare, managed by the Future Warfare Center (FWC), Modeling and Simulation Division (MSD), US Army Space and Missile Defense Command (SMDC)
- **Joint Warfare System—JWARS:** a constructive simulation system that provides operational, planning, system trade analysis, and execution support in joint settings

Table A2.1 shows a rough categorization of the previously mentioned simulation systems. The table is not meant to be a comprehensive list as there are many more systems that did not make it to this review.

FINAL REMARKS

Simulation systems are powerful tools that assist the military in their training, analysis, mission rehearsal, acquisition, and experimentation needs. However, there are some important considerations worth mentioning. First is that of interoperability. Although systems seem to be interacting with one another, it does not mean that the output of this interaction is consistent, as previously mentioned. This is more an issue of modeling than of simulating as the modeler needs to address polymorphism, data availability multi-scope and multi-granularity issues, among other things, during the modeling of the federation. A second consideration is that of achieving training objectives. Designing training objectives is a major challenge as a federation is used to train different people, at different levels, at the same time, and individuals have different training needs. Despite these systems saving money, they are by no means inexpensive and their cost/benefit effectiveness relies on achieving those training objectives. Notice that the training objective is tightly tied to the interoperability consideration. As more than one simulation is required to achieve the training objectives, consistency is key for the training to be effective and avoid cost overruns by solving inconsistencies due to poor modeling. Finally, when referring to cost reduction, the idea of reusability comes to mind. When reusing existing simulations one expects to reduce costs by using simulations to achieve training objectives they were not designed for. In this sense, to establish reuse, both previously mentioned considerations need to be maintained.

Table A2.1 Categorization of Simulation Systems

System	Type		Purpose					Main user	Other users	Level of support			Military unit (up to)
	C	V	T	E	A	M	Q			S	O	T	
JSAF	×		×	×				US Joint	Navy, Marines, Joint, Coalition, UK	×	×	×	Brigade
OneSAF	×		×	×			×	US Army	Navy, Marines, Joint, Canada, UK	×	×	×	Brigade
JCATS	×		×	×		×	×	US Joint	Navy, Marines, DHS, DOE, NATO			×	Battalion
VR Forces	×		×	×				US Army	Spain, Norway, the Netherlands			×	Platoon
VBS2	×	×	×			×		US Army	Secret Service, Marines, UK,			×	Squad
AFSERS	×	×	×			×		US Air Force	Army, Joint, South Korea			×	Battalion
AWSIM	×		×	×		×		US Air Force	NATO, South Korea, NASA Langley		×	×	Wing
RESA	×		×				×	US Navy	South Korea		×	×	–
MTWS	×		×		×			US Marines	Joint, South Korea			×	–
CATS TYR	×		×	×				Sweden AF	United Nations	×	×	×	Battalion
KORA	×		×	×				German AF	Training Academies of German AF	×	×	×	–
CBS	×		×					US Army	Joint			×	–
JTLS	×		×			×		US Joint	Coalition		×	×	Theater
ModSAF	×		×	×	×			US Army	Joint		×	×	–
WARSIM	×		×			×		US Army	Joint, Coalition		×	×	–
EADSIM	×		×	×	×			US Air Force	Joint, Coalition	×	×	×	Theater
JWARS	×		×		×			US Joint	Navy		×		Theater

Type: C, constructive; V, virtual. Purpose: T, training; E, experimentation; A, analysis; M, mission rehearsal; Q, acquisition. Level of support: S, strategic; O, operational; T, tactical.

ACKNOWLEDGEMENTS

The author is grateful to a special group of individuals that made this annex possible: Edmund Spinella, Pierre Hollis, Max Karlström, Staffan Löf, Karl Heinz Neumann, and Eckehard Neugebauer.

REFERENCES

Bohemia Interactive Simulations. VBS Worlds Demos. http://www.bisimulations.com (last accessed 4 July 2011).

Hardy D, Allen E, Adams K, Peters C and Peterson L (2001). Advanced distributed simulation: decade in review and future challenges. *Space and Naval Warfare Systems Center, San Diego, Biennial Review*, San Diego, US Navy, pp. 165–175.

IABG (2010) *GUPPIS/KORA:* Model Profile. IABG, Germany.

JCATS: Simulation User's Guide, Version 4.1.0. (2003) Lawrence Livermore National Laboratory.

JSAF Quick Guide. JCATS: CA. http://www.i-mef.usmc.mil/external/wss/deployable_virtual_trianing_environment/dvte_handouts/can/jsaf/jsaf_qg.pdf (last accessed 4 July 2011).

JSAF User's Guide (2003) *Volume 1, Introduction and Basic Controls.* Joint Semi-Automated Forces.

Karlström M (2011). *Personal Communication*, July 14, 2011.

Lenoir T and Lowood H. *Theaters of War: the Military–Entertainment Complex.* Stanford University, Palo Alto, CA. http://www.stanford.edu/class/sts145/Library/Lenoir-Lowood_Theaters OfWar.pdf (last accessed 4 July 2011).

McClung S and Jones J. *Up and Away, Equipment, Training and Support News.* http://www.metavr.com/aboutus/articles/ETSspring02muse_up%2Baway.pdf (last accessed 4 July 2011).

Neyland D (1997) *Virtual Combat: A Guide to Distributed Interactive Simulation.* Satckpole Books, Mechanicsburg, PA.

OTB SAF Overview. University of Pittsburgh, Pittsburgh, PA. http://usl.sis.pitt.edu/wjj/otbsaf/USER_VOL1.HTML#OTB%20SAF%20Overview (last accessed 4 July 2011).

ShimamotoF (2000) Simulating warfare is no video game. *Sci Tech Rev*, 4–11.

Training Transformation Defined (2006) National Defense University Press, Joint Force Quarterly 42, 55–56.

VT MÄK (2009) VR-Forces: Getting Started Guide. VT-MÄK, Cambridge, MA.

Xinhua News Agency (2010) *S Korea Conducts Anti-Terrorism Drills as Part of UFG Exercises.* http://news.xinhuanet.com/english2010/ world/2010-08/17/c_13449108.htm (last accessed 4 July 2011).

Index

absolute breakpoint rule, 162
accreditation, 14, 70, 211, 243, 245–6, 262, 264, 274, 276–9, 281, 285–6, 293–4, 349, 447
accreditation agents, 276–7, 282
accuracy, 14, 44, 48, 64, 69, 84, 99, 133, 150, 266–9, 278
acoustic sensor, 9, 129
acoustic, 91, 140
acquisition, 35, 71, 131, 142, 145, 222, 253
action domain, 675
activities
  capabilities viewpoint, 781–3
  operational viewpoint, 780–783, 792
activity diagrams, 219–20, 790, 794
actors, non-state, 649, 706–7
adaptation, 124, 126, 258, 363, 545, 673–4, 767–8, 802, 805
adaptation domain, 675
adaptive atomic components of thought or adaptive character of thought (ACT-R), 684
adaptive communication environment (ace), 466–7
aerospace, 77, 142
agent, 10, 18, 168–9, 172–5, 276–7, 281–4, 292, 298
agent architecture-generic, 674
agent based simulation for non-kinetic decision support systems, 677, 679, 681, 683, 685, 687, 689, 691, 693, 695, 697, 699, 701, 703, 705

agent based simulation, 670–672, 707
agent directed simulation (ads), 10, 18, 126, 292, 379, 669–71
agent directed Simulation for combat modeling, 670, 672, 674, 676, 678, 680, 682, 684, 686, 690, 692, 694, 696, 698, 702
agent frameworks, 679–80, 707–8, 710
agent heterogeneity, 694
agent metaphor, 17–18, 669
agent paradigm, 169, 173, 669–70
agent simulation, 671
agent supported simulation, 672, 676, 706
agent support simulation for combat decision support systems, 676–7
agents for simulation, 670
agents, 10, 18, 168–9, 172, 175, 646–7, 658–61, 663, 669–77, 679–92, 694–8, 700–710, 712–13, 716, 743
aggregate, 9–10, 63, 65–6, 85, 96, 113–14, 122–5, 128, 135, 137–42, 148, 167, 174, 199–200, 232
aggregate level engagement and attrition models, 153, 155, 157, 159, 161
aggregate level simulation protocol (ALSP), 77, 345, 355, 416–19, 443–5, 447, 554
aggregate level simulations, 65–6, 345, 834
aggregate models, 9, 113, 118, 124–5, 137, 146, 614

Printed in the United States
By Bookmasters